VOLUME ONE HUNDRED

VITAMINS AND HORMONES

Vitamin D Hormone

VOLUME ONE HUNDRED

VITAMINS AND HORMONES

Vitamin D Hormone

Series Editor

GERALD LITWACK, PhD

Toluca Lake, North Hollywood, California

ELSEVIER

AMSTERDAM • BOSTON • HEIDELBERG • LONDON
NEW YORK • OXFORD • PARIS • SAN DIEGO
SAN FRANCISCO • SINGAPORE • SYDNEY • TOKYO
Academic Press is an imprint of Elsevier

Cover photo credit:
Belorusova, A.Y., Rochel, N.
Structural Studies of Vitamin D Nuclear Receptor Ligand-Binding Properties
Vitamins and Hormones (2016) **100**, pp. 83–116

Academic Press is an imprint of Elsevier
50 Hampshire Street, 5th Floor, Cambridge, MA 02139, USA
525 B Street, Suite 1800, San Diego, CA 92101-4495, USA
The Boulevard, Langford Lane, Kidlington, Oxford OX5 1GB, UK
125 London Wall, London, EC2Y 5AS, UK

First edition 2016

ISBN: 978-0-12-804824-5
ISSN: 0083-6729

For information on all Academic Press publications
visit our website at store.elsevier.com

CONTENTS

CONTRIBUTORS

Jungmi Ahn
Department of Molecular Medicine/Institute of Biotechnology, The University of Texas Health Science Center at San Antonio, Texas Research Park, San Antonio, Texas, USA

Leggy A. Arnold
Department of Chemistry and Biochemistry, Milwaukee Institute for Drug Discovery (MIDD), University of Wisconsin-Milwaukee, Milwaukee, Wisconsin, USA

Jorge N. Artaza
Department of Health & Life Sciences, Charles R. Drew University of Medicine and Science, and Department of Medicine, David Geffen School of Medicine at UCLA, Los Angeles, California, USA

Anna Y. Belorusova
Department of Integrative Structural Biology, Institut de Génétique et de Biologie Moléculaire et Cellulaire (IGBMC), Institut National de Santé et de Recherche Médicale (INSERM) U964, Centre National de Recherche Scientifique (CNRS) UMR 7104, Université de Strasbourg, Illkirch, France

Nancy A. Benkusky
Department of Biochemistry, University of Wisconsin–Madison, Madison, Wisconsin, USA

Alakesh Bera
Department of Molecular Medicine/Institute of Biotechnology, The University of Texas Health Science Center at San Antonio, Texas Research Park, San Antonio, Texas, USA

Megan Campbell
Department of Microbiology, Biochemistry and Molecular Genetics, Rutgers, The State University of New Jersey, New Jersey Medical School, Newark, New Jersey, USA

Carsten Carlberg
Institute of Biomedicine, School of Medicine, University of Eastern Finland, Kuopio, Finland

Alex Carlson
Department of Biochemistry, University of Wisconsin–Madison, Madison, Wisconsin, USA

Bandana Chatterjee
Department of Molecular Medicine/Institute of Biotechnology, The University of Texas Health Science Center at San Antonio, Texas Research Park, and South Texas Veterans Health Care System, Audie L Murphy VA Hospital, San Antonio, Texas, USA

Tai C. Chen
Clinical Translational Science Institute and Department of Medicine, Boston University School of Medicine, Boston, Massachusetts, USA

Sylvia Christakos
Department of Microbiology, Biochemistry and Molecular Genetics, Rutgers, The State University of New Jersey, New Jersey Medical School, Newark, New Jersey, USA

Hector F. DeLuca
Department of Biochemistry, University of Wisconsin-Madison, Madison, Wisconsin, USA

Puneet Dhawan
Department of Microbiology, Biochemistry and Molecular Genetics, Rutgers, The State University of New Jersey, New Jersey Medical School, Newark, New Jersey, USA

Rosemary A. Fricker
School of Medicine and Institute for Science and Technology in Medicine, Keele University, Keele, United Kingdom

Carol A. Haussler
Department of Basic Medical Sciences, University of Arizona College of Medicine, Phoenix, Arizona, USA

Mark R. Haussler
Department of Basic Medical Sciences, University of Arizona College of Medicine, Phoenix, Arizona, USA

Teikichi Ikura
Medical Research Institute, Tokyo Medical and Dental University, Tokyo, Japan

Nobutoshi Ito
Medical Research Institute, Tokyo Medical and Dental University, Tokyo, Japan

Candace S. Johnson
Department of Pharmacology and Therapeutics, Roswell Park Cancer Institute, Buffalo, New York, USA

Peter W. Jurutka
Department of Basic Medical Sciences, University of Arizona College of Medicine, Phoenix, and School of Mathematical and Natural Sciences, Arizona State University, Glendale, Arizona, USA

G. Kerr Whitfield
Department of Basic Medical Sciences, University of Arizona College of Medicine, Phoenix, Arizona, USA

Zainab Khan
School of Mathematical and Natural Sciences, Arizona State University, Glendale, Arizona, USA

Irene M. Kim
Department of Health & Life Sciences, Charles R. Drew University of Medicine and Science, Los Angeles, California, USA

Atsushi Kittaka
Faculty of Pharmaceutical Sciences, Teikyo University, Tokyo, Japan

M. Kyle Hadden
Department of Pharmaceutical Sciences, University of Connecticut, Storrs, Connecticut, USA

Seong Min Lee
Department of Biochemistry, University of Wisconsin–Madison, Madison, Wisconsin, USA

Wei Luo
Department of Pharmacology and Therapeutics, Roswell Park Cancer Institute, Buffalo, New York, USA

Kamil Lupicki
Department of Microbiology, Biochemistry and Molecular Genetics, Rutgers, The State University of New Jersey, New Jersey Medical School, Newark, New Jersey, USA

Yingyu Ma
Department of Pharmacology and Therapeutics, Roswell Park Cancer Institute, Buffalo, New York, USA

Mark B. Meyer
Department of Biochemistry, University of Wisconsin–Madison, Madison, Wisconsin, USA

Charlotte Middleditch
School of Medicine and Institute for Science and Technology in Medicine, Keele University, Keele, United Kingdom

Eiji Munetsuna
Department of Biotechnology, Faculty of Engineering, Toyama Prefectural University, Toyama, and Department of Biochemistry, Fujita Health University for Medical Science, Toyoake, Japan

Keith C. Norris
Department of Medicine, David Geffen School of Medicine at UCLA, Los Angeles, California, USA

Melda Onal
Department of Biochemistry, University of Wisconsin–Madison, Madison, Wisconsin, USA

Rowan P. Orme
School of Medicine and Institute for Science and Technology in Medicine, Keele University, Keele, and Cardiff Institute of Infection and Immunity, School of Medicine, Cardiff, United Kingdom

Sulgi Park
Department of Molecular Medicine/Institute of Biotechnology, The University of Texas Health Science Center at San Antonio, Texas Research Park, San Antonio, Texas, USA

J. Wesley Pike
Department of Biochemistry, University of Wisconsin–Madison, Madison, Wisconsin, USA

Lori A. Plum
Department of Biochemistry, University of Wisconsin-Madison, Madison, Wisconsin, USA

Natacha Rochel
Department of Integrative Structural Biology, Institut de Génétique et de Biologie Moléculaire et Cellulaire (IGBMC), Institut National de Santé et de Recherche Médicale (INSERM) U964, Centre National de Recherche Scientifique (CNRS) UMR 7104, Université de Strasbourg, Illkirch, France

Marya S. Sabir
School of Mathematical and Natural Sciences, Arizona State University, Glendale, Arizona, USA

Hiroshi Saitoh
Teijin Institute for Bio-medical Research, Teijin Pharma Ltd., Tokyo, Japan

Toshiyuki Sakaki
Department of Biotechnology, Faculty of Engineering, Toyama Prefectural University, Toyama, Japan

Ruby Sandoval
School of Mathematical and Natural Sciences, Arizona State University, Glendale, Arizona, USA

Sohel Shamsuzzaman
Department of Biochemistry, University of Wisconsin–Madison, Madison, Wisconsin, USA

Chung Seog Song
Department of Molecular Medicine/Institute of Biotechnology, The University of Texas Health Science Center at San Antonio, Texas Research Park, San Antonio, Texas, USA

Hillary St. John
Department of Biochemistry, University of Wisconsin–Madison, Madison, Wisconsin, USA

Masashi Takano
Faculty of Pharmaceutical Sciences, Teikyo University, Tokyo, Japan

Kelly A. Teske
Department of Chemistry and Biochemistry, Milwaukee Institute for Drug Discovery (MIDD), University of Wisconsin-Milwaukee, Milwaukee, Wisconsin, USA

Donald L. Trump
Department of Medicine, Roswell Park Cancer Institute, Buffalo, New York, and Inova Dwight and Martha Schar Cancer Institute, Falls Church, Virginia, USA

Vaishali Veldurthy
Department of Microbiology, Biochemistry and Molecular Genetics, Rutgers, The State University of New Jersey, New Jersey Medical School, Newark, New Jersey, USA

Lauren Waite
School of Medicine and Institute for Science and Technology in Medicine, Keele University, Keele, United Kingdom

Ran Wei
Department of Microbiology, Biochemistry and Molecular Genetics, Rutgers, The State University of New Jersey, New Jersey Medical School, Newark, New Jersey, USA

Olivia Yu
Department of Chemistry and Biochemistry, Milwaukee Institute for Drug Discovery (MIDD), University of Wisconsin-Milwaukee, Milwaukee, Wisconsin, USA

Baltazar Zuniga
Department of Molecular Medicine/Institute of Biotechnology, The University of Texas Health Science Center at San Antonio, Texas Research Park, San Antonio, and The University of Texas at Austin, Austin, Texas, USA

PREFACE

Ultraviolet radiation in sunlight causes skin cells to convert a precursor, 7-dehydrocholesterol, to a precursor of the active form of vitamin D, 1,25-dihydroxy cholecalciferol (dihydroxyvitamin D3 or calcitriol). The precursor is translocated to the liver where a hydroxyl group is added in the 25 position and, after further translocation to the kidney, a second hydroxyl group in added in the 1-position producing the active form of the vitamin that can bind to and activate the vitamin D receptor. Many essential activities are produced by the activated vitamin D receptor including the production of proteins that transport calcium ion, mineralization of bone, enhancement of the immune system, and many others. Focus on this vitamin that acts as a hormone in the body is important because a large portion of the population is deficient in vitamin D. Deficiency can result from living in a region of limiting sunlight, insufficient ingestion of sources of the vitamin by vegan diets, or because of intestinal diseases and as a consequence of aging where the elderly utilize sunlight less efficiently than younger persons.

In this volume are described the history of vitamin D research; the basic science on the vitamin and related structures, some of which are used clinically; as well as the structure of the vitamin D receptor and its interactions with coregulator and interactions of the vitamin D system with many types of cells in normal systems and in diseases.

In the opening chapter, H.F. DeLuca presents the history of vitamin research in "Vitamin D: historical overview." This is followed by a series of chapters on the basic science of the vitamin D system initiated by "Genomic determinants of vitamin D-regulated gene expression" by J. W. Pike, M.B. Meyer, N.A. Benkusky, S.-M. Lee, H. St. John, A. Carlson. M. Onal, and S. Shamsuzzaman. K.A. Teske, O. Yu, and L. A. Arnold present "Vitamin D receptor–coregulator interaction." A. Y. Belorusova and N. Rochel report on "Structural studies of vitamin D receptor ligand-binding properties." "Crystal structure of the vitamin D receptor ligand-binding domain with lithocholic acids" is described by T. Ikura and N. Ito. V. Veldurthy, R. Wei, M. Campbell, K. Lupicki, P. Dhawan, and S. Christakos offer "25-hydroxyvitamin D-24-hydroxylase (CYP24A1): a key regulator of 1,25(OH)2D3 catabolism and calcium homeostasis." H.F. DeLuca and L.A. Plum relate "Analogs of

1α,25-dihydroxyvitamin D3 in clinical use." M.R. Haussler, G.K. Whitfield, C. A. Haussler, M.S. Sabir, Z. Kahn, R. Sandoval, and P.W. Jurutka author "1,25-dihydroxyvitamin D and Klotho: a tale of two renal hormones coming of age." M.K. Hadden writes on "Hedgehog and vitamin D signaling pathways in development and disease." "Molecular approaches for optimizing vitamin D supplementation" is the contribution of C. Carlberg.

In a group of chapters focusing on the effects of vitamin D on various cell types including cancer, R.P Orme, C. Middleditch, L. Waite, and R.A. Fricker report on "The Role of Vitamin D3 in the Development and Neuroprotection of Midbrain dopamine neurons." This is followed by a report from I.M. Kim, K.C. Norris, and J.N. Artaza on "Vitamin D and cardiac differentiation." J. Ahn, S. Park, B. Zuniga, A. Bera, C. S. Song, and B. Chatterjee describe "Vitamin D in prostate cancer." E. Munetsuna, A. Kittaka, T.C. Chen, and T. Sakaki discuss "Metabolism and action of 25-hydroxy-19-nor-vitamin D3 in human prostate cells." A. Kittaka, M. Takano, and H. Saitoh write on "Vitamin D analogs with nitrogen atom at C2 substitution and effect on bone formation." In the final two chapters: Y. Ma, C.S. Johnson, and D.L. Trump review "Mechanistic insights of vitamin D anticancer effects" and W. Luo, C.S. Johnson, and D.L. Trump report on "Vitamin D signaling modulators in cancer therapy."

The cover illustration is reproduced from Fig. 2 of Chapter 4 by A. Y. Belorusova and N. Rochel showing the crystal structure of the ligand-binding domain of the vitamin D receptor.

Helene Kabes of Elsevier, Oxford, UK, and the group from Reed Elsevier of Chennai, India, were instrumental in the production of this volume.

Special thanks are due to Professor Hector DeLuca of the University of Wisconsin for his contributions to this book.

Gerald Litwack
North Hollywood, California
September 28, 2015

CHAPTER ONE

Vitamin D: Historical Overview

Hector F. DeLuca[1]
Department of Biochemistry, University of Wisconsin-Madison, Madison, Wisconsin, USA
[1]Corresponding author: e-mail address: deluca@biochem.wisc.edu

Contents

Abstract

A history of vitamin D has been provided, dating from the earliest description of rickets, the disease resulting from vitamin D deficiency, to a current understanding of vitamin D metabolism and the mechanism of action of its hormonal form in regulating gene expression in target organs. Vitamin D is produced in skin by impact of 280–310 nm light on 7-dehydrocholesterol. The vitamin D is then converted in the liver to a circulating form, 25-hydroxyvitamin D that is converted largely, if not exclusively, in the kidney to the final hormone, 1α,25-dihydroxyvitamin D. This hormone functions through a nuclear receptor that regulates expression of key genes in target organs. Among its many resulting functions are increased intestinal calcium and phosphate absorption, bone calcium mobilization, and renal reabsorption of calcium. The resultant increase in serum calcium and phosphate supports bone mineralization, curing rickets, and osteomalacia. There are many other functions of vitamin D that remain to be described that contribute to its health supporting role.

1. THE DISCOVERY OF VITAMIN D

It seems likely that rickets appeared in the human population following the migration from Africa into the Northern Hemisphere. However, as far as this author is concerned, the first documentation of the deficiency disease of vitamin D is that of Whistler (1645), who clearly described the disease

Vitamins and Hormones, Volume 100
ISSN 0083-6729
http://dx.doi.org/10.1016/bs.vh.2015.11.001

in which the skeleton was poorly mineralized and deformed. Very likely, this disease occurred only rarely until urbanization of the human race took place in areas where marginal sunlight occurred. The disease became an epidemic especially in England and in Northern Europe during the nineteenth century. In fact, the disease was called the "English Disease" because the incidence in England and Scotland among children was as high as 70–80% (Hess, 1929).

At about the same time, the biological scientists of the nineteenth century were considering that the essential components of a diet were protein, carbohydrate, fat, and minerals. The idea of essential micronutrients had not yet surfaced. Magendie (1816), Funk (1911), Hopkins (1912), and others attempted to grow experimental animals on a diet of purified proteins, carbohydrates, fat, and salts but the animals failed to thrive and survive. Clearly, something was missing from these diets, which later proved to be organic micronutrients known as the vitamins. One of the first indications of the necessity of these extra micronutrients was the experiment of Eijkman (1897) followed by a Dutch colleague, Grijns (1901). Eijkman noted that prisoners suffering beri-beri in the Dutch East Indies were fed dehulled rice. The beri-beri could be corrected by the provision of the hulls, which Eijkman suggested neutralized toxins present in polished rice. His colleague, Grijns (1901), however, realized that it was an essential nutrient found in hulls that corrected the beri-beri. Similarly, the widespread disease, scurvy among British sailors on long voyages, could be corrected by providing limes or in the case of Dutch and German sailors, sauerkraut. Of course, we now know that sauerkraut and limes contain ascorbic acid or vitamin C (Drummond, 1920).

The existence of vitamins was first established by the work of McCollum and Davis (1913) at the University of Wisconsin in which rats that were fed a highly purified diet, developed an eye disease, xerophthalmia that could be cured with butter fat or cod liver oil. McCollum was very quick to realize that this was an accessory food factor, which he termed "vitamin A" (McCollum, Simmonds, & Pitz, 1916). Somewhat later, McCollum's group found that a water-soluble substance prevented the disease, neuritis. This he designated as "vitamin B" (McCollum et al., 1916). Almost identical experiments were conducted by Osborne and Mendel (1917) clearly demonstrating the existence of a fat-soluble micronutrient termed "vitamin A" and a water-soluble nutrient termed "vitamin B."

The discovery of the vitamins was undoubtedly an important factor that stimulated Sir Edward Mellanby in Great Britain to carry out a classical experiment. He was able to produce the disease rickets in dogs by providing them the diet consumed by the Scotts, a population suffering the highest incidence of rickets, who at that time subsisted largely on a diet of oatmeal (Mellanby, 1919). Mellanby unknowingly maintained the dogs in indoor kennels, depriving them of sunlight. Mellanby was well aware of the experiments of McCollum and of Osborne and Mendel and, therefore, used cod liver oil to treat his animals. Unknown to Mellanby, however, cod liver oil contained not only the vitamin A discovered by McCollum but also contained vitamin D. In any case, Mellanby surmised that this was another property of vitamin A, namely the prevention of the skeletal rachitic symptoms.

Quite early, Steenbock and Hart (1913) at the University of Wisconsin were studying the retention of "lime," i.e., calcium, in a group of experimental goats. He reported that goats kept indoors in the winter lost calcium from their bones, whereas they retained calcium when put out in sunlit pastures in the summer. Undoubtedly, this experience was to catalyze his later investigation into the role of sunlight in the treatment of rickets.

McCollum was attracted to Johns Hopkins University and left the University of Wisconsin. McCollum being aware of the work of Mellanby questioned whether the vitamin A in cod liver oil was responsible for curing rickets, as suggested by Mellanby, in addition to xerophthalmia. To test this, McCollum very cleverly destroyed the vitamin A activity by both bubbling oxygen through cod liver oil and heating it. Although this procedure destroyed the ability to cure xerophthalmia, it did not destroy its ability to cure rickets, McCollum correctly reasoned that cod liver oil contains another micronutrient, which he termed vitamin D (McCollum, Simmonds, Becker, & Shipley, 1922; Fig. 1).

While Sir Edward Mellanby (1919) was carrying out his monumental experiments in rachitic dogs, a physician, Hulshinsky (1919) in Vienna, began to realize that sunlight might also be a curative factor for the disease rickets. Thus, he demonstrated in children that exposure to sunlight or artificial ultraviolet light could cure rickets (Hulshinsky, 1919). A similar discovery was made by Chick and her associates (Chick, Palzell, & Hume, 1923). Thus, the strange circumstance then appeared, namely that sunlight equals cod liver oil in its ability to cure rickets. This dichotomy stimulated Steenbock to irradiate the animals, their food, as well as their cages. The irradiation of their food or the animals themselves cured or prevented rickets (Steenbock & Black, 1924; Fig. 1). At about the same time, a similar finding was reported by Hess and Weinstock (1924). By continuing to follow where the antirachitic activity could be found, Steenbock and Black (1925) demonstrated that following irradiation, the antirachitic activity appeared in the nonsaponifiable fraction of fat components of the diet. This discovery ended the dilemma that cod liver oil equals ultraviolet light in healing rickets in that

Figure 1 Scientists who played a major role in the discovery of vitamin D and its availability for the treatment of rickets and osteomalacia. Left: E.V. McCollum, middle: H. Steenbock, and Right: F. Windaus.

sunlight induced a change in the nonsaponifiable fraction of the lipid component of diet to become the antirachitic factor (Steenbock & Black, 1925). The monumental discovery of Steenbock essentially eliminated rickets as a major medical problem worldwide. Although nutritional rickets was eliminated, vitamin D-resistant rickets and other vitamin D-resistant diseases remained, albeit in very small numbers. Because sterol enriched fats could be irradiated to produce vitamin D, a major effort to identify vitamin D was launched by two groups. The first endeavor was that of Windaus and his colleagues who identified what he termed "vitamin D_1" but was unable to determine its exact structure (Windaus & Linsert, 1928). Unfortunately, vitamin D_1 proved to be an adduct of tachysterol, an irradiation side product of vitamin D_2 and vitamin D_2 itself; thus, vitamin D_1 was actually an error in identification. Clear identification of vitamin D_2 or ergocalciferol was by Askew and his colleagues who determined the structure from the irradiation of plant sterols (Askew, Bourdillon, Bruce, Jenkins, & Webster, 1931). The Windaus group also confirmed the structure of vitamin D_2 by their own identification (Windaus, Linsert, Luttringhaus, & Weidlich, 1932). The identification of vitamin D_3 which is the form of vitamin D produced in skin by sunlight was carried out by Windaus and Bock who isolated the 7-dehydrocholesterol from skin and provided the evidence that vitamin D_3 is produced by the impact of ultraviolet light on 7-dehydrocholesterol (Windaus & Bock, 1937; Fig. 1). However, vitamin D_3 was never isolated from the irradiation of skin itself until Esvelt et al. chemically identified it as the product produced in skin by ultraviolet irradiation (Esvelt, Schnoes, &

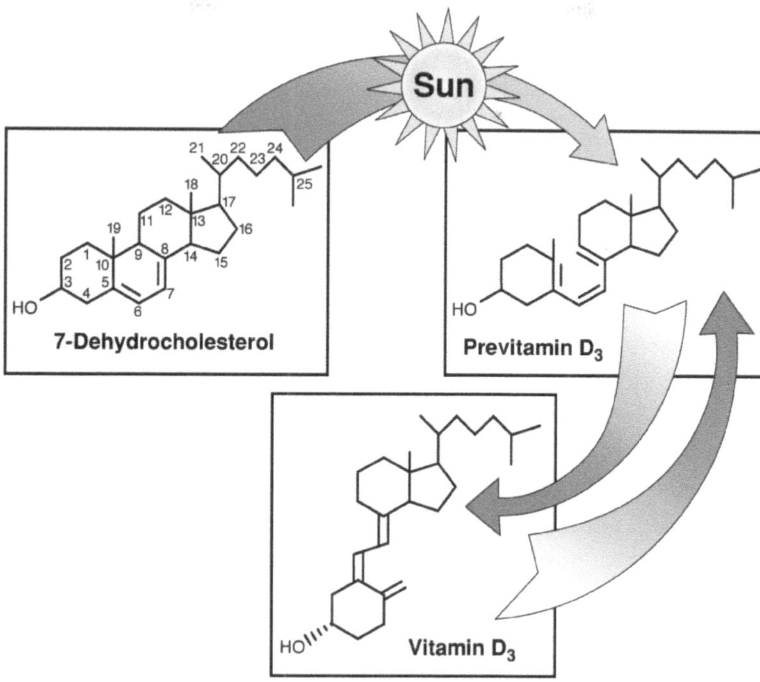

Figure 2 Conversion of 7-dehydrocholesterol to previtamin D_3 by ultraviolet light (280–310 nm) from sunlight or artificial light. Previtamin D_3 isomerizes to vitamin D_3 and is always in a equilibrium mixture of 5.94% previtamin D_3 and 94.06% vitamin D3 at body temperature. (See the color plate.)

DeLuca, 1978). Holick et al. provided evidence that the first intermediate in the production of vitamin D_3 in skin was previtamin D_3 (Holick et al., 1980; Fig. 2). As soon as a relatively inexpensive vitamin D_3 could be produced by the irradiation of animal sterols and also vitamin D_2 from the irradiation of plant sterols, vitamin D_3 and vitamin D_2 replaced irradiation as a means of fortifying foods for the prevention of rickets. This eliminated any changes in taste of the foods brought about by irradiation.

2. THE DISCOVERY OF THE PHYSIOLOGICAL FUNCTIONS OF VITAMIN D

Although the mineralization of the skeleton was the endpoint of vitamin D administration, exactly how this occurred remained relatively unknown. The first insight into the mechanism was by Orr, Holt, Wilkins, and Boone (1923) and later by Kletzien, Templin, Steenbock and Thomas (1932). Both groups determined that vitamin D plays an important role in the utilization of calcium from diet. However, it was the group of Nicolaysen (1937) who clearly demonstrated that vitamin D stimulates the absorption of calcium and independently of phosphorus from diet. Nicolaysen and his group added the additional observation that animals on a low calcium diet showed higher rates of calcium absorption than those animals maintained on a high calcium diet. Nicolaysen then postulated the existence of an endogenous factor that would inform the intestine of the need for calcium (Nicolaysen, 1943; Nicolaysen, Eeg-Larsen, & Malm, 1953). As it turned out, Nicolaysen's endogenous factor proved to be the vitamin D endocrine system (described in a later section). The explanation of the Nicolaysen endogenous factor came from the work of Boyle (Boyle, Gray, & DeLuca, 1971a, 1971b; Boyle, Gray, Omdahl, & DeLuca, 1972) and was finally proved by Ribovich and DeLuca (1975, 1976, 1978), who showed that animals maintained on a constant level of the active form of vitamin D did not show any adaptation of calcium absorption to low calcium diets, whereas animals maintained on vitamin D_3 itself clearly did.

During the 1930–1960 era, a major advance was the finding that vitamin D increases the absorption of calcium from the diet (Nicolaysen, 1937; Nicolaysen et al., 1953). Since the endpoint of vitamin D administration is the calcification of bone that actually cured rickets, for many years, it was believed that vitamin D might play a role directly in the skeleton by stimulating the calcification process. Early work by Shipley, Kramer, and Howland (1925) and Shipley, Kramer, and Howland (1926) showed that slices of bone taken from rachitic animals would calcify normally when incubated in serum from animals given vitamin D, whereas they were unable to mineralize then they were incubated in serum from vitamin D-deficient animals. The more recent work by Lamm and Neuman (1958), revealing that to calcify skeleton, serum must be supersaturated with calcium and phosphorus, suggested that the defect in mineralization in rickets was an inadequate supply of calcium and phosphorus to the calcification sites. The final conclusive evidence was carried out by Underwood and DeLuca, who were able to show by infusion of calcium and phosphorus into the bloodstream of vitamin D-deficient animals to maintain normal levels of calcium and phosphorus therein, resulted in fully mineralized bone illustrating that there was no defect in the mineralization process in vitamin D deficiency and that vitamin D produced the calcification of bone by elevating plasma calcium and phosphorus (Underwood & DeLuca, 1984). This then redirected the focus on how vitamin D elevates plasma calcium and phosphorus to facilitate the mineralization of the skeleton. Clearly, improvement of intestinal absorption of calcium demonstrated by Nicolaysen was one important factor. However, mineralization of new skeleton occurs even in animals that are placed on a diet that is virtually devoid of calcium (Steenbock & Herting, 1955). An important discovery was made by Carlsson (1952) and Bauer, Carlsson, and Lindquist (1955), who realized and demonstrated that a major function of vitamin D is to cause the mobilization of calcium from bone when required. Thus, animals given the low calcium diet will resorb calcium from their formed skeleton and cause it deposition in newly forming bone. The mobilization of calcium from bone is very likely an osteoclastic-mediated process, which involves the RANK ligand secretion induced by both parathyroid hormone and 1,25-$(OH)_2D_3$. RANK ligand stimulates osteoclasts to become activated and resorb bone (Pike, Lee, & Meyer, 2014; Takahashi, Udagawa, & Suda, 2014). Furthermore, RANK ligand stimulates the formation of giant osteoclasts from precursors.

It is well known from the work of Collip (1925) that parathyroid hormone, a peptide hormone that is produced by the parathyroid glands, stimulates the mobilization of calcium from bone in order to prevent hypocalcemic tetany. However, the work of Harrison and Harrison (1964) and Rasmussen, DeLuca, Arnaud, Hawker, and von Stedingk (1963) showed that in vitamin D-deficient animals, the parathyroid hormone is unable to mobilize calcium from bone. Thus, vitamin D-deficient animals are immune to even large amounts of parathyroid hormone in terms of mobilization of skeleton to increase serum calcium. Similarly, vitamin D, unless given in very large amounts, is unable to elevate serum calcium in the absence of parathyroid hormone (Arnaud & Kolb, 1983; Rasmussen et al., 1963). These two calcium homeostatic agents require each other to mobilize calcium from bone (Holick, Garabedian, & DeLuca, 1972; Holick, Schnoes, et al., 1972).

Another important site of action of vitamin D and parathyroid hormone is the distal renal tubule. The first clear demonstration that vitamin D plays an important role in the renal conservation of calcium came from the work of Yamamoto et al. (1984). Thus, the proximal tubule reabsorbs 99% of filtered calcium even in the absence of vitamin D. The remaining 1% of the filtered urinary calcium comes under control of both vitamin D and parathyroid hormone in the distal tubule. Thus, when calcium is required and both the active form of vitamin D and the parathyroid hormone are elevated, these agents work in concert to increase renal reabsorption of calcium as well as the mobilization of calcium from bone (Fig. 3).

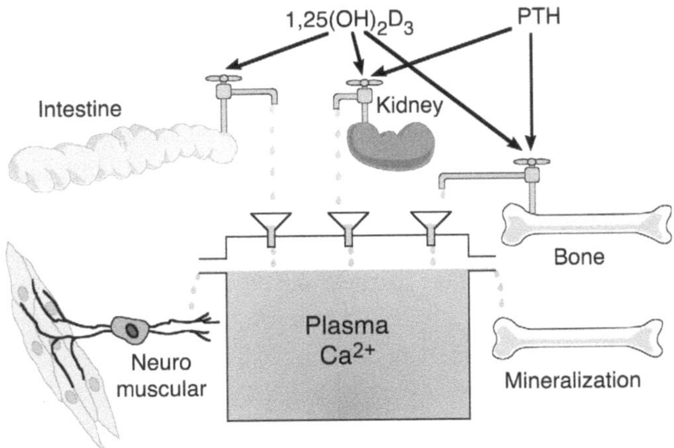

Figure 3 A diagrammatic representation of the mineralization of bone brought about by vitamin D after its conversion to the hormonal form, 1,25-$(OH)_2D_3$. Note that hypocalcemic tetany is also prevented by the action of vitamin D and parathyroid hormone (PTH). (See the color plate.)

3. THE DISCOVERY OF THE HORMONAL FORM OF VITAMIN D

The very key to understanding the vitamin D endocrine system was an understanding of vitamin D metabolism. The pioneer in vitamin D metabolism was Egan Kodicek of Cambridge, England. For 20 years, he investigated what happens to the vitamin D molecule. To carry out these studies, he prepared ^{14}C vitamin D_2. Vitamin D is required in such small amounts that its metabolism could not be studied with ^{14}C-labeled vitamin D at physiologic doses. Nevertheless, Kodicek studied the distribution of ^{14}C vitamin D_2, given in supra physiologic doses, and after 10 years of investigation concluded that vitamin D was active directly without further metabolism and that all metabolism was a degradation pathway (Kodicek, 1960). All metabolites investigated by Kodicek and colleagues were found biologically inactive. This conclusion was also reached by Haussler and Norman (1967) in which they found that vitamin D_3 itself was responsible for intestinal calcium transport. The chemical synthesis of vitamin D_3 labeled with ^{3}H in the 1- and 2-position by Neville and DeLuca (1966) resulted in a high-specific activity form of vitamin D that could be truly studied at physiologic doses. Providing 10 units of ^{3}H vitamin D_3 to vitamin D-deficient animals, Lund and DeLuca (1966) clearly found that it is rapidly converted to polar metabolites, which upon their isolation proved to be more biologically active than vitamin D itself. Furthermore, the metabolites caused a more rapid response of intestinal calcium transport than did vitamin D_3 itself (Morii, Lund, Neville, & DeLuca, 1967). Blunt, DeLuca, and Schnoes (1968), using improved chromatography and several steps of chromatography, isolated the first biologically active metabolite from the plasma of pigs given large doses of vitamin D_3. This metabolite was unequivocally identified as 25-hydroxyvitamin D_3 (25-OH-D_3) (Blunt et al., 1968), which seemed to be the metabolically active form of vitamin D. From this early work, it became clear that vitamin D_3 is in fact converted to more biologically active forms *in vivo*.

4. THE ISOLATION OF THE FINAL ACTIVE FORM OF VITAMIN D

To confirm the identified plasma metabolite as 25-OH-D_3, chemical synthesis of this metabolite was carried out by Blunt and DeLuca (1969). This synthesis allowed for the introduction of tritium to the side chain of the metabolite. When radiolabeled 25-OH-D_3 was administered to vitamin D-deficient animals, it also disappeared and the label appeared in more polar metabolites (Cousins, DeLuca, & Gray, 1970). The DeLuca group isolated three metabolites from the plasma and identified two of them but both had less biological activity than 25-OH-D_3 (Suda, DeLuca, & Hallick, 1971; Suda, DeLuca, Schnoes, Ponchon, et al., 1970; Suda, DeLuca, Schnoes, Tanaka, & Holick, 1970). To determine the final active form of vitamin D, the DeLuca group turned its attention to a target organ. From 1600 vitamin D-deficient chickens given radiolabeled vitamin D_3, 2 μg of the metabolite found in intestinal mucosa was isolated in pure form and the structure clearly identified as 1α,25-dihydroxyvitamin D_3 (1,25-$(OH)_2D_3$) (Holick et al., 1972; Holick, Schnoes, & DeLuca, 1971; Holick, Schnoes, DeLuca, Suda, & Cousins, 1971). A year later, chemical synthesis confirmed the structure (Semmler, Holick, Schnoes, & DeLuca, 1972) and *in vivo* experiments, described below, proved it to be the metabolically active form of vitamin D_3 in intestine and bone (Garabedian, Tanaka, Holick, & DeLuca, 1974). The activation of vitamin D_3 to its final active form, 1,25-$(OH)_2D_3$, is shown in Fig. 4.

In the meantime, the Kodicek group had reconsidered their earlier conclusions and began working on a metabolite called peak P (Lawson, Wilson, & Kodicek, 1969). Myrtle, Haussler, and Norman (1970) had also realized

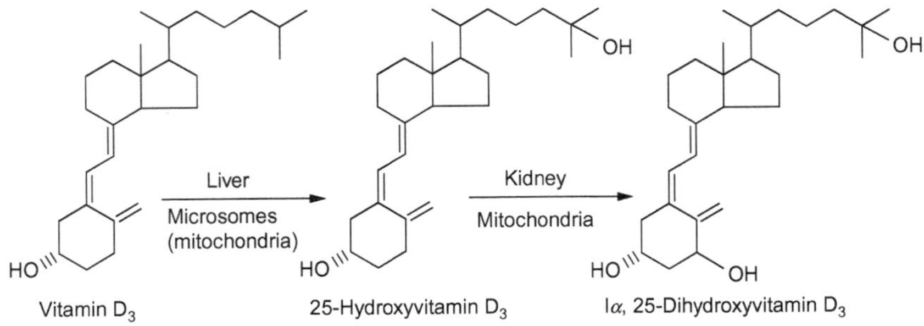

Figure 4 The activation of vitamin D_3 to its final active form, 1,25-dihydroxyvitamin D_3 (1,25-$(OH)_2D_3$).

that vitamin D was metabolized to metabolically active forms and began to work on a metabolite called peak 4B. However, the Wisconsin group provided unambiguous proof that the active form of vitamin D was, in fact, 1α,25-$(OH)_2D_3$ (Holick, Schnoes, & DeLuca, 1971; Holick, Schnoes, DeLuca, Suda, et al., 1971). The Wisconsin group then chemically synthesized both 1α,25-$(OH)_2D_3$ and 1β,25-$(OH)_2D_3$ and proved unequivocally that the natural metabolite was actually 1α,25-$(OH)_2D_3$ (Semmler et al., 1972). In preparation for the synthesis of 1α,25-$(OH)_2D_3$, the Wisconsin group also synthesized 1α-OH-D_3, an important analog that became a pharmaceutical product (Holick, Semmler, Schnoes, & DeLuca, 1973).

One of the key discoveries that proved to be of great importance in the elucidation of the vitamin D endocrine system was that of Fraser and Kodicek (1970), who demonstrated that the kidney, when removed from rats, prevented the production of their peak P metabolite which the Wisconsin group proved was 1α,25-$(OH)_2D_3$ The kidney is the site where the 1α-hydroxyl was added to the vitamin D molecule. This then set the stage for demonstrating that 1,25-$(OH)_2D_3$ and not any of its precursors is the metabolically active form of vitamin D. By using anephric rats, it could be easily shown that 25-OH-D_3 when given a physiologic dose could not stimulate intestinal calcium transport, while 1,25-$(OH)_2D_3$ was clearly able to induce intestinal calcium transport (Holick et al., 1972). Similar experiments (Garabedian et al., 1974) were carried out in regard to the mobilization of calcium from bone. These results proved that 1,25-$(OH)_2D_3$ or perhaps a further metabolite of 1,25-$(OH)_2D_3$ is the active form and not any of its precursors. Ultimately, evidence by Fraser et al. (1973) that this mandatory pathway occurs in humans came with an understanding of vitamin D-dependency rickets type I. Although this disease could be healed by large amounts of vitamin D or 25-OH-D, it could also be healed with very small and physiological amounts of 1,25-$(OH)_2D_3$ (Fraser et al., 1973). Undoubtedly, 25-OH-D at high concentrations acts as an analog of 1,25-$(OH)_2D_3$ (Eisman & DeLuca, 1977). These patients have a defect in the 1α-hydroxylase, which was clearly demonstrated later by an accumulation of patients suffering from this disease (Glorieux, Edouard, & St-Arnaud, 2011).

5. DISCOVERY OF THE VITAMIN D ENDOCRINE SYSTEM (FIG. 5)

The work of Boyle et al. provided key evidence that led to an understanding of the vitamin D endocrine system (Boyle et al., 1971a, 1971b, 1972). By following the conversion of 25-OH-D_3 to 1,25-$(OH)_2D_3$ in animals on a low calcium diet versus those on a high calcium diet, Boyle and colleagues found a large production of 1,25-$(OH)_2D_3$ when a low calcium diet was fed and only small amounts produced when a high calcium diet was fed. An alternate metabolite, 24,25-$(OH)_2D_3$, was produced with a high calcium diet and very little of it was produced with a low calcium diet. This discovery led to understanding the Nicolaysen endogenous factor as described earlier (Nicolaysen et al., 1953). As it turns out, the 24-hydroxylation reaction is the first signal of degradation of the vitamin molecule. When the active form is not needed, it is degraded by 24-hydroxylation leading ultimately to calcitroic acid (Beckman et al., 1996; Esvelt, Schnoes, & DeLuca, 1979). Although there have been many efforts to

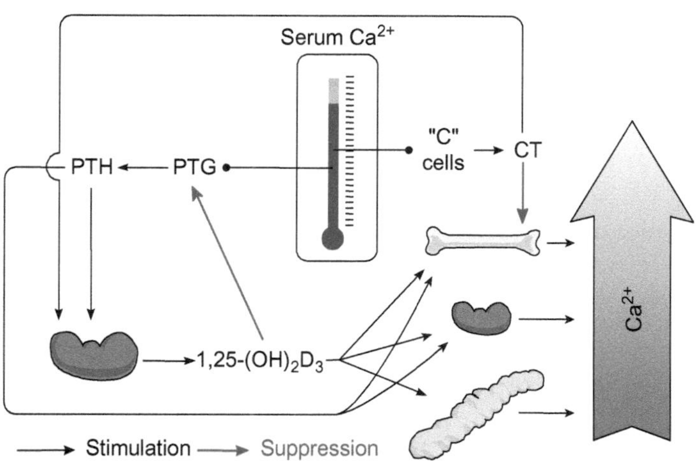

Figure 5 A diagrammatic representation of the regulation of serum calcium concentration by the actions of 1,25-$(OH)_2D_3$, PTH, and calcitonin (CT). (See the color plate.)

ascribe hormonal activity to 24,25-$(OH)_2D_3$, those have ultimately failed, and it is now quite well accepted that 24-hydroxylation is the first step in degradation not only of 25-OH-D_3 but also of the hormone itself (Beckman et al., 1996).

Another key was the finding that the parathyroid hormone is the signal that is released following even slight hypocalcemia that stimulates production of the enzyme, 1α-hydroxylase or CYP24B1 (Garabedian, Holick, DeLuca, & Boyle, 1972). The key role of the parathyroid hormone in this signaling pathway came at the hands of Garabedian et al. (1972) who demonstrated that the parathyroid hormone markedly stimulates 1α-hydroxylation of 25-OH-D_3, a finding confirmed by Fraser and Kodicek (1973). The parathyroids sense hypocalcemia and in turn secrete the parathyroid hormone that cause the 1α-hydroxylase in the kidneys to produce the major calcium mobilizing hormone, 1,25-$(OH)_2D_3$ (DeLuca, 1974). This hormone from vitamin D functions in kidney, bone, and intestine to elevate plasma calcium to levels that then suppress PTH secretion resulting in a shut down of 1α-hydroxylation. The key elements of the vitamin D endocrine system were clearly demonstrated by 1974 and were provided in a presentation to the American Federation of the Biological Scientists (DeLuca, 1974).

Of considerable importance is that rickets in rats is produced not by a low calcium diet but by a low phosphorus diet (Tanaka, Frank, & DeLuca, 1973a, 1973b). These animals are hypophosphatemic and fail to calcify their skeleton because the calcium phosphorus product is too low despite serum calcium being in the normal range. The stimulation of intestinal phosphate transport is a key element in healing rickets in this case (Chen, Castillo, Korycka-Dahl, & DeLuca, 1974). 1,25-Dihydroxyvitamin D_3 has an important function in stimulating intestinal phosphate transport as well. Hypophosphatemia also, in an unknown fashion, stimulates 1α-hydroxylation of 1,25-$(OH)_2D_3$ (Tanaka & DeLuca, 1973). Likely, this is through the FGF-23 pathway of signaling but so far as not yet been elucidated. Nevertheless, the vitamin D system plays an important role in phosphate homeostasis as well.

6. OTHER METABOLISM OF VITAMIN D

Some 33 metabolites of vitamin D have been identified (DeLuca & Schnoes, 1983). Almost all of them are either devoid of biological activity or intermediates in the pathway of degradation. There are analog forms of vitamin D_2 and of vitamin D_3 thus multiplying the number of active vitamin D compounds. Originally, an alternate metabolite in the plasma of animals given vitamin D was identified 21,25-$(OH)_2D_3$ (Suda, DeLuca, Schnoes, Ponchon, et al., 1970; Suda, DeLuca, Schnoes, Tanaka, et al., 1970), which proved to be an error that was later corrected to 24,25-$(OH)_2D_3$ (Holick et al., 1972). A minor metabolite is also 25,26-$(OH)_2D_3$ and exactly how this is produced is not known at this time. It is, nevertheless, quite weak in biological activity and likely represents a minor pathway of degradation.

The major effort in metabolites other than the 1,25-$(OH)_2D_3$ pathway is 24,25-$(OH)_2D_3$ which has been identified since 1972 (Holick et al., 1972). In an extensive effort, Boyle, Omdahl, Gray, and DeLuca (1973) attempted to decipher a possible functional role of 24,25-$(OH)_2D_3$ and came to the conclusion that it is likely enroute to degradation. Reports that 24,25-$(OH)_2D_3$ is a hatching hormone in birds and has a function in bone still persisted (Henry & Norman, 1978). However, through the use of 24,24-difluoro-25-OH-D_3 and 24,24-difluoro-1,25-$(OH)_2D_3$, the role of 24-hydroxylation could be examined in detail. Rats maintained on 24,24-difluoro-25-OH-D_3 as its sole source of vitamin D for two generations revealed no biological defects and were able to reproduce and grow normally and perform normal vitamin D functions (Brommage & DeLuca, 1985). These results suggest that 24-hydroxylation is not a functional pathway. 24,24-Dihydroxyvitamin D_3 and 1,24,25-trihydroxyvitamin D_3 are degraded to the corresponding C-23 acid. Thus, 24-hydroxylation appears to be the initial signal for degradation of vitamin D to its 23-carboxylic acid excretion product (Esvelt et al., 1979).

7. DISCOVERY OF THE VITAMIN D RECEPTOR

The first clear demonstration of the existence of the vitamin D receptor (VDR) was provided by Brumbaugh and Haussler (1973), followed closely by Kream, Reynolds, Knutson, Eisman, and DeLuca (1976). The receptor was found highly specific for 1,25-$(OH)_2D_3$ (Eisman & DeLuca, 1977). Although several attempts to isolate the VDR were made by the Haussler group and the DeLuca group, a homogenous or pure protein preparation was not accomplished. However, using the cloning techniques, Burmester, Maeda, and DeLuca (1988), Burmester, Wiese, Maeda, and DeLuca (1988), and Baker et al. (1988) determined the primary structure of rat and human VDR. The genes encoding the VDR were subsequently

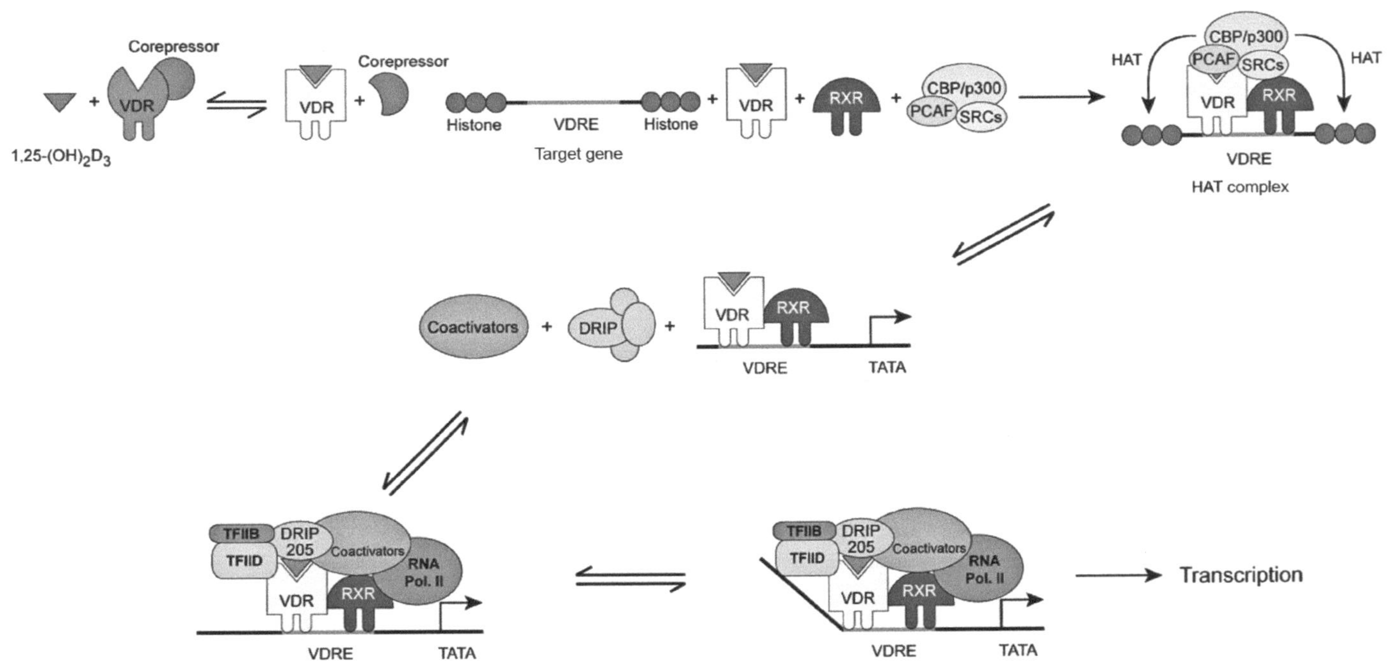

Figure 6 A current model of gene expression regulated by the vitamin D hormone interacting with the vitamin D receptor (VDR) that requires a participation of the retinoid X receptor (RXR). These protein factors bind to the vitamin D responsive elements throughout the target gene including those in the promoter region. These elements are imperfect hexamers separated by three unspecified nucleotides. The factors involved are many including histone acetylase (HAT), Drip protein, CBP/p300, PCAF, and SRCs. (See the color plate.)

isolated and their structure elucidated including the promoter (Jehan & DeLuca, 1997, 2000; Miyamoto et al., 1997). The VDR does not contain a TATA box and is SP-1 driven (Jehan & DeLuca, 2000), while the human gene appears to have alternate promoters (Miyamoto et al., 1997). Brooks et al. (1978) first described vitamin D-dependency rickets type II, which is the result of defects in the receptor gene (Hughes et al., 1988; Wiese et al., 1993). These discoveries provided proof that the functions of vitamin D in humans are via the nuclear VDR. The preparation of VDR KO animals (Li et al., 1997; Yoshizawa et al., 1997) allowed extensive work defining the essential nature of VDR for all of the functions of vitamin D.

The number of genes responsive to 1,25-$(OH)_2D_3$ bound to the receptor is quite large with an estimate of more than 50 that are upregulated and 50 more that are downregulated (Kutuzova & DeLuca, 2004). Responsive elements for VDR were discovered and are imperfect hexamer repeats interrupted by three nonspecified bases (Umesono, Murakami, Thompson, & Evans, 1991). Although originally found in the promoters (Demay, Gerardi, DeLuca, & Kronenberg, 1990; Noda et al., 1990), recent reports have shown the responsive element within the genes distant from the promoter (Pike et al., 2014). Exactly how the expression or repression of genes is accomplished by 1,25-$(OH)_2D_3$ through its receptor remains largely unknown, although working models have been proposed, one of which is provided in Fig. 6.

REFERENCES

Arnaud, C. D., & Kolb, F. O. (1983). The calciotropic hormones and metabolic bone disease. In F. S. Greenspan & P. H. Forsham (Eds.), *Basic and clinical endocrinology* (pp. 187–257). Los Altos, CA: Lange Medical Publications.

Askew, F. A., Bourdillon, R. B., Bruce, H. M., Jenkins, R. G. C., & Webster, T. A. (1931). The distillation of vitamin D. *Proceedings of the Royal Society B, 107*, 76–90.

Baker, A. R., McDonnell, D. P., Hughes, M., Crisp, T. M., Mangelsdorf, D. J., Haussler, M. R., et al. (1988). Cloning and expression of full-length cDNA encoding human vitamin D receptor. *Proceedings of the National Academy of Sciences of the United States of America, 85*, 3294–3298.

Bauer, G. C. H., Carlsson, A., & Lindquist, B. (1955). Evaluation of accretion, resorption and exchange reactions in the skeleton. *Kungl Fysiografen Sallskap I, Lund. Forhandlingar, 25*, 3–18.

Beckman, M. J., Tadikona, P., Werner, E., Prahl, J., Yamada, S., & DeLuca, H. F. (1996). The human 25-hydroxyvitamin D_3-24-hydroxylase, a multicatalytic enzyme. *Biochemistry, 35*, 8465–8472.

Blunt, J. W., & DeLuca, H. F. (1969). The synthesis of 25-hydroxycholecalciferol. A biologically active metabolite of vitamin D_3. *Biochemistry, 8*, 671–675.

Blunt, J. W., DeLuca, H. F., & Schnoes, H. K. (1968). 25-Hydroxycholecalciferol. A biologically active metabolite of vitamin D_3. *Biochemistry*, 7, 3317–3322.

Boyle, I. T., Gray, R. W., & DeLuca, H. F. (1971a). Regulation by calcium of in vivo synthesis of 1,25-dihydroxycholecalciferol and 21,25-dihydroxycholecalciferol. *Proceedings of the National Academy of Sciences of the United States of America, 68*, 2131–2134.

Boyle, I. T., Gray, R. W., & DeLuca, H. F. (1971b). Regulation by calcium of in vivo synthesis of 1,25-dihydroxycholecalciferol and 21,25-dihydroxycholecalciferol. *Proceedings of the National Academy of Sciences of the United States of America, 68*, 2131–2134.

Boyle, I. T., Gray, R. W., Omdahl, J. L., & DeLuca. (1972). Calcium control of the in vivo biosynthesis of 1,25-dihydroxyvitamin D3: Nicolaysen's endogenous factor. In S. Taylor (Ed.), *Endocrinology 1971.Proceedings of the third international symposium* (pp. 468–476) London: William Heineman Medical Books Ltd.

Boyle, I. T., Omdahl, J. L., Gray, R. W., & DeLuca, H. F. (1973). The biological activity and metabolism of 24,25-dihydroxyvitamin D_3. *Journal of Biological Chemistry, 248*, 4174–4180.

Brommage, R., & DeLuca, H. F. (1985). Evidence that 1,25-dihydroxyvitamin D_3 is the physiologically active metabolite of vitamin D_3. *Endocrine Reviews, 6*, 491–511.

Brooks, M. H., Bell, N. H., Love, L., Stern, P. H., Orfei, E., Queener, S. F., et al. (1978). Vitamin D-dependent rickets type II. Resistance of target organs to 1,25-dihydroxyvitamin D. *New England Journal of Medicine, 298*, 997–999.

Brumbaugh, P. F., & Haussler, M. R. (1973). 1α,25-Dihydroxyvitamin D_3 receptor: Competitive binding of vitamin D analogs. *Life Sciences, 13*, 1737–1746.

Burmester, J. K., Maeda, N., & DeLuca, H. F. (1988). Isolation and expression of rat 1,25-dihydroxyvitamin D_3 receptor cDNA. *Proceedings of the National Academy of Sciences of the United States of America, 85*, 1005–1009.

Burmester, J. K., Wiese, R. J., Maeda, N., & DeLuca, H. F. (1988). Structure and regulation of the rat 1,25-dihydroxyvitamin D_3 receptor. *Proceedings of the National Academy of Sciences of the United States of America, 85*, 9499–9502.

Carlsson, A. (1952). Tracer experiment on the effect of vitamin D on the skeletal metabolism of calcium and phosphorus. *Acta Physiologica Scandinavica, 26*, 212–220.

Chen, T. C., Castillo, L., Korycka-Dahl, M., & DeLuca, H. F. (1974). Role of vitamin D metabolites in phosphate transport of rat intestine. *Journal of Nutrition, 104*, 1056–1060.

Chick, H., Palzell, E. J., & Hume, E. M. (1923). *Studies of rickets in Vienna 1919–1922: Medical research council, special report no. 77.*

Collip, J. B. (1925). The extraction of a parathyroid hormone which will prevent or control parathyroid tetany and which regulates the level of blood calcium. *Journal of Biological Chemistry, 63*, 395–438.

Cousins, R. J., DeLuca, H. F., & Gray, R. W. (1970). Metabolism of 25-hydroxycholecalciferol in target and nontarget tissues. *Biochemistry, 9*, 3649–3652.

DeLuca, H. F. (1974). Vitamin D: The vitamin and the hormone. *Federation Proceedings, 33*, 2211–2219.

DeLuca, H. F., & Schnoes, H. K. (1983). Vitamin D: Recent advances. *Annual Review of Biochemistry, 52*, 411–439.

Demay, M. B., Gerardi, J. M., DeLuca, H. F., & Kronenberg. (1990). DNA sequences in the rat osteocalcin gene that bind the 1,25-dihydroxyvitamin D_3 receptor and confer responsiveness to 1,25-dihydroxyvitamin D_3. *Proceedings of the National Academy of Sciences of the United States of America, 87*, 369–373.

Drummond, J. C. (1920). The nomenclature of the so-called accessory food factors (vitamins). *Biochemical Journal, 14*, 660.

Eijkman, C. (1897). Ein beri-beri anliche der huhner. *Virchow's Archiv, 148*, 523–527.

Eisman, J. A., & DeLuca, H. F. (1977). Intestinal 1,25-dihydroxyvitamin D_3 binding protein: Specificity of binding. *Steroids, 30*, 245–257.

Esvelt, R. P., Schnoes, H. K., & DeLuca, H. F. (1978). Vitamin D_3 from rat skins irradiated *in vitro* with ultraviolet light. *Archives of Biochemistry and Biophysics, 188*, 282–286.

Esvelt, R. P., Schnoes, H. K., & DeLuca, H. F. (1979). Isolation and characterization of 1α-hydroxy-23-carboxytetranorvitamin D: A major metabolite of 1,25-dihydroxyvitamin D_3. *Biochemistry, 18*, 3977–3983.
Fraser, D. R., & Kodicek, E. (1970). Unique biosynthesis by kidney of a biologically active vitamin D metabolite. *Nature, 228*, 764–766.
Fraser, D. R., & Kodicek, E. (1973). Regulation of 25-hydroxycholecalciferol-1-hydroxylase activity in kidney by parathyroid hormone. *Nature New Biology, 241*, 163–166.
Fraser, D., Kooh, S. W., Kind, H. P., Holick, M. F., Tanaka, Y., & DeLuca, H. F. (1973). Pathogenesis of hereditary vitamin D dependent rickets. An inborn error of vitamin D metabolism involving defective conversion of 25-hydroxyvitamin D to 1α,25-dihydroxyvitamin D. *New England Journal of Medicine, 289*, 817–822.
Funk, C. (1911). The chemical nature of the substance that cures polyneuritis in birds produced by a diet of polished rice. *Journal of Physiology (London), 43*, 395–402.
Garabedian, M., Holick, M. F., DeLuca, H. F., & Boyle, I. T. (1972). Control of 25-hydroxycholecalciferol metabolism by the parathyroid glands. *Proceedings of the National Academy of Sciences of the United States of America, 69*, 1673–1676.
Garabedian, M., Tanaka, Y., Holick, M. F., & DeLuca, H. F. (1974). Response of intestinal calcium transport and bone calcium mobilization to 1,25-dihydroxyvitamin D_3 in thyroparathyroidectomized rats. *Endocrinology, 94*, 1022–1027.
Glorieux, F. H., Edouard, T., & St-Arnaud, R. (2011). Pseudo-vitamin D deficiency. In D. Feldman, J. W. Pike, & J. S. Adams (Eds.), *Vitamin D* (3rd ed., pp. 1187–1196). San Diego: Academic Press. Chapter 64.
Grijns, G. (1901). Over polyneuritis gallinarum. *Geneeskundig Tijdschrift voor Nederlandsch-Indie, 41*, 3–41. and cited In: E.V.McCollum, & H.B. Glass (Eds.), A history of nutrition (p. 216). Boston: Houghton Mifflin.
Harrison, H. E., & Harrison, H. C. (1964). The interaction of vitamin D and parathyroid hormone on calcium phosphorus and magnesium homeostasis in the rat. *Metabolism, 13*, 952–958.
Haussler, M. R., & Norman, A. W. (1967). The subcellular distribution of physiological doses of vitamin D_3. *Archives of Biochemistry and Biophysics, 118*, 145–153.
Henry, H. L., & Norman, A. W. (1978). Vitamin D: Two dihydroxylated metabolites are required for normal chicken egg hatchability. *Science, 201*, 835–837.
Hess, A. (1929). The history of rickets. In *Rickets, including osteomalacia and tetany* (pp. 22–37). Philadelphia: Lea & Febiger.
Hess, A., & Weinstock, M. (1924). Antirachitic properties imparted to lettuce and to growing wheat by ultraviolet irradiation. *Proceedings of the Society of Experimental Biology & Medicine, 22*, 5–6.
Holick, M. F., Garabedian, M., & DeLuca, H. F. (1972). 1,25-Dihydroxycholecalciferol: Metabolite of vitamin D_3 active on bone in anephric rats. *Science, 176*, 1146–1147.
Holick, M. F., MacLaughlin, J. A., Clark, M. B., Holick, S. A., Potts, J. T., Jr., Anderson, R. R., et al. (1980). Photosynthesis of previtamin D_3 in human skin and the physiologic consequences. *Science, 210*, 203–205.
Holick, M. F., Schnoes, H. K., & DeLuca, H. F. (1971). Identification of 1,25-dihydroxycholecalciferol, a form of vitamin D_3 metabolically active in the intestine. *Proceedings of the National Academy of Sciences of the United States of America, 68*, 803–804.
Holick, M. F., Schnoes, H. K., DeLuca, H. F., Gray, R. W., Boyle, I. T., & Suda, T. (1972). Isolation and identification of 24,25-dihydroxycholecalciferol: A metabolite of vitamin D_3 made in the kidney. *Biochemistry, 11*, 4252–4255.
Holick, M. F., Schnoes, H. K., DeLuca, H. F., Suda, T., & Cousins, R. J. (1971). Isolation and identification of 1,25-dihydroxycholecalciferol. A metabolite of vitamin D active in intestine. *Biochemistry, 10*, 2799–2804.

Holick, M. F., Semmler, E. J., Schnoes, H. K., & DeLuca, H. F. (1973). 1α-Hydroxy derivative of vitamin D_3: A highly potent analog of 1α,25-dihydroxyvitamin D_3. *Science, 180*, 190–191.

Hopkins, G. (1912). Feeding experiments illustrating the importance of accessory food factors in normal dietaries. *Journal of Physiology, 44*, 425–460.

Hughes, M. R., Malloy, P. J., Kieback, D. G., Kesterson, R. A., Pike, J. W., Feldman, D., et al. (1988). Point mutations in the human vitamin D receptor gene associated with hypocalcemic rickets. *Science, 242*, 1702–1706.

Hulshinsky, K. (1919). Heilung von rachitis durch kunstlich hohen-sonne. *Deutsche Medizinische Wochenschrif, 45*, 712–713.

Jehan, F., & DeLuca, H. F. (1997). Cloning and characterization of the mouse vitamin D receptor promoter. *Proceedings of the National Academy of Sciences of the United States of America, 94*, 10138–10143.

Jehan, F., & DeLuca, H. F. (2000). The mouse vitamin D receptor is mainly expressed through an Sp1-driven promoter in vivo. *Archives of Biochemistry and Biophysics, 377*, 273–283.

Kletzien, S. W. F., Templin, V. M., Steenbock, H., & Thomas, B. H. (1932). Vitamin D and the conservation of calcium in the adult. *Journal of Biological Chemistry, 97*, 265–280.

Kodicek, K. (1960). The metabolism of vitamin D. In W. Umbreit H. Molitor (Eds.), *Proceedings of the fourth international congress of biochemistry: Vol. 11* (pp. 198–208). London: Pergamon Press.

Kream, B. E., Reynolds, R. D., Knutson, J. C., Eisman, J. A., & DeLuca, H. F. (1976). Intestinal cytosol binders of 1,25-dihydroxyvitamin D_3 and 25-hydroxyvitamin D_3. *Archives of Biochemistry and Biophysics, 176*, 779–787.

Kutuzova, G. D., & DeLuca, H. F. (2004). Gene expression profiles in rat intestine identify pathways for 1,25-dihydroxyvitamin D_3 stimulated calcium absorption and clarify its immunomodulatory properties. *Archives of Biochemistry and Biophysics, 432*, 152–166.

Lamm, M., & Neuman, W. F. (1958). On the role of vitamin D in calcification. *Archives of Pathology, 66*, 204–209.

Lawson, D. E. M., Wilson, P. W., & Kodicek. (1969). Metabolism of vitamin D. A new cholecalciferol metabolite, involving loss of hydrogen at C-1, in chicken intestinal nuclei. *Biochemical Journal, 115*, 269–277.

Li, Y. C., Pirro, A. E., Amling, M., Delling, D. G., Bronson, R. R., & Demay, M. B. (1997). Targeted ablation of the vitamin D receptor—An animal model of vitamin D-dependent rickets type II with alopecia. *Proceedings of the National Academy of Sciences of the United States of America, 94*, 9831–9835.

Lund, J., & DeLuca, H. F. (1966). Biologically active metabolite of vitamin D_3 from bone, liver, and blood serum. *Journal of Lipid Research, 7*, 739–744.

Magendie, F. (1816). Annales de chimie et de physique. In H. B. Glass (Ed.), *Vol. 3. A history of nutrition by McCollum EV (1957)* (pp. 86–87). Cambridge: The Riverside Press.

McCollum, E. V., & Davis, M. (1913). The necessity of certain lipids in the diet during growth. *Journal of Biological Chemistry, 25*, 167–175.

McCollum, E. V., Simmonds, N., Becker, J. E., & Shipley, P. G. (1922). An experimental demonstration of the existence of a vitamin which promotes calcium deposition. *Journal of Biological Chemistry, 53*, 293–298.

McCollum, E. V., Simmonds, N., & Pitz, W. (1916). The relation of the unidentified dietary factors, the fat-soluble A and water-soluble B of the diet to the growth promoting properties of milk. *Journal of Biological Chemistry, 27*, 33–38.

Mellanby, E. (1919). An experimental investigation on rickets. *Lancet, 1*, 407–412.

Miyamoto, K., Kesterson, R. A., Yamamoto, H., Nishiwaki, E., Tatsumi, S., Taketani, Y., et al. (1997). Structural organization of the human vitamin D receptor chromosomal gene and its promoter. *Molecular Endocrinology, 11*, 1165–1179.

Morii, H., Lund, J., Neville, P. F., & DeLuca, H. F. (1967). Biological activity of a vitamin D metabolite. *Archives of Biochemistry and Biophysics, 120*, 508–512.

Myrtle, J. F., Haussler, M. R., & Norman, A. W. (1970). Evidence for the biologically active form of cholecalciferol in the intestine. *Journal of Biological Chemistry, 245*, 1190–1196.

Neville, P. F., & DeLuca, H. F. (1966). The synthesis of [1,2-^{3}H] vitamin D_3 and the tissue localization of a 0.25 μg (10 IU) dose per rat. *Biochemistry, 5*, 2201–2207.

Nicolaysen, R. (1937). Studies upon the mode of action of vitamin D. III. The influence of vitamin D on the absorption of calcium and phosphorus in the rat. *Biochemical Journal, 31*, 122–129.

Nicolaysen, R. (1943). The absorption of calcium as a function of body saturation with calcium. *Acta Physiologica Scandinavica, 6*, 201–209.

Nicolaysen, R., Eeg-Larsen, N., & Malm, O. J. (1953). Physiology of calcium metabolism. *Physiologic Reviews, 33*, 424–444.

Noda, M., Vogel, R. L., Craig, A. M., Prahl, J., DeLuca, H. F., & Denhardt, D. T. (1990). Identification of a DNA sequence responsible for binding of the 1,25-dihydroxyvitamin D_3 receptor and 1,25-dihydroxyvitamin D_3 enhancement of mouse secreted phosphoprotein 1 (Spp1, osteopontin) gene expression. *Proceedings of the National Academy of Sciences of the United States of America, 87*, 9995–9999.

Orr, W. J., Holt, L. E., Jr., Wilkins, L., & Boone, F. H. (1923). The calcium and phosphorus metabolism in rickets, with special reference to ultraviolet rat therapy. *American Journal of Diseases of Children, 26*, 362–372.

Osborne, T. B., & Mendel, L. B. (1917). The role of vitamins in the diet. *Journal of Biological Chemistry, 31*, 149–163.

Pike, J. W., Lee, S. M., & Meyer, M. B. (2014). Regulation of gene expression by 1,25-dihydroxyvitamin D_3 in bone cells: Exploiting new approaches and defining new mechanisms. *BoneKEy Reports, 3*, 482. 1–9.

Rasmussen, R., DeLuca, H., Arnaud, C., Hawker, C., & von Stedingk, M. (1963). The relationship between vitamin D and parathyroid hormone. *Journal of Clinical Investigation, 42*, 1940–1946.

Ribovich, M. L., & DeLuca, H. F. (1975). The influence of dietary calcium and phosphorus on intestinal calcium transport in rats given vitamin D metabolites. *Archives of Biochemistry and Biophysics, 170*, 529–535.

Ribovich, M. L., & DeLuca, H. F. (1976). Intestinal calcium transport: Parathyroid hormone and adaptation to dietary calcium. *Archives of Biochemistry and Biophysics, 175*, 256–261.

Ribovich, M. L., & DeLuca, H. F. (1978). Adaptation of intestinal calcium absorption: Parathyroid hormone and vitamin D metabolism. *Archives of Biochemistry and Biophysics, 188*, 157–163.

Semmler, E. J., Holick, M. F., Schnoes, H. K., & DeLuca, H. F. (1972). The synthesis of 1α,25-dihydroxycholecalciferol—A metabolically active form of vitamin D_3. *Tetrahedron Letters, 40*, 4147–4150.

Shipley, P. G., Kramer, B., & Howland, J. (1925). Calcification of rachitic bones *in vitro*. *American Journal of Diseases of Children, 30*, 37–39.

Shipley, P. G., Kramer, B., & Howland, J. (1926). Studies upon calcification *in vitro*. *Biochemical Journal, 20*, 379–387.

Steenbock, H., & Black, A. (1924). Fat-soluble vitamins. XVII. The induction of growth-promoting and calcifying properties in a ration by exposure to ultraviolet light. *Journal of Biological Chemistry, 61*, 405–422.

Steenbock, H., & Black, A. (1925). Fat-soluble vitamins. XXIII. The induction of growth-promoting and calcifying properties in fats and their unsaponifiable constituents by exposure to light. *Journal of Biological Chemistry, 64*, 263–298.

Steenbock, H., & Hart, E. B. (1913). The influence of function on the lime requirements of animals. *Journal of Biological Chemistry, 14*, 59–73.

Steenbock, H., & Herting, D. C. (1955). Vitamin D and growth. *Journal of Nutrition, 57*, 449–468.
Suda, T., DeLuca, H. F., & Hallick, T. N. (1971). Synthesis of [26,27-^{3}H]-25-hydroxycholecalciferol. *Analytical Biochemistry, 43*, 139–146.
Suda, T., DeLuca, H. F., Schnoes, H. K., Ponchon, G., Tanaka, Y., & Holick, M. F. (1970a). 21,25-Dihydroxycholecalciferol, a form of vitamin D_3 metabolically active in the intestine. *Biochemistry, 9*, 2917–2922.
Suda, T., DeLuca, H. F., Schnoes, H. K., Tanaka, Y., & Holick, M. F. (1970b). 25,26-dihydroxycholecalciferol, a metabolite of vitamin D_3 with intestinal calcium transport activity. *Biochemistry, 9*, 4776–4780.
Takahashi, N., Udagawa, N., & Suda, T. (2014). Vitamin D endocrine system and osteoclasts. *BoneKEy Reports, 3*(495), 1–9.
Tanaka, Y., & DeLuca, H. F. (1973). The control of 25-hydroxyvitamin D metabolism by inorganic phosphorus. *Archives of Biochemistry and Biophysics, 154*, 566–574.
Tanaka, Y., Frank, H., & DeLuca, H. F. (1973a). Biologically activity of 1,25-dihydroxyvitamin D_3 in the rat. *Endocrinology, 92*, 417–422.
Tanaka, Y., Frank, H., & DeLuca, H. F. (1973b). Intestinal calcium transport: Stimulation by low phosphorus diets. *Science, 181*, 564–566.
Umesono, K., Murakami, K. K., Thompson, C. C., & Evans, R. M. (1991). Direct repeats as selective response elements for the thyroid hormone, retinoic acid, and vitamin D_3 receptors. *Cell, 65*, 1255–1266.
Underwood, J. L., & DeLuca, H. F. (1984). Vitamin D is not directly necessary for bone growth and mineralization. *American Journal of Physiology, 246*, E493–E498.
Whistler, D. (1645). De morbopuerli anglorum, quem patrio ideiomate indigenase vocant "the rickets" (lugduni, batavorum 1645). As cited by Smerdon, GT, Daniel Whistler and the English Disease. A translation and biographical note. *Journal of the History of Medicine and Allied Sciences, 1950*(5), 397–415.
Wiese, R. J., Goto, H., Prahl, J. M., Marx, S. J., Thomas, M., Al-Aqeel, A., et al. (1993). Vitamin D-dependency rickets Type II: Truncated vitamin D receptor in three kindreds. *Molecular and Cellular Endocrinology, 90*, 197–201.
Windaus, A., & Bock, F. (1937). Uber das provitamin aus dem sterin der schweineschwarte. *Hoppe-Seyler's Zeitschrift fur Physiologische Chemie, 245*, 168–170.
Windaus, A., & Linsert, O. (1928). Vitamin D_1. *Annnalen der Chemie, 465*, 148.
Windaus, A., Linsert, O., Luttringhaus, A., & Weidlich, G. (1932). Crystalline-vitamin D_2. *Annalen der Chemie, 492*, 226–241.
Yamamoto, M., Kawanobe, Y., Takahashi, H., Shimazawa, E., Kimura, S., & Ogata, E. (1984). Vitamin D deficiency and renal calcium transport in the rat. *Journal of Clinical Investigation, 74*, 507–513.
Yoshizawa, T., Handa, Y., Uematsu, Y., Takeda, S., Sekine, K., Yoshihara, Y., et al. (1997). Mice lacking the vitamin D receptor exhibit impaired bone formation, uterine hypoplasia and growth retardation after weaning. *Nature Genetics, 16*, 391–396.

CHAPTER TWO

Genomic Determinants of Vitamin D-Regulated Gene Expression

J. Wesley Pike[1], Mark B. Meyer, Nancy A. Benkusky, Seong Min Lee, Hillary St. John, Alex Carlson, Melda Onal, Sohel Shamsuzzaman
Department of Biochemistry, University of Wisconsin–Madison, Madison, Wisconsin, USA
[1]Corresponding author: e-mail address: pike@biochem.wisc.edu

Contents

Abstract

Insight into mechanisms that link the actions of 1,25-dihydroxyvitamin D_3 ($1,25(OH)_2D_3$) to the regulation of gene expression has evolved extensively since the initial discovery of a nuclear protein known as the vitamin D receptor (VDR). Perhaps most important was the molecular cloning of this receptor which enabled its inclusion within the nuclear receptor gene family and further studies of both its structure and regulatory function. Current studies are now refocused on the vitamin D hormone's action at the genome, where VDR together with other transcription factors coordinates the recruitment of chromatin active coregulatory complexes that participate directly in the modification of gene output. These studies highlight the role of chromatin in the expression of genes and the dynamic impact of the epigenetic landscape that contextualizes individual gene loci thus influencing the VDR's transcriptional actions. In this chapter, we summarize advances made over the past few years in understanding vitamin D action on a genome-wide scale, focusing on overarching principles that have

Vitamins and Hormones, Volume 100
ISSN 0083-6729
http://dx.doi.org/10.1016/bs.vh.2015.10.011

emerged at this level. Of particular significance is the finding that dynamic changes that occur to the genome during cellular differentiation at both genetic and epigenetic levels profoundly alter the ability of $1,25(OH)_2D_3$ and its receptor to regulate gene expression. We address the broad impact of differentiation on specific epigenetic histone modifications that occur across the genome and the ability of the VDR to influence this activity at selected gene loci as well. These studies advance our understanding of not only vitamin D action but also of the complex and dynamic role played by the genome itself as a major determinant of VDR activity.

1. INTRODUCTION

A binding protein eventually designated the vitamin D receptor (VDR) was discovered first in the intestine (Brumbaugh & Haussler, 1974a, 1974b) and then in other tissues including the parathyroid glands, kidney, and bone (Pike, 1991). This protein's biochemical features, including its retention on chromatin (Haussler & Norman, 1969) and subsequently its ability to bind to DNA (Pike & Haussler, 1979), suggested that it was similar to that of other steroid hormone receptors and that it might play a role in transcriptional regulation. The DNA-binding capacity of the VDR enabled investigators to purify the protein, to generate both polyclonal and monoclonal antibodies useful in the protein's characterization (Dame, Pierce, Prahl, Hayes, & DeLuca, 1986; Pike, Donaldson, Marion, & Haussler, 1982), and beginning in 1987 to clone the chicken (McDonnell, Mangelsdorf, Pike, Haussler, & O'Malley, 1987) and subsequently the human (Baker et al., 1988) and rat (Burmester, Maeda, & DeLuca, 1988) genes as well. These latter achievements and the domain structure that was revealed from subsequent studies (McDonnell, Pike, & O'Malley, 1988; McDonnell, Scott, Kerner, O'Malley, & Pike, 1989) confirmed that $1,25(OH)_2D_3$ was a true steroid hormone and that the receptor was a *bona fide* member of the steroid receptor gene family (Evans, 1988). They also enabled subsequent studies in patients with hypocalcemic rickets and other clinical features that identified a series of mutations within the VDR gene itself that was causative for hereditary $1,25(OH)_2D_3$-resistant rickets (HVDRR) (Feldman & Malloy, 1990; Forghani et al., 2010; Hughes et al., 1988). This syndrome was first identified by Bell and colleagues in 1978 (Brooks et al., 1978) and proved to be due to defects in the VDR protein (Eil, Liberman, Rosen, & Marx, 1981; Marx et al., 1978; Pike et al., 1984). This discovery, the first for any member of the nuclear receptor family, confirmed the integral and essential role for the VDR as the mediator of the activities of the vitamin D hormone. Importantly, the human phenotype of this disease has been recapitulated in mice through genetic deletion of key elements of the VDR gene from the mouse genome (Bouillon et al., 2008; Li et al., 1997; Yoshizawa et al., 1997). More recently, advanced studies of the VDR gene from both mouse and humans have defined the genetic loci spanning these two genes and determined the locations of key regulatory elements that function to modulate VDR gene output in response to hormones such as the glucocorticoids, retinoic acid, and $1,25(OH)_2D_3$ itself (Zella, Kim, Shevde, & Pike, 2006; Zella et al., 2010). The ability of a transgene that contained either the mouse or the human version of the VDR gene to recapitulate the tissue-specific expression of the VDR in the mouse and to rescue the phenotype of the VDR-null mouse has provided final confirmation of the role of the VDR in $1,25(OH)_2D_3$ action (Lee, Bishop, Goellner, O'Brien, & Pike, 2014). It has also enabled the creation of a humanized mouse model that replicates a particular syndromic subset of HVDRR patients, wherein a VDR molecule incapable of binding $1,25(OH)_2D_3$ is able to prevent the development of alopecia that is seen in mice that do not express the VDR; this activity appears to be $1,25(OH)_2D_3$ independent (Lee, Goellner, O'Brien, & Pike, 2014). These up-to-date studies conclusively demonstrate the importance of the VDR as the mediator of all of the known actions of $1,25(OH)_2D_3$ in disease.

2. GENOME-WIDE ANALYSIS REVEALS NEW CONCEPTS IN VITAMIN D ACTION

Traditional studies over several decades using reporter plasmid analyses facilitated the conclusion that $1,25(OH)_2D_3$ is capable of regulating many vitamin D target genes including osteocalcin (Kerner, Scott, & Pike, 1989; Ozono, Liao, Kerner, Scott, & Pike, 1990), osteopontin (Nilsson et al., 2005), and *Cyp24a1* (Ohyama et al., 1994; Zierold, Darwish, & DeLuca, 1995). These studies coupled with the extensive use of electrophoretic mobility shift and other analyses identified key features of the VDR's DNA-binding sites, termed vitamin D response elements or VDREs, and the participation of RXR as a heterodimer partner essential for adequate DNA-binding affinity (Pike & Meyer, 2014; Pike et al., 2010). However, despite the fact that many of these interactions at target genes have been confirmed via the application of chromatin immunoprecipitation analysis (ChIP) (Kim, Shevde, & Pike, 2005), this latter technique was unable to provide sought after confirmation for many gene promoters and more importantly failed to identify mechanisms that mediated regulation for genes such as *Tnfsf11* (RANKL), *Vdr*, and numerous others as well. Accordingly, it was the development of genome-wide methods such as ChIP-chip (tiled microarrays) and then ChIP-seq (DNA sequencing) analyses that extended the technical reach of ChIP analysis to resolve many of the unknown mechanisms, eventually enabling the quantification of transcription factor-binding sites across entire cellular genomes. Importantly, these techniques were also used to acquire genome-wide data sets for coregulatory factors, chromatin modifiers, and for the presence of epigenetic modifications on both DNA and histones as well. Indeed, any feature for which an antibody could be developed was a potential target. This largely unbiased approach to transcriptional regulation has fundamentally revolutionized our approach to the study of genetic and epigenetic components that are essential for gene regulation, simultaneously revealing an abundance of new insights. Indeed, we have used ChIP-chip and subsequently ChIP-seq analyses to gain a genome-wide perspective through which $1,25(OH)_2D_3$ and its receptor mediate the regulation of cellular transcriptomes in numerous cell types (Lee et al., 2015; Meyer, Benkusky, Lee, & Pike, 2014; Meyer, Benkusky, & Pike, 2014; Meyer, Goetsch, & Pike, 2010b, 2012; Meyer & Pike, 2013; St John, Bishop, et al., 2014). A list of the overarching principles that have emerged is presented in Table 1 (Pike, Lee, & Meyer, 2014). These studies indicated that between 2000 and 8000 VDR-binding sites are detected following activation by $1,25(OH)_2D_3$ in a cell-type-dependent quantitation and that these sites are highly enriched for a DNA sequence found previously in representative genes such as osteocalcin and osteopontin. Furthermore, the majority of these sites are co-occupied by RXR, thereby confirming this principle on a genome-wide scale. Interestingly, we also discovered that while the DNA-binding activity of the VDR at these cellular genomes was largely dependent upon the presence of $1,25(OH)_2D_3$, a significant number of sites were fully occupied by the receptor even in the absence of ligand. The basis for this type of DNA binding is unknown, but does not appear to be due to the absence of RXR, which frequently occupied most of the VDR-binding sites regardless of the presence of $1,25(OH)_2D_3$. This latter finding was accompanied by the discovery that most regulatory regions were located within introns and intergenic regions highly distal to the genes they regulate. These observations will be discussed in the next sections, but provide both

Table 1 Overarching Principles of 1,25(OH)$_2$D$_3$-Mediated Gene Regulation in Target Cells

VDR-binding sites (the cistrome): 2000–8000 1,25(OH)$_2$D$_3$-sensitive binding sites/genome whose number and location are determined by cell type
Active transcription unit: The VDR/RXR heterodimer
Distal-binding site location: dispersed in cis-regulatory modules (CRMs or enhancers) across the genome; located in a cell-type-specific manner near promoters, but predominantly within introns and distal intergenic regions; frequently located in clusters of elements
VDR/RXR-binding site sequence (VDRE): induction mediated by classic hexameric half-sites (AGGTCA) separated by three base pairs; repression mediated by divergent sites
Mode of DNA binding: predominantly, but not exclusively, 1,25(OH)$_2$D$_3$ dependent
Modular features: CRMs contain binding sites for multiple transcription factors that facilitate both independent or synergistic interaction
Epigenetic CRM signatures: defined by the dynamically regulated posttranslational histone H3 and H4 modifications and selectively regulated by 1,25(OH)$_2$D$_3$
VDR cistromes are highly dynamic: cistromes change during cell differentiation, maturation, and disease activation and thus have consequential effects on gene expression

Principles in bold represent those previously defined and now confirmed by genome-wide analysis. Principles in italics represent newly defined genome-wide features of vitamin D action.

novel insight into vitamin D action and explain the difficulties that emerged early on in identifying gene regulatory mechanisms when the focus was limited technically to regions located near gene promoters. The summary provided in Table 1 is supported in part by additional studies that have been conducted by other investigators in the vitamin D field (Ding et al., 2013; Heikkinen et al., 2011; Ramagopalan et al., 2010; Sherman et al., 2014).

3. NOVEL PRINCIPLES OF VITAMIN D ACTION

The observation that DNA binding of the VDR occurs in both a 1,25(OH)$_2$D$_3$-dependent and -independent fashion and that the sites to which the receptor binds occur most frequently many kilobases distal to genetic start sites has profound implications for vitamin D action. Perhaps the most interesting principle has been the discovery that the VDR cistrome is highly dynamic during the course of cellular differentiation resulting in a striking change in the gene expression profile that characterizes the more mature cell. We consider these three issues in the context of vitamin D action in the next sections.

3.1 Modes of DNA Binding and Implications for the Regulatory Activity of the VDR

As indicated above, ChIP-seq analysis revealed that while VDR-binding site occupancy on cellular genomes is highly dependent upon $1,25(OH)_2D_3$, a number of sites were found to contain prebound VDR and RXR prior to ligand activation (Meyer, Benkusky, Lee, et al., 2014; Meyer et al., 2012). This novel observation highlights at least two modes of VDR DNA binding and raises questions as to both the nature of the underlying mechanism through which the VDR binds to DNA in the absence of activation by the hormone and whether this interaction is capable of regulating transcriptional activity independent of $1,25(OH)_2D_3$. Although the mechanism remains obscure, emerging evidence suggests that the VDR may regulate the expression of specific genes and their associated biology through at least two different mechanisms that likely involve VDR DNA binding yet are independent of $1,25(OH)_2D_3$. It is worth noting that despite several unique biochemical properties inherent to the VDR, this latter type of activity is not unexpected given the complex nature of the transcriptional actions of other nuclear receptors.

3.1.1 Ligand-Independent Function of the VDR in the Hair Cycle

As discussed earlier, the syndrome of HVDRR in humans is due to a wide variety of mutations within the VDR gene that results in the production of a receptor that is unable to regulate gene expression (Malloy, Pike, & Feldman, 1999; Malloy et al., 2014). These molecular defects lead to a broad disease phenotype that is particularly evident at the skeleton. Only a subset of these patients display alopecia, however, and this feature was ultimately linked to VDR gene mutations that compromise the overall expression of the VDR rather than to mutations that alter its functional capability to bind to and interact with either $1,25(OH)_2D_3$ or DNA, or to recruit coregulatory proteins essential for gene regulation. Further studies using mouse models supported these observations in humans; VDR-null mice become alopecic whereas *Cyp27b1*-deleted mice unable to produce $1,25(OH)_2D_3$ are unaffected, prompting the emerging hypothesis that control of the hair cycle in the skin could be the result of a $1,25(OH)_2D_3$-independent function of the VDR (Cianferotti, Cox, Skorija, & Demay, 2007; Demay et al., 2007; Li et al., 1997; Panda et al., 2001). While it is known that this biological activity of the VDR involves interaction with components of the Wnt-β catenin and Hedgehog signaling pathways, the mechanism of this regulation remains unclear (Demay et al., 2007). To address this issue directly, we recently developed genetic models in which mice express either the human wild-type VDR or a mutant form that is incapable of binding $1,25(OH)_2D_3$ from transgenes that are comprised of the natural human VDR gene locus (Lee, Bishop, et al., 2014; Lee, Goellner, et al., 2014). Following introduction into the VDR-null mouse background, both transgenes recapitulate expression of the endogenous mouse VDR gene in tissues in which the VDR is known to be expressed. However, while the wild-type human VDR was able to rescue both the deranged skeletal phenotype of the VDR-null mouse and to prevent alopecia, the $1,25(OH)_2D_3$-binding-deficient form of the human VDR could only rescue the alopecia (Lee, Goellner, et al., 2014; Fig. 1). This study confirmed a previous observation involving the expression of a mutant VDR in keratinocytes and established the fundamental paradigm that the VDR is capable of $1,25(OH)_2D_3$-independent actions

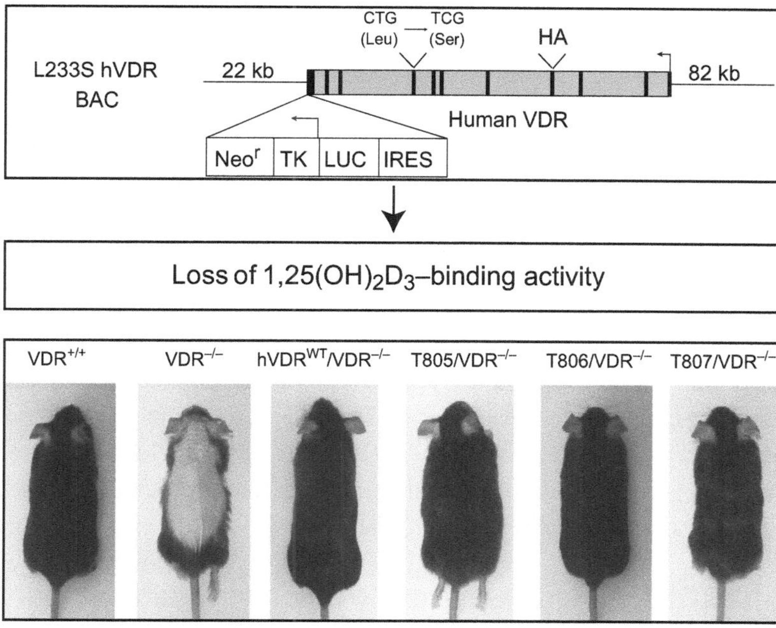

Figure 1 A mutant human VDR deficient in $1,25(OH)_2D_3$-binding activity restores hair follicle cycling in VDR-null mice and prevents alopecia. Minigenes containing the human VDR gene locus were introduced into a VDR-null mouse background to create humanized mice expressing either wild-type (hVDR) or mutant (hVDR-L233S) VDR proteins. Both hVDR-WT and hVDR-L233S rescue the alopecia observed in VDR-null mice. VDR+/+, unmodified normal mice; hVDR-L233S 805, 806, and 807 represent three rescued mouse strains that express increasing concentrations of VDR protein in all appropriate tissues. See Lee, Goellner, et al. (2014). (See the color plate.)

(Chen, Sakai, & Demay, 2001; Skorija et al., 2005). Interestingly, the VDR displays numerous 1,25(OH)$_2$D$_3$-dependent activities in the skin as well suggesting that both processes can occur simultaneously (Bikle, 2004, 2012). It is perhaps of relevance that loss of the VDR in mice potentiates tumor development in the skin whereas loss 1,25(OH)$_2$D$_3$ production does not.

3.1.2 Ligand-Independent Suppression of Gene Expression by the VDR

Is there evidence that the VDR functions in a ligand-independent manner in additional biologic processes? Interestingly, the expression of a 1,25(OH)$_2$D$_3$-binding-defective VDR in mice as described above has provided additional circumstantial evidence. For example, while the mutant receptor is unable to rescue systemic features of altered mineral homeostasis, several of these parameters such as the level of PTH appear to be exaggerated relative of their VDR-null mouse counterparts and certain features of the skeletal phenotype appear to be exacerbated as well (Lee, Goellner, et al., 2014). Interestingly, these skeletal abnormalities were documented early on in the *Cyp27b1*-null mouse. Whether these aberrations are due to reverse VDR activity (suppression) in the absence of ligand remains to be determined. Interestingly, recent studies suggest that in the absence of 1,25(OH)$_2$D$_3$, the VDR may suppress the expression of genes that are normally induced by the receptor in the presence of ligand (Lee, Goellner, et al., 2014). This may indicate that the unliganded VDR can exert transcriptional effects on target genes that are opposite to those identified following 1,25(OH)$_2$D$_3$ activation. Much additional research will be required to prove this hypothesis, however, as at present virtually all of the fundamental mechanistic support for this potential set of VDR activities remains to be delineated. Nevertheless, the role of the VDR in the hair follicle to regulate the hair cycle provides strong conceptual framework that the VDR may regulate gene expression in the absence of 1,25(OH)$_2$D$_3$.

3.2 1,25(OH)$_2$D$_3$ Regulates Transcription via Multiple Enhancers Located at Sites Distal to Gene Promoters

A striking result of extensive unbiased genome-wide ChIP-seq analysis of transcription factor localization across multiple genomes has been the discovery that many if not most genes are regulated by multiple enhancers that are not positioned near promoter regions, but rather within intronic and/or intergenic regions 10s if not 100s of kilobases distal to their transcriptional start sites (TSSs) (Meyer, Benkusky, Lee, et al., 2014; Meyer et al., 2010b,

2012). Indeed, it has been estimated that most genes are regulated by an average of 10 enhancers and that the average distance of an enhancer from its promoter target is greater than 250 kb (Dunham et al., 2012; Gerstein et al., 2012). These findings suggest that the identification of promoter proximal elements near genes based upon the transfected plasmid approach is highly biased, at best incomplete, and often incorrect. A major consequence of results emerging from extensive ChIP-seq analyses is that it is no longer reasonable to explore regulatory mechanisms based upon these earlier molecular biologic approaches with any expectation that an understanding of the regulatory features of a gene will be forthcoming. Significant additional problems with the traditional approaches have also emerged; in the absence of the entire gene locus as well as an appropriate chromatin context, the contribution and interaction of multiple enhancers and the myriad of chromatin regulatory proteins that impact the architecture of the gene locus through both DNA and histone modifications are largely negated. As a consequence, while the results of unbiased ChIP-seq analyses have provided a better understanding of the mechanisms that regulate the output of individual genes, they have at the same time made it much more difficult to define the genetic and epigenetic contributors to such regulation.

3.2.1 Defining Gene Regulation from Distal Sites

Cistromic analysis in many cell types has now revealed the presence of multiple VDR-binding sites within genetic loci that are frequently dispersed to distal intronic and intergenic sites. Each enhancer site may contain one or more VDREs that can be located within a few nucleotides of each other or more than 200 bp (> a nucleosome) from each other. While hundreds of vitamin D target genes are configured in this manner, specific examples include *Cyp24a1*, *Vdr*, *Cbs*, *Tnfsf11*, *c-FOS*, *Spp1*, *Runx2*, *Cdon*, *Mmp13*, *Col2a*, *Trpv6*, *S100g*, and others as well. In the case of *Cyp24a1*, while the promoter proximal element defined in earlier studies (Zierold et al., 1995) was confirmed, a downstream cluster of regulatory elements has also been identified more recently (Meyer, Goetsch, & Pike, 2010a). In genes such as the *Vdr* (Zella et al., 2006, 2010), *Mmp13* (Meyer, Benkusky, Lee, et al., 2014; Meyer, Benkusky, & Pike, 2015), and many others, the presence of proximal elements in earlier reports was not confirmed, although more distal sites of VDR actions have been delineated. In other cases, VDR-binding sites at genes for which the regulatory activity of $1{,}25(OH)_2D_3$ had previously been unknown were clarified, frequently located many kilobases distal to the gene's TSS. These studies of the VDR support the concept that has emerged for most transcription factors that the enhancers to which they bind are often found not only distal to genetic start sites but may contain unregulated genes that are dispersed between the enhancer and the gene it actually regulates.

3.2.2 Confirming Enhancer Function at Target Genes

A major consequence of the above arrangement of regulatory enhancers is that while enhancers can be readily located, it is no longer possible to identify the genes to which they are linked without direct experimentation conducted in an endogenous gene-like chromatin context. Accordingly, we have examined enhancer/target gene relationships by stably integrating recombinantly modified bacterial artificial chromosomes (BAC clones) containing extended gene loci for the *Vdr*, T*nfsf11*, and *Cyp24a1* genes, for example, into the genomes of cells *in vitro* and comparing $1,25(OH)_2D_3$ regulatory activities of both wild-type BAC and mutant clones that contain specific enhancer deletions (Lee, Goellner, et al., 2014; Meyer et al., 2010a; Onal et al., 2014; Zella et al., 2006, 2010). Most importantly, these minigenes have also been inserted into the mouse genome as well, verifying that the clones contain sufficient genetic information to direct not only appropriate tissue-specific expression of the gene but a level of expression capable of rescuing the phenotype of mice that are null for the gene product complimented through the transgene (Lee, Bishop, et al., 2014; Onal et al., 2014). A distinct advantage of this *in vivo* approach is that both the basal and the regulatory features of the transgene of interest can be examined in a wide variety of tissues known to express the gene rather than from a single cell type in culture. A successful rescue permits the conclusion that the transgene contains the fully complement of features that control expression of the gene in its original endogenous setting. The consequence of enhancer deletion on the regulated expression of the transgene as well as the expression of mutant forms of the transcriptional product can then be evaluated as well, the latter highlighted for the human VDR gene in the above section which essentially humanizes the mouse (Lee, Bishop, et al., 2014). We have also deleted putative gene enhancers within the mouse genome and evaluated the consequence of this action on subsequent target gene activity in selected tissues as well (Galli et al., 2008). This has been particularly informative for the mouse *Tnfsf11* gene, which contains at least 10 distinct regulatory regions that differentially control the expression of RANKL in osteoblast-lineage cells and in hematopoietic immune cells (Bishop, Coy, Nerenz, Meyer, & Pike, 2011; Bishop, Meyer, & Pike, 2009; Bishop et al., 2014;

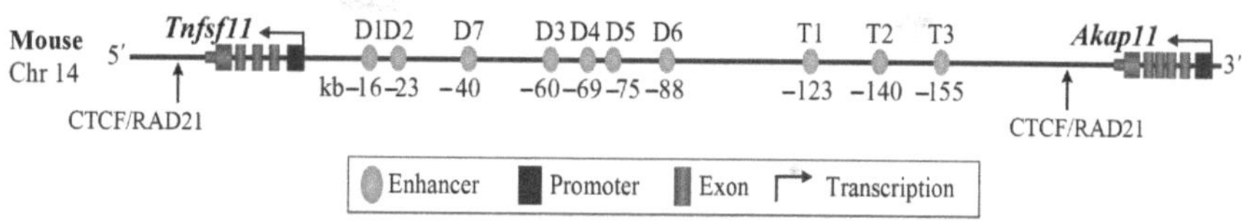

Figure 2 Schematic linear structure of the mouse *Tnfsf11* gene locus with associated regulatory enhancers located on chromosome 14. The gene spans over 220 kb and is defined by two CTCF/RAD21 sites that serve as boundary elements for the gene. The *Tnfsf11* exons are defined as brown rectangles, and enhancers are defined by orange ovals. Arrows define the direction of transcription. Both the *Tnfsf11* and neighboring *Akap11* genes are located on the reverse strand. D1–D7 represent enhancers that regulate *Tnfsf11* expression in osteoblast-lineage cells, whereas T1–T3 represent enhancers that regulate *Tnfsf11* expression in hematopoietic B and T cells. See Onal et al. (2014). (See the color plate.)

Kim, Yamazaki, Shevde, & Pike, 2007; Kim, Yamazaki, Zella, Shevde, & Pike, 2006; Fig. 2). Finally, we have begun to explore the role of enhancers in diseased tissue as well, such as in the atherosclerotic plaques that emerge in ApoE-null mice fed a high-fat diet. In this case, mice containing enhancer deletions that compromise the expression of specific genes are crossed into the ApoE-null background and the consequence of enhancer-mediated gene misexpression examined in the high-fat diet-induced atherosclerotic plaques. In another vein, assessing enhancer/target gene relationship for genes that are suppressed by $1,25(OH)_2D_3$ is particularly complex as many features that are routinely apparent and responsible for gene activation frequently do not apply. For example, while VDR-binding sites can be found near a subset of genes that are repressed by $1,25(OH)_2D_3$, many repressed genes do not contain these adjacent VDR-binding sites (Meyer, Benkusky, Lee, et al., 2014; St John, Bishop, et al., 2014). Thus, the mechanisms associated with repression continue to remain obscure. The distal nature of regulatory elements also suggests a requirement for DNA looping as assessed by chromosome conformation capture (3C) technology to bring multiple distal sites into proximity with the active transcriptional centers associated with individual genes.

3.2.3 Application of the CRISPR/Cas9 Method to the Study of Gene Regulation

Current studies are now focused on using genome-editing methods employing CRISPR (clustered regularly interspaced short palindromic repeats)/Cas9 to establish the relevance of specific regulatory elements both in cells in culture and in mice *in vivo* (Cong et al., 2013). Indeed, a detailed examination of the regulation of *Mmp13* expression by $1{,}25(OH)_2D_3$ in osteoblast-lineage cells has revealed the presence of three enhancers located −10, −20, and −30 kb upstream of the *Mmp13* TSS (Meyer et al., 2015; Fig. 3). While the promoter region of the *Mmp13* manifested no vitamin D-sensitive activity in regulating *Mmp13* mRNA transcript production, the enhancer located at −10 kb not only bound the VDR but mediated all of the actions of $1{,}25(OH)_2D_3$ on *Mmp13* mRNA expression. The more distal enhancers mediated the regulation of this gene by other transcription

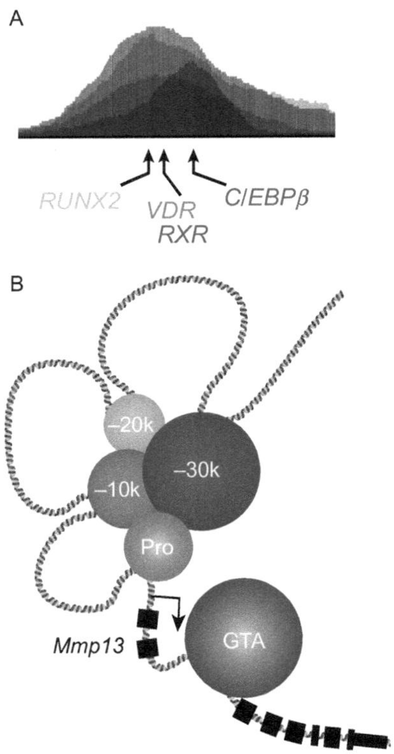

Figure 3 Organization of osteoblast enhancer complexes (OEC) that bind RUNX2, C/EBPβ, and the VDR. (A) Consolidated OEC-binding arrangement for RUNX2, C/EBPβ, and VDR at the *Spp1* (osteopontin) gene as defined by ChIP-seq analysis. (B) Dispersed OEC-binding arrangement for RUNX2, C/EBPβ, and VDR at the *Mmp13* gene locus as defined by ChIP-seq and other functional analyses. Schematic depiction of the individual RUNX2 (blue), C/EBPβ (yellow), and VDR (green) enhancers shown clustered and interacting upstream of the *Mmp13* gene locus with the promoter proximal region (gray) and the general transcription assembly (brown) located near the transcriptional start site. See Meyer, Benkusky, and Pike (2015). (See the color plate.)

factors including RUNX2 and C/EBPβ. Further dissection of the activities of these enhancers using deletion directed by the CRISPR/Cas9 system revealed unique activities inherent to each enhancer in the osteoblast and the fact that despite their linear distances from each other, the loss of key regulatory capability derived from the most distal enhancer (−30 kb) conferred a profound impact on the binding activities of the other transcription factors at the remaining sites and at the promoter. These activities via genetic dissection by the CRISPR/Cas9 system are likely to provide the preferred methodological approach to dissecting transcriptional regulation and enhancer function in the near future in exquisite detail. Indeed, we have already deleted a series of enhancers from vitamin D target genes in mice *in vivo* and shown that these mutations have a striking impact of the expression of the genes to which they are linked.

4. THE INFLUENCE OF CELLULAR DIFFERENTIATION ON VITAMIN D ACTIVITY

Perhaps the most important observation made regarding the actions of $1,25(OH)_2D_3$ on a genome-wide scale has been the discovery that cellular differentiation exerts a dramatic quantitative and qualitative impact on the hormone's ability to regulate gene expression (Meyer, Benkusky, Lee, et al., 2014; Meyer, Benkusky, & Pike, 2014; St John, Bishop, et al., 2014; St John, Meyer, et al., 2014). It has been known since early times that the effects of $1,25(OH)_2D_3$ on osteoblast-lineage cells differ significantly depending upon the cellular state of differentiation, although these conclusions were drawn largely from studies of single genes such as osteocalcin (Lian et al., 1989). In recent studies that have highlighted this concept more broadly, we observed that the treatment of early osteoblasts with $1,25(OH)_2D_3$ resulted in a transcriptome that was more expansive and strikingly different than that observed in mineralizing osteoblasts following differentiation (Meyer, Benkusky, Lee, et al., 2014). While an extensive overlap of genes was noted in both cell types, the transcriptome itself was quantitatively reduced in more mature cells, in part because expression of the VDR itself was downregulated. Indeed, some genes were no longer responsive to $1,25(OH)_2D_3$. The most striking observation, however, was the finding that despite a significant reduction in VDR expression, the basal levels of many vitamin D target genes were altered due to differentiation and the response to $1,25(OH)_2D_3$ was qualitatively altered as well. In some cases, the gene exhibited an increase in response to $1,25(OH)_2D_3$ and in others a decrease; in some cases, the directionality of regulation was reversed. These findings highlight the highly dynamic nature of the biologic activities of $1,25(OH)_2D_3$ in cells that are closely related and raise the important underlying question of how the differentiation process is able to impact response to $1,25(OH)_2D_3$ mechanistically.

4.1 Differentiation Is Accompanied by Direct Alterations in the VDR Cistrome

A comparison of the VDR/RXR cistrome in early precursors and late mineralizing osteoblasts using ChIP-seq analysis revealed a striking reduction in the number of binding sites for the VDR in more mature cells and a significant reduction in their location as well (Meyer, Benkusky, Lee, et al., 2014). As expected, VDR binding was frequently lost at sites near genes that were no longer responsive to $1,25(OH)_2D_3$. In contrast, however, although frequently reduced following differentiation, VDR binding was retained at many sites near genes that remained responsive or showed increased response to $1,25(OH)_2D_3$, although in many cases the level of receptor binding activity was significantly reduced as assessed by ChIP-seq analysis. This observation suggests the possibility that transcription factors other than the VDR that are either recruited to or enriched at sites near these genes following differentiation may contribute in enhancing response to $1,25(OH)_2D_3$. It is also possible that these factors may influence the epigenetic landscape that defines functional features of the target gene, facilitating further the ease with which $1,25(OH)_2D_3$ and its receptor may be able to modify the gene's expression. This hypothesis is supported by the observation that the basal levels of expression of many of these genes are often changed following cellular differentiation.

4.2 The Impact of Osteoblast Differentiation on Master Regulatory Factor Distribution, Histone Modifying Activity, and Response to $1,25(OH)_2D_3$

To explore the idea elaborated above, we first used ChIP-seq analysis to determine the distribution of binding sites for two key osteoblast-lineage determining factors RUNX2 and C/EBPβ across the osteoblast genome before and after differentiation (Meyer, Benkusky, & Pike, 2014). Importantly, while the level of expression of these individual factors changed only modestly following differentiation, both the level of their occupancy at existing sites as well as their accumulation (or loss thereof) at new sites of action was significantly altered. An examination of the effects of $1,25(OH)_2D_3$ on these cistromes revealed that in addition to the hormone's known direct inhibitory role on RUNX2 and C/EBPβ expression, the hormone also caused a modest redistribution of RUNX2 and C/EBPβ DNA binding in early osteoblasts suggesting that the hormone was likely capable of affecting the presence of RUNX2 and C/EBPβ concentrations at selected sites as well (Meyer, Benkusky, Lee, et al., 2014). Given the role of these factors in the recruitment of chromatin regulatory modifiers, these results suggest further that significant quantitative and/or qualitative changes in the levels of epigenetic modification might also be detectable at genes whose expression levels were altered as a function of osteoblast differentiation both in the absence and presence of 1,25$(OH)_2D_3$. Interestingly, we found that while the epigenetic landscape was generally unchanged when examined using all genes that were expressed in both precursor cell and their differentiated counterpart, significant changes were observed when the analysis was restricted to genes whose expression patterns were altered as a result of the differentiation process (Meyer, Benkusky, & Pike, 2014; St John, Bishop, et al., 2014). Of particular importance were the changes observed at H3K4me1, H3K4me2, H3K9ac, H3K27ac, and H4K5ac, modifications that denote the locations of regulatory enhancers or that highlight variations in chromatin decondensation and accessibility. Numerous changes were also noted at H3K4me3, a mark that specifies the location of gene promoters, and at H3K36me3, H4K20me1, and H4K5ac, marks that identify genomic regions spanning the transcription units (exons and introns) of genes. Interestingly, bioinformatic examination also revealed that most of the changes in the levels of signature marks at enhancers were quantitative (St John, Bishop, et al., 2014). Thus, only a few changes in histone modification that denoted the novel appearance of a newly minted enhancer (or its loss) could uniquely regulate the expression of the gene to which it was linked were observed. These results suggest broadly that programmed creation of the vast majority of regulatory enhancers in cells of the osteoblast lineage likely occurs early in the mesenchymal progression, and that alterations in transcription factor expression and activity (i.e., RUNX2 and C/EBPβ) may be dominant at influencing the epigenetic landscape surrounding regulated genes, could influence gene output, and likely modulate response to 1,25$(OH)_2D_3$ and other systemic regulatory hormones.

4.3 Identification and Structure of the Osteoblast Enhancer Complex

The potential influence of both RUNX2 and C/EBPβ activity on cellular response to 1,25$(OH)_2D_3$ together with a preliminary bioinformatic analysis that suggested the enriched presence of DNA-binding motifs for these latter two factors at sites of VDR binding on a genome-wide scale prompted us to examine the potential relationship between the VDR cistrome and those for RUNX2 and C/EBPβ directly. This analysis confirmed that either RUNX2 or C/EBPβ or both were frequent occupants at active VDR DNA-binding sites (Meyer, Benkusky, Lee, et al., 2014). Further inspection revealed that 70% of the 4174 VDR/RXR-binding sites that were identified in early osteoblasts also contained RUNX2, while 42% contain both RUNX2 and C/EBPβ. A more detailed examination identified an even closer physical relationship between RUNX2, C/EBPβ, and VDR/RXR, prompting its description as a consolidated "osteoblast enhancer complex" (Fig. 3A). Indeed, we found that RUNX2 and C/EBPβ bind bidirectionally 8 and 9 bp on average from the VDR/RXR peak center, respectively. As RUNX2 and C/EBPβ are independently active in the regulation of gene expression in osteoblast-lineage cells, these findings support the idea that enhancers of this type are likely capable of mediating both the independent actions of these specific transcription factors and of the actions of the VDR, and perhaps in some cases of integrating the actions of all three as well. Interestingly, other transcription factor arrangements for VDR/RXR, RUNX2, and C/EBPβ are also apparent. Thus, many genes including *Mmp13* are regulated by set of dispersed enhancers that bind RUNX2, C/EBPβ, or the VDR individually (Meyer, Benkusky, Lee, et al., 2014; Meyer et al., 2015). Given the linear distances between each enhancer in these examples, we speculate that the activities of each of these regulatory modules are likely integrated collectively at the promoters for target genes via complex DNA looping in a manner that is reminiscent of that seen for the consolidated osteoblast enhancer complex. Our recent studies of the *Mmp13* gene support this concept through the demonstration that targeted CRISPR/Cas9-mediated deletion of several of the genes enhancers impacts not only the overall expression of *Mmp13* but influences binding activity of transcription factors such as RUNX2, C/EBPβ, and the VDR at the enhancers that remain in the *Mmp13* locus (Meyer et al., 2015; Fig. 3B). The prebound nature of both RUNX2 and C/EBPβ on DNA and their broad master regulatory properties in osteoblasts suggest that they may play an instrumental role in establishing and maintaining enhancers for genes that are not only relevant to the osteoblast lineage but that their actions at these enhancers may facilitate the availability of sites to which the VDR and other secondary regulators can be recruited. If this hypothesis is correct, while the VDR is a primary determinant of vitamin D action, both RUNX2 and

C/EBPβ and likely others that operate in a lineage-dependent fashion are also determinants of the quantitative and qualitative nature of the response, in part by contributing to processes such as histone modification that control the output of gene expression.

5. VDR MODULATES HISTONE ACETYLATION AT TARGET GENES

Initial studies of vitamin D action revealed that VDR binding at the proximal elements associated with *Spp1* and *Cyp24a1* regulation resulted in a differential increase in the level of histone H3 and H4 acetylation at these genes, suggesting the existence of a chromatin response to the actions of 1,25 $(OH)_2D_3$ that might be gene-selective (Kim et al., 2005). Subsequent studies of the genes for *Vdr*, *Tnfsf11*, and others support this view (Kim et al., 2006; Zella et al., 2006). Consistent with these observations, we subsequently discovered that the effects of VDR binding on a genome-wide scale in osteoblasts and osteocytes also reflect this premise (Meyer, Benkusky, Lee, et al., 2014; St John, Bishop, et al., 2014). Accordingly, while acetylation levels of H3K9, H4K5, and H3K27 were increased at sites of VDR action near many genes, sites in other genes were unaffected. It has long been known that one of the functions of the VDR in gene activation is to initiate the recruitment of coregulatory factors that include CBP, p300, and the SRC family of histone acetyltransferases and several histone deacetyltransferases as well (Sutton & MacDonald, 2003). It is clear that the actions of these enzymes at the histones associated with many genes likely account for the changes in acetylation that are observed, although the mechanism that underlies this site-selectivity is not understood. It seems likely that the requirement for gene activation differs among individual genes, perhaps based upon the nature of the residual expression level of the gene in question and the presence of additional transcription factors that contribute to this level of expression.

Acetylation levels represent a hallmark of chromatin decondensation and transcription factor accessibility to binding sites on DNA, particularly if access to those sites is restricted due to nucleosome positioning (Shahbazian & Grunstein, 2007). Alternative explanations as to the role of increased acetylation as well as methylation include the possibility that site-specific increases lead to the recruitment of additional chromatin regulators that are necessary for nucleosomal redistribution, eviction, or exchange, thereby enabling enhancer/promoter engagement through

DNA reorganization (Berger, 2007). Separate studies using 3C analysis have shown, for example, that the presence of estrogen and the estrogen receptor at distal enhancers facilitates this type of DNA reorganization (Pan et al., 2008). Interestingly, while our studies of this event at the *Tnfsf11* (RANKL) and *Cyp24a1* genes have shown linkage between the promoters for these genes and their associated distal enhancers, they do not appear to be influenced by $1,25(OH)_2D_3$ (Bishop et al., 2011; Meyer et al., 2010a). Increased methylation at specific sites on histones likewise precipitates changes in gene output, likely due in this case to the selective recruitment of chromatin regulators known as "readers" whose downstream actions are currently being characterized (Calo & Wysocka, 2013; Ruthenburg, Allis, & Wysocka, 2007). Future studies will be required to delineate the consequence of increased acetylation and methylation by VDR at the molecular level and identify the specific players that are involved. Nevertheless, the observation that activated VDR initiates enhanced expression of specific chromatin regulators as well as their recruitment to genes provides an initial starting point.

6. SUMMARY

Recent studies have revealed a set of general overarching principles through which $1,25(OH)_2D_3$ and its receptor regulate the expression of genes in cellular targets. These findings confirm and extend many of the general features that have been identified for vitamin D action over the past several decades, but have also revealed important new concepts as well. These include the finding that while VDR binds to DNA largely in response to activation by $1,25(OH)_2D_3$, a subset of genetic targets contain prebound VDR even in the absence of the ligand. This finding highlights the possibility that the VDR may have a more ubiquitous $1,25(OH)_2D_3$-independent function in gene regulation, an activity that appears to be illustrated by VDR actions in the hair cycle. An additional discovery is the finding that regulatory enhancers for the VDR as well as for most other transcription factors are often multiple at single gene loci and more importantly located at sites that are frequently remote relative to the genes they regulate. This finding highlights the likely role of DNA looping that enables these distal enhancers to contact the transcriptional machinery near the promoter and to impact transcriptional output. Finally, our results suggest that the process of differentiation is capable of altering the transcription factor milieu at important target genes and directly affecting the epigenetic landscape surrounding these genes as well. This combination of effects is capable of altering the DNA-binding activity of secondary regulatory factors such as the VDR and influencing in both a positive and negative manner the protein's transcriptional activity as well. These results illustrate the highly dynamic nature of VDR action within tissues thereby adding a new dimension to our understanding of the role of the genome in defining cellular context that ultimately impacts hormonal response. This molecular and regulatory complexity and the certain consequence of disease on these events at the genomic levels are likely to be significant *in vivo*.

ACKNOWLEDGMENTS

We acknowledge the helpful contributions of members of the Pike laboratory to this review and financial support from National Institutes of Health grants from NIDDK (DK-072281, DK-074993) and NIAMS (AR-045173 and AR-062442) to J.W.P.

Disclosure: The authors declare no conflict of interest.

REFERENCES

Baker, A. R., McDonnell, D. P., Hughes, M., Crisp, T. M., Mangelsdorf, D. J., Haussler, M. R., et al. (1988). Cloning and expression of full-length cDNA encoding human vitamin D receptor. *Proceedings of the National Academy of Sciences of the United States of America*, *85*(10), 3294–3298.

Berger, S. L. (2007). The complex language of chromatin regulation during transcription. *Nature*, *447*(7143), 407–412. http://dx.doi.org/10.1038/nature05915.

Bikle, D. D. (2004). Vitamin D regulated keratinocyte differentiation. *Journal of Cellular Biochemistry*, *92*(3), 436–444.

Bikle, D. D. (2012). Vitamin D and the skin: Physiology and pathophysiology. *Reviews in Endocrine & Metabolic Disorders*, *13*(1), 3–19. http://dx.doi.org/10.1007/s11154-011-9194-0.

Bishop, K. A., Coy, H. M., Nerenz, R. D., Meyer, M. B., & Pike, J. W. (2011). Mouse Rankl expression is regulated in T cells by c-Fos through a cluster of distal regulatory enhancers designated the T cell control region. *The Journal of Biological Chemistry*, *286*(23), 20880–20891. http://dx.doi.org/10.1074/jbc.M111.231548.

Bishop, K. A., Meyer, M. B., & Pike, J. W. (2009). A novel distal enhancer mediates cytokine induction of mouse RANKl gene expression. *Molecular Endocrinology*, *23*(12), 2095–2110. doi:me.2009-0209 [pii]10.1210/me.2009-0209.

Bishop, K. A., Wang, X., Coy, H. M., Meyer, M. B., Gumperz, J. E., & Pike, J. W. (2014). Transcriptional regulation of the human TNFSF11 gene in t cells via a cell type-selective set of distal enhancers. *Journal of Cellular Biochemistry*, *116*(2), 320–330. http://dx.doi.org/10.1002/jcb.24974.

Bouillon, R., Carmeliet, G., Verlinden, L., van Etten, E., Verstuyf, A., Luderer, H. F., et al. (2008). Vitamin D and human health: Lessons from vitamin D receptor null mice. *Endocrine Reviews*, *29*(6), 726–776. doi:er.2008-0004 [pii] 10.1210/er.2008-0004.

Brooks, M. H., Bell, N. H., Love, L., Stern, P. H., Orfei, E., Queener, S. F., et al. (1978). Vitamin-D-dependent rickets type II. Resistance of target organs to 1,25-dihydroxyvitamin D. *The New England Journal of Medicine*, *298*(18), 996–999. http://dx.doi.org/10.1056/NEJM197805042981804.

Brumbaugh, P., & Haussler, M. (1974a). 1 Alpha,25-dihydroxycholecalciferol receptors in intestine. I. Association of 1 alpha,25-dihydroxycholecalciferol with intestinal mucosa chromatin. *The Journal of Biological Chemistry*, *249*(4), 1251–1257.

Brumbaugh, P. F., & Haussler, M. R. (1974b). 1a,25-dihydroxycholecalciferol receptors in intestine. II. Temperature-dependent transfer of the hormone to chromatin via a specific cytosol receptor. *The Journal of Biological Chemistry*, *249*(4), 1258–1262.

Burmester, J. K., Maeda, N., & DeLuca, H. F. (1988). Isolation and expression of rat 1,25-dihydroxyvitamin D3 receptor cDNA. *Proceedings of the National Academy of Sciences of the United States of America*, *85*(4), 1005–1009.

Calo, E., & Wysocka, J. (2013). Modification of enhancer chromatin: What, how, and why? *Molecular Cell*, *49*(5), 825–837. http://dx.doi.org/10.1016/j.molcel.2013.01.038.

Chen, C. H., Sakai, Y., & Demay, M. B. (2001). Targeting expression of the human vitamin D receptor to the keratinocytes of vitamin D receptor null mice prevents alopecia. *Endocrinology*, *142*(12), 5386–5389.

Cianferotti, L., Cox, M., Skorija, K., & Demay, M. B. (2007). Vitamin D receptor is essential for normal keratinocyte stem cell function. *Proceedings of the National Academy of Sciences of the United States of America*, *104*(22), 9428–9433. http://dx.doi.org/10.1073/pnas.0702884104.

Cong, L., Ran, F. A., Cox, D., Lin, S., Barretto, R., Habib, N., et al. (2013). Multiplex genome engineering using CRISPR/Cas systems. *Science*, *339*(6121), 819–823. http://dx.doi.org/10.1126/science.1231143.

Dame, M., Pierce, E., Prahl, J., Hayes, C., & DeLuca, H. (1986). Monoclonal antibodies to the porcine intestinal receptor for 1,25-dihydroxyvitamin D3: Interaction with distinct receptor domains. *Biochemistry*, *25*(16), 4523–4534.

Demay, M. B., MacDonald, P. N., Skorija, K., Dowd, D. R., Cianferotti, L., & Cox, M. (2007). Role of the vitamin D receptor in hair follicle biology. *The Journal of Steroid Biochemistry and Molecular Biology*, *103*(3–5), 344–346. http://dx.doi.org/10.1016/j.jsbmb.2006.12.036.

Ding, N., Yu, R. T., Subramaniam, N., Sherman, M. H., Wilson, C., Rao, R., et al. (2013). A vitamin D receptor/SMAD genomic circuit gates hepatic fibrotic response. *Cell*, *153*(3), 601–613. http://dx.doi.org/10.1016/j.cell.2013.03.028.

Dunham, I., Kundaje, A., Aldred, S. F., Collins, P. J., Davis, C. A., Doyle, F., et al. (2012). An integrated encyclopedia of DNA elements in the human genome. *Nature*, *489*(7414), 57–74. http://dx.doi.org/10.1038/nature11247.

Eil, C., Liberman, U. A., Rosen, J. F., & Marx, S. J. (1981). A cellular defect in hereditary vitamin-D-dependent rickets type II: Defective nuclear uptake of 1,25-dihydroxyvitamin D in cultured skin fibroblasts. *The New England Journal of Medicine*, *304*(26), 1588–1591. http://dx.doi.org/10.1056/NEJM198106253042608.

Evans, R. (1988). The steroid and thyroid hormone receptor superfamily. *Science*, *240*(4854), 889–895.

Feldman, D., & Malloy, P. (1990). Hereditary 1,25-dihydroxyvitamin D resistant rickets: Molecular basis and implications for the role of 1,25(OH) 2D3 in normal physiology. *Molecular and Cellular Endocrinology*, *72*(3), C57–C62.

Forghani, N., Lum, C., Krishnan, S., Wang, J., Wilson, D., Blackett, P., et al. (2010). Two new unrelated cases of hereditary 1,25-dihydroxyvitamin D-resistant rickets with alopecia resulting from the same novel nonsense mutation in the vitamin D receptor gene. *Journal of Pediatric Endocrinology & Metabolism*, *23*(8), 843–850.

Galli, C., Zella, L., Fretz, J., Fu, Q., Pike, J., Weinstein, R., et al. (2008). Targeted deletion of a distant transcriptional enhancer of the receptor activator of nuclear factor-kappaB ligand gene reduces bone remodeling and increases bone mass. *Endocrinology*, *149*(1), 146–153.

Gerstein, M. B., Kundaje, A., Hariharan, M., Landt, S. G., Yan, K. K., Cheng, C., et al. (2012). Architecture of the human regulatory network derived from ENCODE data. *Nature*, *489*(7414), 91–100. http://dx.doi.org/10.1038/nature11245.

Haussler, M. R., & Norman, A. W. (1969). Chromosomal receptor for a vitamin D metabolite. *Proceedings of the National Academy of Sciences of the United States of America*, *62*(1), 155–162.

Heikkinen, S., Väisänen, S., Pehkonen, P., Seuter, S., Benes, V., & Carlberg, C. (2011). Nuclear hormone 1α,25-dihydroxyvitamin D3 elicits a genome-wide shift in the locations of VDR chromatin occupancy. *Nucleic Acids Research*, *39*(21), 9181–9193. http://dx.doi.org/10.1093/nar/gkr654.

Hughes, M. R., Malloy, P. J., Kieback, D. G., Kesterson, R. A., Pike, J. W., Feldman, D., et al. (1988). Point mutations in the human vitamin D receptor gene associated with hypocalcemic rickets. *Science*, *242*(4886), 1702–1705.

Kerner, S. A., Scott, R. A., & Pike, J. W. (1989). Sequence elements in the human osteocalcin gene confer basal activation and inducible response to hormonal vitamin D3. *Proceedings of the National Academy of Sciences of the United States of America*, *86*(12), 4455–4459.

Kim, S., Shevde, N., & Pike, J. (2005). 1,25-Dihydroxyvitamin D3 stimulates cyclic vitamin D receptor/retinoid X receptor DNA-binding, co-activator recruitment, and histone acetylation in intact osteoblasts. *Journal of Bone and Mineral Research*, *20*(2), 305–317.

Kim, S., Yamazaki, M., Shevde, N. K., & Pike, J. W. (2007). Transcriptional control of receptor activator of nuclear factor-kappaB ligand by the protein kinase A activator forskolin and the transmembrane glycoprotein 130-activating cytokine, oncostatin M, is exerted through multiple distal enhancers. *Molecular Endocrinology*, *21*(1), 197–214.

Kim, S., Yamazaki, M., Zella, L. A., Shevde, N. K., & Pike, J. W. (2006). Activation of receptor activator of NF-kappaB ligand gene expression by 1,25-dihydroxyvitamin D3 is mediated through multiple long-range enhancers. *Molecular and Cellular Biology*, *26*(17), 6469–6486.

Lee, S. M., Bishop, K. A., Goellner, J. J., O'Brien, C. A., & Pike, J. W. (2014). Mouse and human BAC transgenes recapitulate tissue-specific expression of the vitamin D receptor in mice and rescue the VDR-null phenotype. *Endocrinology*, *155*(6), 2064–2076. http://dx.doi.org/10.1210/en.2014-1107.

Lee, S. M., Goellner, J. J., O'Brien, C. A., & Pike, J. W. (2014). A humanized mouse model of hereditary 1,25-dihydroxyvitamin D-resistant rickets without alopecia. *Endocrinology*, *155*(11), 4137–4148. http://dx.doi.org/10.1210/en.2014-1417.

Lee, S. M., Riley, E. M., Meyer, M. B., Benkusky, N. A., Plum, L. A., DeLuca, H. F., et al. (2015). 1,25-dihydroxyvitamin D3 controls a cohort of vitamin D receptor target genes in the proximal intestine that is enriched for calcium-regulating components. *The Journal of Biological Chemistry*, *290*(29), 18199–18215. http://dx.doi.org/10.1074/jbc.M115.665794.

Li, Y. C., Pirro, A. E., Amling, M., Delling, G., Baron, R., Bronson, R., et al. (1997). Targeted ablation of the vitamin D receptor: An animal model of vitamin D-dependent rickets type II with alopecia. *Proceedings of the National Academy of Sciences of the United States of America*, *94*(18), 9831–9835.

Lian, J., Stewart, C., Puchacz, E., Mackowiak, S., Shalhoub, V., Collart, D., et al. (1989). Structure of the rat osteocalcin gene and regulation of vitamin D-dependent expression. *Proceedings of the National Academy of Sciences of the United States of America*, *86*(4), 1143–1147.

Malloy, P. J., Pike, J. W., & Feldman, D. (1999). The vitamin D receptor and the syndrome of hereditary 1,25-dihydroxyvitamin D-resistant rickets. *Endocrine Reviews*, *20*(2), 156–188.

Malloy, P. J., Tasic, V., Taha, D., Tütüncüler, F., Ying, G. S., Yin, L. K., et al. (2014). Vitamin D receptor mutations in patients with hereditary 1,25-dihydroxyvitamin D-resistant rickets. *Molecular Genetics and Metabolism*, *111*(1), 33–40. http://dx.doi.org/10.1016/j.ymgme.2013.10.014.

Marx, S. J., Spiegel, A. M., Brown, E. M., Gardner, D. G., Downs, R. W., Attie, M., et al. (1978). A familial syndrome of decrease in sensitivity to 1,25-dihydroxyvitamin D. *The Journal of Clinical Endocrinology and Metabolism*, *47*(6), 1303–1310.

McDonnell, D. P., Mangelsdorf, D. J., Pike, J. W., Haussler, M. R., & O'Malley, B. W. (1987). Molecular cloning of complementary DNA encoding the avian receptor for vitamin D. *Science*, *235*(4793), 1214–1217.

McDonnell, D., Pike, J., & O'Malley, B. (1988). The vitamin D receptor: A primitive steroid receptor related to thyroid hormone receptor. *Journal of Steroid Biochemistry*, *30*(1–6), 41–46.

McDonnell, D., Scott, R., Kerner, S., O'Malley, B., & Pike, J. (1989). Functional domains of the human vitamin D3 receptor regulate osteocalcin gene expression. *Molecular Endocrinology*, *3*(4), 635–644.

Meyer, M. B., Benkusky, N. A., Lee, C. H., & Pike, J. W. (2014). Genomic determinants of gene regulation by 1,25-dihydroxyvitamin D3 during osteoblast-lineage cell differentiation. *The Journal of Biological Chemistry*, *289*(28), 19539–19554. http://dx.doi.org/10.1074/jbc.M114.578104.

Meyer, M. B., Benkusky, N. A., & Pike, J. W. (2014). The RUNX2 cistrome in osteoblasts: Characterization, down-regulation following differentiation, and relationship to gene expression. *The Journal of Biological Chemistry*, *289*(23), 16016–16031. http://dx.doi.org/10.1074/jbc.M114.552216.

Meyer, M. B., Benkusky, N. A., & Pike, J. W. (2015). Selective distal enhancer control of the Mmp13 gene identified through clustered regularly interspaced short palindromic repeat (CRISPR) genomic deletions. *The Journal of Biological Chemistry*, *290*(17), 11093–11107. http://dx.doi.org/10.1074/jbc.M115.648394.

Meyer, M. B., Goetsch, P. D., & Pike, J. W. (2010a). A downstream intergenic cluster of regulatory enhancers contributes to the induction of CYP24A1 expression by 1alpha,25-dihydroxyvitamin D3. *The Journal of Biological Chemistry*, *285*(20), 15599–15610. http://dx.doi.org/10.1074/jbc.M110.119958.

Meyer, M. B., Goetsch, P. D., & Pike, J. W. (2010b). Genome-wide analysis of the VDR/RXR cistrome in osteoblast cells provides new mechanistic insight into the actions of the vitamin D hormone. *The Journal of Steroid Biochemistry and Molecular Biology*, *121*(1–2), 136–141. doi:S0960-0760(10)00065-8 [pii]10.1016/j.jsbmb.2010.02.011.

Meyer, M. B., Goetsch, P. D., & Pike, J. W. (2012). VDR/RXR and TCF4/β-catenin cistromes in colonic cells of colorectal tumor origin: Impact on c-FOS and c-MYC gene expression. *Molecular Endocrinology*, *26*(1), 37–51. doi:me.2011-1109 [pii]10.1210/me.2011-1109.

Meyer, M. B., & Pike, J. W. (2013). Corepressors (NCoR and SMRT) as well as coactivators are recruited to positively regulated 1α,25-dihydroxyvitamin D3-responsive genes. *The Journal of Steroid Biochemistry and Molecular Biology*, *136*, 120–124. http://dx.doi.org/10.1016/j.jsbmb.2012.08.006.

Nilsson, S. K., Johnston, H. M., Whitty, G. A., Williams, B., Webb, R. J., Denhardt, D. T., et al. (2005). Osteopontin, a key component of the hematopoietic stem cell niche and regulator of primitive hematopoietic progenitor cells. *Blood*, *106*(4), 1232–1239. doi:2004-11-4422 [pii]10.1182/blood-2004-11-4422.

Ohyama, Y., Ozono, K., Uchida, M., Shinki, T., Kato, S., Suda, T., et al. (1994). Identification of a vitamin D-responsive element in the 5′-flanking region of the rat 25-hydroxyvitamin D3 24-hydroxylase gene. *The Journal of Biological Chemistry*, *269*(14), 10545–10550.

Onal, M., Bishop, K. A., St John, H. C., Danielson, A. L., Riley, E. M., Piemontese, M., et al. (2014). A DNA segment spanning the mouse Tnfsf11 transcription unit and its upstream regulatory domain rescues the pleiotropic biologic phenotype of the RANKL null mouse. *Journal of Bone and Mineral Research*, *30*(5), 855–868. http://dx.doi.org/10.1002/jbmr.2417.

Ozono, K., Liao, J., Kerner, S. A., Scott, R. A., & Pike, J. W. (1990). The vitamin D-responsive element in the human osteocalcin gene. Association with a nuclear proto-oncogene enhancer. *The Journal of Biological Chemistry*, *265*(35), 21881–21888.

Pan, Y. F., Wansa, K. D., Liu, M. H., Zhao, B., Hong, S. Z., Tan, P. Y., et al. (2008). Regulation of estrogen receptor-mediated long range transcription via evolutionarily conserved distal response elements. *The Journal of Biological Chemistry*, *283*(47), 32977–32988. http://dx.doi.org/10.1074/jbc.M802024200.

Panda, D. K., Miao, D., Tremblay, M. L., Sirois, J., Farookhi, R., Hendy, G. N., et al. (2001). Targeted ablation of the 25-hydroxyvitamin D 1alpha -hydroxylase enzyme: Evidence for skeletal, reproductive, and immune dysfunction. *Proceedings of the National Academy of Sciences of the United States of America*, *98*(13), 7498–7503. http://dx.doi.org/10.1073/pnas.131029498.

Pike, J. (1991). Vitamin D3 receptors: Structure and function in transcription. *Annual Review of Nutrition*, *11*, 189–216.

Pike, J., Dokoh, S., Haussler, M., Liberman, U., Marx, S., & Eil, C. (1984). Vitamin D3-resistant fibroblasts have immunoassayable 1,25-dihydroxyvitamin D3 receptors. *Science*, *224*(4651), 879–881.

Pike, J., Donaldson, C., Marion, S., & Haussler, M. (1982). Development of hybridomas secreting monoclonal antibodies to the chicken intestinal 1 alpha,25-dihydroxyvitamin D3 receptor. *Proceedings of the National Academy of Sciences of the United States of America*, *79*(24), 7719–7723.

Pike, J. W., & Haussler, M. R. (1979). Purification of chicken intestinal receptor for 1,25-dihydroxyvitamin D. *Proceedings of the National Academy of Sciences of the United States of America*, *76*(11), 5485–5489.

Pike, J. W., Lee, S. M., & Meyer, M. B. (2014). Regulation of gene expression by 1,25-dihydroxyvitamin D3 in bone cells: Exploiting new approaches and defining new mechanisms. *Bonekey Report*, *3*, 482. http://dx.doi.org/10.1038/bonekey.2013.216.

Pike, J. W., & Meyer, M. B. (2014). Fundamentals of vitamin D hormone-regulated gene expression. *The Journal of Steroid Biochemistry and Molecular Biology*, *144PA*, 5–11. http://dx.doi.org/10.1016/j.jsbmb.2013.11.004.

Pike, J. W., Meyer, M. B., Martowicz, M. L., Bishop, K. A., Lee, S. M., Nerenz, R. D., et al. (2010). Emerging regulatory paradigms for control of gene expression by 1,25-dihydroxyvitamin D3. *The Journal of Steroid Biochemistry and Molecular Biology*, *121*(1–2), 130–135. http://dx.doi.org/10.1016/j.jsbmb.2010.02.036.

Ramagopalan, S. V., Heger, A., Berlanga, A. J., Maugeri, N. J., Lincoln, M. R., Burrell, A., et al. (2010). A ChIP-seq defined genome-wide map of vitamin D receptor binding: Associations with disease and evolution. *Genome Research*, *20*(10), 1352–1360. http://dx.doi.org/10.1101/gr.107920.110.

Ruthenburg, A. J., Allis, C. D., & Wysocka, J. (2007). Methylation of lysine 4 on histone H3: Intricacy of writing and reading a single epigenetic mark. *Molecular Cell*, *25*(1), 15–30. doi:S1097-2765(06)00872-0 [pii]10.1016/j.molcel.2006.12.014.

Shahbazian, M. D., & Grunstein, M. (2007). Functions of site-specific histone acetylation and deacetylation. *Annual Review of Biochemistry*, *76*, 75–100. http://dx.doi.org/10.1146/annurev.biochem.76.052705.162114.

Sherman, M. H., Yu, R. T., Engle, D. D., Ding, N., Atkins, A. R., Tiriac, H., et al. (2014). Vitamin D receptor-mediated stromal reprogramming suppresses pancreatitis and enhances pancreatic cancer therapy. *Cell*, *159*(1), 80–93. http://dx.doi.org/10.1016/j.cell.2014.08.007.

Skorija, K., Cox, M., Sisk, J. M., Dowd, D. R., MacDonald, P. N., Thompson, C. C., et al. (2005). Ligand-independent actions of the vitamin D receptor maintain hair follicle homeostasis. *Molecular Endocrinology*, *19*(4), 855–862. http://dx.doi.org/10.1210/me.2004-0415.

St John, H. C., Bishop, K. A., Meyer, M. B., Benkusky, N. A., Leng, N., Kendziorski, C., et al. (2014). The osteoblast to osteocyte transition: Epigenetic changes and response to the vitamin D3 hormone. *Molecular Endocrinology*, *28*(7), 1150–1165. http://dx.doi.org/10.1210/me.2014-1091.

St John, H. C., Meyer, M. B., Benkusky, N. A., Carlson, A. H., Prideaux, M., Bonewald, L. F., et al. (2014). The parathyroid hormone-regulated transcriptome in osteocytes: Parallel actions with 1,25-dihydroxyvitamin D3 to oppose gene expression changes during differentiation and to promote mature cell function. *Bone*, *72C*, 81–91. http://dx.doi.org/10.1016/j.bone.2014.11.010.

Sutton, A. L., & MacDonald, P. N. (2003). Vitamin D: More than a "bone-a-fide" hormone. *Molecular Endocrinology*, *17*(5), 777–791.

Yoshizawa, T., Handa, Y., Uematsu, Y., Takeda, S., Sekine, K., Yoshihara, Y., et al. (1997). Mice lacking the vitamin D receptor exhibit impaired bone formation, uterine hypoplasia and growth retardation after weaning. *Nature Genetics*, *16*(4), 391–396. http://dx.doi.org/10.1038/ng0897-391.

Zella, L. A., Kim, S., Shevde, N. K., & Pike, J. W. (2006). Enhancers located within two introns of the vitamin D receptor gene mediate transcriptional autoregulation by 1,25-dihydroxyvitamin D3. *Molecular Endocrinology*, *20*(6), 1231–1247. http://dx.doi.org/10.1210/me.2006-0015.

Zella, L. A., Meyer, M. B., Nerenz, R. D., Lee, S. M., Martowicz, M. L., & Pike, J. W. (2010). Multifunctional enhancers regulate mouse and human vitamin D receptor gene transcription. *Molecular Endocrinology*, *24*(1), 128–147. doi:me.2009-0140 [pii] 10.1210/me.2009-0140.

Zierold, C., Darwish, H. M., & DeLuca, H. F. (1995). Two vitamin D response elements function in the rat 1,25-dihydroxyvitamin D 24-hydroxylase promoter. *The Journal of Biological Chemistry*, *270*(4), 1675–1678.

CHAPTER THREE

Inhibitors for the Vitamin D Receptor–Coregulator Interaction

Kelly A. Teske, Olivia Yu, Leggy A. Arnold[1]
Department of Chemistry and Biochemistry, Milwaukee Institute for Drug Discovery (MIDD), University of Wisconsin-Milwaukee, Milwaukee, Wisconsin, USA
[1]Corresponding author: e-mail address: arnold2@uwm.edu

Contents

Abstract

The vitamin D receptor (VDR) belongs to the superfamily of nuclear receptors and is activated by the endogenous ligand 1,25-dihydroxyvitamin D_3. The genomic effects mediated by VDR consist of the activation and repression of gene transcription, which includes the formation of multiprotein complexes with coregulator proteins. Coregulators bind many nuclear receptors and can be categorized according to their role as coactivators (gene activation) or corepressors (gene repression). Herein, different approaches to develop compounds that modulate the interaction between VDR and coregulators are summarized. This includes coregulator peptides that were identified by creating phage display libraries. Subsequent modification of these peptides including the introduction of a tether or nonhydrolyzable bonds resulted in the first direct VDR–coregulator inhibitors. Later, small molecules that inhibit VDR–coregulator inhibitors were identified using rational drug design and high-throughput screening. Early on, allosteric inhibition of VDR–coregulator interactions was achieved with VDR

Vitamins and Hormones, Volume 100
ISSN 0083-6729
http://dx.doi.org/10.1016/bs.vh.2015.10.002

antagonists that change the conformation of VDR and modulate the interactions with coregulators. A detailed discussion of their dual agonist/antagonist effects is given as well as a summary of their biological effects in cell-based assays and *in vivo* studies.

1. INTRODUCTION

1,25-Dihydroxyvitamin D_3 (1,25-$(OH)_2D_3$) is a highly active metabolic product of vitamin D_3, which is produced from cholesterol via a light-induced rearrangement in skin cells or absorbed from various food sources in the intestine. The phase I metabolism of vitamin D3 is mediated by P_{450} enzymes, especially CYP2R1 (Cheng, Levine, Bell, Mangelsdorf, & Russell, 2004) and CYP27B1 (Takeyama et al., 1997), which are mainly expressed in the liver and kidney. The hydroxylation of vitamin D_3 to 1,25-$(OH)_2D_3$ increases its affinity for the vitamin D receptor (VDR), which is a member of the nuclear receptor superfamily. VDR, like many nuclear receptors, binds DNA as a heterodimer with retinoid X receptor (RXR), which is promoted by 1,25-$(OH)_2D_3$ (Thompson, Jurutka, Haussler, Whitfield, & Haussler, 1998). Although ligand binding is essential for gene regulation, many other proteins have been identified that mediate the process of gene transcription. These include coregulatory proteins or coregulators that bridge the gap between the nuclear receptor dimer and RNA polymerase II. Currently more than 250 coregulators have been identified, but their exact functions and interactions with DNA-bound VDR are only partially understood (Lonard & O'Malley, 2012). Coregulators can be categorized as either coactivators or corepressors depending on their influence on gene expression. Coactivators may stabilize the transcriptional machinery, have endogenous histone acetyl transferase (HAT) activity or other enzymatic functions that alter chromatin structure, or are involved in mRNA maturation (Fig. 1). In contrast, corepressors may reduce the basal activity of nuclear receptor-mediated transcription by means of disrupting the DNA-bound multiprotein transcriptional complex or by recruitment of histone deacetylases (HDAC) to induce a tighter chromatin structure. Herein, we will discuss coregulators relevant to VDR and small molecules that can inhibit these interactions, with an emphasis on VDR–coactivator interactions. We will also summarize the physiological effects of these small molecules *in vitro* and *in vivo*.

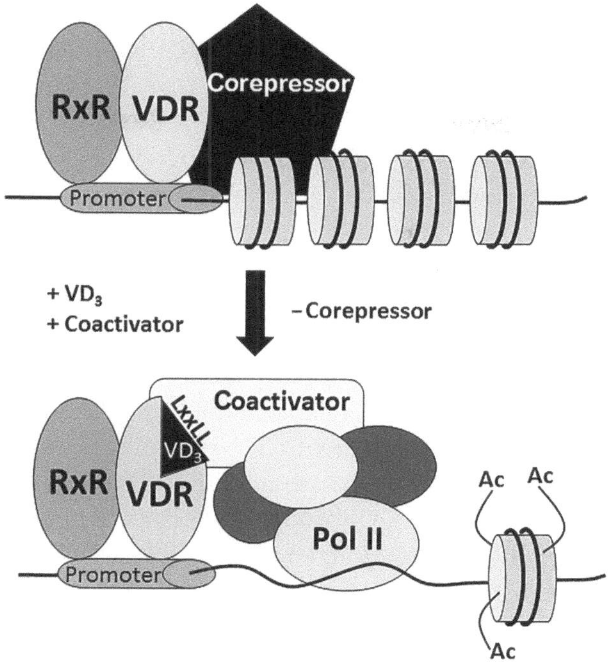

Figure 1 Simplified cartoon of VDR-mediated transcription. Repressed basal transcription of VDR as a heterodimer in the absence of ligand and presence of corepressor followed by the activation of transcription in the presence of 1,25-$(OH)_2D_3$ (VD3) and coactivator leading to histone acetylation and recruitment of RNA polymerase II.

1.1 VDR Coactivators

The binding between VDR and 1,25-$(OH)_2D_3$ induces a conformational change that enables the interaction between VDR and coactivators; thus, agonist binding is important for most VDR–coactivator interactions. On the molecular level, this includes the repositioning of helix 12 (or the C-terminus of VDR) to form a new hydrophobic pocket that enables coactivator interaction. The molecular interaction between VDR and steroid coactivator 1 (SRC1) has been elucidated by X-ray crystallography with a SRC1 peptide that bears the nuclear interaction domain (NID) with a central LXXLL motif (L = leucine and X = any amino acid) (Vanhooke, Benning, Bauer, Pike, & DeLuca, 2004). Although coactivator mutation studies have identified essential amino acids and a minimal amino acid sequence for nuclear receptor binding, crystal structures confirmed the presence of hydrogen bonding with the E420 and K246 residues of VDR, as well as hydrophobic interactions with NID leucine residues. SRC1 and its related coactivators SRC2 and SRC3 exhibit three NIDs (Xu, Wu, & O'Malley, 2009). Binding studies with different SRC NIDs have demonstrated the preference of VDR for the third NID of all SRCs (Teichert et al., 2009). Once bound to the DNA/VDR/RXR complex, SRCs are able to transfer acetyl groups to histones to weaken the interaction between negatively charged DNA phosphate backbone and the usually positively charged histone (Chen et al., 1997; Spencer et al., 1997). In addition, SRCs recruit other proteins with HAT activity such as CREB-binding protein/p300 (Ogryzko, Schiltz, Russanova, Howard, & Nakatani, 1996) and p300/CBP-associated factor (Yang, Ogryzko, Nishikawa, Howard, & Nakatani, 1996). Further relaxation of chromatin may be achieved with the recruitment of methyltransferase 1 and protein arginine *N*-methyltransferase 1 (Chen et al., 1999; Koh, Chen, Lee, & Stallcup, 2001). The vitamin D receptor-interacting protein (DRIP) complex includes coactivator DRIP205 (Rachez et al., 1998), which directly interacts with VDR and is important for the recruitment of RNA polymerase II (Chiba, Suldan, Freedman, & Parvin, 2000). DRIP205 and other partners of this coactivator complex lack HAT activity (Rachez et al., 1999), supporting the model of a sequential recruitment of DRIP205 after the dissociation of VDR and SRCs (Sharma & Fondell, 2002). Other coactivators have been identified that interact with VDR, such as PGC-1 alpha (Savkur, Bramlett, Stayrook, Nagpal, & Burris, 2005), NCoA62 (Baudino et al., 1998), Smad3 (Lekanne Deprez et al., 1995), Ets-1 (Tolon, Castillo, Jimenez-Lara, & Aranda, 2000), WINAC (Kitagawa et al., 2003), and CCAAT displacement protein (Ochiai et al., 2010).

1.2 VDR Corepressors

Similar to other nuclear receptors, VDR interacts with corepressors in the absence of ligand or in the presence of antagonists (Perissi, Jepsen, Glass, & Rosenfeld, 2010). Crystal structures reveal that this interaction includes a different orientation of helix 12 and requires multiple and longer NIDs (Xu et al., 2002). The most studied corepressors for VDR are the nuclear receptor corepressor (NCoR) (Horlein et al., 1995) and the silencing mediator for retinoid or thyroid-hormone receptors (SMRT) (Chen & Evans, 1995). In complex with VDR, they recruit HDAC3 which in turn improves chromatin packing close to the promotor site and represses transcription (Perissi et al., 2010). Other corepressors interacting with VDR are Hairless (Xie, Chang, Oda, & Bikle, 2006) and Alien (Polly et al., 2000).

2. PEPTIDE-BASED INHIBITORS OF THE VDR–COREGULATOR INTERACTION

The identification of the central coactivator LXXLL motif as essential to meditating nuclear receptor binding prompted investigations to develop peptide-based VDR–coactivator inhibitors to evaluate the function of this protein–protein interaction. Pioneered by McDonnell et al. for the estrogen receptor, a phage display library of synthetic LXXLL peptides was generated and screened with two-hybrid assays against a panel of nuclear receptors including VDR (Chang et al., 1999; Hall, Chang, & McDonnell, 2000; McDonnell, Chang, & Norris, 2000). Two peptides, C33 and D47, were identified to bind VDR (Fig. 2). Further investigations of this library identified two more peptides, EBIP41 and EBIP44, with moderate affinity for VDR. Importantly, when C33, D47, EBIP41, and EBIP44 peptides were expressed as Gal4 DBD (DNA-binding domain) fusions in cells, they inhibited the VDR-mediated transcription in a reporter assay under control of an osteocalcin promoter (Pathrose et al., 2002). In addition, RXR-selective peptide F6 was able to inhibit VDR-mediated transcription, demonstrating transactivation between RXR and VDR. A more exhaustive phage display library identified three more LXXLL peptides (Fig. 2, compounds 3, 4, and 5) that not only bind VDR in two-hybrid assays but also inhibit VDR-mediated transcription when expressed in cells (Zella, Chang, McDonnell, & Pike, 2007). These peptides exhibit a consensus sequence of (H/F)P(L/M)LXXLL. Importantly, the binding of these peptides to VDR was more pronounced in the presence of VDR agonists than VDR antagonists. However, the limitation of these peptides for further investigation is their inability to regulate endogenous VDR target genes. Overall, phage display studies contributed to the validation of VDR–coactivator interactions as a target to modulate transcription. In addition, they produced the first

																					Target
C33		H	V	E	M	H	P	L	L	M	G	L	L	M	E	S	Q	W	G	A	VDR
D47		H	V	Y	Q	H	P	L	L	L	S	L	L	S	S	E	H	E	S	G	VDR
EBIP41		R	R	D	D	F	P	L	L	I	S	L	L	K	D	G	A	L	S	Q	VDR
EBIP44		Y	G	L	K	M	S	L	L	E	S	L	L	R	E	D	I	S	T	V	VDR
F6		G	H	E	P	L	T	L	L	E	R	L	L	S	G	T	S	V	A	E	RxR
3		L	S	E	T	H	P	L	L	W	T	L	L	S	S	E	G	D	S	M	VDR
4		M	Q	E	R	F	P	M	L	W	D	L	L	D	L	P	S	P	T	S	VDR
5		L	G	E	S	H	P	L	L	M	Q	L	L	T	E	N	V	G	T	H	VDR

Figure 2 Sequences of coactivator peptides that inhibit the interaction between VDR and coactivators.

potent and selective VDR–coactivator inhibitors. Currently, coregulator peptides are used as fluorescent probes for high throughput assays to identify new nuclear receptor modulators (Arnold et al., 2006).

Different strategies have been developed to overcome the limitations of peptide reagents in cell-based assays, such as inactivity when transfected as a fusion peptide or limited cell permeability and stability (Caboni & Lloyd, 2013). For the thyroid receptor and estrogen receptor–coactivator interactions, cyclic LXXLL peptides were generated by means of a lactam bridge, resulting in inhibitors with IC_{50} values of less than 100 nM (Geistlinger & Guy, 2001, 2003). Improvements on these peptide inhibitors were made by introducing a nonredox-sensitive thioether bridge (Galande, Bramlett, Burris, Wittliff, & Spatola, 2004) and unnatural amino acids such as neopentyl glycine instead of leucine. The most active inhibitors, known as PERMS (peptidomimetic estrogen receptor modulators) had IC_{50} values as low as 70 pM (Galande et al., 2005). Cellular activity for these peptide inhibitors was not reported until Carraz, Zwart, Phan, Michalides, and Brunsveld (2009) developed them with a nona-arginine tag (R_9) that was earlier introduced by Futaki et al. (2001) as an intracellular delivery tag. PERMs with a R_9 tag were reported recently with cellular activities as low as 3 μM for the regulation of the pS2 gene expression in MCF-7 cells (Nagakubo et al., 2014). Recently, new cyclic peptide VDR–coactivator inhibitors were introduced by Demizu et al. (2013) and Misawa, Demizu, Kawamura, Yamagata, and Kurihara (2015) bearing hydrocarbon linkers that have been reported to increase stability and oral bioavailability (Bird et al., 2010). A dramatic IC_{50} value change from 220 to 3.2 μM was observed by changing the nonfunctional linker (Fig. 3, DPI-06) to a functionalized linker (Fig. 3, DPI-07). Prior to this study, it was reported that all hydrocarbon linkers can result in multiple binding modes of cyclic peptides and, in the case of stapled peptide ER–coactivator inhibitors, they did not significantly increase their potency (Phillips et al., 2011). The strong influence of the linker on VDR binding makes the design of these molecules unpredictable and without cell-based data, uncertain of their permeability and stability.

3. SMALL-MOLECULE INHIBITORS OF THE VDR–COREGULATOR INTERACTION

During the last decade, many direct inhibitors of various nuclear receptor–coregulator interactions have been developed (Caboni & Lloyd, 2013;

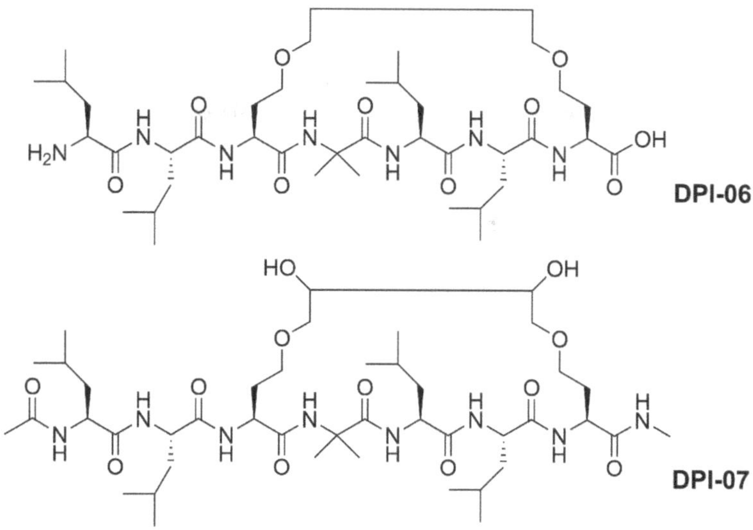

Figure 3 Structures of cyclic peptide-based VDR–coactivator inhibitors.

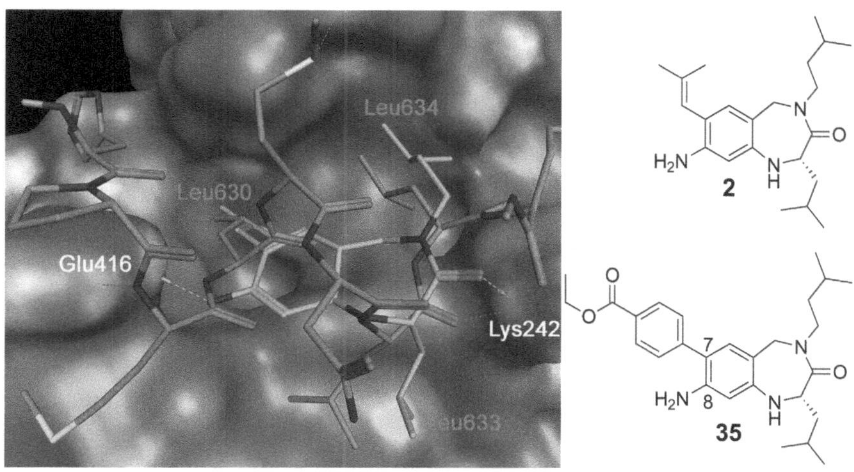

Figure 4 Overlay between a crystal structure of VDR and coactivator peptide DRIP205 and docked conformation of compound 2 as well as structures of compounds 2 and 35. (See the color plate.)

Moore, Mayne, & Katzenellenbogen, 2010). In regard to VDR, Mita et al. (2010) reported the first reversible inhibitor of the VDR–coactivator interaction in 2010. Using a rational design approach, a benzodiazepine scaffold was substituted with branched hydrophobic groups to mimic the i, $i+3$, and $i+4$ positions of leucine in coactivator DRIP205 (Fig. 4; Vanhooke et al., 2004). Docking studies revealed that compound 2 might form hydrogen bonds with rat VDR clamp residues Glu416 and Lys242. The inhibition activity of compound 2 (IC_{50} = 17 μM) was recapitulated in cells with a reporter gene assay. In addition, compound 2 inhibited estrogen receptor β-mediated transcription at a similar concentration, whereas ERα-mediated transcription was not modulated. Prompted by these results, a more exhaustive structure–activity relationship (SAR) study was reported in 2013 by the same group (Mita, Dodo, Noguchi-Yachide, Hashimoto, & Ishikawa, 2013). Despite the large number of analogs with various substituents in the 7- and 8-position, only marginal improvement (IC_{50} = 14 μM) was observed for compound 35 (Fig. 4). However, the aniline function in the 8-position was confirmed to be important for binding, probably interacting with Glu417 of VDR.

The first irreversible VDR–coactivator inhibitors were identified in 2012 using high-throughput screening (Nandhikonda et al., 2012). Among 275,000 compounds, 140 inhibitors with cellular activity were identified, including a group of 3-indolylmethanamines. A comprehensive SAR study around the 3-indolylmethanamine scaffold identified compound 31B as the most active VDR–coactivator inhibitor in cells (IC_{50} = 4.2 μM [Fig. 5A]). In addition, a linear free-energy relationship between inhibition rates of 3-indolylmethanamines bearing different electronic substituents confirmed irreversibility. Due to the unique mode of binding, a high selectivity of 31B toward VDR with respect to other nuclear receptors was observed. In addition, 31B is selective toward the interaction between VDR and coregulator peptide SRC2-3 above other LXXLL coregulator peptides. Importantly, downregulation of VDR target gene *TRPV6* by 31B was observed in the presence of $1,25(OH)_2D_3$ for DU145 cancer cells as well as antiproliferation at higher concentration. Inhibition of VDR-mediated transcription and

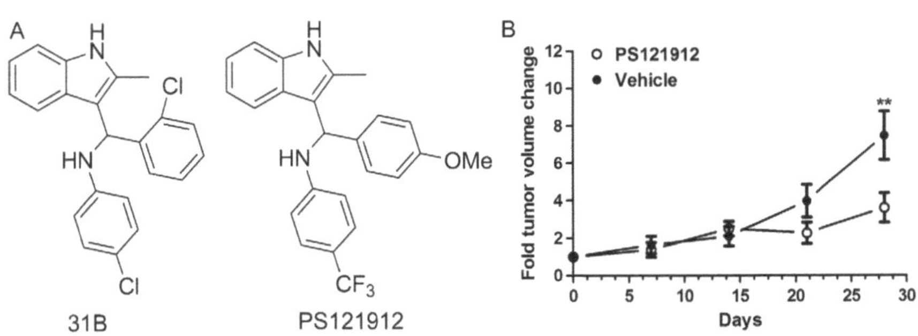

Figure 5 (A) Structures of 31B and PS121912; (B) antiproliferative effect of PS121912 in a HL60 xenograft model.

antiproliferation in the presence of 31B was also observed for ovarian cancer cells OVCAR8 and SKOV3 and endometrial cancer cells ECC-1. In *cis*-platinum resistant SKOV3 cells, other biomarkers of antiproliferation and apoptosis were upregulated in the presence of 31B, such as activation of caspase 3, phosphorylation of MAP kinases p38 and SAPN/JNK, upregulation of P21, and cell-cycle arrest. In a cisplatin-resistant SKOV3 xenograft tumor model, 31B treatment delivered five times a week at a dose of 5 mg/kg led to suppressed tumor growth after 2 weeks (Guthrie et al., 2015). In addition, reduced tumor formation was partially caused by a compromised *de novo* production of fatty acids due to lower expression of FASN in the tumor. Further SAR studies resulted in the discovery of 3-indolylmethanamine PS121912, a VDR–coactivator inhibitor that inhibited VDR-mediated transcription with an IC_{50} of 590 nM (Fig. 5A; Sidhu, Nassif, et al., 2014). Similar to 31B, PS121912 is selective toward VDR and has a preference for the interaction between VDR and coregulator peptide SRC2-3. Importantly, ChIP studies revealed that in HL60 leukemia cells PS121912 was able to reduce the DNA occupancy of VDR and binding of SRC2. However, PS121912 promoted the recruitment of NCoR to the VDR–DNA complex (Sidhu, Teske, et al., 2014). PS121912 reversed the regulation of VDR target genes in the presence of 1,25-$(OH)_2D_3$ at a concentration of 500 nM and modulated the transcription of many genes affiliated with the cell-cycle control. Elevated levels of P21 protein levels were observed for the PS121912 in the presence and absence of 1,25-$(OH)_2D_3$ in HL60 cells as well as increased levels of pro-apoptotic serine protease HTRA. In a mouse HL60 xenograft model at 3 mg/kg five times a week, a significant change in tumor volume was observed after 3 weeks of treatment (Fig. 5B). The blood calcium levels and animal weight did not differ from the control group.

4. VDR ANTAGONISTS OR ALLOSTERIC INHIBITION OF THE VDR–COREGULATORS INTERACTION

The synthesis of new synthetic analogs of 1,25-$(OH)_2D_3$ resulted in the identification of new VDR ligands that initiate the recruitment of coactivators much like 1,25-$(OH)_2D_3$. However, a different class of VDR ligands was discovered that binds VDR and only weakly promotes VDR–coactivator interactions. Usually, the biological effects of these antagonists have been determined in the presence of agonists like 1,25-$(OH)_2D_3$, giving results similar to the vehicle control. Interestingly, the degree of coactivator recruitment by VDR depends on the chemical structure of

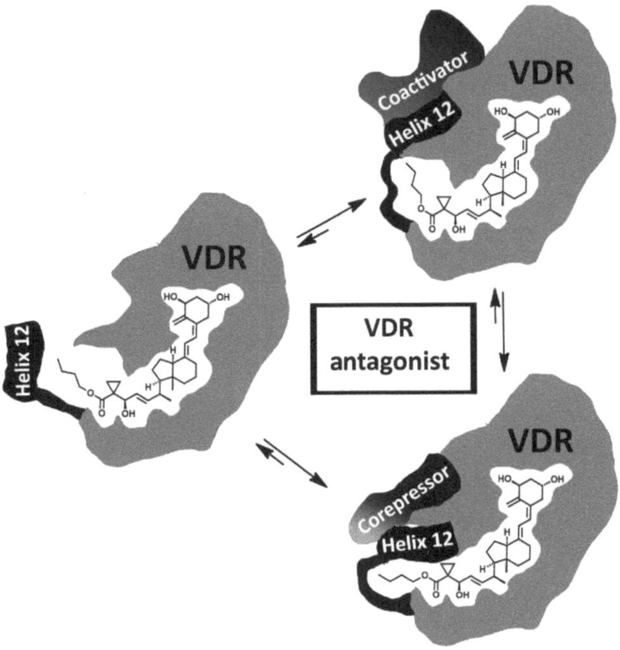

Figure 6 Possible equilibrium structures of VDR in the presence of antagonist.

the VDR antagonist. Thus, the quality of a VDR antagonist can be defined by its residual agonistic activity. On the molecular level, this behavior is believed to be caused by the orientation of helix 12 (Fig. 6). Depending on the structure, VDR antagonists may influence the equilibrium of VDR bound to coactivators, corepressors, or neither. Crystal structures of all three possible complexes have been reported for nuclear receptors. However, VDR prefers to crystallize solely with an agonist arrangement. Recently, VDR-antagonist structures showed some significant differences in their overall structure in comparison with the VDR-agonist complex. However, it is believed that these high energy structures are accompanied by less-ordered VDR-antagonist structures that do not crystallize. Herein, we will discuss the biological consequences of VDR antagonists in the presence and absence of 1,25-$(OH)_2D_3$.

4.1 TEI-9647

Early identification of antagonist effects of VDR ligands was based on their ability to inhibit the differentiation of promyelocytic leukemia cells. In the presence of 1,25-$(OH)_2D_3$, HL60 cells transition to monocytes, which is believed to be mediated by genomic effects of VDR, including the recruitment of coregulators (Lee, Inaba, DeLuca, & Mellon, 1989). In the contrary, NB4 cell differentiation in the presence of 1,25-$(OH)_2D_3$ is considered a model for nongenomic 1,25-$(OH)_2D_3$-mediated effects (Bhatia, Kirkland, & Meckling-Gill, 1995). TEI-9647 and its diastereomer TEI-9648 (Fig. 7) inhibited HL60 differentiation but not NB4 differentiation in the presence of 1,25-$(OH)_2D_3$ (Miura, Manabe, Gao, Norman, & Ishizuka, 1999). The differentiation was quantified by the reduction of a redox-sensitive dye (nitro blue tetrazolium chloride or NBT) and modulation of cell surface marker expression of CD71 and CD11b (Miura, Manabe, Ozono, et al., 1999). Cell differentiation can also be induced by retinoic acids (Breitman, Selonick, & Collins, 1980), promoting the transition of HL60 cells into granulocytes. TEI-9647 and TEI-9648 were not able to inhibit this transition in the presence of 1,25-$(OH)_2D_3$; thus both compounds are not interacting with RXR (Miura, Manabe, Ozono, et al., 1999). TEI-9647 and TEI-9648 bind VDR 10-fold less than 1,25-$(OH)_2D_3$ (Miura, Manabe, Ozono, et al., 1999). TEI-9647 reduced the expression of VDR target genes *CYP24A1* and *P21* in HL60 cells treated with 1,25-$(OH)_2D_3$ (Ishizuka et al., 2000; Ishizuka, Miura, Ozono, Saito et al., 2001). Reporter assays carried out with different cells lines confirmed transcriptional inhibition by TEI-9747 in the presence of 1,25-$(OH)_2D_3$—although to a different degree. In HeLa cells (human cervical carcinoma) TEI-9647 behaved like a very weak antagonist, which was partially caused by the inability of TEI-9647 to block VDR–RXR binding (Ozono et al., 1999). However, TEI-9647 did not influence the VDR–DNA binding shown by gel shift. Importantly, TEI-9647 reduced the binding between VDR and coactivator SRC1 in Saos-2 cells but did not influence the effect

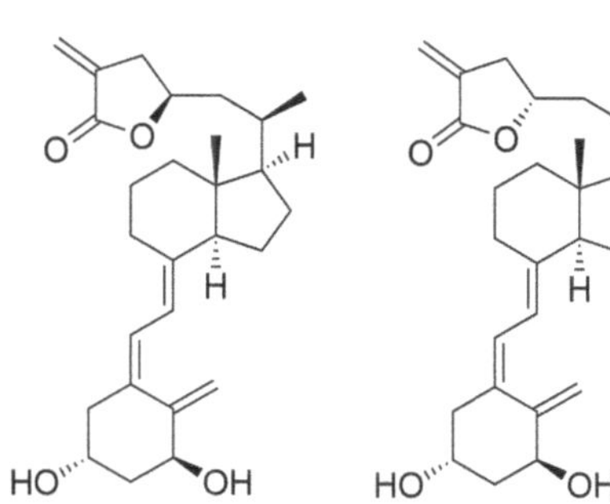

TEI-9648 TEI-9647

Figure 7 Structures of TEI-9647 and TEI-9648.

of 1,25-$(OH)_2D_3$ with respect to translocation to the nucleus as shown with a GFP-labeled VDR in COS7 cells. The partial antagonist/agonist effect of TEI-9647 was also observed by (Toell et al., 2001) using gel shift experiments to quantify the binding between VDR and RXR. However, the antagonist activity of TEI-9647 was much greater than its agonist activity with an IC_{50} value of 2.5 nM. TEI-9647 induces the release of coactivator SRC2 and corepressor NCoR *in vitro*, which was observed in the presence and absence of RXR (Ochiai et al., 2005; Toell et al., 2001). In addition, TEI-9647 inhibited the recruitment of SRC3 and DRIP205 (Yamaoka et al., 2006).

An osteocalcin VDRE reporter assay in COS7 cells was used to investigate TEI-9647 in regard to a possible conjugate addition with VDR's nucleophilic residues H397 or H305 because analogs of TEI-9647 lacking the exocyclic double bond are very weak VDR binders (Fig. 8; Bula, Bishop, Ishizuka, & Norman, 2000). Although TEI-9647 was fully reversible with 1,25-$(OH)_2D_3$, limited proteolysis assays showed a new 30 kDa band of VDR-LBD in the presence of TEI-9647. Further mechanistic studies using X-ray crystallography and mass spectrometry demonstrated that rat VDR but not human VDR showed a second mass peak after incubation with TEI9647 (Kakuda et al., 2010). The formation of a covalent adduct between VDR and TEI-9647 is supported by the fact that in contrast to rat VDR, human VDR has two cysteine residues (C403 and C410) in proximity to the unsaturated lactone of TEI-9647 (Fig. 8). However, an agonist conformation with respect to the orientation of VDR's helix 12 was observed for crystal structures with TEI-9647 and rVDRwt, hVDR H305F, and hVDRH305F/H397F mutants. Although 1,25-$(OH)_2D_3$ interacts with both VDR histidine residues, TEI-9647 prefers H309 and interacts with H395 if H305 is mutated. The authors suggested that the antagonist action of TEI-9647 involves the interaction with H395 that positions the Michael acceptor (TEI-9647) toward C403 followed by alkylation. Further analysis of TEI-9647 using binding studies and molecular modeling demonstrated that TEI-9647 is a potent VDR agonist for hVDRH305F and hVDRH305F/H397F, reducing the ability of TEI-9647 to covalently

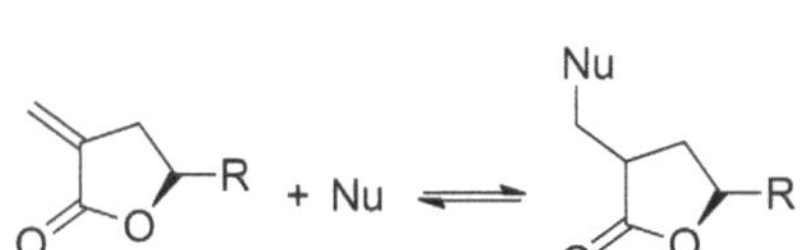

Figure 8 Nucleophilic addition between TEI-9647 and VDR (Nu).

interact with VDR (Mizwicki et al., 2009). Therefore, TEI-9647 behaved like an antagonist in human cells but as a weak agonist in rat cells (Perakyla, Molnar, & Carlberg, 2004). Mutation of hVDR C403 and C410 residues eliminated the antagonistic effects of TEI-9647, and incorporation of cysteine residues into rVDR increased the antagonist activity of TEI-9647 (Ochiai et al., 2005). Interestingly, the affinity of TEI-9647 for hVDR and hVDR C403S/C410N was similar, whereas the efficacy for hVDR was significantly higher for the double mutant. Importantly, TEI-9747 bound to hVDR was not able to recruit coactivator SRC2, in contrast to the hVDR double mutant.

Not surprisingly, the first analogs of TEI-9647, HLV, and GC-3, behaved as agonists in a rat ROS17/2.8 osteoblast transcription assay and were able to increase calcium transport *in vivo* in D-deficient rats without antagonizing this action in the presence of 1,25-$(OH)_2D_3$ (Fig. 9; Chiellini et al., 2008). However, in COS7 monkey kidney cells, both compounds behaved as transcriptional antagonists in the presence of 1,25-$(OH)_2D_3$ (Inaba et al., 2007; Yoshimoto et al., 2008). Interestingly, only GC-3 was able to promote the binding between VDR and RXR, although both compounds reduced the VDR–RXR interaction in the presence of 1,25-$(OH)_2D_3$. In addition, both compounds reduced the recruitment of coactivators SRC1, SRC2, and SRC3 to VDR in the presence of 1,25-$(OH)_2D_3$ and inhibited the interaction with corepressors NCoR and SMRT in the absence of 1,25-$(OH)_2D_3$. The mRNA levels of VDR target genes such as *CYP3A4*, *E-Cadherin*, and *CaT1* were repressed in the presence of 1,25-$(OH)_2D_3$ in SW480 cells (Inaba et al., 2007). Further analogs of TEI-9647 include compounds with different substituents at the 2-position

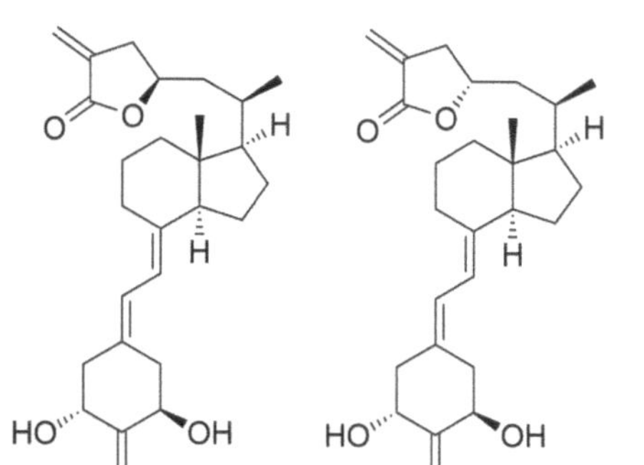

HLV
LAC67a

GC-3
LAC67b

Figure 9 Structures of HLV and GC-3.

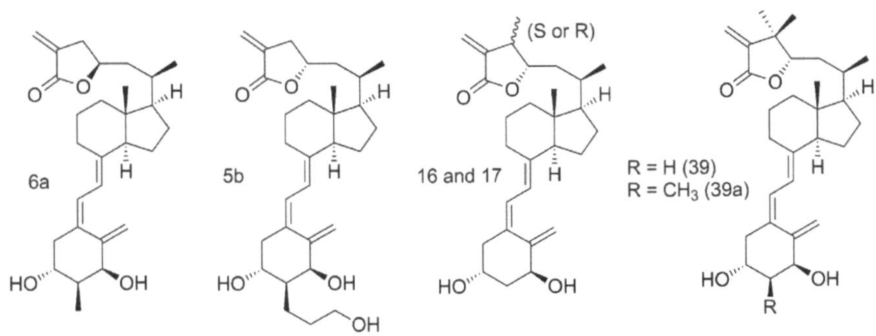

Figure 10 Structures of TEI-9647 analogs.

(Fig. 10; Saito & Kittaka, 2006; Saito, Matsunaga, et al., 2003; Saito et al., 2006b; Saito, Saito, et al., 2003; Takenouchi et al., 2004). Compound 6a bearing a methyl substituent had three times greater affinity toward VDR than TEI-9647; however, the ability to differentiate HL60 cells was only 32% of that of TEI-9647 (Saito & Kittaka, 2006). A 28-fold improvement of antagonist activity (HL60 differentiation) was observed for 5b in comparison to TEI-9647 exhibiting the same VDR affinity. The introduction of substituents in the 24-position resulted in epimers 16 and 17 that exhibited more than a twofold improvement of binding and antagonist activity. Further improvements were achieved with the introduction of two methyl groups at the 24-position, increasing the antagonist activity of 39 more than 13-fold. Finally, compound 39a has an IC_{50} value of 93 pM and a VDR affinity of 67% in comparison with 1,25-$(OH)_2D_3$. QSAR analysis of these TEI-9647 analogs using a comparative molecular field analysis and a comparative similarity indices analysis resulted in the identification of the optimal spatial arrangement of substituents at the 2 and 24 position of TEI-9647 analogs (Saito et al., 2004, 2006a; Wang et al., 2010).

In contrast to healthy patients, endogenous 1,25-$(OH)_2D_3$ strongly induced the formation of osteoclasts in patient with Paget's disease, leading to hyper-bone resorption (Menaa et al., 2000). In bone marrow cells from patients with Paget's disease, TEI-9647 reduced the formation of osteoclasts in the presence and absence of 1,25-$(OH)_2D_3$ and in turn inhibited bone resorption induced by 1,25-$(OH)_2D_3$ (Ishizuka et al., 2004). In addition, TEI-9647 inhibited the upregulation of coregulator TAF_{II}-17 in the presence of 1,25-$(OH)_2D_3$, which is believed to mediate the hyper-responsiveness in Paget's disease patients toward 1,25-$(OH)_2D_3$ (Ishizuka et al., 2005). TEI-9647 was also investigated in rats that were fed a vitamin

D-deficient, low calcium diet (−D rats) (Ishizuka, Miura, Ozono, Chokki et al., 2001). The effect of 1,25-$(OH)_2D_3$ on intestinal calcium transport is biphasic with an earlier independent phase (Bikle, Zolock, Morrissey, & Herman, 1978) and a later genomic driven process (Spielvogel, Farley, & Norman, 1972). TEI-9647 did not modulate the first phase but rather inhibited the effect of 1,25-$(OH)_2D_3$ in the later genomic response. Parathyroid hormone (PTH) is a negatively controlled gene by 1,25-$(OH)_2D_3$ (Silver, Naveh-Many, Mayer, Schmelzer, & Popovtzer, 1986). In −D rats, different doses of TEI-9647 acted as a weak agonist by decreasing PTH levels, stimulating calcium absorption, and mobilizing bone calcium. However, in combination with 1,25-$(OH)_2D_3$, TEI-9647 antagonized the action of the natural vitamin D hormone. In renal tissue, it was found that TEI-9647 inhibited the 1,25-$(OH)_2D_3$-induced relaxation of renal arteries from hypertensive patients (Dong et al., 2012). Similar results were observed in *ex vivo* rat renal arteries treated with U46619, a thromboxane A2 agonist. Here, TEI9647 reduced the phosphorylation of endothelial nitric oxide synthase and levels of NO that were elevated in the presence of 1,25-$(OH)_2D_3$ (Dong et al., 2013). In respect to cancer, TEI9647 reversed the downregulation of *CYP21A2* in the presence of 1,25-$(OH)_2D_3$ in a reporter assay using mouse and human adrenocortical carcinoma cells (Lundqvist, Wikvall, & Norlin, 2012). In MCF-7 cells, TEI-9647 reversed the 1,25-$(OH)_2D_3$-induced inhibition of aromatase enzyme, which is responsible for the conversion of testosterone to estradiol (Lundqvist, Norlin, & Wikvall, 2011). In primary ERα-negative breast cancer cells, TEI-9647 reduced the induction of mRNA levels of ERα in the presence of 1,25-$(OH)_2D_3$ (Santos-Martinez et al., 2014). In human acute lymphoblastic leukemia cells, TEI-9647 reduced the induction of proliferation by 1,25-$(OH)_2D_3$ in the presence of dexamethasone (Antony et al., 2012). Furthermore, in trophoblasts, TEI-9747 reversed the downregulation of the *IL-10* mRNA in the presence of 1,25-$(OH)_2D_3$ (Barrera et al., 2012) and antagonized the expression of *TNF-α*, *IL-6*, and *IFN-γ* in cultured trophoblasts in the presence of 1,25-$(OH)_2D_3$ (Diaz et al., 2009). TEI-9647 also reversed the upregulation of *prolactin* gene in human peripheral blood mononuclear cells (PBMCs) in the presence of 1,25$(OH)_2D_3$ (Diaz et al., 2011). In primary T cells, TEI-9647 reversed the effects of 1,25-$(OH)_2D_3$ that induced IL-31 and oncostatin M production and reduced IL-22 expression. TEI-9647 also reduced the surface expression of VDR-target gene CCR10 in IL-21-induced terminal differentiation human B cells in the presence of 1,25-$(OH)_2D_3$. In human keratinocytes and neutrophils in the presence of 1,25-$(OH)_2D_3$, TEI-9647 inhibited the production of IL-37 as well as the induction of mRNA levels of cathelicidin (Kanda, Hau, Tada, Sato, & Watanabe, 2012).

4.2 ZK159222

ZK159222 exhibited a subnanomolar affinity for the VDR-RXR-VDRE (Herdick, Steinmeyer, & Carlberg, 2000a; Toell et al., 2001; Fig. 11). In the presence of 1,25-$(OH)_2D_3$, ZK159222 inhibited VDR-mediated transcription with an IC_{50} value of 300 nM. SDS-PAGE demonstrated three different conformations of VDR–ZK159222 complex (Perakyla et al., 2004). These conformations may be responsible for the dissociation between liganded VDR and coactivators SRC1, SRC2, SCR3, and DRIP205 (Yamaoka et al., 2006; Zella et al., 2007). In addition, the interaction between corepressor NCoR and VDR was inhibited (Toell et al., 2001). The inhibitory effect was more pronounced in MCF-7 (breast) than HeLa (cervix) cells partially due to different expression levels of coactivators. ZK159222 also inhibited the interaction between VDR and corepressor SMRT as demonstrated with a pull-down assay (Sanchez-Martinez, Zambrano, Castillo, & Aranda, 2008). Furthermore, ZK159222 acted as a weak agonist (20% efficacy) in the absence of 1,25-$(OH)_2D_3$, which might be partially mediated by RXR. Indeed, gel shift assays have shown that 9-*cis*-RA was able to increase the agonist activity of ZK159222, enabling recruitment of SRC1, SRC2, and SRC3 (Sanchez-Martinez, Castillo, Steinmeyer, & Aranda, 2006). In the presence of 1,25-$(OH)_2D_3$, ZK159222 was able to inhibit the gene regulation mediated by RAR in the presence of retinoic acid (Castillo et al., 2006). However, in serum-depleted media the antagonist effect of ZK159222 was not observed at a concentration of 100 nM. For rVDR,

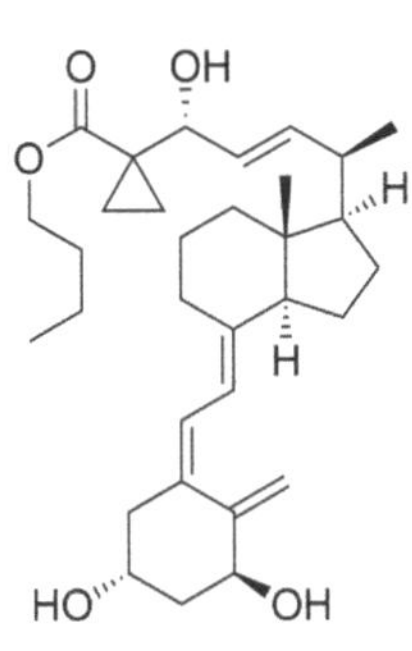

ZK159222

Figure 11 Structure of ZK159222.

ZK159222 inhibited the interaction between VDR and SRC1 and DRIP205 but not between VDR and SRC2 or SRC3 at a concentration of 100 nM. For both rVDR and hVDR, ZK159222 antagonized the transcriptional activation by 1,25-$(OH)_2D_3$ in a reporter assay (Castillo et al., 2006; Ochiai et al., 2005). Mutation studies and molecular dynamics (MD) studies revealed that ZK159222 disturbed the H397–F422 interaction, while C403 and C410 VDR mutants did not influence the antagonistic effect of ZK159222, thus excluding covalent interactions as observed for TEI-9647 (Perakyla et al., 2004). In osteoblastic MC3T3-E1 cells, *osteopontin* (*OPN*) but not *CYP24A1* mRNA levels were upregulated in the presence of ZK159222. ZK159222 induced a weak recruitment of VDR and RXR to the *CYP24A1* promoter but a more pronounced recruitment to the *OPN* promoter. Further studies revealed that the agonistic effect of ZK159222 at the *OPN* promoter was probably caused by fully acetylated histones that facilitate transcription. Coactivator recruitment was not promoted by ZK159222 at both promoter sites (Kim, Shevde, & Pike, 2005). In human fetal osteoblastic cells (SV-HFO) in the presence of 1,25-$(OH)_2D_3$, ZK159222 inhibited the expression of *osteocalcin*, alkaline phosphatase activity, and calcium contents (van Driel et al., 2006). In osteoblastic ST2 cells, ZK159222 blocked the activation of the mRLD5 region of *mRANKL* in the presence of 1,25-$(OH)_2D_3$ (Kim, Yamazaki, Zella, Shevde, & Pike, 2006). In addition, ZK159222 inhibited the upregulation of *creatine kinase* in the presence of VDR agonist and estradiol in osteoblast-like ROS 17/2.8 cells (Somjen, Waisman, Lee, Posner, & Kaye, 2001). The calcemic activity of ZK159222 was 0.02% of that of 1,25-$(OH)_2D_3$ in mice after 5 days of 10 μg/kg/d (Castillo et al., 2006).

ZK159222 also inhibited the differentiation of HL60 cells in the presence of 1,25-$(OH)_2D_3$ at a concentration of 6 nM (Fujishima, Kojima, Azumaya, Kittaka, & Takayama, 2003). The process involves the upregulation of *kinase suppressor of Ras-2* gene (*KRS-2*), which was demonstrated to be inhibited by ZK159222 (Wang, Wang, White, & Studzinski, 2007). In addition, ZK159222 inhibited the phosphorylation of Raf-1 (Studzinski et al., 2005; Wang, Wang, White, & Studzinski, 2006) and the expression of *pRb* and *c/EBPβ* in the presence of 1,25-$(OH)_2D_3$ (Ji & Studzinski, 2004). ZK159222 inhibited the phosphorylation of phosphoinositide and Akt-mediated by phosphatidylinositol 3-kinase in the presence of 1,25-$(OH)_2D_3$ (Hughes, Lee, Reiner, & Brown, 2008). Furthermore, ZK159222 inhibited the steroid sulfatase activity in the presence of 1,25-$(OH)_2D_3$ in HL60 cells (Hughes, Steinmeyer, Chandraratna, & Brown, 2005). Interestingly, this effect was not observed in NB4 cells

(Hughes & Brown, 2006). In trophoblasts, ZK159222-reduced mRNA expression of *CYP24A1* in the presence of 1,25-$(OH)_2D_3$ and attenuated slightly the expression of *CYP27B1*. In the absence of 1,25-$(OH)_2D_3$, upregulation of *CYP27B1* was more pronounced. ZK159222 was able to increase the concentration of cellular cAMP by itself. This effect was additive in the presence of 1,25-$(OH)_2D_3$. Similar results were obtained for hCG (human chorionic gonadotrophin) (Avila et al., 2007). Weak induction of *calbindin-D28K*, a cytosolic calcium-binding protein, and *VDR* itself was observed in the presence of ZK159222 in choriocarcinoma-derived cells (JEG-3). In the presence of 1,25-$(OH)_2D_3$, ZK159222 exhibited strong antagonist effects in these cells (Belkacemi, Zuegel, Steinmeyer, Dion, & Lafond, 2005).

Antimicrobial response in monocytes pretreated with ZK159222 reduced the elevated *cathelicidin* mRNA levels when stimulated with a synthetic 19-kD *M. tuberculosis*-derived lipopeptide (Liu et al., 2006). In addition, ZK159222 reduced the upregulation of *CYP24A1*, *CCL22*, and *CD300LF* in the presence of 1,25-$(OH)_2D_3$ in $CD14^+$ monocytes (Szeles et al., 2009). Furthermore, pretreatment with ZK159222 reduced the elevated *DEFB4* mRNA levels when stimulated with rIFN-γ and reversed the upregulation of *CD14* and downregulation of (toll-like receptors) *TRL2* and *TRL4* in the presence of 1,25-$(OH)_2D_3$ (Sadeghi et al., 2006). In differentiated macrophages from human blood, addition of the VDR antagonist ZK159222 inhibited the induction of *human cationic antimicrobial protein-18* in the presence of 25-vitamin D_3 and TLR2/6 ligand Pam2CSK4 (Li et al., 2013). ZK159222 also reversed the effect of 1,25-$(OH)_2D_3$ in *M. tuberculosis*-stimulated PBMCs, which had decreased the protein and mRNA expression of *TLR2*, *TRL4*, *Dectin-1*, and *mannose receptor* (Khoo et al., 2011). ZK159222 also reduced the production of IL-10 in stimulated B cells (Heine et al., 2008) and inhibited the differentiation of Langerhans cells (Gobel et al., 2009).

In primary HUVEC cells, ZK159222 reduced the production of nitric oxide in the presence of 1,25-$(OH)_2D_3$ and reduced the phosphorylation of eNOS, p38, Akt, and ERK1/2 promoted by 1,25-$(OH)_2D_3$ (Molinari et al., 2011). Overall, ZK159222 inhibited the growth of HUVEC cells in the presence of pro-proliferative 1,25-$(OH)_2D_3$ (Pittarella et al., 2015). In addition, ZK159222-reduced 3D cell migration stimulated by 1,25-$(OH)_2D_3$ and reduced mRNA levels of *MMP-2*. In keratinocytes, ZK159222 weakly activated the expression of *cathelicidin* in a luciferase reporter assay and slightly increased the production of cathelicidin in primary human keratinocytes (Peric et al., 2009). When stimulated with 25-vitamin D_3 and TGF-β1, ZK159222 inhibited the expression of *cathelicidin*, *CD14*, and *TLR2* (Schauber et al., 2007). In cancer cells such as cervical carcinoma cells, ZK159222 restored mRNA levels of *human ether a-go-go-1* (*EAG1*) that was downregulated by 1,25-$(OH)_2D_3$ (Avila et al., 2010). In HeLa cells, a reporter assay under control of a human immunodeficiency virus type I long terminal repeat HIV-1 LTR promoter demonstrated the antagonist effects of ZK159222 in the presence of 1,25-$(OH)_2D_3$ (Nevado, Tenbaum, Castillo, Sanchez-Pacheco, & Aranda, 2007). ZK159222 also reduced the production of P-glycoprotein, a member of the ABC transporter family, in LS174T colon adenocarcinoma cells in the presence of 1,25-$(OH)_2D_3$ (Kota, Allen, & Roufogalis, 2011). In LNCaP cells, ZK159222 reduced the mRNA and protein levels of the prostate-derived factor stimulated 1,25-$(OH)_2D_3$ at 10 nM (Lambert et al., 2006). Finally, ZK159222 blocked the stimulatory effects of a VDR agonist on DNA synthesis in epithelia E304 cells (Somjen, Kohen, Amir-Zaltsman, Knoll, & Stern, 2000) and prevented the decrease of *ceramide kinase* expression elicited by 1,25-$(OH)_2D_3$ in neuroblast-like SH-SY5Y cells (Bini et al., 2012).

4.3 ZK168281

ZK168281 has a picomolar affinity for the VDR–RXR–VDRE complex (Fig. 12; Bury, Steinmeyer, & Carlberg, 2000). In addition, ZK168281 is less agonistic and three times more potent as an antagonist than ZK159222. Limited digestion studies demonstrated the formation of different VDR structures in the presence of ZK168281 in comparison to 1,25-$(OH)_2D_3$. The residual agonist effect of ZK168281 was 5% that of 1,25-$(OH)_2D_3$ (Herdick, Steinmeyer, & Carlberg, 2000b). ZK168281

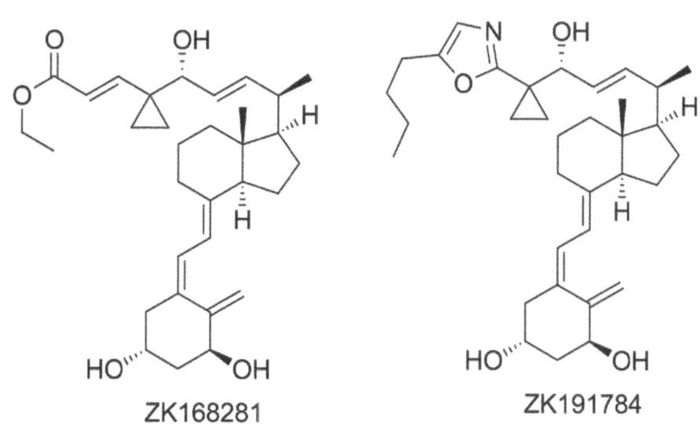

Figure 12 Structures of ZK168281 and ZK191784.

promoted the recruitment of NCoR (Lempiainen, Molnar, Macias Gonzalez, Perakyla, & Carlberg, 2005) and inhibited the interaction with DRIP205 and SRC1 in two-hybrid assays (Zella et al., 2007). ZK168281 behaved like a pure antagonist in five different cell lines (Perakyla et al., 2004; Shah et al., 2006), and C403S/C410N mutation did not alter the binding of ZK168281. Importantly, ZK168281 did not promote the complexation between VDR–RXR–SRC2–VDRE (Vaisanen, Perakyla, Karkkainen, Steinmeyer, & Carlberg, 2002). In contrast to ZK159222, ZK168281 antagonist activity was not influenced by truncation of the VDR helix 12 or RXR helix 12. Mutation studies and MD studies revealed that ZK168281 severely disturbs the H397 and F422 interaction of VDR (Vaisanen et al., 2002). In addition, the agonistic effects of ZK168281 increased with VDR H302A and H397A mutations (Yamamoto et al., 2006). Finally, ZK168281 inhibited the phosphorylation of phosphoinositide mediated by phosphatidylinositol 3-kinase in the presence of 1,25-$(OH)_2D_3$ (Hughes et al., 2008).

4.4 ZK191784

ZK191784 has a structure similar to ZK159222 with a bioisosteric replacement of the ester functionality (Fig. 12). The relative VDR binding is 33% that of 1,25-$(OH)_2D_3$ (Zugel, Steinmeyer, Giesen, & Asadullah, 2002). ZK191784 inhibited the differentiation of HL60 cells in the presence of 1,25-$(OH)_2D_3$ and exerted only weak agonist effects for inducing the expression of *CD14*. Like 1,25-$(OH)_2D_3$, ZK191784 reduced the proliferation of stimulated lymphocytes and inhibited the expression of surface maker HLA-DR, B7.1, and ICAM-1, although to a lesser degree than 1,25-$(OH)_2D_3$. In addition, IL-12, IL-10, and TNFα secretion were reduced by ZK191784. In mice, ZK191784 reduced the hyperresponsiveness induced by 2,4-dimethylfluorobenzene without increasing the concentration of calcium in urine. For chronic intestinal inflammation, ZK191784 reduced the expression IFN-γ and IL-6 in mesenteric lymph node cells and lowered the numbers of activated CD11c + dendritic-cells in the colon (Strauch et al., 2007). In TRPV5$^{-/-}$ mice, ZK191784 normalized the calcium hyperabsorption and expression of intestinal calcium transport proteins. Furthermore, ZK191784 reduced 1,25-$(OH)_2D_3$-dependent calcium uptake by Caco-2 cells (intestine). In WT mice, ZK191784 increased renal *TRPV5* and *calbindin-D28K* expression and decreased urine calcium excretion. In rat osteosarcoma cells, ZK191784 and 1,25-$(OH)_2D_3$

enhanced bone *TRPV6* mRNA levels and secretion of osteocalcin (Nijenhuis et al., 2006). However, *CYP24A1* was downregulated in femoral bone for mice treated with ZK191784. The phosphate homeostasis was unaffected by ZK191784 (van der Eerden et al., 2013). For HUVEC cells, ZK191784 increased the formation of nitric oxide and cell viability in the presence of 1,25-$(OH)_2D_3$ in the presence and absence of hydrogen peroxide. ZK191784 improved the expression of *beclin 1* and the phosphorylation of ERK1/2 increased as well as the repression of *BAX* in the presence of 1,25-$(OH)_2D_3$. Furthermore, 1,25-$(OH)_2D_3$ alone or in combination with ZK191784 was able to prevent the loss of mitochondrial potential and the consequent cytochrome C release and caspase activation (Uberti et al., 2014). In keratinocytes, ZK191784 exhibited weak agonist behavior, reducing *HBD2* expression and increasing the expression of *cathelicidin* and *hCAP18* (Peric et al., 2009). In DU145 and LNCaP cells, ZK191784 reduced *MMP-9* and *MMP-2* as well as the surface expression of ICAM-1 (Stio et al., 2011).

4.5 Amide-Based VDR Antagonists

ML-3-452 binds VDR with an EC_{50} of 107 nM (Fig. 13; Lamblin et al., 2010). In CCS25 cells, ML-3-452 did not increase the expression of *CYP24A1* and *TSLP* but reduced their expression in the presence of 1,25-$(OH)_2D_3$. ML-3-452 reduced the recruitment of SRC3 to the VDR–DNA complex and increased the interaction between VDR and NCoR on the *CYP24A1* promoter.

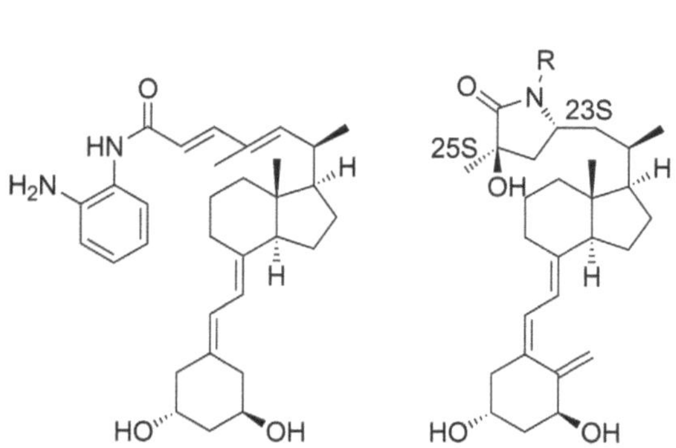

ML 3-452

R = Me DLAM-01
R = Bn DLAM-1P
R = Bn-3,5-$(OEt)_2$ DLAM-1P-3,5$(OEt)_2$

Figure 13 Structures of ML-3-452, DLAM-01, DLAM-1P, DLAM-1P-3,5$(OEt)_2$.

The first compound in the lactam series was DLAM-01 exhibiting a VDR binding affinity of 10.2 nM (Fig. 13). DLAM-1P-binding affinity was 1.9 nM and inhibited the differentiation of HL60 cell at 1 μM (Kato et al., 2004). DLAM-1P inhibited VDR-mediated transcription in a reporter assay in the nanomolar range, without showing any agonist activity in the absence of 1,25-$(OH)_2D_3$ (Nakano et al., 2006). In contrast to TEI-9647, hVDR and rVDR were inhibited by DLAM-1P at the same concentration; thus hVDR cysteine residues did not influence this binding. DLAM-1P also suppressed the differentiation of bone marrow cells induced by 1,25-$(OH)_2D_3$ as well as the upregulation of *RANKL* mRNA in mouse primary osteoblast (Inada et al., 2008). In addition, calcium reabsorption induced by 1,25-$(OH)_2D_3$ in mouse calvarial organ culture was reduced by DLAM-1P. For LNCaP cells, DLAM-1P promoted proliferation reduced by 1,25-$(OH)_2D_3$ and further decreased the upregulation of *P21* (Takita et al., 2008). The introduction of substituents in the meta-position resulted in DLAM-1P-3,5$(OEt)_2$ with a 3.5 times higher affinity to VDR than 1,25-$(OH)_2D_3$ and an IC_{50} of 90 nM for the inhibition of differentiation of HL60 cells (Cho et al., 2008; Ishizuka et al., 2008). DLAM-2P, bearing an ethylbenzene amide substituent reversed the decreased induction of *DMP-1* mRNA levels in the presence of 1,25-$(OH)_2D_3$ in cementoblasts and osteocytes-like cells (Nociti et al., 2014).

4.6 Adamantane-Based VDR Antagonists

AD47 exhibited weak agonist activity in reporter assays employing an *osteopontin*, *CYP24A1*, and repeated VDRE as promoters at 10 nM (Fig. 14; Inaba et al., 2007; Yamamoto et al., 2006). However, in the presence of 1,25-$(OH)_2D_3$, AD47 inhibited transcription at 100 nM. In the

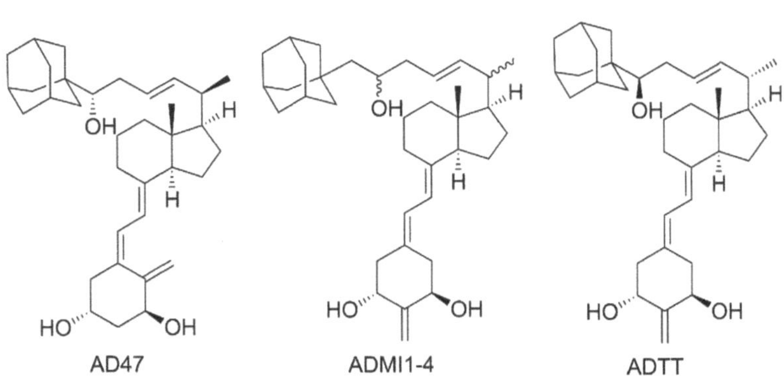

Figure 14 Structures of AD47, ADI1-4, and ADTT.

absence of 1,25-$(OH)_2D_3$, a weak interaction between VDR and RXR was observed. In the presence of 1,25-$(OH)_2D_3$, this interaction was significantly reduced by AD47. Similarly, AD47 was able to weakly promote the interaction between VDR and coactivator SRC1, but inhibited this interaction in the presence 1,25-$(OH)_2D_3$. Interestingly, AD47 strongly promoted the interaction of VDR with DRIP205 and no inhibition was observed in the presence of 1,25-$(OH)_2D_3$. In regard to corepressors, AD47 inhibited the recruitment of NCoR and SMRT but did not suppress the dissociation of the VDR–corepressor complex induced by 1,25-$(OH)_2D_3$. For reporter assays carried out with different cell lines, AD47 reduced the expression of *CYP24A1* in the presence of 1,25-$(OH)_2D_3$, except for HCT116 intestine cells. In addition, AD47 behaved as an agonist in the absence of 1,25-$(OH)_2D_3$ except for HEK293 kidney cells. Other genes such as *CaT1*, *CYP3A4*, and *E-cadherin* were upregulated by AD47 in SW480 intestine cells and reduced in the presence of 1,25-$(OH)_2D_3$. Four analogs, called ADMI1-4 with different stereochemical configurations, were all able to bind VDR and inhibited transcription in the presence of 1,25-$(OH)_2D_3$ in a luciferase reporter assay (Fig. 14; Igarashi et al., 2007; Nakabayashi et al., 2008). *CYP24A1* expression was reduced in the presence of 1,25-$(OH)_2D_3$ for ADMI1 and ADMI3. ADTT exhibited a weak agonistic activity and did not promote the dimerization of VDR and RXR in the absence of 1,25-$(OH)_2D_3$ in HEK293 cells (Fig. 14; Choi, Yamada, & Makishima, 2011). However, in the absence of 1,25-$(OH)_2D_3$, ADTT induced a weak VDR recruitment of SRC1 and dissociation of SMRT. Similar to AD47, ADTT behaved like an agonist in various cells except HEK293 kidney cells. In those cells, ADTT reduced VDR–DNA binding. In HCT116 colon cancer cells, however, ADTT induced recruitment of RXR and surprisingly SMRT to the DNA-bound VDR. In THP-1 cells, ADTT and ADMI3 reduced the mRNA levels of *CYP1A1* in the presence of 1,25-$(OH)_2D_3$ and benzo[a]pyrene (an aryl hydrocarbon agonist) (Matsunawa et al., 2012). A recent series of ADTK1-4 analogs exhibited significantly less antagonist activity than ADTT (Kudo et al., 2014).

4.7 Branched VDR Antagonists

Further development of VDR agonist GEMINI (Norman et al., 2000) resulted in VDR super-agonists (Yamamoto et al., 2007) with higher affinity toward VDR than endogenous ligand 1,25-$(OH)_2D_3$ and VDR antagonists 4, 6, 7, 8, and 10 (Fig. 15; Inaba et al., 2009). The agonistic activity of these

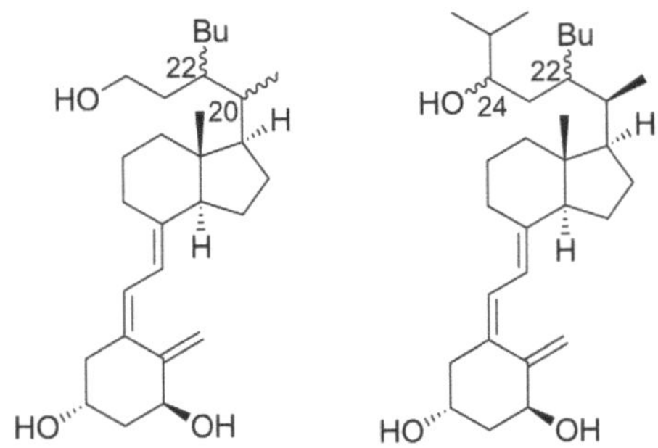

20R,22R = 3
20R,22S = 4
20S,22S = 5
20S,22R = 6

24R,22R = 7
24S,22R = 8
24R,22S = 9
24S,22S = 10

Figure 15 Structures of branched VDR antagonists.

ligands in the presence of 1,25-$(OH)_2D_3$ was very weak when measured by luciferase reporter assay. *CYP24A1* expression in the presence of 1,25-$(OH)_2D_3$ was reduced as well. No change in differentiation of HL60 cells was observed in the absence of 1,25-$(OH)_2D_3$, but inhibition of cell differentiation in the presence of 1,25-$(OH)_2D_3$ was demonstrated for compound 4 in conjunction with occupancy of a new binding pocket for the butyl substituent. Compound 4 was weakly able to support the binding between VDR and RXR but not between VDR and SRC1. However, in the presence of 1,25-$(OH)_2D_3$, both interactions were inhibited by compound 4 at a concentration of 1 μM (Inaba et al., 2010). Elongation of compound 4's carbon chain by one CH_2 unit resulted in a slightly more potent antagonist (Sakamaki, Inaba, Yoshimoto, & Yamamoto, 2010). Introduction of a methyl or an ethyl substituent in the 24 position increased the agonist activity of these ligands determined by a luciferase reporter assay (Yoshimoto et al., 2012). The substitution of a butyl by an ethyl substituent resulted in strong antagonistic behavior of VDR ligands that bear the hydroxyl functionality in the 24 position or have no alkyl substituents in the 25 position (Anami, Itoh, Egawa, Yoshimoto, & Yamamoto, 2014).

High-throughput screening also identified PPARδ agonist GW0742 as a novel VDR antagonist (Fig. 16; Nandhikonda et al., 2013). GW0742 did not only bind VDR but also other nuclear receptors. In the presence of GW0742, nuclear receptor target gene mRNA levels were suppressed in the presence of their endogenous ligands. The conversion of GW0742 acid functionality to an alcohol increased its agonistic activity in cells (Teske

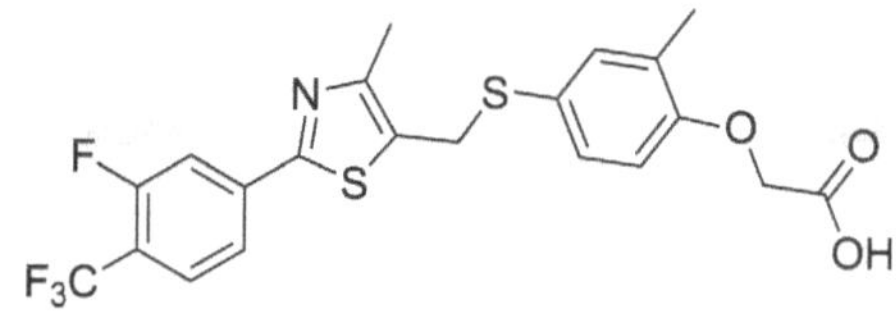

Figure 16 Structure of GW0742.

et al., 2014a). Virtual screening using a nuclear receptor ligand database in conjunction with a stringent pharmacophore model for 1,25-$(OH)_2D_3$ identified many nuclear receptor ligands that might interact with VDR (Teske et al., 2014b). For selected NR ligands such as H6036 VDR–coactivator inhibition was demonstrated.

5. CONCLUSION AND FUTURE DIRECTIONS

The development of VDR–coregulator inhibitors has allowed us to identify different biological functions of VDR. VDR antagonists in particular have been invaluable to verify the involvement of VDR for biological effects induced by 1,25-$(OH)_2D_3$. However, the development of VDR antagonists as novel therapeutics has been met with caution partially because of the fear that antagonizing VDR will cause dysfunction and partially because VDR antagonists exhibit various degrees of agonism. ZK191784 is an excellent example of a situation where these obstacles can be overcome by careful ligand design. The regulation of VDR–coregulator interactions with VDR antagonists has been more challenging because recent discoveries have shown many factors that influence nuclear receptor-mediated transcription, such as position of DNA promoter and activation sites, cell type and expression of coregulators, the differentiation state of cells and their circadian rhythms, and many other factors. However, the development of metabolically stable VDR antagonists will enable us to characterize the *in vivo* effects mediated by VDR due to the presence of endogenous 1,25-$(OH)_2D_3$. Direct small-molecule inhibitors of the VDR–coregulators interaction have been very challenging to develop as well, due to the fact that the interaction between VDR and coregulators is relatively weak and represents protein–protein interactions with large interaction surfaces. Nevertheless, the first scaffolds have been identified by rational design and high-throughput screening. The optimization of these inhibitors in respect to affinity and selectivity among different nuclear receptors and their coregulators is still at an early stage. Small-molecule VDR–coregulator inhibitors do not only represent research tools to dissect the transcriptional multiprotein complex that governs transcription but also drug candidates for diseases that deregulate VDR due to abnormal coregulators binding.

ACKNOWLEDGMENTS

This work was supported by the University of Wisconsin-Milwaukee, the Milwaukee Institute for Drug Discovery, the UWM Research Growth Initiative, NIH R03DA031090, the UWM Research Foundation, the Lynde and Harry Bradley Foundation, and the Richard and Ethel Herzfeld Foundation.

REFERENCES

Anami, Y., Itoh, T., Egawa, D., Yoshimoto, N., & Yamamoto, K. (2014). A mixed population of antagonist and agonist binding conformers in a single crystal explains partial agonism against vitamin D receptor: Active vitamin D analogues with 22R-alkyl group. *Journal of Medicinal Chemistry*, *57*(10), 4351–4367. http://dx.doi.org/10.1021/Jm500392t.

Antony, R., Sheng, X., Ehsanipour, E. A., Ng, E., Pramanik, R., Klemm, L., et al. (2012). Vitamin D protects acute lymphoblastic leukemia cells from dexamethasone. *Leukemia Research*, *36*(5), 591–593. http://dx.doi.org/10.1016/j.leukres.2012.01.011.

Arnold, L. A., Estebanez-Perpina, E., Togashi, M., Shelat, A., Ocasio, C. A., McReynolds, A. C., et al. (2006). A high-throughput screening method to identify small molecule inhibitors of thyroid hormone receptor coactivator binding. *Science's STKE*, *2006*(341), pl3. http://dx.doi.org/10.1126/stke.3412006pl3.

Avila, E., Diaz, L., Barrera, D., Halhali, A., Mendez, I., Gonzalez, L., et al. (2007). Regulation of vitamin D hydroxylases gene expression by 1,25-dihydroxyvitamin D3 and cyclic AMP in cultured human syncytiotrophoblasts. *The Journal of Steroid Biochemistry and Molecular Biology*, *103*(1), 90–96. http://dx.doi.org/10.1016/j.jsbmb.2006.07.010.

Avila, E., Garcia-Becerra, R., Rodriguez-Rasgado, J. A., Diaz, L., Ordaz-Rosado, D., Zugel, U., et al. (2010). Calcitriol down-regulates human ether a go-go 1 potassium channel expression in cervical cancer cells. *Anticancer Research*, *30*(7), 2667–2672.

Barrera, D., Noyola-Martinez, N., Avila, E., Halhali, A., Larrea, F., & Diaz, L. (2012). Calcitriol inhibits interleukin-10 expression in cultured human trophoblasts under normal and inflammatory conditions. *Cytokine*, *57*(3), 316–321. http://dx.doi.org/10.1016/j.cyto.2011.11.020.

Baudino, T. A., Kraichely, D. M., Jefcoat, S. C., Jr., Winchester, S. K., Partridge, N. C., & MacDonald, P. N. (1998). Isolation and characterization of a novel coactivator protein, NCoA-62, involved in vitamin D-mediated transcription. *The Journal of Biological Chemistry*, *273*(26), 16434–16441.

Belkacemi, L., Zuegel, U., Steinmeyer, A., Dion, J. P., & Lafond, J. (2005). Calbindin-D28k (CaBP28k) identification and regulation by 1,25-dihydroxyvitamin D3 in human choriocarcinoma cell line JEG-3. *Molecular and Cellular Endocrinology*, *236*(1–2), 31–41. http://dx.doi.org/10.1016/j.mce.2005.03.002.

Bhatia, M., Kirkland, J. B., & Meckling-Gill, K. A. (1995). Monocytic differentiation of acute promyelocytic leukemia cells in response to 1,25-dihydroxyvitamin D3 is independent of nuclear receptor binding. *The Journal of Biological Chemistry*, *270*(27), 15962–15965.

Bikle, D. D., Zolock, D. T., Morrissey, R. L., & Herman, R. H. (1978). Independence of 1,25-dihydroxyvitamin D3-mediated calcium transport from de novo RNA and protein synthesis. *The Journal of Biological Chemistry*, *253*(2), 484–488.

Bini, F., Frati, A., Garcia-Gil, M., Battistini, C., Granado, M., Martinesi, M., et al. (2012). New signalling pathway involved in the anti-proliferative action of vitamin D(3) and its analogues in human neuroblastoma cells. A role for ceramide kinase. *Neuropharmacology*, *63*(4), 524–537. http://dx.doi.org/10.1016/j.neuropharm.2012.04.026.

Bird, G. H., Madani, N., Perry, A. F., Princiotto, A. M., Supko, J. G., He, X., et al. (2010). Hydrocarbon double-stapling remedies the proteolytic instability of a lengthy peptide therapeutic. *Proceedings of the National Academy of Sciences of the United States of America*, *107*(32), 14093–14098. http://dx.doi.org/10.1073/pnas.1002713107.

Breitman, T. R., Selonick, S. E., & Collins, S. J. (1980). Induction of differentiation of the human promyelocytic leukemia cell line (HL-60) by retinoic acid. *Proceedings of the National Academy of Sciences of the United States of America*, 77(5), 2936–2940.

Bula, C. M., Bishop, J. E., Ishizuka, S., & Norman, A. W. (2000). 25-Dehydro-1alpha-hydroxyvitamin D3-26,23S-lactone antagonizes the nuclear vitamin D receptor by mediating a unique noncovalent conformational change. *Molecular Endocrinology*, *14*(11), 1788–1796. http://dx.doi.org/10.1210/mend.14.11.0552.

Bury, Y., Steinmeyer, A., & Carlberg, C. (2000). Structure activity relationship of carboxylic ester antagonists of the vitamin D(3) receptor. *Molecular Pharmacology*, *58*(5), 1067–1074.

Caboni, L., & Lloyd, D. G. (2013). Beyond the ligand-binding pocket: Targeting alternate sites in nuclear receptors. *Medicinal Research Reviews*, *33*(5), 1081–1118. http://dx.doi.org/10.1002/med.21275.

Carraz, M., Zwart, W., Phan, T., Michalides, R., & Brunsveld, L. (2009). Perturbation of estrogen receptor alpha localization with synthetic nona-arginine LXXLL-peptide coactivator binding inhibitors. *Chemistry & Biology*, *16*(7), 702–711. http://dx.doi.org/10.1016/j.chembiol.2009.06.009.

Castillo, A. I., Sanchez-Martinez, R., Jimenez-Lara, A. M., Steinmeyer, A., Zugel, U., & Aranda, A. (2006). Characterization of vitamin D receptor ligands with cell-specific and dissociated activity. *Molecular Endocrinology*, *20*(12), 3093–3104. http://dx.doi.org/10.1210/me.2006-0215.

Chang, C., Norris, J. D., Gron, H., Paige, L. A., Hamilton, P. T., Kenan, D. J., et al. (1999). Dissection of the LXXLL nuclear receptor-coactivator interaction motif using combinatorial peptide libraries: Discovery of peptide antagonists of estrogen receptors alpha and beta. *Molecular and Cellular Biology*, *19*(12), 8226–8239.

Chen, J. D., & Evans, R. M. (1995). A transcriptional co-repressor that interacts with nuclear hormone receptors. *Nature*, *377*(6548), 454–457. http://dx.doi.org/10.1038/377454a0.

Chen, H., Lin, R. J., Schiltz, R. L., Chakravarti, D., Nash, A., Nagy, L., et al. (1997). Nuclear receptor coactivator ACTR is a novel histone acetyltransferase and forms a multimeric activation complex with P/CAF and CBP/p300. *Cell*, *90*(3), 569–580.

Chen, D., Ma, H., Hong, H., Koh, S. S., Huang, S. M., Schurter, B. T., et al. (1999). Regulation of transcription by a protein methyltransferase. *Science*, *284*(5423), 2174–2177.

Cheng, J. B., Levine, M. A., Bell, N. H., Mangelsdorf, D. J., & Russell, D. W. (2004). Genetic evidence that the human CYP2R1 enzyme is a key vitamin D 25-hydroxylase. *Proceedings of the National Academy of Sciences of the United States of America*, *101*(20), 7711–7715. http://dx.doi.org/10.1073/pnas.0402490101.

Chiba, N., Suldan, Z., Freedman, L. P., & Parvin, J. D. (2000). Binding of liganded vitamin D receptor to the vitamin D receptor interacting protein coactivator complex induces interaction with RNA polymerase II holoenzyme. *The Journal of Biological Chemistry*, *275*(15), 10719–10722.

Chiellini, G., Grzywacz, P., Plum, L. A., Barycki, R., Clagett-Dame, M., & DeLuca, H. F. (2008). Synthesis and biological properties of 2-methylene-19-nor-25-dehydro-1alpha-hydroxyvitamin D(3)-26,23-lactones—Weak agonists. *Bioorganic & Medicinal Chemistry*, *16*(18), 8563–8573. http://dx.doi.org/10.1016/j.bmc.2008.08.011.

Cho, K., Uneuchi, F., Kato-Nakamura, Y., Namekawa, J. I., Ishizuka, S., Takenouchi, K., et al. (2008). Structure-activity relationship studies on vitamin D lactam derivatives as vitamin D receptor antagonist. *Bioorganic & Medicinal Chemistry Letters*, *18*(15), 4287–4290. http://dx.doi.org/10.1016/j.bmcl.2008.06.095.

Choi, M., Yamada, S., & Makishima, M. (2011). Dynamic and ligand-selective interactions of vitamin D receptor with retinoid X receptor and cofactors in living cells. *Molecular Pharmacology*, *80*(6), 1147–1155. http://dx.doi.org/10.1124/mol.111.074138.

Demizu, Y., Nagoya, S., Shirakawa, M., Kawamura, M., Yamagata, N., Sato, Y., et al. (2013). Development of stapled short helical peptides capable of inhibiting vitamin D receptor (VDR)-coactivator interactions. *Bioorganic & Medicinal Chemistry Letters*, *23*(15), 4292–4296. http://dx.doi.org/10.1016/j.bmcl.2013.06.002.

Diaz, L., Martinez-Reza, I., Garcia-Becerra, R., Gonzalez, L., Larrea, F., & Mendez, I. (2011). Calcitriol stimulates prolactin expression in non-activated human peripheral blood mononuclear cells: Breaking paradigms. *Cytokine*, *55*(2), 188–194. http://dx.doi.org/10.1016/j.cyto.2011.04.013.

Diaz, L., Noyola-Martinez, N., Barrera, D., Hernandez, G., Avila, E., Halhali, A., et al. (2009). Calcitriol inhibits TNF-alpha-induced inflammatory cytokines in human trophoblasts. *Journal of Reproductive Immunology*, *81*(1), 17–24. http://dx.doi.org/10.1016/j.jri.2009.02.005.

Dong, J., Wong, S. L., Lau, C. W., Lee, H. K., Ng, C. F., Zhang, L., et al. (2012). Calcitriol protects renovascular function in hypertension by down-regulating angiotensin II type 1 receptors and reducing oxidative stress. *European Heart Journal*, *33*(23), 2980–2990. http://dx.doi.org/10.1093/eurheartj/ehr459.

Dong, J., Wong, S. L., Lau, C. W., Liu, J., Wang, Y. X., Dan He, Z., et al. (2013). Calcitriol restores renovascular function in estrogen-deficient rats through downregulation of cyclooxygenase-2 and the thromboxane-prostanoid receptor. *Kidney International*, *84*(1), 54–63. http://dx.doi.org/10.1038/ki.2013.12.

Fujishima, T., Kojima, Y., Azumaya, I., Kittaka, A., & Takayama, H. (2003). Design and synthesis of potent vitamin D receptor antagonists with a-ring modifications: Remarkable effects of 2alpha-methyl introduction on antagonistic activity. *Bioorganic & Medicinal Chemistry*, *11*(17), 3621–3631.

Futaki, S., Suzuki, T., Ohashi, W., Yagami, T., Tanaka, S., Ueda, K., et al. (2001). Arginine-rich peptides. An abundant source of membrane-permeable peptides having potential as carriers for intracellular protein delivery. *The Journal of Biological Chemistry*, *276*(8), 5836–5840. http://dx.doi.org/10.1074/jbc.M007540200.

Galande, A. K., Bramlett, K. S., Burris, T. P., Wittliff, J. L., & Spatola, A. F. (2004). Thioether side chain cyclization for helical peptide formation: Inhibitors of estrogen receptor-coactivator interactions. *The Journal of Peptide Research*, *63*(3), 297–302. http://dx.doi.org/10.1111/j.1399-3011.2004.00152.x.

Galande, A. K., Bramlett, K. S., Trent, J. O., Burris, T. P., Wittliff, J. L., & Spatola, A. F. (2005). Potent inhibitors of LXXLL-based protein-protein interactions. *Chembiochem*, *6*(11), 1991–1998. http://dx.doi.org/10.1002/cbic.200500083.

Geistlinger, T. R., & Guy, R. K. (2001). An inhibitor of the interaction of thyroid hormone receptor beta and glucocorticoid interacting protein 1. *Journal of the American Chemical Society*, *123*(7), 1525–1526.

Geistlinger, T. R., & Guy, R. K. (2003). Novel selective inhibitors of the interaction of individual nuclear hormone receptors with a mutually shared steroid receptor coactivator 2. *Journal of the American Chemical Society*, *125*(23), 6852–6853. http://dx.doi.org/10.1021/ja0348391.

Gobel, F., Taschner, S., Jurkin, J., Konradi, S., Vaculik, C., Richter, S., et al. (2009). Reciprocal role of GATA-1 and vitamin D receptor in human myeloid dendritic cell differentiation. *Blood, 114*(18), 3813–3821. http://dx.doi.org/10.1182/blood-2009-03-210484.

Guthrie, M. L., Sidhu, P. S., Hill, E. K., Horan, T. C., Nandhikonda, P., Teske, K. A., et al. (2015). Antitumor activity of 3-indolylmethanamines 31B and PS121912. *Anticancer Research, 35*(11), 6001–6007.

Hall, J. M., Chang, C. Y., & McDonnell, D. P. (2000). Development of peptide antagonists that target estrogen receptor beta-coactivator interactions. *Molecular Endocrinology, 14*(12), 2010–2023. http://dx.doi.org/10.1210/mend.14.12.0561.

Heine, G., Niesner, U., Chang, H. D., Steinmeyer, A., Zugel, U., Zuberbier, T., et al. (2008). 1,25-dihydroxyvitamin D(3) promotes IL-10 production in human B cells. *European Journal of Immunology, 38*(8), 2210–2218. http://dx.doi.org/10.1002/eji.200838216.

Herdick, M., Steinmeyer, A., & Carlberg, C. (2000a). Antagonistic action of a 25-carboxylic ester analogue of 1alpha, 25-dihydroxyvitamin D3 is mediated by a lack of ligand-induced vitamin D receptor interaction with coactivators. *The Journal of Biological Chemistry, 275*(22), 16506–16512. http://dx.doi.org/10.1074/jbc.M910000199.

Herdick, M., Steinmeyer, A., & Carlberg, C. (2000b). Carboxylic ester antagonists of 1alpha,25-dihydroxyvitamin D(3) show cell-specific actions. *Chemistry & Biology,* 7(11), 885–894.

Horlein, A. J., Naar, A. M., Heinzel, T., Torchia, J., Gloss, B., Kurokawa, R., et al. (1995). Ligand-independent repression by the thyroid hormone receptor mediated by a nuclear receptor co-repressor. *Nature, 377*(6548), 397–404. http://dx.doi.org/10.1038/377397a0.

Hughes, P. J., & Brown, G. (2006). 1Alpha,25-dihydroxyvitamin D3-mediated stimulation of steroid sulphatase activity in myeloid leukaemic cell lines requires VDRnuc-mediated activation of the RAS/RAF/ERK-MAP kinase signalling pathway. *Journal of Cellular Biochemistry, 98*(3), 590–617. http://dx.doi.org/10.1002/jcb.20787.

Hughes, P. J., Lee, J. S., Reiner, N. E., & Brown, G. (2008). The vitamin D receptor-mediated activation of phosphatidylinositol 3-kinase (PI3K alpha) plays a role in the 1 alpha,25-dihydroxyvitamin D3-stimulated increase in steroid sulphatase activity in myeloid leukaemic cell lines. *Journal of Cellular Biochemistry, 103*(5), 1551–1572. http://dx.doi.org/10.1002/Jcb.21545.

Hughes, P. J., Steinmeyer, A., Chandraratna, R. A., & Brown, G. (2005). 1alpha,25-dihydroxyvitamin D3 stimulates steroid sulphatase activity in HL60 and NB4 acute myeloid leukaemia cell lines by different receptor-mediated mechanisms. *Journal of Cellular Biochemistry, 94*(6), 1175–1189. http://dx.doi.org/10.1002/jcb.20377.

Igarashi, M., Yoshimoto, N., Yamamoto, K., Shimizu, M., Ishizawa, M., Makishima, M., et al. (2007). Identification of a highly potent vitamin D receptor antagonist: (25S)-26-adamantyl-25-hydroxy-2-methylene-22,23-didehydro-19,27-dinor-20-epi-vita min D3 (ADMI3). *Archives of Biochemistry and Biophysics, 460*(2), 240–253. http://dx.doi.org/10.1016/j.abb.2006.11.026.

Inaba, Y., Nakabayashi, M., Itoh, T., Yoshimoto, N., Ikura, T., Ito, N., et al. (2010). 22S-butyl-1 alpha,24R-dihydroxyvitamin D-3: Recovery of vitamin D receptor agonistic activity. *Journal of Steroid Biochemistry and Molecular Biology, 121*(1–2), 146–150. http://dx.doi.org/10.1016/j.jsbmb.2010.02.033.

Inaba, Y., Yamamoto, K., Yoshimoto, N., Matsunawa, M., Uno, S., Yamada, S., et al. (2007). Vitamin D3 derivatives with adamantane or lactone ring side chains are cell type-selective vitamin D receptor modulators. *Molecular Pharmacology, 71*(5), 1298–1311. http://dx.doi.org/10.1124/mol.106.032318.

Inaba, Y., Yoshimoto, N., Sakamaki, Y., Nakabayashi, M., Ikura, T., Tamamura, H., et al. (2009). A New class of vitamin D analogues that induce structural rearrangement of the ligand-binding pocket of the receptor. *Journal of Medicinal Chemistry, 52*(5), 1438–1449. http://dx.doi.org/10.1021/Jm8014348.

Inada, M., Tsukamoto, K., Hirata, M., Takita, M., Nagasawa, K., & Miyaura, C. (2008). Novel vitamin D3 analogs, 1alpha, 25(OH)2D(3)-26, 23-lactam (DLAMs), antagonize bone resorption via suppressing RANKL expression in osteoblasts. *Biochemical and Biophysical Research Communications*, *372*(3), 434–439. http://dx.doi.org/10.1016/j.bbrc.2008.05.041.

Ishizuka, S., Kurihara, N., Hiruma, Y., Miura, D., Namekawa, J., Tamura, A., et al. (2008). 1alpha,25-dihydroxyvitamin D(3)-26,23-lactam analogues function as vitamin D receptor antagonists in human and rodent cells. *The Journal of Steroid Biochemistry and Molecular Biology*, *110*(3–5), 269–277. http://dx.doi.org/10.1016/j.jsbmb.2007.11.007.

Ishizuka, S., Kurihara, N., Miura, D., Takenouchi, K., Cornish, J., Cundy, T., et al. (2004). Vitamin D antagonist, TEI-9647, inhibits osteoclast formation induced by 1alpha,25-dihydroxyvitamin D3 from pagetic bone marrow cells. *The Journal of Steroid Biochemistry and Molecular Biology*, *89–90*(1–5), 331–334. http://dx.doi.org/10.1016/j.jsbmb.2004.03.025.

Ishizuka, S., Kurihara, N., Reddy, S. V., Cornish, J., Cundy, T., & Roodman, G. D. (2005). (23S)-25-dehydro-1{alpha}-hydroxyvitamin D3-26,23-lactone, a vitamin D receptor antagonist that inhibits osteoclast formation and bone resorption in bone marrow cultures from patients with Paget's disease. *Endocrinology*, *146*(4), 2023–2030. http://dx.doi.org/10.1210/en.2004-1140.

Ishizuka, S., Miura, D., Eguchi, H., Ozono, K., Chokki, M., Kamimura, T., et al. (2000). Antagonistic action of novel 1alpha,25-dihydroxyvitamin D(3)-26, 23-lactone analogs on 25-hydroxyvitamin-D(3)-24-hydroxylase gene expression induced by 1alpha,25-dihydroxy-vitamin D(3) in human promyelocytic leukemia (HL-60) cells. *Archives of Biochemistry and Biophysics*, *380*(1), 92–102. http://dx.doi.org/10.1006/abbi.2000.1902.

Ishizuka, S., Miura, D., Ozono, K., Chokki, M., Mimura, H., & Norman, A. W. (2001). Antagonistic actions in vivo of (23S)-25-dehydro-1alpha-hydroxyvitamin D(3-) 26,23-lactone on calcium metabolism induced by 1alpha,25-dihydroxyvitamin D(3). *Endocrinology*, *142*(1), 59–67. http://dx.doi.org/10.1210/endo.142.1.7925.

Ishizuka, S., Miura, D., Ozono, K., Saito, M., Eguchi, H., Chokki, M., et al. (2001). 23S)- and (23R)-25-dehydro-1alpha-hydroxyvitamin D(3)-26,23-lactone function as antagonists of vitamin D receptor-mediated genomic actions of 1alpha,25-dihydroxyvitamin D (3. *Steroids*, *66*(3–5), 227–237.

Ji, Y., & Studzinski, G. P. (2004). Retinoblastoma protein and CCAAT/enhancer-binding protein beta are required for 1,25-dihydroxyvitamin D3-induced monocytic differentiation of HL60 cells. *Cancer Research*, *64*(1), 370–377.

Kakuda, S., Ishizuka, S., Eguchi, H., Mizwicki, M. T., Norman, A. W., & Takimoto-Kamimura, M. (2010). Structural basis of the histidine-mediated vitamin D receptor agonistic and antagonistic mechanisms of (23S)-25-dehydro-1alpha-hydroxyvitamin D3-26,23-lactone. *Acta Crystallographica. Section D, Biological Crystallography*, *66*(Pt. 8), 918–926. http://dx.doi.org/10.1107/S0907444910020810.

Kanda, N., Hau, C. S., Tada, Y., Sato, S., & Watanabe, S. (2012). Decreased serum LL-37 and vitamin D3 levels in atopic dermatitis: Relationship between IL-31 and oncostatin M. *Allergy*, *67*(6), 804–812. http://dx.doi.org/10.1111/j.1398-9995.2012.02824.x.

Kato, Y., Nakano, Y., Sano, H., Tanatani, A., Kobayashi, H., Shimazawa, R., et al. (2004). Synthesis of 1 alpha 25-dihydroxyvitamin D-3-26,23-lactams (DLAMs), a novel series of 1 alpha 25,-dihydroxyvitamin D-3 antagonist. *Bioorganic & Medicinal Chemistry Letters*, *14*(10), 2579–2583. http://dx.doi.org/10.1016/j.bmcl.2004.02.076.

Khoo, A. L., Chai, L. Y., Koenen, H. J., Oosting, M., Steinmeyer, A., Zuegel, U., et al. (2011). Vitamin D(3) down-regulates proinflammatory cytokine response to mycobacterium tuberculosis through pattern recognition receptors while inducing protective cathelicidin production. *Cytokine*, *55*(2), 294–300. http://dx.doi.org/10.1016/j.cyto.2011.04.016.

Kim, S., Shevde, N. K., & Pike, J. W. (2005). 1,25-dihydroxyvitamin D3 stimulates cyclic vitamin D receptor/retinoid X receptor DNA-binding, co-activator recruitment, and histone acetylation in intact osteoblasts. *Journal of Bone and Mineral Research*, *20*(2), 305–317. http://dx.doi.org/10.1359/JBMR.041112.

Kim, S., Yamazaki, M., Zella, L. A., Shevde, N. K., & Pike, J. W. (2006). Activation of receptor activator of NF-kappa B ligand gene expression by 1,25-dihydroxyvitamin D-3 is mediated through multiple long-range enhancers. *Molecular and Cellular Biology*, *26*(17), 6469–6486. http://dx.doi.org/10.1128/Mcb.00353-06.

Kitagawa, H., Fujiki, R., Yoshimura, K., Mezaki, Y., Uematsu, Y., Matsui, D., et al. (2003). The chromatin-remodeling complex WINAC targets a nuclear receptor to promoters and is impaired in Williams syndrome. *Cell*, *113*(7), 905–917.

Koh, S. S., Chen, D., Lee, Y. H., & Stallcup, M. R. (2001). Synergistic enhancement of nuclear receptor function by p160 coactivators and two coactivators with protein methyltransferase activities. *The Journal of Biological Chemistry*, *276*(2), 1089–1098. http://dx.doi.org/10.1074/jbc.M004228200.

Kota, B. P., Allen, J. D., & Roufogalis, B. D. (2011). The effect of vitamin D3 and ketoconazole combination on VDR-mediated P-gp expression and function in human colon adenocarcinoma cells: Implications in drug disposition and resistance. *Basic & Clinical Pharmacology & Toxicology*, *109*(2), 97–102. http://dx.doi.org/10.1111/j.1742-7843.2011.00693.x.

Kudo, T., Ishizawa, M., Maekawa, K., Nakabayashi, M., Watarai, Y., Uchida, H., et al. (2014). Combination of triple bond and adamantane ring on the vitamin D side chain produced partial agonists for vitamin D receptor. *Journal of Medicinal Chemistry*, *57*(10), 4073–4087. http://dx.doi.org/10.1021/jm401989c.

Lambert, J. R., Kelly, J. A., Shim, M., Huffer, W. E., Nordeen, S. K., Baek, S. J., et al. (2006). Prostate derived factor in human prostate cancer cells: Gene induction by vitamin D via a p53-dependent mechanism and inhibition of prostate cancer cell growth. *Journal of Cellular Physiology*, *208*(3), 566–574. http://dx.doi.org/10.1002/jcp.20692.

Lamblin, M., Spingarn, R., Wang, T. T., Burger, M. C., Dabbas, B., Moitessier, N., et al. (2010). An o-aminoanilide analogue of 1alpha,25-dihydroxyvitamin D(3) functions as a strong vitamin D receptor antagonist. *Journal of Medicinal Chemistry*, *53*(20), 7461–7465. http://dx.doi.org/10.1021/jm1007159.

Lee, Y., Inaba, M., DeLuca, H. F., & Mellon, W. S. (1989). Immunological identification of 1,25-dihydroxyvitamin D3 receptors in human promyelocytic leukemic cells (HL-60) during homologous regulation. *The Journal of Biological Chemistry*, *264*(23), 13701–13705.

Lekanne Deprez, R. H., Riegman, P. H., Groen, N. A., Warringa, U. L., van Biezen, N. A., Molijn, A. C., et al. (1995). Cloning and characterization of MN1, a gene from chromosome 22q11, which is disrupted by a balanced translocation in a meningioma. *Oncogene*, *10*(8), 1521–1528.

Lempiainen, H., Molnar, F., Macias Gonzalez, M., Perakyla, M., & Carlberg, C. (2005). Antagonist- and inverse agonist-driven interactions of the vitamin D receptor and the constitutive androstane receptor with corepressor protein. *Molecular Endocrinology*, *19*(9), 2258–2272. http://dx.doi.org/10.1210/me.2004-0534.

Li, D., Wang, X., Wu, J. L., Quan, W. Q., Ma, L., Yang, F., et al. (2013). Tumor-produced versican V1 enhances hCAP18/LL-37 expression in macrophages through activation of TLR2 and vitamin D3 signaling to promote ovarian cancer progression in vitro. *Plos One*, *8*(2), e56616. http://dx.doi.org/10.1371/journal.pone.0056616.

Liu, P. T., Stenger, S., Li, H. Y., Wenzel, L., Tan, B. H., Krutzik, S. R., et al. (2006). Toll-like receptor triggering of a vitamin D-mediated human antimicrobial response. *Science*, *311*(5768), 1770–1773. http://dx.doi.org/10.1126/science.1123933.

Lonard, D. M., & O'Malley, B. W. (2012). Nuclear receptor coregulators: Modulators of pathology and therapeutic targets. *Nature Reviews. Endocrinology*, *8*(10), 598–604. http://dx.doi.org/10.1038/nrendo.2012.100.

Lundqvist, J., Norlin, M., & Wikvall, K. (2011). 1alpha,25-dihydroxyvitamin D3 exerts tissue-specific effects on estrogen and androgen metabolism. *Biochimica et Biophysica Acta*, *1811*(4), 263–270. http://dx.doi.org/10.1016/j.bbalip.2011.01.004.

Lundqvist, J., Wikvall, K., & Norlin, M. (2012). Vitamin D-mediated regulation of CYP21A2 transcription - a novel mechanism for vitamin D action. *Biochimica et Biophysica Acta*, *1820*(10), 1553–1559. http://dx.doi.org/10.1016/j.bbagen.2012.04.017.

Matsunawa, M., Akagi, D., Uno, S., Endo-Umeda, K., Yamada, S., Ikeda, K., et al. (2012). Vitamin D receptor activation enhances benzo[a]pyrene metabolism via CYP1A1 expression in macrophages. *Drug Metabolism and Disposition*, *40*(11), 2059–2066. http://dx.doi.org/10.1124/dmd.112.046839.

McDonnell, D. P., Chang, C. Y., & Norris, J. D. (2000). Development of peptide antagonists that target estrogen receptor-cofactor interactions. *The Journal of Steroid Biochemistry and Molecular Biology*, *74*(5), 327–335.

Menaa, C., Barsony, J., Reddy, S. V., Cornish, J., Cundy, T., & Roodman, G. D. (2000). 1,25-dihydroxyvitamin D3 hypersensitivity of osteoclast precursors from patients with Paget's disease. *Journal of Bone and Mineral Research*, *15*(2), 228–236. http://dx.doi.org/10.1359/jbmr.2000.15.2.228.

Misawa, T., Demizu, Y., Kawamura, M., Yamagata, N., & Kurihara, M. (2015). Structural development of stapled short helical peptides as vitamin D receptor (VDR)-coactivator interaction inhibitors. *Bioorganic & Medicinal Chemistry*, *23*(5), 1055–1061. http://dx.doi.org/10.1016/j.bmc.2015.01.007.

Mita, Y., Dodo, K., Noguchi-Yachide, T., Hashimoto, Y., & Ishikawa, M. (2013). Structure-activity relationship of benzodiazepine derivatives as LXXLL peptide mimetics that inhibit the interaction of vitamin D receptor with coactivators. *Bioorganic & Medicinal Chemistry*, *21*(4), 993–1005. http://dx.doi.org/10.1016/j.bmc.2012.11.042. S0968-0896(12)00935-2 [pii].

Mita, Y., Dodo, K., Noguchi-Yachide, T., Miyachi, H., Makishima, M., Hashimoto, Y., et al. (2010). LXXLL peptide mimetics as inhibitors of the interaction of vitamin D receptor with coactivators. *Bioorganic & Medicinal Chemistry Letters*, *20*(5), 1712–1717. http://dx.doi.org/10.1016/j.bmcl.2010.01.079. S0960-894X(10)00091-0 [pii].

Miura, D., Manabe, K., Gao, Q., Norman, A. W., & Ishizuka, S. (1999). 1alpha,25-dihydroxyvitamin D(3)-26,23-lactone analogs antagonize differentiation of human leukemia cells (HL-60 cells) but not of human acute promyelocytic leukemia cells (NB4 cells). *FEBS Letters*, *460*(2), 297–302.

Miura, D., Manabe, K., Ozono, K., Saito, M., Gao, Q., Norman, A. W., et al. (1999). Antagonistic action of novel 1alpha,25-dihydroxyvitamin D3-26, 23-lactone analogs on differentiation of human leukemia cells (HL-60) induced by 1alpha,25-dihydroxyvitamin D3. *The Journal of Biological Chemistry*, *274*(23), 16392–16399.

Mizwicki, M. T., Bula, C. M., Mahinthichaichan, P., Henry, H. L., Ishizuka, S., & Norman, A. W. (2009). On the mechanism underlying (23S)-25-dehydro-1alpha (OH)-vitamin D3-26,23-lactone antagonism of hVDRwt gene activation and its switch to a superagonist. *The Journal of Biological Chemistry*, *284*(52), 36292–36301. http://dx.doi.org/10.1074/jbc.M109.042069.

Molinari, C., Uberti, F., Grossini, E., Vacca, G., Carda, S., Invernizzi, M., et al. (2011). 1alpha,25-dihydroxycholecalciferol induces nitric oxide production in cultured endothelial cells. *Cellular Physiology and Biochemistry*, *27*(6), 661–668. http://dx.doi.org/10.1159/000330075.

Moore, T. W., Mayne, C. G., & Katzenellenbogen, J. A. (2010). Minireview: Not picking pockets: Nuclear receptor alternate-site modulators (NRAMs). *Molecular Endocrinology*, *24*(4), 683–695. http://dx.doi.org/10.1210/me.2009-0362.

Nagakubo, T., Demizu, Y., Kanda, Y., Misawa, T., Shoda, T., Okuhira, K., et al. (2014). Development of cell-penetrating R7 fragment-conjugated helical peptides as inhibitors of estrogen receptor-mediated transcription. *Bioconjugate Chemistry*, *25*(11), 1921–1924. http://dx.doi.org/10.1021/bc500480e.

Nakabayashi, M., Yamada, S., Yoshimoto, N., Tanaka, T., Igarashi, M., Ikura, T., et al. (2008). Crystal structures of rat vitamin D receptor bound to adamantyl vitamin D analogs: Structural basis for vitamin D receptor antagonism and partial agonism. *Journal of Medicinal Chemistry*, *51*(17), 5320–5329. http://dx.doi.org/10.1021/jm8004477.

Nakano, Y., Kato, Y., Imai, K., Ochiai, E., Namekawa, J., Ishizuka, S., et al. (2006). Practical synthesis and evaluation of the biological activities of 1 alpha,25-dihydroxyvitamin D-3 antagonists, 1 alpha,25-dihydroxyvitamin D-3-26,23-lactams. Designed on the basis of the helix 12-folding inhibition hypothesis. *Journal of Medicinal Chemistry*, *49*(8), 2398–2406. http://dx.doi.org/10.1021/Jm050738x.

Nandhikonda, P., Lynt, W. Z., McCallum, M. M., Ara, T., Baranowski, A. M., Yuan, N. Y., et al. (2012). Discovery of the first irreversible small molecule inhibitors of the interaction between the vitamin D receptor and coactivators. *Journal of Medicinal Chemistry*, *55*(10), 4640–4651. http://dx.doi.org/10.1021/jm300460c.

Nandhikonda, P., Yasgar, A., Baranowski, A. M., Sidhu, P. S., McCallum, M. M., Pawlak, A. J., et al. (2013). Peroxisome proliferation-activated receptor delta agonist GW0742 interacts weakly with multiple nuclear receptors, including the vitamin D receptor. *Biochemistry*, *52*(24), 4193–4203. http://dx.doi.org/10.1021/bi400321p.

Nevado, J., Tenbaum, S. P., Castillo, A. I., Sanchez-Pacheco, A., & Aranda, A. (2007). Activation of the human immunodeficiency virus type I long terminal repeat by 1 alpha,25-dihydroxyvitamin D3. *Journal of Molecular Endocrinology*, *38*(6), 587–601. http://dx.doi.org/10.1677/JME-06-0065.

Nijenhuis, T., van der Eerden, B. C., Zugel, U., Steinmeyer, A., Weinans, H., Hoenderop, J. G., et al. (2006). The novel vitamin D analog ZK191784 as an intestine-specific vitamin D antagonist. *The FASEB Journal*, *20*(12), 2171–2173. http://dx.doi.org/10.1096/fj.05-5515fje.

Nociti, F. H., Jr., Foster, B. L., Tran, A. B., Dunn, D., Presland, R. B., Wang, L., et al. (2014). Vitamin D represses dentin matrix protein 1 in cementoblasts and osteocytes. *Journal of Dental Research*, *93*(2), 148–154. http://dx.doi.org/10.1177/0022034513516344.

Norman, A. W., Manchand, P. S., Uskokovic, M. R., Okamura, W. H., Takeuchi, J. A., Bishop, J. E., et al. (2000). Characterization of a novel analogue of 1alpha,25(OH)(2)-vitamin D(3) with two side chains: Interaction with its nuclear receptor and cellular actions. *Journal of Medicinal Chemistry*, *43*(14), 2719–2730.

Ochiai, E., Kitagawa, H., Takada, I., Fujiyama, S., Sawatsubashi, S., Kim, M. S., et al. (2010). CDP/cut is an osteoblastic coactivator of the vitamin D receptor (VDR). *Journal of Bone and Mineral Research*, *25*(5), 1157–1166. http://dx.doi.org/10.1359/jbmr.091105.

Ochiai, E., Miura, D., Eguchi, H., Ohara, S., Takenouchi, K., Azuma, Y., et al. (2005). Molecular mechanism of the vitamin D antagonistic actions of (23S)-25-dehydro-1alpha-hydroxyvitamin D3-26,23-lactone depends on the primary structure of the carboxyl-terminal region of the vitamin d receptor. *Molecular Endocrinology*, *19*(5), 1147–1157. http://dx.doi.org/10.1210/me.2004-0234.

Ogryzko, V. V., Schiltz, R. L., Russanova, V., Howard, B. H., & Nakatani, Y. (1996). The transcriptional coactivators p300 and CBP are histone acetyltransferases. *Cell*, *87*(5), 953–959.

Ozono, K., Saito, M., Miura, D., Michigami, T., Nakajima, S., & Ishizuka, S. (1999). Analysis of the molecular mechanism for the antagonistic action of a novel 1alpha,25-dihydroxyvitamin D(3) analogue toward vitamin D receptor function. *The Journal of Biological Chemistry*, *274*(45), 32376–32381.

Pathrose, P., Barmina, O., Chang, C. Y., McDonnell, D. P., Shevde, N. K., & Pike, J. W. (2002). Inhibition of 1,25-dihydroxyvitamin D3-dependent transcription by synthetic

LXXLL peptide antagonists that target the activation domains of the vitamin D and retinoid X receptors. *Journal of Bone and Mineral Research*, *17*(12), 2196–2205. http://dx.doi.org/10.1359/jbmr.2002.17.12.2196.
Perakyla, M., Molnar, F., & Carlberg, C. (2004). A structural basis for the species-specific antagonism of 26,23-lactones on vitamin D signaling. *Chemistry & Biology*, *11*(8), 1147–1156. http://dx.doi.org/10.1016/j.chembiol.2004.05.023.
Peric, M., Koglin, S., Dombrowski, Y., Gross, K., Bradac, E., Buchau, A., et al. (2009). Vitamin D analogs differentially control antimicrobial peptide/'alarmin' expression in psoriasis. *Plos One*, *4*(7), e6340. http://dx.doi.org/10.1371/journal.pone.0006340.
Perissi, V., Jepsen, K., Glass, C. K., & Rosenfeld, M. G. (2010). Deconstructing repression: Evolving models of co-repressor action. *Nature Reviews. Genetics*, *11*(2), 109–123. http://dx.doi.org/10.1038/nrg2736.
Phillips, C., Roberts, L. R., Schade, M., Bazin, R., Bent, A., Davies, N. L., et al. (2011). Design and structure of stapled peptides binding to estrogen receptors. *Journal of the American Chemical Society*, *133*(25), 9696–9699. http://dx.doi.org/10.1021/ja202946k.
Pittarella, P., Squarzanti, D. F., Molinari, C., Invernizzi, M., Uberti, F., & Reno, F. (2015). NO-dependent proliferation and migration induced by vitamin D in HUVEC. *The Journal of Steroid Biochemistry and Molecular Biology*, *149C*, 35–42. http://dx.doi.org/10.1016/j.jsbmb.2014.12.012.
Polly, P., Herdick, M., Moehren, U., Baniahmad, A., Heinzel, T., & Carlberg, C. (2000). VDR-alien: A novel, DNA-selective vitamin D(3) receptor-corepressor partnership. *The FASEB Journal*, *14*(10), 1455–1463.
Rachez, C., Lemon, B. D., Suldan, Z., Bromleigh, V., Gamble, M., Naar, A. M., et al. (1999). Ligand-dependent transcription activation by nuclear receptors requires the DRIP complex. *Nature*, *398*(6730), 824–828. http://dx.doi.org/10.1038/19783.
Rachez, C., Suldan, Z., Ward, J., Chang, C. P. B., Burakov, D., Erdjument-Bromage, H., et al. (1998). A novel protein complex that interacts with the vitamin D-3 receptor in a ligand-dependent manner and enhances VDR transactivation in a cell-free system. *Genes & Development*, *12*(12), 1787–1800. http://dx.doi.org/10.1101/gad.12.12.1787.
Sadeghi, K., Wessner, B., Laggner, U., Ploder, M., Tamandl, D., Friedl, J., et al. (2006). Vitamin D3 down-regulates monocyte TLR expression and triggers hyporesponsiveness to pathogen-associated molecular patterns. *European Journal of Immunology*, *36*(2), 361–370. http://dx.doi.org/10.1002/eji.200425995.
Saito, N., & Kittaka, A. (2006). Highly potent vitamin D receptor antagonists: Design, synthesis, and biological evaluation. *Chembiochem*, 7(10), 1479–1490. http://dx.doi.org/10.1002/cbic.200600054.
Saito, N., Masuda, M., Matsunaga, T., Saito, H., Anzai, M., Takenouchi, K., et al. (2004). 24,24-dimethylvitamin D-3-26,23-lactones and their 2 alpha-functionalized analogues as highly potent VDR antagonists. *Tetrahedron*, *60*(36), 7951–7961. http://dx.doi.org/10.1016/j.tet.2004.05.113.
Saito, N., Matsunaga, T., Fujishima, T., Anzai, M., Saito, H., Takenouchi, K., et al. (2003). Remarkable effect of 2[small alpha]-modification on the VDR antagonistic activity of 1small alpha-hydroxyvitamin D3-26,23-lactones. *Organic & Biomolecular Chemistry*, *1*(24), 4396–4402. http://dx.doi.org/10.1039/b311107e.
Saito, N., Matsunaga, T., Saito, H., Anzai, M., Takenouchi, K., Miura, D., et al. (2006a). Synthesis and 2 alpha-modification of 24-phenylvitamin D-3 lactones: Effects on VDR antagonistic activity. *Heterocycles*, *67*(1), 311–336.
Saito, N., Matsunaga, T., Saito, H., Anzai, M., Takenouchi, K., Miura, D., et al. (2006b). Further synthetic and biological studies on vitamin D hormone antagonists based on C24-alkylation and C2alpha-functionalization of 25-dehydro-1alpha-hydroxyvitamin D(3)-26,23-lactones. *Journal of Medicinal Chemistry*, *49*(24), 7063–7075. http://dx.doi.org/10.1021/jm060797q.

Saito, N., Saito, H., Anzai, M., Yoshida, A., Fujishima, T., Takenouchi, K., et al. (2003). Dramatic enhancement of antagonistic activity on vitamin D receptor: A double functionalization of 1alpha-hydroxyvitamin D3 26,23-lactones. *Organic Letters*, *5*(25), 4859–4862. http://dx.doi.org/10.1021/ol035922w.

Sakamaki, Y., Inaba, Y., Yoshimoto, N., & Yamamoto, K. (2010). Potent antagonist for the vitamin D receptor: Vitamin D analogues with simple side chain structure. *Journal of Medicinal Chemistry*, *53*(15), 5813–5826. http://dx.doi.org/10.1021/Jm100649d.

Sanchez-Martinez, R., Castillo, A. I., Steinmeyer, A., & Aranda, A. (2006). The retinoid X receptor ligand restores defective signalling by the vitamin D receptor. *Embo Reports*, 7(10), 1030–1034. http://dx.doi.org/10.1038/sj.embor.7400776.

Sanchez-Martinez, R., Zambrano, A., Castillo, A. I., & Aranda, A. (2008). Vitamin D-dependent recruitment of corepressors to vitamin D/retinoid X receptor heterodimers. *Molecular and Cellular Biology*, *28*(11), 3817–3829. http://dx.doi.org/10.1128/MCB.01909-07.

Santos-Martinez, N., Diaz, L., Ordaz-Rosado, D., Garcia-Quiroz, J., Barrera, D., Avila, E., et al. (2014). Calcitriol restores antiestrogen responsiveness in estrogen receptor negative breast cancer cells: A potential new therapeutic approach. *BMC Cancer*, *14*, 230. http://dx.doi.org/10.1186/1471-2407-14-230.

Savkur, R. S., Bramlett, K. S., Stayrook, K. R., Nagpal, S., & Burris, T. P. (2005). Coactivation of the human vitamin D receptor by the peroxisome proliferator-activated receptor gamma coactivator-1 alpha. *Molecular Pharmacology*, *68*(2), 511–517. http://dx.doi.org/10.1124/mol.105.012708.

Schauber, J., Dorschner, R. A., Coda, A. B., Buchau, A. S., Liu, P. T., Kiken, D., et al. (2007). Injury enhances TLR2 function and antimicrobial peptide expression through a vitamin D-dependent mechanism. *The Journal of Clinical Investigation*, *117*(3), 803–811. http://dx.doi.org/10.1172/JCI30142.

Shah, S., Islam, M. N., Dakshanamurthy, S., Rizvi, I., Rao, M., Herrell, R., et al. (2006). The molecular basis of vitamin D receptor and beta-catenin crossregulation. *Molecular Cell*, *21*(6), 799–809. http://dx.doi.org/10.1016/j.molcel.2006.01.037.

Sharma, D., & Fondell, J. D. (2002). Ordered recruitment of histone acetyltransferases and the TRAP/mediator complex to thyroid hormone-responsive promoters in vivo. *Proceedings of the National Academy of Sciences of the United States of America*, *99*(12), 7934–7939. http://dx.doi.org/10.1073/pnas.122004799.

Sidhu, P. S., Nassif, N., McCallum, M. M., Teske, K., Feleke, B., Yuan, N. Y., et al. (2014). Development of novel vitamin D receptor-coactivator inhibitors. *ACS Medicinal Chemistry Letters*, *5*(2), 199–204. http://dx.doi.org/10.1021/ml400462j.

Sidhu, P. S., Teske, K., Feleke, B., Yuan, N. Y., Guthrie, M. L., Fernstrum, G. B., et al. (2014). Anticancer activity of VDR-coregulator inhibitor PS121912. *Cancer Chemotherapy and Pharmacology*, *74*(4), 787–798. http://dx.doi.org/10.1007/s00280-014-2549-y.

Silver, J., Naveh-Many, T., Mayer, H., Schmelzer, H. J., & Popovtzer, M. M. (1986). Regulation by vitamin D metabolites of parathyroid hormone gene transcription in vivo in the rat. *The Journal of Clinical Investigation*, *78*(5), 1296–1301. http://dx.doi.org/10.1172/JCI112714.

Somjen, D., Kohen, F., Amir-Zaltsman, Y., Knoll, E., & Stern, N. (2000). Vitamin D analogs modulate the action of gonadal steroids in human vascular cells in vitro. *American Journal of Hypertension*, *13*(4 Pt. 1), 396–403.

Somjen, D., Waisman, A., Lee, J. K., Posner, G. H., & Kaye, A. M. (2001). A non-calcemic analog of 1 alpha,25 dihydroxy vitamin D(3) (JKF) upregulates the induction of creatine kinase B by 17 beta estradiol in osteoblast-like ROS 17/2.8 cells and in rat diaphysis. *The Journal of Steroid Biochemistry and Molecular Biology*, 77(4–5), 205–212.

Spencer, T. E., Jenster, G., Burcin, M. M., Allis, C. D., Zhou, J., Mizzen, C. A., et al. (1997). Steroid receptor coactivator-1 is a histone acetyltransferase. *Nature*, *389*(6647), 194–198. http://dx.doi.org/10.1038/38304.

Spielvogel, A. M., Farley, R. D., & Norman, A. W. (1972). Studies on the mechanism of action of calciferol. V. Turnover time of chick intestinal epithelial cells in relation to the intestinal action of vitamin D. *Experimental Cell Research*, *74*(2), 359–366.

Stio, M., Martinesi, M., Simoni, A., Zuegel, U., Steinmeyer, A., Santi, R., et al. (2011). The novel vitamin D analog ZK191784 inhibits prostate cancer cell invasion. *Anticancer Research*, *31*(12), 4091–4098.

Strauch, U. G., Obermeier, F., Grunwald, N., Dunger, N., Rath, H. C., Scholmerich, J., et al. (2007). Calcitriol analog ZK191784 ameliorates acute and chronic dextran sodium sulfate-induced colitis by modulation of intestinal dendritic cell numbers and phenotype. *World Journal of Gastroenterology*, *13*(48), 6529–6537.

Studzinski, G. P., Wang, X. N., Ji, Y., Wang, Q., Zhang, Y. Y., Kutner, A., et al. (2005). The rationale for deltanoids in therapy for myeloid leukemia: Role of KSR-MAPK-C/EBP pathway. *Journal of Steroid Biochemistry and Molecular Biology*, *97*(1–2), 47–55. http://dx.doi.org/10.1016/j.jsbmb.2005.06.010.

Szeles, L., Keresztes, G., Torocsik, D., Balajthy, Z., Krenacs, L., Poliska, S., et al. (2009). 1,25-dihydroxyvitamin D-3 is an autonomous regulator of the transcriptional changes leading to a tolerogenic dendritic cell phenotype. *Journal of Immunology*, *182*(4), 2074–2083. http://dx.doi.org/10.4049/jimmunol.0803345.

Takenouchi, K., Sogawa, R., Manabe, K., Saitoh, H., Gao, Q., Miura, D., et al. (2004). Synthesis and structure-activity relationships of TEI-9647 derivatives as vitamin D3 antagonists. *The Journal of Steroid Biochemistry and Molecular Biology*, *89–90*(1–5), 31–34. http://dx.doi.org/10.1016/j.jsbmb.2004.03.046.

Takeyama, K., Kitanaka, S., Sato, T., Kobori, M., Yanagisawa, J., & Kato, S. (1997). 25-Hydroxyvitamin D3 1alpha-hydroxylase and vitamin D synthesis. *Science*, *277*(5333), 1827–1830.

Takita, M., Hirata, M., Tsukamoto, K., Nagasawa, K., Miyaura, C., & Inada, M. (2008). 1 alpha,25-Dihydroxyvitamin D(3)-26,23-lactam, a novel vitamin D(3) analog, acts as a vitamin D(3) antagonist in human prostate cancer cells. *Journal of Health Science*, *54*(4), 497–502. http://dx.doi.org/10.1248/Jhs.54.497.

Teichert, A., Arnold, L. A., Otieno, S., Oda, Y., Augustinaite, I., Geistlinger, T. R., et al. (2009). Quantification of the vitamin D receptor-coregulator interaction. *Biochemistry*, *48*(7), 1454–1461. http://dx.doi.org/10.1021/Bi801874n.

Teske, K., Nandhikonda, P., Bogart, J. W., Feleke, B., Sidhu, P., Yuan, N., et al. (2014a). Modulation of transcription mediated by the vitamin D receptor and the peroxisome proliferator-activated receptor delta in the presence of GW0742 analogs. *Journal of Biomolecular Research & Therapeutics*, *3*(1). http://dx.doi.org/10.4172/2167-7956.1000111.

Teske, K., Nandhikonda, P., Bogart, J. W., Feleke, B., Sidhu, P., Yuan, N., et al. (2014b). Identification of Vdr antagonists among nuclear receptor ligands using virtual screening. *Nuclear Receptor Research*, *1*, 101076. http://dx.doi.org/10.11131/2014/101076.

Thompson, P. D., Jurutka, P. W., Haussler, C. A., Whitfield, G. K., & Haussler, M. R. (1998). Heterodimeric DNA binding by the vitamin D receptor and retinoid X receptors is enhanced by 1,25-dihydroxyvitamin D3 and inhibited by 9-cis-retinoic acid. Evidence for allosteric receptor interactions. *The Journal of Biological Chemistry*, *273*(14), 8483–8491.

Toell, A., Gonzalez, M. M., Ruf, D., Steinmeyer, A., Ishizuka, S., & Carlberg, C. (2001). Different molecular mechanisms of vitamin D(3) receptor antagonists. *Molecular Pharmacology*, *59*(6), 1478–1485.

Tolon, R. M., Castillo, A. I., Jimenez-Lara, A. M., & Aranda, A. (2000). Association with Ets-1 causes ligand- and AF2-independent activation of nuclear receptors. *Molecular and Cellular Biology*, *20*(23), 8793–8802.

Uberti, F., Lattuada, D., Morsanuto, V., Nava, U., Bolis, G., Vacca, G., et al. (2014). Vitamin D protects human endothelial cells from oxidative stress through the autophagic and survival pathways. *The Journal of Clinical Endocrinology and Metabolism*, *99*(4), 1367–1374. http://dx.doi.org/10.1210/jc.2013-2103.

Vaisanen, S., Perakyla, M., Karkkainen, J. I., Steinmeyer, A., & Carlberg, C. (2002). Critical role of helix 12 of the vitamin D(3) receptor for the partial agonism of carboxylic ester antagonists. *Journal of Molecular Biology*, *315*(2), 229–238. http://dx.doi.org/10.1006/jmbi.2001.5225.

van der Eerden, B. C., Fratzl-Zelman, N., Nijenhuis, T., Roschger, P., Zugel, U., Steinmeyer, A., et al. (2013). The vitamin D analog ZK191784 normalizes decreased bone matrix mineralization in mice lacking the calcium channel TRPV5. *Journal of Cellular Physiology*, *228*(2), 402–407. http://dx.doi.org/10.1002/jcp.24144.

van Driel, M., Koedam, M., Buurman, C. J., Roelse, M., Weyts, F., Chiba, H., et al. (2006). Evidence that both 1 alpha,25-dihydroxyvitamin D-3 and 24-hydroxylated D-3 enhance human osteoblast differentiation and mineralization. *Journal of Cellular Biochemistry*, *99*(3), 922–935. http://dx.doi.org/10.1002/Jcb.20875.

Vanhooke, J. L., Benning, M. M., Bauer, C. B., Pike, J. W., & DeLuca, H. F. (2004). Molecular structure of the rat vitamin D receptor ligand binding domain complexed with 2-carbon-substituted vitamin D3 hormone analogues and a LXXLL-containing coactivator peptide. *Biochemistry*, *43*(14), 4101–4110. http://dx.doi.org/10.1021/bi036056y.

Wang, J. H., Tang, K., Hou, Q. Q., Cheng, X. L., Dong, L. H., Liu, Y. J., et al. (2010). 3D-QSAR studies on C24-monoalkylated vitamin D-3 26,23-lactones and their C2 alpha-modified derivatives with inhibitory activity to vitamin D receptor. *Molecular Informatics*, *29*(8–9), 621–632. http://dx.doi.org/10.1002/minf.201000071.

Wang, X., Wang, T. T., White, J. H., & Studzinski, G. P. (2006). Induction of kinase suppressor of RAS-1(KSR-1) gene by 1, alpha25-dihydroxyvitamin D3 in human leukemia HL60 cells through a vitamin D response element in the 5'-flanking region. *Oncogene*, *25*(53), 7078–7085. http://dx.doi.org/10.1038/sj.onc.1209697.

Wang, X., Wang, T. T., White, J. H., & Studzinski, G. P. (2007). Expression of human kinase suppressor of Ras 2 (hKSR-2) gene in HL60 leukemia cells is directly upregulated by 1,25-dihydroxyvitamin D(3) and is required for optimal cell differentiation. *Experimental Cell Research*, *313*(14), 3034–3045. http://dx.doi.org/10.1016/j.yexcr.2007.05.021.

Xie, Z., Chang, S., Oda, Y., & Bikle, D. D. (2006). Hairless suppresses vitamin D receptor transactivation in human keratinocytes. *Endocrinology*, *147*(1), 314–323. http://dx.doi.org/10.1210/en.2005-1111.

Xu, H. E., Stanley, T. B., Montana, V. G., Lambert, M. H., Shearer, B. G., Cobb, J. E., et al. (2002). Structural basis for antagonist-mediated recruitment of nuclear co-repressors by PPARalpha. *Nature*, *415*(6873), 813–817. http://dx.doi.org/10.1038/415813a.

Xu, J., Wu, R. C., & O'Malley, B. W. (2009). Normal and cancer-related functions of the p160 steroid receptor co-activator (SRC) family. *Nature Reviews. Cancer*, *9*(9), 615–630. http://dx.doi.org/10.1038/nrc2695.

Yamamoto, K., Abe, D., Yoshimoto, N., Choi, M., Yamagishi, K., Tokiwa, H., et al. (2006). Vitamin D receptor: Ligand recognition and allosteric network. *Journal of Medicinal Chemistry*, *49*(4), 1313–1324. http://dx.doi.org/10.1021/jm050795q.

Yamamoto, K., Inaba, Y., Yoshimoto, N., Choi, M., DeLuca, H. F., & Yamada, S. (2007). 22-alkyl-20-epi-1 alpha,25-dihydroxyvitamin D-3 compounds of superagonistic activity: Syntheses, biological activities and interaction with the receptor. *Journal of Medicinal Chemistry*, *50*(5), 932–939. http://dx.doi.org/10.1021/Jm060889fd.

Yamaoka, K., Kim, M. S., Takada, I., Takeyama, K., Kamimura, T., & Kato, S. (2006). Culture serum-induced conversion from agonist to antagonist of a vitamin D analog, TEI-9647. *The Journal of Steroid Biochemistry and Molecular Biology*, *100*(4-5), 177–183. http://dx.doi.org/10.1016/j.jsbmb.2006.04.008.

Yang, X. J., Ogryzko, V. V., Nishikawa, J., Howard, B. H., & Nakatani, Y. (1996). A p300/CBP-associated factor that competes with the adenoviral oncoprotein E1A. *Nature*, *382*(6589), 319–324. http://dx.doi.org/10.1038/382319a0.

Yoshimoto, N., Inaba, Y., Yamada, S., Makishima, M., Shimizu, M., & Yamamoto, K. (2008). 2-methylene 19-nor-25-dehydro-1alpha-hydroxyvitamin D3 26,23-lactones: Synthesis, biological activities and molecular basis of passive antagonism. *Bioorganic & Medicinal Chemistry*, *16*(1), 457–473. http://dx.doi.org/10.1016/j.bmc.2007.09.017.

Yoshimoto, N., Sakamaki, Y., Haeta, M., Kato, A., Inaba, Y., Itoh, T., et al. (2012). Butyl pocket formation in the vitamin D receptor strongly affects the agonistic or antagonistic behavior of ligands. *Journal of Medicinal Chemistry*, *55*(9), 4373–4381. http://dx.doi.org/10.1021/Jm300230a.

Zella, L. A., Chang, C. Y., McDonnell, D. P., & Pike, J. W. (2007). The vitamin D receptor interacts preferentially with DRIP205-like LxxLL motifs. *Archives of Biochemistry and Biophysics*, *460*(2), 206–212. http://dx.doi.org/10.1016/j.abb.2006.12.016.

Zugel, U., Steinmeyer, A., Giesen, C., & Asadullah, K. (2002). A novel immunosuppressive 1alpha,25-dihydroxyvitamin D3 analog with reduced hypercalcemic activity. *The Journal of Investigative Dermatology*, *119*(6), 1434–1442. http://dx.doi.org/10.1046/j.1523-1747.2002.19623.x.

CHAPTER FOUR

Structural Studies of Vitamin D Nuclear Receptor Ligand-Binding Properties

Anna Y. Belorusova, Natacha Rochel[1]

Department of Integrative Structural Biology, Institut de Génétique et de Biologie Moléculaire et Cellulaire (IGBMC), Institut National de Santé et de Recherche Médicale (INSERM) U964, Centre National de Recherche Scientifique (CNRS) UMR 7104, Université de Strasbourg, Illkirch, France

[1]Corresponding author: e-mail address: rochel@igbmc.fr

Contents

Abstract

The vitamin D nuclear receptor (VDR) and its natural ligand, 1α,25-dihydroxyvitamin D3 hormone ($1,25(OH)_2D_3$, or calcitriol), classically regulate mineral homeostasis and

Vitamins and Hormones, Volume 100
ISSN 0083-6729
http://dx.doi.org/10.1016/bs.vh.2015.10.003

metabolism but also much broader range of biological functions, such as cell growth, differentiation, antiproliferation, apoptosis, adaptive/innate immune responses. Being widely expressed in various tissues, VDR represents an important therapeutic target in the treatment of diverse disorders. Since ligand binding is a key step in VDR-mediated signaling, numerous 1,25$(OH)_2D_3$ analogs have been synthesized in order to selectively modulate the receptor activity. Most of the synthetic analogs have been developed by modification of a parental compound and some of them mimic 1,25$(OH)_2D_3$ scaffold without being structurally related to it. The ability of ligands that have different size and conformation to bind to VDR and to demonstrate biological effects is intriguing, and therefore, ligand-binding properties of the receptor have been extensively investigated using a variety of biochemical, biophysical, and computational methods. In this chapter, we describe different aspects of the structure–function relationship of VDR in complex with natural and synthetic ligands coming from structural analysis. With the emphasis on the binding modes of the most promising compounds, such as secosteroidal agonists and 1,25$(OH)_2D_3$ mimics, we also highlight the action of VDR antagonists and the evidence for the existence of an alternative ligand-binding site within the receptor. Additionally, we describe the crystal structures of VDR mutants associated with hereditary vitamin D-resistant rickets that display impaired ligand-binding function.

1. INTRODUCTION

The vitamin D nuclear receptor (VDR) is a transcription factor binding with high affinity its natural ligand, the 1α,25-dihydroxyvitamin D3 (1,25 $(OH)_2D_3$), also known as calcitriol. VDR and its ligand regulate various biological functions, such as cell growth, differentiation, antiproliferation, apoptosis, adaptive/innate immune responses, bone mineralization, and calcium/phosphate homeostasis (Haussler, Jurutka, Mizwicki, & Norman, 2011). VDR is a promiscuous nuclear receptor (NR) which is found in prostate, ovary, breast, and skin, and also in brain, heart, pancreas, kidney, intestine, and colon. Consequently, deregulation of VDR function may lead to severe diseases such as cancers, psoriasis, rickets, renal osteodystrophy, and autoimmunity disorders (multiple sclerosis, rheumatoid arthritis, inflammatory bowel diseases, type I diabetes), and, therefore, VDR represents an important drug target (Bouillon et al., 2006; Holick, 2003; Pinette, Yee, Amegadzie, & Nagpal, 2003). Several synthetic analogs of calcitriol are used for the standard topical treatment of psoriasis (Leyssens, Verlinden, & Verstuyf, 2014), a chronic skin disease characterized by hyperproliferation of keratinocytes, but their implication in the systemic treatment is limited as they induce hypercalcemia. Therefore, the development of new vitamin D analogs that possess desired calcitriol activities with reduced calcemic effect is strongly required. Calcitriol and its analogs also have potential use in the treatment of neurodegenerative and autoimmune diseases.

VDR mediates the biological effects of its ligand by regulating the transcription of target genes: it forms a heterodimer with the retinoid X nuclear receptors (RXRs) that binds specific DNA motif of controlled genes, so-called VDR response element (VDRE). Steric constraints of the VDR-RXR complex determine the optimal heterodimer binding site within VDRE as a direct repeat of the sequence RGKTSA (R=A or G, K=G or T, S=C or G) separated by three nucleotides and therefore called DR3. Recent genome-wide studies on VDR binding have revealed that only few dozens of VDREs are located within 10 kb from the TSS of $1,25(OH)_2D_3$ target genes (Haussler et al., 2013), while thousands of additional VDR loci spread over the whole genome (Bernstein et al., 2012). In a ligand-dependent manner, the DNA-bound heterodimer recruits various coregulators of transcription: typically, in the absence of ligands or in the presence of antagonists corepressors are recruited to the target genes, while agonist ligands induce a change in the structure of the NR that allows interaction with coactivators. Recruitment of coactivators results first in histone acetylation which prepares target gene promoters through decondensation of the chromatin and, further, in a link with the basal transcriptional machinery.

VDR shares the main structural characteristics of NRs, which is modular and consists of a highly conserved DNA-binding domain (DBD) and a ligand-binding domain (LBD) linked by a hinge region (Fig. 1). Like the

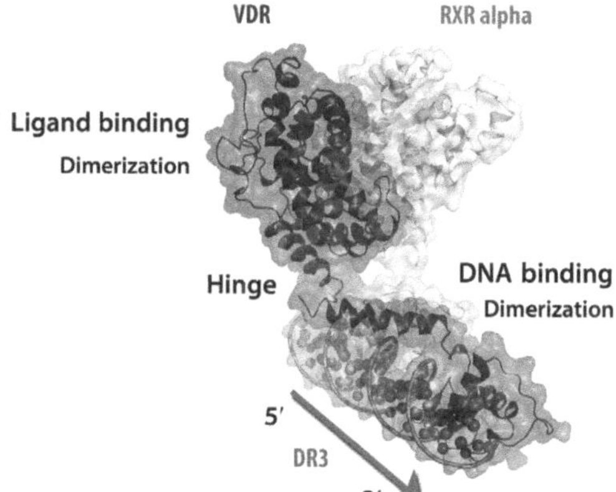

Figure 1 Structural organization of the VDR–RXRα heterodimer bound to canonical DR3 response element as suggested by solution studies. (See the color plate.)

other NRs, VDR is a multifunctional protein interacting simultaneously with molecules of different size and origin (hormone, DNA, RXR, coregulators of transcription, and various protein complexes). Structural studies of VDR/RXR DNA-bound heterodimer revealed a relative orientation of the LBDs and DBDs: the LBDs are situated in an asymmetric manner on the 5′-end of the response element, shifted away from the center of the two half-sites (Orlov, Rochel, Moras, & Klaholz, 2012; Rochel et al., 2011). However, even with the relatively separated positioning, there is an evidence of long-range allosteric connections between the domains (Zhang et al., 2011). Other NRs comprise also a variable N-terminal A/B domain displaying a ligand-independent activation function; however, the A/B domain of VDR is very short and its function is not completely understood.

The LBD of the NRs harbors a ligand-dependent activation function or AF-2, a major interface for homo- and heterodimerization and an interface for coactivators as well as corepressors. Therefore, the binding of ligand is a key step of NR signaling that has been intensively investigated due to the high pharmaceutical potential of NR ligands. In case of VDR, since the first crystal structure of the hVDR LBD–1,25$(OH)_2D_3$ complex, 102 crystal structures of VDR LBD–ligand complexes have been deposited in the protein data bank (PDB) (reviewed in Carlberg, Molnár, & Mouriño, 2012; Molnár, 2014; Rochel & Moras, 2011; Yamada & Makishima, 2014). These crystal structures explain most features of VDR ligand binding, superagonism, mutual adaptability of the ligands and of the pocket, provide clues for the antagonism mechanism and indicate the presence of an alternative binding site in the VDR LBD.

2. CRYSTAL STRUCTURES OF VDR LBD IN COMPLEX WITH 1,25$(OH)_2D_3$

2.1 Overall Organization of Human, Rat, and Zebrafish VDR–LBD Complexes

Detailed information on the molecular mechanism of action of 1,25 $(OH)_2D_3$ has been obtained by the elucidation of the crystal structure of its complex with the VDR LBD. Structural characterization of VDR–ligand interactions has been investigated using either human (h) or rat (r) VDR LBD (Rochel, Wurtz, Mitschler, Klaholz, & Moras, 2000; Vanhooke, Benning, Bauer, Pike, & DeLuca, 2004), or the LBD from zebrafish (z) VDR (Ciesielski, Rochel, & Moras, 2007). The crystal structure of

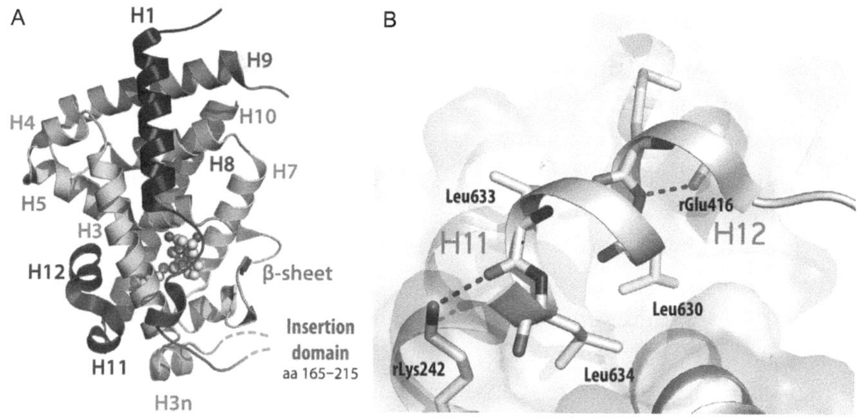

Figure 2 Overall structure of the VDR LBD. (A) The VDR LBD bound to $1,25(OH)_2D_3$ is composed of 13 helices (H1, 2, 2n, 3–12) (PDB ID: 1DB1). VDR members present an additional insertion domain (ID) between helices H1 and H2n. (B) Binding mode of the DRIP205 coactivator peptide LXXLL (PDB ID: 1RK3). (See the color plate.)

the hVDR–LBD complex (Rochel et al., 2000) has revealed a conformation similar to other NR LBDs crystal structures (Fig. 2A). The general fold of VDR LBD consists of a three-layered α-helical sandwich composed of 12 helices (H1–H12) and a three-stranded β-sheet. The ligand-binding site is located on the bottom of the LBD and is surrounded by helices H2, H3, H5, H6, H7, H10, and H12. The residues of each of β-sheet strands also form contacts with a ligand.

For the crystallization of the hVDR–$1,25(OH)_2D_3$ complex, a truncated form of the hVDR LBD was used: 50 amino acid residues from a large insertion domain at the N-terminal part of the LBD connecting helices H1–H3 was deleted. This region is characterized by poor sequence conservation between VDR family members and predicted disordered state. No clear biological function has been assigned to the insertion region (Rochel et al., 2001). The truncation of the hVDR construct is positioned just before H3n (Fig. 2). In case when rVDR LBD is used for the crystallization, the same truncation is present. For the zVDR LBD, a wild-type protein is used; however, the insertion region is not visible in the electron density maps. The differences observed between the structures of h, r, and z VDR LBDs are small and primarily involve the loops. The ligand-binding pocket is conserved both in sequence and structure, and the ligand conformation and interactions observed in all crystal structures are very similar.

2.2 Activation Function-2

The first crystal structures of apo-RXRa and holo-RARg LBDs have revealed that LBDs undergo major conformational changes upon ligand binding that has been called the "mousetrap mechanism" (Renaud et al., 1995). Upon ligand binding, helix H11 is repositioned in the continuity of helix H10, and helix H12 swings in to seal the binding cavity while the ω-loop flips over underneath helix H6 bringing along the N-terminal part of helix H3. Helix H12, also referred to as the activating domain of the AF-2 function, stabilizes ligand binding by contributing to the hydrophobic environment, making, in some cases, additional contacts with the ligand. Functionally, ligands act as switches for coregulator binding by triggering major movements of helix H12. The agonist conformation of H12 allows interactions with coactivators from the p160 family of steroid receptor coactivators (SRCs) that remodel chromatin (Oñate, Tsai, Tsai, & O'Malley, 1995), or of the DRIP/TRAP mediator complexes that interact with the basal transcriptional machinery and help recruit the RNA polymerase to the transcription start site (Rachez et al., 1999). Only one agonist active conformation is observed for all NRs–agonist complexes that permits coactivator binding, while several other conformations have been shown allowing corepressor binding (Gallastegui, Mackinnon, Fletterick, & Estébanez-Perpiñá, 2015). In the absence of ligand, helix H12 can adopt a conformation leading to an autorepressed conformation of the NR, which blocks coactivator interaction, to a conformation where H12 is freely moving in solution.

2.3 Recognition of Coactivator Peptide

All the crystal structures of VDR LBDs of different species display an active conformation. Nevertheless, there is a major difference between the crystallization conditions that are typically used for h, r, and z VDR LBDs: hVDR LBD is cocrystallized with ligands in the absence of coactivator peptide, while in case of the rVDR and the zVDR LBDs short coactivator peptides are added in order to stabilize the complex and promote crystal growth. For the rVDR complexes, the second LXXLL motif of DRIP205 has been used (Vanhooke et al., 2004) and for the zVDR the second NR box of SRC-1 or SRC-2 (Ciesielski, Rochel, Mitschler, Kouzmenko, & Moras, 2004; Huet et al., 2011).

The LXXLL peptides observed in the crystal structures of z and rVDR are bound to a groove formed by helices H3, H4, and H12 of the LBD (Fig. 2B).

For the LXXLL motif of DRIP205 (625–636 KNHPMLMNLLKD), this interaction buries about 507 $Å^2$ of the receptor's surface. The side chains of Leu630, 633, and 634 of a peptide are buried within the pocket and surrounded by hydrophobic residues, while Met631 and Asn632 are oriented outward from the solvent. Coactivator peptide is additionally locked through the hydrogen bonds between Glu416 from H12 (hGlu420) with the backbone amide nitrogen of Leu630 and Leu633. At the other end, Lys242 from H3 (hLys246) forms a hydrogen bond to the main chain oxygen of Leu634. The same binding is observed for the zVDR LBD–peptide–ligand complex. These interactions are similar to those described for other NRs (Darimont et al., 1998; Nolte et al., 1998; Shiau et al., 1998).

2.4 VDR Ligand-Binding Pocket

The A-ring of 1,25$(OH)_2D_3$ in its VDR LBP adopts a chair B conformation with the 19-methylene "up" and the 1α-OH and 3β-OH groups in equatorial and axial orientations, respectively. The aliphatic chain at position 17 of the D-ring adopts an extended conformation (Fig. 3A). Conformation of the ligand is maintained through the interactions of two types: first, with the hydrophobic residues lining the LBP and, second, through the hydrogen bonds formed between polar residues and three hydroxyl groups of the ligand. Main anchoring interaction points are: for the 1-OH group with Ser237 (H3) and Arg274 (H5), for the 3-OH group with Ser278 (H5) and Tyr143 (loop H1–H2), and for the 25-OH group with His305 (loop H6–H7) and His397 (H11) (Fig. 3B). The triene connecting the A- and C-ring is tightly sandwiched in a hydrophobic channel between Ser275 (loop H5-β) and Trp286 (β1) on one side and Leu233 (H3) on the other side. The C-ring contacts Trp-286, while the C18 methyl group points toward Val234 (H3). The side chain is surrounded by hydrophobic residues (Fig. 3C).

In order to investigate structurally and functionally important amino acids within the LBP of the wild-type hVDR in the presence of several synthetic vitamin D analogs, several site-directed alanine mutagenesis studies have been performed. Väisänen et al. (2002) have shown that structurally different agonists have distinct ligand–receptor interactions and that residues His229, Asp232, Glu269, Phe279, and Tyr295 are critical for the agonistic properties of ligands. Using a two-dimensional alanine scanning mutational analysis, Choi et al. (2003) have studied 18 alanine mutants of residues forming the LBP. These residues form either hydrogen bonds with the ligands

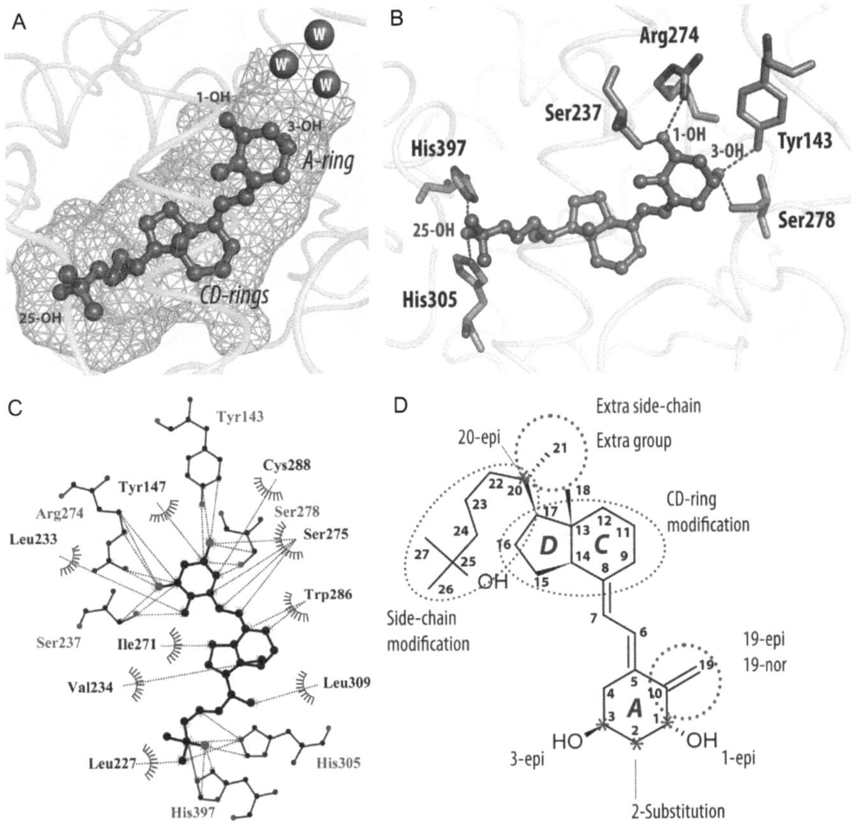

Figure 3 Atomic details of the 1,25$(OH)_2D_3$ interactions to the human VDR LBD (PDB ID: 1DB1). (A) Binding mode of 1,25$(OH)_2D_3$ in the VDR LBP. The volume of the LBP is shown as a gray surface. (B) Specific H-bonds anchoring the three hydroxyl groups of the 1,25 $(OH)_2D_3$. (C) Electrostatic and hydrophobic interactions between 1,25$(OH)_2D_3$ and LBP residues presented as 2D diagram. Obtained using LigPlot+ (Laskowski & Swindells, 2011). (D) Chemical structure and regions of 1,25$(OH)_2D_3$ subjected to chemical modifications. (See the color plate.)

(Y143, S237, R274, S278, H305, H397) or hydrophobic interactions (D144, L233, V234, I238, I268, I271, S275, C288, W286, V300, Q400, Y401). The transactivation potencies of these mutants with the natural hormone and 11 other ligands were evaluated through functional studies based on a luciferase reporter assay. Eight residues (Y143, D144, L233, I271, R274, W286, H397, Y401) have been shown to be essential for transactivation by vitamin D3 and agonistic ligands. These results are in agreement with the ligand-binding mode observed in the different crystal structures.

The crystal structures of VDR LBDs in complex with the 1,25$(OH)_2D_3$ and with its various synthetic analogs revealed two main features of VDR LBP. First, the ligand-binding cavity of hVDR is large (697 Å^3) with the 1,25$(OH)_2D_3$ occupying only 56% of its volume. This provides an additional space for fitting of the modified moieties of the hormone, particularly at the position C2 of A-ring and around the aliphatic chain. A channel of water molecules near the position 2 of the A-ring creates an extra cavity; additional space around the aliphatic chain is also observed. Several examples of ligand modifications aiming to create tighter contacts with the LBP in these positions are described in Section 2. Second, mutual adaptability and flexibility of the ligand and the amino acids lining the LBP were observed in several crystal structures. In some cases, the LBP adaptability affects the ligand agonistic/antagonistic properties as described in more details in Section 3 upon the discussion on 22-alkyl and Gemini analogs.

3. SECOSTEROIDAL DERIVATIVES OF 1,25$(OH)_2D_3$

More than 3000 synthetic analogs of 1,25$(OH)_2D_3$ have been synthesized with the goal to increase the physiological potency and specificity of the natural ligand. Most of these analogs carry modifications of the secosteroidal 1,25$(OH)_2D_3$ ligand on the A-ring, the CD-rings, or the side chain (Fig. 3D). For the A-ring, the modifications comprise additional functional groups on C2 (Hourai et al., 2006; Vanhooke et al., 2004), isomers of the 1-OH or 3-OH groups (Molnár et al., 2011) or functional groups on C1, and C19 epimer or removal (Vanhooke et al., 2004). For the CD-rings, modifications can be on the aromatic rings (Verstuyf et al., 2010). For the side chain, C20 epimerization (Tocchini-Valentini, Rochel, Wurtz, Mitschler, & Moras, 2001), additional side chains (Ciesielski et al., 2007; Inaba et al., 2009), and addition of heteroatoms or functional groups (introduction of aromatic rings (Ciesielski et al., 2012), extra-alkyls (Anami, Itoh, Egawa, Yoshimoto, & Yamamoto, 2014; Yoshimoto et al., 2012) or adamantly (Nakabayashi et al., 2008)). Most of the analogs are VDR agonists or partial agonists, and few of them are antagonists. In the reported crystal structures of the VDR LBD in complex with synthetic agonists, all complexes adopt a similar active conformation. All compounds are anchored by the same residues in the LBP, with the hydroxyls of the A-ring and of the side chain located in identical positions and forming the same hydrogen bonds. Subtle changes are seen as differential or additional contacts mediated by some analogs.

3.1 Superagonists of VDR

Superagonist analogs are at least 10 times more potent in transactivation assays and act as antiproliferative agents with a magnitude of several orders higher than 1,25$(OH)_2D_3$ *in vitro*. The elucidation of the crystal structures of some of the superagonist ligands in complexes with VDR have not only helped to understand their biological potency but also provided with important clues for the rational synthesis of a new generation of vitamin D3 analogs.

3.1.1 20-Epi Derivatives

One of best studied class of VDR superagonists is the 20-epi compounds that have an inverted stereochemistry at position C20 in the flexible aliphatic chain. 20-epi compounds induce VDR-mediated transcription at concentrations of at least 100-fold lower than the natural ligand and provoke antiproliferative activity several orders of magnitude higher than 1,25$(OH)_2D_3$ (Binderup et al., 1991; Liu, Collins, Norman, & Peleg, 1997; Peleg, Sastry, Collins, Bishop, & Norman, 1995; Yang & Freedman, 1999). Biochemical data and cellular assays suggested that the enhanced transactivation ability of 20-epi-analogs correlated with the ability of these compounds to promote coactivator interaction (Yang & Freedman, 1999). The crystal structures of the hVDR LBD in complex with MC1288 and KH1060 have shown that, contrary to a belief that synthetic analogs would induce a different active conformation, the overall organization and especially the position of helix H12 are strictly maintained in all VDR–ligand complexes (Tocchini-Valentini et al., 2001). Furthermore, the ligand-binding cavity is unique and conserved. The specific interactions observed in the ligand–protein complexes with MC1288 and KH1060 involve the hydrophobic contacts of the 17β-aliphatic chains. For the MC1288, the main difference is the positioning of the methyl group C21 which results in different contacts with Val300, Leu309, and His397 (Fig. 4A). In the case of KH1060, the major differences observed are tighter and more numerous ligand–protein contacts. The methyl groups C26a and C27a, specific to KH1060, form additional contacts with H3, loop 6–7, H11, and H12 (Fig. 4B).

Additional interactions through a different conformation of the side chain and additional methyl groups of 20-epi compounds have been shown to stabilize the active conformation of VDR, therefore, explaining the superagonistic nature of KH1060. A longer half-life of VDR–superagonist complexes is responsible for a more potent complex of the receptor with coactivator and subsequent transcriptional activation. On the basis of the

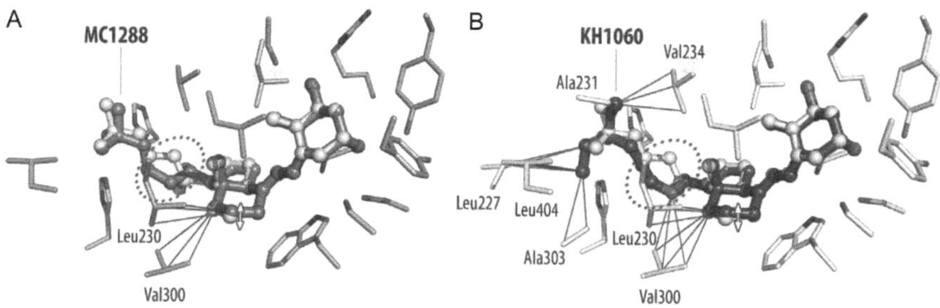

Figure 4 Crystal structures of hVDR LBD in complex with 20-epi-analogs. (A) hVDR–LBD complex with KH1060 (PDB ID: 1IE8). (B) hVDR–LBD complex with MC1288 (PDB ID: 1IE9). Specific hydrophobic interactions are presented as lines. Different positioning of the aliphatic chain in comparison with 1,25$(OH)_2D_3$ is highlighted. (See the color plate.)

crystal structures of VDR complexes, new superagonist analogs bearing an entropically favored oxolane ring in the side chain at position C20, and a 2α-methyl group on the A-ring were developed (Antony et al., 2010; Hourai et al., 2008). These compounds were able to stimulate VDR-mediated transcription at lower doses than 1,25$(OH)_2D_3$.

3.1.2 C2-Substituted Analogs

Several water molecules forming a channel near the position 2 of the A-ring can be observed in the different crystal structures of VDR in complex with 1,25$(OH)_2D_3$ (Fig. 5A). The space surrounding the water molecules channel is considered as an extra cavity that can accommodate ligands with additional moieties at position 2. Indeed, the fourfold increase in binding affinity of the 2α-methyl analogs is in agreement with this observation (Fujishima et al., 1998). Several crystal structures of the VDR LBD bound to selected C2-substituted analogs that fill this water channel have been described (Fujishima et al., 1998; Hourai et al., 2006; Vanhooke et al., 2004). In case of C2-substituted analogs bearing methyl, propyl, propoxy, hydroxypropyl, and hydroxypropoxy groups on C2 position, the water molecules are displaced or their location is affected (Fig. 5B). The exception here is a methyl substituent, which is small enough not to affect the water molecules network while providing additional van der Waals contacts that explain the higher binding affinity of this analog (Hourai et al., 2006; Fig. 5C). The affinities of the C2-substituted analogs for VDR result from the balance between the loss of water-mediated H-bonds, additional van der Waals contacts, and entropic effect. Several C2-substituents bearing additional modifications act as VDR superagonists: 2MD which presents a 19-nor configuration, a

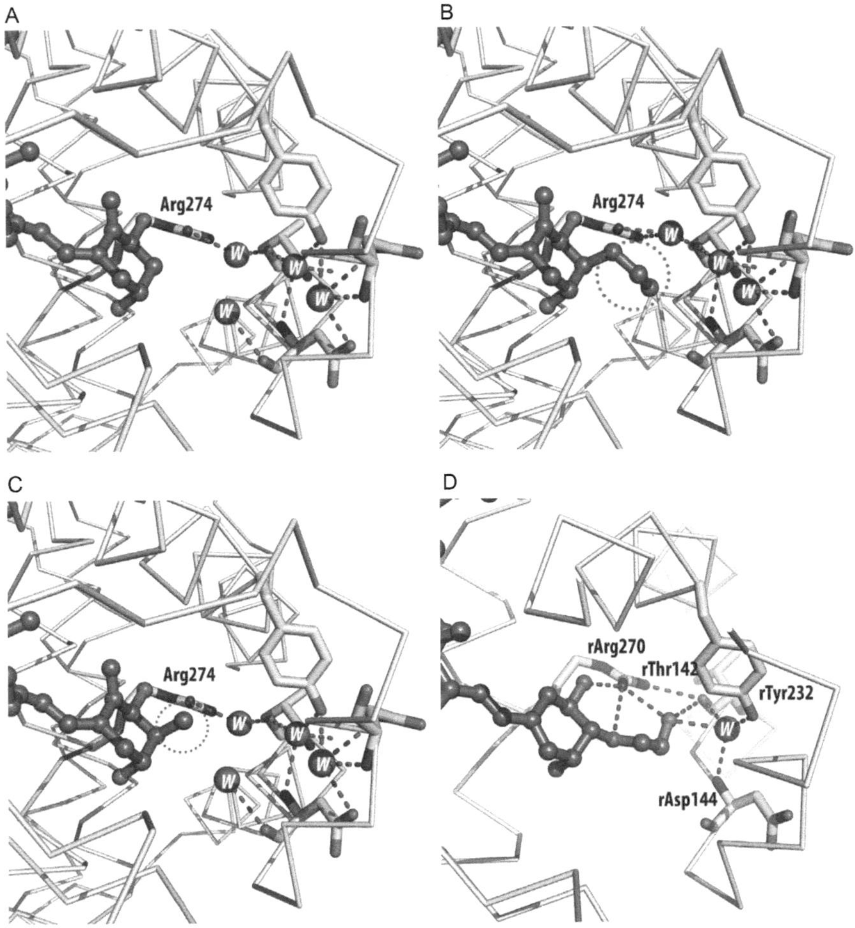

Figure 5 Effect of C2-substituted analogs on a water channel formation. (A) hVDR–LBD complex with $1,25(OH)_2D_3$ (PDB ID: 1DB1). (B) hVDR–LBD complex with C2α-propyl analog (PDB ID: 2HAM). (C) hVDR–LBD complex with C2α-methyl analog (PDB ID: 2HB8). (D) rVDR–LBD complex with the 20S analog bearing a 2β-hydroxyethoxy group (PDB ID: 2ZLA).

methylene moiety on C2α, and a 20-epi-configuration on C20 (Vanhooke et al., 2004); 20S analog bearing 2β-hydroxyethoxy group together with other modifications (16ene-22thia-19nor and C26 and C27 methylation) (Shimizu et al., 2008) and others (reviewed in Sibilska, Szybinski, Sicinski, Plum, and Deluca (2013). The 2β-hydroxyethoxy group in the 20S complex shows stabilizing interactions via hydrogen bonds between the terminal OH moiety of the 2-substituent and both rArg270 (hArg274) and a water molecule (Fig. 5D). These interactions may stabilize the active receptor conformation and explain its increased potency.

3.1.3 14-Epi Analogs

Two 14-epi-analogs TX527 and TX522 show a strongly enhanced antiproliferative action (at least 10-fold) coupled to markedly lower calcemic effects (50 and 400 times less than $1,25(OH)_2D_3$, respectively) (Verlinden et al., 2000). Both ligands significantly enhance interaction between VDR and different coactivators, including SRC-1, TIF2, and the DRIP205 compared to the natural hormone (Eelen et al., 2005). The crystal structure of the hVDR LBD in complex with TX522 revealed modified contacts of C12 and C22 of the ligand with the LBP (Eelen et al., 2005). As a consequence of the epi-configuration of C14, the CD-rings are shifted by 0.5 Å. The C12 shift induces a closer contact of this atom to Val300 (H6) in the VDR-TX522 (Fig. 6). Due to the rigidity of the side chain of TX522, another pathway of the chain is taken in the pocket and makes an additional contact with the Ile268 (H5). This study suggests that the enhanced potency to induce VDR–coactivator interactions may be the basis for the superagonistic profile of TX522 and TX527.

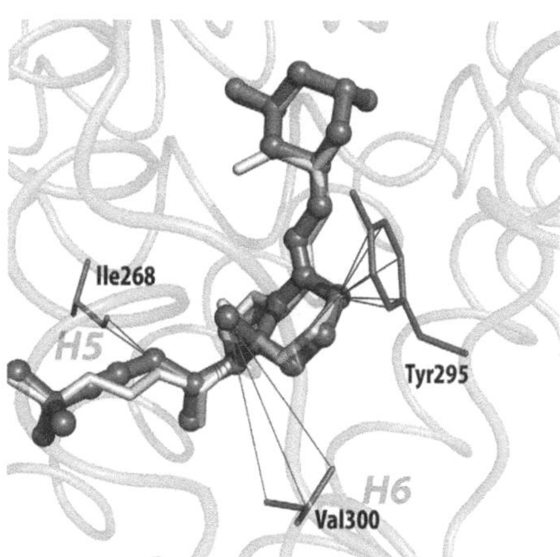

Figure 6 Binding mode of 14-epi-analog TX522. Overlay of hVDR–LBD complexes with 14-epi-analog TX522 (PDB ID: 1TXI) and $1,25(OH)_2D_3$ (PDB ID: 1DB1; light gray). Hydrophobic interactions specific for TX522 are presented as lines.

3.2 Synthetic Compounds Expanding the VDR LBP

A second tendency in the rational development of 1,25(OH_2)D_3 analogs is to include bulky groups (aromatic rings, heterocyclic and adamantyl groups, additional side chains, etc.) in order to drastically change the VDR LBP–ligand interactions.

3.2.1 Gemini Analogs

The analogs with an additional side chain, such as the Gemini analogs, cause expansion of the LBP in order to accommodate the additional moieties (Ciesielski et al., 2007; Huet et al., 2011). Interestingly, the Gemini analogs display the agonistic properties. The crystal structure of Gemini bound to the zVDR LBD revealed an intriguing ability of the LBP to expand in order to adapt the second side chain of the compound (Ciesielski et al., 2007). The increase of the LBP volume is caused by the repositioning of zLeu337 (hLeu309) at the beginning of H7 (Fig. 7). The hydrogen bonds anchoring Gemini are similar to those formed with 1,25$(OH)_2D_3$ with an additional stabilization effect from zHis333 which interacts with hydroxyl groups of both side chains. Due to the presence of two side chains, a larger number of interactions are formed between Gemini and amino acids lining the LBP: 76 interactions at 4 Å cut-off versus 71 for 1,25$(OH)_2D_3$. New Gemini compounds with increased agonistic properties and resistance to metabolic degradation have been developed (Huet et al., 2011; Maehr,

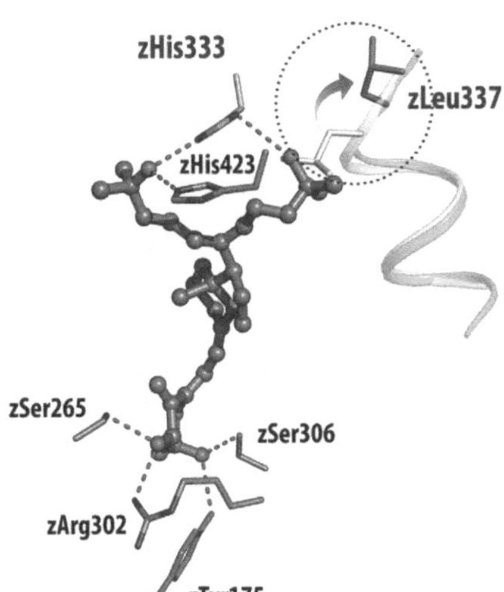

Figure 7 Crystal structure of zVDR–LBD complex with Gemini (PDB ID: 2HCD). H-bonds are shown as dashed lines. Conformational change of the zLeu337 is highlighted.

Rochel, Lee, Suh, & Uskokovic, 2013). 19-nor compounds Gemini-0072 and Gemini-0097 bear modifications on the side chains: deuterated geminal methyl groups, trifluoromethyl groups introduced to the second side chain, and unsaturated C-23. The side-chain fluorine atoms stabilize helices H3, H11, and H12, therefore, explaining the superagonistic properties of these novel Gemini derivatives.

3.2.2 22-Alkyl Derivatives

Similarly to Gemini and its derivatives, 22-alkyl analogs cause the formation of an extra cavity within the LBP (Fig. 8). In the case of compounds of this type, the synergy between the pocket expansion and the ligand interaction with the VDR C-terminus determines whether the compound demonstrates agonistic/partial agonistic or antagonistic properties (Anami et al., 2014; Yoshimoto et al., 2012). When the ligand is weakly bound to the helices H11 and H12, repositioning of the hLeu309 (rLeu305) leads to the further destabilization of the LBD secondary structure which explains the antagonistic nature of these ligands.

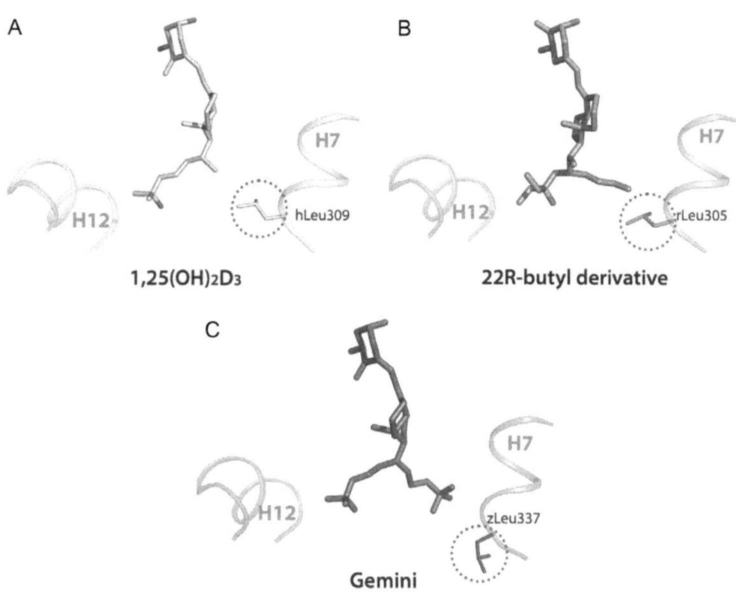

Figure 8 Adaptability of the VDR LBP caused by repositioning of hLeu309 (zLeu337, rLeu305). (A) hVDR–LBD complex with 1,25(OH)$_2$D$_3$ (PDB ID: 1DB1). (B) zVDR–LBD complex with Gemini (PDB ID: 2HCD). (C) rVDR–LBD complex with 22R-butyl derivative (PDB ID: 3WT7).

3.2.3 Adamantyl-Containing Compounds

The compounds with adamantyl group and 19-nor-1,25$(OH)_2D_3$ derivatives have a bulky adamantyl side chain that expands the LBP near the C-terminus of H11, loop 11–12, the N-terminus of H3, and loop 6–7 (Fig. 9; Nakabayashi et al., 2008). As a result of the LBD secondary structure perturbation, adamantyl analogs demonstrate weak agonistic or antagonistic properties, although the VDR LBD adopts the active conformation in all cocrystallized complexes. One of the main advantages of these compounds is their cell-selective VDR modulation activity, which explains their further investigation and derivatization. Newly synthesized adamantyl analogs with higher VDR agonistic activities have been recently described (Kudo et al., 2014). Compounds ADTK1-4 have more rigid side chains due to the presence of a triple bond. When bound to VDR, the side chain of the adamantyl compounds is significantly repositioned in comparison with that of the natural hormone. In this way, the adamantyl group takes the place of the 26,27-dimethyl group of 1,25$(OH)_2D_3$. Importantly, the adamantyl moiety forms hydrophobic interactions with Phe418 and Val414, residues of H12 of the rVDR LBD, and these contacts are tighter than those observed with the 1,25 $(OH)_2D_3$.

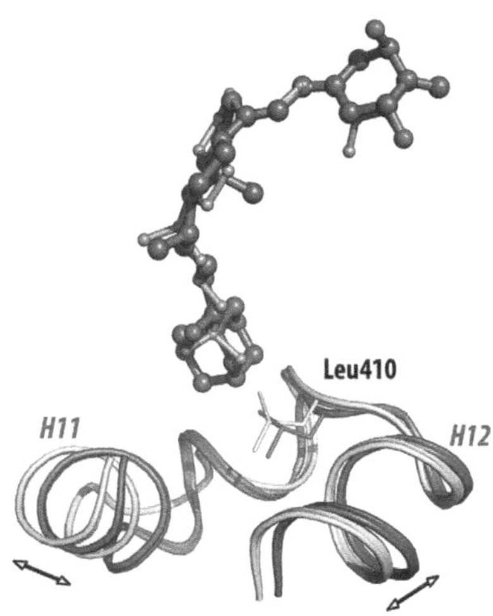

Figure 9 Crystal structure of rVDR–LBD complex with adamantyl compound ADTT. Overlay of rVDR–LBD complex with ADTT (PDB ID: 2ZMI) and hVDR–LBD complex with 1,25$(OH)_2D_3$ (PDB ID: 1DB1; light gray). Repositioning of helices 11 and 12 is shown by arrows. (See the color plate.)

4. SYNTHETIC MIMICS OF $1,25(OH)_2D_3$

Most of the $1,25(OH)_2D_3$ analogs bear minor modifications compared to the natural ligand, sometimes with altered A-, C-, or D-rings. Another type of ligands are the nonsecosteroidal molecules that mimic 1,25 $(OH)_2D_3$ without being structurally related to it, and hundreds of them have already been described. There are currently three classes of synthetic nonsecosteroidal mimetics of $1,25(OH)_2D_3$: diphenylmethane LG190178 and its derivatives (Asano et al., 2013; Boehm et al., 1999; Demizu et al., 2011; Fischer et al., 2012; Kakuda et al., 2008; Swann, Bergh, Farach-Carson, Ocasio, & Koh, 2002); bis- and tris-aromatic derivatives (Ciesielski et al., 2012; Peräkylä, Malinen, Herzig, & Carlberg, 2005); and compounds with a *p*-carborane group (Fujii et al., 2011). Only some of the diphenylmethane derivatives and of the bis- and tris-aromatic derivatives have been shown to be potent VDR agonists. Some of these nonsecosteroidal ligands possess less calcium mobilization activity and are attractive therapeutics against psoriasis, osteoporosis, and cancer. Nonsteroidal compounds for other steroid NRs are currently used in cancer treatment such as raloxifene for estrogen receptor (ER) (Fink, Mortensen, Stauffer, Aron, & Katzenellenbogen, 1999) or flutamide for androgene receptor (AR) (Yin et al., 2003). Ligands of all three classes, including different derivatives of LG190178 (carboxylic and hydroxamic acid derivatives and phenyl-pyrrolyl pentane derivatives) are described in details in the recent review of Yamada and Makishima (2014). In case of all nonsecosteroidal ligands cocrystallized with the VDR LBD, these compounds take a similar position and a conformation to that of the natural ligand.

4.1 Derivatives of LG190178

For the first class of $1,25(OH)_2D_3$ mimics, the YR301 compound which is a stereoisomer of LG190178 has been crystallized in complex with the rVDR LBD (Kakuda et al., 2008). This compound has been shown to exhibit potent transcriptional activity *in vitro*. The X-ray crystallographic analysis revealed that the 2-hydroxy-3,3-dimethylbutyl group acts as a mimic for the side-chain of $1,25(OH)_2D_3$, while the 2,3-hydroxypropyl group replaces the A-ring and its diethyl-methyl-group takes the same spatial position as the CD-rings of the natural compound. YR301 possesses two functional hydroxyl groups: the first forms hydrogen bonds with His301 and His393 (C2) and the second with Ser233 and Arg270 (C2′). C2 and C2′

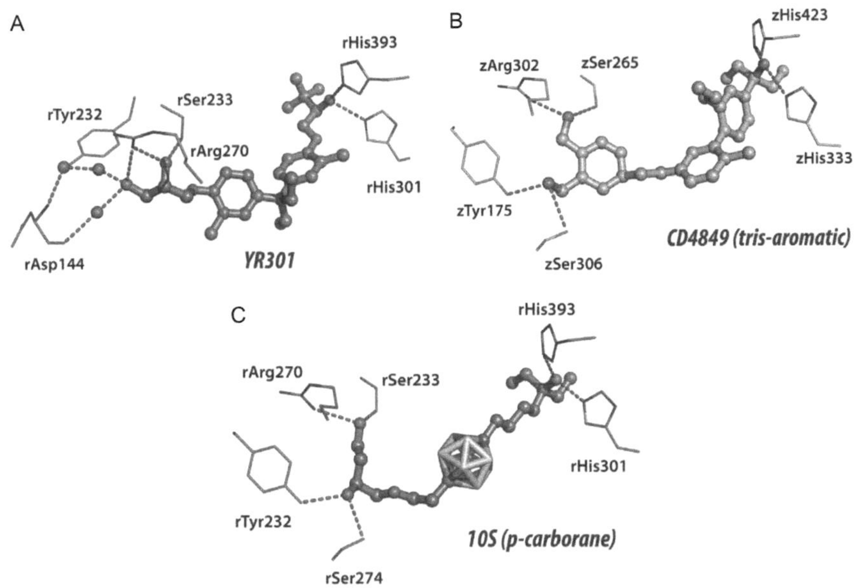

Figure 10 Synthetic mimics of 1,25(OH)$_2$D$_3$. (A) Crystal structure of rVDR–LBD complex with YR301 (PDB ID: 2ZFX), a derivative of LG190178. (B) Crystal structure of zVDR–LBD complex with tris-aromatic derivative CD4528 (PDB ID: 4G2H). (C) Crystal structure of rVDR–LBD complex with 10S carborane derivative (PDB ID: 3VJS). H-bonds are shown as dashed lines.

groups replace the 1-OH and 25-OH groups of the natural ligand, respectively. The terminal hydroxyl group of YR301 forms two direct hydrogen bonds with Arg270 and interacts with Tyr232 and Asp144 via water molecules (Fig. 10A). Additional derivatization with a hydroxyl group in YR335 allows this hydroxyl group to form H-bonds with Ser274 and Tyr139, similarly to the 3-OH group of 1,25(OH)$_2$D$_3$, but without improving its transcriptional potency (Demizu et al., 2011). In an attempt to build a bifunctional molecule with combined VDR agonistic and histone deacetylase activities, hydroxamic acid was introduced to diphenylmethane derivatives (Fischer et al., 2012).

4.2 Bis- and Tris-Aromatic Compounds

The second class of 1,25(OH)$_2$D$_3$ mimics are the bis- and tris-aromatic derivatives. A dibenzyl alcohol is a common moiety in these ligands. The two phenyl groups are either linked by an ether (CD3938, CD4720, CD4742, and CD4802) or an alkyl (CD4528 and CD4849) chain (Peräkylä et al., 2005). For the bis-aromatic derivatives, a dienyl alcohol is branched to the second phenyl, while the methyl moieties of the dienyl alcohol in CD4528 and CD4420 are replaced by tri-fluoromethyls. In the tris-aromatic derivatives, a third phenyl group is included in the branched side chain replacing the dienyl alcohol of the bis-aromatic analogs. Like $1,25(OH)_2D_3$, bis- and tris-aromatic derivatives stimulate transcription at nanomolar concentrations (Ciesielski et al., 2012). Examination of the protein–ligand interactions reveals that these compounds have a position and an H-bond network within the LBP similar to the natural ligand (Fig. 10B). Bis- and tris-aromatic analogs are better accommodated within the VDR LBP than the other nonsecosteroidal ligands as the intrinsic hydrophobic van der Waals contacts of the VDR–$1,25(OH)_2D_3$ complex are maintained in these new structures. In case of CD4528, additional interactions are mediated through the trifluoromethyl groups. For the tris-aromatic analogs, a larger fraction of the LBP is occupied by the ligand that results in the formation of additional stabilizing contacts. The length of the alkyl group attached to the second phenyl ring induces some rearrangement of the VDR LBP in the region of helix H6 without affecting the agonistic behavior of the complex. These extended pockets may be important to achieve selectivity and dissociated biological profiles without affecting the agonistic activity of the ligands. The nature of the linker (oxygen vs. carbon) between the two first phenyl rings plays a crucial role in VDR binding affinity through the desolvation cost of the ether linker upon VDR binding.

4.3 Compounds with *p*-Carborane Core

The third class of $1,25(OH)_2D_3$ mimics consists of ligands with *p*-carborane (1,12-diarba-closo-dodecarborane) as hydrophobic core structure and flexible acyclic triols. Although first reported carborane ligands have a low binding affinity to VDR, they induce the active receptor conformation (Fujii et al., 2011). The carborane group is located at the same place as the CD-rings of the natural ligand and functions as a hydrophobic anchor in binding to the VDR. Three hydroxyl groups of the flexible acyclic triol allow its suitable positioning in the LBP to form hydrogen bonds similarly to the natural ligand (Fig. 10C). The same authors reported design and synthesis of ω-hydroxyalkoxy derivatives of the carborane-containing compound. The new ligands display increased affinity to VDR and are highly potential (Fujii et al., 2014).

The structural studies on different nonsecosteroidal analogs have provided the molecular basis for their transcriptional activity. A significant effect of the nonsecosteroidal ligands on VDR activity confirms the fact that the secosteroid backbone of $1,25(OH)_2D_3$ is less important than the positioning of the three anchoring hydroxyl groups. Structural scaffolds of non-secosteroidal ligands different from that of the natural ligand and their potent VDR agonistic activities with low calcemic actions make these mimics potential drugs for clinical applications.

5. CRYSTAL STRUCTURES OF VDR LBD WITH LITHOCHOLIC ACID

Over a decade ago, the secondary bile acids including lithocholic acid (LCA) were identified as selective low-potency agonists for VDR (Adachi et al., 2005; Makishima, Lu, Xie, Haussler, & Mangelsdorf, 2002). Interestingly, LCA activates VDR in target organs, intestine and kidneys, without causing hypercalcemia (Ishizawa et al., 2008). As the scaffold of this compound is significantly different from that of $1,25(OH)_2D_3$, it was important to obtain structural information on the mechanism of selective activation of VDR by LCA.

5.1 Crystal Structures of rVDR LBD with LCA and Its Derivatives

In 2013, the crystal structures of the rat VDR LBD with LCA and its derivatives were published (Masuno et al., 2013). The ligands in cocrystallized complexes adapt the opposite orientation to $1,25(OH)_2D_3$, their A-ring is positioned at the top of the LBP, whereas the acyclic tail is located at the bottom of the LBP. LCA as well as its derivatives form similar H-bond networks and hydrophobic interactions as $1,25(OH)_2D_3$. Only two direct H-bonds are formed between rTyr143 (H1) and rSer274 (H5) and the 24-carboxyl group of the LCA. The other anchoring residues of $1,25(OH)_2D_3$, rSer233 (H3), and rArg270 (H5), interact with the 24-carboxyl group of LCA through water molecules. Residues rHis301 (loop 6–7) and rHis393 (H11) interact with the hydroxyl group on the A-ring of LCA (Fig. 11A). The shorter length of LCA over $1,25(OH)_2D_3$ (14 vs. 15.1 Å) leads to weakening of interactions with the C-terminal part of VDR. That, and water-mediated anchoring hydrogen bonds, explain the weak agonistic properties of this ligand. LCA and its 3-substituents derivatives (3-keto, acetate, and proponiate) form similar interactions with rVDR although through water-mediated H-bonds for the LCA and 3-ketoLCA and direct H-bonds for the two other derivatives, with rHis301 (loop 6–7) and rHis393 (H11) (Masuno et al., 2013).

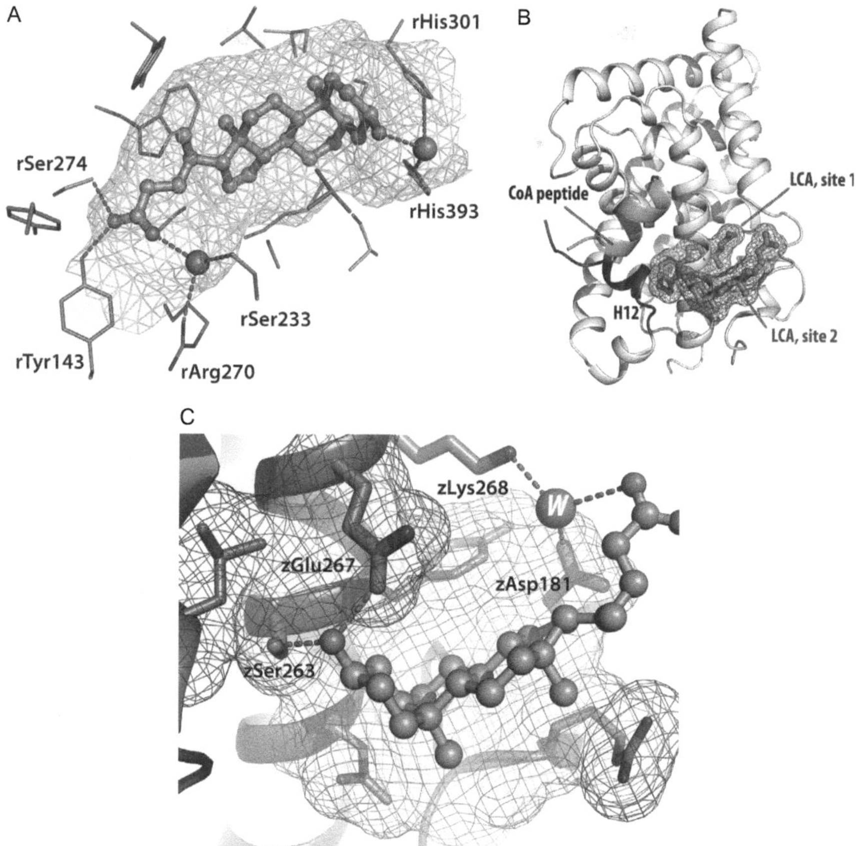

Figure 11 Crystal structure of VDR–LBD complex with LCA. (A) Binding mode of LCA in the canonical rVDR LBP (PDB ID: 3W5P). The volume of the LBP is shown as a gray surface. The interacting residues are shown. (B) Overall structure of zVDR–LBD complex with LCA showing the location of the alternative binding site within LBD (PDB ID: 4Q0A). (C) Details of interactions in the second LCA-binding site.

5.2 Crystal Structure of zVDR LBD with LCA: Alternative LCA-Binding Site

In the crystal structure of the zVDR LBD in complex with LCA recently obtained in our group, an alternative second binding site for LCA is observed (Belorusova et al., 2014). In addition to the first ligand molecule occupying the classical LBP, a second LCA molecule is not fully buried but is rather weakly attached to the surface of VDR close to the loop 1–3 and H3 (Fig. 11B). This second LCA molecule is primarily stabilized by nonpolar interactions with zAsp181, zArg184, zPhe185 (H2), zAsp260, zSer263, zTyr264, zGln267 (H3), and zLeu443 (H12). The hydroxyl group on the A-ring forms two direct H-bonds with zSer263 (H3) and zGln267 (H3) (Fig. 11C). The side chain is more loosely positioned, and its 24-carboxyl group interacts through a water molecule with zAsp181 (H2′) and zLys268 (H3). Biophysical characterization confirmed a multiple LCA binding by the VDR LBD in solution, while the structural analysis and molecular dynamics simulation suggest that the binding of an additional ligand has a stabilizing effect on active receptor conformation, and, importantly, on a bound coactivator peptide. Binding residues forming both the canonical LBP and the second binding site are conserved between VDR family members, suggesting a common mechanism of their activation by LCA.

This crystal structure was the first one that revealed two ligands bound to VDR. At the same time, other NRs have been shown to be able to bind two ligands: ERβ (Wang et al., 2006), thyroid receptor (TR) (Estébanez-Perpiñá, Arnold, Jouravel, et al., 2007), AR (Estébanez-Perpiñá, Arnold, Arnold, et al., 2007), peroxisome proliferator-activated receptors alpha (Bernardes et al., 2013), and gamma (Hughes et al., 2014). The rising number of reports on multiple ligand-binding events clearly indicates that ligand recognition by NRs is far more complex than had initially been postulated. It is clear that the cooperativity between ligand-binding sites needs to be further investigated in order to better understand the mechanism of specific actions of diverse NR modulators.

6. CRYSTAL STRUCTURES OF HVDRR-ASSOCIATED VDR MUTANTS

Hereditary vitamin D-resistant rickets (HVDRR) is a rare genetic disease caused by mutations in the VDR gene that result in the complete or partial loss of receptor functions, including DNA binding, heterodimerization with RXR-alpha, coregulator recruitment, and ligand binding. Up to date, around 50 different mutations have been identified within the VDR (reviewed in Feldman & Malloy, 2014). Twenty-three missense or nonsense mutations affecting the LBD of VDR have been described (Table 1), five of them leading to the substitution of amino acids forming the LBP (Fig. 12A). These LBP-localized mutations affect the ability of VDR to bind 1,25 $(OH)_2D_3$ to different degrees: while the patients with Leu227Pro, Ile268Thr, Arg274His, and His305Gln mutations display response to the treatment with unphysiologically high concentrations of $1,25(OH)_2D_3$ (Aljubeh et al., 2011;

Table 1 Missense and Nonsense Mutations Affecting the VDR LBD

Mutation	References
K141L, T142W, and insertion of A143	Malloy, Xu, Cattani, Reyes, and Feldman (2004)
T146I	Song, Yoon, Shim, and Bae (2011)
R158C	Song et al. (2011)
L227P[a]	Huang, Malloy, Feldman, and Pitukcheewanont (2013)
deltaK246	Zhou, Wang, Malloy, Dolezel, and Feldman (2009)
F251C	Malloy, Zhu, Zhao, Pehling, and Feldman (2001)
Q259E, Q259P	Cockerill et al. (1997) and Macedo et al. (2008)
L263R	Nguyen et al. (2006)
I268T[a]	Malloy, Xu, Peng, et al. (2004)
G319V	Macedo et al. (2008)
R274L[a], R274H[a]	Aljubeh, Wang, Al-Remeithi, Malloy, and Feldman (2011) and Kristjansson, Rut, Hewison, O'Riordan, and Hughes (1993)
N276Y	Malloy et al. (2014)
W286R[a]	Nguyen et al. (2002)
H305Q[a]	Malloy et al. (1997)
I314S	Whitfield et al. (1996)
E329K	Miller et al. (2001)
V346M	Arita et al. (2008)
R391C, R391S	Nguyen et al. (2006) and Whitfield et al. (1996)
E420A, E420K	Malloy, Xu, Peng, Clark, and Feldman (2002) and Malloy, Zhou, Wang, Hiort, and Feldman (2011)

[a]Mutations of amino acid residues forming the LBP.

Huang et al., 2013; Malloy, Xu, Peng, et al., 2004; Malloy et al., 1997), there is almost no response to the hormone in case of Arg274Leu and Trp286Arg substitutions (Kristjansson et al., 1993; Nguyen et al., 2002). It is of high importance to develop the mutant-specific agonists that could be used for the treatment of HDVRR patients.

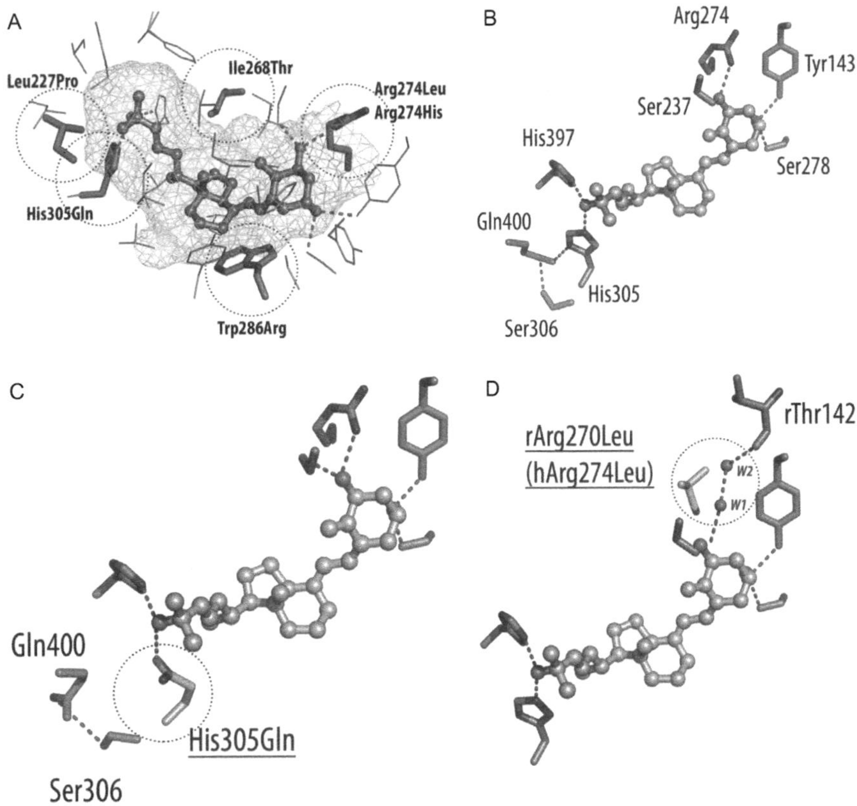

Figure 12 HVDRR-associated VDR mutations. (A) Location of HVDRR-associated mutations within the VDR LBD. (B) Hydrogen bonds anchoring the three hydroxyl-groups of the $1,25(OH)_2D_3$. (C) Crystal structure of hVDR His305Gln complex with $1,25(OH)_2D_3$ (PDB ID: 3M7R). (D) Crystal structure of rVDR Arg270Leu complex with $1,25(OH)_2D_3$ (PDB ID: 3VT3).

By comparing the crystal structures of wild-type and mutated VDR LBDs, clues for the design of selective VDR agonists can be found. Crystal structures of three HVDRR-associated mutants have been reported: structure of the hVDR H305Q mutant in complex with $1,25(OH)_2D_3$ (Rochel, Hourai, & Moras, 2010) and rVDR LBD R270L and W282R mutants (hR274L and hW286R, accordingly) in complexes with natural and synthetic ligands (Nakabayashi et al., 2013).

6.1 His305Gln VDR Mutant

His305Gln mutation is associated with a 10-fold decrease in $1,25(OH)_2D_3$-dependent transactivation and an eightfold lower affinity for calcitriol than the normal VDR (Malloy et al., 1997). His305 is one of the residues forming direct hydrogen bond with calcitriol and, subsequently, with Gln400 and Ser306 (Fig. 12B). Mutation of this residue would lead to the destabilization of the hydrogen bond network anchoring the 25-hydroxyl group of the hormone. Indeed, in the crystal structure of the His305Gln mutant in complex with 1,25$(OH)_2D_3$, the interactions between Gln400 and Ser306 are no more observed (Rochel et al., 2010; Fig. 12C). Being not fixed by Gln400, Gln305 adopts one of two different conformations, as was observed from the corresponding electron density. Although the mutated residue is still able to form a hydrogen bond with the 25-hydroxyl group of 1,25$(OH)_2D_3$, the destabilization of surrounding amino acid interconnections results in the significant loss of receptor affinity to the ligand.

6.2 Arg274Leu VDR Mutant

Arg274Leu was the first missense mutation identified in the LBD of VDR; it has been described in early 1990s (Kristjansson et al., 1993) but the detailed structural information about this mutant became available recently (Nakabayashi et al., 2013). Rat VDR LBD Arg270Leu (hArg274Leu) structure shows that a direct hydrogen bond between 1α-OH group of vitamin D3 and rArg270 disappears (Fig. 12D). Instead, two water molecules occupy a vacant space in the absence of long charged arginine side chain and mediate a connection of 1α-OH group with rThr142. As water-mediated bonds are generally weaker than direct, the mutation results in a significant decrease of 1,25$(OH)_2D_3$ binding.

Additionally, the interaction of rArg270Leu mutant with several modified 19-nor VDR superagonists has been described. One of the analogs was able to form an additional direct hydrogen bond through its terminal hydroxyl group (modification of 1,25$(OH)_2D_3$ at 2-position) with Asp144, therefore, possessing a stronger stabilization of the ligand compare to the natural hormone. Earlier, in the absence of a detailed structural information about the hArg274Leu mutant, a docking simulation has been used for a rational design of nonsecosteroidal agonists with the aim to compensate effects of mutated VDR (Swann et al., 2002). Several Arg274Leu-selective analogs have been developed by adding hydrophobic moieties oriented toward Leu274 to the bisphenol skeleton of a parental compound (structure of this compound is similar to the one of diphenylmethane LG190178 discussed above).

6.3 Trp286Arg VDR Mutant

Together with the structure of rArg270Leu mutant, the same researchers described the crystal structure of rTrp282Arg (hTrp286Arg). This VDR mutant is not able to be activated by 1,25(OH)$_2$D$_3$ of even micromolar range (Nguyen et al., 2002). It should be noted that Trp286 is one of key residues forming the tight hydrophobic channel around the conjugated triene between the A- and C-rings of 1,25(OH)$_2$D$_3$; therefore, arginine substitution would destabilize the LBP. Despite the abolished ability to bind 1,25(OH)$_2$D$_3$, rTrp282Arg was cocrystallized in the complex with natural hormone. An important feature of this crystal structure is the disordered state of the LBD region corresponding to a β-sheet. Thus, the hTrp286Arg mutation drastically affects the LBD folding that results in a loss of receptor function.

7. DYNAMIC PROCESS OF LIGAND BINDING

X-ray crystallography has provided detailed information on the molecular mechanism of VDR LBD action. However, the crystal structures give only a static view of the ligand binding process. The dynamics of ligand binding process by VDR has been investigated through the complementary to X-ray crystallography methods such as NMR (Singarapu et al., 2011) and hydrogen/deuterium exchange (HDX) coupled with mass spectrometry (MS) (Zhang et al., 2010).

HDX-MS analysis has been used to probe the conformational dynamics upon ligand binding to VDR. Within the apo form VDR LBD, the entire C-terminal part has been shown to be very dynamic with 80% of amide hydrogen exchange (Zhang et al., 2010). The region forming the LBP also showed high exchange rate, while the central layer of the α-helical sandwich and helix H10 appeared protected. As expected, binding of 1,25(OH)$_2$D$_3$ or the synthetic analog ED-71 led to a significant protection from hydrogen amide exchange. Interestingly, regions remote from the LBP, such as loop 7–8 and H10, displayed altered HDX profile upon ligand binding. H10 forms an interface of heterodimerization with RXR; therefore, the destabilization of this region in the holo-form of VDR is intriguing. Contrary to the VDR LBD, in the context of a full-length VDR/RXR heterodimer, this region appeared to be stabilized in a ligand-induced way. The HDX profile of the ED-71 complex was similar to that of the 1,25(OH)$_2$D$_3$ complex except for helix H1: in agreement with a crystal structure (Hourai et al.,

2006), ED-71 that exhibits a hydroxypropoxy group at position C2 of the A-ring partially destabilizes H1 by displacing the water molecules and therefore disrupting the ability of Asp-144 in H1 to form a hydrogen bond with bound water. Impaired protection from the amide hydrogen exchange was observed for the VDR complex with alfacalcidol (a precursor of 1,25 $(OH)_2D_3$ lacking a 25-OH group). In case of "weak" ligands helices, H7 and H12 are significantly more mobile. Formation of a less stable complex explains partial agonistic behavior of alfacalcidol.

Another type of experiment following the structural changes within LBD upon the ligand binding in solution is NMR. Singarapu et al. (2011) reported assigned NMR chemical shifts of rVDR–LBD in presence of either one of two agonists used (1,25$(OH)_2D_3$ and 2MD) or an antagonist (OU-72). The comparison of apo- and holo-forms of VDR LBD revealed significant conformational changes upon the ligand binding for the exception of OU-72 complex where the AF-2 function remained structurally dynamic. The fact that the antagonist did not stabilize the active VDR conformation was further supported by comparison of ternary complexes between rVDR LBD, ligands, and LXXLL motif of DRIP205: binding of this peptide to rVDR LBD complexed with 1,25$(OH)_2D_3$ or 2MD resulted in significant chemical shift changes, whereas the rVDR–LBD complex with OU-72 failed to exhibit chemical shift changes upon the same conditions.

While the X-ray crystallography provides the atomic details of VDR–ligand interactions, HDX-MS and NMR are the techniques classically used for exploring the conformational dynamics of protein upon ligand binding in solution. Although here we do not describe molecular dynamics simulation experiments performed on VDR, there are also computational approaches that allow registering of amino acid movements upon simulation of ligand binding. Altogether, these methods are extensively used for the investigation of detailed mechanism of ligand agonism/antagonism, as well as for the rational design of new selective modulators of VDR activity.

8. CONCLUSION AND PERSPECTIVES

The crystal structures of the VDR LBD explain ligand agonism and superagonism, adaptability of the ligands, as well as of the LBP, and provide clues on the mechanism of VDR antagonism. Data on the dynamics of ligand binding by VDR complements our understanding of mechanism of the receptor action. In the past years, new classes of 1,25$(OH)_2D_3$ analogs and mimics were developed. Here, we described the structural information on the most important compounds, information that was widely used for the further design of active analogs of 1,25$(OH)_2D_3$, e.g., additional moieties stabilizing the ligand–protein contacts or incorporation of heterogeneous groups, and also highlighted most of the recently published VDR ligands. Despite the constantly rising number of reports on VDR-ligand recognition, the major problem still remains: there is no clear correlation between the structure of ligands and their function *in vivo*. Attempts to synthesize selective analogs of 1,25$(OH)_2D_3$ with low calcemic activities have been only partially successful. Although a significant effect of the nonsecosteroidal ligands on VDR activity confirms the fact that the secosteroid backbone of 1,25$(OH)_2D_3$ is less important than the positioning of the three anchoring hydroxyl groups, the mechanism of the action *in vivo* of these compounds remains yet to be elucidated. At the same time, ligand binding should be more extensively investigated in the context of the full-length receptor in different functional states and in presence of interaction partners.

ACKNOWLEDGMENTS

This work was supported by the Centre National pour la Recherche Scientifique (CNRS), the Institut National de la Santé et de la Recherche Médicale (INSERM), the Agence Nationale de Recherche (ANR-13-BSV8-0024-01), the Fondation ARC pour la Recherche sur le Cancer, the French Infrastructure for Integrated Structural Biology (FRISBI), and Instruct as part of the European Strategy Forum on Research Infrastructures (ESFRI). A.Y.B. was supported by a grant from the Fondation FRM pour la Recherche Médicale (FDT20140930978).

REFERENCES

Adachi, R., Honma, Y., Masuno, H., Kawana, K., Shimomura, I., Yamada, S., et al. (2005). Selective activation of vitamin D receptor by lithocholic acid acetate, a bile acid derivative. *Journal of Lipid Research*, *46*(1), 46–57.

Aljubeh, J. M., Wang, J., Al-Remeithi, S. S., Malloy, P. J., & Feldman, D. (2011). Report of two unrelated patients with hereditary vitamin D resistant rickets due to the same novel mutation in the vitamin D receptor. *Journal of Pediatric Endocrinology & Metabolism*, *24*(9–10), 793–799.

Anami, Y., Itoh, T., Egawa, D., Yoshimoto, N., & Yamamoto, K. (2014). A mixed population of antagonist and agonist binding conformers in a single crystal explains partial agonism against vitamin D receptor: Active vitamin D analogues with 22R-alkyl group. *Journal of Medicinal Chemistry*, *57*(10), 4351–4367.

Antony, P., Sigüeiro, R., Huet, T., Sato, Y., Ramalanjaona, N., Rodrigues, L. C., et al. (2010). Structure–function relationships and crystal structures of the vitamin D receptor bound 2 alpha-methyl-(20S,23S)- and 2 alpha-methyl-(20S,23R)-epoxymethano-1 alpha,25-dihydroxyvitamin D_3. *Journal of Medicinal Chemistry*, *53*(3), 1159–1171.

Arita, K., Nanda, A., Wessagowit, V., Akiyama, M., Alsaleh, Q. A., & McGrath, J. A. (2008). A novel mutation in the VDR gene in hereditary vitamin D-resistant rickets. *The British Journal of Dermatology, 158*(1), 168–171.

Asano, L., Ito, I., Kuwabara, N., Waku, T., Yanagisawa, J., Miyachi, H., et al. (2013). Structural basis for vitamin D receptor agonism by novel non-secosteroidal ligands. *FEBS Letters, 587*(7), 957–963.

Belorusova, A. Y., Eberhardt, J., Potier, N. N., Stote, R. H., Dejaegere, A., & Rochel, N. (2014). Structural insights into the molecular mechanism of vitamin D receptor activation by lithocholic acid involving a new mode of ligand recognition. *Journal of Medicinal Chemistry, 57*(11), 4710–4719.

Bernardes, A., Souza, P. C. T., Muniz, J. R. C., Ricci, C. G., Ayers, S. D., Parekh, N. M., et al. (2013). Molecular mechanism of peroxisome proliferator-activated receptor α activation by WY14643: A new mode of ligand recognition and receptor stabilization. *Journal of Molecular Biology, 425*(16), 2878–2893.

Bernstein, B. E., Birney, E., Dunham, I., Green, E. D., Gunter, C., & Snyder, M. (2012). An integrated encyclopedia of DNA elements in the human genome. *Nature, 489*(7414), 57–74.

Binderup, L., Latini, S., Binderup, E., Bretting, C., Calverley, M., & Hansen, K. (1991). 20-epi-vitamin D3 analogues: A novel class of potent regulators of cell growth and immune responses. *Biochemical Pharmacology, 42*(8), 1569–1575.

Boehm, M. F., Fitzgerald, P., Zou, A., Elgort, M. G., Bischoff, E. D., Mere, L., et al. (1999). Novel nonsecosteroidal vitamin D mimics exert VDR-modulating activities with less calcium mobilization than 1,25-dihydroxyvitamin D3. *Chemistry & Biology, 6*(5), 265–275.

Bouillon, R., Eelen, G., Verlinden, L., Mathieu, C., Carmeliet, G., & Verstuyf, A. (2006). Vitamin D and cancer. *The Journal of Steroid Biochemistry and Molecular Biology, 102*(1–5), 156–162.

Carlberg, C., Molnár, F., & Mouriño, A. (2012). Vitamin D receptor ligands: The impact of crystal structures. *Expert Opinion on Therapeutic Patents, 22*(4), 417–435.

Choi, M., Yamamoto, K., Itoh, T., Makishima, M., Mangelsdorf, D. J., Moras, D., et al. (2003). Interaction between vitamin D receptor and vitamin D ligands: Two-dimensional alanine scanning mutational analysis. *Chemistry & Biology, 10*(3), 261–270.

Ciesielski, F., Rochel, N., Mitschler, A., Kouzmenko, A., & Moras, D. (2004). Structural investigation of the ligand binding domain of the zebrafish VDR in complexes with $1\alpha,25(OH)_2D_3$ and Gemini: Purification, crystallization and preliminary X-ray diffraction analysis. *The Journal of Steroid Biochemistry and Molecular Biology, 89–90*, 55–59.

Ciesielski, F., Rochel, N., & Moras, D. (2007). Adaptability of the Vitamin D nuclear receptor to the synthetic ligand Gemini: Remodelling the LBP with one side chain rotation. *The Journal of Steroid Biochemistry and Molecular Biology, 103*(3), 235–242.

Ciesielski, F., Sato, Y., Chebaro, Y., Moras, D., Dejaegere, A., & Rochel, N. (2012). Structural basis for the accommodation of bis- and tris-aromatic derivatives in vitamin D nuclear receptor. *Journal of Medicinal Chemistry, 55*(19), 8440–8449.

Cockerill, F. J., Hawa, N. S., Yousaf, N., Hewison, M., O'Riordan, J. L. H., & Farrow, S. M. (1997). Mutations in the vitamin D receptor gene in three kindreds associated with hereditary vitamin D resistant rickets. *The Journal of Clinical Endocrinology and Metabolism, 82*(9), 3156–3160.

Darimont, B. D., Wagner, R. L., Apriletti, J. W., Stallcup, M. R., Kushner, P. J., Baxter, J. D., et al. (1998). Structure and specificity of nuclear receptor–coactivator interactions. *Genes & Development, 12*(21), 3343–3356.

Demizu, Y., Takahashi, T., Kaneko, F., Sato, Y., Okuda, H., Ochiai, E., et al. (2011). Design, synthesis and X-ray crystallographic study of new nonsecosteroidal vitamin D receptor ligands. *Bioorganic & Medicinal Chemistry Letters, 21*(20), 6104–6107.

Eelen, G., Verlinden, L., Rochel, N., Claessens, F., De Clercq, P., Vandewalle, M., et al. (2005). Superagonistic action of 14-epi-analogs of 1,25-dihydroxyvitamin D explained by vitamin D receptor-coactivator interaction. *Molecular Pharmacology*, *67*(5), 1566–1573.

Estébanez-Perpiñá, E., Arnold, L. A., Arnold, A. A., Nguyen, P., Rodrigues, E. D., Mar, E., et al. (2007a). A surface on the androgen receptor that allosterically regulates coactivator binding. *Proceedings of the National Academy of Sciences of the United States of America*, *104*(41), 16074–16079.

Estébanez-Perpiñá, E., Arnold, L. A., Jouravel, N., Togashi, M., Blethrow, J., Mar, E., et al. (2007b). Structural insight into the mode of action of a direct inhibitor of coregulator binding to the thyroid hormone receptor. *Molecular Endocrinology*, *21*(12), 2919–2928.

Feldman, D., & Malloy, P. J. (2014). Mutations in the vitamin D receptor and hereditary vitamin D-resistant rickets. *BoneKEy Reports*, *3*, 510.

Fink, B. E., Mortensen, D. S., Stauffer, S. R., Aron, Z. D., & Katzenellenbogen, J. A. (1999). Novel structural templates for estrogen-receptor ligands and prospects for combinatorial synthesis of estrogens. *Chemistry & Biology*, *6*(4), 205–219.

Fischer, J., Wang, T.-T., Kaldre, D., Rochel, N., Moras, D., White, J. H., et al. (2012). Synthetically accessible non-secosteroidal hybrid molecules combining vitamin D receptor agonism and histone deacetylase inhibition. *Chemistry & Biology*, *19*(8), 963–971.

Fujii, S., Kano, A., Masuno, H., Songkram, C., Kawachi, E., Hirano, T., et al. (2014). Design and synthesis of tetraol derivatives of 1,12-dicarba-closo-dodecaborane as non-secosteroidal vitamin D analogs. *Bioorganic & Medicinal Chemistry Letters*, *24*(18), 4515–4519.

Fujii, S., Masuno, H., Taoda, Y., Kano, A., Wongmayura, A., Nakabayashi, M., et al. (2011). Boron cluster-based development of potent nonsecosteroidal vitamin D receptor ligands: Direct observation of hydrophobic interaction between protein surface and carborane. *Journal of the American Chemical Society*, *133*(51), 20933–20941.

Fujishima, T., Liu, Z., Miura, D., Chokki, M., Ishizuka, S., Konno, K., et al. (1998). Synthesis and biological activity of 2-methyl-20-epi analogues of 1 alpha,25-dihydroxyvitamin D3. *Bioorganic & Medicinal Chemistry Letters*, *8*(16), 2145–2148.

Gallastegui, N., Mackinnon, J. A., Fletterick, R. J., & Estébanez-Perpiñá, E. (2015). Advances in our structural understanding of orphan nuclear receptors. *Trends in Biochemical Sciences*, *40*(1), 25–35.

Haussler, M. R., Jurutka, P. W., Mizwicki, M., & Norman, A. W. (2011). Vitamin D receptor (VDR)-mediated actions of 1α,25$(OH)_2$vitamin D_3: Genomic and nongenomic mechanisms. *Best Practice & Research. Clinical Endocrinology & Metabolism*, *25*(4), 543–559.

Haussler, M. R., Whitfield, G. K., Kaneko, I., Haussler, C. A., Hsieh, D., Hsieh, J. C., et al. (2013). Molecular mechanisms of vitamin D action. *Calcified Tissue International*, *92*(2), 77–98.

Holick, M. F. (2003). Vitamin D: A millenium perspective. *Journal of Cellular Biochemistry*, *88*(2), 296–307.

Hourai, S., Fujishima, T., Kittaka, A., Suhara, Y., Takayama, H., Rochel, N., et al. (2006). Probing a water channel near the A-ring of receptor-bound 1α,25-dihydroxyvitamin D3 with selected 2α-substituted analogues. *Journal of Medicinal Chemistry*, *49*(17), 5199–5205.

Hourai, S., Rodrigues, L. C., Antony, P., Reina-San-Martin, B., Ciesielski, F., Magnier, B. C., et al. (2008). Structure-based design of a superagonist ligand for the vitamin D nuclear receptor. *Chemistry & Biology*, *15*(4), 383–392.

Huang, K., Malloy, P., Feldman, D., & Pitukcheewanont, P. (2013). Enteral calcium infusion used successfully as treatment for a patient with hereditary vitamin D resistant rickets (HVDRR) without alopecia: A novel mutation. *Gene, 512*(2), 554–559.

Huet, T., Maehr, H., Lee, H. J., Uskokovic, M. R., Suh, N., Moras, D., et al. (2011). Structure-function study of gemini derivatives with two different side chains at C-20, Gemini-0072 and Gemini-0097. *Medicinal Chemistry Communications, 2*(5), 424–429.

Hughes, T. S., Giri, P. K., de Vera, I. M. S., Marciano, D. P., Kuruvilla, D. S., Shin, Y., et al. (2014). An alternate binding site for PPARγ ligands. *Nature Communications, 5*, 3571.

Inaba, Y., Yoshimoto, N., Sakamaki, Y., Nakabayashi, M., Ikura, T., Tamamura, H., et al. (2009). A new class of vitamin D analogues that induce structural rearrangement of the ligand-binding pocket of the receptor. *Journal of Medicinal Chemistry, 52*(5), 1438–1449.

Ishizawa, M., Matsunawa, M., Adachi, R., Uno, S., Ikeda, K., Masuno, H., et al. (2008). Lithocholic acid derivatives act as selective vitamin D receptor modulators without inducing hypercalcemia. *Journal of Lipid Research, 49*(4), 763–772.

Kakuda, S., Okada, K., Eguchi, H., Takenouchi, K., Hakamata, W., Kurihara, M., et al. (2008). Structure of the ligand-binding domain of rat VDR in complex with the nonsecosteroidal vitamin D3 analogue YR301. *Acta Crystallographica. Section F, Structural Biology and Crystallization Communications, 64*(Pt. 11), 970–973.

Kristjansson, K., Rut, A. R., Hewison, M., O'Riordan, J. L., & Hughes, M. R. (1993). Two mutations in the hormone binding domain of the vitamin D receptor cause tissue resistance to 1,25 dihydroxyvitamin D3. *The Journal of Clinical Investigation, 92*(1), 12–16.

Kudo, T., Ishizawa, M., Maekawa, K., Nakabayashi, M., Watarai, Y., Uchida, H., et al. (2014). Combination of triple bond and adamantane ring on the vitamin D side chain produced partial agonists for vitamin D receptor. *Journal of Medicinal Chemistry, 57*(10), 4073–4087.

Laskowski, R. A., & Swindells, M. B. (2011). LigPlot+: Multiple ligand-protein interaction diagrams for drug discovery. *Journal of Chemical Information and Modeling, 51*(10), 2778–2786.

Leyssens, C., Verlinden, L., & Verstuyf, A. (2014). The future of vitamin D analogs. *Frontiers in Physiology, 5*, 122.

Liu, Y. Y., Collins, E. D., Norman, A. W., & Peleg, S. (1997). Differential interaction of 1alpha,25-dihydroxyvitamin D3 analogues and their 20-epi homologues with the vitamin D receptor. *The Journal of Biological Chemistry, 272*(6), 3336–3345.

Macedo, L. C., Soardi, F. C., Ananias, N., Belangero, V. M. S., Rigatto, S. Z. P., De-Mello,-M. P., et al. (2008). Mutations in the vitamin D receptor gene in four patients with hereditary 1,25-dihydroxyvitamin D-resistant rickets. *Arquivos Brasileiros de Endocrinologia e Metabologia, 52*(8), 1244–1251.

Maehr, H., Rochel, N., Lee, H. J., Suh, N., & Uskokovic, M. R. (2013). Diastereotopic and deuterium effects in Gemini. *Journal of Medicinal Chemistry, 56*(10), 3878–3888.

Makishima, M., Lu, T. T., Xie, W., Haussler, M. R., & Mangelsdorf, D. J. (2002). Vitamin D receptor as an intestinal bile acid sensor. *Science, 296*(5571), 1313–1316.

Malloy, P. J., Eccleshall, T. R., Gross, C., Van Maldergem, L., Bouillon, R., & Feldman, D. (1997). Hereditary vitamin D resistant rickets caused by a novel mutation in the vitamin D receptor that results in decreased affinity for hormone and cellular hyporesponsiveness. *The Journal of Clinical Investigation, 99*(2), 297–304.

Malloy, P. J., Tasic, V., Taha, D., Tütüncüler, F., Ying, G. S., Yin, L. K., et al. (2014). Vitamin D receptor mutations in patients with hereditary 1,25-dihydroxyvitamin D-resistant rickets. *Molecular Genetics and Metabolism, 111*(1), 33–40.

Malloy, P. J., Xu, R., Cattani, A., Reyes, M. L., & Feldman, D. (2004). A unique insertion/substitution in helix H1 of the vitamin D receptor ligand binding domain in a patient with hereditary 1,25-dihydroxyvitamin D-resistant rickets. *Journal of Bone and Mineral Research: The Official Journal of the American Society for Bone and Mineral Research, 19*(6), 1018–1024.

Malloy, P. J., Xu, R., Peng, L., Clark, P. A., & Feldman, D. (2002). A novel mutation in helix 12 of the vitamin D receptor impairs coactivator interaction and causes hereditary 1,25-dihydroxyvitamin D-resistant rickets without alopecia. *Molecular Endocrinology, 16*(11), 2538–2546.

Malloy, P. J., Xu, R., Peng, L., Peleg, S., Al-Ashwal, A., & Feldman, D. (2004). Hereditary 1,25-dihydroxyvitamin D resistant rickets due to a mutation causing multiple defects in vitamin D receptor function. *Endocrinology, 145*(11), 5106–5114.

Malloy, P. J., Zhou, Y., Wang, J., Hiort, O., & Feldman, D. (2011). Hereditary vitamin D-resistant rickets (HVDRR) owing to a heterozygous mutation in the vitamin D receptor. *Journal of Bone and Mineral Research: The Official Journal of the American Society for Bone and Mineral Research, 26*(11), 2710–2718.

Malloy, P. J., Zhu, W., Zhao, X. Y., Pehling, G. B., & Feldman, D. (2001). A novel inborn error in the ligand-binding domain of the vitamin D receptor causes hereditary vitamin D-resistant rickets. *Molecular Genetics and Metabolism, 73*(2), 138–148.

Masuno, H., Ikura, T., Morizono, D., Orita, I., Yamada, S., Shimizu, M., et al. (2013). Crystal structures of complexes of vitamin D receptor ligand-binding domain with lithocholic acid derivatives. *Journal of Lipid Research, 54*(8), 2206–2213.

Miller, J., Djabali, K., Chen, T., Liu, Y., Ioffreda, M., Lyle, S., et al. (2001). Atrichia caused by mutations in the vitamin D receptor gene is a phenocopy of generalized atrichia caused by mutations in the hairless gene. *The Journal of Investigative Dermatology, 117*(3), 612–617.

Molnár, F. (2014). Structural considerations of vitamin D signaling. *Frontiers in Physiology, 5*, 191.

Molnár, F., Sigüeiro, R., Sato, Y., Araujo, C., Schuster, I., Antony, P., et al. (2011). 1α,25(OH)2-3-epi-vitamin D3, a natural physiological metabolite of vitamin D3: Its synthesis, biological activity and crystal structure with its receptor. *PLoS One, 6*(3), e18124.

Nakabayashi, M., Tsukahara, Y., Iwasaki-Miyamoto, Y., Mihori-Shimazaki, M., Yamada, S., Inaba, S., et al. (2013). Crystal structures of hereditary vitamin D-resistant rickets-associated vitamin D receptor mutants R270L and W282R bound to 1,25-dihydroxyvitamin D3 and synthetic ligands. *Journal of Medicinal Chemistry, 56*(17), 6745–6760.

Nakabayashi, M., Yamada, S., Yoshimoto, N., Tanaka, T., Igarashi, M., Ikura, T., et al. (2008). Crystal structures of rat vitamin D receptor bound to adamantyl vitamin D analogs: Structural basis for vitamin D receptor antagonism and partial agonism. *Journal of Medicinal Chemistry, 51*(17), 5320–5329.

Nguyen, T. M., Adiceam, P., Kottler, M. L., Guillozo, H., Rizk-Rabin, M., Brouillard, F., et al. (2002). Tryptophan missense mutation in the ligand-binding domain of the vitamin D receptor causes severe resistance to 1,25-dihydroxyvitamin D. *Journal of Bone and Mineral Research: The Official Journal of the American Society for Bone and Mineral Research, 17*(9), 1728–1737.

Nguyen, M., d'Alesio, A., Pascussi, J. M., Kumar, R., Griffin, M. D., Dong, X., et al. (2006). Vitamin D-resistant rickets and type 1 diabetes in a child with compound heterozygous mutations of the vitamin D receptor (L263R and R391S): Dissociated responses of the CYP-24 and rel-B promoters to 1,25-dihydroxyvitamin D3. *Journal of Bone and Mineral Research: The Official Journal of the American Society for Bone and Mineral Research, 21*(6), 886–894.

Nolte, R. T., Wisely, G. B., Westin, S., Cobb, J. E., Lambert, M. H., Kurokawa, R., et al. (1998). Ligand binding and co-activator assembly of the peroxisome proliferator-activated receptor-gamma. *Nature, 395*(6698), 137–143.

Oñate, S. A., Tsai, S. Y., Tsai, M. J., & O'Malley, B. W. (1995). Sequence and characterization of a coactivator for the steroid hormone receptor superfamily. *Science (New York, NY)*, *270*(5240), 1354–1357.

Orlov, I., Rochel, N., Moras, D., & Klaholz, B. P. (2012). Structure of the full human RXR/VDR nuclear receptor heterodimer complex with its DR3 target DNA. *The EMBO Journal*, *31*(2), 291–300.

Peleg, S., Sastry, M., Collins, E. D., Bishop, J. E., & Norman, A. W. (1995). Distinct conformational changes induced by 20-epi analogues of 1 alpha,25-dihydroxyvitamin D3 are associated with enhanced activation of the vitamin D receptor. *The Journal of Biological Chemistry*, *270*(18), 10551–10558.

Peräkylä, M., Malinen, M., Herzig, K.-H., & Carlberg, C. (2005). Gene regulatory potential of nonsteroidal vitamin D receptor ligands. *Molecular Endocrinology*, *19*(8), 2060–2073.

Pinette, K. V., Yee, Y. K., Amegadzie, B. Y., & Nagpal, S. (2003). Vitamin D receptor as a drug discovery target. *Mini Reviews in Medicinal Chemistry*, *3*(3), 193–204.

Rachez, C., Lemon, B. D., Suldan, Z., Bromleigh, V., Gamble, M., Näär, A. M., et al. (1999). Ligand-dependent transcription activation by nuclear receptors requires the DRIP complex. *Nature*, *398*(6730), 824–828.

Renaud, J. P., Rochel, N., Ruff, M., Vivat, V., Chambon, P., Gronemeyer, H., et al. (1995). Crystal structure of the RAR-gamma ligand-binding domain bound to all-trans retinoic acid. *Nature*, *378*(6558), 681–689.

Rochel, N., Ciesielski, F., Godet, J., Moman, E., Roessle, M., Peluso-Iltis, C., et al. (2011). Common architecture of nuclear receptor heterodimers on DNA direct repeat elements with different spacings. *Nature Structural & Molecular Biology*, *18*(5), 564–570.

Rochel, N., Hourai, S., & Moras, D. (2010). Crystal structure of hereditary vitamin D-resistant rickets—Associated mutant H305Q of vitamin D nuclear receptor bound to its natural ligand. *The Journal of Steroid Biochemistry and Molecular Biology*, *121*(1–2), 84–87.

Rochel, N., & Moras, D. (2011). Structural basis for ligand activity in VDR. In D. Feldman, J. W. Pike, & J. S. Adams (Eds.), *Vitamin D* (3rd ed., pp. 171–191). Elsevier Inc.

Rochel, N., Tocchini-Valentini, G., Egea, P. F., Juntunen, K., Garnier, J. M., Vihko, P., et al. (2001). Functional and structural characterization of the insertion region in the ligand binding domain of the vitamin D nuclear receptor. *European Journal of Biochemistry/FEBS*, *268*(4), 971–979.

Rochel, N., Wurtz, J. M., Mitschler, A., Klaholz, B., & Moras, D. (2000). The crystal structure of the nuclear receptor for vitamin D bound to its natural ligand. *Molecular Cell*, *5*(1), 173–179.

Shiau, A. K., Barstad, D., Loria, P. M., Cheng, L., Kushner, P. J., Agard, D. A., et al. (1998). The structural basis of estrogen receptor/coactivator recognition and the antagonism of this interaction by tamoxifen. *Cell*, *95*(7), 927–937.

Shimizu, M., Miyamoto, Y., Takaku, H., Matsuo, M., Nakabayashi, M., Masuno, H., et al. (2008). 2-Substituted-16-ene-22-thia-1alpha,25-dihydroxy-26,27-dimethyl-19-norvitamin D3 analogs: Synthesis, biological evaluation, and crystal structure. *Bioorganic & Medicinal Chemistry*, *16*(14), 6949–6964.

Sibilska, I. K., Szybinski, M., Sicinski, R. R., Plum, L. A., & Deluca, H. F. (2013). Highly potent 2-methylene analogs of 1α,25-dihydroxyvitamin D3: Synthesis and biological evaluation. *The Journal of Steroid Biochemistry and Molecular Biology*, *136*, 9–13.

Singarapu, K. K., Zhu, J., Tonelli, M., Rao, H., Assadi-Porter, F. M., Westler, W. M., et al. (2011). Ligand-specific structural changes in the vitamin D receptor in solution. *Biochemistry*, *50*(51), 11025–11033.

Song, J. K., Yoon, K. S., Shim, K. S., & Bae, C.-W. (2011). Novel compound heterozygous mutations in the vitamin D receptor gene in a Korean girl with hereditary vitamin D resistant rickets. *Journal of Korean Medical Science*, *26*(8), 1111–1114.

Swann, S. L., Bergh, J., Farach-Carson, M. C., Ocasio, C. A., & Koh, J. T. (2002). Structure-based design of selective agonists for a rickets-associated mutant of the vitamin D receptor. *Journal of the American Chemical Society*, *124*(46), 13795–13805.

Tocchini-Valentini, G., Rochel, N., Wurtz, J. M., Mitschler, A., & Moras, D. (2001). Crystal structures of the vitamin D receptor complexed to superagonist 20-epi ligands. *Proceedings of the National Academy of Sciences of the United States of America*, *98*(10), 5491–5496.

Väisänen, S., Ryhänen, S., Saarela, J. T. A., Peräkylä, M., Andersin, T., & Mäenpää, P. H. (2002). Structurally and functionally important amino acids of the agonistic conformation of the human vitamin D receptor. *Molecular Pharmacology*, *62*(4), 788–794.

Vanhooke, J. L., Benning, M. M., Bauer, C. B., Pike, J. W., & DeLuca, H. F. (2004). Molecular structure of the rat vitamin D receptor ligand binding domain complexed with 2-carbon-substituted vitamin D3 hormone analogues and a LXXLL-containing coactivator peptide. *Biochemistry*, *43*(14), 4101–4110.

Verlinden, L., Verstuyf, A., Van Camp, M., Marcelis, S., Sabbe, K., Zhao, X. Y., et al. (2000). Two novel 14-Epi-analogues of 1,25-dihydroxyvitamin D3 inhibit the growth of human breast cancer cells *in vitro* and *in vivo*. *Cancer Research*, *60*(10), 2673–2679.

Verstuyf, A., Verlinden, L., Van Etten, E., Shi, L., Wu, Y., D'Halleweyn, C., et al. (2010). Biological activity of CD-ring modified 1α,25-dihydroxyvitamin D analogues: C-ring and five-membered D-ring analogues. *Journal of Bone and Mineral Research*, *15*(2), 237–252.

Wang, Y., Chirgadze, N. Y., Briggs, S. L., Khan, S., Jensen, E. V., & Burris, T. P. (2006). A second binding site for hydroxytamoxifen within the coactivator-binding groove of estrogen receptor beta. *Proceedings of the National Academy of Sciences of the United States of America*, *103*(26), 9908–9911.

Whitfield, G. K., Selznick, S. H., Haussler, C. A., Hsieh, J. C., Galligan, M. A., Jurutka, P. W., et al. (1996). Vitamin D receptors from patients with resistance to 1,25-dihydroxyvitamin D3: Point mutations confer reduced transactivation in response to ligand and impaired interaction with the retinoid X receptor heterodimeric partner. *Molecular Endocrinology*, *10*(12), 1617–1631.

Yamada, S., & Makishima, M. (2014). Structure–activity relationship of nonsecosteroidal vitamin D receptor modulators. *Trends in Pharmacological Sciences*, *305*, 1–14.

Yang, W., & Freedman, L. P. (1999). 20-Epi analogues of 1,25-dihydroxyvitamin D3 are highly potent inducers of DRIP coactivator complex binding to the vitamin D3 receptor. *The Journal of Biological Chemistry*, *274*(24), 16838–16845.

Yin, D., He, Y., Perera, M. A., Hong, S. S., Marhefka, C., Stourman, N., et al. (2003). Key structural features of nonsteroidal ligands for binding and activation of the androgen receptor. *Molecular Pharmacology*, *63*(1), 211–223.

Yoshimoto, N., Sakamaki, Y., Haeta, M., Kato, A., Inaba, Y., Itoh, T., et al. (2012). Butyl pocket formation in the vitamin D receptor strongly affects the agonistic or antagonistic behavior of ligands. *Journal of Medicinal Chemistry*, *55*(9), 4373–4381.

Zhang, J., Chalmers, M. J., Stayrook, K. R., Burris, L. L., Garcia-Ordonez, R. D., Pascal, B. D., et al. (2010). Hydrogen/deuterium exchange reveals distinct agonist/partial agonist receptor dynamics within vitamin D receptor/retinoid X receptor heterodimer. *Structure*, *18*(10), 1332–1341.

Zhang, J., Chalmers, M. J., Stayrook, K. R., Burris, L. L., Wang, Y., Busby, S. A., et al. (2011). DNA binding alters coactivator interaction surfaces of the intact VDR–RXR complex. *Nature Structural & Molecular Biology*, *18*(5), 556–563.

Zhou, Y., Wang, J., Malloy, P. J., Dolezel, Z., & Feldman, D. (2009). Compound heterozygous mutations in the vitamin D receptor in a patient with hereditary 1,25-dihydroxyvitamin D-resistant rickets with alopecia. *Journal of Bone and Mineral Research: The Official Journal of the American Society for Bone and Mineral Research*, *24*(4), 643–651.

CHAPTER FIVE

Crystal Structure of the Vitamin D Receptor Ligand-Binding Domain with Lithocholic Acids

Teikichi Ikura, Nobutoshi Ito[1]
Medical Research Institute, Tokyo Medical and Dental University, Tokyo, Japan
[1]Corresponding author: e-mail address: ito.str@tmd.ac.jp

Contents

Abstract

The secondary bile acid lithocholic acid (LCA) and its derivatives act as selective modulators of the vitamin D receptor (VDR), although their structures fundamentally differ from that of the natural hormone 1α,25-dihydroxyvitamin D_3 (1,25$(OH)_2D_3$). The complexes of the ligand-binding domain of rat VDR (VDR-LBD) with LCA and its derivatives revealed that the ligands bound to the same ligand-binding pocket (LBP) of VDR-LBD that 1,25$(OH)_2D_3$ binds to, but in the opposite orientation; their A-ring was positioned at the top of the LBP, whereas their acyclic tail was located at the bottom of the LBP. However, most of the hydrophobic and hydrophilic interactions observed in the complex with 1,25$(OH)_2D_3$ were reproduced in the complexes with LCA and its derivatives. Additional interactions between VDR-LBD and the C-3 substituents of the A-ring were also observed in the complexes, probably related to the observed difference in the potency among the LCA-type ligands. Recently, zebrafish VDR has been reported to have the second LBP on the outside of the canonical LBP, although its physiological function is unclear.

Vitamins and Hormones, Volume 100
ISSN 0083-6729
http://dx.doi.org/10.1016/bs.vh.2015.10.004

1. INTRODUCTION

The active metabolite of vitamin D_3, 1,25-dihydroxyvitamin D_3 (1,25 $(OH)_2D_3$), regulates calcium homeostasis (Haussler et al., 1998). It also promotes cellular differentiation, inhibits cellar proliferation, and regulates the immune system (Abe et al., 1981; DeLuca, 2004; Hosomi, Hosoi, Abe, Suda, & Kuroki, 1983; Lemire, 1992; Smith, Walworth, & Holick, 1986; Tanaka et al., 1982). It has been used clinically to treat renal osteodystrophy, vitamin D-dependent rickets type I, and X-linked hypophosphatemic rickets, among other conditions (Bortman, Folgueira, Katayama, Snitcovsky, & Brentani, 2002; Fraser et al., 1973; Glorieux, Marie, Pettifor, & Delvin, 1980; Hayes, 2000; Konety & Getzenberg, 2002; Lamberg-Allardt, 1991; Langner, Verjans, Stapor, Mol, & Fraczykowska, 1993). Most of its effects are mediated by its specific binding to the vitamin D receptor (VDR), which is a member of nuclear receptor (NR) super family (Yamada, Shimizu, & Yamamoto, 2003). When 1,25 $(OH)_2D_3$ is bound to VDR, it activates VDR by inducing conformational changes. The activated complex, VDR/1,25$(OH)_2D_3$, binds as a heterodimer with the retinoid X receptor (RXR) to vitamin D response elements (VDREs) located in the promoter region of the target genes. Recruitment of coactivator proteins to this heterodimer is also essential to the transactivation.

The amino acid sequence of the VDR is highly conserved across species: mammalian VDRs share about 90% identity between rat and mouse and more than 80% between either rat or mouse, and human, and more than 70% with even *Xenopus laevis* VDR (Jehan & DeLuca, 1997; Li, Bergwitz, Juppner, & Demay, 1997). The VDR is constituted of several domains or regions common among proteins in NR super family; as the major ones, the amino terminal DNA-binding domain (DBD), the carboxyl terminal ligand-binding domain (LBD), and the hinge region between the two domains (Fig. 1A) (Allegretto, Pike, & Haussler, 1987; Green et al., 1986). The DBD consists of two zinc-coordinated finger modules, which are different from the well-known zinc finger motif (Shaffer & Gewirth, 2002). Each module consists of two helices perpendicular to each other. A zinc ion, coordinated by four conserved cysteines, holds the base of a loop at the N-terminus of each helix. The two modules appear highly related to each other structurally, but they are not equivalent topologically and their function in DNA binding is also different. Whereas the N-terminal module

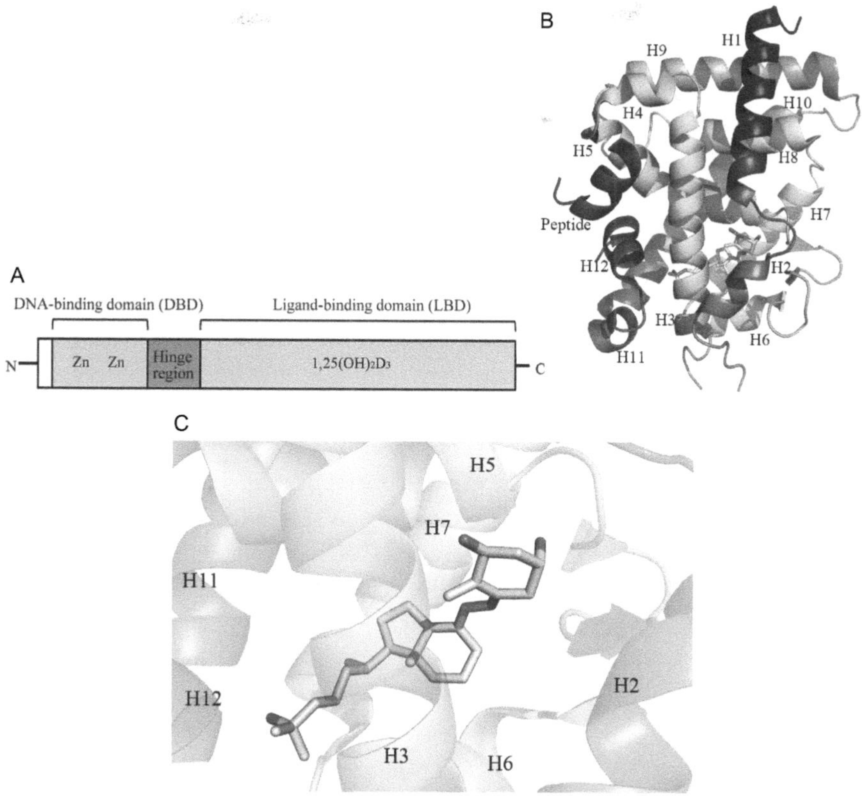

Figure 1 Structure of rVDR in complex with the ligand 1,25(OH)$_2$D$_3$. (A) Schematic of domain structures of rVDR. Two zinc molecules bind to the N-terminal DNA-binding domain (DBD), while the ligand binds to the C-terminal ligand-binding domain (LBD). (B) Tertiary structure of the LBD of rVDR in ternary complex with the ligand 1,25(OH)$_2$D$_3$ and the peptide derived from the coactivator MED1 (PDB code: 2ZLC). Twelve helices of the LDB are numbered after those of the human RXR. (C) Close-up view of the ligand 1,25(OH)$_2$D$_3$ in the ternary complex. The helices are drawn transparently. (See the color plate.)

directs specific DNA-binding site, the C-terminal one and its adjacent several amino acid residues of the hinge region serve as a dimerization interface for interaction with partner proteins like RXR. The hinge region forms a long helix, and the geometric consequences of this structure are expected to explain the requirement of the hinge region for transcriptional activity (Shaffer & Gewirth, 2002; Shaffer, McDonnell, & Gewirth, 2005). In the deletion studies on the hinge region, however, VDRs with different lengths of the hinge region, even if those that did not activate transcription, bound VDREs with equal and high affinity (Shaffer et al., 2005). The ability of VDR to bind VRDEs may not be sufficient for its transcriptional activity.

The LBD is connected with the DBD through the hinge region, and exerts absolute regulatory control over the DNA binding as well as transcription-modifying properties of VDR. The three-dimensional structure of the LBD of human VDR in complex with 1,25 $(OH)_2D_3$ was first determined by Rochel, Wurtz, Mitschler, Klaholz, and Moras (2000). Then, the complex structures of the LBD of rat VDR and human VDR with the natural ligand as well as other ligands were determined one after another (Shimizu et al., 2008; Tocchini-Valentini, Rochel, Wurtz, Mitschler, & Moras, 2001; Tocchini-Valentini, Rochel, Wurtz, & Moras, 2004; Vanhooke, Benning, Bauer, Pike, & DeLuca, 2004). These structures revealed the presence of 12 α-helices and 3 β-strands that together form a highly hydrophobic ligand-binding pocket (LBP) at its core (Fig. 1B). The ligand $1,25(OH)_2D_3$ binds in this pocket in an extended configuration with the A-ring in the β-chair conformation and the 1α-OH group equatorial. It is anchored by six hydrogen bonds and several hydrophobic interactions (Fig. 1C). While two helices H9 and H10 provide a binding surface for RXR, three helices H2, H4, and H12 form the ligand-dependent activation function (AF-2) domain together the loop 3–4. The domain also interacts with the LXXLL motif of MED1 known as the NR box (Fig. 1B). Several residues in H12 also make direct contact with the ligand. Thus, the LBD plays the most important role in the functional activation of VDR.

Clinical use of $1,25(OH)_2D_3$ is limited because therapeutic doses can give rise to significant hypercalciuria and hypercalcemia (Bouillon, Okamura, & Norman, 1995). A number of synthetic ligands to VDR have been developed for medical use; however, most of them can also cause similar problems because they are derived from $1,25(OH)_2D_3$. Several synthetic compounds without the vitamin D_3 scaffold have been reported to bind to VDR and have VDR-modulating activities, including growth inhibition of cancer cells and keratinocytes and induction of leukemic cell differentiation, with less calcium mobilization side effects than $1,25(OH)_2D_3$ (Boehm et al., 1999). Therefore, these synthetic compounds are expected to be therapeutics for cancer, leukemia, and psoriasis. Subsequently, Makishima et al. discovered that secondary bile acids, including lithocholic acid (LCA) and its derivatives, also behaved as VDR agonists (Adachi et al., 2005; Ishizawa et al., 2008; Makishima et al., 2002). LCA acts as a detergent to stabilize fats for absorption, but it has been implicated in human and experimental animal carcinogenesis (Bernstein et al., 2011). The agonistic behavior of LCA as a ligand recognized by VDR, however, was not common knowledge because the structure of LCA is completely different from that of vitamin D_3.

Additional studies showed that VDR had dual functions as a metabolic sensor of bile acids and as an endocrine receptor for $1,25(OH)_2D_3$. Both functions are closely related to colon cancer suppression (Degirolamo, Modica, Palasciano, & Moschetta, 2011). Although theoretical studies were performed to elucidate how LCA binds to VDR, the agonistic mechanism of VDR by LCA was still unclear (Adachi et al., 2005; Choi et al., 2003).

Recently, the complex structures of VDR with LCA were determined by us and Belorusova et al. (Belorusova et al., 2014; Masuno et al., 2013). We determined the crystal structures of the LBD of rat VDR (rVDR) in ternary complexes with a synthetic peptide containing the target sequence of MED1 and the ligands LCA, 3-keto LCA, LCA acetate, and LCA propionate (Masuno et al., 2013), while Belorusova et al. determined those of the LBD of zebrafish VDR (zVDR) in ternary complexes with a peptide containing the target sequence of nuclear receptor coactivator 2 (SRC-2) and LCA (Belorusova et al., 2014). The structures reveal that LCA and its derivatives bind to the same LBP of VDR that $1,25(OH)_2D_3$ binds to, but in the opposite orientation (Ciesielski, Rochel, Mitschler, Kouzmenko, & Moras, 2004; Eelen et al., 2005; Hourai et al., 2006; Rochel et al., 2007, 2000; Shimizu et al., 2008; Tocchini-Valentini et al., 2001, 2004; Vanhooke et al., 2004). Furthermore, the second LCA molecule was found to bind to zVDR with the low affinity in the completely different manner (Belorusova et al., 2014).

2. LITHOCHOLIC ACIDS

LCAs are included in secondary bile acids. Bile acids are synthesized from cholesterol in the liver, then conjugated with glycine or taurine, and secreted into the bile and delivered to the lumen of the small intestine. They are the major catabolic detergents that are required for the ingestion and intestinal absorption of hydrophobic nutrients such as cholesterol, triacylglycerol, and fat-soluble vitamins. The solubilized nutrients are incorporated into lipoproteins, which are delivered to the liver and metabolized (Russell, 2003). Other physiological functions of bile acids are also reported as follows (Hofmann, 1999). Bile acids stimulate bile flow and biliary phospholipid secretion. The bile flow is induced by their osmotic properties, while they promote the transfer of phospholipids from the canalicular membrane into bile. Bile acids negatively regulate biosynthesis of bile acids themselves and cholesterol. The concentration of bile acids in the hepatocyte seems to act as a signal: when high, bile acid synthesis is low; when low, bile acid synthesis increases up to 15-fold. Because bile acids are synthesized from cholesterol, cholesterol synthesis undergoes a parallel increase. There are additional functions of bile acids in the intestine. Bile acids solubilize polyvalent metals such as iron and calcium in the duodenum, promoting their absorption. They also seem to stimulate the release of motilin, which coordinates the interdigestive migrating motility complex. They have bacteriostatic effects and stimulate mucin secretion; these actions are likely to affect intestinal flora of the small intestine. They affect the absorption of water and electrolytes by the colonic mucosa, and affect colonic motility. Bile acids, however, are cytotoxic when present in abnormally high concentrations. This may occur intracellularly, in the hepatocyte in cholestasis, or occur extracellularly, in the colon in patients with bile acid malabsorption.

On the other hand, secondary bile acids, such as LCA and deoxycholic acids, are converted from bile acids by bacteria in the intestine (Fig. 2). They also keep functions similar to those which bile acids have, but unfortunately increase cytotoxicity as cancer-promoting substances and carcinogen for colon cancer (Bernstein et al., 2011). They also induces caspase-dependent colonocyte apoptosis at physiological relevant concentrations (Bernstein, Bernstein, Payne, Dvorakova, & Garewal, 2005; Glinghammar, Inoue, & Rafter, 2002). Oxidative DNA damage is likely to contribute to the proapoptotic activity of the secondary bile acids.

In consideration of such cytotoxic activity of secondary bile acids, their specific binding to VDR suggests that VDR may act as a (secondary) bile acid sensor as well as an endocrine receptor for vitamin D signaling, although the physiological role of VDR in bile acid metabolism is still unclear (Nehring, Zierold, & DeLuca, 2007; Nishida, Ozeki, & Makishima, 2009). Secondary bile acids have not been demonstrated to regulate calcium

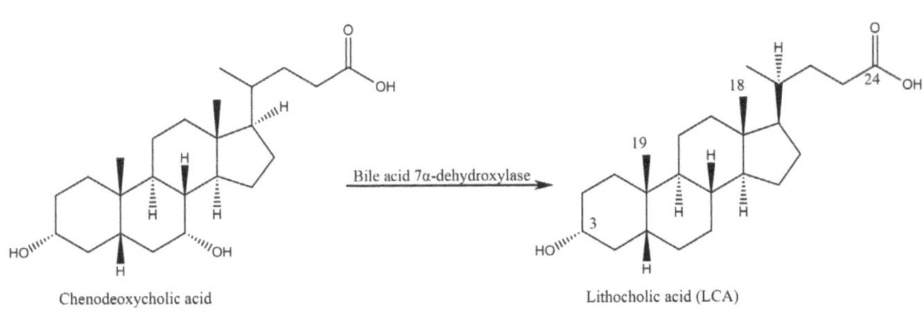

Figure 2 Schematic of lithocholic acid (LCA) production. LCA is converted from chenodeoxycholic acid by bile acid 7α-dehydroxylase of bacteria in the intestine. The derivatives of LCA described in this chapter are substituted at C-3 position of the A-ring.

metabolism, although they decrease vitamin D absorption directly. Accordingly, they may be available to induce partial activities of VDR except for hypercalcemia (Choi & Makishima, 2009).

3. STRUCTURES OF THE VDR LBD WITH LCAs

3.1 Structures of the Ternary Complexes

Recently, the crystal structures of the complexes of the LBD of rat VDR (rVDR-LBD) with LCA and its three derivatives, 3-keto LCA, LCA acetate, and LCA propionate (Fig. 3A–D), and the complex of the LDB of zebrafish (zVDR-LDB) with LCA (Fig. 3E) were determined by X-ray crystallography (Belorusova et al., 2014; Masuno et al., 2013). The overall structures of VDR-LBD in the five complexes are nearly identical to each others, and furthermore, to the proteins in the $1,25(OH)_2D_3$ complexes. Therefore, no significant structural differences are found among the proteins in the complexes with LCA, its derivatives, and $1,25(OH)_2D_3$ (Shimizu et al., 2008; Vanhooke et al., 2004).

The AF-2 domain of the VDR-LBD consists of helices 3, 4, and 12 and loop 3–4, and interacts with the LXXLL motif of the coactivator known as the NR box. How do the LCA and its derivatives affect coactivator binding? The structure of the peptide in each of the five complexes is nearly identical, although the peptides derived from different coactivators, MED1 and SRC-2, are applied for the rVDR and zVDR complexes, respectively. The structure of the peptide in the LCA complex is also compared with that in the $1,25(OH)_2D_3$ complex, indicating no significant structural differences among the peptides in the ternary complexes with LCA, its derivatives, and $1,25(OH)_2D_3$ (Fig. 4A) (Shimizu et al., 2008; Vanhooke et al., 2004). The peptide forms an α-helix, and binds to the AF-2 domain to that in the complex with $1,25(OH)_2D_3$. The MED1 peptide forms a little kink at Pro628 in each complex, and the deviation from the typical α-helix is also observed at Lys689 of the SRC-2. This suggests that the kink is not caused by the restricted structure of Pro628 but by the inherent interaction between the peptide and VDR.

The complex between MED1 and rVDR shows a typical example of the interactions between the coactivator and the AF-2 domain of the LCA complex (Fig. 4B). The AF-2 domain forms a shallow pit consisting of five hydrophobic residues, Ile238 in helix 3, Ile256 in helix 4/5, Leu259 in helix 4/5, Leu413 in helix 12, and Val417 in helix 12. The peptide binds to this pit through the hydrophobic interactions between the LXXLL motif of the

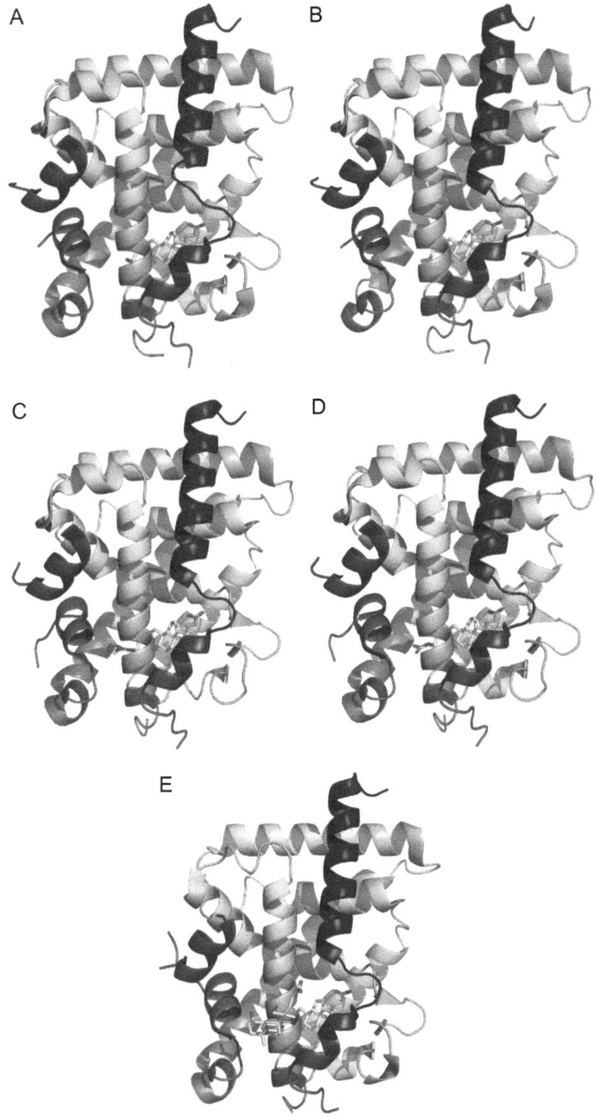

Figure 3 Overall structures of the ternary complexes of VDR-LBD with LCA and its derivatives: (A) rVDR-LBD with LCA (PDB code: 3W5P), (B) rVDR-LBD with 3-keto LCA (PDB code: 3W5Q), (C) rVDR-LBD with LCA acetate (PDB code: 3W5R), (D) rVDR-LBD with LCA propionate (PDB code: 3W5T), and (E) zVDR-LBD with LCA (PDB code: 4Q0A). (See the color plate.)

peptide and the complementary pit of the protein. The polar side chains of Lys242 in helix 3 and Glu416 in helix 12 also facilitate the binding of the peptide by clamping it on the both edges of the AF-2 domain (a charge clamping). Three hydrogen bonds are formed between the oxygen atoms

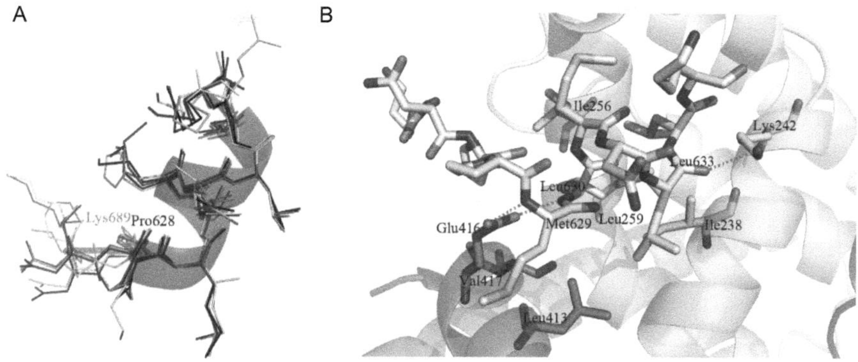

Figure 4 Structures of the coactivator peptides in VDR complexes. (A) Superposition of MED1 peptides in rVDR complexes with 1,25$(OH)_2D_3$ (black), LCA (red), 3-keto LCA (green), LCA acetate (blue), LCA propionate (yellow), and SRC-2 peptide in zVDR complex with LCA (cyan). Pro628 of MED1 peptide corresponds to Lys689 of SRC-2 peptide. Secondary structure of the peptide in rVDR complex with LCA is also drawn with red. (B) Interactions between MED1 and rVDR-LBD complex with LCA. Hydrogen bonds are drawn with magenta dashed lines. (See the color plate.)

of the side chain carboxyl group of Glu416 and the amide nitrogen atoms of Met629 and Leu630, and the nitrogen atom of the side chain amino group of Lys242 and the carbonyl oxygen of Leu633. All these interactions observed in the LCA complexes are the same as those seen in the complex with 1,25 $(OH)_2D_3$. Therefore, these results indicate that the interactions between the coactivator MED1 and the AF-2 domain are well conserved in the ternary complexes with LCA and its derivatives.

3.2 Structures of the Ligands and Their Interactions with VDR-LDB

Proteins in the NR super family have a common LBP. Residues in helices 1, 3, 5, 11, and 12, all β-turns, and loops 6–7 and 11–12 form the framework for the LBP of VDR. The natural hormone 1,25$(OH)_2D_3$ is accommodated in the LBP (Fig. 1C). We observed clear electron density in the rVDR-LBP, as was previously reported in the complex with 1,25$(OH)_2D_3$ (Shimizu et al., 2008), and crystallographic refinement allowed us unambiguous determination of the structure of LCA (Fig. 5A) and its derivatives in complex.

Except for their respective substituents, LCA and its three derivatives are accommodated in the LBP of rVDR-LBD with almost identical structures. However, their orientation is opposite to that of 1,25$(OH)_2D_3$ in both the horizontal and vertical planes (Fig. 5B–D). The 24-carboxyl group is

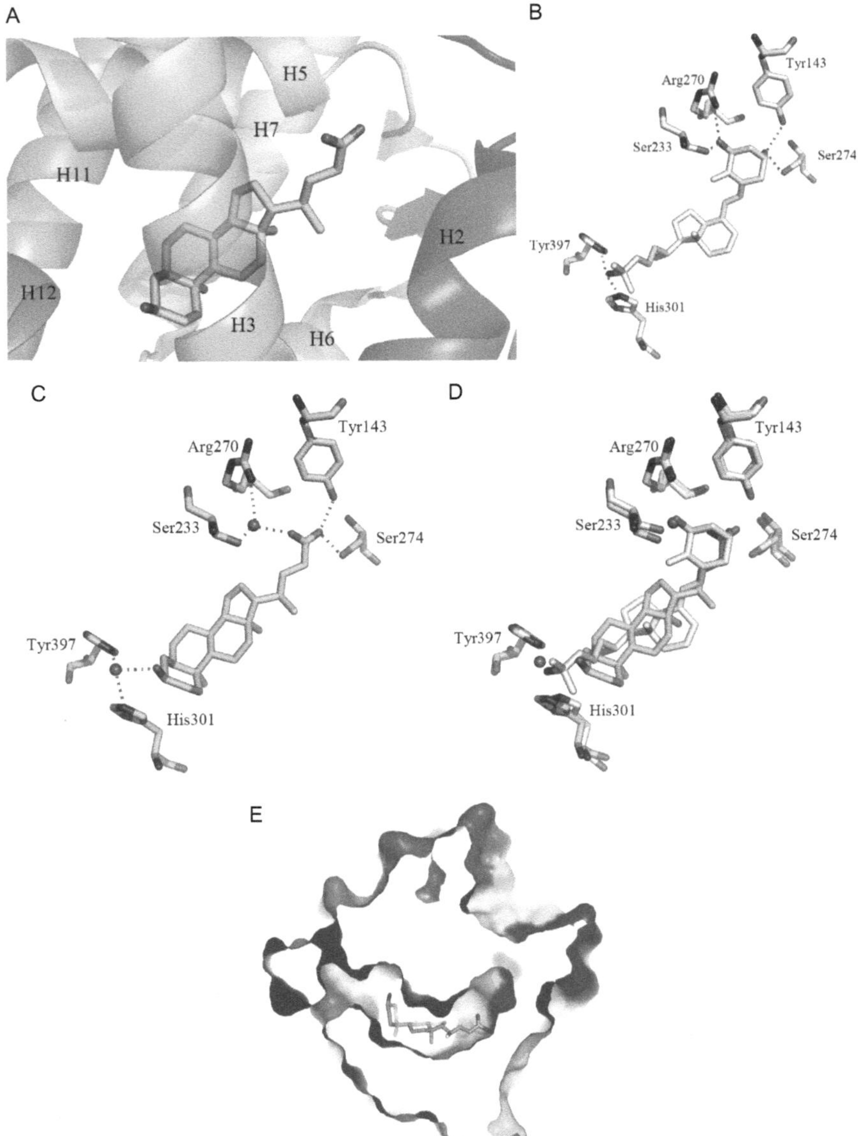

Figure 5 Structures of the ligands in the ternary complexes and interactions between them and rVDR-LBD. (A) Structure of LCA in the LBP of rVDR-LBD. Its location is comparable with that of $1,25(OH)_2D_3$ shown in Fig. 1C, but its orientation is opposite. (B) Interactions between $1,25(OH)_2D_3$ and rVDR-LBD shown in Fig. 1C. (C) Interactions between LCA and rVDR-LBD. (D) Superposition of $1,25(OH)_2D_3$ (light gray) and LCA (green) in the ternary complexes of rVDR-LDB. (E) Location of LCA in the LBP of rVDR-LBD, which shows its hydrophobic interactions with the alkyl and aromatic groups of the surrounding residues. Hydrogen bonds are drawn with magenta dashed lines in panels (B) and (C). Red spheres stand for water molecules in panels (C) and (D). (See the color plate.)

directed toward the β-turns and the β-face of the steroid is directed toward helix 7 in the bottom of the LBP, while the A-ring faces helix 12 (Fig. 5A). LCA forms three hydrogen bonds at the carboxyl group. One oxygen atom of the carboxyl group directly interacts with the hydroxyl groups in the side chains of Tyr143 in helix 1 (the distance between the oxygen atoms of the carboxyl group and the hydroxyl group is 2.46 Å) and Ser274 in helices 4/5 (the distance between the oxygen atoms of the carboxyl group and the hydroxyl group is 2.76 Å) through hydrogen bonds, while the other oxygen atom of the carboxyl group interacts via a water molecule (the distance between the oxygen atoms of the water and the carboxyl group is 2.69 Å), with the hydroxyl group in the side chain of Ser233 in helix 3 (the distance between the oxygen atoms of the water and the hydroxyl group is 2.92 Å) and the guanidinium group in the side chain of Arg270 in helix 4/5 (the distance between the oxygen atom of the water and the nitrogen atom of the guanidinium group is 3.01 Å) (Fig. 5C). These hydrogen bonds are also observed in the other three complexes. The four rings of the steroid in each of the complexes interact with hydrophobic residues in the LBP through hydrophobic interactions. There are 12 residues, Leu226, Leu229, Val230, Ile264, Ile267, Met268, Trp282, Val296, Ala299, Leu305, Ile306, and Leu309, distributed within 4.3 Å from the rings (Fig. 5E). Such hydrophobic interactions are also conserved in the other complexes.

Hydrogen bonds between VDR and the ligands are also observed at the other end of the ligands, the C-3 position of the A-ring. The four ligands differ in their substituents at this position. The hydroxyl group of LCA, the carbonyl group of 3-keto LCA, the propionyl group of LCA propionate, and the acetyl group of LCA acetate interact with residues in helix 6, loop 6–7, and helix 11 (Fig. 6A–E). In the complexes with LCA and 3-keto LCA, the oxygen atoms of the respective hydroxyl and carbonyl groups of the substituents interact via a water-mediated hydrogen bond with the nitrogen atoms of imidazole rings of His301 in helix 6 and His393 in helix 11 (Fig. 6B and C). In contrast, in the complex with LCA acetate, the oxygen atom of the acetyl group directly forms a hydrogen bond with the nitrogen atom of the imidazole ring of His301 (Fig. 6D). In the complex with LCA propionate, the oxygen atom of the propionyl group directly forms two hydrogen bonds with the nitrogen atoms of the imidazole ring of His301 and His393 (Fig. 6E). Furthermore, the alkyl parts of the two substituents interact with the aromatic rings of Tyr397 in helix 11 and Phe418 in helix 12 and the side chains of Leu410 and Val414 in helix 12, stabilizing the

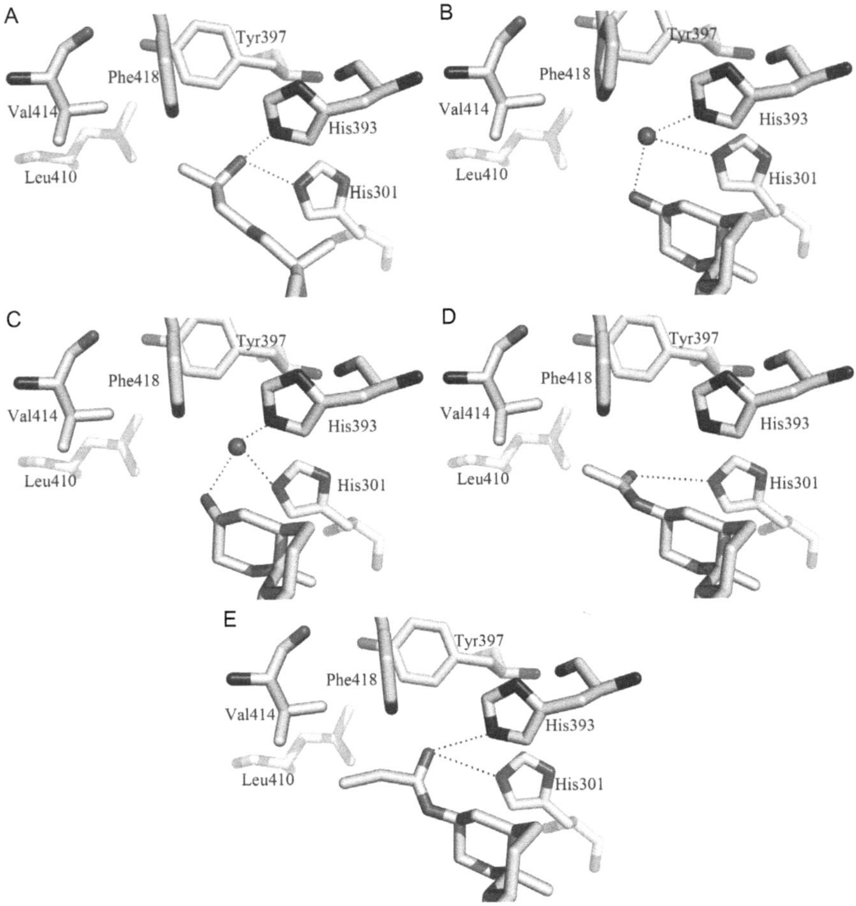

Figure 6 Interactions of the ligands with residues in helix 6, loop 6–7, and helix 11 of rVDR-LDB. (A) Interactions between $1,25(OH)_2D_3$ and its surrounding residues. (B) Interactions between LCA and its surrounding residues. (C) Interactions between 3-keto LCA and its surrounding residues. (D) Interactions between LCA acetate and its surrounding residues. (E) Interactions between LCA propionate and its surrounding residues. Hydrogen bonds are drawn with magenta dashed lines in all panels. Red spheres stand for water molecules in panels (B) and (C). (See the color plate.)

binding of the two derivatives to the LBP of VDR-LBD (Fig. 6D and E). From this viewpoint, LCA propionate may be the most effective of the four ligands because it has the longest alkyl part in the substituent. This is consistent with recent experimental results showing that LCA propionate was the most effective agonist of the four (Adachi et al., 2005; Ishizawa et al., 2008).

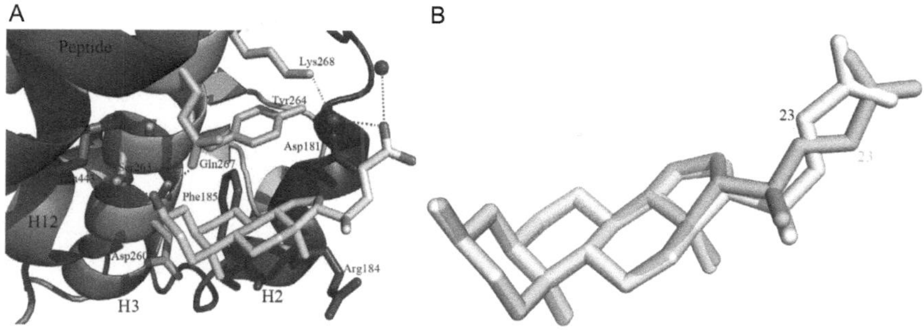

Figure 7 Feature of the second LCA-binding site. (A) Interactions between LCA and its surrounding residues in the second LCA-binding site of zVDR-LBD. Hydrogen bonds are drawn with magenta dashed lines. Red spheres stand for water molecules. (B) Superposition of LCA molecules in the first canonical site (green) and the second site (gray). The orientation of the C-23 carbon of the second LCA is different from that of the first one. (See the color plate.)

According to Belorusova et al., the LBD of zVDR possesses the second LCA-binding site, which is located on the surface of zVDR-LBD close to the loop 1–3 and H3 (Fig. 7A) (Belorusova et al., 2014). The structure of the second LCA is similar to that of the first one except for the orientation of the carbon atom at C-23 position (Fig. 6B), but the interactions between the zVDR-LBD and LCA in the second site are completely different from those observed in the canonical first site (Fig. 7A). The second LCA is partially exposed to the solvent, but primarily stabilized by hydrophobic interactions with Asp181, Arg184, and Phe185 in helix 2 and Leu443 in helix 12. The hydroxyl group of the A-ring forms two hydrogen bonds with the hydroxyl group in the side chain of Ser263 and the amide group in the side chain of Gln267 in helix 3. The 24-carboxyl group interacts via a water-mediated hydrogen bond with the side chain carboxyl group of Asp181 in helix 2 and the side chain amino group of Lys268 in helix 3. The amino acid residues forming the second LCA-binding site are well conserved among many species of VDRs, although Arg184 of zVDR-LDB is substituted by Asp152 in rat VDR-LBD or Gln152 in human VDR-LBD (hVDR-LBD). This suggests that the second LCA-binding site may exist in rat and human VDR-LBD. Actually, Belorusova and her colleagues showed that hVDR-LBD has at least two LCA-binding sites by isothermal titration calorimetry, and the interaction between LCA and the second binding site was about 30 times weaker than that between LCA and the first (canonical) one (Belorusova et al., 2014). However, low solubility of LCA may undermine the reliability of the results. We also tried to crystallize the complex of rVDR-LDB with two LCA molecules using several techniques, but have not obtained it yet (data not shown). The function of the second LCA-binding site is still unclear, but the second LCA molecule may contribute to structural stability of VDR-LBD as indicated by molecular dynamics simulations (Belorusova et al., 2014).

4. MECHANISM OF AGONIST ACTIVITIES OF LCA TO VDR

Since the discovery of its function as an agonist of VDR, LCA has been expected to be used a vitamin D alternative, especially because LCA appears to activate VDR without causing hypercalcemia. However, because the functional mechanism of LCA is still unclear, LCA derivatives with higher activities have been found mainly by trial and error. As shown in the previous section, the structures of ternary complexes of VDR-LBD with LCA and its derivatives elucidates how they bind to VRD-LDB.

LCA and its derivatives bind to the same canonical LBP that $1,25(OH)_2D_3$ binds. LCA and its derivatives are similar in size (14–16 Å × 5 Å × 3 Å, length × width × thickness) to natural $1,25(OH)_2D_3$ (approximately 15 Å × 5 Å × 2 Å) and, although their shapes are somewhat different, the nonplanar *cis* A/B rings of the steroid skeleton of LCA mimic the curvature of the 9,10-secosteroid portion of $1,25(OH)_2D_3$ (Fig. 5D). Often, the size seems to be critical to the agonistic activity of a ligand, since several larger ligands do not have the agonistic activity, but instead have the antagonistic activity (Inaba et al., 2010; Nakabayashi et al., 2008).

The orientation of LCA and its derivatives, however, is opposite to that of $1,25(OH)_2D_3$ (Fig. 5D). Its A-ring was set on the inlet of the LBP, while its 24-carboxyl group wedged itself into the LBP. Interestingly, this orientation is the same as that of a bile acid analog, 6α-ethyl-chenodeoxycholic acid (6α-ethyl-CDCA), in the farnesoid X receptor (FXR) (Fig. 8) (Mi et al., 2003). FXR plays a key role in regulation of bile acid transport in the enterohepatic system. Activation of liver FXR protects hepatocytes from toxic bile acid accumulation by stimulating bile acid secretion at the canalicular membrane and limiting bile acid uptake from the portal circulation. Thus, this orientation of bile acid may be optimized to increase its affinity in receptor binding.

On the other hand, despite their opposite orientation, LCA and its derivatives can reproduce the interactions between VDR and $1,25(OH)_2D_3$, which consists of a hydrophobic secosteroid framework and three polar groups. The polar groups are located at the both ends of the ligand and

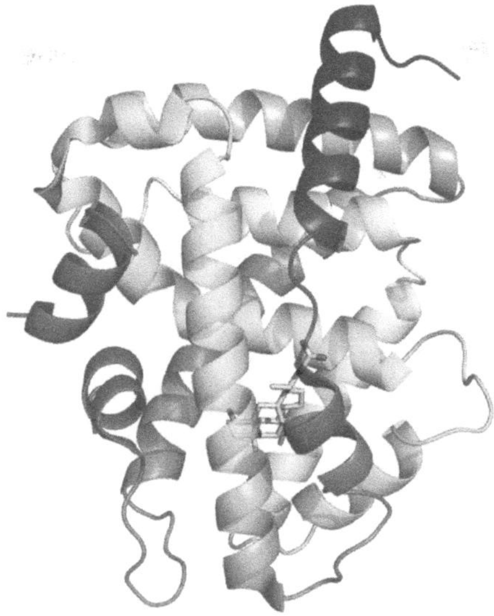

Figure 8 Tertiary structure of the LBD of rat FXR in ternary complex with 6α-ethyl-CDCA and the peptide derived from the coactivator GRIP-1 (PDB code: 1OSV). The orientation of the ligand is the same as that of LCA bound to the VDR. (See the color plate.)

stabilize the ligand binding through several hydrogen bonds, while the hydrophobic secosteroid frame just fits the hydrophobic tunnel of the LBP. In the complex with $1,25(OH)_2D_3$, the A-ring of $1,25(OH)_2D_3$ deeply wedges itself into the LBP and its 25-hydroxyl group is set on the inlet of the LBP. There are two polar groups at the C-1 and C-3 positions of the A-ring of $1,25(OH)_2D_3$, whose oxygen atoms form two pairs of bifurcated hydrogen bonds with the side chains of Tyr143 and Ser274, and Ser233 and Arg270, respectively. Furthermore, the 25-hydroxyl group of $1,25(OH)_2D_3$ also forms a bifurcated hydrogen bond with the nitrogen atoms of the imidazole rings of His301 and His393. In contrast, LCA and its derivatives consist of the hydrophobic steroid framework and two polar groups at the both ends (Fig. 2). However, they maintain most of the hydrogen bonds between the ligand and the protein by exchanging the hydrogen-bonding partners between the A-ring and the carboxyl group. Two oxygen atoms of the 24-carboxyl group form, with the help of a water molecule, two pairs of bifurcated hydrogen bonds directly with the side chains of Tyr143 and Ser274, and indirectly with Ser233 and Arg270, respectively. The hydrogen bonds with the nitrogen atoms of the imidazole rings of His301 and His393 are partially formed, depending on the substituents at the C-3 position of the A-ring. Therefore, the hydrogen-bonding network between the ligand and the protein is essentially the same for both LCA and 1,25(OH)$_2$D$_3$, even though their orientations are opposite. Adachi et al. also indicated the importance of the hydrogen-bonding network for agonistic activity by showing that esterification of the carboxyl group of LCA weakened its agonistic activity (Adachi et al., 2005).

Although LCA mimicked the dimensions and chemical properties of 1,25(OH)$_2$D$_3$ as discussed above, the hydrophobic surface area of LCA differs markedly from that of 1,25(OH)$_2$D$_3$ because their shapes differ (Fig. 5D). The accessible surface areas of the hydrophobic surface of LCA and 1,25(OH)$_2$D$_3$ are 371 and 442 Å^2, respectively, suggesting that the LBP accommodates LCA somewhat loosely. Such a feature allows several water molecules to penetrate the LBP. One of the water molecules is involved in the hydrogen-bond network around the 24-carboxyl group. Another water molecule is also incorporated in the hydrogen-bond network around the C-3 position of the A-ring in the complex with LCA or 3-keto LCA. However, interactions mediated by hydrogen bonding seem to be less important for stabilizing the complex than the hydrophobic interactions. In fact, the binding affinity of LCA for VDR is much lower than that of 1,25 (OH)$_2$D$_3$ (Makishima et al., 2002).

The substituents at the C-3 position of the A-ring of LCA also affect the activation of human VDR, and LCA acetate and LCA propionate are more potent agonists than LCA and 3-keto LCA. Our crystal structures show that the C-3 substituents of the former derivatives interact directly with VDR via a hydrogen bond, while the latter ones require a water molecule to form indirect hydrogen bonds with VDR. In addition, the alkyl part of these acyl groups interacts with the hydrophobic residues of VDR. These interactions are probably the reason for the higher potency of these ligands. These additional interactions at the C-3 position may also explain some of the mutation analyses; the human VDR-S275A and S278A mutations (corresponding to S271A and S274A of rat VDR, respectively) almost completely abolished the activity of LCA, whereas they are still activated by LCA acetate (Adachi et al., 2005). The extra interactions at the C-3 position compensate for the loss of the hydrogen bond due to the mutation to some extent.

Some of the other results from the mutation analyses are more difficult to explain (Adachi et al., 2005; Choi et al., 2003; Ishizawa et al., 2008; Sato et al., 2008). Although VDR-H305A (H301A of rat VDR) had a significant effect on LCA activity, this mutation has little effect on activation by LCA acetate or 3-keto LCA. The VDR-S237M mutation (S233M of rat VDR) weakly diminishes LCA acetate and LCA activity, and VDR-S278V

(S274V of rat VDR) drastically decreases LCA acetate activity. However, it is difficult to elucidate the relationship between the character of the ligand in the ternary complex and the mutational response of VDR, even though there is a small structural difference in a portion of helix 11, which interacts with the substituent of LCA.

5. CONCLUSIONS AND FUTURE DIRECTIONS

The VDR-LDB selectively binds to the secondary bile acid, LCA and its derivatives, as well as its natural hormone 1,25$(OH)_2D_3$, although their structures fundamentally differ from that of the natural hormone. The crystal structures have shown the similarities and differences between LCA and 1,25 $(OH)_2D_3$ in their interactions with VDR. The reason for the functional differences between LCA and 1,25$(OH)_2D_3$, such as induction of hypercalcemia, is still unclear, especially because no significant differences in the interactions between VDR and the coactivator peptide are detected. To rationally elucidate the underlying reason for the difference in the agonistic activity between the ligands, it will be necessary to analyze the higher ordered complexes using full-length VDR, and the complex structure will provide a sound basis for the design of new ligands based on LCA, hopefully with better pharmaceutical features.

REFERENCES

Abe, E., Miyaura, C., Sakagami, H., Takeda, M., Konno, K., Yamazaki, T., et al. (1981). Differentiation of mouse myeloid leukemia cells induced by 1 alpha,25-dihydroxyvitamin D3. *Proceedings of the National Academy of Sciences of the United States of America*, *78*(8), 4990–4994.

Adachi, R., Honma, Y., Masuno, H., Kawana, K., Shimomura, I., Yamada, S., et al. (2005). Selective activation of vitamin D receptor by lithocholic acid acetate, a bile acid derivative. *Journal of Lipid Research*, *46*(1), 46–57. http://dx.doi.org/10.1194/jlr.M400294-JLR200.

Allegretto, E. A., Pike, J. W., & Haussler, M. R. (1987). Immunochemical detection of unique proteolytic fragments of the chick 1,25-dihydroxyvitamin D3 receptor. Distinct 20-kDa DNA-binding and 45-kDa hormone-binding species. *The Journal of Biological Chemistry*, *262*(3), 1312–1319.

Belorusova, A. Y., Eberhardt, J., Potier, N., Stote, R. H., Dejaegere, A., & Rochel, N. (2014). Structural insights into the molecular mechanism of vitamin D receptor activation by lithocholic acid involving a new mode of ligand recognition. *Journal of Medicinal Chemistry*, *57*(11), 4710–4719. http://dx.doi.org/10.1021/jm5002524.

Bernstein, H., Bernstein, C., Payne, C. M., Dvorakova, K., & Garewal, H. (2005). Bile acids as carcinogens in human gastrointestinal cancers. *Mutation Research*, *589*(1), 47–65. http://dx.doi.org/10.1016/j.mrrev.2004.08.001.

Bernstein, C., Holubec, H., Bhattacharyya, A. K., Nguyen, H., Payne, C. M., Zaitlin, B., et al. (2011). Carcinogenicity of deoxycholate, a secondary bile acid. *Archives of Toxicology*, *85*(8), 863–871. http://dx.doi.org/10.1007/s00204-011-0648-7.

Boehm, M. F., Fitzgerald, P., Zou, A., Elgort, M. G., Bischoff, E. D., Mere, L., et al. (1999). Novel nonsecosteroidal vitamin D mimics exert VDR-modulating activities with less calcium mobilization than 1,25-dihydroxyvitamin D3. *Chemistry & Biology*, *6*(5), 265–275. http://dx.doi.org/10.1016/S1074-5521(99)80072-6.

Bortman, P., Folgueira, M. A., Katayama, M. L., Snitcovsky, I. M., & Brentani, M. M. (2002). Antiproliferative effects of 1,25-dihydroxyvitamin D3 on breast cells: A mini review. *Brazilian Journal of Medical and Biological Research = Revista Brasileira de Pesquisas Médicas e Biológicas/Sociedade Brasileira de Biofísica ... [et al]*, *35*(1), 1–9.

Bouillon, R., Okamura, W. H., & Norman, A. W. (1995). Structure-function relationships in the vitamin D endocrine system. *Endocrine Reviews*, *16*(2), 200–257.

Choi, M., & Makishima, M. (2009). Therapeutic applications for novel non-hypercalcemic vitamin D receptor ligands. *Expert Opinion on Therapeutic Patents*, *19*(5), 593–606. http://dx.doi.org/10.1517/13543770902877717.

Choi, M., Yamamoto, K., Itoh, T., Makishima, M., Mangelsdorf, D. J., Moras, D., et al. (2003). Interaction between vitamin D receptor and vitamin D ligands: Two-dimensional alanine scanning mutational analysis. *Chemistry & Biology*, *10*(3), 261–270.

Ciesielski, F., Rochel, N., Mitschler, A., Kouzmenko, A., & Moras, D. (2004). Structural investigation of the ligand binding domain of the zebrafish VDR in complexes with 1alpha,25(OH)2D3 and Gemini: Purification, crystallization and preliminary X-ray diffraction analysis. *The Journal of Steroid Biochemistry and Molecular Biology*, *89–90*(1–5), 55–59. http://dx.doi.org/10.1016/j.jsbmb.2004.03.109.

Degirolamo, C., Modica, S., Palasciano, G., & Moschetta, A. (2011). Bile acids and colon cancer: Solving the puzzle with nuclear receptors. *Trends in Molecular Medicine*, *17*(10), 564–572. http://dx.doi.org/10.1016/j.molmed.2011.05.010.

DeLuca, H. F. (2004). Overview of general physiologic features and functions of vitamin D. *The American Journal of Clinical Nutrition*, *80*(6 Suppl.), 1689S–1696S.

Eelen, G., Verlinden, L., Rochel, N., Claessens, F., De Clercq, P., Vandewalle, M., et al. (2005). Superagonistic action of 14-epi-analogs of 1,25-dihydroxyvitamin D explained by vitamin D receptor-coactivator interaction. *Molecular Pharmacology*, *67*(5), 1566–1573. http://dx.doi.org/10.1124/mol.104.008730.

Fraser, D., Kooh, S. W., Kind, H. P., Holick, M. F., Tanaka, Y., & DeLuca, H. F. (1973). Pathogenesis of hereditary vitamin-D-dependent rickets. An inborn error of vitamin D metabolism involving defective conversion of 25-hydroxyvitamin D to 1 alpha,25-dihydroxyvitamin D. *The New England Journal of Medicine*, *289*(16), 817–822. http://dx.doi.org/10.1056/NEJM197310182891601.

Glinghammar, B., Inoue, H., & Rafter, J. J. (2002). Deoxycholic acid causes DNA damage in colonic cells with subsequent induction of caspases, COX-2 promoter activity and the transcription factors NF-kB and AP-1. *Carcinogenesis*, *23*(5), 839–845.

Glorieux, F. H., Marie, P. J., Pettifor, J. M., & Delvin, E. E. (1980). Bone response to phosphate salts, ergocalciferol, and calcitriol in hypophosphatemic vitamin D-resistant rickets. *The New England Journal of Medicine*, *303*(18), 1023–1031. http://dx.doi.org/10.1056/NEJM198010303031802.

Green, S., Walter, P., Kumar, V., Krust, A., Bornert, J. M., Argos, P., et al. (1986). Human oestrogen receptor cDNA: Sequence, expression and homology to v-erb-A. *Nature*, *320*(6058), 134–139. http://dx.doi.org/10.1038/320134a0.

Haussler, M. R., Whitfield, G. K., Haussler, C. A., Hsieh, J. C., Thompson, P. D., Selznick, S. H., et al. (1998). The nuclear vitamin D receptor: Biological and molecular regulatory properties revealed. *Journal of Bone and Mineral Research: The Official Journal of the American Society for Bone and Mineral Research*, *13*(3), 325–349. http://dx.doi.org/10.1359/jbmr.1998.13.3.325.

Hayes, C. E. (2000). Vitamin D: A natural inhibitor of multiple sclerosis. *The Proceedings of the Nutrition Society*, *59*(4), 531–535.

Hofmann, A. F. (1999). The continuing importance of bile acids in liver and intestinal disease. *Archives of Internal Medicine, 159*(22), 2647–2658.

Hosomi, J., Hosoi, J., Abe, E., Suda, T., & Kuroki, T. (1983). Regulation of terminal differentiation of cultured mouse epidermal cells by 1 alpha,25-dihydroxyvitamin D3. *Endocrinology, 113*(6), 1950–1957.

Hourai, S., Fujishima, T., Kittaka, A., Suhara, Y., Takayama, H., Rochel, N., et al. (2006). Probing a water channel near the A-ring of receptor-bound 1 alpha,25-dihydroxyvitamin D3 with selected 2 alpha-substituted analogues. *Journal of Medicinal Chemistry, 49*(17), 5199–5205. http://dx.doi.org/10.1021/jm0604070.

Inaba, Y., Nakabayashi, M., Itoh, T., Yoshimoto, N., Ikura, T., Ito, N., et al. (2010). 22S-butyl-1alpha,24R-dihydroxyvitamin D3: Recovery of vitamin D receptor agonistic activity. *The Journal of Steroid Biochemistry and Molecular Biology, 121*(1–2), 146–150. http://dx.doi.org/10.1016/j.jsbmb.2010.02.033.

Ishizawa, M., Matsunawa, M., Adachi, R., Uno, S., Ikeda, K., Masuno, H., et al. (2008). Lithocholic acid derivatives act as selective vitamin D receptor modulators without inducing hypercalcemia. *Journal of Lipid Research, 49*(4), 763–772. http://dx.doi.org/10.1194/jlr.M700293-JLR200.

Jehan, F., & DeLuca, H. F. (1997). Cloning and characterization of the mouse vitamin D receptor promoter. *Proceedings of the National Academy of Sciences of the United States of America, 94*(19), 10138–10143.

Konety, B. R., & Getzenberg, R. H. (2002). Vitamin D and prostate cancer. *The Urologic Clinics of North America, 29*(1), 95–106, ix.

Lamberg-Allardt, C. (1991). Is there a role for vitamin D in osteoporosis? *Calcified Tissue International, 49*(Suppl.), S46–S49.

Langner, A., Verjans, H., Stapor, V., Mol, M., & Fraczykowska, M. (1993). Topical calcitriol in the treatment of chronic plaque psoriasis: A double-blind study. *The British Journal of Dermatology, 128*(5), 566–571.

Lemire, J. M. (1992). Immunomodulatory role of 1,25-dihydroxyvitamin D3. *Journal of Cellular Biochemistry, 49*(1), 26–31. http://dx.doi.org/10.1002/jcb.240490106.

Li, Y. C., Bergwitz, C., Juppner, H., & Demay, M. B. (1997). Cloning and characterization of the vitamin D receptor from *Xenopus laevis*. *Endocrinology, 138*(6), 2347–2353. http://dx.doi.org/10.1210/endo.138.6.5210.

Makishima, M., Lu, T. T., Xie, W., Whitfield, G. K., Domoto, H., Evans, R. M., et al. (2002). Vitamin D receptor as an intestinal bile acid sensor. *Science, 296*(5571), 1313–1316. http://dx.doi.org/10.1126/science.1070477.

Masuno, H., Ikura, T., Morizono, D., Orita, I., Yamada, S., Shimizu, M., et al. (2013). Crystal structures of complexes of vitamin D receptor ligand-binding domain with lithocholic acid derivatives. *Journal of Lipid Research, 54*(8), 2206–2213. http://dx.doi.org/10.1194/jlr.M038307.

Mi, L. Z., Devarakonda, S., Harp, J. M., Han, Q., Pellicciari, R., Willson, T. M., et al. (2003). Structural basis for bile acid binding and activation of the nuclear receptor FXR. *Molecular Cell, 11*(4), 1093–1100.

Nakabayashi, M., Yamada, S., Yoshimoto, N., Tanaka, T., Igarashi, M., Ikura, T., et al. (2008). Crystal structures of rat vitamin D receptor bound to adamantyl vitamin D analogs: Structural basis for vitamin D receptor antagonism and partial agonism. *Journal of Medicinal Chemistry, 51*(17), 5320–5329. http://dx.doi.org/10.1021/jm8004477.

Nehring, J. A., Zierold, C., & DeLuca, H. F. (2007). Lithocholic acid can carry out *in vivo* functions of vitamin D. *Proceedings of the National Academy of Sciences of the United States of America, 104*(24), 10006–10009. http://dx.doi.org/10.1073/pnas.0703512104.

Nishida, S., Ozeki, J., & Makishima, M. (2009). Modulation of bile acid metabolism by 1alpha-hydroxyvitamin D3 administration in mice. *Drug Metabolism and Disposition, 37*(10), 2037–2044. http://dx.doi.org/10.1124/dmd.109.027334.

Rochel, N., Hourai, S., Perez-Garcia, X., Rumbo, A., Mourino, A., & Moras, D. (2007). Crystal structure of the vitamin D nuclear receptor ligand binding domain in complex with a locked side chain analog of calcitriol. *Archives of Biochemistry and Biophysics*, *460*(2), 172–176. http://dx.doi.org/10.1016/j.abb.2007.01.031.

Rochel, N., Wurtz, J. M., Mitschler, A., Klaholz, B., & Moras, D. (2000). The crystal structure of the nuclear receptor for vitamin D bound to its natural ligand. *Molecular Cell*, *5*(1), 173–179.

Russell, D. W. (2003). The enzymes, regulation, and genetics of bile acid synthesis. *Annual Review of Biochemistry*, *72*, 137–174. http://dx.doi.org/10.1146/annurev.biochem.72.121801.161712.

Sato, H., Macchiarulo, A., Thomas, C., Gioiello, A., Une, M., Hofmann, A. F., et al. (2008). Novel potent and selective bile acid derivatives as TGR5 agonists: Biological screening, structure-activity relationships, and molecular modeling studies. *Journal of Medicinal Chemistry*, *51*(6), 1831–1841. http://dx.doi.org/10.1021/jm7015864.

Shaffer, P. L., & Gewirth, D. T. (2002). Structural basis of VDR-DNA interactions on direct repeat response elements. *The EMBO Journal*, *21*(9), 2242–2252. http://dx.doi.org/10.1093/emboj/21.9.2242.

Shaffer, P. L., McDonnell, D. P., & Gewirth, D. T. (2005). Characterization of transcriptional activation and DNA-binding functions in the hinge region of the vitamin D receptor. *Biochemistry*, *44*(7), 2678–2685. http://dx.doi.org/10.1021/bi0477182.

Shimizu, M., Miyamoto, Y., Takaku, H., Matsuo, M., Nakabayashi, M., Masuno, H., et al. (2008). 2-Substituted-16-ene-22-thia-1alpha,25-dihydroxy-26,27-dimethyl-19-norvita min D3 analogs: Synthesis, biological evaluation, and crystal structure. *Bioorganic & Medicinal Chemistry*, *16*(14), 6949–6964. http://dx.doi.org/10.1016/j.bmc.2008.05.043.

Smith, E. L., Walworth, N. C., & Holick, M. F. (1986). Effect of 1 alpha,25-dihydroxyvitamin D3 on the morphologic and biochemical differentiation of cultured human epidermal keratinocytes grown in serum-free conditions. *The Journal of Investigative Dermatology*, *86*(6), 709–714.

Tanaka, H., Abe, E., Miyaura, C., Kuribayashi, T., Konno, K., Nishii, Y., et al. (1982). 1 alpha,25-Dihydroxycholecalciferol and a human myeloid leukaemia cell line (HL-60). *The Biochemical Journal*, *204*(3), 713–719.

Tocchini-Valentini, G., Rochel, N., Wurtz, J. M., Mitschler, A., & Moras, D. (2001). Crystal structures of the vitamin D receptor complexed to superagonist 20-epi ligands. *Proceedings of the National Academy of Sciences of the United States of America*, *98*(10), 5491–5496. http://dx.doi.org/10.1073/pnas.091018698.

Tocchini-Valentini, G., Rochel, N., Wurtz, J. M., & Moras, D. (2004). Crystal structures of the vitamin D nuclear receptor liganded with the vitamin D side chain analogues calcipotriol and seocalcitol, receptor agonists of clinical importance. Insights into a structural basis for the switching of calcipotriol to a receptor antagonist by further side chain modification. *Journal of Medicinal Chemistry*, *47*(8), 1956–1961. http://dx.doi.org/10.1021/jm0310582.

Vanhooke, J. L., Benning, M. M., Bauer, C. B., Pike, J. W., & DeLuca, H. F. (2004). Molecular structure of the rat vitamin D receptor ligand binding domain complexed with 2-carbon-substituted vitamin D3 hormone analogues and a LXXLL-containing coactivator peptide. *Biochemistry*, *43*(14), 4101–4110. http://dx.doi.org/10.1021/bi036056y.

Yamada, S., Shimizu, M., & Yamamoto, K. (2003). Vitamin D receptor. *Endocrine Development*, *6*, 50–68.

CHAPTER SIX

25-Hydroxyvitamin D_3 24-Hydroxylase: A Key Regulator of $1,25(OH)_2D_3$ Catabolism and Calcium Homeostasis

Vaishali Veldurthy, Ran Wei, Megan Campbell, Kamil Lupicki, Puneet Dhawan, Sylvia Christakos[1]
Department of Microbiology, Biochemistry and Molecular Genetics, Rutgers, The State University of New Jersey, New Jersey Medical School, Newark, New Jersey, USA
[1]Corresponding author: e-mail address: christak@njms.rutgers.edu

Contents

Abstract

One of the most pronounced effects of the hormonally active form of vitamin D, 1,25-dihydroxyvitamin D_3 ($1,25(OH)_2D_3$), is increased synthesis of 25-hydroxyvitamin D_3 24-hydroxylase (CYP24A1), the enzyme responsible for the catabolism of $1,25(OH)_2D_3$. Thus, $1,25(OH)_2D_3$ regulates its own metabolism, protecting against hypercalcemia and limiting the levels of $1,25(OH)_2D_3$ in cells. This chapter summarizes the catalytic properties of CYP24A1, the recent data related to the crystal structure of CYP24A1, the findings obtained from the generation of mice deficient for the *Cyp24a1* gene as well as recent data identifying a causal role of a genetic defect in *CYP24A1* in certain patients with idiopathic infantile hypercalcemia. This chapter also reviews the regulation of renal and placental CYP24A1 as well as the genomic mechanisms, including coactivators, repressors, and epigenetic modification, involved in modulating $1,25(OH)_2D_3$ regulation of CYP24A1. We conclude with future research directions related to this key regulator of $1,25(OH)_2D_3$ catabolism and calcium homeostasis.

Vitamins and Hormones, Volume 100
ISSN 0083-6729
http://dx.doi.org/10.1016/bs.vh.2015.10.005

1. INTRODUCTION AND CATALYTIC PROPERTIES

25-Hydroxyvitamin D_3 24-hydroxylase (CYP24A1) is a mitochondrial inner membrane P450 enzyme that utilizes adrenodoxin and adrenodoxin reductase for electron transfer from NADPH. CYP24A1 is responsible for the catabolism of calcitriol, $1,25(OH)_2D_3$, the physiologically active form of vitamin D (Jones, Prosser, & Kaufmann, 2012; St-Arnaud, 2011). CYP24A1 initiates the C24 oxidation pathway that results in the catabolism of $1,25(OH)_2D_3$ in five enzymatic steps to produce calcitroic acid which is excreted in the bile (Jones et al., 2012; St-Arnaud, 2011). CYP24A1 can also catalyze the C23 oxidation pathway resulting in the formation of $1,25(OH)_2D_3$-26,23 lactone from the initial substrate $1,25(OH)_2D_3$ (Jones et al., 2012; St-Arnaud, 2011). CYP24A1 can also hydroxylate $25(OH)D_3$, the precursor of $1,25(OH)_2D_3$, resulting in the formation of $24,25(OH)_2D_3$ which can be further converted to different metabolites and finally to calcitroic acid (Jones et al., 2012; St-Arnaud, 2011). $25(OH)D_3$ can also be converted by CYP24A1 via C23 hydroxylation to $25(OH)D_3$-26,23 lactone (Jones et al., 2012; St-Arnaud, 2011). It has been suggested that $1,25(OH)_2D_3$ is the preferred substrate for CYP24A1 (Shinki et al., 1992). CYP24A1 thus limits the amount of $1,25(OH)_2D_3$ by catalyzing the conversion of $1,25(OH)_2D_3$ to hydroxylated products targeted for excretion or by converting $25(OH)D_3$ to $24,25(OH)_2D_3$ or $23,25(OH)_2D_3$ which are further converted to additional metabolic products, thus decreasing the pool available for 1-hydroxylation. It has been reported that the C23 oxidation product $1,25(OH)_2D_3$-26,23-lactone acts as a vitamin D receptor (VDR) antagonist (Ishizuka et al., 2005). The C23 oxidation pathway may provide a redundant alternative mechanism to attenuate the level of physiologically active $1,25(OH)_2D_3$, protecting against excessive VDR activation. It should be noted that species specificity has been reported for the C24 and C23 hydroxylation pathways (Hamamoto et al., 2006; Prosser, Kaufmann, O'Leary, Byford, & Jones, 2007). Rat CYP24A1 primarily uses the C24 hydroxylation pathway, while human CYP24A1 uses both pathways (Hamamoto et al., 2006). CYP24A1 is present in all cells containing the VDR and is strongly induced by $1,25(OH)_2D_3$. Thus, $1,25(OH)_2D_3$ self-induces its deactivation protecting against hypercalcemia and limiting the levels of $1,25(OH)_2D_3$ in cells.

2. *CYP24A1* GENE AND CRYSTAL STRUCTURE OF CYP24A1

The human *CYP24A1* gene has been localized to chromosome 20q13.2-q13.3 (Hahn et al., 1993). This gene is highly conserved across species. The rat *Cyp24a1* gene spans approximately 15 kb, comprises 12 exons, and is present as a single copy (Ohyama et al., 1993). The *Cyp24a1* gene product CYP24A1 is a 50–55-kDa polypeptide which possesses about 500 amino acids similar to other cytochrome proteins (Jones et al., 2012). In 2010, Annalora et al. reported the crystal structure of rat CYP24A1 (Annalora et al., 2010). The CYP24A1 structure exhibits 12 α-helices (A–L) and 4 β-sheet systems (β1–β4) as well as additional helices (A′, B′, G′ on the distal surface and K′ and K″ between β2 and the heme-binding motif) (Fig. 1).

Hydrophobic sequences surrounding helices A′ and G′ were identified as membrane insertion sequences, and key residues important for substrate binding and catalysis were identified that are consistent with results obtained

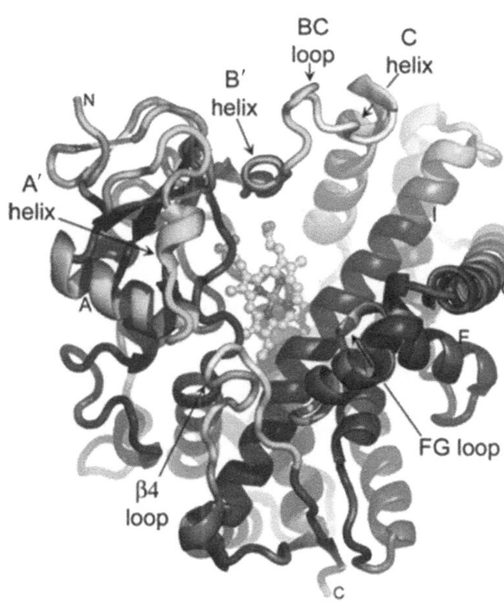

Figure 1 The crystal structure of CYP24A1. α-Helices, the β4 loop as well as BC loop and FG loop regions and the N and C terminus are labeled. *From Andrew J. Annalora and G. David Stout with permission.* (See the color plate.)

by homology modeling (Annalora et al., 2010). In addition, four basic residues from helices K and L involved in electron chain partner interactions were identified. These findings are important for determining functional consequences of mutations in CYP24A1 in patients and in the design of vitamin D analogs and CYP24A1 inhibitors.

3. *CYP24A1*-NULL MICE

St. Arnaud et al., by disrupting exons encoding the core heme-binding domain, generated *Cyp24a1*-null mice, which provided the first direct *in vivo* evidence supporting a catabolic role for CYP24A1 (St-Arnaud et al., 2000). About 50% of homozygous mutant mice died in the perinatal period. The surviving homozygous mice were fertile, failed to clear exogenously administered $1,25(OH)_2D_3$, and exhibited defects in intramembranous ossification. When *Cyp24a1*-null mice were crossed with *Vdr*-null mice, the bone defect was rescued, demonstrating that it is the elevated $1,25(OH)_2D_3$ in the *Cyp24a1*-null mice, acting through the VDR, which resulted in the observed defects in intramembranous ossification. Although it has been reported that ablation of *Cyp24a1* also delays bone fracture repair and that this delay can be rescued by treatment with $24,25(OH)_2D_3$ (St-Arnaud, 2010; St-Arnaud & Naja, 2011), further studies are needed to determine the mechanisms and pathways mediating these effects of $24,25(OH)_2D_3$.

4. GENETIC DEFECT IN *CYP24A1*: A CAUSE OF IDIOPATHIC INFANTILE HYPERCALCEMIA

It has recently been discovered that several infants who presented with idiopathic infantile hypercalcemia have a genetic mutation in *CYP24A1*, thus reinforcing the findings using the *Cyp24a1*-null mouse of a critical role for CYP24A1 in inactivating $1,25(OH)_2D_3$ (Schlingmann et al., 2011). The patients' symptoms include severe hypercalcemia, failure to thrive, vomiting, dehydration, and, in some cases, nephrocalcinosis (Schlingmann et al., 2011). Five different CYP24A1 mutations, all affecting residues of structural importance, were reported [E143del (in-frame deletion of E143), E322K, R396W, L409S, and R159Q] (Schlingmann et al., 2011). Molecular modeling studies have reported that the mutation-induced loss of function occurs through different pathways (Ji & Shen, 2011). For example, the R159Q and the R396W mutations may weaken the binding of heme to

CYP24A1, while the L409S mutation would weaken the binding of 1,25$(OH)_2D_3$ to CYP24A1. The findings of Schlingmann et al. (2011) indicate a causal role of a genetic defect in *CYP24A1* in certain patients with idiopathic infantile hypercalcemia and suggest the need for careful monitoring of infants who receive prophylactic vitamin D to avoid vitamin D toxicity. *CYP24A1* mutations have been identified not only in infants but also in adults (Colussi et al., 2014; Dinour et al., 2013, 2015; Jacobs et al., 2014; Tebben et al., 2012). The findings in adults suggest that mutations in *CYP24A1* should be considered in the diagnosis of unexplained hypercalcemia associated with elevated 1,25$(OH)_2D_3$ and nephrocalcinosis.

5. REGULATION OF CYP24A1

1,25$(OH)_2D_3$ regulates its own biosynthesis by inducing the expression of CYP24A1. CYP24A1 functions in cooperation with CYP27B1. CYP27B1, 25(OH)D 1α hydroxylase, is a cytochrome P450 enzyme which metabolizes 25(OH)D to 1,25$(OH)_2D_3$ and is present predominantly in the kidney. Together, CYP24A1 and CYP27B1 tightly regulate the bioavailability of 1,25$(OH)_2D_3$ depending on the physiologic state (Bikle, 2014; Jones, Prosser, & Kaufmann, 2014; Plum & DeLuca, 2010). Low dietary calcium results in hypocalcemia and elevated PTH (parathyroid hormone). PTH positively regulates CYP27B1 and inhibits CYP24A1, thus stimulating 1,25$(OH)_2D_3$ synthesis, 1,25$(OH)_2D_3$-mediated increased intestinal calcium absorption, and an increase in serum calcium. PTH directly stimulates *Cyp27b1* transcription (Zierold, Nehring, & DeLuca, 2007). PTH has been reported to negatively regulate renal CYP24A1 by destabilizing *Cyp24a1* mRNA (Zierold, Mings, & DeLuca, 2001). Increased levels of 1,25$(OH)_2D_3$ in turn inhibit PTH and CYP27B1 and stimulate CYP24A1, thus protecting against hypercalcemia (Bikle, 2014; Jones et al., 2014; Plum & DeLuca, 2010). In addition to PTH and 1,25$(OH)_2D_3$, fibroblast growth factor 23 (FGF23) can also modulate vitamin D metabolism. FGF23 is a ~32-kDa glycoprotein that belongs to the FGF19 subfamily and acts to promote renal phosphate excretion, at least in part, by inhibiting the renal sodium-dependent Pi cotransporter, Npt2a (Hu, Shiizaki, Kuro-o, & Moe, 2013). FGF23 is produced in bone, most prominently in osteocytes (Bonewald & Wacker, 2013; Hu et al., 2013; Perwad & Portale, 2011). Chronically elevated serum phosphate and 1,25$(OH)_2D_3$ independently stimulate FGF23 production in osteocytes. αKlotho, which is highly expressed in the distal tubule of the kidney, is an obligate coreceptor for

FGF23. αKlotho forms a complex with FGF receptors 1, 3, and 4, and together FGF23 and klotho inhibit CYP27B1 and induce CYP24A1 resulting in a reduction in 1,25$(OH)_2D_3$ levels (Hu et al., 2013). Mice lacking FGF23 (*Fgf23−/−* mice) or klotho (*Klotho−/−* mice) exhibit similar phenotypes including hyperphosphatemia, hypercalcemia, elevated 1,25$(OH)_2D_3$ levels, ectopic calcification, and aging-like features (skin atrophy, osteoporosis, and atherosclerosis), further indicating the requirement of klotho for FGF23-mediated regulation of phosphate homeostasis and vitamin D metabolism (Kuro-o et al., 1997; Shimada et al., 2004). FGF23 is a causative factor for several hereditary hypophosphatemic disorders including X-linked hypophosphatemia (XLH), the most common form of inherited rickets (Martin et al., 2011; Pettifor & Thandrayen, 2012). XLH is due to an inactivating mutation in the *PHEX* (phosphate-regulating gene with homologies to Endopeptidase on the X chromosome) gene ("A gene (PEX) with homologies to endopeptidases is mutated in patients with X-linked hypophosphatemic rickets. The HYP Consortium," 1995). XLH is characterized by renal Pi wasting, low 1,25$(OH)_2D_3$ production, and rickets. The mechanism of the inhibiting effect of PHEX on FGF23 remains to be determined. In the Hyp mouse, which also harbors an inactivating mutation in *Phex* and exhibits similar features of human XLH, the loss the *Phex* results in an increase in FGF23 which results in hypophosphatemia and disordered regulation of vitamin D metabolism (decreased renal synthesis and increased renal catabolism of 1,25$(OH)_2D_3$) (Liu et al., 2006; Sitara et al., 2004). It has been reported that the suppression of 1,25$(OH)_2D_3$ production by FGF23 depends on activation of the mitogen-activated protein kinase signaling pathway (Perwad, Zhang, Tenenhouse, & Portale, 2007). Although it is clear that FGF23 (in addition to PTH and 1,25$(OH)_2D_3$) plays an essential role in regulation of both CYP27B1 and CYP24A1, the mechanisms involved in induction of CYP24A1 by FGF23 remain to be determined.

6. AGING AND CYP24A1

In aging, there is a decline in the ability of the kidney to synthesize 1,25$(OH)_2D_3$ and a decrease in intestinal calcium absorption (Armbrecht, Zenser, Bruns, & Davis, 1979; Armbrecht, Zenser, & Davis, 1980; Tsai, Heath, Kumar, & Riggs, 1984; van Abel et al., 2006). We and others have shown an increase in *Cyp24a1* gene expression with aging which may be one factor leading to age-related alterations in calcium homeostasis

(Johnson et al., 1995; Matkovits & Christakos, 1995; Tsai et al., 1984). A greater increase in renal *Cyp24a1* mRNA in aging females compared to aging males was observed suggesting that an increase in CYP24A1 with age contributes to age- and gender-related bone loss (Matkovits & Christakos, 1995). Further studies are needed to determine the mechanisms that result in dysregulation of calcium homeostasis with age including changes in epigenetic status and changes in VDR signaling that occur with age.

7. PLACENTAL CYP24A1

Maternal 1,25$(OH)_2D_3$, which is derived in part from decidua and placental cells, is significantly elevated during pregnancy (Delvin, Arabian, Glorieux, & Mamer, 1985; Kovacs & Kronenberg, 1997). Elevated maternal 1,25$(OH)_2D_3$ levels are accompanied by increased CYP27B1 in placenta (Zehnder et al., 2002). In addition to increased 1,25$(OH)_2D_3$ synthesis in placenta, suppression of CYP24A1 expression, due to methylation of the *CYP24A1* gene, has been reported (Novakovic et al., 2009). Methylation of the *CYP24A1* gene was reported to be specific for placenta (*CYP24A1* methylation in other human tissues tested was not detected), and methylation of *VDR* and *CYP27B1* genes in placenta was not observed. *CYP24A1* promoter methylation was accompanied by downregulation of basal promoter activity as well as by a lack of responsiveness of the *CYP24A1* promoter to induction by 1,25$(OH)_2D_3$ (Novakovic et al., 2009). These findings suggest that methylation of *CYP24A1*, resulting in suppression of CYP24A1 in the placenta, contributes to increased bioavailability of 1,25$(OH)_2D_3$ during pregnancy.

8. GENOMIC MECHANISMS MEDIATING 1,25$(OH)_2D_3$ REGULATION OF CYP24A1

1,25$(OH)_2D_3$ is the primary regulator of CYP24A1. Thus, 1,25 $(OH)_2D_3$, by inducing its own deactivation, limits the 1,25$(OH)_2D_3$ response and protects cells against excessive activation of the VDR signaling pathway. In general, the genomic actions of 1,25$(OH)_2D_3$ are mediated by the VDR which heterodimerizes with RXR and interacts with vitamin D response elements (VDREs) in and around target genes and mediates their transcription. 1,25$(OH)_2D_3$/VDR-mediated transcription is a multifactorial process that involves the recruitment by 1,25$(OH)_2D_3$ of transcriptional coactivators (Christakos, 2008; Pike & Meyer, 2010). The p160 coactivators steroid receptor activator 1, 2, and 3 (SRC-1, SRC-2, and SRC-3) that have histone acetyltransferase (HAT) activity are primary coactivators that bind to VDR and recruit secondary coactivators such as CBP/p300 (which also has HAT activity) and CARM1 and G9a (histone methyltransferases) (Christakos, 2008; Christakos et al., 2007; Pike & Meyer, 2010). Acetylation and methylation of core histones, by disrupting histone/DNA binding, play a fundamental role in VDR-mediated transcription. VDR-mediated transcription is also mediated by TFIIB, several TATA-binding protein-associated factors as well as mediator complex (a protein complex which regulates RNA polymerase II activity) (Christakos, 2008; Pike & Meyer, 2010). A number of additional factors that cooperate with VDR in the transcriptional regulation of *Cyp24a1* have been identified. It has been reported that the Ras-activated transcription factors Ets-1 and Ets-2 are essential for $1{,}25(OH)_2D_3$ induction of rat *Cyp24a1* transcription (Dwivedi, Omdahl, Kola, Hume, & May, 2000). In the absence of $1{,}25(OH)_2D_3$, *Cyp24a1* transcription is unaffected by Ets proteins. In addition to Ets proteins, functional cooperation between VDR and C/EBPβ in $1{,}25(OH)_2D_3$-mediated induction of *Cyp24a1* transcription has also been reported (Dhawan et al., 2005). Similar to the Ets transcription factors, C/EBPβ is unable to affect *Cyp24a1* transcription in the absence of $1{,}25(OH)_2D_3$. It has been suggested that the inability of Ets and C/EBP transcription factors to function in the absence of $1{,}25(OH)_2D_3$ may be due to repression of *Cyp24a1* transcription by unliganded VDR/RXR (Dhawan et al., 2005; Dwivedi et al., 2000). $1{,}25(OH)_2D_3$-mediated recruitment of coactivators such as CBP/p300 may be needed to relieve the inhibition. The SWI/SNF complex that remodels chromatin using the energy of ATP hydrolysis has also been shown to contribute to VDR-mediated transcriptional activation of *Cyp24a1* (Seth-Vollenweider, Joshi, Dhawan, Sif, & Christakos, 2014). BRG1, an ATPase that is a component of the SWI/SNF complex, was found to associate with C/EBPβ and to cooperate with C/EBPβ and VDR in regulating *Cyp24a1* transcription (Seth-Vollenweider et al., 2014; Fig. 2). Factors involved in the repression of VDR-mediated *Cyp24a1* transcription have also been identified. Ying Yang transcription factor (YY1) was found to repress *Cyp24a1* transcription, at least in part, by sequestering activator proteins including CBP and C/EBPβ (Dhawan et al., 2005; Raval-Pandya, Dhawan, Barletta, & Christakos, 2001). In addition, protein arginine methyl transferase 5 (PRMT5) has been reported to silence VDR-mediated *Cyp24a1* transcription through its methylation of H3R8 and H4R3 (Seth-Vollenweider et al., 2014; Fig. 2).

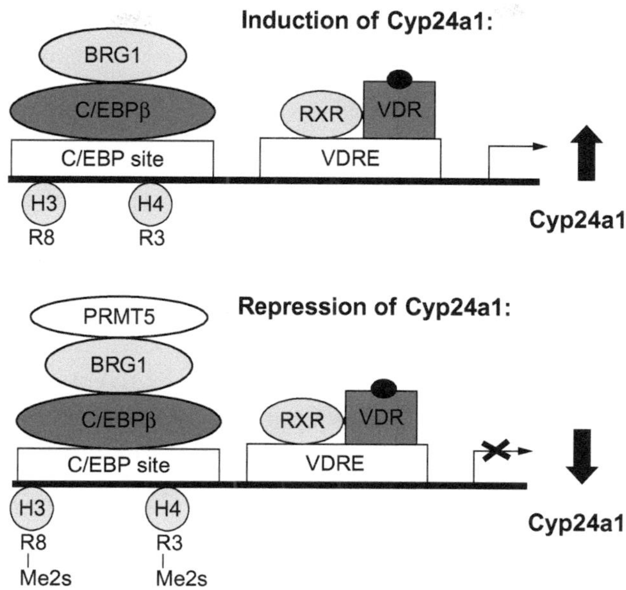

Figure 2 Mechanistic model depicting induction and repression of *Cyp24a1* transcription. $1,25(OH)_2D_3$ induction of *Cyp24a1* involves cooperation among C/EBPβ, BRG1, and VDR. PRMT5, which is recruited to the C/EBP site by BRG1, represses *Cyp24a1* transcription via symmetrical dimethylation of H3R8 and H4R3. VDRE, vitamin D response element; RXR, retinoid X receptor. *Seth-Vollenweider et al. (2014), Copyright (2014)*

Since VDR coregulatory proteins are master regulators of $1,25(OH)_2D_3$ action, the factors involved in repression of CYP24A1 may prevent activation at times that do not require catabolism of $1,25(OH)_2D_3$. Genome-wide studies using ChIP-chip and ChIP-seq have confirmed the VDREs previously defined in the mouse *Cyp24a1* gene at −160 and −265. In addition, a novel intergenic VDRE was identified at +35 and +37 kb (Meyer, Goetsch, & Pike, 2010). Genome-wide studies also confirmed $1,25(OH)_2D_3$-mediated occupancy of C/EBPβ at −345 (Pike & Meyer, 2012). Thus, VDR-mediated *Cyp24a1* transcription involves many different factors and mechanisms including coactivators, repressors, and epigenetic modification that modulate CYP24A1 expression and thus affect regulation of $1,25(OH)_2D_3$ metabolism and the maintenance of calcium homeostasis.

9. CONCLUSION AND FUTURE DIRECTIONS

Understanding vitamin D metabolism is critical to our understanding of the mechanisms involved in maintenance of calcium homeostasis.

CYP24A1 has an important role in the tight regulation of 1,25$(OH)_2D_3$ and thus in controlling intracellular 1,25$(OH)_2D_3$ degradation and in the maintenance of normal calcemia and bone health. Increased catabolism of 1,25 $(OH)_2D_3$ has been suggested to contribute to age-related bone loss (Matkovits & Christakos, 1995; Tsai et al., 1984) and has been implicated in chronic kidney disease (Petkovich & Jones, 2011) and in certain types of cancer (Hobaus et al., 2013; Narvaez et al., 2014). In addition, inactivating mutations in *CYP24A1* have been identified in patients with idiopathic infantile hypercalcemia (Schlingmann et al., 2011). Future clinical studies may identify CYP24A1 dysregulation in other diseases associated with hypercalcemia. The crystal structure of CYP24A1 will allow future studies related to the design of 1,25$(OH)_2D_3$ analogs with altered sensitivity to 1,25$(OH)_2D_3$ catabolism. In addition, studies related to the epigenetic regulation of CYP24A1, currently in progress, may suggest a new class of drug targets for therapeutic intervention.

ACKNOWLEDGMENT

S.C. receives funding from the National Institutes of Health (AI-100379, AG044552).

REFERENCES

Annalora, A. J., Goodin, D. B., Hong, W. X., Zhang, Q., Johnson, E. F., & Stout, C. D. (2010). Crystal structure of CYP24A1, a mitochondrial cytochrome P450 involved in vitamin D metabolism. *Journal of Molecular Biology*, *396*(2), 441–451. http://dx.doi.org/10.1016/j.jmb.2009.11.057.

Armbrecht, H. J., Zenser, T. V., Bruns, M. E., & Davis, B. B. (1979). Effect of age on intestinal calcium absorption and adaptation to dietary calcium. *The American Journal of Physiology*, *236*(6), E769–E774.

Armbrecht, H. J., Zenser, T. V., & Davis, B. B. (1980). Effect of age on the conversion of 25-hydroxyvitamin D3 to 1,25-dihydroxyvitamin D3 by kidney of rat. *The Journal of Clinical Investigation*, *66*(5), 1118–1123. http://dx.doi.org/10.1172/JCI109941.

Bikle, D. D. (2014). Vitamin D metabolism, mechanism of action, and clinical applications. *Chemistry & Biology*, *21*(3), 319–329. http://dx.doi.org/10.1016/j.chembiol.2013.12.016.

Bonewald, L. F., & Wacker, M. J. (2013). FGF23 production by osteocytes. *Pediatric Nephrology*, *28*(4), 563–568. http://dx.doi.org/10.1007/s00467-012-2309-3.

Christakos, S. (2008). Vitamin D gene regulation. In J. Bilezikian, L. Raisz, & T. J. Martin (Eds.), *Principles of bone biology* (pp. 779–794). New York, NY: Academic Press.

Christakos, S., Dhawan, P., Benn, B., Porta, A., Hediger, M., Oh, G. T., et al. (2007). Vitamin D: Molecular mechanism of action. *Annals of the New York Academy of Sciences*, *1116*, 340–348. http://dx.doi.org/10.1196/annals.1402.070.

Colussi, G., Ganon, L., Penco, S., De Ferrari, M. E., Ravera, F., Querques, M., et al. (2014). Chronic hypercalcaemia from inactivating mutations of vitamin D 24-hydroxylase (CYP24A1): Implications for mineral metabolism changes in chronic renal failure. *Nephrology, Dialysis, Transplantation*, *29*(3), 636–643. http://dx.doi.org/10.1093/ndt/gft460.

Delvin, E. E., Arabian, A., Glorieux, F. H., & Mamer, O. A. (1985). In vitro metabolism of 25-hydroxycholecalciferol by isolated cells from human decidua. *The Journal of Clinical Endocrinology and Metabolism*, *60*(5), 880–885. http://dx.doi.org/10.1210/jcem-60-5-880.

Dhawan, P., Peng, X., Sutton, A. L., MacDonald, P. N., Croniger, C. M., Trautwein, C., et al. (2005). Functional cooperation between CCAAT/enhancer-binding proteins and the vitamin D receptor in regulation of 25-hydroxyvitamin D3 24-hydroxylase. *Molecular and Cellular Biology*, *25*(1), 472–487. http://dx.doi.org/10.1128/MCB.25.1.472-487.2005.

Dinour, D., Beckerman, P., Ganon, L., Tordjman, K., Eisenstein, Z., & Holtzman, E. J. (2013). Loss-of-function mutations of CYP24A1, the vitamin D 24-hydroxylase gene, cause long-standing hypercalciuric nephrolithiasis and nephrocalcinosis. *The Journal of Urology*, *190*(2), 552–557. http://dx.doi.org/10.1016/j.juro.2013.02.3188.

Dinour, D., Davidovits, M., Aviner, S., Ganon, L., Michael, L., Modan-Moses, D., et al. (2015). Maternal and infantile hypercalcemia caused by vitamin-D-hydroxylase mutations and vitamin D intake. *Pediatric Nephrology*, *30*(1), 145–152. http://dx.doi.org/10.1007/s00467-014-2889-1.

Dwivedi, P. P., Omdahl, J. L., Kola, I., Hume, D. A., & May, B. K. (2000). Regulation of rat cytochrome P450C24 (CYP24) gene expression. Evidence for functional cooperation of Ras-activated Ets transcription factors with the vitamin D receptor in 1,25-dihydroxyvitamin D(3)-mediated induction. *The Journal of Biological Chemistry*, *275*(1), 47–55.

A gene (PEX) with homologies to endopeptidases is mutated in patients with X-linked hypophosphatemic rickets. The HYP Consortium. (1995). *Nature Genetics*, *11*(2), 130–136. http://dx.doi.org/10.1038/ng1095-130.

Hahn, C. N., Baker, E., Laslo, P., May, B. K., Omdahl, J. L., & Sutherland, G. R. (1993). Localization of the human vitamin D 24-hydroxylase gene (CYP24) to chromosome 20q13.2–>q13.3. *Cytogenetics and Cell Genetics*, *62*(4), 192–193.

Hamamoto, H., Kusudo, T., Urushino, N., Masuno, H., Yamamoto, K., Yamada, S., et al. (2006). Structure-function analysis of vitamin D 24-hydroxylase (CYP24A1) by site-directed mutagenesis: Amino acid residues responsible for species-based difference of CYP24A1 between humans and rats. *Molecular Pharmacology*, *70*(1), 120–128. http://dx.doi.org/10.1124/mol.106.023275.

Hobaus, J., Hummel, D. M., Thiem, U., Fetahu, I. S., Aggarwal, A., Mullauer, L., et al. (2013). Increased copy-number and not DNA hypomethylation causes overexpression of the candidate proto-oncogene CYP24A1 in colorectal cancer. *International Journal of Cancer*, *133*(6), 1380–1388. http://dx.doi.org/10.1002/ijc.28143.

Hu, M. C., Shiizaki, K., Kuro-o, M., & Moe, O. W. (2013). Fibroblast growth factor 23 and Klotho: Physiology and pathophysiology of an endocrine network of mineral metabolism. *Annual Review of Physiology*, *75*, 503–533. http://dx.doi.org/10.1146/annurev-physiol-030212-183727.

Ishizuka, S., Kurihara, N., Reddy, S. V., Cornish, J., Cundy, T., & Roodman, G. D. (2005). (23S)-25-Dehydro-1{alpha}-hydroxyvitamin D3-26,23-lactone, a vitamin D receptor antagonist that inhibits osteoclast formation and bone resorption in bone marrow cultures from patients with Paget's disease. *Endocrinology*, *146*(4), 2023–2030. http://dx.doi.org/10.1210/en.2004-1140.

Jacobs, T. P., Kaufman, M., Jones, G., Kumar, R., Schlingmann, K. P., Shapses, S., et al. (2014). A lifetime of hypercalcemia and hypercalciuria, finally explained. *The Journal of Clinical Endocrinology and Metabolism*, *99*(3), 708–712. http://dx.doi.org/10.1210/jc.2013-3802.

Ji, H. F., & Shen, L. (2011). CYP24A1 mutations in idiopathic infantile hypercalcemia. *The New England Journal of Medicine*, *365*(18), 1741. http://dx.doi.org/10.1056/NEJMc1110226#SA1. author reply 1742–1743.

Johnson, J. A., Beckman, M. J., Pansini-Porta, A., Christakos, S., Bruns, M. E., Beitz, D. C., et al. (1995). Age and gender effects on 1,25-dihydroxyvitamin D3-regulated gene expression. *Experimental Gerontology*, *30*(6), 631–643.

Jones, G., Prosser, D. E., & Kaufmann, M. (2012). 25-Hydroxyvitamin D-24-hydroxylase (CYP24A1): Its important role in the degradation of vitamin D. *Archives of Biochemistry and Biophysics*, *523*(1), 9–18. http://dx.doi.org/10.1016/j.abb.2011.11.003.

Jones, G., Prosser, D. E., & Kaufmann, M. (2014). Cytochrome P450-mediated metabolism of vitamin D. *Journal of Lipid Research*, *55*(1), 13–31. http://dx.doi.org/10.1194/jlr.R031534.

Kovacs, C. S., & Kronenberg, H. M. (1997). Maternal-fetal calcium and bone metabolism during pregnancy, puerperium, and lactation. *Endocrine Reviews*, *18*(6), 832–872. http://dx.doi.org/10.1210/edrv.18.6.0319.

Kuro-o, M., Matsumura, Y., Aizawa, H., Kawaguchi, H., Suga, T., Utsugi, T., et al. (1997). Mutation of the mouse klotho gene leads to a syndrome resembling ageing. *Nature*, *390*(6655), 45–51. http://dx.doi.org/10.1038/36285.

Liu, S., Zhou, J., Tang, W., Jiang, X., Rowe, D. W., & Quarles, L. D. (2006). Pathogenic role of Fgf23 in Hyp mice. *American Journal of Physiology. Endocrinology and Metabolism*, *291*(1), E38–E49. http://dx.doi.org/10.1152/ajpendo.00008.2006.

Martin, A., Liu, S., David, V., Li, H., Karydis, A., Feng, J. Q., et al. (2011). Bone proteins PHEX and DMP1 regulate fibroblastic growth factor Fgf23 expression in osteocytes through a common pathway involving FGF receptor (FGFR) signaling. *The FASEB Journal*, *25*(8), 2551–2562. http://dx.doi.org/10.1096/fj.10-177816.

Matkovits, T., & Christakos, S. (1995). Variable in vivo regulation of rat vitamin D-dependent genes (osteopontin, Ca, Mg-adenosine triphosphatase, and 25-hydroxyvitamin D3 24-hydroxylase): Implications for differing mechanisms of regulation and involvement of multiple factors. *Endocrinology*, *136*(9), 3971–3982. http://dx.doi.org/10.1210/endo.136.9.7649106.

Meyer, M. B., Goetsch, P. D., & Pike, J. W. (2010). A downstream intergenic cluster of regulatory enhancers contributes to the induction of CYP24A1 expression by 1alpha,25-dihydroxyvitamin D3. *The Journal of Biological Chemistry*, *285*(20), 15599–15610. http://dx.doi.org/10.1074/jbc.M110.119958.

Narvaez, C. J., Matthews, D., LaPorta, E., Simmons, K. M., Beaudin, S., & Welsh, J. (2014). The impact of vitamin D in breast cancer: Genomics, pathways, metabolism. *Frontiers in Physiology*, *5*, 213. http://dx.doi.org/10.3389/fphys.2014.00213.

Novakovic, B., Sibson, M., Ng, H. K., Manuelpillai, U., Rakyan, V., Down, T., et al. (2009). Placenta-specific methylation of the vitamin D 24-hydroxylase gene: Implications for feedback autoregulation of active vitamin D levels at the fetomaternal interface. *The Journal of Biological Chemistry*, *284*(22), 14838–14848. http://dx.doi.org/10.1074/jbc.M809542200.

Ohyama, Y., Noshiro, M., Eggertsen, G., Gotoh, O., Kato, Y., Bjorkhem, I., et al. (1993). Structural characterization of the gene encoding rat 25-hydroxyvitamin D3 24-hydroxylase. *Biochemistry*, *32*(1), 76–82.

Perwad, F., & Portale, A. A. (2011). Vitamin D metabolism in the kidney: Regulation by phosphorus and fibroblast growth factor 23. *Molecular and Cellular Endocrinology*, *347*(1–2), 17–24. http://dx.doi.org/10.1016/j.mce.2011.08.030.

Perwad, F., Zhang, M. Y., Tenenhouse, H. S., & Portale, A. A. (2007). Fibroblast growth factor 23 impairs phosphorus and vitamin D metabolism in vivo and suppresses 25-hydroxyvitamin D-1alpha-hydroxylase expression in vitro. *American Journal of Physiology Renal Physiology*, *293*(5), F1577–F1583. http://dx.doi.org/10.1152/ajprenal.00463.2006.

Petkovich, M., & Jones, G. (2011). CYP24A1 and kidney disease. *Current Opinion in Nephrology and Hypertension*, *20*(4), 337–344. http://dx.doi.org/10.1097/MNH.0b013e3283477a7b.

Pettifor, J. M., & Thandrayen, K. (2012). Hypophosphatemic rickets: Unraveling the role of FGF23. *Calcified Tissue International, 91*(5), 297–306. http://dx.doi.org/10.1007/s00223-012-9651-0.

Pike, J. W., & Meyer, M. B. (2010). The vitamin D receptor: New paradigms for the regulation of gene expression by 1,25-dihydroxyvitamin D(3). *Endocrinology and Metabolism Clinics of North America, 39*(2), 255–269. http://dx.doi.org/10.1016/j.ecl.2010.02.007.

Pike, J. W., & Meyer, M. B. (2012). Regulation of mouse Cyp24a1 expression via promoter-proximal and downstream-distal enhancers highlights new concepts of 1,25-dihydroxyvitamin D(3) action. *Archives of Biochemistry and Biophysics, 523*(1), 2–8. http://dx.doi.org/10.1016/j.abb.2011.12.003.

Plum, L. A., & DeLuca, H. F. (2010). Vitamin D, disease and therapeutic opportunities. *Nature Reviews. Drug Discovery, 9*(12), 941–955. http://dx.doi.org/10.1038/nrd3318.

Prosser, D. E., Kaufmann, M., O'Leary, B., Byford, V., & Jones, G. (2007). Single A326G mutation converts human CYP24A1 from 25-OH-D3-24-hydroxylase into -23-hydroxylase, generating 1alpha,25-(OH)2D3-26,23-lactone. *Proceedings of the National Academy of Sciences of the United States of America, 104*(31), 12673–12678. http://dx.doi.org/10.1073/pnas.0702093104.

Raval-Pandya, M., Dhawan, P., Barletta, F., & Christakos, S. (2001). YY1 represses vitamin D receptor-mediated 25-hydroxyvitamin D(3)24-hydroxylase transcription: Relief of repression by CREB-binding protein. *Molecular Endocrinology, 15*(6), 1035–1046. http://dx.doi.org/10.1210/mend.15.6.0651.

Schlingmann, K. P., Kaufmann, M., Weber, S., Irwin, A., Goos, C., John, U., et al. (2011). Mutations in CYP24A1 and idiopathic infantile hypercalcemia. *The New England Journal of Medicine, 365*(5), 410–421. http://dx.doi.org/10.1056/NEJMoa1103864.

Seth-Vollenweider, T., Joshi, S., Dhawan, P., Sif, S., & Christakos, S. (2014). Novel mechanism of negative regulation of 1,25-dihydroxyvitamin D3-induced 25-hydroxyvitamin D3 24-hydroxylase (Cyp24a1) transcription: Epigenetic modification involving cross-talk between protein-arginine methyltransferase 5 and the SWI/SNF complex. *The Journal of Biological Chemistry, 289*(49), 33958–33970. http://dx.doi.org/10.1074/jbc.M114.583302.

Shimada, T., Kakitani, M., Yamazaki, Y., Hasegawa, H., Takeuchi, Y., Fujita, T., et al. (2004). Targeted ablation of Fgf23 demonstrates an essential physiological role of FGF23 in phosphate and vitamin D metabolism. *The Journal of Clinical Investigation, 113*(4), 561–568. http://dx.doi.org/10.1172/JCI19081.

Shinki, T., Jin, C. H., Nishimura, A., Nagai, Y., Ohyama, Y., Noshiro, M., et al. (1992). Parathyroid hormone inhibits 25-hydroxyvitamin D3-24-hydroxylase mRNA expression stimulated by 1 alpha,25-dihydroxyvitamin D3 in rat kidney but not in intestine. *The Journal of Biological Chemistry, 267*(19), 13757–13762.

Sitara, D., Razzaque, M. S., Hesse, M., Yoganathan, S., Taguchi, T., Erben, R. G., et al. (2004). Homozygous ablation of fibroblast growth factor-23 results in hyperphosphatemia and impaired skeletogenesis, and reverses hypophosphatemia in Phex-deficient mice. *Matrix Biology, 23*(7), 421–432. http://dx.doi.org/10.1016/j.matbio.2004.09.007.

St-Arnaud, R. (2010). CYP24A1-deficient mice as a tool to uncover a biological activity for vitamin D metabolites hydroxylated at position 24. *The Journal of Steroid Biochemistry and Molecular Biology, 121*(1-2), 254–256. http://dx.doi.org/10.1016/j.jsbmb.2010.02.002.

St-Arnaud, R. (2011). Cyp24A1: Structure, function and physiological role. In D. Feldman, J. W. Pike, & J. S. Adams (Eds.), *Vitamin D* (3rd ed., pp. 43–56): New York, NY: Academic Press.

St-Arnaud, R., Arabian, A., Travers, R., Barletta, F., Raval-Pandya, M., Chapin, K., et al. (2000). Deficient mineralization of intramembranous bone in vitamin D-24-hydroxylase-ablated mice is due to elevated 1,25-dihydroxyvitamin D and not to the absence of 24,25-dihydroxyvitamin D. *Endocrinology*, *141*(7), 2658–2666. http://dx.doi.org/10.1210/endo.141.7.7579.

St-Arnaud, R., & Naja, R. P. (2011). Vitamin D metabolism, cartilage and bone fracture repair. *Molecular and Cellular Endocrinology*, *347*(1-2), 48–54. http://dx.doi.org/10.1016/j.mce.2011.05.018.

Tebben, P. J., Milliner, D. S., Horst, R. L., Harris, P. C., Singh, R. J., Wu, Y., et al. (2012). Hypercalcemia, hypercalciuria, and elevated calcitriol concentrations with autosomal dominant transmission due to CYP24A1 mutations: Effects of ketoconazole therapy. *The Journal of Clinical Endocrinology and Metabolism*, *97*(3), E423–E427. http://dx.doi.org/10.1210/jc.2011-1935.

Tsai, K. S., Heath, H., 3rd., Kumar, R., & Riggs, B. L. (1984). Impaired vitamin D metabolism with aging in women. Possible role in pathogenesis of senile osteoporosis. *The Journal of Clinical Investigation*, *73*(6), 1668–1672. http://dx.doi.org/10.1172/JCI111373.

van Abel, M., Huybers, S., Hoenderop, J. G., van der Kemp, A. W., van Leeuwen, J. P., & Bindels, R. J. (2006). Age-dependent alterations in Ca2+ homeostasis: Role of TRPV5 and TRPV6. *American Journal of Physiology. Renal Physiology*, *291*(6), F1177–F1183. http://dx.doi.org/10.1152/ajprenal.00038.2006.

Zehnder, D., Evans, K. N., Kilby, M. D., Bulmer, J. N., Innes, B. A., Stewart, P. M., et al. (2002). The ontogeny of 25-hydroxyvitamin D(3) 1alpha-hydroxylase expression in human placenta and decidua. *The American Journal of Pathology*, *161*(1), 105–114.

Zierold, C., Mings, J. A., & DeLuca, H. F. (2001). Parathyroid hormone regulates 25-hydroxyvitamin D(3)-24-hydroxylase mRNA by altering its stability. *Proceedings of the National Academy of Sciences of the United States of America*, *98*(24), 13572–13576. http://dx.doi.org/10.1073/pnas.241516798.

Zierold, C., Nehring, J. A., & DeLuca, H. F. (2007). Nuclear receptor 4A2 and C/EBPbeta regulate the parathyroid hormone-mediated transcriptional regulation of the 25-hydroxyvitamin D3-1alpha-hydroxylase. *Archives of Biochemistry and Biophysics*, *460*(2), 233–239. http://dx.doi.org/10.1016/j.abb.2006.11.028.

CHAPTER SEVEN

Analogs of 1α,25-Dihydroxyvitamin D_3 in Clinical Use

Hector F. DeLuca[1], Lori A. Plum
Department of Biochemistry, University of Wisconsin-Madison, Madison, Wisconsin, USA
[1]Corresponding author: e-mail address: deluca@biochem.wisc.edu

Contents

Abstract

Biologically active metabolites of vitamin D that have been successfully developed for the clinical market are described. Their properties that resulted in their success in the clinic are also provided. Precursors of the metabolically active 1α,25-dihydroxyvitamin D have been prepared and successfully marketed not only for renal failure patients but also for a variety of patients having metabolic bone disorders. Finally, successful analogs of 1α,25-dihydroxyvitamin D in use in the clinic worldwide are presented including properties that have contributed to their success.

Vitamins and Hormones, Volume 100
ISSN 0083-6729
http://dx.doi.org/10.1016/bs.vh.2015.11.002

1. DISCOVERY OF THE VITAMIN D ENDOCRINE SYSTEM

With the discovery of the vitamin D-based endocrine system (1974) and the demonstration that 1α,25-dihydroxyvitamin D_3 (1,25-$(OH)_2D_3$) and not vitamin D_3, 25-hydroxyvitamin D_3 (25-OH-D_3) or other known metabolites are the active form of vitamin D in maintaining calcium homeostasis and bone, came a rush of synthetic analogs in an attempt to improve selectivity over the natural hormone, 1,25-$(OH)_2D_3$ (Binderup, Binderup, Godtfredsen, & Kissmeyer, 2005; Uskokovic et al., 2005). This review will address only those analogs that have been successfully introduced into the commercial market and will not attempt to review the numerous analogs that have been prepared. Many of these lack significant biological activity data to permit speculation on their potential. Reviews of some of the analogs are available in the series *Vitamin D* by Feldman, Pike, and Glorieux (2005).

2. VITAMIN D METABOLITES

2.1 25-Hydroxyvitamin D_3

The first vitamin D metabolite to achieve commercialization was 25-hydroxyvitamin D_3 (25-OH-D_3). It was discovered in 1968 (Blunt, DeLuca, & Schnoes, 1968) and chemically synthesized in 1969 (Blunt & DeLuca, 1969). It became clear that 25-OH-D_3 is further metabolized in the kidney to the final active form, 1α,25-dihydroxyvitamin D_3 (1,25-$(OH)_2D_3$) as discussed below. Clearly, 25-OH-D is the circulating form of vitamin D and is found in blood at a concentration of 30–40 ng/ml or 75–100 nM, and is the form measured to assess vitamin D status of a subject (Eisman, Shepard, & DeLuca, 1977; Hollis, 2005; Lips, 2005). Its measurement is widely used and represents a significant industry. In the United States, 25-OH-D was first marketed in oral capsules by Upjohn Company and later by Organon under the trade name of Calderol® for renal failure patients and vitamin D-resistant conditions. It was and still is marketed as Dedrogyl® in southern Europe. Marketing of Calderol® was discontinued with the development of improved therapies for the treatment of secondary hyperparathyroidism of renal failure. It is likely that 25-OH-D_3 will be reintroduced into the clinical market as an alternative to vitamin D_3 for correction of vitamin D deficiency (Fig. 1).

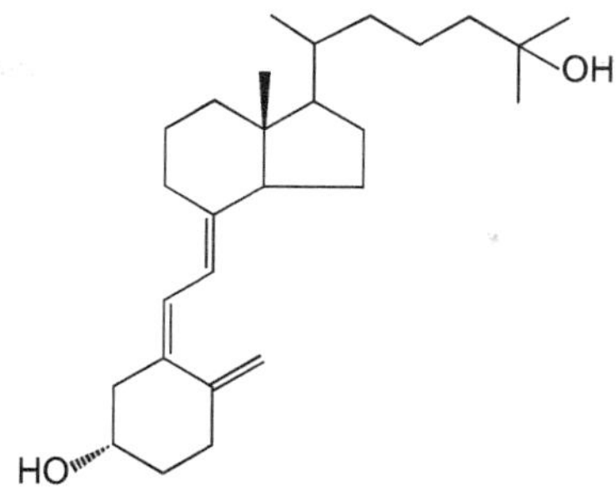

Figure 1 Structure of 25-hydroxyvitamin D_3 (25-OH-D_3; calcidiol; Calderol®; Dedrogyl®; Vectical®).

2.2 1α,25-Dihydroxyvitamin D_3

Holick et al. isolated and identified 1,25-$(OH)_2D_3$ from the intestines of vitamin D-deficient chicks given a physiological dose of radiolabeled vitamin D_3 and unequivocally determined its structure (Holick, Schnoes, & DeLuca, 1971; Holick, Schnoes, DeLuca, Suda, & Cousins, 1971; Holick et al., 1976). Chemical synthesis of the 1α-hydroxy and 1β-hydroxy forms by Semmler, Holick, Schnoes, and DeLuca (1972) proved that the active hormone is 1α,25-$(OH)_2D_3$ and not the 1β-hydroxy derivative. More facile syntheses allowed the preparation of commercial quantities (Baggiolini et al., 1986) and its introduction to the clinic as Rocaltrol® in the oral form (Gallagher, 1990). Oral 1,25-$(OH)_2D_3$ or calcitriol preferentially activates intestinal absorption of calcium (first-pass effect) and also induces intestinal CYP24A1 resulting in partial metabolic inactivation, reducing its overall biological impact (Frolik & DeLuca, 1971, 1972, 1973).

Oral calcitriol was originally approved for use in hemodialysis patients but because of its first-pass effect on intestine, hypercalcemia limited its use (Slatopolsky et al., 1984). This problem was avoided by an intravenous preparation given at dialysis that proved quite successful (Andress, Norris, Coburn, Slatopolsky, & Sherrard, 1989). However, there is a narrow window of therapy between suppression of parathyroid hormone (PTH) synthesis, secretion, and hypercalcemia (Finch, Brown, & Slatopolsky, 1999). Its use in this population has largely been replaced with vitamin D compounds having a wider therapeutic window. Unfortunately, currently the vitamin D compound used in dialysis patients is often on a cost basis and not on the basis of therapeutic safety and efficacy.

Oral calcitriol has been approved for the treatment of osteoporosis in some countries but not in the United States (Eastell & Riggs, 2005). The

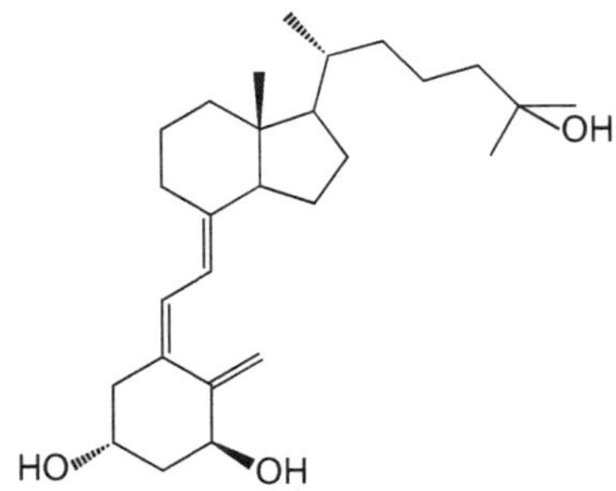

Figure 2 Structure of 1α,25-dihydroxyvitamin D_3 (1,25-$(OH)_2D_3$; calcitriol; Calcijex®; Rocaltrol®).

U.S. FDA requires very large and expensive fracture reduction trials that have not been performed. Tilyard, Spears, Thomson, and Dovey (1992) demonstrated a reduction in fractures in 400 osteoporosis patients by the daily administration of 0.5 μg. Further, application of oral calcitriol for vitamin D dependency rickets (Fraser et al., 1973) and other vitamin D-resistant conditions is well known.

Topical calcitriol named Silkis® and Vectical® is marketed for plaque psoriasis by Galderma (Fig. 2).

3. PRECURSORS TO 1,25-DIHYDROXYVITAMIN D_3

3.1 1α-Hydroxyvitamin D_3 (1α-OH-D_3, Alfacalcidol)

This analog was prepared in a synthetic exercise in developing chemical methods for introduction of the 1α-hydroxyl group in the vitamin D molecule (Holick, Semmler, Schnoes, & DeLuca, 1973). Improved methods of synthesis increased interest in this analog, especially for patients lacking functional kidneys. This compound was clearly an improvement over oral 1,25-$(OH)_2D_3$ because it is not destroyed by CYP24A1 during absorption and also lacks the first-pass effect of 1,25-$(OH)_2D_3$ (Frolik & DeLuca, 1971, 1972). In vitamin D-deficient animals, 1α-OH-D_3 is two to five times more active than vitamin D_3, and in nephrectomized rats, it is active while vitamin D is not (Holick, Kasten-Schraufrogel, Tavela, & DeLuca, 1975; Sjoden, Lindgren, & DeLuca, 1984). Its ease of synthesis coupled with the absence of metabolism during oral absorption and absence of first-pass effect on intestinal calcium absorption make it an excellent oral active vitamin D pharmaceutical. As an intravenous drug, it has little advantage over 1,25-$(OH)_2D_3$. The fact that it must be hydroxylated on C-25 before it is either

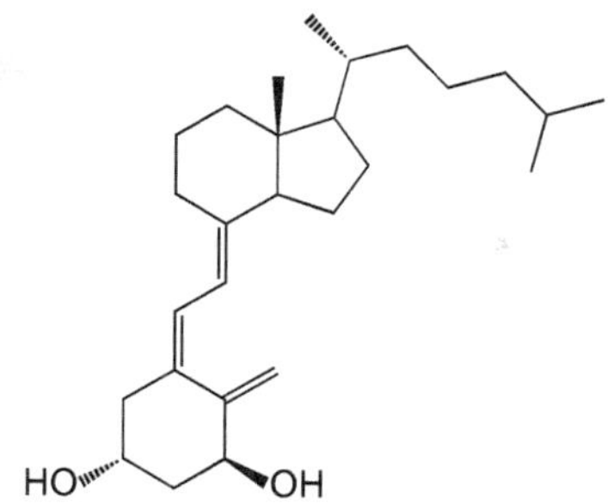

Figure 3 Structure of 1α-hydroxyvitamin D_3 (1α-OH-D_3, alfacalcidol; Alfarol®, One-Alfa®).

active or metabolized gives it a longer lifetime than 1,25-$(OH)_2D_3$ (Holick et al., 1975). It has not been developed and marketed in the United States. However, it is widely used elsewhere for both renal failure patients and osteoporosis (Hamdy et al., 1995; Orimo et al., 1994; Orimo, Shiraki, Hayashi, & Nakamura, 1987; Shiraki et al., 1999; Fig. 3).

3.2 1α-Hydroxyvitamin D_2 (1α-OH-D_2, Doxercalciferol)

This derivative was first prepared by Lam, Schnoes, and DeLuca (1977) in small amounts by a process that is commercially not feasible. Paaren, DeLuca, and Schnoes (1980) developed a direct 1α-hydroxylation method making it a commercially feasible pharmaceutical. Bone Care International first attempted to develop it as a possible treatment for osteoporosis (Gallagher, Bishop, Knutson, Mazess, & DeLuca, 1994; Gallagher, Bishop, Kyllo, DeLuca, & Mazess, 1993). Because of the prohibitive cost of the clinical trials required to enter this market and a failure of a small clinical trial, its development was shifted to the renal osteodystrophy field (secondary hyperparathyroidism). Thus, 1α-OH-D_2 successfully entered the market as one of the major therapies for this disorder in the United States (Frazao et al., 2000). Its activity is similar to 1α-OH-D_3. Its usefulness is largely based on its longer lifetime than 1,25-$(OH)_2D_3$, its lack of first-pass effect as an oral drug, and its survival during intestinal absorption. As an intravenous drug, it has no known advantage over 1α-OH-D_3 (Fig. 4).

4. ANALOGS OF 1,25-DIHYDROXYVITAMIN D_3

4.1 22-Oxa-1α,25-Dihydroxyvitamin D_3 (Oxacalcitriol)

Chugai Company of Japan first licensed 1α-OH-D_3 from the Wisconsin Alumni Research Foundation (WARF) and developed it for market both

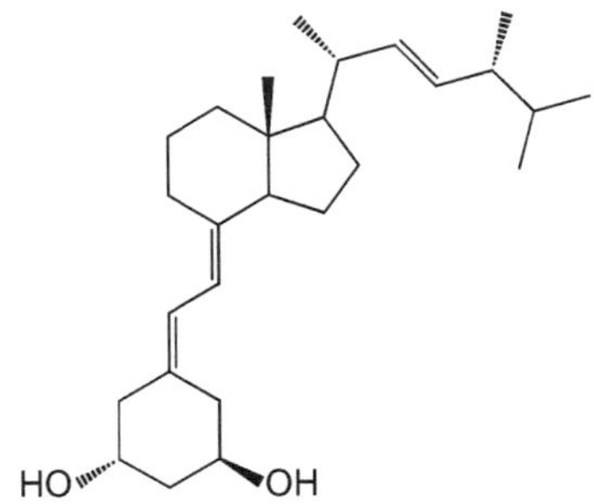

Figure 4 Structure of 1α-hydroxyvitamin D_2 (doxercalciferol; Hectorol®).

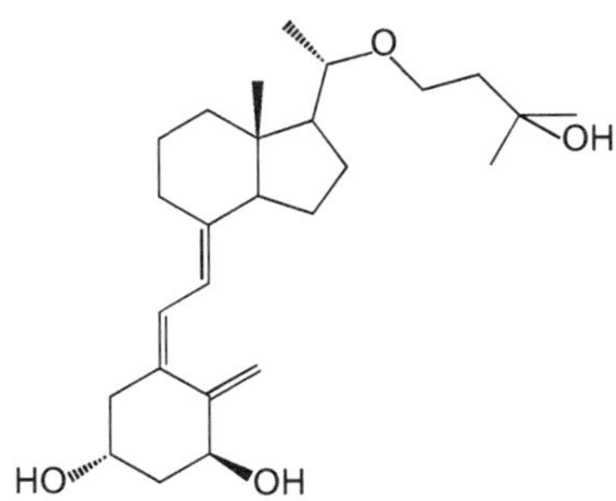

Figure 5 Structure of 22-oxa-1,25-$(OH)_2D_3$ (oxacalcitriol; Oxarol®).

for osteoporosis and for renal failure patients. The success of this endeavor gave birth to analog preparation within the Chugai Company. One of the successes was the development of 22-oxa-1,25-$(OH)_2D_3$ (Kubodera, Miyamoto, Ochi, & Matsunaga, 1986). On a weight basis, this compound has much less activity in raising serum calcium while retaining activity in suppressing serum PTH levels. It has little activity on mineralizing bone so it was applied to dialysis patients to suppress iPTH. It is clearly successful in this regard and is marketed in Japan for dialysis patients. It is rapidly metabolized having a lifetime measured in minutes (Kubodera, 2005) which likely accounts for its low activity in mineralizing bone. It is also quite applicable to psoriasis as a topical treatment for plaque psoriasis. If it enters the blood stream, it is rapidly eliminated, providing safety as well as efficacy for this disease (Fig. 5).

4.2 MC903 or Calcipotriol

An ambitious program of 1,25-$(OH)_2D_3$ analog synthesis was initiated by Leo Pharmaceuticals following the early success of 1α-OH-D_3 from the

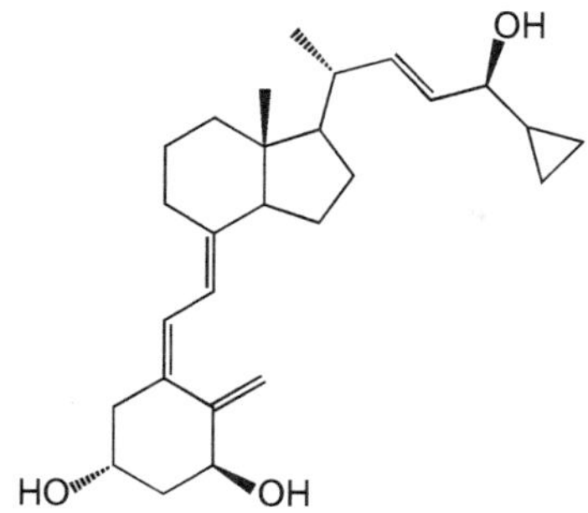

Figure 6 Structure of calcipotriol (calcipotriene; Dovonex®).

licensed DeLuca patents from WARF (Binderup et al., 2005). From their impressive list of analogs prepared by a sizeable group of chemists has emerged Calverley's MC903 (Dovonex® or calcipotriol; Calverley, 1987), a topical treatment of plaque psoriasis. This compound binds to the VDR comparably to 1,25-$(OH)_2D_3$, but when provided *in vivo*, it is 100–200 times less active in raising serum calcium (Kissmeyer & Binderup, 1991). This low calcemic activity results from rapid metabolism and excretion when provided orally or intravenously. Similar to the 22-oxa compound (mentioned above), its lifetime *in vivo* is about 10 min (Kissmeyer & Binderup, 1991). Because *in vitro* it is very effective in suppressing keratinocyte growth and differentiation, it was applied to the hyperproliferative keratinocyte disease, plaque psoriasis (Binderup et al., 2005, p. 1504, table VIII). It is successful in 80% of patients suffering this disorder (Kragballe, Fogh, & Sogaard, 1991). It has not been successfully applied to any other disease including many cancers. The mechanism whereby calcipotriol suppresses proliferation of overactive keratinocytes remains unknown (Fig. 6).

4.3 [2β-(3-Hydroxypropoxy)-1α,25-Dihydroxyvitamin D_3] (ED-71)

Similar to the European development of analogs of 1,25-$(OH)_2D_3$, Chugai Pharmaceutical Company launched its development of analogs after licensing of 1α-OH-D_3 patents from the Wisconsin group. Chugai impressively developed 1α-OH-D_3 (Alfarol®) for renal failure as well as osteoporosis. As described above, it successfully introduced 22-oxa-1,25-$(OH)_2D_3$ (OCT) for the topical treatment of psoriasis and secondary hyperparathyroidism in dialysis patients. Chugai extensively investigated analogs for the treatment of osteoporosis. The result of many years of development was the introduction of eldecalcitol, ED-71, for this disease (Tsurukami et al., 1994). ED-71 is at least equal and more often more effective than 1,25-$(OH)_2D_3$ in mineralizing rachitic bone or in elevating serum calcium. Its increased activity in this respect may be related to its much longer lifetime (70 h vs. 4 h in rats) than 1,25-$(OH)_2D_3$ (Kubodera, 2005). When tested in ovariectomized female rats, ED-71 is very effective in increasing bone mass. ED-71 has been successfully developed for the treatment of osteoporosis in Japan. Although it was first believed its success resulted from increased bone synthesis (Kubodera, Tsuji, Uchiyama, & Endo, 2003), it seems likely that the increased bone mass following treatment with ED-71 results from a suppression of bone resorption (Kubodera et al., 2003; Noguchi, Kawate, Nomura, & Takayanagi, 2013). To date, no vitamin D compound has been developed for osteoporosis that functions by increasing bone synthesis despite a clear demonstration that 1,25-$(OH)_2D_3$ increases bone synthesis in PTH and CYP27B1 knockout mice (Xue, Karaplis, Hendy, Goltzman, & Miao, 2006; Fig. 7).

4.4 26,27-Hexafluoro-1α,25-Dihydroxyvitamin D_3 (Falecalcitriol)

Sumitomo Dainippon Pharma and Taisho Pharmaceutical entered the Japanese renal market with another synthetic modification of the natural hormone. Fluorination of 25-OH-D_3 was originally done to provide evidence that 24-hydroxylation was primarily a route of catabolism (Ameenuddin, Sunde, DeLuca, Ikekawa, & Kobayashi, 1982). Placing fluoro groups in various locations of the natural hormone helped delineate the important structural groups needed for biological activity (Brommage & DeLuca, 1985;

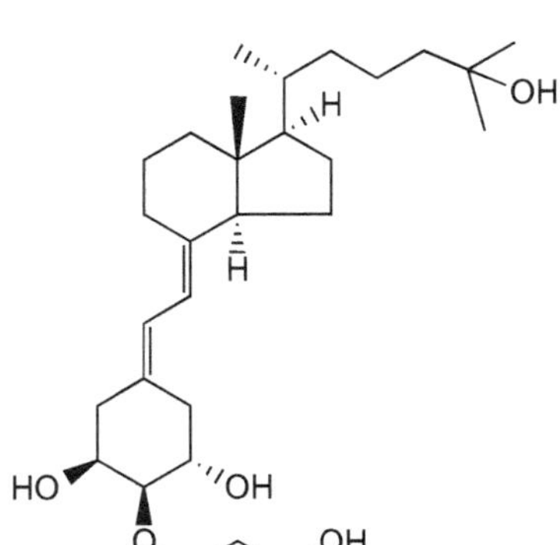

Figure 7 Structure of [2β-(3-hydroxypropoxy)-1α,25-dihydroxyvitamin D_3] (ED-71; Edirol®).

Nakada, Tanaka, DeLuca, Kobayashi, & Ikekawa, 1985; Napoli, Fivizzani, Schnoes, & DeLuca, 1978; Ohshima et al., 1984; Okamoto, Tanaka, DeLuca, Kobayashi, & Ikekawa, 1983; Tanaka, Wichmann, & DeLuca, 1983). Falecalcitriol is a hexafluoro derivative of $1,25(OH)_2D_3$ that is converted to a 23S-hydroxylated form which has strong biological activity (Komuro, Sato, & Kanamaru, 2003; Tanaka, DeLuca, Kobayashi, & Ikekawa, 1984). The 23S-hydroxylated form is resistant to further metabolism and therefore has an extended half-life and greater potency than $1,25\text{-}(OH)_2D_3$. Clinical trials conducted in patients with moderate-to-severe secondary hyperparathyroidism show the increased potency of falecalcitriol leads to increased efficacy over a less potent, marketed oral vitamin D therapy (Akiba et al., 1998; Fig. 8).

4.5 19-Nor-1α,25-Dihydroxyvitamin D_2 (Paricalcitol)

The most recently introduced analog of $1,25\text{-}(OH)_2D$ is 19-nor-1α,25-dihydroxyvitamin D_2 known commercially as Zemplar®. The elimination of a methylene group from carbon 10 decreased the hypercalcemic activity of $1,25\text{-}(OH)_2D$ while maintaining the ability to suppress circulating PTH, providing an improved therapeutic window over that of $1,25\text{-}(OH)_2D_3$ (Slatopolsky et al., 1995). Unknowingly, this compound is a poor substrate for the CYP24A1 enzyme that inactivates $1,25\text{-}(OH)_2D_3$. As a result, it has improved survival during intestinal absorption and has a longer half-life than $1,25\text{-}(OH)_2D_3$ of 16 h compared to 2–4 h in dialysis patients (Bailie & Johnson, 2002). It clearly has been associated with improved survival of dialysis patients (Teng et al., 2003). It certainly ranks as one of the most successful vitamin D analogs in the treatment of secondary hyperparathyroidism of stage 5 renal failure patients (Coyne et al., 2002; Llach et al., 1998; Martin et

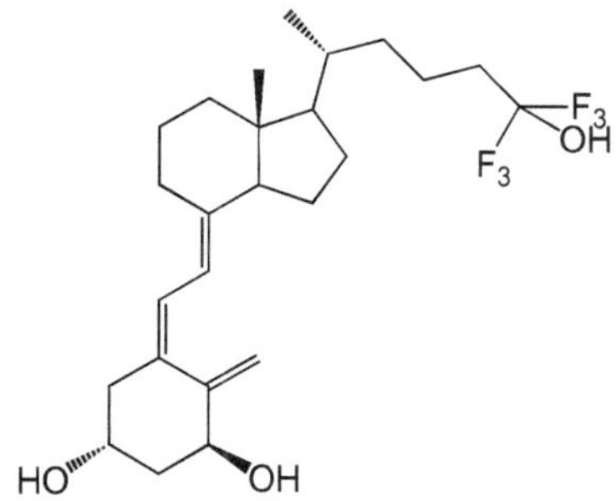

Figure 8 Structure of 26,27-hexafluoro-1α,25-dihydroxyvitamin D_3 (falecalcitriol; Hornel®).

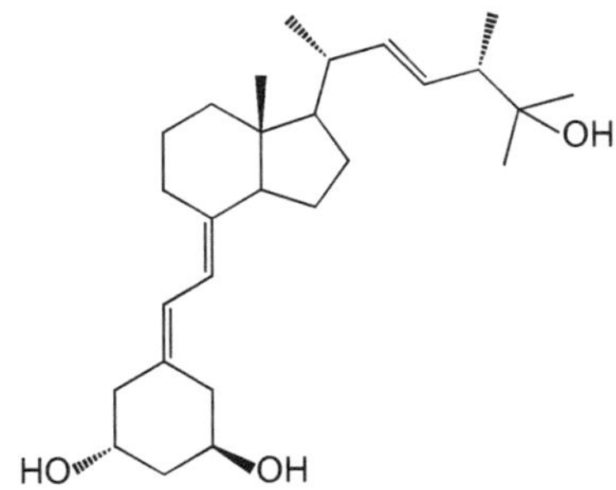

Figure 9 Structure of 19-nor-1α,25-dihydroxyvitamin D_2 (paricalcitol; Zemplar®).

al., 1998). Currently in the United States, it and 1α-hydroxyvitamin D_2 (Hectorol®) are the leaders for this application. Both oral and intravenous preparations are available with the intravenous preparation most widely used. The dosage range is from 1 to 9 μg/3 times a week (Llach et al., 1998; Martin et al., 1998; Fig. 9).

5. FUNCTIONS OF VITAMIN D BEYOND BONE, PARATHYROID, CALCIUM, AND PHOSPHORUS

It is likely that vitamin D has more subtle functions yet to be discovered. This is largely based on the presence of the VDR in tissues not involved with calcium phosphorus, bone, and PTH. Among these sites are keratinocytes of skin, various cells of the immune system, neural structures, reproductive organs, stomach, and pituitary (Plum & DeLuca, 2010; Stumpf, Sar, & DeLuca, 1981; Stumpf, Sar, Reid, Tanaka, & DeLuca, 1979; Wang, Zhu, & DeLuca, 2012, 2014). Although many claims have been made, a clear function at these sites has yet to be described.

Furthermore, new and more selective analogs will likely be developed. Thus, this chapter can only be considered a progress report from the discovery of the vitamin D-based endocrine system to date.

REFERENCES

Akiba, T., Marumo, F., Owada, A., Kurihara, S., Inboue, A., Chida, Y., et al. (1998). Controlled trial of falecalcitriol versus alfacalcidol in suppression of parathyroid hormone in hemodialysis patients with secondary hyperparathyroidism. *American Journal of Kidney Diseases*, *32*, 238–246.

Ameenuddin, S., Sunde, M., DeLuca, H. F., Ikekawa, N., & Kobayashi, Y. (1982). 24-Hydroxylation of 25-hydroxyvitamin D_3: Is it required for embryonic development in chicks? *Science*, *217*, 451–452.

Andress, D. L., Norris, K. C., Coburn, J. W., Slatopolsky, E. A., & Sherrard, D. J. (1989). Intravenous calcitriol in the treatment of refractory osteitis fibrosa of chronic renal failure. *New England Journal of Medicine*, *321*, 274–279.

Baggiolini, E. G., Iacobeli, J. A., Hennessy, B. M., Batchoo, A. D., Sereno, J. F., & Uskokovic, M. R. (1986). Stereocontrolled total synthesis of 1α,25-dihydroxycholecalciferol and 1α,25-dihydroxyergocalciferol. *Journal of Organic Chemistry*, *51*, 3098–3108.

Bailie, G. R., & Johnson, C. A. (2002). Comparative review of the pharmacokinetics of vitamin D analogues. *Seminars in Dialysis*, *15*, 352–357.

Binderup, L., Binderup, E., Godtfredsen, W. O., & Kissmeyer, A. M. (2005). Development of new vitamin D analogs. In D. Feldman, J. W. Pike, & F. H. Glorieux (Eds.), *Vitamin D: Vol. 2.* (2nd ed, pp. 1489–1510). Burlington, MA: Elsevier Academic Press. chapter 84.

Blunt, J. W., & DeLuca, H. F. (1969). The synthesis of 25-hydroxycholecalciferol. A biologically active metabolite of vitamin D_3. *Biochemistry*, *8*, 671–675.

Blunt, J. W., DeLuca, H. F., & Schnoes, H. K. (1968). 25-Hydroxycholecalciferol. A biologically active metabolite of vitamin D_3. *Biochemistry*, *7*, 3317–3322.

Brommage, R., & DeLuca, H. F. (1985). Evidence that 1,25-dihydroxyvitamin D_3 is the physiologically active metabolite of vitamin D_3. *Endocrine Reviews*, *6*, 491–511.

Calverley, J. J. (1987). Synthesis of MC 903 a biologically active vitamin D analogue. *Tetrahedron*, *43*, 4609–4619.

Coyne, D. W., Grieff, M., Ahya, S. N., Giles, K., Norwood, K., & Slatopolsky, E. (2002). Differential effects of acute administration of 19-nor-1,25-dihydroxyvitamin D_2 and 1,25-dihydroxyvitamin D_3 on serum calcium and phosphorus in hemodialysis patients. *American Journal of Kidney Diseases*, *40*, 1283–1288.

Eastell, R., & Riggs, B. L. (2005). Vitamin D and osteoporosis. In D. Feldman, J. W. Pike, & F. H. Glorioeux (Eds.), *Vitamin D: Vol. 2.* (2nd ed, pp. 1101–1120). San Diego: Elsevier Academic Press.

Eisman, J. A., Shepard, R. M., & DeLuca, H. F. (1977). Determination of 25-hydroxyvitamin D_2 and 25-hydroxyvitamin D_3 in human plasma using high-pressure liquid chromatography. *Analytical Biochemistry*, *80*, 298–305.

Feldman, D., Pike, J. W., & Glorieux, F. H. (2005). *Vitamin D* (Vols. 1&2) (1st and 2nd eds.). Burlington, MA: Elsevier Academic Press.

Finch, J. L., Brown, A. J., & Slatopolsky, E. (1999). Differential effects of 1,25-dihydroxyvitamin D_3 and 19-nor-1,25-dihydroxyvitamin D_2 on calcium and phosphorus resorption in bone. *Journal of the American Society of Nephrology*, *10*, 980–985.

Fraser, D., Kooh, S. W., Kind, H. P., Holick, M. F., Tanaka, Y., & DeLuca, H. F. (1973). Pathogenesis of hereditary vitamin D dependent rickets. An inborn error of vitamin D metabolism involving defective conversion of 25-hydroxyvitamin D to 1α,25-dihydroxyvitamin D. *New England Journal of Medicine*, *289*, 817–822.

Frazao, J. M., Elangovan, L., Maung, H. M., Chesney, R. W., Acchiardo, S. R., Bower, J. D., et al. (2000). Intermittent doxercalciferol (1alpha-hydroxyvitamin D_2) therapy for secondary hyperparathyroidism. *American Journal of Kidney Diseases*, *36*, 550–561.

Frolik, C. A., & DeLuca, H. F. (1971). 1,25-Dihydroxycholecalciferol: The metabolite of vitamin D responsible for increased intestinal calcium transport. *Archives of Biochemistry and Biophysics*, *147*, 143–147.

Frolik, C. A., & DeLuca, H. F. (1972). Metabolism of 1,25-dihydroxycholecalciferol in the rat. *Journal of Clinical Investigation*, *51*, 2900–2906.

Frolik, C. A., & DeLuca, H. F. (1973). The stimulation of 1,25-dihydroxycholecalciferol metabolism in vitamin D-deficient rats by 1,25-dihydroxycholecalciferol treatment. *Journal of Clinical Investigation*, *52*, 543–548.

Gallagher, J. C. (1990). Metabolic effects of synthetic calcitriol (Rocaltrol) in the treatment of postmenopausal osteoporosis. *Metabolism*, *39*(4 Suppl. 1), 27–29.

Gallagher, J. C., Bishop, C. W., Knutson, J. C., Mazess, R. B., & DeLuca, H. F. (1994). Effects of increasing doses of 1α-hydroxyvitamin D_2 on calcium homeostasis in postmenopausal osteopenic women. *Journal of Bone and Mineral Research*, *9*, 607–614.

Gallagher, J. C., Bishop, C. W., Kyllo, D. M., DeLuca, H. F., & Mazess, R. B. (1993). Long term evaluation of the safety of 1αOHD$_2$ in post-menopausal women. *In Journal of Bone and Mineral Research. 15th annual meeting of the American Society for Bone and Mineral Research, Tampa, FL, September 18–22 Abstract.*

Hamdy, N. A. T., Kannis, J. A., Beneton, M. N. C., Brown, C. B., Juttman, R., Jordans, J. G. M., et al. (1995). Effect of alfacalcidol on natural course of renal bone disease in mild to moderate renal failure. *British Medical Journal*, *310*, 358–363.

Holick, M. F., Kasten-Schraufrogel, P., Tavela, T., & DeLuca, H. F. (1975). Biological activity of 1α-hydroxyvitamin D_3 in the rat. *Archives of Biochemistry and Biophysics*, *166*, 63–66.

Holick, M. F., Schnoes, H. K., & DeLuca, H. F. (1971). Identification of 1,25-dihydroxycholecalciferol, a form of vitamin D_3 metabolically active in the intestine. *Proceedings of the National Academy of Sciences of the United States of America*, *68*, 803–804.

Holick, M. F., Schnoes, H. K., DeLuca, H. F., Suda, T., & Cousins, R. J. (1971). Isolation and identification of 1,25-dihydroxycholecalciferol. A metabolite of vitamin D active in intestine. *Biochemistry*, *10*, 2799–2804.

Holick, M. F., Semmler, E. J., Schnoes, H. K., & DeLuca, H. F. (1973). 1α-Hydroxy derivative of vitamin D_3. A highly potent analog of 1α,25-dihydroxyvitamin D_3. *Science*, *180*, 190–191.

Holick, M. F., Tavela, T. E., Holick, S. A., Schnoes, H. K., DeLuca, H. F., & Gallagher, B. M. (1976). Synthesis of 1α-hydroxy[6-^{3}H]vitamin D_3 and its metabolism to 1α,25-dihydroxy[6-^{3}H]vitamin D_3 in the rat. *The Journal of Biological Chemistry*, *251*, 1020–1028.

Hollis, B. W. (2005). Detection of vitamin D and its major metabolites. In D. Feldman, J. W. Pike, & F. H. Glorieux (Eds.), *Vitamin D: Vol. 2*. (2nd ed, pp. 931–950). Burlington, MA: Elsevier Academic Press. chapter 58.

Kissmeyer, A. M., & Binderup, L. (1991). Calcipotriol (MC 903): Pharmacokinetics in rats and biological activities of metabolites. A comparative study with 1,25$(OH)_2D_3$. *Biochemical Pharmacology*, *41*, 1601–1606.

Komuro, S., Sato, M., & Kanamaru, H. (2003). Disposition and metabolism of F6-1α,25 $(OH)_2$ vitamin D_3 and 1α,25$(OH)_2$ vitamin D_3 in the parathyroid glands of rats dosed with tritium-labeled compounds. *Drug Metabolism and Disposition*, *31*(8), 973–978.

Kragballe, K., Fogh, K., & Sogaard, H. (1991). Long-term efficacy and tolerability of topical calcipotriol in psoriasis. Results of an open study. *Acta Dermato-Venereologica*, *71*, 475–478.

Kubodera, N. (2005). Development of OCT and ED-71. In D. Feldman, J. W. Pike, & F. H. Glorieux (Eds.), *Vitamin D: Vol. 2*. (2nd ed, pp. 1525–1541). Burlington, MA: Elsevier Academic Press.

Kubodera, N., Miyamoto, K., Ochi, K., & Matsunaga, I. (1986). Synthetic studies of vitamin D analogs. VII. Synthesis of 20-oxa-21-norvitamin D_3 analogs. *Chemical & Pharmaceutical Bulletin*, *34*, 2286–2289.

Kubodera, N., Tsuji, N., Uchiyama, Y., & Endo, K. (2003). A new active vitamin D analog, ED-71, causes increase in bone mass with preferential effects on bone in osteoporotic patients. *Journal of Cellular Biochemistry*, *88*, 286–289.

Lam, H.-Y., Schnoes, H. K., & DeLuca, H. F. (1977). Synthesis of 1α-hydroxyergocalciferol. *Steroids*, *30*, 671–677.

Lips, P. (2005). How to define normal values for serum concentrations of 25-hydroxyvitamin D? An overview. In D. Feldman, J. W. Pike, & F. H. Glorieux (Eds.), *Vitamin D: Vol. 2*. (2nd ed, pp. 1020–1028). Burlington, MA: Elsevier Academic Press. chapter 62.

Llach, F., Keshav, G., Goldblat, M. V., Lindberg, J. S., Sadler, R., Delmez, J., et al. (1998). Suppression of parathyroid hormone secretion in hemodialysis patients by a novel vitamin D analogue: 19-nor-1,25-dihydroxyvitamin D_3. *American Journal of Kidney Diseases*, *32*, S48–S54.

Martin, K. J., Gonzalez, E. A., Gellens, M., Hamm, L. L., Abboud, H., & Lindbert, J. (1998). 19-Nor-1α,25-dihydroxyvitamin D_2 (paricalcitol) safely and effectively reduces the levels of intact parathyroid hormone in patients on hemodialysis. *Journal of the American Society of Nephrology*, *9*, 1427–1432.

Nakada, M., Tanaka, Y., DeLuca, H. F., Kobayashi, Y., & Ikekawa, N. (1985). Biological activities and binding properties of 23,23-difluoro-25-hydroxyvitamin D_3 and its 1α-hydroxy derivative. *Archives of Biochemistry and Biophysics*, *241*, 173–178.

Napoli, J. L., Fivizzani, M. A., Schnoes, H. K., & DeLuca, H. F. (1978). 1α-Hydroxy-25-fluorovitamin D_3: A potent analogue of 1α,25-dihydroxyvitamin D_3. *Biochemistry*, *17*, 2387–2392.

Noguchi, Y., Kawate, H., Nomura, M., & Takayanagi, R. (2013). Eldecalcitol for the treatment of osteoporosis. *Clinical Interventions in Aging*, *8*, 1313–1321.

Ohshima, E., Sai, H., Takatsuto, S., Ikekawa, N., Kobayashi, Y., Tanaka, Y., et al. (1984). Synthesis and biological activity of 1α-fluoro-25-hydroxyvitamin D_3. *Chemical & Pharmaceutical Bulletin*, *32*, 3525–3531.

Okamoto, S., Tanaka, Y., DeLuca, H. F., Kobayashi, Y., & Ikekawa, N. (1983). Biological activity of 24,24-difluoro-1,25-dihydroxyvitamin D_3. *American Journal of Physiology*, *244*, E159–E163.

Orimo, H., Shiraki, M., Hayashi, Y., Hoshino, T., Onaya, T., Miyazaki, S., et al. (1994). Effects of 1-hydroxyvitamin D_3 on lumbar bone mineral density and vertebral fractures in patients with postmenopausal osteoporosis. *Calcified Tissue International*, *54*, 370–376.

Orimo, H., Shiraki, M., Hayashi, T., & Nakamura, T. (1987). Reduced occurrence of vertebral crush fractures in senile osteoporosis treated with 1 alpha (OH)-vitamin D_3. *Bone and Mineral*, *47*, 52.

Paaren, H. E., DeLuca, H. F., & Schnoes, H. K. (1980). Direct (1) hydroxylation of vitamin D_3 and related compounds. *Journal of Organic Chemistry*, *45*, 3253–3258.

Plum, L. A., & DeLuca, H. F. (2010). Vitamin D, disease and therapeutic opportunities. *Nature Reviews. Drug Discovery*, *9*, 941–955.

Semmler, E. J., Holick, M. F., Schnoes, H. K., & DeLuca, H. F. (1972). The synthesis of 1α,25-dihydroxycholecalciferol—A metabolically active form of vitamin D_3. *Tetrahedron Letters*, *40*, 4147–4150.

Shiraki, M., Kushida, K., Fukunaga, M., Kishimoto, H., Taga, M., Nakamura, T., et al. (1999). A double-masked multicenter comparative study between alendronate and alfacalcidol in Japanese patients with osteoporosis. The Alendronate Phase III Osteoporosis Treatment Research Group. *Osteoporosis International*, *10*, 183–192.

Sjoden, G., Lindgren, J. U., & DeLuca, H. F. (1984). Antirachitic activity of 1α-hydroxyergocalciferol and 1α-hydroxycholecalciferol in rat. *Journal of Nutrition*, *114*, 2043–2046.

Slatopolsky, E., Finch, J., Ritter, C., Denda, M., Morrissey, J., Brown, A., et al. (1995). A new analog of calcitriol, 19-nor-1,25-$(OH)_2D_2$, suppresses parathyroid hormone secretion in uremic rats in the absence of hypercalcemia. *American Journal of Kidney Diseases*, *26*, 852–860.

Slatopolsky, E., Weerts, C., Thielan, J., Horst, R., Harter, H., & Martin, K. J. (1984). Marked suppression of secondary hyperparathyroidism by intravenous administration of 1,25-dihydroxycholecalciferol in uremic patients. *Journal of Clinical Investigation*, *74*, 2138–2143.

Stumpf, W. E., Sar, M., & DeLuca, H. F. (1981). Sites of action of 1,25$(OH)_2$ vitamin D_3 identified by thaw-mount autoradiography. In D. V. Cohn, R. V. Talmage, &

J. L. Matthews (Eds.), *Hormonal control of calcium metabolism, proceedings of the 7th international conference on calcium regulating hormones* (pp. 222–229). Amsterdam: Excerpta Medica.

Stumpf, W. E., Sar, M., Reid, F. A., Tanaka, Y., & DeLuca, H. F. (1979). Target cells for 1,25-dihydroxyvitamin D_3 in intestinal tract, stomach, kidney, skin, pituitary and parathyroid. *Science, 206*, 1188–1190.

Tanaka, Y., DeLuca, H. F., Kobayashi, Y., & Ikekawa, N. (1984). 26,26,26,27,27,27-Hexafluoro-1,25-dihydroxyvitamin D_3: A highly potent, long-lasting analog of 1,25-dihydroxyvitamin D_3. *Archives of Biochemistry and Biophysics, 229*, 348–354.

Tanaka, Y., Wichmann, J. K., & DeLuca, H. F. (1983). Metabolism and binding properties of 24,24-difluoro-25-hydroxyvitamin D_3. *Archives of Biochemistry and Biophysics, 225*, 649–655.

Teng, M., Wolf, M., Lowrie, E., Ofsthun, N., Lazarus, S. M., & Thadani, R. (2003). Survival of patients undergoing hemodialysis with paricalcitol or calcitriol therapy. *New England Journal of Medicine, 349*, 446–456.

Tilyard, M. W., Spears, G. F. S., Thomson, J., & Dovey, S. (1992). Treatment of postmenopausal osteoporosis with calcitriol or calcium. *The New England Journal of Medicine, 326*, 357–362.

Tsurukami, H., Nakamura, T., Suzuki, K., Sato, K., Higuchi, Y., & Nishii, Y. (1994). A novel synthetic vitamin D analogue, 2β-(3-hydroxypropoxy)1α,25-dihydroxyvitamin D_3 (ED-71), increases bone mass by stimulating the bone formation in normal and ovariectomized rats. *Calcified Tissue International, 54*, 142–149.

Uskokovic, M. R., Maehr, H., Reddy, S. G., Li, Y. C., Adorini, L., & Holick, M. F. (2005). Gemini: The 1,25-dihydroxy vitamin D analogs with two side-chains. In D. Feldman, J. W. Pike, & F. H. Glorieux (Eds.), *Vitamin D: Vol. 2.* (2nd ed, pp. 1511–1524). Burlington, MA: Elsevier Academic Press. chapter 85.

Wang, Y., Zhu, J., & DeLuca, H. F. (2012). Where is the vitamin D receptor? *Archives of Biochemistry and Biophysics, 523*, 123–133.

Wang, Y., Zhu, J., & DeLuca, H. F. (2014). Identification of the vitamin D receptor in osteoblasts and chondrocytes but not osteoclasts in mouse bone. *Journal of Bone and Mineral Research, 29*, 685–692.

Xue, Y., Karaplis, A. C., Hendy, G. N., Goltzman, D., & Miao, D. (2006). Exogenous 1,25-dihydroxyvitamin D_3 exerts a skeletal anabolic effect and improves mineral ion homeostasis in mice that are homozygous for both the 1α-hydroxylase and parathyroid hormone null alleles. *Endocrinology, 147*, 4801–4810.

CHAPTER EIGHT

1,25-Dihydroxyvitamin D and Klotho: A Tale of Two Renal Hormones Coming of Age

Mark R. Haussler*[,1], G. Kerr Whitfield*, Carol A. Haussler*, Marya S. Sabir[†], Zainab Khan[†], Ruby Sandoval[†], Peter W. Jurutka*[,†]

*Department of Basic Medical Sciences, University of Arizona College of Medicine, Phoenix, Arizona, USA
[†]School of Mathematical and Natural Sciences, Arizona State University, Glendale, Arizona, USA
[1]Corresponding author: e-mail address: haussler@email.arizona.edu

Contents

Abstract

1,25-Dihydroxyvitamin D_3 (1,25D) is the renal metabolite of vitamin D that signals through binding to the nuclear vitamin D receptor (VDR). The ligand–receptor complex transcriptionally regulates genes encoding factors stimulating calcium and phosphate absorption plus bone remodeling, maintaining a skeleton with reduced risk of age-related osteoporotic fractures. 1,25D/VDR signaling exerts feedback control of Ca/PO_4 via regulation of FGF23, klotho, and CYP24A1 to prevent age-related, ectopic calcification, fibrosis, and associated pathologies. Vitamin D also elicits xenobiotic detoxification, oxidative stress reduction, neuroprotective functions, antimicrobial defense,

Vitamins and Hormones, Volume 100
ISSN 0083-6729
http://dx.doi.org/10.1016/bs.vh.2015.11.005

immunoregulation, anti-inflammatory/anticancer actions, and cardiovascular benefits. Many of the healthspan advantages conferred by 1,25D are promulgated by its induction of klotho, a renal hormone that is an anti-aging enzyme/coreceptor that protects against skin atrophy, osteopenia, hyperphosphatemia, endothelial dysfunction, cognitive defects, neurodegenerative disorders, and impaired hearing. In addition to the high-affinity 1,25D hormone, low-affinity nutritional VDR ligands including curcumin, polyunsaturated fatty acids, and anthocyanidins initiate VDR signaling, whereas the longevity principles resveratrol and SIRT1 potentiate VDR signaling. 1,25D exerts actions against neural excitotoxicity and induces serotonin mood elevation to support cognitive function and prosocial behavior. Together, 1,25D and klotho maintain the molecular signaling systems that promote growth (p21), development (Wnt), antioxidation (Nrf2/FOXO), and homeostasis (FGF23) in tissues crucial for normal physiology, while simultaneously guarding against malignancy and degeneration. Therefore, liganded-VDR modulates the expression of a "fountain of youth" array of genes, with the klotho target emerging as a major player in the facilitation of health span by delaying the chronic diseases of aging.

1. INTRODUCTION

The kidneys are known to elaborate four hormones/enzymes vital to normal physiology and long-term survival in mammals. These four principles are: (A) 1,25-dihydroxyvitamin D_3 (1,25D) (synthesized in the proximal convoluted tubule), (B) erythropoietin (produced by peritubular capillary endothelial cells in the proximal convoluted tubule), (C) renin (secreted by granular cells of the juxtaglomerular apparatus), and (D) klotho (synthesized and secreted by the distal convoluted tubule). Each of these hormones has multifaceted actions. For example, 1,25D, via its nuclear vitamin D receptor (VDR), modulates the expression of approximately 1500 genes in numerous cell types, contributing to many of life's most fundamental processes. 1,25D maintains the molecular signaling systems that promote growth (p21), development (Wnt), antioxidation (Nrf2/FOXO), and homeostasis (FGF23) in crucial tissues, while at the same time guarding against malignancy and degeneration. A second hormone, erythropoietin, is essential for production of red blood cells in bone marrow, especially under hypoxic conditions, but also promotes hematopoietic cell survival by attenuating apoptosis. Further, in rats, pretreatment with erythropoietin protects neurons during induced cerebral hypoxia (Siren et al., 2001). Other studies suggest that erythropoietin improves memory, mood, neuronal plasticity, and memory-related neural networks (Adamcio et al., 2008). Thus, erythropoietin appears to affect brain function independent of oxygen delivery for metabolism. Similarly, 1,25D, traditionally considered as a renal hormone acting on intestine, bone, and kidney to recruit calcium for the musculoskeletal system, may also influence the CNS and behavior independent of calcium delivery (Patrick & Ames, 2015). A third kidney principle, the enzyme renin, participates in the renin–angiotensin–aldosterone system (RAAS) that is known to control extracellular volume and arterial vasoconstriction. However, the RAAS is also present in many extrarenal sites (Dzau, 1988), including brain, where it stimulates thirst, as well as promotes secretion of antidiuretic hormone from the neurohypophysis to conserve blood volume. Finally, and somewhat analogously to renin, α-klotho (hereafter referred to as klotho) is an enzymatically active protein with a circulating form that is expressed primarily in the kidneys and brain. Klotho is a multifunctional protein that regulates the metabolism of phosphate, calcium, and 25-hydroxyvitamin D. Klotho also may act as a peptide hormone, although an α-klotho receptor has not been identified to date. Point mutations of the *klotho* (*KL*) gene in humans are associated with hypertension and kidney disease, indicating that klotho may be essential to the maintenance of normal renal function (Xu & Sun, 2015). Finally, *KL* is an aging-suppressor gene, and may also be a tumor suppressor gene. The fact that *KL* expression is affected by erythropoietin, 1,25D, and the RAAS is noteworthy, suggesting that it may be the ultimate integrator of renal hormone action (Berridge, 2015).

The present chapter focuses on two of these renal hormones, 1,25D and klotho, that oppose each other in a closed endocrine loop. However, although 1,25D and klotho counteract one another with respect to phosphate and vitamin D metabolism, the two principles appear to cooperate locally to mediate the molecular signaling systems that promote growth, development, and antioxidation, as well as maintaining mechanisms that effect calcium transient stabilization and prevent phosphate-precipitated senescence that leads to CNS, cardiovascular, and renal decline. 1,25D exerts actions against neural excitotoxicity (L-type calcium channels) and induces serotonin as a mood elevator to support cognitive function and prosocial behavior. Herein, we develop, and provide supporting data for, a vitamin D/klotho healthspan premise, akin to the "vitamin D phenotypic stability hypothesis" recently proposed by Berridge (2015). His view of 1,25D/klotho signaling is calcium-centric, yet incorporates the antioxidation, anti-inflammatory, and antiproliferation functions of 1,25D and klotho, concepts we introduced in a previous treatise (Jurutka et al., 2013). Berridge (2015) extends his theory to include vitamin D in the prevention of neuropsychiatric disorders, a notion that we have also recently addressed (Kaneko, Sabir, et al., 2015), and update herein. Thus, we present in this chapter a related, renal endocrine-centric premise, arguing that the kidneys are the nexus of health span well beyond their obvious function to eliminate nitrogenous waste and balance electrolytes and water. We contend that renal hormones, including 1,25D and klotho, help prevent fibrosis, ectopic calcification, inflammation, and neoplastic transformation, support central nervous system health, as well as benefit the cardiovascular and cerebrovascular systems to lessen the risk of myocardial infarction and ischemic stroke.

2. 1,25-DIHYDROXYVITAMIN D

2.1 Synthesis and Degradation

The hormone precursor, vitamin D_3, can be obtained from the diet or synthesized from 7-dehydrocholesterol in skin in a UV light-dependent reaction (Fig. 1). Vitamin D_3 then circulates to the liver, where it is converted to 25-hydroxyvitamin D_3 (25D), the major circulating form that is assayed to quantitate clinical vitamin D status. The final step in the production of the hormonal form occurs mainly, but not exclusively, in the kidney, via a tightly regulated 1α-hydroxylation reaction catalyzed by mitochondrial CYP27B1 (Fig. 1). The major inducer of CYP27B1 in kidney is parathyroid hormone (PTH) that is secreted during hypocalcemia (Hughes, Brumbaugh, Haussler, Wergedal, & Baylink, 1975). When 1,25D levels then rise, PTH synthesis in the parathyroid glands is suppressed by a direct action of 1,25D-liganded VDR on gene transcription (DeMay, Kiernan, DeLuca, & Kronenberg, 1992). This negative feedback loop (not shown in Fig. 1) is vital to curtail the bone-resorbing effects of PTH in anticipation of 1,25D-mediated increases in both intestinal calcium absorption and bone resorption, thus preventing hypercalcemia. The major repressor of CYP27B1 in kidney is FGF23, the phosphaturic peptide hormone secreted during hyperphosphatemia (Bergwitz & Juppner, 2010). We (Kolek et al., 2005) and others (Quarles, 2008) proved that 1,25D induces FGF23 release from bone osteocytes in a process that is independently stimulated by high circulating phosphate levels. Thus, PTH is repressed by 1,25D and calcium, whereas FGF23 is induced by 1,25D and phosphate, protecting against hypercalcemia and hyperphosphatemia, respectively, either of which can elicit ectopic calcification. Finally, a second inducer of CYP27B1 is

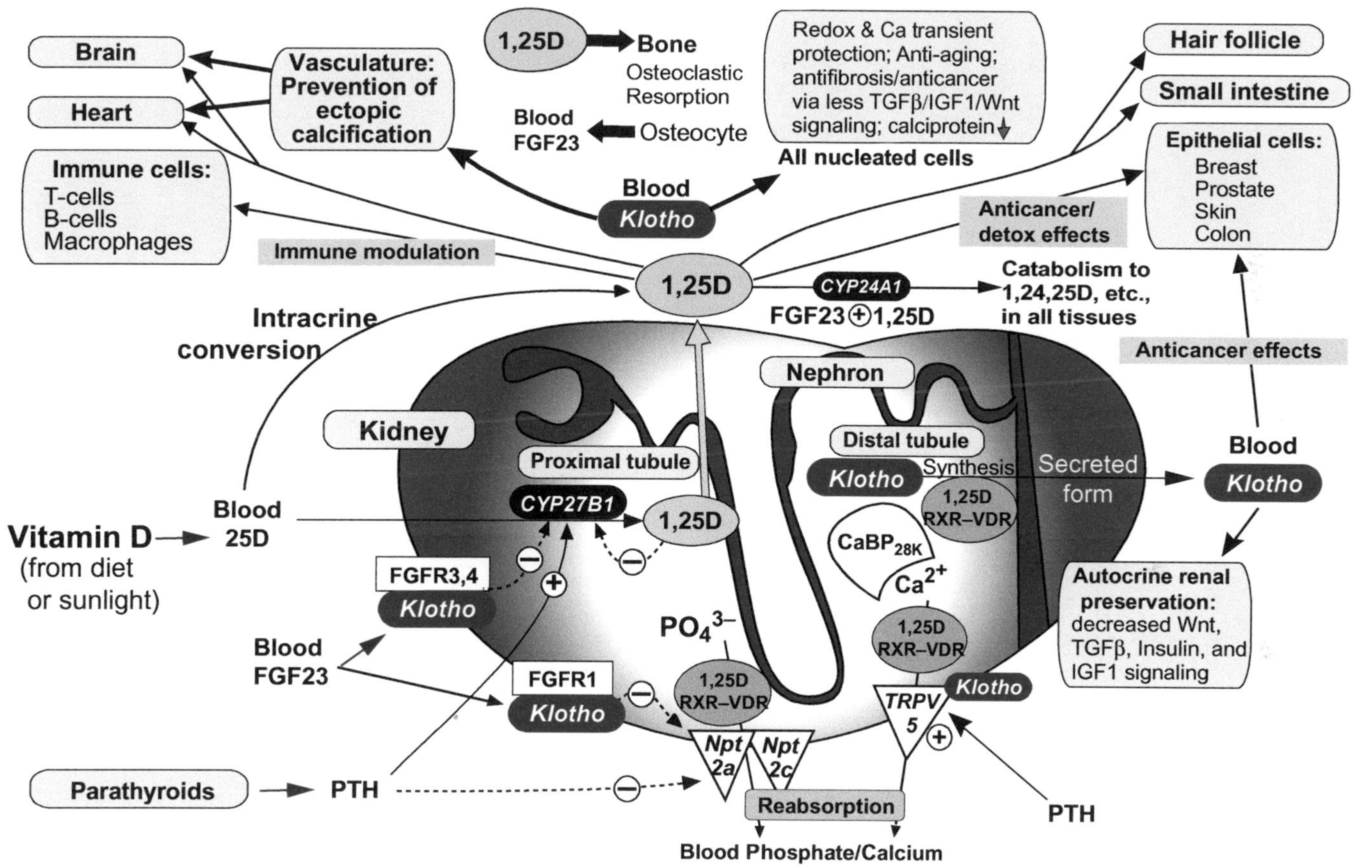

Figure 1 See figure legend on next page

phosphate depletion (Hughes et al., 1975), a phenomenon we now understand to be mediated by relief of FGF23-mediated suppression of CYP27B1, since FGF23 is no longer secreted under low phosphate conditions.

Also illustrated in Fig. 1 (center right) is the mechanism that initiates the process of 1,25D catabolism in all target cells, namely the action of CYP24A1 (St-Arnaud, 2010). The CYP24A1 gene is transcriptionally activated by 1,25D (Ohyama et al., 1994; Zierold, Darwish, & DeLuca, 1994), as well as by FGF23 (Shimada, Hasegawa, et al., 2004). Thus, the vitamin D endocrine system is elegantly choreographed by feedback controls that interpret bone mineral ion status to prevent bone mineral excess as well as hypervitaminosis D. The vitamin D intracrine system, in contrast, appears to be more dependent on the availability of ample 25D substrate to generate 1,25D locally in order to maintain healthy epithelial, immune, cardiovascular, and nervous systems.

2.2 Nuclear Receptor-Mediated Mechanism of Ligand Action

The hormonal 1,25D metabolite acts as a classic nuclear receptor ligand that binds VDR with high affinity to control the transcription of a multitude of genes (Haussler et al., 2013). Liganded VDR binds to one of the retinoid X receptors (RXRs) to form a heterodimer that recognizes vitamin D responsive elements (VDREs) in the vicinity of target genes and recruits comodulator complexes that modify chromatin to effect either induction or repression of transcription (Haussler, Jurutka, Mizwicki, & Norman, 2011; Pike, Lee, & Meyer, 2014; Pike & Meyer, 2014). Two major functional regions of VDR highlighted in Fig. 2 are the N-terminal zinc finger DNA-binding domain (DBD), and the C-terminal ligand-binding (LBD)/heterodimerization domain. When the X-ray crystallographic structure of the VDR LBD consisting of 12 α-helices (Rochel, Wurtz, Mitschler, Klaholz, & Moras, 2000) is compared to that of its closest relative in the

---

Figure 1 The kidney is the nexus of healthful aging. The kidney responds to 1,25D, FGF23, and PTH to regulate vitamin D bioactivation and calcium/phosphate reabsorption, and serves as an endocrine source of 1,25D and klotho. Thus, the kidney is the endocrine nexus of health by conserving calcium, eliminating phosphate, and producing 1,25D and klotho "fountain of youth" hormones. Renal hormones 1,25D (shaded in light blue) and klotho (shaded in dark blue) play crucial roles in bone mineral homeostasis to prevent osteomalacia and osteoporosis, but reach beyond these traditional roles to delay chronic disorders of aging such as ectopic calcification, fibrosis, vascular stiffening, heart and kidney function decline, epithelial cell cancers, autoimmune disease, hair loss, and neuropsychiatric conditions. (See the color plate.)

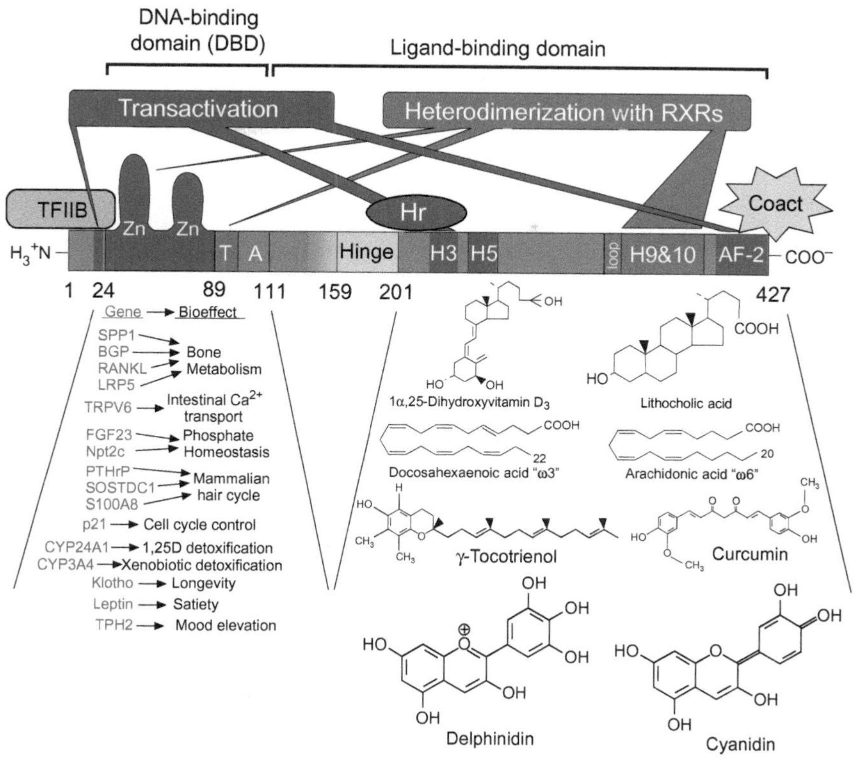

Figure 2 Functional domains in human VDR. Highlighted at the left is the human VDR zinc finger DNA-binding domain that, in cooperation with the corresponding domain in the RXR heteropartner, mediates direct association with the target genes listed at the lower left, leading to the indicated physiological effects. The official gene symbol for bone Gla protein (BGP) is *BGLAP*, for RANKL is *TNFSF11*, for Npt2c is *SLC34A3*, for PTHrP is *PTHLH*, and for klotho is *KL*. Below the ligand-binding domain (at the right) are illustrated selected VDR ligands, including several novel ligands discussed in the text. Also shown above the schematic are the proposed interactions with TFIIB, hairless (Hr), and coactivators (coact), the latter of which associate with the activation function-2 (AF-2) domain. (See the color plate.)

nuclear receptor superfamily, the pregnane X receptor (PXR) (Watkins et al., 2001), it is noted that the LBDs of both receptors have an expansive ligand binding pocket (approx. 700–1150 Å^3) that can potentially accommodate a variety of ligands. Known ligands for PXR are in fact very diverse and include endogenous steroids, the secondary bile acid lithocholic acid (LCA), the antibiotic rifampicin, as well as xenobiotics such as hyperforin, the active ingredient of St. John's Wort (Moore et al., 2002). Similarly, it is now emerging that VDR can accommodate several nutritional lipids as low affinity ligands (Fig. 2, lower right) including the dietarily essential polyunsaturated fatty acids, docosahexaenoic acid (DHA, an ω3 fatty acid), and arachidonic acid (an ω6 fatty acid), the vitamin E derivative γ-tocotrienol, curcumin (Bartik et al., 2010), which is a turmeric-derived polyphenol found in curry, and the anthocyanidins delphinidin and cyanidin found in pigmented fruits and vegetables (Austin et al., 2014; Hoss et al., 2013). VDR has thus retained its evolutionarily ancient, PXR-like ability to bind diverse ligands (Whitfield et al., 2003), yet VDR appears to have evolved as a "specialty" regulator of intestinal calcium absorption and hair growth in terrestrial animals, providing both a mineralized skeleton in a calcium-scarce environment, and physical protection against the harmful UV radiation of the sun.

The structure of 1,25D-occupied hVDR, heterodimerized with full-length RXRα, docked on a VDRE, and bound with a single coactivator, has been determined in solution via small angle X-ray scattering and fluorescence resonance energy transfer techniques (Rochel et al., 2011) and more recently by cryo-electron microscopy (Orlov, Rochel, Moras, & Klaholz, 2012). These, and other structural studies of the VDR–RXR heterodimer, e.g. (Zhang et al., 2011), allow us to visualize how the DBD and the ligand binding/heterodimerization domains are arranged relative to one another, and how their binding to ligand, DNA, and coactivators influence one another. Figure 3A illustrates in schematic fashion the spatial arrangement of a liganded VDR–RXR heterodimer bound to a generic VDRE element. The right side of Fig. 3A is adapted from Orlov et al. (2012). A key event in VDR-mediated gene activation is the binding of a ligand, which results in a dramatic conformational change in the position of helix 12 at the C-terminus of VDR, bringing it to the "closed" position to serve as part of a platform for binding the LxxLL domains of coactivators (Jurutka et al., 1997; Masuyama, Brownfield, St-Arnaud, & MacDonald, 1997). The binding of coactivator, in turn, likely stabilizes the VDR–RXR heterodimer on the VDRE, and may even assist in heterodimerization by conformationally inducing the VDR LBD to face the open side of the DNA helix at the 5′ end of the VDRE (see right portion of Fig. 3A). The model of Orlov et al. (2012) also portrays the C-terminal extension (CTE) of the VDR DBD as facing the open side of the DNA helix opposite the side occupied by the RXR DBD, suggesting that the CTE may participate in coactivator recruitment (Hsieh et al., 1999). Finally, it should be noted that the DNA sequence of a positive VDRE may itself allosterically influence the stability of helix 12 for coactivator binding (Zhang et al., 2011).

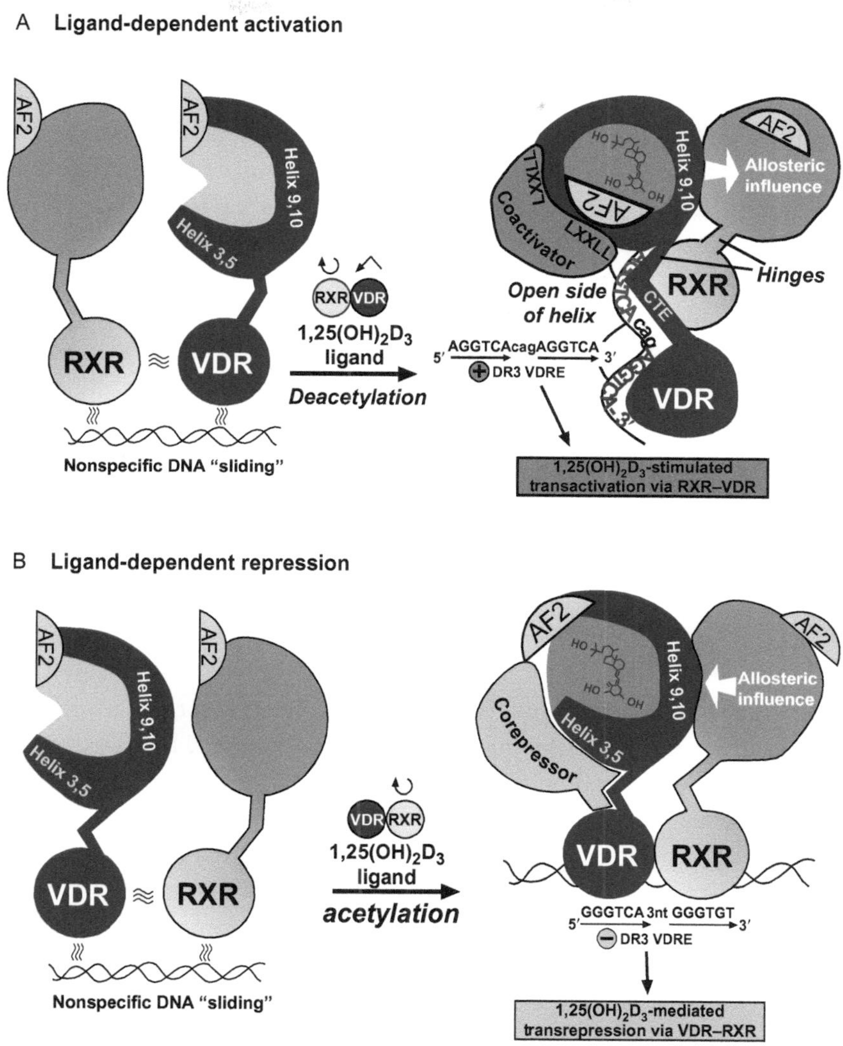

Figure 3 Proposed mechanisms of gene induction and repression by VDR. (A) Allosteric model of RXR–VDR activation after binding 1,25D and coactivator, deacetylation, and docking on a generic positive VDRE with the sequence acAGGTCAcagAGGTCActc (Orlov et al., 2012). See text for explanation. (B) Allosteric model for VDR–RXR inactivation after binding 1,25D and corepressor, acetylation, and docking in reverse polarity on a high-affinity negative VDRE (chicken PTH). See text for explanation. (See the color plate.)

The structural specifics of ligand-dependent gene repression by VDR–RXR are still not well characterized. However, repression may be mediated, in part, by VDR and RXR binding on the DR3 in the opposite orientation (i.e., VDR on the 5′ rather than the 3′ half element), as shown in

Fig. 3B. Another key feature of repression presumably involves the recruitment of nuclear receptor corepressor(s) to alter the architecture of chromatin in the vicinity of the target gene. This chromatin restructuring is likely catalyzed by histone deacetylases (HDACs) and/or demethylases attracted to the VDR- (or RXR-)tethered corepressor. We postulate that the information driving the opposite orientation of the heterodimer as well as the binding of repressor rather than coactivator is intrinsic to the negative VDRE DNA sequence (Whitfield et al., 2005). For both induction (Fig. 3A) and repression (Fig. 3B), we propose that the process is further regulated by deacetylation/acetylation of the VDR protein (Dampf-Stone et al., 2015), as has been shown for other nuclear receptors (Popov et al., 2007). The models depicted in Fig. 3 utilize 1,25D as the VDR ligand. However, alternative VDR ligands, such as those depicted in Fig. 2, may trigger the formation of similar complexes. Further, there exists the exciting possibility that individual alternative ligands may serve as selective VDR modulators to drive VDRE- or comodulator-specific target gene regulation.

2.3 Target Genes

Table 1 provides a compilation of VDREs that are known to be recognized by the combined DBD zinc fingers of the two receptors. The VDR–RXR controlled genes located near these VDREs (and listed with them) encode proteins that determine bone growth and remodeling, bone mineral homeostasis, detoxification, the mammalian hair cycle, cell proliferation/differentiation, apoptosis, lipid metabolism, immune and CNS function, and longevity. In general, VDREs possess either a direct repeat of two hexanucleotide half-elements with a spacer of three nucleotides (DR3) or an everted repeat of two half-elements with a spacer of six nucleotides (ER6), with DR3s being the most common. Single VDREs may occur in the proximal promoter of vitamin D-regulated genes; however, studies of 1,25D-controlled cytochrome P450 (CYP) genes introduced the concepts of multiplicity and remoteness to VDREs. Examples of VDR target genes possessing at least two VDREs are the human CYP24A1 and CYP3A4 genes, with the 5′ CYP3A4 DR3 being located 7.7 kb upstream of the proximal ER6 VDRE. Further examples have been revealed by ChIP or ChIP scanning (Barthel et al., 2007; Fretz et al., 2006; Kim, Yamazaki, Shevde, & Pike, 2007; Meyer et al., 2006) of genomic DNA surrounding the transient-receptor potential vanilloid type 6 (TRPV6), LRP5, and receptor activator of nuclear factor κB ligand (RANKL) genes, which have

Table 1 VDREs in Genes Directly Modulated in Their Expression by 1,25D and Possibly Other VDR Ligands

Gene	Bioeffect	Type	Location	5′-Half	Spacer	3′-Half	References	Group
rBGP	Bone metabolism	Positive	−456	GGGTGA	atg	AGGACA	Terpening et al. (1991)	Bone
mBGP	Bone metabolism	Negative	−444	GGGCAA	atg	AGGACA	Lian et al. (1997)	Bone
hBGP	Bone metabolism	Positive	−485	GGGTGA	acg	GGGGCA	Kerner, Scott, and Pike (1989)	Bone
mSPP1	Bone metabolism	Positive	−757	GGTTCA	cga	GGTTCA	Noda et al. (1990)	Bone
mSPP1	Bone metabolism	Positive	−2000	GGGTCA	tat	GGTTCA	Pike et al. (2007)	Bone
mLRP5	Bone anabolism	Positive	+656	GGGTCA	ctg	GGGTCA	Barthel et al. (2007)	Bone
mLRP5	Bone anabolism	Positive	+19 kb	GGGTCA	tgc	AGGTTC	Fretz, Zella, Kim, Shevde, and Pike (2006)	Bone
rRUNX2	Bone anabolism	Negative	−78	AGTACT	gtg	AGGTCA	Drissi et al. (2002)	Bone
mRANKL	Bone resorption	Positive	−22.7 kb	TGACCT	cctttg	GGGTCA	Haussler et al. (2008)	Bone
mRANKL	Bone resorption	Positive	−76 kb	GAGTCA	ccg	AGTTGT	Kim, Yamazaki, Zella, Shevde, and Pike (2006)	Bone
mRANKL	Bone resorption	Positive	−76 kb	GGTTGC	ctg	AGTTCA	Kim et al. (2006)	Bone
cIntegrin-beta3	Bone resorption, platelet aggregation	Positive	−756	GAGGCA	gaa	GGGAGA	Cao et al. (1993)	Bone
cCarbonic anhydrase II	Bone resorption, brain function	Positive	−39	AGGGCA	tgg	AGTTCG	Quelo, Machuca, and Jurdic (1998)	Bone

Continued

Table 1 VDREs in Genes Directly Modulated in Their Expression by 1,25D and Possibly Other VDR Ligands—cont'd

Gene	Bioeffect	Type	Location	5′-Half	Spacer	3′-Half	References	Group
cPTH	Mineral homeostasis	Negative	−60	GGGTCA	gga	GGGTGT	Liu, Koszewski, et al. (1996)	Mineral
mVDR	Autoregulation of VDR	Positive	+8467	GGGTTA	gag	AGGACA	Zella, Kim, Shevde, and Pike (2007)	Mineral
hTRPV6	Intestinal Ca^{2+} transport	Positive	−1270	AGGTCA	ttt	AGTTCA	Meyer, Watanuki, Kim, Shevde, and Pike (2006)	Mineral
hTRPV6	Intestinal Ca^{2+} transport	Positive	−2100	GGGTCA	gtg	GGTTCG	Meyer et al. (2006)	Mineral
hTRPV6	Intestinal Ca^{2+} transport	Positive	−2155	AGGTCT	tgg	GGTTCA	Meyer et al. (2006)	Mineral
hTRPV6	Intestinal Ca^{2+} transport	Positive	−4287	GGGGTA	gtg	AGGTCA	Meyer et al. (2006)	Mineral
hTRPV6	Intestinal Ca^{2+} transport	Positive	−4337	CAGTCA	ctg	GGTTCA	Meyer et al. (2006)	Mineral
hNPT2a	Renal phosphate reabsorption	Positive	−1963	GGGGCA	gca	AGGGCA	Taketani et al. (1998)	Mineral
hNpt2c	Renal phosphate reabsorption	Positive	−556	AGGTCA	gag	GGTTCA	Barthel et al. (2007)	Mineral
hFGF23	Renal phosphate elimination	Positive	−35.7 kb	GGGAGA	atg	AGGGCA	Haussler et al. (2011) and Saini et al. (2013)	Mineral

hFGF23	Renal phosphate elimination	Positive	−32.9 kb	TGAACT	caaggg	AGGGCA	Haussler et al. (2011) and Saini et al. (2013)	Mineral
hFGF23	Renal phosphate elimination	Positive	−16.2 kb	TAACCC	tgcttt	AGTTCA	Haussler et al. (2011) and Saini et al. (2013)	Mineral
hFGF23	Renal phosphate elimination	Positive	+8.6 kb	AGGGCA	gga	AGGACA	Haussler et al. (2011) and Saini et al. (2013)	Mineral
mFGF23	Renal phosphate elimination	Positive	−334	AGTGGG	gac	AGGTCA	Kaneko, Saini, et al. (2015)	Mineral
hklotho	Renal phosphate elimination	Positive	−31 kb	AGTTCA	aga	AGTTCA	Forster et al. (2011)	Mineral
hklotho	Renal phosphate elimination	Positive	−46 kb	GGTTCG	tag	AGTTCA	Forster et al. (2011)	Mineral
mklotho	Renal phosphate elimination	Positive	−35 kb	AGGTCA	gag	AGTTCA	Forster et al. (2011)	Mineral
rCYP24A1	1,25D detoxification	Positive	−151	AGGTGA	gtg	AGGGCG	Ohyama et al. (1994)	Detox
rCYP24A1	1,25D detoxification	Positive	−238	GGTTCA	gcg	GGTGCG	Zierold et al. (1994)	Detox
hCYP24A1	1,25D detoxification	Positive	−164	AGGTGA	gcg	AGGGCG	Zou, Elgort, and Allegretto (1997)	Detox
hCYP24A1	1,25D detoxification	Positive	−285	AGTTCA	ccg	GGTGTG	Zou et al. (1997)	Detox

Continued

Table 1 VDREs in Genes Directly Modulated in Their Expression by 1,25D and Possibly Other VDR Ligands—cont'd

Gene	Bioeffect	Type	Location	5′-Half	Spacer	3′-Half	References	Group
hCYP3A4	Xenobiotic detoxification	Positive	−169	TGAACT	caaagg	AGGTCA	Thummel et al. (2001) and Thompson et al. (2002)	Detox
hCYP3A4	Xenobiotic detoxification	Positive	−7.7 kb	GGGTCA	gca	AGTTCA	Makishima et al. (2002)	Detox
rCYP3A23	Xenobiotic detoxification	Positive	−120	AGTTCA	tga	AGTTCA	Barwick et al. (1996) and Thompson et al. (2002)	Detox
hMDR1	P-glycoprotein, drug resistance	Positive	−7863	AGTTCA	atg	AGGTAA	Saeki, Kurose, Tohkin, and Hasegawa (2008)	Detox
hMDR1	P-glycoprotein, drug resistance	Positive	−7853	AGGTCA	agtt	AGTTCA	Saeki et al. (2008)	Detox
hp21	Cell cycle control	Positive	−765	AGGGAG	att	GGTTCA	Liu, Lee, Cohen, Bommakanti, and Freedman (1996)	Cell life
hFOXO1	Cell cycle control	Positive	−2856	GGGTCA	cca	AGGTGA	Wang et al. (2005)	Cell life
hIGFBP-3	Cell proliferation/ apoptosis	Positive	−3282	GGTTCA	ccg	GGTGCA	Peng, Malloy, and Feldman (2004)	Cell life
hInvolucrin	Skin barrier function	Positive	−2083	GGCAGA	tct	GGCAGA	Bikle et al. (2002)	Cell life
hPLD1	Keratinocyte differentiation	Positive	−246	GGGTGA	tgc	GGTCGA	Kikuchi et al. (2007)	Cell life

hCCR10	Homing of T-cells to skin	Positive	−110	GGGTCT	acg	GGGTCA	Shirakawa et al. (2008)	Cell life
rPTHrP	Mammalian hair cycle	Negative	−805	AGGTTA	ctc	AGTGAA	Falzon (1996)	Cell life
hSOSTDC1	Mammalian hair cycle	Negative	−6215	AGGACA	gca	GGGACA	Hsieh et al. (2014)	Cell life
hHairless	Mammalian hair cycle	Positive	−7269	TGGTGA	gtg	AGGTCA	Hsieh et al. (2014)	Cell life
rVEGF	Angiogenesis	Positive	−2730	AGGTGA	ctc	AGGGCA	Cardus et al. (2009)	Cell life
hMIS	Müllerian-inhibiting substance	Positive	−381	GGGTGA	gca	GGGACA	Malloy, Peng, Wang, and Feldman (2009)	Cell life
hHLA-DRB1	Major histocompatibility complex	Positive	−1	GGGTGG	agg	GGTTCA	Ramagopalan et al. (2009)	Immune
hCAMP	Antimicrobial peptide	Positive	−615	GGTTCA	atg	GGTTCA	Gombart, Borregaard, and Koeffler (2005)	Immune
hKSR-1	Monocytic differentiation	Positive	−8156	GGTGCA	tat	AGGTCA	Wang, Wang, White, and Studzinski (2006)	Immune
hKSR-2	Monocytic differentiation	Positive	−2501	AGTTCA	gca	TGGTCA	Wang, Wang, White, and Studzinski (2007)	Immune

Continued

Table 1 VDREs in Genes Directly Modulated in Their Expression by 1,25D and Possibly Other VDR Ligands—cont'd

Gene	Bioeffect	Type	Location	5′-Half	Spacer	3′-Half	References	Group
hKSR-2	Monocytic differentiation	Positive	+3185	GGTTCA	aac	AGTTCT	Wang et al. (2007)	Immune
mInsig-2	Regulation of lipid synthesis	Positive	−2470	AGGGTA	acg	AGGGCA	Lee, Lee, Choi, and Lee (2005)	Metabolism
hPCFT	Intestinal folate transporter	Positive	−1680	AGGTTA	ttc	AGTTCA	Eloranta et al. (2009)	Metabolism
hCystathionine β synthase	Homocysteine clearance	Positive	+5923	GGGTTG	atg	AGTTCA	Kriebitzsch et al. (2011)	Metabolism
hTryptophan hydroxylase 2	Serotonin synthesis	Positive	−9971 −7059	TGGTCA AGGTCA	att att	AGTTCA TGGTCA	Patrick and Ames (2014)	Metabolism CNS

uncovered VDREs at considerable distances from the transcription start site (Table 1). Indeed, a genome-wide study of the VDR/RXR cistrome found that 98% of VDR/RXR binding sites in LS180 cells are located >500 bp upstream or downstream from the transcriptional start site of the nearest gene (Meyer, Goetsch, & Pike, 2012). Strikingly, it is thought that functional enhancers in the human genome often occur over hundreds of kb 5′ or 3′ of the target gene transcription start site (Mifsud et al., 2015). The most attractive model to explain such remoteness is that distant VDREs are juxtapositioned with more proximal VDREs via chromatin looping, creating a single platform that supports the transcriptional machinery (Saini et al., 2013).

2.4 Impact on Disease

2.4.1 Bone Mineral Metabolism (Osteoporosis)

Intestinal calcium absorption is mediated by 1,25D-VDR induction of TRPV6 (Barthel et al., 2007; Meyer et al., 2006). Indeed, TRPV6 null mice have 60% decreased intestinal calcium absorption, decreased bone mineral density and, strikingly, 20% of animals exhibit alopecia and dermatitis (Bianco et al., 2007) similar to VDR knockout mice (Li et al., 1997). Since the skin phenotype in VDR-null mice is not ameliorated by the high calcium rescue diet (Amling et al., 1999), we speculate that TRPV6 may mediate calcium entry into keratinocytes to elicit differentiation and hair cycling. Although 1,25D also enhances intestinal phosphate absorption via the induction of Npt2b (Katai et al., 1999), because phosphate is abundant in the diet and constitutively absorbed by the small intestine, the phosphate absorption effect of 1,25D may not be as physiologically important as the profound effect of 1,25D to trigger calcium transport. Further, because calcium is protective against colon cancer (Garland et al., 1985), while hair along with an intact skin barrier reduces UV-elicited cancer, VDR-induced TRPV6 could also function in colon and skin to lower the risk of neoplasia in these two epithelia (Fig. 2).

In the osteoblast, RANKL constitutes one of the most dramatically 1,25D-upregulated bone genes, the product of which effects bone resorption through osteoclastogenesis (Fig. 2). We have shown that RANKL is induced over 5000-fold by 1,25D in mouse ST-2 stromal cells in culture (Haussler et al., 2010). OPG, the soluble decoy receptor for RANKL that tempers its activity, is simultaneously repressed by 86% (Haussler et al., 2010) to amplify the bioeffect of RANKL. Thus, like PTH, 1,25D is a potent bone-resorbing, hypercalcemic hormone and, although chronic excess of either hormone elicits a severe osteopenic pathology, physiologic bone remodeling can be argued to strengthen the skeleton and make it less susceptible to fractures and the eventual ravages of senile osteoporosis.

2.4.2 Detoxification and Antioxidation (Healthspan/Senescence)

A common theme for both VDR and PXR is the induction of CYP enzymes that participate in xenobiotic detoxification. A major target for both receptors in humans is CYP3A4 (Makishima et al., 2002; Thompson et al., 2002; Thummel et al., 2001) for which the detoxification substrates include LCA (Araya & Wikvall, 1999), a toxic and potentially carcinogenic secondary bile acid produced by gut bacteria. Thus, by acting as VDR ligands, several natural agonists for VDR, including the high-affinity 1,25D hormonal metabolite, and the lower affinity, nutritionally modulated bile acids (or potentially any other VDR ligand) can promote detoxification of LCA and possibly other intestinal endo- or xenobiotics, with the end result likely being a reduction in colon cancer incidence. Additionally, both VDR and PXR induce SULT2A, an enzyme that detoxifies sterols via 3α-sulfation (Echchgadda, Song, Roy, & Chatterjee, 2004). Moreover, in the realm of cardiovascular disease (Guilliams, 2004) and neurodegenerative disorders of aging such as Alzheimer's disease (AD) (Seshadri et al., 2002), liganded VDR can induce cystathionine β-synthase (Kriebitzsch et al., 2011) to catalyze the metabolic elimination of homocysteine, an excess of which is a risk factor because of the vascular and CNS toxicity of this methionine metabolite. Finally, as discussed above, upregulation of CYP24A1 by 1,25D and by FGF23 is a key feedback link whereby circulating and local 1,25D levels are maintained optimally to prevent ectopic calcification and other pro-aging actions that could result from an excess of the potent 1,25D hormone. We contend that this CYP24A1 induction (Shimada, Hasegawa, et al., 2004) maintains intracrine as well as endocrine 1,25D homeostasis. Thus, *CYP24A1* constitutes, beyond *Klotho*, a second anti-aging gene. Indeed, mice with ablation of the *CYP24A1* gene have a reduced lifespan because of 1,25D toxicity (Masuda et al., 2005).

Liganded-VDR has been shown to oppose oxidative damage, the leading candidate for the cause of aging (Lin & Beal, 2003). In one example, 1,25D-bound VDR induces *FOXO3* (Eelen et al., 2013), a major human longevity gene that is involved in cellular metabolism and the response to oxidative stress (Morris, Willcox, Donlon, & Willcox, 2015). In an experiment from our laboratory, HEK-293 cells were transfected with the natural superoxide dismutase (SOD) promoter containing multiple FOXO responsive elements. The results (Fig. 4) illustrate that 1,25D stimulates, and β-catenin inhibits, FOXO3 activity. As a second example of VDR actions to oppose oxidative damage, 1,25D has been reported to induce the expression of nuclear factor-erythroid-2-related factor 2 (Nrf2) (Nakai et al., 2014; Wang et al., 2005). Via a complex mechanism involving the cytoplasmic protein Keap1, Nrf2 accumulates in the nucleus during conditions of oxidative stress, where it heterodimerizes with the Maf transcription factor and binds to the antioxidant response element enhancer present in numerous stress response genes, inducing transcription of their cognate mRNAs. These antioxidative genes include: glutamate cysteine ligase, which catalyzes the rate-limiting step in the synthesis of glutathione (GSH); sulfiredoxin1 and thioredoxin reductase 1, which support the action of peroxiredoxins to detoxify highly reactive peroxides; glutathione-*S*-transferase, which catalyzes the conjugation of GSH with endogenous and xenobiotic electrophiles; UDP-glucuronosyltransferase, which catalyzes the conjugation of glucuronic acid to a variety of endogenous and exogenous substrates, rendering them more water soluble; and multidrug resistance-associated proteins, which are membrane transporters that efflux cytotoxic compounds. Thus, the Nrf2–Keap1 system is well recognized as a major cellular

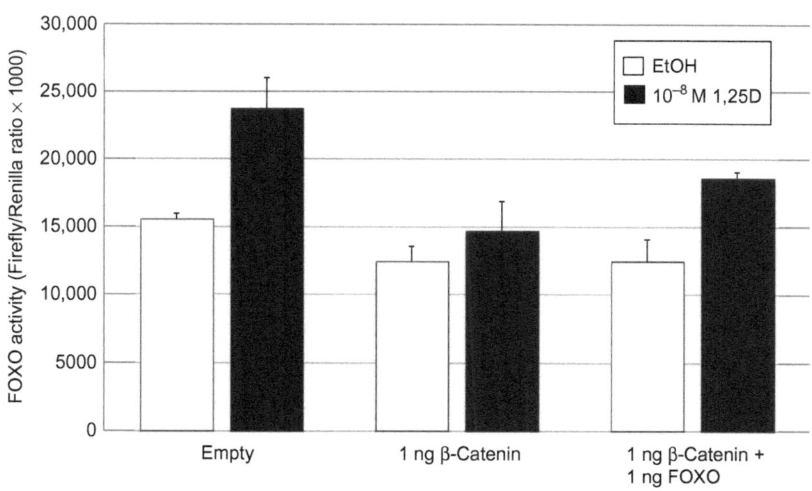

Figure 4 1,25D stimulates, and β-catenin inhibits, FOXO3 activity in HEK-293 human kidney cells transfected with a Firefly luciferase reporter construct containing the natural promoter from the superoxide dismutase (SOD-2) gene that contains multiple FOXO responsive elements, along with the indicated expression plasmids for β-catenin and FOXO3. Firefly luciferase values were normalized to expression of Renilla luciferase and are expressed as the ratio of Firefly/Renilla values × 1000 ± STDEV.

defense mechanism against oxidative and xenobiotic stresses (Suzuki & Yamamoto, 2015).

The effectiveness of 1,25D in potentially reversing aging-associated pathologies caused by oxidative, xenobiotic, and glucose stress is illustrated by the recent studies of Manna and Jain (2015a) who reported that treatment of 3T3L1 adipocytes with 1,25D and high glucose decreased reactive oxygen species (ROS) production (31%), NOX4 protein expression (71%), and NF-κB phosphorylation, and increased protein expression of Nrf2 (78%) and thioredoxin (30%) in high glucose-treated cells. They also showed that 1,25D upregulated SIRT1 protein expression and AMPK phosphorylation, and stimulated the IRS1/PI3K/PIP3/PKCζ/λ signaling cascade, GLUT4 translocation (44%), and glucose uptake (34%) in high glucose-treated cells (Manna & Jain, 2012; Manna & Jain, 2015b). Data from our laboratory (Fig. 10A) extend this conclusion by revealing that SIRT1 and its resveratrol activator cooperate with 1,25D to enhance VDR signaling as monitored by induction of *CYP24A1*, an action that is presumably reflected in the augmented expression of other genes such as klotho. These results suggest that 1,25D prevents oxidative stress through multiple mechanisms, including modulation of NOX4/Nrf2/thioredoxin redox signaling and the induction of the SIRT1-mediated AMPK/IRS1/GLUT4 antiglucose toxicity pathway, and, relevant to this chapter, the generation of endocrine klotho (see Section 3 and Fig. 10A and C). Utilizing these mechanisms, together with upregulating CYPs, FOXO3, and the Nrf2 antioxidative/detoxification cascade and klotho, the vitamin D hormone appears to deter aging-associated pathologies caused by oxidative, xenobiotic, and glucose stress and delay senescence while preventing renal and liver toxicity, respiratory distress, Type 2 diabetes mellitus, cardiovascular disease, inflammation, and carcinogenesis.

Not surprisingly, deficiency in vitamin D has been linked to a set of human diseases resembling those discussed above. As noted by Berridge (2015), changes in gene transcription and organelle function occur during aging, including a decline in the maintenance of mitochondrial energy metabolism and antioxidant defenses. Further, there is an age-dependent decline in vitamin D and klotho levels as aging progresses that, along with a diminution of Nrf2 function, may impair the stability of ROS and Ca^{2+} signaling. Vitamin D deficiency also may contribute to the pathophysiology of age-related disease through dysregulation of mTORC1, the mechanistic target of rapamycin (Berridge, 2015). mTORC1 is activated by a number of signaling pathways such as growth factors, nutrients, and the energy status of the cell. Activation of mTORC1, in turn, generates multiple output signals that regulate protein and lipid synthesis, autophagy, inflammation, and glycolysis. Excessive activation of mTORC1 has been associated with aging and with many diseases such as AD, cancer, obesity, and type II diabetes (Laplante & Sabatini, 2012). It is worth noting that both vitamin D and klotho downregulate the activity of the PI3K/Akt/mTOR pathway (Halicka et al., 2012). The inhibitory activity of 1,25D is mediated by enhanced expression of PTEN, a tumor suppressor, and induction of DNA-damage-inducible transcript 4 (DDIT4), which suppresses the activity of mTORC1 (Lisse et al., 2011). The ability of both 1,25D and klotho to harness mTORC1 activity could therefore be part of the explanation for why vitamin D deficiency is associated with the aging and disease states discussed above. Finally, when evaluated in aging rats, vitamin D potentiated hippocampal synaptic function and appeared to prevent a decline in cognition (Latimer et al., 2014), lending support to the notion that a deficiency in vitamin D could presage the decline in human cognition that transpires with aging (Toffanello et al., 2014).

2.4.3 Neuropsychiatric Disorders (ADHD, Autism, Bipolar Disorder, Depression, Schizophrenia; Antisocial, Obsessive-Compulsive, and Suicidal Behaviors)

VDR is expressed broadly in brain, including in both neurons and glial cells (Harms, Burne, Eyles, & McGrath, 2011). The most intense immunochemical signal is present in the hypothalamus (paraventricular and supraoptic nuclei as well as the lateral/ventromedial regions), and in the dopaminergic (DA) neurons of the substantia nigra (Eyles, Smith, Kinobe, Hewison, & McGrath, 2005), but VDR is also detected in prefrontal cortex, cingulate gyrus, and CA2 region of the hippocampus. Despite the demonstrated presence of VDR, we are only recently gaining information on the function(s) of VDR and its 1,25D ligand in these brain regions. Data are emerging that 1,25D influences neuronal cell differentiation and exerts neuroprotective actions against cytotoxicity (Harms et al., 2011). Specifically, 1,25D treatment is reported to increase NGF production and neurite outgrowth in cultured hippocampal neurons (Brown, Bianco, McGrath, & Eyles, 2003), with concomitant protection against excitotoxicity in association with downregulation of L-type calcium channel expression (Brewer et al., 2001). In an intriguing study, daily injections of vitamin D_3 significantly reduced the cortical infarction volume in rats with middle cerebral artery ligation, in concert with upregulation of glial cell derived neurotrophic factor (GDNF) (Wang et al., 2000). Further experiments in which cultured rat neocortical neurons were briefly exposed to harmful doses of glutamate (Glu) showed that chronic vitamin D_3 treatment protected against neurotoxicity, a phenomenon occurring in parallel with upregulation of *VDR* mRNA expression (Taniura et al., 2006). Finally, recent work indicates that 1,25D modulates γ-aminobutyric acid (GABA) and GABAergic function. Specifically, Jiang et al. (2014) observed a significant rise of Glu and glutamine (Gln) concentrations in the prefrontal cortex of rats after chronic 1,25D administration. Consistent with this finding, Groves et al. (2013) reported decreased levels of Glu and Gln in vitamin D-deficient BALB/c mice. Dysfunction of the Glu system in the prefrontal cortex is associated with schizophrenia, and the possibility that 1,25D modulates Glu neurotransmission is suggestive of possible neuropsychiatric benefits of vitamin D centered on supporting proper GABA neurotransmission (Cieslak et al., 2014; Yuksel et al., 2014). The schizophrenic changes in brain rhythms affecting processes such as perception, memory, and consciousness are caused by defects, some of which are developmental, in fast-spiking parvalbumin-expressing GABAergic inhibitory interneurons (Steullet et al., 2014). One of the etiologies of these defects appears to be inflammation-elicited oxidative stress, possibly resulting from viral infections during pregnancy. Increased oxidation attenuates the activity of NMDA receptors (NMDARs), and NMDAR hypofunction, in turn, is responsible for the onset of schizophrenia (Behrens & Sejnowski, 2009). As outlined by Berridge (2015), consistent with the notion that oxidation is involved in the etiology of schizophrenia, depressed levels of the antioxidant GSH are observed in the prefrontal cortex from patients with schizophrenia and other psychiatric diseases such as major depressive and bipolar disorders (Gawryluk, Wang, Andreazza, Shao, & Young, 2011). Further, there are epidemiological studies that have directly linked low vitamin D levels to schizophrenia (Harms et al., 2011) and this association has been supported by animal models (Kesby et al., 2010). Low maternal vitamin D levels in rats impair brain development, resulting in large scale structural changes that resemble those seen in schizophrenia (Eyles, Burne, & McGrath, 2013). The possibility of a mechanistic connection between low vitamin D and low GSH is suggested by the ability of 1,25D-liganded VDR, as discussed above, to induce expression of Nrf2, which in turn upregulates antioxidation factors, including thioredoxin and GSH. A potential role of 1,25D/VDR as a guardian of ROS signaling pathways is not only consistent with what is known about the signaling defects responsible for schizophrenia (Berridge, 2013, 2014) but also raises the possibility that maintaining optimal vitamin D concentrations in pregnancy and in neonates might prevent some cases of schizophrenia (Berridge, 2015).

A second neurotransmitter that we hypothesize is controlled in part by 1,25D is serotonin. Dysregulation of serotonergic neurotransmission is implicated in a number of psychiatric disorders, including ADHD, autism, depression, as well as antisocial, obsessive-compulsive, and suicidal behaviors (Lesch, 2005; Wrase, Reimold, Puls, Kienast, & Heinz, 2006). Adequate production of serotonin is predominantly controlled by tryptophan hydroxylase (TPH), which catalyzes the rate-limiting step in serotonin biosynthesis (Lovenberg, Jequier, & Sjoerdsma, 1967). Interestingly, two genes with distinct patterns of expression encode different TPH isoforms, namely *TPH1* and *TPH2*. In nonneuronal serotonergic cells, *TPH1* is the primary isoform expressed. Conversely, serotonergic neurons in the adult CNS express predominantly *TPH2* (Zhang, Beaulieu, Sotnikova, Gainetdinov, & Caron, 2004). Expression of *TPH2* has been studied in differentiated serotonergic rat medullary raphe RN46A B-14 cells and shown to be transcriptionally regulated by calcium (Remes Lenicov, Lemonde, Czesak, Mosher, & Albert, 2007) as well as by estradiol-17β (Hiroi & Handa, 2013). More pertinent to the current discussion, Patrick and Ames (2014) identified, *in silico,* four candidate VDREs in the 5′ flanking regions of human *TPH* genes, two in *TPH1* and two in *TPH2,* and hypothesized that 1,25D regulates the synthesis of serotonin. In a recent study, we evaluated the activity of these four candidate VDREs, with particular focus on the two VDREs near the *TPH2* gene, and analyzed whether 1,25D modulated *TPH2* mRNA in a number of cell types, including serotonergic RN46A B-14 cells. Figure 5A is a photomicrograph of RN46A-B14 cells in culture, demonstrating their fusiform stellate morphology and neurite interconnections. Strikingly, our real-time PCR results (Kaneko, Sabir, et al., 2015) revealed that *TPH2* mRNA was 26- to 86-fold upregulated by 10 nM 1,25D in these serotonergic rat brain cells (Fig. 5B–D). Interestingly, the response to 1,25D was biphasic, reminiscent of the "U-shaped" curves generated when plotting circulating 25-hydroxyvitamin D levels versus disease and death in humans (Keisala et al., 2009). In other words, there appears to exist an optimal level of circulating vitamin D for health, and this may correspond empirically to 10 nM 1,25D in the case of RN46A-B14 cell treatment. If one assumes that there is a corresponding optimum level of 1,25D for *TPH2* expression, and resulting serotonin, in the human brain, then this effective level could be argued to elicit normal neurotransmission and development. A further conclusion from these data is that, although the VDRE sequences themselves are not conserved between human and rat, the fact that rat *TPH2* is so potently induced by 1,25D in serotonergic cells suggests that the pair of VDRE(s) positioned at approximately −7 and −10 kb

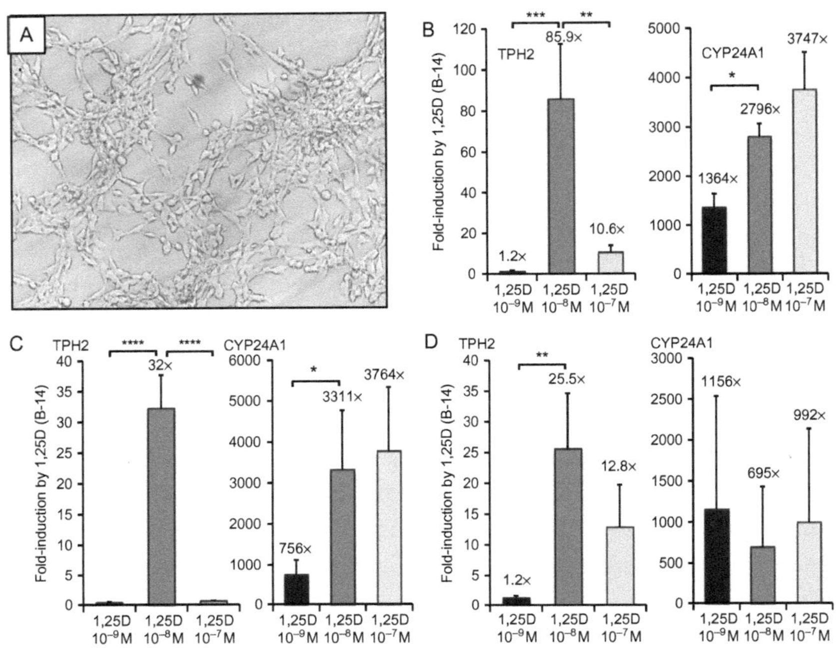

Figure 5 The rat *TPH2* gene is induced biphasically by 1,25D in embryonic medullary raphe neuronal (RN46A-B14) cells. (A) RN46A-B14 cells exhibit a fusiform stellate morphology with neurite interconnections. (B–D) RN46A-B14 cells were treated with 1,25D for 24 h at the indicated concentrations, and mRNA levels determined by real-time PCR as described (Kaneko, Sabir, et al., 2015). Passage numbers are 3 for panel (B), 8 for panel C, and 30 for panel (D). Fold induction by 1,25D is shown on the ordinate of each panel for *TPH2* or *CYP24A1* as indicated. Statistical significance is shown in the body of each panel: $*p<0.05$; $**p<0.01$; $***p<0.001$; $****p<0.00001$. *Kaneko, Saini, et al. (2015), Copyright (2015) Federation of Societies for Experimental Biology, USA.* (See the color plate.)

in the human *TPH2* gene and shown in Table 1, may be relevant to *TPH2* expression in human serotonergic cells, and *in vivo*, although proof of this will require further testing. Encouragingly, Jiang et al. (2014) demonstrated a significant induction of *TPH2* mRNA in the prefrontal cortex of rats after chronic 1,25D administration. This finding validates, *in vivo*, the data on *TPH2* induction by 1,25D in cultured cells (Fig. 5). However, this study also raises an important caveat, because Jiang et al. (2014) did not observe an increase in steady-state serotonin levels, apparently due to the fact that serotonin-metabolizing enzymes were also induced in prefrontal cortex, either in a primary or secondary fashion, by chronic 1,25D treatment (Jiang et al., 2014). Nevertheless, one could argue that the flux of serotonin in certain neurons is amplified by 1,25D. Therefore, we conclude that the vitamin D hormone is capable of governing serotonin concentrations in relevant regions of the brain where both *VDR* and *TPH2* are expressed.

Expanding further on the significance of *TPH2* induction by 1,25D, Patrick and Ames have published two papers (Patrick & Ames, 2014, 2015) in which they explore the observed association between vitamin D insufficiency and a plethora of neuropsychiatric disorders. In brief, they illuminate aberrant serotonin concentrations, both during development and in the adult, as a common denominator in a broad range of behavioral illnesses including ADHD, autism, bipolar disorder, depression, and schizophrenia, as well as antisocial, obsessive-compulsive, and suicidal behaviors. Their proposal is founded on evidence that executive function, sensory gating, and prosocial behavior are all regulated by serotonin and that serotonin levels are often suboptimal in disorders of these actions. They further hypothesize that a fundamental mechanism connecting vitamin D deficiency, low central serotonin concentrations, and neuropsychiatric disease is the extrinsic property of the vitamin D hormone, 1,25D, to upregulate *TPH2* expression, and that adequate 1,25D is in fact required for serotonin synthesis in serotonergic neurons. The results in Fig. 5 support the vitamin D–serotonin hypothesis both experimentally and mechanistically. Taking all of these data together, it seems highly likely that vitamin D can amplify serotonin synthesis in the central nervous system, making vitamin D a candidate for the prevention and treatment of neuropsychiatric disorders in which vitamin D and/or serotonin are implicated. Low levels of vitamin D are common in ASD, ADHD, bipolar disorder, schizophrenia, and impulsive behavior (Anglin, Samaan, Walter, & McDonald, 2013; Belvederi Murri et al., 2013; Goksugur et al., 2014; Gracious, Finucane, Friedman-Campbell, Messing, & Parkhurst, 2012; Grudet, Malm, Westrin, & Brundin, 2014; Patrick & Ames, 2014; Rylander & Verhulst, 2013). For this reason, many individuals at risk or already diagnosed with any of these disorders may benefit from a vitamin D supplement. In this regard, a recent case report (Jia et al., 2015) on a 32-month-old boy with vitamin D-deficiency and autism spectrum disorder who improved markedly in his core symptoms when treated with vitamin D_3, is pertinent. Moreover, vitamin D supplementation during the first year of life was shown in one study to decrease the incidence of schizophrenia by 77% (McGrath et al., 2004). This is particularly relevant because vitamin D insufficiency is rampant (up to 91%) among pregnant women in the United States and varies according to state, perhaps because of differences in sun exposure (Hossein-nezhad & Holick, 2012). Even 50% of mothers taking prenatal vitamins and their neonates had insufficient levels of vitamin D, whereas supplementation with 4000 IU/day was safe and effective in achieving adequate vitamin D concentrations without toxicity (Hollis, Johnson, Hulsey, Ebeling, & Wagner, 2011). Vitamin D supplementation has also been shown to improve inattention, hyperactivity, and impulsivity in children and adults with ADHD (Rucklidge, Frampton, Gorman, & Boggis, 2014).

In addition to the neuropsychiatric disorders listed above, an increasing number of studies indicate that a deficiency in vitamin D may contribute to the onset of neuro*degenerative* diseases such as AD and Parkinson's disease (PD) (Garcion, Wion-Barbot, Montero-Menei, Berger, & Wion, 2002). Similarly to the psychiatric disorders, the role of vitamin D as a guardian of pathways involving Ca^{2+} and antioxidation may be significant, since dysregulation of Ca^{2+} and ROS is a feature of these neurodegenerative diseases (Eyles et al., 2013). As one example highlighted by Berridge (2015), AD is a progressive neurodegenerative disorder characterized by memory loss, neuronal cell death, and dementia. These changes are initiated by extracellular β-amyloid (Aβ) deposits that impede neuronal signaling to blunt cognition. Pathologic Aβ enhances the level of Ca^{2+}, which then feeds back to escalate the formation of Aβ (Bezprozvanny & Mattson, 2008), creating a vicious cycle that is based on spiraling Ca^{2+}. Given the role of 1,25D in modulating Ca^{2+} signaling, vitamin D deficiency and the resulting dysregulation of Ca^{2+} may then trigger this positive feedback loop to drive the onset and progression of AD. Aβ-induced elevation in Ca^{2+} may contribute further to AD pathogenesis by activating mitochondrial generation of ROS (Butterfield, Swomley, & Sultana, 2013), rendering the actions of vitamin D, as detailed above, in upregulating Nrf2 to maintain levels of GSH, thioredoxin, and NMDAR doubly important in the prevention or attenuation of AD. Notably, bexarotene, an RXR-selective agonist that has been shown to reverse AD pathology and cognitive deficits in AD animal models (Cramer et al., 2012) is also known to suppress ROS (Tang et al., 2014). Moreover, there are several studies directly implicating a deficiency in vitamin D in the onset of AD (DeLuca, Kimball, Kolasinski, Ramagopalan, & Ebers, 2013; Tuohimaa, Keisala, Minasyan, Cachat, & Kalueff, 2009). The level of vitamin D in AD patients is demonstrably lower than that in controls and enhanced dietary vitamin D intake lowers the risk of developing AD in a study of older women (Annweiler, Llewellyn, & Beauchet, 2013). Therefore, as elaborated by Berridge (2015), 1,25D may protect the brain by regulating toolkit components that maintain ROS and Ca^{2+} levels. Further contributions by vitamin D to intracellular Ca^{2+} regulation include the upregulation of Ca^{2+} pumps and Ca^{2+} buffers such as calbindin-D_{9K} (CB) and parvalbumin. In support of this notion, neuronal levels of CB are known to be reduced in AD (Sutherland et al., 1992). Further, the level of Nrf2 is markedly reduced in the brain of patients with AD (Ramsey et al., 2007) and cognition in AD transgenic mice was strikingly improved following vector-mediated expression of Nrf2 in the hippocampus (Kanninen et al., 2009). Finally, recent studies (Laczmanski et al., 2015; Lee, Kim, & Song, 2014) have identified associations between VDR polymorphisms and AD susceptibility. Taken together, these and other studies provide a tantalizing glimpse into the role for vitamin D deficiency in the onset and/or progression of AD; however, more work clearly needs to be performed to prove the direct involvement of 1,25D/VDR and its mechanism of action.

PD is another neurodegenerative disorder with vitamin D connections. PD is caused by loss of DA neurons situated in the substantia nigra pars compacta (SNc). There is emerging evidence that PD is associated with vitamin D deficiency (DeLuca et al., 2013; Knekt et al., 2010). VDR is intensely expressed in DA SNc neurons which experience repetitive surges in Ca^{2+} during the operation of the DA neuronal pacemaker mechanism, which involves repeated activation of L-type Ca^{2+} channels (Cali, Ottolini, & Brini, 2014). Because the expression of these channels is reduced by 1,25D (Brewer et al., 2001), the L-type channels are particularly active when vitamin D is deficient. As noted by Berridge (2015), given that the plasma membrane Ca^{2+} pump (PMCA), the Na^{+}/Ca^{2+} exchanger (NCX), and the cytosolic buffers such as calbindin-D_{28K}, as well as mitochondria (*per se*), are present in SNc neurons at very low concentrations relative to other neurons (Surmeier & Schumacker, 2013), the mitochondria are subjected to undue pressure to accommodate repetitive Ca^{2+} uptake and associated increases in mitochondrial ROS generation. Consequently, when vitamin D levels are low, Ca^{2+} and ROS levels will rise, leading to the death of SNc neurons (Berridge, 2015). Moreover, oxidative stress induces nuclear export of the orphan receptor Nurr1 (Garcia-Yague, Rada, Rojo, Lastres-Becker, & Cuadrado, 2013), and there is emerging evidence that reduced Nurr1 function can contribute to the pathogenesis of PD (Decressac, Volakakis, Bjorklund, & Perlmann, 2013). Significantly, oxidative stress in the locus coeruleus and SNc is reported to be reduced by a local infusion of 1,25D (Chen, Lin, & Chiu, 2003), presumably via the functions of 1,25D to maintain the normal low resting levels of ROS and Ca^{2+}, thus reducing pressure on the mitochondria (Berridge, 2015). If these beneficial effects of 1,25D can be confirmed in the setting of human PD, then opportunities may emerge for the development of VDR ligands or other small molecule pharmaceuticals for the prevention or mitigation of PD.

2.4.4 Skin (Hair Loss and Psoriasis)

VDR is abundantly expressed in the skin and VDR-null mice display a prominent skin phenotype including dermal cysts and alopecia which is not ameliorated with a high calcium, lactose, and phosphate rescue diet. Surprisingly, however, mice that are unable to synthesize 1,25D are not alopecic, suggesting that VDR action in the hair follicle is independent of the 1,25D ligand. Along with VDR (Cianferotti, Cox, Skorija, & Demay, 2007), β-catenin is also required for mammalian hair cycling (Huelsken, Vogel, Erdmann, Cotsarelis, & Birchmeier, 2001), and it has been suggested that VDR interacts with the Wnt/β-catenin signaling pathway in mediating this process (Beaudoin, Sisk, Coulombe, & Thompson, 2005; Cianferotti et al., 2007). As illustrated schematically in Fig. 6 and explained in the legend, the regulation of mammalian hair cycling involves at least two signaling pathways, BMP and Wnt. Unoccupied VDR is thought to promote β-catenin-Lef1/TCF action by cooperating physically and/or functionally with Lef1/TCF (Luderer, Gori, & Demay, 2011), with several scenarios being possible (see Fig. 6 right side and bottom center). In support of a role for VDR, Lisse et al. (2014) have shown that unliganded VDR is required for induction of both the cWnt and hedgehog signaling pathways during anagen. Thus, VDR and Lef1/TCF transcription factors are recruited to identical regions in the regulatory sequences of murine *shh* and *gli1* genes (Lisse et al., 2014), suggesting that cooperation between VDR and Lef1/TCF, potentiated by β-catenin, is required for induction of these pathways during postmorphogenic hair cycles. Figure 7 lists relevant upstream sequences in the mouse *shh* and *gli1* genes, highlighting the proximity of apparent VDRE and Lef1/TCF enhancer elements as determined via ChIP-seq analysis (Lisse et al., 2014; Luderer et al., 2011). One unique feature of these apparent *cis*-regulatory modules (CRMs) in the mouse *shh* and *gli1* genes is the absence of classic DR3 VDREs (Fig. 7). Instead, a DR4 or, more commonly, solitary consensus VDRE half-elements are observed, often juxtapositioned to a Lef1/TCF element. One possibility is that unliganded VDR binds with Lef1/TCF on a composite element, perhaps excluding the traditional RXR heteropartner of VDR (Fig. 6). However, RXRα is also absolutely required for hair cycling (Li et al., 2000), arguing for an as-yet-undefined role in the association of VDR with CRMs in the *shh* and *gli1* genes. Moreover, 1,25D is known to act through VDR–RXR

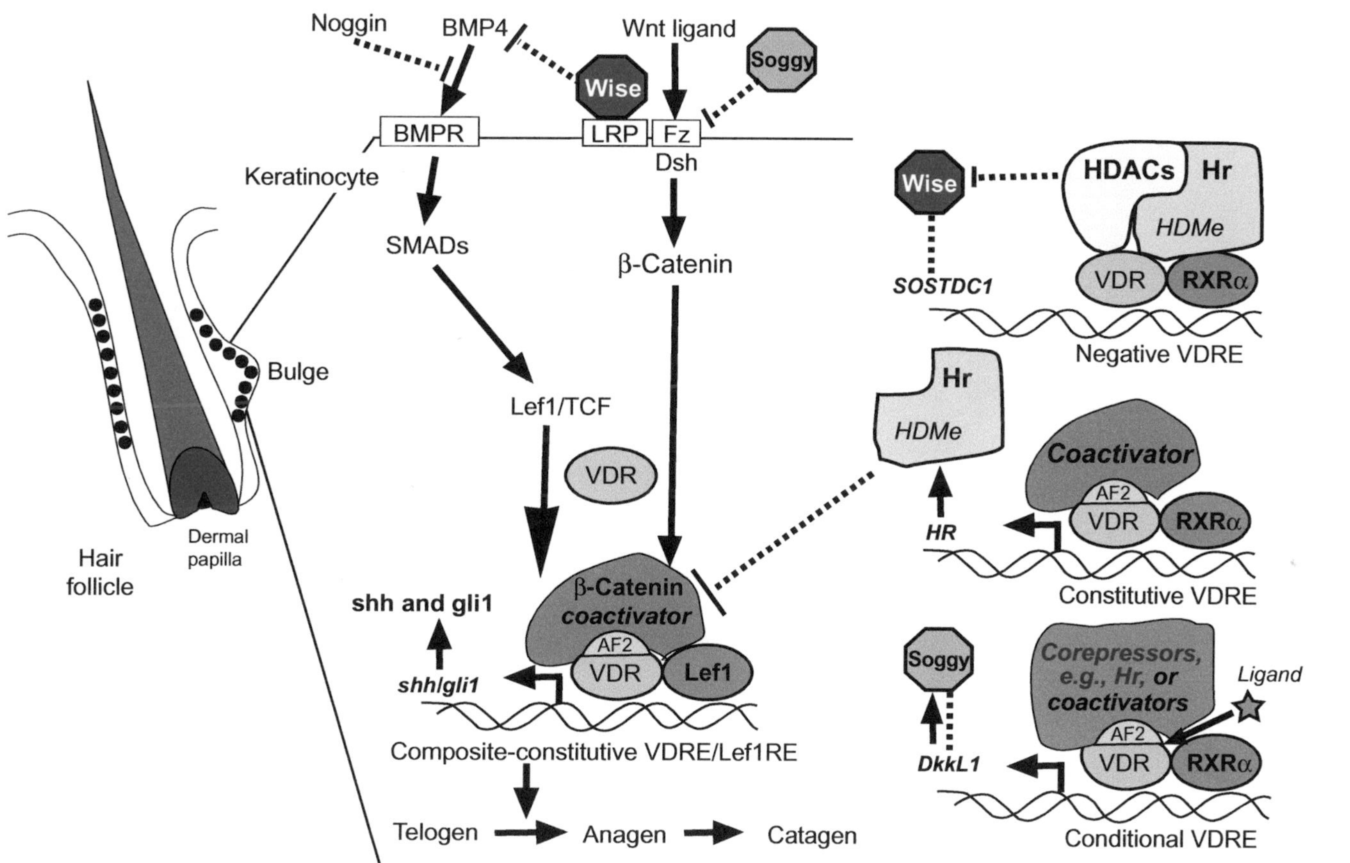

Figure 6 See figure legend on next page

to induce LRP5 (Table 1; Barthel et al., 2007; Fretz et al., 2006), the coreceptor of frizzled in the binding of Wnt ligands (Fig. 6), as well as the Boc and Cdon coreceptors that are active in hedgehog signaling in osteocytes (St John et al., 2014). The contribution of RXR to hair cycling also could involve the coregulation with VDR of genes other than those in the Wnt and hedgehog pathways.

Gene ablation studies point to the role of another player, namely the corepressor, hairless (Hr) (Thompson, Sisk, & Beaudoin, 2006), for which a (*hr*) gene knockout produces an alopecic phenocopy of the VDR-null mouse (Zarach, Beaudoin, Coulombe, & Thompson, 2004). Hr colocalizes with VDR in the outer root sheath of the hair follicle (Hsieh et al., 2003) and we have demonstrated a direct protein–protein interaction between Hr and VDR (Hsieh et al., 2003). The exact role of Hr in the regulation of the hair cycle remains to be defined, but we and other laboratories have postulated that Hr and VDR may cooperate to mediate transcriptional repression of target genes relevant to the hair cycle. This notion is based on the following considerations. First, Hr is already known to function as a transcriptional

Figure 6 Model for control of the hair cycle by unliganded VDR and β-catenin. Regulation of *HR, SOSTDC1*, and *DkkL1* expression by VDR may contribute to epidermal cell functions and hair cycling by upregulating *HR,* and in turn repressing *SOSTDC1* (Wise) and *DkkL1* (Soggy) expression. Signaling in the mammalian hair cycling is complex, consisting of the convergence of two signaling pathways, BMP and Wnt. Noggin from the dermal papilla initially antagonizes BMP4 signaling in bulge keratinocytes, allowing for the accumulation of Lef1/TCF, a DNA-binding protein that targets genes and controls their expression via transcriptional coactivator partners such as β-catenin. Cessation of Noggin signaling reinstates BMP signal transduction via SMADs provided that Wise (encoded by *SOSTDC1*), which antagonizes both BMP and Wnt pathways (Lintern, Guidato, Rowe, Saldanha, & Itasaki, 2009), has also been repressed, either directly by VDR (upper right), indirectly through VDR induction of Hr (right center), or by a combination thereof. Wnt ligand (e.g., Wnt 10b) signaling leads to accumulation of β-catenin, which cooperates with unliganded VDR and Lef1/TCF to induce genes encoding factors such as sonic hedgehog (*shh*), and *gli1* that trigger the hair cycle to transition from telogen (resting) to anagen (growth). See text for additional discussion. "Constitutive" VDREs refers to those *cis*-elements that dock unliganded VDR (shown as a pink oval), often in combination with RXR or potentially another transcription factor such as Lef1/TCF. "Conditional" VDREs signifies that the direction of gene control (positive vs. negative) may be cell context specific as well as differentiation stage selective. Abbreviations not defined in the text are: HDMe, histone demethylase; Wnt, ortholog of *Drosophila* wingless and mouse int-1; Lef1, lymphoid enhancer factor-1; TCF, T cell-specific factor; and Fz, frizzled. Factors that are membrane receptors or transporters are boxed. Solid arrows indicate activation and dotted lines ending in a solid perpendicular line denote inhibition. (See the color plate.)

```
shh:
aacccatttccagctccagtcatacgtgcatggagtttcagaaagtctcg
atttggctgggagattggcagcctggaaatctcaaaggaggtgggatggg
aagagaggctcgtgctttgctttgccgtctgtccatccaccctcgtcgcc
gaatatttattcgcttttaattcttatgcaagcaggttaaaaattaaagc
gattgcaaagccagcaagttccaagtccctcccctaaggtaccgcggct
ctggagaaatgaggagcatccttaaagaaatatcaatacattctctgacc
```

```
gli1:
gaggaggtcatagagtaaggtcagagcctctgaaagtggatttatgagag
atcgtgagccatcatgtggttgctgggaattgagctcaggacctctggaa
gagcagtccttgctcttagccactgagccatcttgccagcccctgattgg
atgattgctttgaaacagagtttctccatgttggtcctggttgacctgg
aactttctatgtagatcaggctggccttgaacttacagagatgcccttgc
ttctgtcttcccagtgctgggattaaaggcatgtactattatgcacatct
```

Figure 7 *Cis*-regulatory modules (CRMs) in the *shh* and *gli1* genes. CRMs are demarcated by green (forward) and red (reverse) highlighted primers employed by Luderer et al. (2011) in ChIP analysis to localize VDR and Lef1/TCF binding regions. The *shh* CRM, located in the first intron of the mouse *shh* gene, contains 3 Lef1/TCF sites (purple highlight, italics) and 6 VDRE half-elements (yellow highlight, except when antisense (pale blue)). The *gli1* CRM, located approximately 2 kb upstream of the mouse *gli1* transcriptional start site, contains 3 Lef1/TCF sites (pale blue or purple/italics highlight) and 5 VDRE half-elements (yellow highlight), with two of the latter configured in a DR4 motif. (See the color plate.)

repressor, and at least one of its molecular functions is recruitment of HDACs to promote a repressive heterochromatin architecture (Potter, Zarach, Sisk, & Thompson, 2002); Hr is also reported to possess intrinsic histone 3 lysine 9 demethylase (HDMe) activity, possibly controlling transcription via the histone code as an epigenetic "eraser" (Liu et al., 2011). Regarding which genes might be targeted by a complex of Hr, VDR (and potentially RXR), Thompson and coworkers (Beaudoin et al., 2005) have defined *SOSTDC1* (encoding Wise) as a gene overexpressed in keratinocytes from Hr-null mice, and Kato and colleagues (Yamamoto et al., 2009) characterized *S100A8* as a gene overexpressed in VDR-null keratinocytes. In examining the effect of activated VDR on the expression of *SOSTDC1* and *S100A8* mRNA levels in human keratinocytes (Haussler et al., 2010), we observed that *SOSTDC1* is strikingly repressed at the level of mRNA production after 18 h of 1,25D treatment of keratinocytes. Further, we (Hsieh et al., 2014) showed that *SOSTDC1* mRNA was repressed 41–59% by 1,25D in KERTr and primary human keratinocytes; a functional (negative) VDRE was located at −6215 bp in the human *SOSTDC1* (Table 1). Because Wise not only antagonizes the Wnt pathway by binding to LRP but also inhibits the BMP pathway through neutralization of BMP4 (Lintern et al., 2009), repression of *SOSTDC1* by VDR could constitute a major event in initiating the mammalian hair cycle (Fig. 6). Similarly, 1,25D rapidly represses expression in human keratinocytes of S100A8 and its obligatory S100A9 heteropartner (Haussler et al., 2010). Two caveats in these conclusions must be acknowledged: first, in cell culture systems, inhibition of both *SOSTDC1* and *S100A8* is 1,25D dependent, but we note that, *in vivo*, the appropriate combination of factors in the hair follicle renders this repression 1,25D independent. Second, this inhibition of S100A8/A9 expression by 1,25D-VDR is in stark contrast to the induction of S100A8/A9 observed in HL-60 promyelocytic leukemia cells when differentiated by 1,25D along the macrophage lineage (Suzuki et al., 2006). We explain this discrepancy with the concept that VDR can regulate S100A8/A9 expression in a cell selective fashion. One additional gene repressed by 1,25D-VDR, namely PTHrP (Falzon, 1996), is already known to encode a suppressor of the telogen to anagen transition in the hair follicle, as well as promote entry into catagen (Cho et al., 2003), providing yet another VDR–RXRα–Hr repressed gene target that participates in hair cycle control (Table 1). In conclusion, VDR is crucial for the regeneration of hair to protect skin and facilitate healthful aging. Based upon its functioning as a corepressor of VDR, we (Hsieh et al., 2014) hypothesized that Hr may target VDR-VDRE signaling and subsequently modulate downstream *SOSTDC1* and *DkkL1* expression. Thus, Hr is postulated to function as a corepressor with VDR/RXR to suppress inhibitors of Wnt signaling to trigger the hair cycle. As a final note, human *HR* is transcriptionally activated by unliganded VDR (Hsieh et al., 2014), and we have discovered an apparent constitutive VDRE at −7269 bp in the human *HR* promoter (Table 1), as well as a novel, ligand-independent thyroid hormone responsive element in the first intron of *HR*. We reasoned that the resulting upregulation of Hr could reciprocally repress *HR* expression, as well as that of *VDR* and *TR* mRNAs, thus establishing a novel inhibitory feedback loop to control the level of all three transcription factors.

Another action of VDR (in this instance liganded with 1,25D) is to inhibit proliferation and promote differentiation in skin. These actions underlie, at least in part, the clinical use of 1,25D and its analogs in the treatment of psoriasis (Kragballe, 1997). Further, the VDR-null mouse is supersensitive to DMBA-induced skin cancer (Zinser, Sundberg, & Welsh, 2002) as well as UV light-induced skin malignancy (Ellison, Smith, Gilliam, & MacDonald, 2008), providing a rationale for 1,25D use in the prevention of skin cancer. Interestingly, photoirradiation of skin produces vitamin D, and the CYPs catalyzing bioactivation to 1,25D are expressed in skin and therefore are able to produce local 1,25D to protect the epithelium against UV-induced photo-damage and malignancy. In addition, 1,25D induces the expression of a number of genes in cultured keratinocytes, the products of which are potential prodifferentiative and structural components as well as detoxification, immunomodulation, and anti-inflammatory/antioxidation principles (Haussler et al., 2013). For example, 1,25D induces caspase-14 in keratinocytes (Haussler et al., 2013) which is crucial for keratinocyte differentiation (Zarach et al., 2004). 1,25D also induces cathelicidin and several defensins in keratinocytes (Haussler et al., 2013), indicating that vitamin D has antibacterial actions in skin. Finally, 1,25D increases the expression in human keratinocytes of a number of keratin-related transcripts (Haussler et al., 2013), as well as certain late cornified envelope (LCE) proteins, namely LCE2B, -3A, -3B, -3C, -3D, and -3E (Austin et al., 2014; Hoss et al., 2013), with the five *LCE3* gene products reportedly playing a role in skin repair after minor injury (Bergboer et al., 2011). Interestingly, a common deletion of *LCE3B* and *-3C* is a risk factor for psoriasis (de Cid et al., 2009), suggesting that vitamin D signaling not only supports the skin structurally and mediates barrier function development but also facilitates repair to prevent inflammation (Austin et al., 2014; Hoss et al., 2013). In summary, unliganded VDR functions to drive the mammalian hair cycle in cooperation with Lef1/TCF, β-catenin, and Hr, primarily via Wnt and hedgehog pathways, and secondarily by modulation of Wnt pathway inhibitors, whereas liganded VDR signals the development, barrier function, and repair of the skin. These 1,25D-dependent activities are likely redundant with other signalers, such as calcium, but nevertheless are important therapeutically in the prevention and treatment of skin diseases such as psoriasis, with the caveat that severe psoriasis appears to be a T cell-centric disease, making therapeutic approaches that target IL-17A and TNF-α pathways more effective strategies for treating severe disease.

2.4.5 Cancer Prevention (Epithelial and Blood Cell Malignancies)

Vitamin D status has been associated with the incidence of numerous types of cancer (Feldman, Krishnan, Swami, Giovannucci, & Feldman, 2014; Heaney, 2008), and mortality rates from colon, breast, and prostate cancer are elevated at latitudes with diminished UV irradiation (Bouillon et al., 2008; Garland, Garland, & Gorham, 1999; Hibler, Molmenti, Lance, Jurutka, & Jacobs, 2014). 1,25D exerts anticancer effects in animal models of breast, ovary, lung, and prostate cancers (Welsh, 2012), and at all stages of carcinogenesis, from tumor cell proliferation, to angiogenesis and metastasis. 1,25D suppresses cell proliferation by repressing cyclin D and by inducing CDK inhibitors such as p21 and p27 (Verlinden et al., 1998). The antiproliferative action of 1,25D also includes suppression of the Wnt/β-catenin signaling pathway (Egan et al., 2010; Palmer et al., 2001). 1,25D induces the expression of DICKKOPF-1 (DKK-1), which, like Wise, inhibits the Wnt pathway (Aguilera et al., 2007), and liganded VDR inhibits the activity of the transcription factor β-catenin (Palmer et al., 2001). 1,25D also induces the expression of growth arrest and DNA-damage-inducible protein 45 (GADD45), which curbs cell growth during intervals of DNA repair (Jiang, Li, Fornace, Nicosia, & Bai, 2003). The ability of curcumin, a low-affinity VDR ligand, to inhibit the proliferation of breast cancer cells, also appears to be dependent on Nrf2 induction (Chen, Zhang, et al., 2014). Finally, the suppression of ROS signaling by 1,25D (Bao, Ting, Hsu, & Lee, 2008), as discussed above, is also antiproliferative because it attenuates growth factor signaling pathways.

As a further anticancer action, 1,25D dampens the proliferation of the endothelial cells required for angiogenesis (Deeb, Trump, & Johnson, 2007) and suppresses the growth of solid tumors in animal models (Eisman, Barkla, & Tutton, 1987). This effect of 1,25D appears to be mediated by repression of vascular endothelial growth factor (Nakagawa, Kawaura, Kato, Takeda, & Okano, 2005). VDR liganded with 1,25D also exerts antimetastatic actions in animal models, such as prostate (Lokeshwar et al., 1999) and lung (Nakagawa et al., 2005) cancers. 1,25D plays an essential role in maintaining cell–cell adhesion by controlling the expression of E-cadherin and other adhesion components such as zonula occludens-1. 1,25D also mitigates the ability of migrating cells to penetrate and generate secondary tumors by inhibiting secretion of matrix metalloproteinases 2 and 9 (MMP-2 and MMP-9), which proteolyze components of the extracellular matrix to allow for cancer cell migration (Nakagawa et al., 2005). 1,25D/VDR represses MMP-13 expression (Meyer, Benkusky, & Pike, 2015), and MMP-13 is considered to be a major factor in the etiology of breast cancer. Finally, genetic polymorphisms in CYP27B1 and CYP24A1, enzymes that execute vitamin D metabolism, result in marked changes in the activity of these CYP enzymes in colon cancer cells. These alterations would likely result in differential 1,25D exposure in colonic cells and render individuals more susceptible to the development of colon cancer (Jacobs et al., 2013).

2.4.6 Metabolism (Obesity)

Epidemiologic studies have concluded that vitamin D levels associate more closely with glucose metabolism and diabetes than with obesity, *per se* (Clemente-Postigo et al., 2015). Our laboratories have focused on the interrelationship between leptin, a hormone that enhances peripheral tissue fatty acid metabolism, and 1,25D. Leptin is also a satiety hormone which can be generated centrally (Morash, Li, Murphy, Wilkinson, & Ur, 1999), although its major endocrine source is adipose tissue. Leptin is controlled primarily at the transcriptional level, and a wide variety of physiological conditions and pharmacological agents have been shown to affect its expression, including fasting and feeding, insulin, glucocorticoids, and thiazolidinediones. Studies of the regulation of leptin expression by vitamin D are conflicting. Initially, it was reported that 1,25D suppressed leptin secretion from human adipose tissue (Menendez et al., 2001). However, a more recent investigation revealed that 1,25D stimulates leptin production in mouse adipose tissue (Kong, Chen, Zhu, Zhao, & Li, 2013). Also, Narvaez, Matthews, Broun, Chan, and Welsh (2009) observed that leptin expression in both white and brown adipose tissue is sharply depressed in VDR-null mice, compared to wild-type littermates. Finally, VDR knockout mice are lean and resistant to diet-induced obesity (Narvaez et al., 2009), likely because of overexpression of uncoupling protein-1, which is repressed by VDR (Malloy & Feldman, 2013). Our own recent study (Kaneko, Sabir, et al., 2015) demonstrates that 1,25D dramatically represses leptin gene expression in mouse 3T3-L1 adipocytes while inducing leptin mRNA expression in human glioblastoma (U87) cells, providing evidence that 1,25D regulates leptin expression differentially in adipose and brain, repressing adipose leptin possibly to decrease fat catabolism, and inducing central leptin to elicit satiety.

In a search for active VDREs near the mouse leptin gene using ChIP-seq data obtained previously in MC3T3-E1 osteoblastic cells (Meyer, Benkusky, Lee, & Pike, 2014; Meyer, Benkusky, & Pike, 2014), we (Kaneko, Sabir, et al., 2015) identified a 1,25D-dependent, VDR-binding CRM located 28 kb upstream of the *Lep* transcriptional start site. The involvement of this CRM in transcriptional activation by 1,25D is further confirmed by the presence of epigenetic histone marks such as H3K4me2 and H4K5ac that represent signature modifications of active gene enhancers. This CRM (Fig. 8) harbors not only three canonical DR3 VDREs but also binding sites for RUNX2 and C/EBPβ, both of which are known to be involved in gene activation by the VDR/RXR heterodimer in osteoblasts

-28141

```
AGGGCAAATGAGTCAGCAGTAGGTTTAAGCAATGAACAGTTGCACCACGGGGTGTTTCTC
AAATTCTTGTCTACTTTTAGAAAGTACGCAATGCCATTCAGTGGGCACCGTGATTCTGAA
ATAGCTGCGAAAATGTTAAAAATGCTGTGGGCAAAGATGTGCCCTCATTGACCTCTTTCT
GAAAACAATAGTCCACAATTGTGTGGCGCGCCCTTCAGCAGCCTGGCTGCTCCCTCTGAT
CCTTGTTGGTGACATGAAGGGAATGCATGAAACTCTCTCTCTGGAGGCTCCTTCCAGCCT
CGAATTCTGATACCTGACACTAGGCCCACATTCCTGATACTCCGGCTGCATCTCTACAGG
ACCAAGACCACCAAGCCAATCGGGTGAGTGTGGACAAAATGTTCTCCCCAATCGATTCTT
TGGCACAGGGTAGTTGAAACAATTAGTTAGCTGGCATTGGTTTTCATTTCACAACTGGCG
GCTCTGGCATCATCGTGACCAGCGGTTTTTCCCACAGCGGTGTCCTAACATGCTGCATGC
CTGGAGCAGTTTTGAGGTGTGTCATAGGCAAGTGGCTACCTAATAACAAACTGTCGAGGC
TTTCGAATGATTGATGATCTTCCTGGTTTTAATCAATTAGAACAGATTCCACATAACCCC
GTAATGATGCCCAGACGGTTGCTTGCATTCTGACATGCCTCTTTGTATGGAAGACGTTTT
GCAACTGGCAGACCAGGTCTGCTTGGAGCTCCAATTCTGCTGGATGCCACGGGAAGAGTG
TCACGAGCCATACGCATGGATATCTGTCATCGGAAGCAGCCGCAGGCTGGCCATCGGAAG
CCCCTGATGCATCCTTAAACATTAAACCACTAAAGATCCAAGCTCTAGAATAACCCGGCC
TAAACAGCCTGCTACCGGCATAGGCTGAGGTTTAGCAAGACAGCTGGTAAACAAGGCAGT
TCAGAAGAACATGGAACAAAGGGCCTGCTCATTATGCTGATCCTTGCTCTAAAAAGAAAC
ATTCCACCATGATCTGTTCATCGATGAGGAAGGGAATGGGGTGTGATGCAGTATAGATTG
AGTATTTGGGTGTCCATTCCTGTACCAATCAATCATGGCCAGTTTAATACTGTTCATGAG
AGACTTTGGGTACTCAGCCTGCGAGCTCGTATTTCCTCCCTCCTGTGGTG
```

-26951

Figure 8 Sequence of a proposed 1,25D/VDR–RXR CRM in the mouse leptin gene. The positioning of the 1190 bp CRM (depicted in sense orientation) in relation to the start of transcription is indicated by numbers bordering the sequence. DR3 VDREs are highlighted yellow when they are sense, and green (italicized typeface) when they occur antisense. The C/EBP site (antisense) is highlighted pale blue and the RUNX2 site (sense) is highlighted light purple. *Kaneko, Saini, et al. (2015), Copyright (2015) Federation of Societies for Experimental Biology, USA.* (See the color plate.)

(Meyer, Benkusky, Lee, et al., 2014). Given that leptin is produced at multiple sites, and modulates physiologic phenomena in tissues as diverse as the anterior pituitary, GnRH/kisspeptin neurons, muscle, adipose and bone, not to mention actions on insulin sensitivity, blood pressure, and immunity, the challenge now is to identify biologically relevant target cells in order to decipher which of the many roles played by leptin are governed by, or dependent upon, vitamin D and its receptor. We postulate that, by controlling leptin, likely through the CRM we identified (Fig. 8), 1,25D may regulate eating behavior, as well as fatty acid metabolism. However, these effects of 1,25D may cancel each other out to render the vitamin D hormone relatively neutral with respect to preventing obesity. Indeed, an epidemiological investigation (Vimaleswaran et al., 2013) generated the conclusion that, whereas obesity may drive down vitamin D levels, a vitamin D deficiency in and of itself may not lead to obesity. Nevertheless, a very recent study suggests that higher concentrations of 1,25D and 25D were both associated with lower risk of elevated triglycerides and metabolic syndrome, suggesting that the interplay between obesity and vitamin D is complex and multifaceted (Bea et al., 2015).

2.4.7 Cardiovascular Disease (Hypertrophy, Fibrosis, and Vascular Calcification)

Hypertension and cardiovascular disease are associated with vitamin D deficiency (Dong, Lau, Wong, & Huang, 2014; Heaney, 2008). To confirm this association, Ni et al. (2014) created an endothelial cell-specific knockout of mouse *VDR* and examined vascular endothelial function, both at baseline and under challenge by angiotensin II. Acetylcholine-induced aortic relaxation was significantly diminished in these mice, accompanied by a reduction in endothelial NO synthase expression and signal transduction. Blood pressure at baseline was comparable at 12 and 24 weeks of age, but the endothelial VDR knockout mice displayed increased sensitivity to the hypertensive effects of angiotensin II compared to wild-type littermate controls. These data not only indicate a role for VDR but also point to the functional properties of the endothelium in preventing vascular stiffness, hypertension, and myocardial hypertrophy. Preservation of the cardiovascular system by 1,25D apparently also includes prevention of cardiac hypertrophy, congestive heart failure, and atrial arrhythmias. These effects occur via multiple mechanisms, such as stabilizing ROS and Ca^{2+} transients, both of which are dysregulated in hypertension, and curbing the renin-angiotensin system to prevent hypertension-related heart disease. For example, VDR liganded with 1,25D prevents cyclic AMP response element-binding protein (CREBP) from docking on its DNA transcriptional enhancer in the renin gene, thus inhibiting the production of renin (Yuan et al., 2007) and its downstream effectors angiotensin II, endothelin-1 (ET-1) and NOX-1. In support of this finding, knockout of either *CYP27B1* or *VDR* in mice elicited activation of the RAAS, leading to hypertension and cardiac hypertrophy (Li et al., 2002; Zhou et al., 2008). The *CYP27B1* knockout mouse could be rescued by 1,25D treatment (Zhou et al., 2008) and, correspondingly, vitamin D supplementation has been reported to prolong the survival of patients with cardiovascular disease (Vacek et al., 2012). The above-discussed ability of VDR to control ROS is also significant in controlling hypertension, and an additional mechanism for this is the ability of 1,25D to reduce ROS levels in the renal arteries of hypertensive patients by inhibiting the NOX enzyme that generates ROS and by increasing the SOD-1 enzyme (Fig. 4) that metabolizes ROS (Dong et al., 2012). Vascular smooth muscle tone is also normalized by 1,25D via a reduction in the entry of Ca^{2+} that is driving the excessive release of contracting factors (Wong, Delansorne, Man, & Vanhoutte, 2008).

Thus, hypertension associated with vitamin D deficiency is driven by an increase in the levels of renin and its effectors angiotensin II and endothelin-1 (ET-1), which are responsible for precipitating cardiac hypertrophy and congestive heart failure (Berridge, 2012). As explained by Berridge, in the presence of increased angiotensin II and ET-1, there are subtle spatial and temporal changes in the individual Ca^{2+} transients that drive contraction. Vitamin D deficiency apparently initiates hypertrophy by increasing Ca^{2+} signaling because of a deficit in the expression of both SERCA and phospholamban, contributing to enhanced Ca^{2+} transient amplitude and a decline in the recovery phase (Choudhury et al., 2014). Such changes are promulgated by elevated angiotensin II and ET-1, plus escalated ROS/Ca^{2+}. ROS act by boosting the activity of ion channels ($Na_V1.5$ sodium channel, $Ca_V1.2$ channels, and RYR2) and pumps (SERCA) to effect the Ca^{2+} cycling events that occur during each heartbeat (Berridge, 2015). These pathologically elevated ROS and Ca^{2+} signaling processes elicit alterations in gene transcription that culminate in hypertrophy (Kohler, Sag, & Maier, 2014). 1,25D reverses cardiac hypertrophy in spontaneously hypertensive rats (Mancuso et al., 2008), and vitamin D supplementation markedly improves the outcome for patients experiencing heart failure (Vacek et al., 2012). There also exists a relationship between vitamin D deficiency and atrial fibrillation (Chen, Liu, et al., 2014; Hanafy et al., 2014). Atrial $InsP_3R2s$ are activated in the presence of ET-1, which is increased during vitamin D deficiency, causing a spike in atrial arrhythmias (Mackenzie et al., 2002). Taken together, all of these observations suggest that vitamin D hormone replacement is a reasonable consideration in the management of patients with cardiac hypertrophy, congestive heart failure, and atrial fibrillation. Furthermore, as discussed in Section 3, klotho assists 1,25D in preventing fibrosis and vascular calcification, two major age-related pathologies that affect healthspan negatively.

3. KLOTHO

3.1 Membrane and Secreted Forms

An aging-suppressor gene, *α-klotho* (hereafter referred to as *klotho* or *KL*), was discovered by Kuro-o et al. (1997), and its disruption in mice was associated with soft tissue calcification, profound hyperphosphatemia, osteoporosis, emphysema, arteriosclerosis, skin atrophy, infertility, hypoglycemia, and a curtailed lifespan. *Klotho* is the only reported single gene mutation that leads to premature aging in the mouse (Kuro-o et al., 1997) and a recessive inactivating mutation in the human *KL* gene elicits a phenotype of severe tumoral calcinosis (Ichikawa et al., 2007). *Klotho*- and *FGF23*-null mice have identical hyperphosphatemic phenotypes of short lifespan/premature aging, ectopic calcification, arteriosclerosis, osteoporosis, muscle atrophy, skin atrophy, and hearing loss (Kuro-o et al., 1997; Shimada, Kakitani, et al., 2004), and the mechanisms underlying some of these deficits will be discussed in more detail below. *Klotho* is expressed primarily in the kidneys and brain choroid plexus (Wang & Sun, 2009a). Multiple klotho protein forms have been characterized: a full-length transmembrane klotho (abbreviated mKL), two truncated soluble klotho forms, and a secreted Klotho (sKL) form (Xu & Sun, 2015). Full-length klotho (130 kDa) contains two extracellular glycosyl hydrolase domains, KL1 and KL2, a 20-amino acid single transmembrane domain, and a short, 9-amino acid intracellular domain. This mKL form is cleaved by proteases ADAM10, ADAM17, and BACE1 (β-APP cleavage enzyme 1), generating soluble forms that possess either the KL1 domain alone (65 kDa), or both KL1 and KL2 domains (130 kDa). After entering the circulation, soluble klothos appear to function as hormones (Chen, Podvin, Gillespie, Leeman, & Abraham, 2007; Kurosu et al., 2005; Wang & Sun, 2009a). Secreted klotho is generated by alternative mRNA splicing and consists of 549-amino acids (65 kDa), including the N-terminal signal peptide followed by the KL1 domain alone. Interestingly, circulating endocrine klotho encompasses both soluble and secreted species, and the short-form soluble (cleaved) klotho and secreted (alternatively spliced) sKL both contain the single KL1 domain and have approximately the same molecular mass of about 65 kDa. These circulating, short klotho forms are reported to employ the catalytic activity of their KL1 domains to regulate the TRPV5 channel (Fig. 1; Chang et al., 2005) and the renal outer medullary potassium channel 1 (ROMK1) (Cha et al., 2009). As illustrated in Fig. 1, other anti-aging actions of circulating klotho species include antifibrogenic and antineoplastic benefits in the vasculature and other target tissues. These endocrine functions of klotho will be discussed in detail in later sections.

3.2 Coreceptor Function of Membrane Klotho: Feedback Control of Phosphate and Vitamin D Metabolism

By far the best-characterized function of full-length, membrane klotho is to act as a renal coreceptor of FGFR isoforms in the feedback control of phosphate and vitamin D metabolism by bone-derived fibroblast growth factor

23 (FGF23) (Fig. 1). FGF23 signals via renal FGFR1/klotho coreceptors to promote phosphaturia (Razzaque, 2009) and, via renal FGFR3,4/klotho (Gattineni, Twombley, Goetz, Mohammadi, & Baum, 2011), to repress CYP27B1 (Perwad, Zhang, Tenenhouse, & Portale, 2007) as well as induce CYP24A1 (Razzaque, 2009; Shimada, Hasegawa, et al., 2004; Fig. 1), with the latter two actions serving to curb 1,25D levels. Remarkably, double knockouts of FGF23 (or its klotho coreceptor) with either VDR (Hesse, Frohlich, Zeitz, Lanske, & Erben, 2007) or CYP27B1 (Renkema, Alexander, Bindels, & Hoenderop, 2008) essentially rescue FGF23 null mice, underscoring the role of FGF23 and klotho as counter-regulatory hormones to 1,25D, which appears to be the key to their health and longevity benefits. Klotho likely possesses systemic anti-aging properties independent of its phosphaturic actions, perhaps through its glycosyl hydrolase enzymatic activity (Cha et al., 2009). Conversely, although FGF23 is anti-aging at the kidney by eliciting phosphate elimination and detoxifying 1,25D, its "off-target" actions could actually be pro-aging in terms of coronary artery disease, as well as potential neoplastic actions in the colon (Jacobs et al., 2011), and it is possible that these off-target FGF23 pathologies are "buffered" by secreted klotho (Bergwitz & Juppner, 2010). Upregulation of klotho by 1,25D (Forster et al., 2011) is thus consistent not only with potentiation of FGF23 signaling in the kidney, but also protection of other cell types (e.g., vascular and colon), in which a secreted form of klotho may exert beneficial actions (Wang & Sun, 2009b).

Regarding the regulation of klotho biosynthesis, Thurston et al. have demonstrated that TNF-α and γ-interferon are suppressors of renal klotho expression (Thurston et al., 2010) and the FGF23 ligand is also a putative repressor of klotho expression (Quarles, 2008). On the other hand, inducers of klotho are incompletely characterized. However, as detailed below, we have reported (Forster et al., 2011) that 1,25D significantly induces *klotho* mRNA expression in human and mouse renal cell lines. Figure 9A illustrates our real-time PCR results (Forster et al., 2011) utilizing RNA isolated from distal convoluted tubule (mpkDCT) cells, the primary expression site for klotho in kidney, with primers designed to capture both alternatively spliced mRNAs for the membrane and secreted forms of klotho. These data demonstrate that 1,25D treatment induces mRNA expression of both the membrane and secreted forms of *klotho* mRNA, suggesting that 1,25D may be capable of both amplifying FGF23 responsiveness and eliciting secretion of circulating klotho hormone. Interestingly, curcumin, an alternative VDR ligand (Bartik et al., 2010), selectively upregulates membrane klotho mRNA in

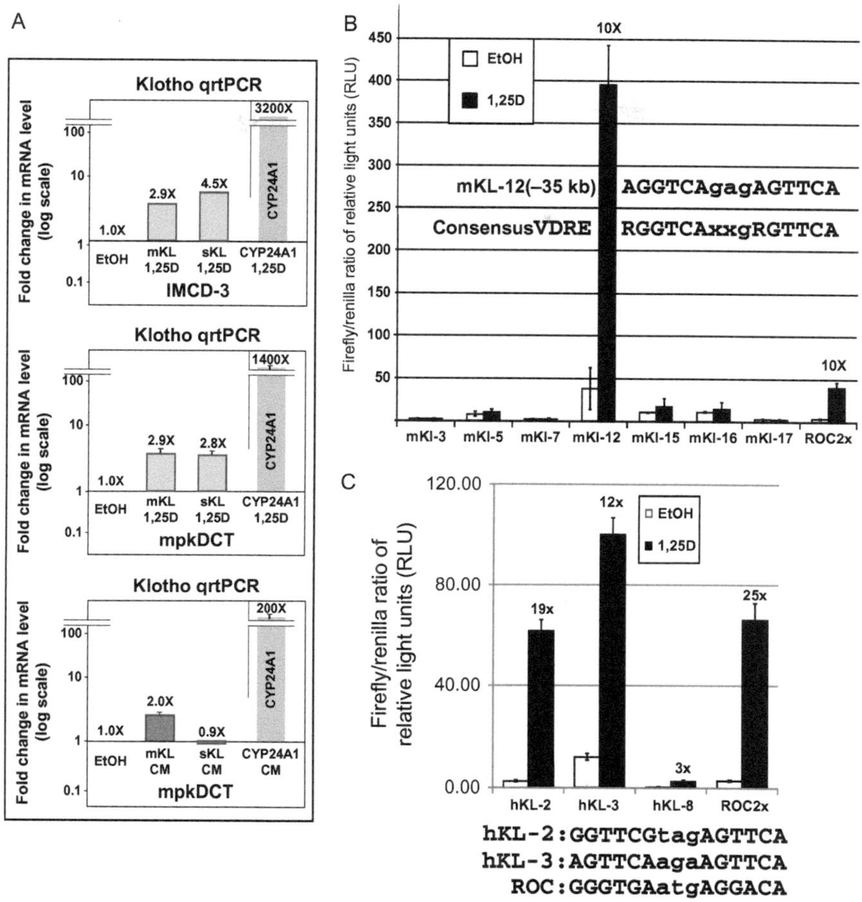

Figure 9 Regulation of membrane or secreted klotho forms (abbreviated mKL or sKL) by 1,25 and activity of candidate mouse and human klotho VDREs. (A) Upregulation of mKL and sKL in mouse intermedullary collecting duct (IMCD-3) cells (upper panel) or mouse distal convoluted (mpkDCT) cells (lower two panels) by either 1,25D (upper two panels), or curcumin (CM, lower panel). Cells were treated with ligand for 24 h prior to RNA isolation and real-time PCR (Forster et al., 2011). (B) Candidate mouse klotho VDREs were cloned into a pLUC-MCS reporter vector, cotransfected into HK-2 human kidney cells along with a pSG5-VDR cDNA expression plasmid and treated with 1,25D (10^{-8} M) for 24 h. Firefly luciferase values were normalized to expression of Renilla luciferase. Data are depicted as a fold effect of 1,25D. Results reveal that only the mouse klotho VDRE located at −35 kb (mKl-12) displays transactivation ability. (C) Transfection of candidate human klotho VDREs demonstrates a striking (>10-fold) 1,25D responsiveness of VDREs corresponding to sequences at −46 kb (hKL-2) and −31 kb (hKL-3), but not +3.2 kb (hKL-8). The sequences of the active VDREs in the mouse and human *Klotho* genes are listed in panels (B) and below (C); they are very similar to proven VDREs, with the mouse VDRE at −35 kb (mKL-12) conforming exactly to the consensus VDRE. (See the color plate.)

mpkDCT cells (Forster et al., 2011), leading to the hypothesis that designer vitamin D analogs or alternative VDR ligands can promote the healthful aging benefits of systemic klotho without accentuating FGF23 action to perhaps elicit hypophosphatemia. In order to mechanistically explain regulation of *KL* by 1,25D, we performed bioinformatic analysis of both the human and mouse klotho genes, which revealed 17 candidate VDREs in the mouse gene and 11 putative VDREs in the human gene (Forster et al., 2011). When assessed for functionality by cotransfection of reporter constructs into human kidney (HK-2) cells (Fig. 9B and C), only one mouse VDRE at −35 kb (mKL-12) and two human VDREs at −46 and −31 kb (hKL-2 and hKL-3) displayed a potency similar to the established rat osteocalcin (ROC) VDRE (Forster et al., 2011). We thus postulate that 1,25D-liganded VDR–RXR induces klotho expression by binding to functional VDREs near both the human and mouse *klotho* genes. The sequences of these VDREs are included in Table 1 and Fig. 9B and C. In combination with the data of Tsujikawa, Kurotaki, Fujimori, Fukuda, and Nabeshima (2003) that 1,25D increases steady-state *klotho* mRNA levels in mouse kidney, *in vivo*, the results shown in Fig. 9 indicate that 1,25D is the first discovered natural inducer of the *klotho* longevity gene.

3.3 Calcium Metabolism and Antioxidation

In addition to its phosphaturic actions as a coreceptor for FGF23, klotho also has actions to promote calcium reabsorption at the kidney. Chang et al. (2005) reported that soluble klotho increases the abundance of TRPV5 on the cell membrane and that this action is abolished by a single glycosylation site mutation. Klotho may also possess sialidase activity to remove a sialylated LacNAc residue from the TRPV5 N-glycosylation branch and expose a galectin-1 binding site. When this galectin-1 site is occupied, TRPV5 endocytosis and removal from the plasma membrane are diminished (Wolf, An, Nie, Bal, & Huang, 2014). The same modification was also discovered by Cha et al. (2009) for ROMK, an ATP-dependent potassium transporter in the cell membrane for which membrane accumulation is similarly dependent on klotho sialidase activity (Cha et al., 2009). Klotho may therefore regulate calcium metabolism via TRPV5, and potassium flux through ROMK. Thus, in indirect fashion, 1,25D, by inducing klotho, appears to elicit calcium and potassium retention at the kidneys.

Klotho also exerts antioxidative effects, some of which are reminiscent of 1,25D actions. For example, klotho suppresses oxidative stress through enhancing the expression of peroxiredoxins (Prx-2 and Prx-3) and thioredoxin reductase 1 (Trxrd-1) that act together to reduce ROS (Zeldich et al., 2014). Klotho also binds to transient-receptor potential canonical Ca^{2+} channel 1 (TRPC-1) through its KL2 domain and regulates TRPC-1-mediated Ca^{2+} entry to maintain endothelial integrity and prevent Ca^{2+}-stimulated NOS formation, which contributes to the formation of the potent $ONOO^-$ species (Kusaba et al., 2010). Thus, the maintenance of Ca^{2+} and redox signaling at a low resting state by 1,25D and its effectors, Nrf2 and klotho, appears to constitute the mechanism whereby 1,25D and klotho prevent the ravages of oxidation.

Another mechanism for the antioxidative actions of klotho involves binding of soluble klotho to presumed cell-surface receptors to signal inhibition of FOXO1 phosphorylation and promote its nuclear translocation (Yamamoto et al., 2005). Unphosphorylated FOXO1 induces expression of the *SOD2* (*MnSOD*) gene, thereby facilitating removal of ROS and conferring resistance to oxidative stress. Consistent with this finding, transgenic mice that overexpress klotho exhibit higher MnSOD expression and lower oxidative stress (Yamamoto et al., 2005). Other studies show that klotho overexpression decreases H_2O_2-induced apoptosis, β-galactosidase activity, mitochondrial DNA fragmentation, superoxide anion generation, lipid peroxidation, and Bax protein expression (Xu & Sun, 2015).

Finally, Wang, Kuro-o, and Sun (2012) reported that klotho downregulates the expression of a catalytic subunit of NADPH oxidase and suppresses angiotensin II-induced superoxide production, oxidative damage, and apoptosis through the cAMP/PKA pathway. *In vivo klotho* gene delivery similarly attenuates NADPH oxidase activity and superoxide production to prevent the progression of spontaneous hypertension and resulting renal damage (Wang & Sun, 2009b). Moreover, both 1,25D (Fig. 4) and sKlotho (data not shown) stimulate FOXO activity to quench ROS via SOD gene induction. Thus, these conclusions and others discussed above lead us to profess that 1,25D and klotho represent a dynamic "one-two punch" in maintaining healthful aging, with intracellular calcium current regulation and mitigation of oxidation being a common theme.

3.4 Effects on Wnt Signaling: Antifibrotic and Anticancer Actions

Klotho is reported to modulate Wnt signaling (Wang & Sun, 2009a), and data from our laboratories show that sKlotho is a potent suppressor of both endogenous and exogenous β-catenin activity in HEK-293 normal human embryonic kidney cells (Fig. 10B) and in HCT-116 colon cancer cells (data not shown), a phenomenon potentiated by 1,25D-VDR but not by

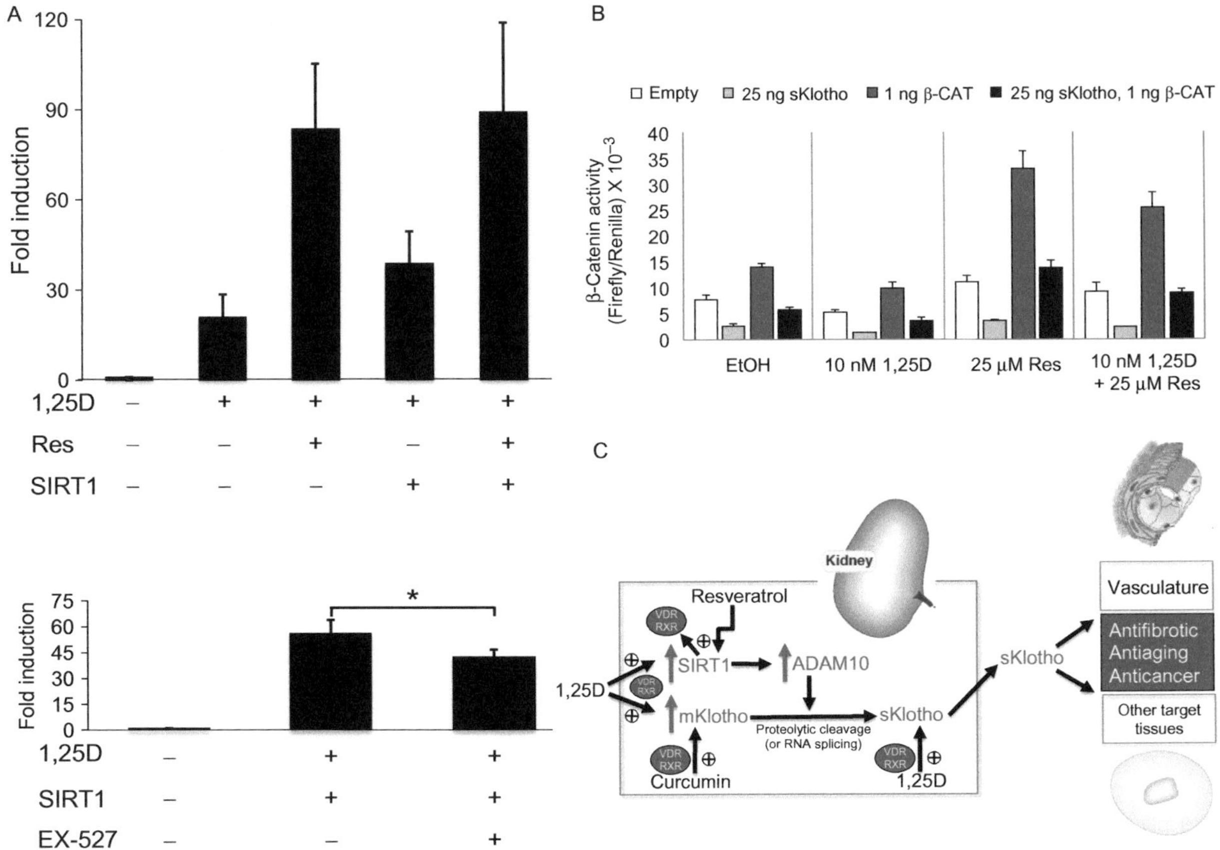

Figure 10 See figure legend on next page

resveratrol. Interestingly, this suppression of β-catenin is specific for secreted klotho in that it does not occur with membrane klotho overexpression (data not shown). Klotho-mediated regulation of Wnt signaling was also shown by Liu et al. (2007), who demonstrated that secreted klotho binds to Wnt ligands to suppress downstream signal transduction, and that *klotho* knockout enhances Wnt signaling in mice. Regarding the disease-related consequences of Wnt signaling suppression by klotho, activated Wnt3 signaling extends the cell cycle by arresting it at the G2/M phase and induces fibrogenic cytokines in mouse kidney, but klotho-treated cells circumvent this phase and are protected against renal fibrosis (Satoh et al., 2012). The loss of klotho may therefore contribute to kidney injury by releasing the inhibition of pathogenic Wnt/β-catenin signaling (Zhou, Li, Zhou, Tan, & Liu, 2013). Indeed, *in vivo* expression of klotho decreases the activation of renal β-catenin and diminishes renal fibrosis in chronic kidney disease (CKD) (Zhou et al., 2013). Conversely, reduced klotho expression aggravates renal interstitial fibrosis (Sugiura et al., 2012), and overexpression of secreted klotho abolishes the fibrogenic effects of TGF-β1 (Doi et al., 2011; Zhou et al., 2013).

Regarding antitumor actions of klotho, Behera et al. (2015) demonstrated a correlation between loss of klotho and a gain in Wnt5A expression, leading to progression of melanoma. Similarly, Abramovitz et al. (2011) have reported that both membrane and secreted klotho serve as tumor

Figure 10 Resveratrol and SIRT1 cooperate with 1,25D to enhance VDR signaling and the production of α-klotho. (A) Upper panel: HEK-293 cells were treated with 1,25D, resveratrol and/or SIRT1, and endogenous *CYP24A1* expression in human embryonic kidney cells was measured by real-time PCR. Lower panel: As in the upper panel, but treatments included the selective SIRT1 inhibitor, EX-527. (B) HEK-293 cells were cotransfected with a Firefly luciferase plasmid containing a β-catenin responsive element along with the indicated expression plasmids encoding soluble klotho (sKlotho) or β-catenin (β-CAT), with Firefly luciferase results normalized to Renilla luciferase. (C) A hypothetical model for resveratrol activation of VDR via stimulation of SIRT1 (Baur, 2010). SIRT1 catalyzes deacetylation of VDR (to increase the capacity of 1,25D binding), RXR, or comodulators. SIRT1 activation also leads to ADAM10 stimulation (Donmez, Wang, Cohen, & Guarente, 2010) to produce soluble klotho (sKlotho) via ADAM10-mediated cleavage of membrane klotho (mKlotho). Curcumin selectively induces mKlotho, while 1,25D stimulates SIRT1 activity (An et al., 2010) as well as expression of both mKlotho and sKlotho. The integration of these regulatory circuits, which are controlled by the levels of nutritionally derived "healthy" lipids (1,25D, curcumin, and resveratrol), culminates in the elaboration of sKlotho from the kidney to exert proposed endocrine anti-aging effects consisting of antifibrogenic and antineoplastic actions in the vasculature and other target tissues. (See the color plate.)

suppressors by inhibiting tumor cell proliferation through regulation of insulin-like growth factor-1 (IGF-1) signaling. Another *in vivo* experiment showed that secreted klotho possesses greater inhibitory effects on tumor cell growth than full-length klotho (Abramovitz et al., 2011). These results, along with data from our laboratories that reveal potent suppression of β-catenin activity by klotho in cancer cells which is further potentiated by 1,25D, raise the possibility that a subset of the antitumor actions of 1,25D could be mediated via upregulation of klotho expression.

3.5 Influence on Insulin/IGF-1 Actions

It has been observed that klotho increases the plasma membrane retention of TRPV2, leading to enhanced glucose-triggered insulin secretion from pancreatic β-cells (Lin & Sun, 2012). Vitamin D has long been known to promote insulin secretion (Norman, Frankel, Heldt, & Grodsky, 1980), meaning that insulin release is yet another example of the dual beneficial effects of 1,25D and klotho. *Klotho* knockout mice exhibit less energy storage and expenditure compared to wild-type mice (Mori et al., 2000), as well as attenuated insulin production and enhanced insulin sensitivity (Utsugi et al., 2000; Wolf et al., 2008). Apparently, klotho suppresses the downstream signaling pathways of both the insulin receptor (mediated by insulin receptor substrate (IRS)), and the insulin-like growth factor 1 receptor (IGF-1R), without directly associating with either of these receptors (Mori et al., 2000; Yamamoto et al., 2005). Instead, klotho likely affects IRS and IGF-1R activity via modulation of the forkhead box proteins (FOXOs). Activated IRS signals in part via phosphorylation of FOXO1, FOXO3a, and FOXO4, which then remain in the cytoplasm rather than traveling to the nucleus. As noted above, klotho modulates FOXO phosphorylation, raising the possibility that klotho may also regulate the ability of FOXO1 to mediate insulin stimulation of glucose production (Nakae, Kitamura, Silver, & Accili, 2001). Klotho also promotes adipogenesis through induction of CCAAT/enhancer-binding proteins (C/EBPs) (Wolf et al., 2008) and PPARγ (Chihara et al., 2006), both of which elicit differentiation of pre-adipocytes (Cristancho & Lazar, 2011).

3.6 Anti-aging and Organ Protection

Inhibition of IGF-1 signaling, whether by IGF-1R haploinsufficiency, a reduction in IGF-1 ligand levels or by specific deletion of the insulin receptor in fat tissue, prolongs lifespan (Katic et al., 2007). Moreover, as detailed above, overexpression of klotho inhibits both insulin and IGF-1 signaling. Thus, klotho could decelerate aging either by directly signaling in all target tissues, or indirectly by inhibiting insulin signaling and inducing a FOXO1-mediated factor in fat tissue. In addition, the many other actions of klotho, such as antioxidation, antifibrosis, antimalignancy, anticalcium transients, and antiphosphatemia, contribute to the anti-aging potential of klotho. Based on all these observations, klotho can be considered an organ protection hormone that promulgates healthful aging by delaying chronic diseases through its beneficial signaling in virtually all cells.

As pointed out by Kuro-o (2012), and discussed herein, high concentrations of extracellular phosphate are toxic to cells, and impaired urinary phosphate excretion increases serum phosphate level to induce a premature-aging phenotype. Urinary phosphate levels are increased by dietary phosphate overload and might induce tubular injury and interstitial fibrosis. Extracellular phosphate exerts its cytotoxic effects by forming insoluble nanoparticles with calcium and fetuin-A. These nanoparticles are referred to by Kuro-o as calciprotein particles and are capable of inducing various cellular responses, including the osteogenic transformation of vascular smooth muscle cells and cell death of vascular endothelial cells and renal tubular epithelial cells. Calciprotein particles can be detected in the serum of animal models of kidney disease and in patients with CKD and probably contribute to the pathogenesis of CKD. This important insight provides a mechanism whereby klotho, by preventing hyperphosphatemia, protects the vascular and renal systems, thereby prolonging lifespan. In addition, 1,25D is thought to cooperate with klotho in retarding vascular calcification by the induction of osteopontin (Table 1), a powerful antimineralization factor (Lau et al., 2012). Ironically, the two renal hormones that are deficient in patients with chronic renal failure because of loss of renal mass, namely 1,25D and klotho, are two vital effectors of renal and vascular health, suggesting a strategy for the prevention and treatment of vascular disease, as well as CKD. We conclude that the future for the dynamic duo of 1,25D and Klotho is bright, and not only have these renal hormones come of (endocrine) age, but they may well be the secret to a long and healthy life.

4. CONCLUSION AND FUTURE DIRECTIONS

1,25D is anti-aging when at optimal levels, but pro-aging when either deficient or in excess. Klotho is a "D"-helper that supports the anti-aging benefits of D/VDR, both by keeping 1,25D in check and being induced by 1,25D to work along side as a "one-two punch" that knocks out vascular calcification, fibrosis, oxidative stress, protracted calcium signaling intracellularly (esp. in the CNS), and excessive Wnt signaling that promotes cancer, as well as insulin/IGF signaling that leads to aging. Summarized in Fig. 10C is the authors' conceptualization of how 1,25D, its nutritional surrogates, and the induction and secretion of klotho, potentiated by bona fide anti-aging principles such as resveratrol and SIRT1 (Fig. 10A), ultimately mediate vascular and tissue protection to effect healthspan. The basic molecular mechanisms outlined herein represent plausible explanations for how 1,25D and klotho enhance healthspan, but ultimately they will need to be supported by evidence-based medicine in rigorous clinical studies that provide proof of concept.

ACKNOWLEDGMENTS

This study was supported by National Institutes of Health grants NIH DK033351 to M.R.H. and CA140285 to P.W.J.

REFERENCES

Abramovitz, L., Rubinek, T., Ligumsky, H., Bose, S., Barshack, I., Avivi, C., et al. (2011). KL1 internal repeat mediates klotho tumor suppressor activities and inhibits BFGF and IGF-I signaling in pancreatic cancer. *Clinical Cancer Research, 17*(13), 4254–4266. http://dx.doi.org/10.1158/1078-0432.CCR-10-2749.

Adamcio, B., Sargin, D., Stradomska, A., Medrihan, L., Gertler, C., Theis, F., et al. (2008). Erythropoietin enhances hippocampal long-term potentiation and memory. *BMC Biology, 6*, 37. http://dx.doi.org/10.1186/1741-7007-6-37.

Aguilera, O., Pena, C., Garcia, J. M., Larriba, M. J., Ordonez-Moran, P., Navarro, D., et al. (2007). The Wnt antagonist dickkopf-1 gene is induced by 1alpha,25-dihydroxyvitamin D3 associated to the differentiation of human colon cancer cells. *Carcinogenesis, 28*(9), 1877–1884. http://dx.doi.org/10.1093/carcin/bgm094.

Amling, M., Priemel, M., Holzmann, T., Chapin, K., Rueger, J. M., Baron, R., et al. (1999). Rescue of the skeletal phenotype of vitamin D receptor-ablated mice in the setting of normal mineral ion homeostasis: Formal histomorphometric and biomechanical analyses. *Endocrinology, 140*(11), 4982–4987.

An, B. S., Tavera-Mendoza, L. E., Dimitrov, V., Wang, X., Calderon, M. R., Wang, H. J., et al. (2010). Stimulation of Sirt1-regulated foxo protein function by the ligand-bound vitamin D receptor. *Molecular and Cellular Biology, 30*(20), 4890–4900. http://dx.doi.org/10.1128/MCB.00180-10.

Anglin, R. E., Samaan, Z., Walter, S. D., & McDonald, S. D. (2013). Vitamin D deficiency and depression in adults: Systematic review and meta-analysis. *British Journal of Psychiatry, 202*, 100–107. http://dx.doi.org/10.1192/bjp.bp.111.106666.

Annweiler, C., Llewellyn, D. J., & Beauchet, O. (2013). Low serum vitamin D concentrations in Alzheimer's disease: A systematic review and meta-analysis. *Journal of Alzheimer's Disease, 33*(3), 659–674. http://dx.doi.org/10.3233/JAD-2012-121432.

Araya, Z., & Wikvall, K. (1999). 6Alpha-hydroxylation of taurochenodeoxycholic acid and lithocholic acid by CYP3A4 in human liver microsomes. *Biochimica et Biophysica Acta, 1438*(1), 47–54.

Austin, H. R., Hoss, E., Batie, S. F., Moffet, E. W., Jurutka, P. W., Haussler, M. R., et al. (2014). Regulation of late cornified envelope genes relevant to psoriasis risk by plant-derived cyanidin. *Biochemical and Biophysical Research Communications, 443*(4), 1275–1279. http://dx.doi.org/10.1016/j.bbrc.2013.12.128.

Bao, B. Y., Ting, H. J., Hsu, J. W., & Lee, Y. F. (2008). Protective role of 1 alpha, 25-dihydroxyvitamin D3 against oxidative stress in nonmalignant human prostate epithelial cells. *International Journal of Cancer, 122*(12), 2699–2706. http://dx.doi.org/10.1002/ijc.23460.

Barthel, T. K., Mathern, D. R., Whitfield, G. K., Haussler, C. A., Hopper, H. A. I., Hsieh, J. C., et al. (2007). 1,25-Dihydroxyvitamin D3/VDR-mediated induction of FGF23 as well as transcriptional control of other bone anabolic and catabolic genes that orchestrate the regulation of phosphate and calcium mineral metabolism. *Journal of Steroid Biochemistry and Molecular Biology, 103*(3–5), 381–388.

Bartik, L., Whitfield, G. K., Kaczmarska, M., Lowmiller, C. L., Moffet, E. W., Furmick, J. K., et al. (2010). Curcumin: A novel nutritionally derived ligand of the vitamin D receptor with implications for colon cancer chemoprevention. *Journal of Nutritional Biochemistry, 21*, 1153–1161. http://dx.doi.org/10.1016/j.jnutbio.2009.09.012.

Barwick, J. L., Quattrochi, L. C., Mills, A. S., Potenza, C., Tukey, R. H., & Guzelian, P. S. (1996). Trans-species gene transfer for analysis of glucocorticoid-inducible transcriptional activation of transiently expressed human CYP3A4 and rabbit CYP3A6 in primary cultures of adult rat and rabbit hepatocytes. *Molecular Pharmacology, 50*(1), 10–16.

Baur, J. A. (2010). Resveratrol, sirtuins, and the promise of a dr mimetic. *Mechanisms of Ageing and Development, 131*(4), 261–269. http://dx.doi.org/10.1016/j.mad.2010.02.007.

Bea, J. W., Jurutka, P. W., Hibler, E. A., Lance, P., Martinez, M. E., Roe, D. J., et al. (2015). Concentrations of the vitamin D metabolite 1,25(OH)2D and odds of metabolic syndrome and its components. *Metabolism: Clinical and Experimental, 64*(3), 447–459. http://dx.doi.org/10.1016/j.metabol.2014.11.010.

Beaudoin, G. M., 3rd, Sisk, J. M., Coulombe, P. A., & Thompson, C. C. (2005). Hairless triggers reactivation of hair growth by promoting Wnt signaling. *Proceedings of the National Academy of Sciences of the United States of America, 102*(41), 14653–14658.

Behera, R., Marchbank, K., Kaur, A., Dang, V., Webster, M., O'Connell, M., et al. (2015). Crosstalk between klotho and Wnt5a drives age-related melanoma progression. *Cancer Research, 75*(14 Suppl.). Abstract A11. http://dx.doi.org/10.1158/1538-7445.MEL2014-A11.

Behrens, M. M., & Sejnowski, T. J. (2009). Does schizophrenia arise from oxidative dysregulation of parvalbumin-interneurons in the developing cortex? *Neuropharmacology, 57*(3), 193–200. http://dx.doi.org/10.1016/j.neuropharm.2009.06.002.

Belvederi Murri, M., Respino, M., Masotti, M., Innamorati, M., Mondelli, V., Pariante, C., et al. (2013). Vitamin D and psychosis: Mini meta-analysis. *Schizophrenia Research, 150*(1), 235–239. http://dx.doi.org/10.1016/j.schres.2013.07.017.

Bergboer, J. G., Tjabringa, G. S., Kamsteeg, M., van Vlijmen-Willems, I. M., Rodijk-Olthuis, D., Jansen, P. A., et al. (2011). Psoriasis risk genes of the late cornified envelope-3 group are distinctly expressed compared with genes of other lce groups. *American Journal of Pathology, 178*(4), 1470–1477. http://dx.doi.org/10.1016/j.ajpath.2010.12.017.

Bergwitz, C., & Juppner, H. (2010). Regulation of phosphate homeostasis by PTH, vitamin D, and FGF23. *Annual Review of Medicine, 61*, 91–104. http://dx.doi.org/10.1146/annurev.med.051308.111339.

Berridge, M. J. (2012). Calcium signalling remodelling and disease. *Biochemical Society Transactions, 40*(2), 297–309. http://dx.doi.org/10.1042/BST20110766.

Berridge, M. J. (2013). Dysregulation of neural calcium signaling in Alzheimer disease, bipolar disorder and schizophrenia. *Prion*, 7(1), 2–13. http://dx.doi.org/10.4161/pri.21767.

Berridge, M. J. (2014). Calcium signalling and psychiatric disease: Bipolar disorder and schizophrenia. *Cell and Tissue Research, 357*(2), 477–492. http://dx.doi.org/10.1007/s00441-014-1806-z.

Berridge, M. J. (2015). Vitamin D cell signalling in health and disease. *Biochemical and Biophysical Research Communications, 460*(1), 53–71. http://dx.doi.org/10.1016/j.bbrc.2015.01.008.

Bezprozvanny, I., & Mattson, M. P. (2008). Neuronal calcium mishandling and the pathogenesis of Alzheimer's disease. *Trends in Neurosciences, 31*(9), 454–463. http://dx.doi.org/10.1016/j.tins.2008.06.005.

Bianco, S. D., Peng, J. B., Takanaga, H., Suzuki, Y., Crescenzi, A., Kos, C. H., et al. (2007). Marked disturbance of calcium homeostasis in mice with targeted disruption of the trpv6 calcium channel gene. *Journal of Bone and Mineral Research, 22*(2), 274–285. http://dx.doi.org/10.1359/jbmr.061110.

Bikle, D. D., Ng, D., Oda, Y., Hanley, K., Feingold, K., & Xie, Z. (2002). The vitamin D response element of the involucrin gene mediates its regulation by 1,25-dihydroxyvitamin D_3. *Journal of Investigative Dermatology, 119*(5), 1109–1113.

Bouillon, R., Carmeliet, G., Verlinden, L., van Etten, E., Verstuyf, A., Luderer, H. F., et al. (2008). Vitamin D and human health: Lessons from vitamin D receptor null mice. *Endocrine Reviews, 29*(6), 726–776. http://dx.doi.org/10.1210/er.2008-0004.

Brewer, L. D., Thibault, V., Chen, K. C., Langub, M. C., Landfield, P. W., & Porter, N. M. (2001). Vitamin D hormone confers neuroprotection in parallel with downregulation of l-type calcium channel expression in hippocampal neurons. *Journal of Neuroscience, 21*(1), 98–108.

Brown, J., Bianco, J. I., McGrath, J. J., & Eyles, D. W. (2003). 1,25-Dihydroxyvitamin D3 induces nerve growth factor, promotes neurite outgrowth and inhibits mitosis in embryonic rat hippocampal neurons. *Neuroscience Letters, 343*(2), 139–143.

Butterfield, D. A., Swomley, A. M., & Sultana, R. (2013). Amyloid beta-peptide (1-42)-induced oxidative stress in Alzheimer disease: Importance in disease pathogenesis and progression. *Antioxidants & Redox Signaling, 19*(8), 823–835. http://dx.doi.org/10.1089/ars.2012.5027.

Cali, T., Ottolini, D., & Brini, M. (2014). Calcium signaling in Parkinson's disease. *Cell and Tissue Research, 357*(2), 439–454. http://dx.doi.org/10.1007/s00441-014-1866-0.

Cao, X., Ross, F. P., Zhang, L., MacDonald, P. N., Chappel, J., & Teitelbaum, S. L. (1993). Cloning of the promoter for the avian integrin b_3 subunit gene and its regulation by 1,25-dihydroxyvitamin D_3. *Journal of Biological Chemistry, 268*(36), 27371–27380.

Cardus, A., Panizo, S., Encinas, M., Dolcet, X., Gallego, C., Aldea, M., et al. (2009). 1,25-Dihydroxyvitamin D3 regulates VEGF production through a vitamin D response element in the VEGF promoter. *Atherosclerosis, 204*(1), 85–89. http://dx.doi.org/10.1016/j.atherosclerosis.2008.08.020.

Cha, S. K., Hu, M. C., Kurosu, H., Kuro-o, M., Moe, O., & Huang, C. L. (2009). Regulation of renal outer medullary potassium channel and renal K(+) excretion by klotho. *Molecular Pharmacology, 76*(1), 38–46. http://dx.doi.org/10.1124/mol.109.055780.

Chang, Q., Hoefs, S., van der Kemp, A. W., Topala, C. N., Bindels, R. J., & Hoenderop, J. G. (2005). The beta-glucuronidase klotho hydrolyzes and activates the TRPV5 channel. *Science, 310*(5747), 490–493. http://dx.doi.org/10.1126/science.1114245.

Chen, K. B., Lin, A. M., & Chiu, T. H. (2003). Systemic vitamin D3 attenuated oxidative injuries in the locus coeruleus of rat brain. *Annals of the New York Academy of Sciences, 993*, 313–324. discussion 345–319.

Chen, W. R., Liu, Z. Y., Shi, Y., Yin da, W., Wang, H., Sha, Y., et al. (2014). Relation of low vitamin D to nonvalvular persistent atrial fibrillation in Chinese patients.

Annals of Noninvasive Electrocardiology, *19*(2), 166–173. http://dx.doi.org/10.1111/anec.12105.

Chen, C. D., Podvin, S., Gillespie, E., Leeman, S. E., & Abraham, C. R. (2007). Insulin stimulates the cleavage and release of the extracellular domain of klotho by ADAM10 and ADAM17. *Proceedings of the National Academy of Sciences of the United States of America*, *104*(50), 19796–19801. http://dx.doi.org/10.1073/pnas.0709805104.

Chen, B., Zhang, Y., Wang, Y., Rao, J., Jiang, X., & Xu, Z. (2014). Curcumin inhibits proliferation of breast cancer cells through Nrf2-mediated down-regulation of Fen1 expression. *Journal of Steroid Biochemistry and Molecular Biology*, *143*, 11–18. http://dx.doi.org/10.1016/j.jsbmb.2014.01.009.

Chihara, Y., Rakugi, H., Ishikawa, K., Ikushima, M., Maekawa, Y., Ohta, J., et al. (2006). Klotho protein promotes adipocyte differentiation. *Endocrinology*, *147*(8), 3835–3842. http://dx.doi.org/10.1210/en.2005-1529.

Cho, Y. M., Woodard, G. L., Dunbar, M., Gocken, T., Jimenez, J. A., & Foley, J. (2003). Hair-cycle-dependent expression of parathyroid hormone-related protein and its type I receptor: Evidence for regulation at the anagen to catagen transition. *Journal of Investigative Dermatology*, *120*(5), 715–727.

Choudhury, S., Bae, S., Ke, Q., Lee, J. Y., Singh, S. S., St-Arnaud, R., et al. (2014). Abnormal calcium handling and exaggerated cardiac dysfunction in mice with defective vitamin D signaling. *PLoS One*, *9*(9), e108382. http://dx.doi.org/10.1371/journal.pone.0108382.

Cianferotti, L., Cox, M., Skorija, K., & Demay, M. B. (2007). Vitamin D receptor is essential for normal keratinocyte stem cell function. *Proceedings of the National Academy of Sciences of the United States of America*, *104*(22), 9428–9433.

Cieslak, K., Feingold, J., Antonius, D., Walsh-Messinger, J., Dracxler, R., Rosedale, M., et al. (2014). Low vitamin D levels predict clinical features of schizophrenia. *Schizophrenia Research*, *159*(2-3), 543–545. http://dx.doi.org/10.1016/j.schres.2014.08.031.

Clemente-Postigo, M., Munoz-Garach, A., Serrano, M., Garrido-Sanchez, L., Bernal-Lopez, M. R., Fernandez-Garcia, D., et al. (2015). Serum 25-hydroxyvitamin D and adipose tissue vitamin D receptor gene expression: Relationship with obesity and type 2 diabetes. *Journal of Clinical Endocrinology and Metabolism*, *100*(4), E591–E595. http://dx.doi.org/10.1210/jc.2014-3016.

Cramer, P. E., Cirrito, J. R., Wesson, D. W., Lee, C. Y., Karlo, J. C., Zinn, A. E., et al. (2012). ApoE-directed therapeutics rapidly clear beta-amyloid and reverse deficits in AD mouse models. *Science*, *335*(6075), 1503–1506. http://dx.doi.org/10.1126/science.1217697.

Cristancho, A. G., & Lazar, M. A. (2011). Forming functional fat: A growing understanding of adipocyte differentiation. *Nature Reviews. Molecular Cell Biology*, *12*(11), 722–734. http://dx.doi.org/10.1038/nrm3198.

Dampf-Stone, A., Batie, S., Sabir, M., Jacobs, E. T., Lee, J. H., Whitfield, G. K., et al. (2015). Resveratrol potentiates vitamin D and nuclear receptor signaling. *Journal of Cellular Biochemistry*, *116*(6), 1130–1143. http://dx.doi.org/10.1002/jcb.25070.

de Cid, R., Riveira-Munoz, E., Zeeuwen, P. L., Robarge, J., Liao, W., Dannhauser, E. N., et al. (2009). Deletion of the late cornified envelope LCE3B and LCE3C genes as a susceptibility factor for psoriasis. *Nature Genetics*, *41*(2), 211–215. http://dx.doi.org/10.1038/ng.313.

Decressac, M., Volakakis, N., Bjorklund, A., & Perlmann, T. (2013). NURR1 in Parkinson disease—From pathogenesis to therapeutic potential. *Nature Reviews. Neurology*, *9*(11), 629–636. http://dx.doi.org/10.1038/nrneurol.2013.209.

Deeb, K. K., Trump, D. L., & Johnson, C. S. (2007). Vitamin D signalling pathways in cancer: Potential for anticancer therapeutics. *Nature Reviews. Cancer*, 7(9), 684–700. http://dx.doi.org/10.1038/nrc2196.

DeLuca, G. C., Kimball, S. M., Kolasinski, J., Ramagopalan, S. V., & Ebers, G. C. (2013). Review: The role of vitamin D in nervous system health and disease. *Neuropathology and Applied Neurobiology*, *39*(5), 458–484. http://dx.doi.org/10.1111/nan.12020.

DeMay, M. B., Kiernan, M. S., DeLuca, H. F., & Kronenberg, H. M. (1992). Sequences in the human parathyroid hormone gene that bind the 1,25-dihydroxyvitamin D_3 receptor and mediate transcriptional repression in response to 1,25-dihydroxyvitamin D_3. *Proceedings of the National Academy of Sciences of the United States of America*, *89*, 8097–8101.

Doi, S., Zou, Y., Togao, O., Pastor, J. V., John, G. B., Wang, L., et al. (2011). Klotho inhibits transforming growth factor-beta1 (TGF-beta1) signaling and suppresses renal fibrosis and cancer metastasis in mice. *Journal of Biological Chemistry*, *286*(10), 8655–8665. http://dx.doi.org/10.1074/jbc.M110.174037.

Dong, J., Lau, C. W., Wong, S. L., & Huang, Y. (2014). Cardiovascular benefits of vitamin D. *Sheng Li Xue Bao*, *66*(1), 30–36.

Dong, J., Wong, S. L., Lau, C. W., Lee, H. K., Ng, C. F., Zhang, L., et al. (2012). Calcitriol protects renovascular function in hypertension by down-regulating angiotensin II type 1 receptors and reducing oxidative stress. *European Heart Journal*, *33*(23), 2980–2990. http://dx.doi.org/10.1093/eurheartj/ehr459.

Donmez, G., Wang, D., Cohen, D. E., & Guarente, L. (2010). SIRT1 suppresses beta-amyloid production by activating the alpha-secretase gene ADAM10. *Cell*, *142*(2), 320–332. http://dx.doi.org/10.1016/j.cell.2010.06.020.

Drissi, H., Pouliot, A., Koolloos, C., Stein, J. L., Lian, J. B., Stein, G. S., et al. (2002). 1,25-$(OH)_2$-Vitamin D_3 suppresses the bone-related Runx2/Cbfa1 gene promoter. *Experimental Cell Research*, *274*(2), 323–333.

Dzau, V. J. (1988). Circulating versus local renin-angiotensin system in cardiovascular homeostasis. *Circulation*, 77(6 Pt. 2), I4–I13.

Echchgadda, I., Song, C. S., Roy, A. K., & Chatterjee, B. (2004). Dehydroepiandrosterone sulfotransferase is a target for transcriptional induction by the vitamin D receptor. *Molecular Pharmacology*, *65*(3), 720–729. http://dx.doi.org/10.1124/mol.65.3.720.

Eelen, G., Verlinden, L., Meyer, M. B., Gijsbers, R., Pike, J. W., Bouillon, R., et al. (2013). 1,25-Dihydroxyvitamin D3 and the aging-related forkhead box O and sestrin proteins in osteoblasts. *Journal of Steroid Biochemistry and Molecular Biology*, *136*(136), 112–119. http://dx.doi.org/10.1016/j.jsbmb.2012.09.011.

Egan, J. B., Thompson, P. A., Vitanov, M. V., Bartik, L., Jacobs, E. T., Haussler, M. R., et al. (2010). Vitamin D receptor ligands, adenomatous polyposis coli, and the vitamin D receptor Foki polymorphism collectively modulate beta-catenin activity in colon cancer cells. *Molecular Carcinogenesis*, *49*(4), 337–352. http://dx.doi.org/10.1002/mc.20603.

Eisman, J. A., Barkla, D. H., & Tutton, P. J. (1987). Suppression of in vivo growth of human cancer solid tumor xenografts by 1,25-dihydroxyvitamin D3. *Cancer Research*, *47*(1), 21–25.

Ellison, T. I., Smith, M. K., Gilliam, A. C., & MacDonald, P. N. (2008). Inactivation of the vitamin D receptor enhances susceptibility of murine skin to UV-induced tumorigenesis. *Journal of Investigative Dermatology*, *128*(10), 2508–2517. http://dx.doi.org/10.1038/jid.2008.131.

Eloranta, J. J., Zair, Z. M., Hiller, C., Hausler, S., Stieger, B., & Kullak-Ublick, G. A. (2009). Vitamin D3 and its nuclear receptor increase the expression and activity of the human proton-coupled folate transporter. *Molecular Pharmacology*, *76*(5), 1062–1071. http://dx.doi.org/10.1124/mol.109.055392.

Eyles, D. W., Burne, T. H., & McGrath, J. J. (2013). Vitamin D, effects on brain development, adult brain function and the links between low levels of vitamin D and neuropsychiatric disease. *Frontiers in Neuroendocrinology*, *34*(1), 47–64. http://dx.doi.org/10.1016/j.yfrne.2012.07.001.

Eyles, D. W., Smith, S., Kinobe, R., Hewison, M., & McGrath, J. J. (2005). Distribution of the vitamin D receptor and 1 alpha-hydroxylase in human brain. *Journal of Chemical Neuroanatomy*, *29*(1), 21–30. http://dx.doi.org/10.1016/j.jchemneu.2004.08.006.

Falzon, M. (1996). DNA sequences in the rat parathyroid hormone-related peptide gene responsible for 1,25-dihydroxyvitamin D_3-mediated transcriptional repression. *Molecular Endocrinology*, *10*, 672–681.

Feldman, D., Krishnan, A. V., Swami, S., Giovannucci, E., & Feldman, B. J. (2014). The role of vitamin D in reducing cancer risk and progression. *Nature Reviews Cancer*, *14*(5), 342–357. http://dx.doi.org/10.1038/nrc3691.

Forster, R. E., Jurutka, P. W., Hsieh, J. C., Haussler, C. A., Lowmiller, C. L., Kaneko, I., et al. (2011). Vitamin D receptor controls expression of the anti-aging klotho gene in mouse and human renal cells. *Biochemical and Biophysical Research Communications*, *414*(3), 557–562. http://dx.doi.org/10.1016/j.bbrc.2011.09.117.

Fretz, J. A., Zella, L. A., Kim, S., Shevde, N. K., & Pike, J. W. (2006). 1,25-Dihydroxyvitamin D3 regulates the expression of low-density lipoprotein receptor-related protein 5 via deoxyribonucleic acid sequence elements located downstream of the start site of transcription. *Molecular Endocrinology*, *20*(9), 2215–2230.

Garcia-Yague, A. J., Rada, P., Rojo, A. I., Lastres-Becker, I., & Cuadrado, A. (2013). Nuclear import and export signals control the subcellular localization of Nurr1 protein in response to oxidative stress. *Journal of Biological Chemistry*, *288*(8), 5506–5517. http://dx.doi.org/10.1074/jbc.M112.439190.

Garcion, E., Wion-Barbot, N., Montero-Menei, C. N., Berger, F., & Wion, D. (2002). New clues about vitamin D functions in the nervous system. *Trends in Biochemical Sciences*, *13*(3), 100–105. http://dx.doi.org/10.1016/S1043-2760(01)00547-1.

Garland, C. F., Garland, F. C., & Gorham, E. D. (1999). Calcium and vitamin D. Their potential roles in colon and breast cancer prevention. *Annals of the New York Academy of Sciences*, *889*, 107–119.

Garland, C., Shekelle, R. B., Barrett-Connor, E., Criqui, M. H., Rossof, A. H., & Paul, O. (1985). Dietary vitamin D and calcium and risk of colorectal cancer: A 19-year prospective study in men. *Lancet*, *1*(8424), 307–309.

Gattineni, J., Twombley, K., Goetz, R., Mohammadi, M., & Baum, M. (2011). Regulation of serum 1,25(OH)2vitamin D3 levels by fibroblast growth factor 23 is mediated by FGF receptors 3 and 4. *American Journal of Physiology. Renal Physiology*, *301*(2), F371–F377. http://dx.doi.org/10.1152/ajprenal.00740.2010.

Gawryluk, J. W., Wang, J. F., Andreazza, A. C., Shao, L., & Young, L. T. (2011). Decreased levels of glutathione, the major brain antioxidant, in post-mortem prefrontal cortex from patients with psychiatric disorders. *International Journal of Neuropsychopharmacology*, *14*(1), 123–130. http://dx.doi.org/10.1017/S1461145710000805.

Goksugur, S. B., Tufan, A. E., Semiz, M., Gunes, C., Bekdas, M., Tosun, M., et al. (2014). Vitamin D status in children with attention-deficit-hyperactivity disorder. *Pediatrics International*, *56*(4), 515–519. http://dx.doi.org/10.1111/ped.12286.

Gombart, A. F., Borregaard, N., & Koeffler, H. P. (2005). Human cathelicidin antimicrobial peptide (CAMP) gene is a direct target of the vitamin D receptor and is strongly up-regulated in myeloid cells by 1,25-dihydroxyvitamin D3. *FASEB Journal*, *19*(9), 1067–1077.

Gracious, B. L., Finucane, T. L., Friedman-Campbell, M., Messing, S., & Parkhurst, M. N. (2012). Vitamin D deficiency and psychotic features in mentally ill adolescents: A cross-sectional study. *BMC Psychiatry*, *12*, 38. http://dx.doi.org/10.1186/1471-244X-12-38.

Groves, N. J., Kesby, J. P., Eyles, D. W., McGrath, J. J., Mackay-Sim, A., & Burne, T. H. (2013). Adult vitamin D deficiency leads to behavioural and brain neurochemical alterations in C57BL/6J and BALB/c mice. *Behavioural Brain Research*, *241*, 120–131. http://dx.doi.org/10.1016/j.bbr.2012.12.001.

Grudet, C., Malm, J., Westrin, A., & Brundin, L. (2014). Suicidal patients are deficient in vitamin D, associated with a pro-inflammatory status in the blood. *Psychoneuroendocrinology*, *50*, 210–219. http://dx.doi.org/10.1016/j.psyneuen.2014.08.016.

Guilliams, T. G. (2004). Homocysteine—A risk factor for vascular diseases: Guidelines for the clinical practice. *Journal of the American Nutraceutical Association*, 7, 11–24.

Halicka, H. D., Zhao, H., Li, J., Lee, Y. S., Hsieh, T. C., Wu, J. M., et al. (2012). Potential anti-aging agents suppress the level of constitutive mTOR- and DNA damage-signaling. *Aging (Albany, NY)*, *4*(12), 952–965.

Hanafy, D. A., Chang, S. L., Lu, Y. Y., Chen, Y. C., Kao, Y. H., Huang, J. H., et al. (2014). Electromechanical effects of 1,25-dihydroxyvitamin D with antiatrial fibrillation activities. *Journal of Cardiovascular Electrophysiology*, *25*(3), 317–323. http://dx.doi.org/10.1111/jce.12309.

Harms, L. R., Burne, T. H., Eyles, D. W., & McGrath, J. J. (2011). Vitamin D and the brain. *Best Practice & Research. Clinical Endocrinology & Metabolism*, *25*(4), 657–669. http://dx.doi.org/10.1016/j.beem.2011.05.009.

Haussler, M. R., Haussler, C. A., Bartik, L., Whitfield, G. K., Hsieh, J. C., Slater, S., et al. (2008). Vitamin D receptor: Molecular signaling and actions of nutritional ligands in disease prevention. *Nutrition Reviews*, *66*(10 Suppl. 2), S98–S112. http://dx.doi.org/10.1111/j.1753-4887.2008.00093.x.

Haussler, M. R., Haussler, C. A., Whitfield, G. K., Hsieh, J. C., Thompson, P. D., Barthel, T. K., et al. (2010). The nuclear vitamin D receptor controls the expression of genes encoding factors which feed the "fountain of youth" to mediate healthful aging. *Journal of Steroid Biochemistry and Molecular Biology*, *121*, 88–97. http://dx.doi.org/10.1016/j.jsbmb.2010.03.019.

Haussler, M. R., Jurutka, P. W., Mizwicki, M., & Norman, A. W. (2011). Vitamin D receptor (VDR)-mediated actions of 1alpha,25(OH)vitamin D: Genomic and non-genomic mechanisms. *Best Practice & Research. Clinical Endocrinology & Metabolism*, *25*(4), 543–559. http://dx.doi.org/10.1016/j.beem.2011.05.010.

Haussler, M. R., Whitfield, G. K., Kaneko, I., Haussler, C. A., Hsieh, D., Hsieh, J. C., et al. (2013). Molecular mechanisms of vitamin D action. *Calcified Tissue International*, *92*(2), 77–98. http://dx.doi.org/10.1007/s00223-012-9619-0.

Heaney, R. P. (2008). Vitamin D in health and disease. *Clinical Journal of the American Society of Nephrology*, *3*(5), 1535–1541. http://dx.doi.org/10.2215/CJN.01160308.

Hesse, M., Frohlich, L. F., Zeitz, U., Lanske, B., & Erben, R. G. (2007). Ablation of vitamin D signaling rescues bone, mineral, and glucose homeostasis in FGF-23 deficient mice. *Matrix Biology*, *26*(2), 75–84. http://dx.doi.org/10.1016/j.matbio.2006.10.003.

Hibler, E. A., Molmenti, C. L., Lance, P., Jurutka, P. W., & Jacobs, E. T. (2014). Associations between circulating 1,25(OH)(2)D concentration and odds of metachronous colorectal adenoma. *Cancer Causes and Control*, *25*(7), 809–817. http://dx.doi.org/10.1007/s10552-014-0382-6.

Hiroi, R., & Handa, R. J. (2013). Estrogen receptor-beta regulates human tryptophan hydroxylase-2 through an estrogen response element in the 5′ untranslated region. *Journal of Neurochemistry*, *127*(4), 487–495. http://dx.doi.org/10.1111/jnc.12401.

Hollis, B. W., Johnson, D., Hulsey, T. C., Ebeling, M., & Wagner, C. L. (2011). Vitamin D supplementation during pregnancy: Double-blind, randomized clinical trial of safety and effectiveness. *Journal of Bone and Mineral Research*, *26*(10), 2341–2357. http://dx.doi.org/10.1002/jbmr.463.

Hoss, E., Austin, H. R., Batie, S. F., Jurutka, P. W., Haussler, M. R., & Whitfield, G. K. (2013). Control of late cornified envelope genes relevant to psoriasis risk: Upregulation by 1,25-dihydroxyvitamin D and plant-derived delphinidin. *Archives for Dermatological Research*, *305*(10), 867–878. http://dx.doi.org/10.1007/s00403-013-1390-1.

Hossein-nezhad, A., & Holick, M. F. (2012). Optimize dietary intake of vitamin D: An epigenetic perspective. *Current Opinion in Clinical Nutrition and Metabolic Care, 15*(6), 567–579. http://dx.doi.org/10.1097/MCO.0b013e3283594978.

Hsieh, J. C., Estess, R. C., Kaneko, I., Whitfield, G. K., Jurutka, P. W., & Haussler, M. R. (2014). Vitamin D receptor-mediated control of soggy, wise, and hairless gene expression in keratinocytes. *Journal of Endocrinology, 220*(2), 165–178. http://dx.doi.org/10.1530/JOE-13-0212.

Hsieh, J.-C., Sisk, J. M., Jurutka, P. W., Haussler, C. A., Slater, S. A., Haussler, M. R., et al. (2003). Physical and functional interaction between the vitamin D receptor and hairless corepressor, two proteins required for hair cycling. *Journal of Biological Chemistry, 278*(40), 38665–38674. http://dx.doi.org/10.1074/jbc.M304886200.

Hsieh, J.-C., Whitfield, G. K., Oza, A. K., Dang, H. T. L., Price, J. N., Galligan, M. A., et al. (1999). Characterization of unique DNA binding and transcriptional activation functions in the carboxyl-terminal extension of the zinc finger region in the human vitamin D receptor. *Biochemistry, 38*, 16347–16358.

Huelsken, J., Vogel, R., Erdmann, B., Cotsarelis, G., & Birchmeier, W. (2001). Beta-catenin controls hair follicle morphogenesis and stem cell differentiation in the skin. *Cell, 105*(4), 533–545. http://dx.doi.org/10.1016/S0092-8674(01)00336-1.

Hughes, M. R., Brumbaugh, P. F., Haussler, M. R., Wergedal, J. E., & Baylink, D. J. (1975). Regulation of serum 1a,25-dihydroxyvitamin D_3 by calcium and phosphate in the rat. *Science, 190*, 578–580.

Ichikawa, S., Imel, E. A., Kreiter, M. L., Yu, X., Mackenzie, D. S., Sorenson, A. H., et al. (2007). A homozygous missense mutation in human klotho causes severe tumoral calcinosis. *Journal of Clinical Investigation, 117*(9), 2684–2691. http://dx.doi.org/10.1172/JCI31330.

Jacobs, E., Martinez, M. E., Buckmeier, J., Lance, P., May, M., & Jurutka, P. (2011). Circulating fibroblast growth factor-23 is associated with increased risk for metachronous colorectal adenoma. *Journal of Carcinogenesis, 10*, 3. http://dx.doi.org/10.4103/1477-3163.76723.

Jacobs, E. T., Van Pelt, C., Forster, R. E., Zaidi, W., Hibler, E. A., Galligan, M. A., et al. (2013). Cyp24a1 and cyp27b1 polymorphisms modulate vitamin D metabolism in colon cancer cells. *Cancer Research, 73*(8), 2563–2573. http://dx.doi.org/10.1158/0008-5472.CAN-12-4134.

Jia, F., Wang, B., Shan, L., Xu, Z., Staal, W. G., & Du, L. (2015). Core symptoms of autism improved after vitamin D supplementation. *Pediatrics, 135*, e196–e198.

Jiang, F., Li, P., Fornace, A. J., Jr., Nicosia, S. V., & Bai, W. (2003). G2/M arrest by 1,25-dihydroxyvitamin D3 in ovarian cancer cells mediated through the induction of GADD45 via an exonic enhancer. *Journal of Biological Chemistry, 278*(48), 48030–48040. http://dx.doi.org/10.1074/jbc.M308430200.

Jiang, P., Zhang, L. H., Cai, H. L., Li, H. D., Liu, Y. P., Tang, M. M., et al. (2014). Neurochemical effects of chronic administration of calcitriol in rats. *Nutrients, 6*(12), 6048–6059. http://dx.doi.org/10.3390/nu6126048.

Jurutka, P. W., Hsieh, J.-C., Remus, L. S., Whitfield, G. K., Thompson, P. D., Haussler, C. A., et al. (1997). Mutations in the 1,25-dihydroxyvitamin D_3 receptor identifying c-terminal amino acids required for transcriptional activation that are functionally dissociated from hormone binding, heterodimeric DNA binding and interaction with basal transcription factor IIB, *in vitro*. *Journal of Biological Chemistry, 272*, 14592–14599.

Jurutka, P. W., Whitfield, G. K., Forster, R., Batie, S., Lee, J., & Haussler, M. R. (2013). Vitamin D: A fountain of youth in gene regulation. In A. F. Gombart (Ed.), *Vitamin D: Oxidative stress, immunity, and aging* (pp. 3–35). Boca Raton: CRC Press.

Kaneko, I., Sabir, M. S., Dussik, C. M., Whitfield, G. K., Karrys, A., Hsieh, J. C., et al. (2015). 1,25-Dihydroxyvitamin D regulates expression of the tryptophan hydroxylase 2 and leptin genes: Implication for behavioral influences of vitamin D. *FASEB Journal*, *29*(9), 4023–4035. http://dx.doi.org/10.1096/fj.14-269811.

Kaneko, I., Saini, R. K., Griffin, K. P., Whitfield, G. K., Haussler, M. R., & Jurutka, P. W. (2015). FGF23 gene regulation by 1,25-dihydroxyvitamin D: Opposing effects in adipocytes and osteocytes. *Journal of Endocrinology*, *226*, 155–166.

Kanninen, K., Heikkinen, R., Malm, T., Rolova, T., Kuhmonen, S., Leinonen, H., et al. (2009). Intrahippocampal injection of a lentiviral vector expressing Nrf2 improves spatial learning in a mouse model of Alzheimer's disease. *Proceedings of the National Academy of Sciences of the United States of America*, *106*(38), 16505–16510. http://dx.doi.org/10.1073/pnas.0908397106.

Katai, K., Miyamoto, K., Kishida, S., Segawa, H., Nii, T., Tanaka, H., et al. (1999). Regulation of intestinal Na+-dependent phosphate co-transporters by a low-phosphate diet and 1,25-dihydroxyvitamin D_3. *Biochemical Journal*, *343*(Pt 3), 705–712.

Katic, M., Kennedy, A. R., Leykin, I., Norris, A., McGettrick, A., Gesta, S., et al. (2007). Mitochondrial gene expression and increased oxidative metabolism: Role in increased lifespan of fat-specific insulin receptor knock-out mice. *Aging Cell*, *6*(6), 827–839. http://dx.doi.org/10.1111/j.1474-9726.2007.00346.x.

Keisala, T., Minasyan, A., Lou, Y. R., Zou, J., Kalueff, A. V., Pyykko, I., et al. (2009). Premature aging in vitamin D receptor mutant mice. *Journal of Steroid Biochemistry and Molecular Biology*, *115*(3–5), 91–97. http://dx.doi.org/10.1016/j.jsbmb.2009.03.007.

Kerner, S. A., Scott, R. A., & Pike, J. W. (1989). Sequence elements in the human osteocalcin gene confer basal activation and inducible response to hormonal vitamin D_3. *Proceedings of the National Academy of Sciences of the United States of America*, *86*, 4455–4459.

Kesby, J. P., Cui, X., O'Loan, J., McGrath, J. J., Burne, T. H., & Eyles, D. W. (2010). Developmental vitamin D deficiency alters dopamine-mediated behaviors and dopamine transporter function in adult female rats. *Psychopharmacology*, *208*(1), 159–168. http://dx.doi.org/10.1007/s00213-009-1717-y.

Kikuchi, R., Sobue, S., Murakami, M., Ito, H., Kimura, A., Iwasaki, T., et al. (2007). Mechanism of vitamin D3-induced transcription of phospholipase d1 in hacat human keratinocytes. *FEBS Letters*, *581*(9), 1800–1804. http://dx.doi.org/10.1016/j.febslet.2007.03.073.

Kim, S., Yamazaki, M., Shevde, N. K., & Pike, J. W. (2007). Transcriptional control of receptor activator of nuclear factor-kappaB ligand by the protein kinase a activator forskolin and the transmembrane glycoprotein 130-activating cytokine, oncostatin M, is exerted through multiple distal enhancers. *Molecular Endocrinology*, *21*(1), 197–214.

Kim, S., Yamazaki, M., Zella, L. A., Shevde, N. K., & Pike, J. W. (2006). Activation of receptor activator of NF-kappaB ligand gene expression by 1,25-dihydroxyvitamin D3 is mediated through multiple long-range enhancers. *Molecular and Cellular Biology*, *26*(17), 6469–6486. http://dx.doi.org/10.1128/MCB.00353-06.

Knekt, P., Kilkkinen, A., Rissanen, H., Marniemi, J., Saaksjarvi, K., & Heliovaara, M. (2010). Serum vitamin D and the risk of Parkinson disease. *Archives of Neurology*, *67*(7), 808–811. http://dx.doi.org/10.1001/archneurol.2010.120.

Kohler, A. C., Sag, C. M., & Maier, L. S. (2014). Reactive oxygen species and excitation-contraction coupling in the context of cardiac pathology. *Journal of Molecular and Cellular Cardiology*, *73*, 92–102. http://dx.doi.org/10.1016/j.yjmcc.2014.03.001.

Kolek, O. I., Hines, E. R., Jones, M. D., Lesueur, L. K., Lipko, M. A., Kiela, P. R., et al. (2005). 1{alpha},25-Dihydroxyvitamin D3 upregulates FGF23 gene expression in bone: The final link in a renal-gastrointestinal-skeletal axis that controls phosphate transport. *American Journal of Physiology—Gastrointestinal and Liver Physiology*, *289*(6), G1036–G1042.

Kong, J., Chen, Y., Zhu, G., Zhao, Q., & Li, Y. C. (2013). 1,25-Dihydroxyvitamin D3 upregulates leptin expression in mouse adipose tissue. *Journal of Endocrinology, 216*(2), 265–271. http://dx.doi.org/10.1530/JOE-12-0344.

Kragballe, K. (1997). The future of vitamin D in dermatology. *Journal of the American Academy of Dermatology, 37*(3 Pt. 2), S72–S76.

Kriebitzsch, C., Verlinden, L., Eelen, G., van Schoor, N. M., Swart, K., Lips, P., et al. (2011). 1,25-Dihydroxyvitamin D3 influences cellular homocysteine levels in murine preosteoblastic MC3T3-E1 cells by direct regulation of cystathionine beta-synthase. *Journal of Bone and Mineral Research, 26*, 2991–3000.

Kuro-o, M. (2012). Klotho in health and disease. *Current Opinion in Nephrology and Hypertension, 21*(4), 362–368. http://dx.doi.org/10.1097/MNH.0b013e32835422ad.

Kuro-o, M., Matsumura, Y., Aizawa, H., Kawaguchi, H., Suga, T., Utsugi, T., et al. (1997). Mutation of the mouse klotho gene leads to a syndrome resembling ageing. *Nature, 390*(6655), 45–51. http://dx.doi.org/10.1038/36285.

Kurosu, H., Yamamoto, M., Clark, J. D., Pastor, J. V., Nandi, A., Gurnani, P., et al. (2005). Suppression of aging in mice by the hormone klotho. *Science, 309*(5742), 1829–1833. http://dx.doi.org/10.1126/science.1112766.

Kusaba, T., Okigaki, M., Matui, A., Murakami, M., Ishikawa, K., Kimura, T., et al. (2010). Klotho is associated with VEGF receptor-2 and the transient receptor potential canonical-1 Ca2+ channel to maintain endothelial integrity. *Proceedings of the National Academy of Sciences of the United States of America, 107*(45), 19308–19313. http://dx.doi.org/10.1073/pnas.1008544107.

Laczmanski, L., Jakubik, M., Bednarek-Tupikowska, G., Rymaszewska, J., Sloka, N., & Lwow, F. (2015). Vitamin D receptor gene polymorphisms in Alzheimer's disease patients. *Experimental Gerontology, 69*, 142–147. http://dx.doi.org/10.1016/j.exger.2015.06.012.

Laplante, M., & Sabatini, D. M. (2012). mTOR signaling in growth control and disease. *Cell, 149*(2), 274–293. http://dx.doi.org/10.1016/j.cell.2012.03.017.

Latimer, C. S., Brewer, L. D., Searcy, J. L., Chen, K. C., Popovic, J., Kraner, S. D., et al. (2014). Vitamin D prevents cognitive decline and enhances hippocampal synaptic function in aging rats. *Proceedings of the National Academy of Sciences of the United States of America, 111*(41), E4359–E4366. http://dx.doi.org/10.1073/pnas.1404477111.

Lau, W. L., Leaf, E. M., Hu, M. C., Takeno, M. M., Kuro-o, M., Moe, O. W., et al. (2012). Vitamin D receptor agonists increase klotho and osteopontin while decreasing aortic calcification in mice with chronic kidney disease fed a high phosphate diet. *Kidney International, 82*(12), 1261–1270. http://dx.doi.org/10.1038/ki.2012.322.

Lee, Y. H., Kim, J. H., & Song, G. G. (2014). Vitamin D receptor polymorphisms and susceptibility to Parkinson's disease and Alzheimer's disease: A meta-analysis. *Neurological Sciences, 35*(12), 1947–1953. http://dx.doi.org/10.1007/s10072-014-1868-4.

Lee, S., Lee, D. K., Choi, E., & Lee, J. W. (2005). Identification of a functional vitamin D response element in the murine insig-2 promoter and its potential role in the differentiation of 3T3-L1 preadipocytes. *Molecular Endocrinology, 19*(2), 399–408. http://dx.doi.org/10.1210/me.2004-0324.

Lesch, K. P. (2005). Serotonergic gene inactivation in mice: Models for anxiety and aggression? In In *Novartis Foundation Symposium: 268.* (pp. 111–140). discussion 140–116, 167–170.

Li, M., Indra, A. K., Warot, X., Brocard, J., Messaddeq, N., Kato, S., et al. (2000). Skin abnormalities generated by temporally controlled RXRalpha mutations in mouse epidermis. *Nature, 407*(6804), 633–636.

Li, Y. C., Kong, J., Wei, M., Chen, Z. F., Liu, S. Q., & Cao, L. P. (2002). 1,25-Dihydroxyvitamin D(3) is a negative endocrine regulator of the renin-angiotensin system. *Journal of Clinical Investigation, 110*(2), 229–238. http://dx.doi.org/10.1172/JCI15219.

Li, Y. C., Pirro, A. E., Amling, M., Delling, G., Baron, R., Bronson, R., et al. (1997). Targeted ablation of the vitamin D receptor: An animal model of vitamin D-dependent rickets type II with alopecia. *Proceedings of the National Academy of Sciences of the United States of America, 94*(18), 9831–9835.

Lian, J. B., Shalhoub, V., Aslam, F., Frenkel, B., Green, J., Hamrah, M., et al. (1997). Species-specific glucocorticoid and 1,25-dihydroxyvitamin D responsiveness in mouse MC3T3-E1 osteoblasts: Dexamethasone inhibits osteoblast differentiation and vitamin D down-regulates osteocalcin gene expression. *Endocrinology, 138*(5), 2117–2127.

Lin, M., & Beal, M. (2003). The oxidative damage theory of aging. *Clinical Neuroscience Research, 2*, 305–315. http://dx.doi.org/10.1016/S1566-2772(03)00007-0.

Lin, Y., & Sun, Z. (2012). Antiaging gene klotho enhances glucose-induced insulin secretion by up-regulating plasma membrane levels of trpv2 in min6 beta-cells. *Endocrinology, 153*(7), 3029–3039. http://dx.doi.org/10.1210/en.2012-1091.

Lintern, K. B., Guidato, S., Rowe, A., Saldanha, J. W., & Itasaki, N. (2009). Characterization of wise protein and its molecular mechanism to interact with both Wnt and bmp signals. *Journal of Biological Chemistry, 284*(34), 23159–23168. http://dx.doi.org/10.1074/jbc.M109.025478.

Lisse, T. S., Liu, T., Irmler, M., Beckers, J., Chen, H., Adams, J. S., et al. (2011). Gene targeting by the vitamin D response element binding protein reveals a role for vitamin D in osteoblast mtor signaling. *FASEB Journal, 25*(3), 937–947. http://dx.doi.org/10.1096/fj.10-172577.

Lisse, T. S., Saini, V., Zhao, H., Luderer, H. F., Gori, F., & Demay, M. B. (2014). The vitamin D receptor is required for activation of cWnt and hedgehog signaling in keratinocytes. *Molecular Endocrinology, 28*(10), 1698–1706. http://dx.doi.org/10.1210/me.2014-1043.

Liu, H., Fergusson, M. M., Castilho, R. M., Liu, J., Cao, L., Chen, J., et al. (2007). Augmented Wnt signaling in a mammalian model of accelerated aging. *Science, 317*(5839), 803–806. http://dx.doi.org/10.1126/science.1143578.

Liu, L., Kim, H., Casta, L. C., Kobayashi, Y., Shapiro, L. S., & Christinao, A. M. (2011). Hairless is a H3K9 histone demethylase. *Journal of Investigative Dermatology, 131*, S69.

Liu, S. M., Koszewski, N., Lupez, M., Malluche, H. H., Olivera, A., & Russell, J. (1996). Characterization of a response element in the 5′-flanking region of the avian (chicken) PTH gene that mediates negative regulation of gene transcription by 1,25-dihydroxyvitamin D_3 and binds the vitamin D_3 receptor. *Molecular Endocrinology, 10*, 206–215.

Liu, M., Lee, M. H., Cohen, M., Bommakanti, M., & Freedman, L. P. (1996). Transcriptional activation of the cdk inhibitor p21 by vitamin D_3 leads to the induced differentiation of the myelomonocytic cell line U937. *Genes and Development, 10*(2), 142–153.

Lokeshwar, B. L., Schwartz, G. G., Selzer, M. G., Burnstein, K. L., Zhuang, S. H., Block, N. L., et al. (1999). Inhibition of prostate cancer metastasis in vivo: A comparison of 1,25-dihydroxyvitamin D (calcitriol) and EB1089. *Cancer Epidemiology, Biomarkers and Prevention, 8*(3), 241–248.

Lovenberg, W., Jequier, E., & Sjoerdsma, A. (1967). Tryptophan hydroxylation: Measurement in pineal gland, brainstem, and carcinoid tumor. *Science, 155*(3759), 217–219.

Luderer, H. F., Gori, F., & Demay, M. B. (2011). Lymphoid enhancer-binding factor-1 (LEF1) interacts with the DNA-binding domain of the vitamin D receptor. *Journal of Biological Chemistry, 286*(21), 18444–18451. http://dx.doi.org/10.1074/jbc.M110.188219.

Mackenzie, L., Bootman, M. D., Laine, M., Berridge, M. J., Thuring, J., Holmes, A., et al. (2002). The role of inositol 1,4,5-trisphosphate receptors in Ca(2+) signalling and the generation of arrhythmias in rat atrial myocytes. *Journal of Physiology, 541*(Pt. 2), 395–409. http://dx.doi.org/10.1113/jphysiol.2001.013411.

Makishima, M., Lu, T. T., Xie, W., Whitfield, G. K., Domoto, H., Evans, R. M., et al. (2002). Vitamin D receptor as an intestinal bile acid sensor. *Science, 296*(5571), 1313–1316.

Malloy, P. J., & Feldman, B. J. (2013). Cell-autonomous regulation of brown fat identity gene UCP1 by unliganded vitamin D receptor. *Molecular Endocrinology, 27*(10), 1632–1642. http://dx.doi.org/10.1210/me.2013-1037.

Malloy, P. J., Peng, L., Wang, J., & Feldman, D. (2009). Interaction of the vitamin D receptor with a vitamin D response element in the Mullerian-inhibiting substance (MIS) promoter: Regulation of MIS expression by calcitriol in prostate cancer cells. *Endocrinology, 150*(4), 1580–1587. http://dx.doi.org/10.1210/en.2008-1555.

Mancuso, P., Rahman, A., Hershey, S. D., Dandu, L., Nibbelink, K. A., & Simpson, R. U. (2008). 1,25-Dihydroxyvitamin-D3 treatment reduces cardiac hypertrophy and left ventricular diameter in spontaneously hypertensive heart failure-prone (cp/+) rats independent of changes in serum leptin. *Journal of Cardiovascular Pharmacology, 51*(6), 559–564. http://dx.doi.org/10.1097/FJC.0b013e3181761906.

Manna, P., & Jain, S. K. (2012). Vitamin D up-regulates glucose transporter 4 (GLUT4) translocation and glucose utilization mediated by cystathionine-gamma-lyase (CSE) activation and H2S formation in 3T3L1 adipocytes. *Journal of Biological Chemistry, 287*(50), 42324–42332. http://dx.doi.org/10.1074/jbc.M112.407833.

Manna, P., & Jain, S. (2015a). Vitamin D (VD) prevents oxidative stress via regulating NOX4/Nrf2/Trx signaling cascade and upregulates SIRT1-mediated AMPK/IRS1/GLUT4 pathway and glucose uptake in high glucose treated 3T3L1 adipocytes. *FASEB Journal, 29*, 253.251.

Manna, P., & Jain, S. K. (2015b). Phosphatidylinositol-3,4,5-triphosphate and cellular signaling: Implications for obesity and diabetes. *Cellular Physiology and Biochemistry, 35*(4), 1253–1275. http://dx.doi.org/10.1159/000373949.

Masuda, S., Byford, V., Arabian, A., Sakai, Y., Demay, M. B., St-Arnaud, R., et al. (2005). Altered pharmacokinetics of 1alpha,25-dihydroxyvitamin D3 and 25-hydroxyvitamin D3 in the blood and tissues of the 25-hydroxyvitamin D-24-hydroxylase (CYP24A1) null mouse. *Endocrinology, 146*(2), 825–834. http://dx.doi.org/10.1210/en.2004-1116.

Masuyama, H., Brownfield, C. M., St-Arnaud, R., & MacDonald, P. N. (1997). Evidence for ligand-dependent intramolecular folding of the AF-2 domain in vitamin D receptor-activated transcription and coactivator interaction. *Molecular Endocrinology, 11*(10), 1507–1517.

McGrath, J., Saari, K., Hakko, H., Jokelainen, J., Jones, P., Jarvelin, M. R., et al. (2004). Vitamin D supplementation during the first year of life and risk of schizophrenia: A Finnish birth cohort study. *Schizophrenia Research, 67*(2–3), 237–245. http://dx.doi.org/10.1016/j.schres.2003.08.005.

Menendez, C., Lage, M., Peino, R., Baldelli, R., Concheiro, P., Dieguez, C., et al. (2001). Retinoic acid and vitamin D(3) powerfully inhibit in vitro leptin secretion by human adipose tissue. *Journal of Endocrinology, 170*(2), 425–431.

Meyer, M. B., Benkusky, N. A., Lee, C. H., & Pike, J. W. (2014a). Genomic determinants of gene regulation by 1,25-dihydroxyvitamin D3 during osteoblast-lineage cell differentiation. *Journal of Biological Chemistry, 289*(28), 19539–19554. http://dx.doi.org/10.1074/jbc.M114.578104.

Meyer, M. B., Benkusky, N. A., & Pike, J. W. (2014b). The RUNX2 cistrome in osteoblasts: Characterization, down-regulation following differentiation, and relationship to gene expression. *Journal of Biological Chemistry, 289*(23), 16016–16031. http://dx.doi.org/10.1074/jbc.M114.552216.

Meyer, M. B., Benkusky, N. A., & Pike, J. W. (2015). Selective distal enhancer control of the Mmp13 gene identified through clustered regularly interspaced short palindromic repeat (CRISPR) genomic deletions. *Journal of Biological Chemistry, 290*(17), 11093–11107. http://dx.doi.org/10.1074/jbc.M115.648394.

Meyer, M. B., Goetsch, P. D., & Pike, J. W. (2012). VDR/RXR and TCF4/beta-catenin cistromes in colonic cells of colorectal tumor origin: Impact on c-FOS and c-MYC gene expression. *Molecular Endocrinology, 26*(1), 37–51. http://dx.doi.org/10.1210/me.2011-1109.

Meyer, M. B., Watanuki, M., Kim, S., Shevde, N. K., & Pike, J. W. (2006). The human transient receptor potential vanilloid type 6 distal promoter contains multiple vitamin D receptor binding sites that mediate activation by 1,25-dihydroxyvitamin D3 in intestinal cells. *Molecular Endocrinology, 20*(6), 1447–1461. http://dx.doi.org/10.1210/me.2006-0031.

Mifsud, B., Tavares-Cadete, F., Young, A. N., Sugar, R., Schoenfelder, S., Ferreira, L., et al. (2015). Mapping long-range promoter contacts in human cells with high-resolution capture Hi-C. *Nature Genetics, 47*(6), 598–606. http://dx.doi.org/10.1038/ng.3286.

Moore, L. B., Maglich, J. M., McKee, D. D., Wisely, B., Willson, T. M., Kliewer, S. A., et al. (2002). Pregnane x receptor (PXR), constitutive androstane receptor (CAR), and benzoate x receptor (BXR) define three pharmacologically distinct classes of nuclear receptors. *Molecular Endocrinology, 16*(5), 977–986. http://dx.doi.org/10.1210/mend.16.5.0828.

Morash, B., Li, A., Murphy, P. R., Wilkinson, M., & Ur, E. (1999). Leptin gene expression in the brain and pituitary gland. *Endocrinology, 140*(12), 5995–5998. http://dx.doi.org/10.1210/endo.140.12.7288.

Mori, K., Yahata, K., Mukoyama, M., Suganami, T., Makino, H., Nagae, T., et al. (2000). Disruption of klotho gene causes an abnormal energy homeostasis in mice. *Biochemical and Biophysical Research Communications, 278*(3), 665–670. http://dx.doi.org/10.1006/bbrc.2000.3864.

Morris, B. J., Willcox, D. C., Donlon, T. A., & Willcox, B. J. (2015). FOXO3: A major gene for human longevity—A mini-review. *Gerontology, 61*(6), 515–525. http://dx.doi.org/10.1159/000375235.

Nakae, J., Kitamura, T., Silver, D. L., & Accili, D. (2001). The forkhead transcription factor Foxo1 (Fkhr) confers insulin sensitivity onto glucose-6-phosphatase expression. *Journal of Clinical Investigation, 108*(9), 1359–1367. http://dx.doi.org/10.1172/JCI12876.

Nakagawa, K., Kawaura, A., Kato, S., Takeda, E., & Okano, T. (2005). 1 alpha,25-Dihydroxyvitamin D(3) is a preventive factor in the metastasis of lung cancer. *Carcinogenesis, 26*(2), 429–440. http://dx.doi.org/10.1093/carcin/bgh332.

Nakai, K., Fujii, H., Kono, K., Goto, S., Kitazawa, R., Kitazawa, S., et al. (2014). Vitamin D activates the Nrf2-Keap1 antioxidant pathway and ameliorates nephropathy in diabetic rats. *American Journal of Hypertension, 27*(4), 586–595. http://dx.doi.org/10.1093/ajh/hpt160.

Narvaez, C. J., Matthews, D., Broun, E., Chan, M., & Welsh, J. (2009). Lean phenotype and resistance to diet-induced obesity in vitamin D receptor knockout mice correlates with induction of uncoupling protein-1 in white adipose tissue. *Endocrinology, 150*(2), 651–661. http://dx.doi.org/10.1210/en.2008-1118.

Ni, W., Watts, S. W., Ng, M., Chen, S., Glenn, D. J., & Gardner, D. G. (2014). Elimination of vitamin D receptor in vascular endothelial cells alters vascular function. *Hypertension, 64*(6), 1290–1298. http://dx.doi.org/10.1161/HYPERTENSIONAHA.114.03971.

Noda, M., Vogel, R. L., Craig, A. M., Prahl, J., DeLuca, H. F., & Denhardt, D. T. (1990). Identification of a DNA sequence responsible for binding of the 1,25-dihydroxyvitamin D_3 receptor and 1,25-dihydroxyvitamin D_3 enhancement of mouse secreted phosphoprotein 1 (SPP-1 or osteopontin) gene expression. *Proceedings of the National Academy of Sciences of the United States of America*, *87*, 9995–9999.

Norman, A. W., Frankel, J. B., Heldt, A. M., & Grodsky, G. M. (1980). Vitamin D deficiency inhibits pancreatic secretion of insulin. *Science*, *209*(4458), 823–825.

Ohyama, Y., Ozono, K., Uchida, M., Shinki, T., Kato, S., Suda, T., et al. (1994). Identification of a vitamin D-responsive element in the 5′-flanking region of the rat 25-hydroxyvitamin D3 24-hydroxylase gene. *Journal of Biological Chemistry*, *269*(14), 10545–10550.

Orlov, I., Rochel, N., Moras, D., & Klaholz, B. P. (2012). Structure of the full human RXR/VDR nuclear receptor heterodimer complex with its DR3 target DNA. *EMBO Journal*, *31*(2), 291–300. http://dx.doi.org/10.1038/emboj.2011.445.

Palmer, H. G., Gonzalez-Sancho, J. M., Espada, J., Berciano, M. T., Puig, I., Baulida, J., et al. (2001). Vitamin D_3 promotes the differentiation of colon carcinoma cells by the induction of E-cadherin and the inhibition of beta-catenin signaling. *Journal of Cell Biology*, *154*(2), 369–387. http://dx.doi.org/10.1083/jcb.200102028.

Patrick, R. P., & Ames, B. N. (2014). Vitamin D hormone regulates serotonin synthesis. Part 1: Relevance for autism. *FASEB Journal*, *28*(6), 2398–2413. http://dx.doi.org/10.1096/fj.13-246546.

Patrick, R. P., & Ames, B. N. (2015). Vitamin D and the omega-3 fatty acids control serotonin synthesis and action, part 2: Relevance for ADHD, bipolar, schizophrenia, and impulsive behavior. *FASEB Journal*, *29*(6), 2207–2222. http://dx.doi.org/10.1096/fj.14-268342.

Peng, L., Malloy, P. J., & Feldman, D. (2004). Identification of a functional vitamin D response element in the human insulin-like growth factor binding protein-3 promoter. *Molecular Endocrinology*, *18*(5), 1109–1119. http://dx.doi.org/10.1210/me.2003-0344.

Perwad, F., Zhang, M. Y., Tenenhouse, H. S., & Portale, A. A. (2007). Fibroblast growth factor 23 impairs phosphorus and vitamin D metabolism in vivo and suppresses 25-hydroxyvitamin D-1alpha-hydroxylase expression in vitro. *American Journal of Physiology Renal Physiology*, *293*(5), F1577–F1583. http://dx.doi.org/10.1152/ajprenal.00463.2006.

Pike, J. W., Lee, S. M., & Meyer, M. B. (2014). Regulation of gene expression by 1,25-dihydroxyvitamin D3 in bone cells: Exploiting new approaches and defining new mechanisms. *BoneKEy Reports*, *3*, 482. http://dx.doi.org/10.1038/bonekey.2013.216.

Pike, J. W., & Meyer, M. B. (2014). Fundamentals of vitamin D hormone-regulated gene expression. *Journal of Steroid Biochemistry and Molecular Biology*, *144*(Pt. A), 5–11. http://dx.doi.org/10.1016/j.jsbmb.2013.11.004.

Pike, J. W., Meyer, M. B., Watanuki, M., Kim, S., Zella, L. A., Fretz, J. A., et al. (2007). Perspectives on mechanisms of gene regulation by 1,25-dihydroxyvitamin D3 and its receptor. *Journal of Steroid Biochemistry and Molecular Biology*, *103*(3-5), 389–395. http://dx.doi.org/10.1016/j.jsbmb.2006.12.050.

Popov, V. M., Wang, C., Shirley, L. A., Rosenberg, A., Li, S., Nevalainen, M., et al. (2007). The functional significance of nuclear receptor acetylation. *Steroids*, *72*(2), 221–230. http://dx.doi.org/10.1016/j.steroids.2006.12.001.

Potter, G. B., Zarach, J. M., Sisk, J. M., & Thompson, C. C. (2002). The thyroid hormone-regulated corepressor hairless associates with histone deacetylases in neonatal rat brain. *Molecular Endocrinology*, *16*(11), 2547–2560.

Quarles, L. D. (2008). Endocrine functions of bone in mineral metabolism regulation. *Journal of Clinical Investigation*, *118*(12), 3820–3828. http://dx.doi.org/10.1172/JCI36479.

Quelo, I., Machuca, I., & Jurdic, P. (1998). Identification of a vitamin D response element in the proximal promoter of the chicken carbonic anhydrase II gene. *Journal of Biological Chemistry*, *273*(17), 10638–10646.

Ramagopalan, S. V., Maugeri, N. J., Handunnetthi, L., Lincoln, M. R., Orton, S. M., Dyment, D. A., et al. (2009). Expression of the multiple sclerosis-associated MHC class II allele HLA-DRB1*1501 is regulated by vitamin D. *PLoS Genetics*, *5*(2), e1000369. http://dx.doi.org/10.1371/journal.pgen.1000369.

Ramsey, C. P., Glass, C. A., Montgomery, M. B., Lindl, K. A., Ritson, G. P., Chia, L. A., et al. (2007). Expression of Nrf2 in neurodegenerative diseases. *Journal of Neuropathology and Experimental Neurology*, *66*(1), 75–85. http://dx.doi.org/10.1097/nen.0b013e31802d6da9.

Razzaque, M. S. (2009). The FGF23-klotho axis: Endocrine regulation of phosphate homeostasis. *Nature Reviews. Endocrinology*, *5*(11), 611–619. http://dx.doi.org/10.1038/nrendo.2009.196.

Remes Lenicov, F., Lemonde, S., Czesak, M., Mosher, T. M., & Albert, P. R. (2007). Cell-type specific induction of tryptophan hydroxylase-2 transcription by calcium mobilization. *Journal of Neurochemistry*, *103*(5), 2047–2057. http://dx.doi.org/10.1111/j.1471-4159.2007.04903.x.

Renkema, K. Y., Alexander, R. T., Bindels, R. J., & Hoenderop, J. G. (2008). Calcium and phosphate homeostasis: Concerted interplay of new regulators. *Annals of Medicine*, *40*(2), 82–91. http://dx.doi.org/10.1080/07853890701689645.

Rochel, N., Ciesielski, F., Godet, J., Moman, E., Roessle, M., Peluso-Iltis, C., et al. (2011). Common architecture of nuclear receptor heterodimers on DNA direct repeat elements with different spacings. *Nature Structural and Molecular Biology*, *18*(5), 564–570. http://dx.doi.org/10.1038/nsmb.2054.

Rochel, N., Wurtz, J. M., Mitschler, A., Klaholz, B., & Moras, D. (2000). The crystal structure of the nuclear receptor for vitamin D bound to its natural ligand. *Molecular Cell*, *5*(1), 173–179. http://dx.doi.org/10.1016/S1097-2765(00)80413-X.

Rucklidge, J. J., Frampton, C. M., Gorman, B., & Boggis, A. (2014). Vitamin-mineral treatment of attention-deficit hyperactivity disorder in adults: Double-blind randomised placebo-controlled trial. *British Journal of Psychiatry*, *204*, 306–315. http://dx.doi.org/10.1192/bjp.bp.113.132126.

Rylander, M., & Verhulst, S. (2013). Vitamin D insufficiency in psychiatric inpatients. *Journal of Psychiatric Practice*, *19*(4), 296–300. http://dx.doi.org/10.1097/01.pra.0000432599.24761.c1.

Saeki, M., Kurose, K., Tohkin, M., & Hasegawa, R. (2008). Identification of the functional vitamin D response elements in the human MDR1 gene. *Biochemical Pharmacology*, *76*(4), 531–542. http://dx.doi.org/10.1016/j.bcp.2008.05.030.

Saini, R. K., Kaneko, I., Jurutka, P. W., Forster, R., Hsieh, A., Hsieh, J. C., et al. (2013). 1,25-Dihydroxyvitamin D(3) regulation of fibroblast growth factor-23 expression in bone cells: Evidence for primary and secondary mechanisms modulated by leptin and interleukin-6. *Calcified Tissue International*, *92*(4), 339–353. http://dx.doi.org/10.1007/s00223-012-9683-5.

Satoh, M., Nagasu, H., Morita, Y., Yamaguchi, T. P., Kanwar, Y. S., & Kashihara, N. (2012). Klotho protects against mouse renal fibrosis by inhibiting Wnt signaling. *American Journal of Physiology. Renal Physiology*, *303*(12), F1641–F1651. http://dx.doi.org/10.1152/ajprenal.00460.2012.

Seshadri, S., Beiser, A., Selhub, J., Jacques, P. F., Rosenberg, I. H., D'Agostino, R. B., et al. (2002). Plasma homocysteine as a risk factor for dementia and Alzheimer's disease. *New England Journal of Medicine*, *346*(7), 476–483. http://dx.doi.org/10.1056/NEJMoa011613.

Shimada, T., Hasegawa, H., Yamazaki, Y., Muto, T., Hino, R., Takeuchi, Y., et al. (2004). FGF-23 is a potent regulator of vitamin D metabolism and phosphate homeostasis. *Journal of Bone and Mineral Research*, *19*(3), 429–435.

Shimada, T., Kakitani, M., Yamazaki, Y., Hasegawa, H., Takeuchi, Y., Fujita, T., et al. (2004). Targeted ablation of FGF23 demonstrates an essential physiological role of FGF23 in phosphate and vitamin D metabolism. *Journal of Clinical Investigation, 113*(4), 561–568.

Shirakawa, A. K., Nagakubo, D., Hieshima, K., Nakayama, T., Jin, Z., & Yoshie, O. (2008). 1,25-Dihydroxyvitamin D3 induces CCR10 expression in terminally differentiating human B cells. *Journal of Immunology, 180*(5), 2786–2795. 180/5/2786 [pii].

Siren, A. L., Fratelli, M., Brines, M., Goemans, C., Casagrande, S., Lewczuk, P., et al. (2001). Erythropoietin prevents neuronal apoptosis after cerebral ischemia and metabolic stress. *Proceedings of the National Academy of Sciences of the United States of America, 98*(7), 4044–4049. http://dx.doi.org/10.1073/pnas.051606598.

St John, H. C., Bishop, K. A., Meyer, M. B., Benkusky, N. A., Leng, N., Kendziorski, C., et al. (2014). The osteoblast to osteocyte transition: Epigenetic changes and response to the vitamin D3 hormone. *Molecular Endocrinology, 28*(7), 1150–1165. http://dx.doi.org/10.1210/me.2014-1091.

St-Arnaud, R. (2010). CYP24A1-deficient mice as a tool to uncover a biological activity for vitamin D metabolites hydroxylated at position 24. *Journal of Steroid Biochemistry and Molecular Biology, 121*(1–2), 254–256. http://dx.doi.org/10.1016/j.jsbmb.2010.02.002.

Steullet, P., Cabungcal, J. H., Monin, A., Dwir, D., O'Donnell, P., Cuenod, M., et al. (2014). Redox dysregulation, neuroinflammation, and NMDA receptor hypofunction: A "central hub" in schizophrenia pathophysiology? *Schizophrenia Research*, http://dx.doi.org/10.1016/j.schres.2014.06.021. (in press).

Sugiura, H., Yoshida, T., Shiohira, S., Kohei, J., Mitobe, M., Kurosu, H., et al. (2012). Reduced klotho expression level in kidney aggravates renal interstitial fibrosis. *American Journal of Physiology. Renal Physiology, 302*(10), F1252–F1264. http://dx.doi.org/10.1152/ajprenal.00294.2011.

Surmeier, D. J., & Schumacker, P. T. (2013). Calcium, bioenergetics, and neuronal vulnerability in Parkinson's disease. *Journal of Biological Chemistry, 288*(15), 10736–10741. http://dx.doi.org/10.1074/jbc.R112.410530.

Sutherland, M. K., Somerville, M. J., Yoong, L. K. K., Bergeron, C., Haussler, M. R., & McLachlan, D. R. C. (1992). Reduction of vitamin D hormone receptor mRNA levels in Alzheimer as compared to Huntington hippocampus: Correlation with calbindin-28k mRNA levels. *Molecular Brain Research, 13*, 239–250.

Suzuki, T., Tazoe, H., Taguchi, K., Koyama, Y., Ichikawa, H., Hayakawa, S., et al. (2006). DNA microarray analysis of changes in gene expression induced by 1,25-dihydroxyvitamin D3 in human promyelocytic leukemia HL-60 cells. *Biomedical Research (Tokyo, Japan), 27*(3), 99–109. JST.JSTAGE/biomedres/27.99 [pii].

Suzuki, T., & Yamamoto, M. (2015). Frbm special issue "Nrf2 regulated redox signaling and metabolism in physiology and medicine" molecular basis of the keap1-Nrf2 system. *Free Radical Biology and Medicine, 88*(Part B), 93–100. http://dx.doi.org/10.1016/j.freeradbiomed.2015.06.006.

Taketani, Y., Segawa, H., Chikamori, M., Morita, K., Tanaka, K., Kido, S., et al. (1998). Regulation of type II renal Na^+-dependent inorganic phosphate transporters by 1,25-dihydroxyvitamin D_3. Identification of a vitamin D-responsive element in the human NaPi-3 gene. *Journal of Biological Chemistry, 273*(23), 14575–14581.

Tang, X. H., Osei-Sarfo, K., Urvalek, A. M., Zhang, T., Scognamiglio, T., & Gudas, L. J. (2014). Combination of bexarotene and the retinoid cd1530 reduces murine oral-cavity carcinogenesis induced by the carcinogen 4-nitroquinoline 1-oxide. *Proceedings of the National Academy of Sciences of the United States of America, 111*(24), 8907–8912. http://dx.doi.org/10.1073/pnas.1404828111.

Taniura, H., Ito, M., Sanada, N., Kuramoto, N., Ohno, Y., Nakamichi, N., et al. (2006). Chronic vitamin D3 treatment protects against neurotoxicity by glutamate in association with upregulation of vitamin D receptor mRNA expression in cultured rat cortical neurons. *Journal of Neuroscience Research, 83*(7), 1179–1189. http://dx.doi.org/10.1002/jnr.20824.

Terpening, C. M., Haussler, C. A., Jurutka, P. W., Galligan, M. A., Komm, B. S., & Haussler, M. R. (1991). The vitamin D-responsive element in the rat bone gla protein gene is an imperfect direct repeat that cooperates with other cis-elements in 1,25-dihydroxyvitamin D3-mediated transcriptional activation. *Molecular Endocrinology*, *5*(3), 373–385.

Thompson, P. D., Jurutka, P. W., Whitfield, G. K., Myskowski, S. M., Eichhorst, K. R., Dominguez, C. E., et al. (2002). Liganded VDR induces CYP3A4 in small intestinal and colon cancer cells via DR3 and ER6 vitamin D responsive elements. *Biochemical and Biophysical Research Communications*, *299*(5), 730–738. http://dx.doi.org/10.1016/S0006-291X(02)02742-0.

Thompson, C. C., Sisk, J. M., & Beaudoin, G. M., 3rd. (2006). Hairless and Wnt signaling: Allies in epithelial stem cell differentiation. *Cell Cycle*, *5*(17), 1913–1917. http://dx.doi.org/10.4161/cc.5.17.3189.

Thummel, K. E., Brimer, C., Yasuda, K., Thottassery, J., Senn, T., Lin, Y., et al. (2001). Transcriptional control of intestinal cytochrome P-4503A by 1alpha,25-dihydroxy vitamin D_3. *Molecular Pharmacology*, *60*(6), 1399–1406.

Thurston, R. D., Larmonier, C. B., Majewski, P. M., Ramalingam, R., Midura-Kiela, M., Laubitz, D., et al. (2010). Tumor necrosis factor and interferon-gamma down-regulate klotho in mice with colitis. *Gastroenterology*, *138*(4), 1384–1394. http://dx.doi.org/10.1053/j.gastro.2009.12.002. e1381–e1382.

Toffanello, E. D., Coin, A., Perissinotto, E., Zambon, S., Sarti, S., Veronese, N., et al. (2014). Vitamin D deficiency predicts cognitive decline in older men and women: The Pro.V.A. Study. *Neurology*, *83*(24), 2292–2298. http://dx.doi.org/10.1212/WNL.0000000000001080.

Tsujikawa, H., Kurotaki, Y., Fujimori, T., Fukuda, K., & Nabeshima, Y. (2003). Klotho, a gene related to a syndrome resembling human premature aging, functions in a negative regulatory circuit of vitamin D endocrine system. *Molecular Endocrinology*, *17*(12), 2393–2403. http://dx.doi.org/10.1210/me.2003-0048.

Tuohimaa, P., Keisala, T., Minasyan, A., Cachat, J., & Kalueff, A. (2009). Vitamin D, nervous system and aging. *Psychoneuroendocrinology*, *34*(Suppl. 1), S278–S286. http://dx.doi.org/10.1016/j.psyneuen.2009.07.003.

Utsugi, T., Ohno, T., Ohyama, Y., Uchiyama, T., Saito, Y., Matsumura, Y., et al. (2000). Decreased insulin production and increased insulin sensitivity in the klotho mutant mouse, a novel animal model for human aging. *Metabolism: Clinical and Experimental*, *49*(9), 1118–1123. http://dx.doi.org/10.1053/meta.2000.8606.

Vacek, J. L., Vanga, S. R., Good, M., Lai, S. M., Lakkireddy, D., & Howard, P. A. (2012). Vitamin D deficiency and supplementation and relation to cardiovascular health. *American Journal of Cardiology*, *109*(3), 359–363. http://dx.doi.org/10.1016/j.amjcard.2011.09.020.

Verlinden, L., Verstuyf, A., Convents, R., Marcelis, S., Van Camp, M., & Bouillon, R. (1998). Action of 1,25(OH)2D3 on the cell cycle genes, cyclin D1, p21 and p27 in MCF-7 cells. *Molecular and Cellular Endocrinology*, *142*(1-2), 57–65.

Vimaleswaran, K. S., Berry, D. J., Lu, C., Tikkanen, E., Pilz, S., Hiraki, L. T., et al. (2013). Causal relationship between obesity and vitamin D status: Bi-directional Mendelian randomization analysis of multiple cohorts. *PLoS Medicine*, *10*(2). e1001383http://dx.doi.org/10.1371/journal.pmed.1001383.

Wang, Y., Chiang, Y. H., Su, T. P., Hayashi, T., Morales, M., Hoffer, B. J., et al. (2000). Vitamin D(3) attenuates cortical infarction induced by middle cerebral arterial ligation in rats. *Neuropharmacology*, *39*(5), 873–880.

Wang, Y., Kuro-o, M., & Sun, Z. (2012). Klotho gene delivery suppresses Nox2 expression and attenuates oxidative stress in rat aortic smooth muscle cells via the cAMP-PKA pathway. *Aging Cell*, *11*(3), 410–417. http://dx.doi.org/10.1111/j.1474-9726.2012.00796.x.

Wang, Y., & Sun, Z. (2009a). Current understanding of klotho. *Ageing Research Reviews*, *8*(1), 43–51. http://dx.doi.org/10.1016/j.arr.2008.10.002.

Wang, Y., & Sun, Z. (2009b). Klotho gene delivery prevents the progression of spontaneous hypertension and renal damage. *Hypertension*, *54*(4), 810–817. http://dx.doi.org/10.1161/HYPERTENSIONAHA.109.134320.

Wang, T. T., Tavera-Mendoza, L. E., Laperriere, D., Libby, E., MacLeod, N. B., Nagai, Y., et al. (2005). Large-scale in silico and microarray-based identification of direct 1,25-dihydroxyvitamin D3 target genes. *Molecular Endocrinology*, *19*(11), 2685–2695.

Wang, X., Wang, T. T., White, J. H., & Studzinski, G. P. (2006). Induction of kinase suppressor of RAS-1 (KSR-1) gene by 1, alpha25-dihydroxyvitamin D3 in human leukemia HL60 cells through a vitamin D response element in the 5′-flanking region. *Oncogene*, *25*(53), 7078–7085. http://dx.doi.org/10.1038/sj.onc.1209697.

Wang, X., Wang, T. T., White, J. H., & Studzinski, G. P. (2007). Expression of human kinase suppressor of ras 2 (hKSR-2) gene in HL60 leukemia cells is directly upregulated by 1,25-dihydroxyvitamin D(3) and is required for optimal cell differentiation. *Experimental Cell Research*, *313*(14), 3034–3045. http://dx.doi.org/10.1016/j.yexcr.2007.05.021.

Watkins, R. E., Wisely, G. B., Moore, L. B., Collins, J. L., Lambert, M. H., Williams, S. P., et al. (2001). The human nuclear xenobiotic receptor PXR: Structural determinants of directed promiscuity. *Science*, *292*(5525), 2329–2333.

Welsh, J. (2012). Cellular and molecular effects of vitamin D on carcinogenesis. *Archives of Biochemistry and Biophysics*, *523*(1), 107–114. http://dx.doi.org/10.1016/j.abb.2011.10.019.

Whitfield, G. K., Dang, H. T. L., Schluter, S. F., Bernstein, R. M., Bunag, T., Manzon, L. A., et al. (2003). Cloning of a functional vitamin D receptor from the lamprey (Petromyzon marinus), an ancient vertebrate lacking a calcified skeleton and teeth. *Endocrinology*, *144*(6), 2704–2716. http://dx.doi.org/10.1210/en.2002-221101.

Whitfield, G. K., Jurutka, P. W., Haussler, C. A., Hsieh, J. C., Barthel, T. K., Jacobs, E. T., et al. (2005). Nuclear vitamin D receptor: Structure-function, molecular control of gene transcription, and novel bioactions. In D. Feldman, J. W. Pike, & F. H. Glorieux (Eds.), *Vitamin D: Vol. 1* (2nd ed., pp. 219–261). Oxford, UK: Elsevier Academic Press.

Wolf, M. T., An, S. W., Nie, M., Bal, M. S., & Huang, C. L. (2014). Klotho up-regulates renal calcium channel transient receptor potential vanilloid 5 (TRPV5) by intra- and extracellular N-glycosylation-dependent mechanisms. *Journal of Biological Chemistry*, *289*(52), 35849–35857. http://dx.doi.org/10.1074/jbc.M114.616649.

Wolf, I., Levanon-Cohen, S., Bose, S., Ligumsky, H., Sredni, B., Kanety, H., et al. (2008). Klotho: A tumor suppressor and a modulator of the IGF-1 and FGF pathways in human breast cancer. *Oncogene*, *27*(56), 7094–7105. http://dx.doi.org/10.1038/onc.2008.292.

Wong, M. S., Delansorne, R., Man, R. Y., & Vanhoutte, P. M. (2008). Vitamin D derivatives acutely reduce endothelium-dependent contractions in the aorta of the spontaneously hypertensive rat. *American Journal of Physiology. Heart and Circulatory Physiology*, *295*(1), H289–H296. http://dx.doi.org/10.1152/ajpheart.00116.2008.

Wrase, J., Reimold, M., Puls, I., Kienast, T., & Heinz, A. (2006). Serotonergic dysfunction: Brain imaging and behavioral correlates. *Cognitive, Affective & Behavioral Neuroscience*, *6*(1), 53–61. http://dx.doi.org/10.3758/CABN.6.1.53.

Xu, Y., & Sun, Z. (2015). Molecular basis of klotho: From gene to function in aging. *Endocrine Reviews*, *36*(2), 174–193. http://dx.doi.org/10.1210/er.2013-1079.

Yamamoto, M., Clark, J. D., Pastor, J. V., Gurnani, P., Nandi, A., Kurosu, H., et al. (2005). Regulation of oxidative stress by the anti-aging hormone klotho. *Journal of Biological Chemistry*, *280*(45), 38029–38034. http://dx.doi.org/10.1074/jbc.M509039200.

Yamamoto, Y., Memezawa, A., Takagi, K., Ochiai, E., Shindo, M., & Kato, S. (2009). A tissue-specific function by unliganded VDR. In *Abstracts from the 14th workshop on vitamin D, Brugge, Belgium, October 4–8* (p. 66).

Yuan, W., Pan, W., Kong, J., Zheng, W., Szeto, F. L., Wong, K. E., et al. (2007). 1,25-Dihydroxyvitamin D3 suppresses renin gene transcription by blocking the activity of the cyclic AMP response element in the renin gene promoter. *Journal of Biological Chemistry*, *282*(41), 29821–29830. http://dx.doi.org/10.1074/jbc.M705495200.

Yuksel, R. N., Altunsoy, N., Tikir, B., Cingi Kuluk, M., Unal, K., Goka, S., et al. (2014). Correlation between total vitamin D levels and psychotic psychopathology in patients with schizophrenia: Therapeutic implications for add-on vitamin D augmentation. *Therapeutic Advances in Psychopharmacology*, *4*(6), 268–275. http://dx.doi.org/10.1177/2045125314553612.

Zarach, J. M., Beaudoin, G. M., 3rd, Coulombe, P. A., & Thompson, C. C. (2004). The co-repressor hairless has a role in epithelial cell differentiation in the skin. *Development*, *131*(17), 4189–4200.

Zeldich, E., Chen, C. D., Colvin, T. A., Bove-Fenderson, E. A., Liang, J., Tucker Zhou, T. B., et al. (2014). The neuroprotective effect of klotho is mediated via regulation of members of the redox system. *Journal of Biological Chemistry*, *289*(35), 24700–24715. http://dx.doi.org/10.1074/jbc.M114.567321.

Zella, L. A., Kim, S., Shevde, N. K., & Pike, J. W. (2007). Enhancers located in the vitamin D receptor gene mediate transcriptional autoregulation by 1,25-dihydroxyvitamin D3. *Journal of Steroid Biochemistry and Molecular Biology*, *103*(3–5), 435–439. http://dx.doi.org/10.1016/j.jsbmb.2006.12.019.

Zhang, X., Beaulieu, J. M., Sotnikova, T. D., Gainetdinov, R. R., & Caron, M. G. (2004). Tryptophan hydroxylase-2 controls brain serotonin synthesis. *Science*, *305*(5681), 217. http://dx.doi.org/10.1126/science.1097540.

Zhang, J., Chalmers, M. J., Stayrook, K. R., Burris, L. L., Wang, Y., Busby, S. A., et al. (2011). DNA binding alters coactivator interaction surfaces of the intact VDR-RXR complex. *Nature Structural and Molecular Biology*, *18*(5), 556–563. http://dx.doi.org/10.1038/nsmb.2046.

Zhou, L., Li, Y., Zhou, D., Tan, R. J., & Liu, Y. (2013). Loss of klotho contributes to kidney injury by derepression of Wnt/beta-catenin signaling. *Journal of the American Society of Nephrology*, *24*(5), 771–785. http://dx.doi.org/10.1681/ASN.2012080865.

Zhou, C., Lu, F., Cao, K., Xu, D., Goltzman, D., & Miao, D. (2008). Calcium-independent and 1,25(OH)2D3-dependent regulation of the renin-angiotensin system in 1alpha-hydroxylase knockout mice. *Kidney International*, *74*(2), 170–179. http://dx.doi.org/10.1038/ki.2008.101.

Zierold, C., Darwish, H. M., & DeLuca, H. F. (1994). Identification of a vitamin D-response element in the rat calcidiol (25-hydroxyvitamin D3) 24-hydroxylase gene. *Proceedings of the National Academy of Sciences of the United States of America*, *91*(3), 900–902.

Zinser, G. M., Sundberg, J. P., & Welsh, J. (2002). Vitamin D_3 receptor ablation sensitizes skin to chemically induced tumorigenesis. *Carcinogenesis*, *23*(12), 2103–2109. http://dx.doi.org/10.1093/carcin/23.12.2103.

Zou, A., Elgort, M. G., & Allegretto, E. A. (1997). Retinoid X receptor (RXR) ligands activate the human 25-hydroxyvitamin D_3-24-hydroxylase promoter via RXR heterodimer binding to two vitamin D-responsive elements and elicit additive effects with 1,25-dihydroxyvitamin D_3. *Journal of Biological Chemistry*, *272*(30), 19027–19034.

CHAPTER NINE

Hedgehog and Vitamin D Signaling Pathways in Development and Disease

M. Kyle Hadden[1]
Department of Pharmaceutical Sciences, University of Connecticut, Storrs, Connecticut, USA
[1]Corresponding author: e-mail address: kyle.hadden@uconn.edu

Contents

Abstract

The classic physiological activity associated with the vitamin D scaffold is the maintenance of calcium and phosphorous homeostasis in the bone. This activity is commonly attributed to direct binding of 1α,25-dihydroxyvitamin D3 [1α,25(OH)$_2$D$_3$, calcitriol], the hormonally active form of vitamin D, to the vitamin D receptor (VDR). More recently, calcitriol and VDR have been shown to control the expression of genes associated with cellular proliferation and differentiation in a wide variety of cells, suggesting more extensive biological activities for the vitamin D system. Recently, calcitriol and several structurally related members of the vitamin D class of *seco*-steroids have demonstrated the ability to regulate the hedgehog (Hh) signaling pathway. Hh signaling plays an essential role with respect to tissue differentiation during embryogenesis and maintains stem cell populations in certain adult tissues. Potential mechanisms of cross talk between the two

Vitamins and Hormones, Volume 100
ISSN 0083-6729
http://dx.doi.org/10.1016/bs.vh.2015.10.006

signaling pathways highlight our evolving understanding of the complexities of calcitriol/VDR signaling and suggest a cooperative role during multiple stages of human development and disease.

1. THE HEDGEHOG SIGNALING PATHWAY

1.1 Hedgehog Signal Transduction

The hedgehog (Hh) pathway is a developmental signaling pathway that plays a key role in directing growth and tissue patterning during embryonic development. In cells regulated by the Hh pathway, signal transmission is controlled through a multistep cascade that includes an Hh ligand [Sonic (SHh), Indian (IHh), or Desert (DHh)]; Patched (Ptch) a 12-transmembrane (TM) domain cell-surface receptor; Smoothened (Smo), a 7TM GPCR-like receptor; Suppressor of Fused (Sufu); and the glioma-associated oncogene (Gli) family of zinc-finger transcription factors. The ultimate result of Hh regulation is a balance between cellular levels of the activator and repressor forms of Gli. When Hh ligand is absent, Ptch suppresses the activity of Smo by preventing its trafficking to the primary cilia (Fig. 1A). Although the exact mechanisms that control Ptch-mediated repression of Smo are poorly understood, it has been hypothesized that Ptch controls the influx or efflux of a small molecule that directly binds and regulates Smo activation/translocation (Amayke, Jagani, & Dorsch, 2013; Hadden, 2014; Sharpe, Wang, Hannoush, & de Sauvage, 2015). When Smo is in the inactive state, Gli and Sufu form a heteroprotein complex in the cytosol that promotes

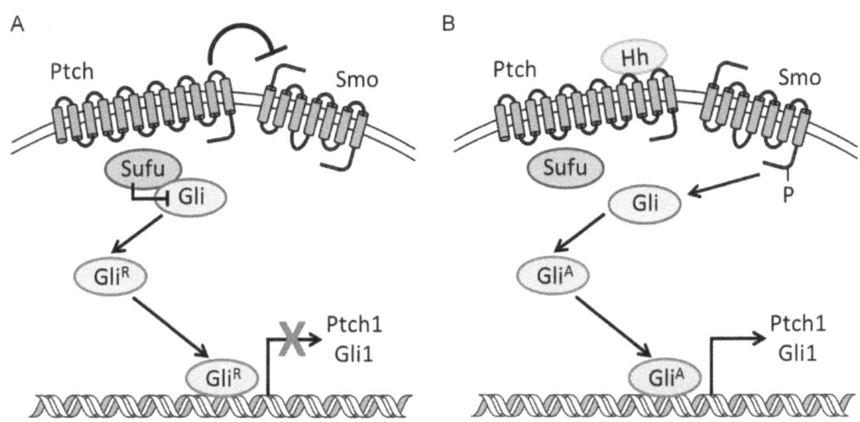

Figure 1 The Hh signaling pathway. *Adapted from Hadden (2014).* (See the color plate.)

proteasome-mediated degradation of Gli into an N-terminal truncated form (Gli^R) that serves as a direct transcriptional repressor of Hh pathway target genes. Upon binding of an Hh ligand to Ptch, the ligand/receptor complex is internalized, and Smo repression is abolished. This allows Smo to accumulate in nonmotile cilia where it is activated through a series of phosphorylations on its intracellular C-terminus. Active, phosphorylated Smo disrupts the Gli–Sufu complex and full-length Gli (Gli^A) translocates to the nucleus where it serves to regulate target gene expression.

As noted above, Hh target gene expression is tightly controlled through the relative levels of Gli^A/Gli^R present in a given cell; transcriptional control that occurs through both direct and indirect mechanisms (Hui & Angers, 2011). For some pathway target genes, a decrease in the nuclear level of Gli^R is sufficient to promote transcription. Other target genes rely on graded increases in Gli^A for expression. The most common target gene associated with the Hh pathway is *Gli1*, which itself controls the expression levels of multiple genes related to Hh signaling, cell cycle progression, cell adhesion, and apoptosis (Hui & Angers, 2011). *Gli1* exerts direct control over several negative regulators of pathway signaling, including *Ptch1*, *Ptch2*, *Boc*, and *Hhip* (Hh-interacting protein), through direct binding interactions with their chromatin sites (Hui & Angers, 2011; Vokes et al., 2007; Yoon et al., 2002). Its ability to directly induce the expression of pathway inhibitors (*Ptch1*, *Ptch2*, and *Hhip*) and inhibit the expression of pathway activators (*Boc*) highlights a precise level of transcriptional control over signal transmission through several negative feedback loops.

1.2 Hh Signaling in Cancer

Dysregulation of Hh signaling results in its constitutive activation and uncontrolled cellular proliferation. Multiple mechanisms through which aberrantly activated Hh signaling contributes to tumor formation, growth, and metastasis have been identified (Fig. 2; Amayke et al., 2013; Hadden, 2014). Cancers generally characterized as Hh-dependent do not require the presence of Hh ligand for constitutive pathway activation. By contrast, these forms of cancer are driven by specific mutations in *Ptch*, *Smo*, or *Sufu*. Hh pathway activation was initially associated with cancer onset following the discovery that germline loss-of-function mutations in *Ptch* caused nevoid basal cell carcinoma syndrome (NBCCS), also known as Gorlin syndrome, a genetic disease that predisposes syndrome patients to develop basal cell carcinomas (BCCs), medulloblastomas (MBs), and rhabdomyosarcomas (RMS;

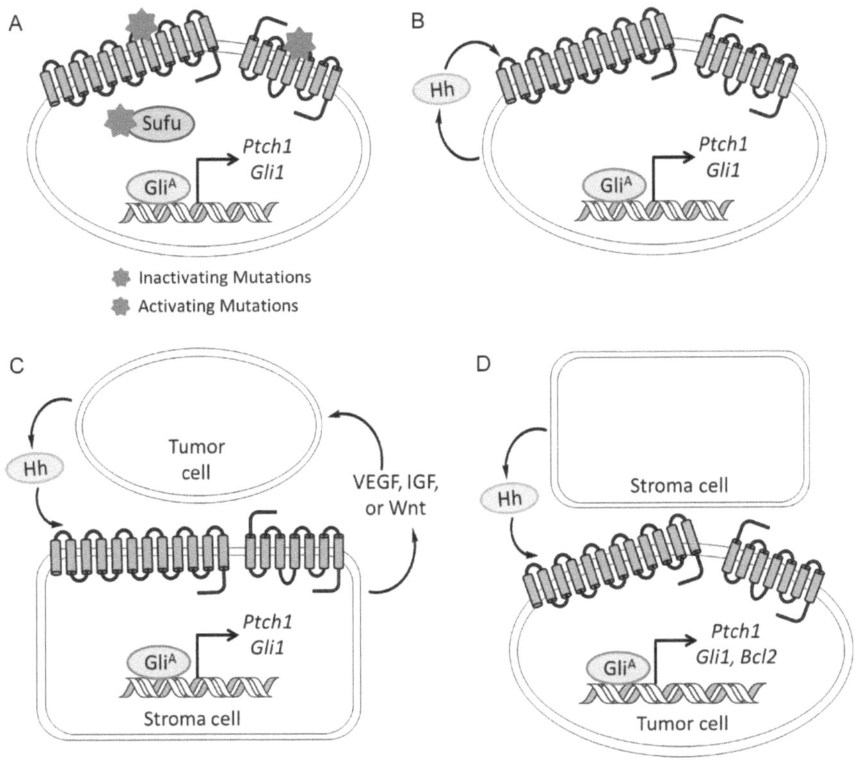

Figure 2 Mechanisms of dysregulated Hh pathway signaling that have been implicated in the development of human cancer. (A) Mutations in key pathway components, (B) autocrine signaling, (C) paracrine signaling, and (D) reverse paracrine signaling. (See the color plate.)

Gorlin, 1995; Hahn et al., 1996). These connections between *Ptch* mutations and the predisposition to several specific forms of cancer provided an important link toward the definitive identification of several Hh-dependent forms of cancer.

BCC is the most commonly diagnosed form of cancer in persons of European ancestry (affecting greater than 3 million people worldwide annually) and it has been estimated that approximately 30% of Caucasians living in areas of high sun exposure will develop a BCC during their lifetime. Genetic analyses have demonstrated that the majority of sporadically occurring BCCs harbor either loss-of-function mutations in *Ptch1* (75%) or *Sufu* (5%) or a gain-of-function mutation in *Smo* (10%) (Ling et al., 2001; Reifenberger et al., 2005). In addition, mutations in each of these proteins have also been implicated as primary contributors to the development of Hh-dependent

MBs, and to a lesser extent RMS (Raffel et al., 1997; Taylor et al., 2002; Tostar et al., 2006; Zibat et al., 2010).

Additional forms of ligand-dependent Hh pathway activation have been implicated in a variety of human cancers; however, it remains unclear whether constitutive pathway activation that is not a direct result of genetic mutation(s) specifically drives oncogenic transformation. Upregulation of Hh pathway components and autocrine activation of Hh signaling has been reported in subsets of tumors originating from lung, breast, gastrointestinal, prostate, or pancreatic tissues (Berman et al., 2003; Karhadkar et al., 2004; Kubo et al., 2004; Thayer et al., 2003; Watkins et al., 2003). A paracrine requirement for oncogenic Hh activation in the tumor microenvironment has been characterized in pancreatic and colorectal cancers that overexpress Hh ligand (Yauch et al., 2008). In this model, tumor epithelial cells produce Hh ligand, which activates pathway signaling in the surrounding stromal cells. Activation of Hh signaling in the stromal cells is thought to increase expression of additional signaling and growth factors in the stroma, which in turn promote tumor cell growth and metastasis (Tian et al., 2009; Yauch et al., 2008). In addition, a ligand-dependent "reverse" paracrine model in which the stromal cells produce the required Hh ligand, which activates Hh signaling and drives oncogenesis in the tumor cells has been implicated as a driving force for several forms of hematologic cancer (Dierks et al., 2007; Hegde et al., 2008).

1.3 Mechanisms of Resistance to Hh Signaling

Several small molecule Hh pathway inhibitors that directly target Smo have demonstrated promising preclinical properties and have progressed into clinical trials; however, the development of resistance to several of these compounds has arisen as a major concern for their continued advancement and clinical success. Vismodegib (Erivedge™) became the first Hh pathway inhibitor that targets Smo to achieve FDA approval in early 2012 when it was approved for the treatment of metastatic BCC (Dlugosz, Agrawal, & Kirkpatrick, 2012). While treatment of an MB patient with vismodegib also demonstrated initial positive results, relapse occurred when a point mutation in Smo (D473H) rendered the patient insensitive to further vismodegib treatment (Dijkgraaf et al., 2011; Rudin et al., 2009; Yauch et al., 2009). Several research groups have recently reported the identification of additional point mutations in Smo that confer resistance to patients receiving vismodegib for the treatment of advanced BCC (Atwood et al., 2015;

Pricl et al., 2015; Sharpe, Pau, et al., 2015). In these studies, Smo mutations that prevent vismodegib binding or confer constitutive reactivation of the pathway were identified in over 50% of patients that experienced a BCC relapse while receiving vismodegib. In addition, the preclinical evaluation of NVP-LDE225, another small molecule Smo antagonist under development by Novartis, in murine models of Hh-dependent MB also demonstrated tumor regrowth following initial regression (Buonamici et al., 2010). In addition to the identification of reactivating point mutations in Smo, treatment with NVP-LDE225 resulted in chromosomal amplification of *Gli2*.

Additional mechanisms of tumor resistance that are independent of canonical Hh signaling have also been reported for Smo antagonists. In a murine model of Hh-dependent MB, resistance to a semi-synthetic derivative of the natural product Hh pathway inhibitor cyclopamine (Cyc) resulted from the induction of P-glycoprotein (P-gp) in the tumor cells and active efflux of the drug (Lee et al., 2012). Components of the phosphoinositide 3-kinase (PI3K) pathway were upregulated in MB tumors resistant to NVP-LDE225 (Buonamici et al., 2010). PI3K signaling was previously shown to enhance Gli-dependent transcription when SHh is present at low levels, suggesting PI3K signaling may partially compensate for inhibition of Hh signaling by Smo antagonists to promote tumor resistance (Riobo, Lu, Ai, Haines, & Emerson, 2006). In esophageal adenocarcinoma, the mTOR pathway component S6 kinase 1 (S6K1) phosphorylates Gli1, resulting in its Smo-independent dissociation from Sufu and activation of Hh signaling (Wang et al., 2012). This mechanism of resistance has not been explored in BCC or MB; however, it provides evidence of pathway activation at the level of Gli and suggests additional resistance mechanisms downstream of Smo may be possible. The inflammatory cytokine osteopontin (OPN) has also been shown to activate Gli-mediated transcription through a nonclassical mechanism by modulation of Akt/GSK3β signaling (Das, Samant, & Shevde, 2013). OPN inactivates GSK3β, which ultimately results in Sufu/Gli dissociation and accumulation of Gli^A within the nucleus. Interestingly, OPN also upregulated P-gp through a Gli-dependent process, suggesting that it may mediate resistance to Smo antagonists through multiple distinct mechanisms.

2. VITAMIN D METABOLISM AND REGULATION

The vitamin D metabolic pathway is a complex, highly regulated multistep process that provides calcitriol from the cholesterol precursor

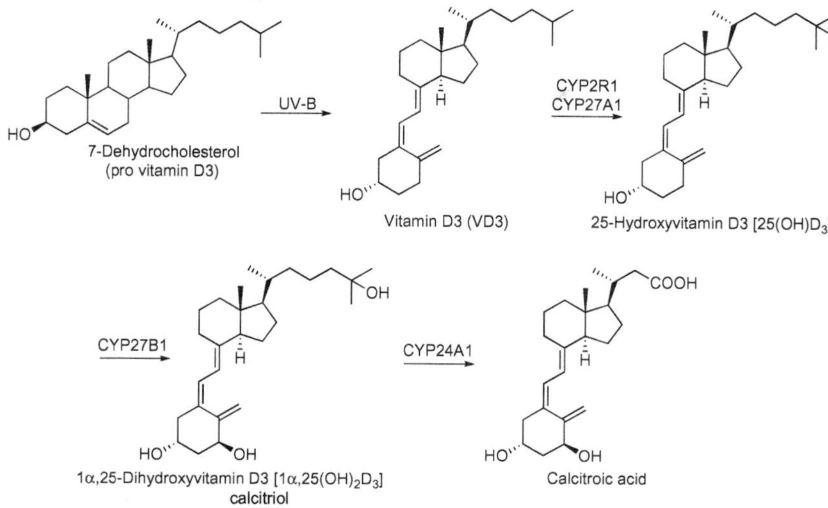

Figure 3 Vitamin D ligand production and metabolism.

7-dehydrocholesterol (7-DHC/provitamin D3, Fig. 3). Vitamin D3 (VD3, cholecalciferol) is either synthesized in the skin from 7-DHC through UV light from the sun or taken in through diet and supplementation. After entering the bloodstream, VD3 is hydroxylated in the liver at C-25 to provide 25-hydroxyvitamin D3 [25(OH)D3]. Several cytochrome P450 (CYPs) enzymes, including CYP2R1, CYP27A1, and CYP3A4, have demonstrated 25-hydroxylase activity; however, studies indicate that CYP2R1 is the primary mediator of hydroxylation on C-25 (Bikle, 2014). Subsequent hydroxylation in the kidney at C-1 to provide calcitriol is mediated by CYP27B1, the only enzyme identified to date that exhibits 1α-hydroxylase activity on the vitamin D scaffold (Bikle, 2014). Metabolism and inactivation of the vitamin D scaffold occurs through initial hydroxylation at C-24 and subsequent oxidative truncation of the side chain to provide calcitroic acid (Bikle, 2014). CYP24A1 is the only established 24-hydroxylase involved in vitamin D metabolism. *Cyp24A1* is a key target gene of the vitamin D receptor (VDR) pathway and its cellular expression is significantly upregulated following activation of VDR, highlighting as its primary function the maintenance of appropriate levels of calcitriol within responding cells.

The physiological effects mediated by vitamin D ligands are traditionally attributed to calcitriol binding to VDR, activation of the receptor, and subsequent regulation of a variety of downstream signaling mechanisms (Bikle,

2014). VDR is a ligand-dependent transcription factor and member of the steroid hormone nuclear receptor family that is most commonly associated with maintaining calcium and phosphorous homeostasis in the bone (Nagpal, Na, & Rathnachalam, 2005). More recently, VDR has been shown to control the expression of genes associated with cellular proliferation and differentiation, autoimmune diseases, diabetes, and blood pressure (Basit, 2013; DeLuca, Kimball, Kolasinski, Ramagopalan, & Ebers, 2013; Mitri, Muraru, & Pittas, 2011; Norman & Powell, 2014; Pike & Meyer, 2010). In addition to its genomic control of many aspects of cellular signaling, VDR activation also results in various "nongenomic" effects, presumably via cross talk with other signaling pathways (Haussler, Jurutka, Mizwicki, & Norman, 2011; Revelli, Massobrio, & Tesarik, 1998).

3. VITAMIN D/Hh PATHWAY REGULATION IN THE SKIN

Cross talk mechanisms between vitamin D/VDR signaling and the Hh pathway are not well defined; however, several lines of evidence exist to suggest their interactions may play an important physiological role, primarily in proper skin homeostasis and the oncogenic development of BCC. These mechanisms include the transcriptional control of Hh pathway components by VDR as well as the ability of vitamin D ligands to directly modulate Hh pathway target genes, details of which are described below and in Section 4.

3.1 Vitamin D Control Over Skin Differentiation

The epidermis, the outermost layers of cells in the skin, is the largest organ in the human body and acts as the major barrier to external pathogens. The primary cell type that constitutes the epidermis is keratinocytes (Bikle, 2004). The bottom layer, the stratum basale, mainly consists of proliferating keratinocytes that begin to differentiate as they travel toward the outermost layer, the stratum corneum. Calcitriol and VDR play an integral role in this differentiation process. To maintain tight control over calcitriol-mediated differentiation, keratinocytes are capable of expressing all the enzymatic machinery necessary to produce and metabolize calcitriol (Bikle, 2004). The levels of active CYP27A1 (25-OHase) and CYP27B1 (1-OHase) in keratinocytes are controlled by multiple factors including calcium levels, calcitriol concentration, UVB radiation, and stage of cellular differentiation, suggesting the levels of calcitriol produced are tightly regulated at multiple stages (Bikle, 2004). In addition to the ability of keratinocytes to directly control intracellular calcitriol levels, the cellular response(s) initiated by VDR activation are also distinct depending on the differentiation status of the cells. Several coactivator proteins that modulate the transcriptional activity of VDR are differentially associated with VDR in proliferating and differentiated keratinocytes, suggesting a distinct transcriptional response depending on differentiation state (Hawker, Pennypacker, Chang, & Bikle, 2007). A reduction in overall VDR levels as keratinocytes differentiate highlights a reduced need for VDR signaling in keratinocytes of the stratum corneum (Bikle, 2004). Taken together, these findings highlight the precise level of control calcitriol and VDR play with respect to keratinocyte homeostasis.

3.2 Tumor Suppressor Properties of Calcitriol/VDR in Skin

Multiple studies have demonstrated the chemopreventative and chemotherapeutic properties of both calcitriol and VDR in skin. Prolonged UVB radiation damages keratinocyte DNA, primarily through the formation of mutagenic cyclobutane pyrimidine dimers (CPDs). Direct topical administration of calcitriol or its analogues protected against CPD formation and increased CPD clearance (Dixon et al., 2005; Wong et al., 2004). Interestingly, these results were seen with calcitriol analogues that only promote the nongenomic actions of VDR, suggesting that genomic activation was not necessary. By contrast, *in vitro* treatment of keratinocytes with calcitriol resulted in the increased expression of DNA damage repair genes XPC (xeroderma pigmentosum complementation group C) and DDB2 (damage-specific DNA-binding protein 2), suggesting a potential model for DNA damage protection/repair that incorporates both genomic and nongenomic actions of VDR (Moll, Sander, Frischauf, & Richter, 2007).

Additional protective properties of VDR have been identified through studies in VDR knockout mice. In one study, approximately 85% of $VDR^{-/-}$ mice, but none of the control animals, developed skin tumors shortly after exposure to the carcinogen 7,12-dimethylbenzanthracene (DMBA; Zinser, Sundberg, & Welsh, 2002). A related study demonstrated that $VDR^{-/-}$ mice were also more susceptible to tumor formation following UVB exposure (Ellison, Smith, Gilliam, & Macdonald, 2008). *In vitro*, keratinocytes lacking VDR are hyper-proliferative and exhibit decreased apoptosis in response to UVB irradiation (Ellison et al., 2008). In addition, evidence suggests that proper VDR function reduces the expression of oncogenic long noncoding RNAs (lncRNAs) while concomitantly upregulating lncRNAs with tumor suppressor properties suggesting its deletion reverses this trend toward a more oncogenic profile (Jiang & Bikle, 2014).

3.3 Vitamin D/VDR in BCC Development

Several lines of evidence from both preclinical and clinical studies have emerged to suggest that vitamin D ligands and the VDR are associated with the onset and progression of BCC; however, these relationships are not well defined and are oftentimes contradictory. A comparison of the vitamin D signaling components in BCC and normal human skin demonstrated that *VDR*, *CYP27B1*, and *CYP24A1* are significantly overexpressed in BCC tissue when compared to normal skin tissue (Mitschele et al., 2004). It is unclear whether the increased expression levels of these pathway components results primarily in an increase in calcitriol synthesis or calcitriol turnover and whether either of these outcomes is a causative factor leading to BCC formation or a result of oncogenic transformation. In addition, several studies have identified an association between genetic variants of *VDR* that result in forms of the receptor that are less active transcriptionally and have been associated with an increased risk of BCC development (Köstner et al., 2012; Lesiak et al., 2011). In contrast to these clinical findings that suggest a reduction in VDR activity correlates to increased BCC development, several studies have demonstrated that increased levels of 25(OH)D3 also correlate to enhanced risk of BCC (Asgari et al., 2010; Reddy, 2013; Van der Pols et al., 2013). In addition, NBCCS patients that are predisposed to developing BCCs during their lifetime had levels of circulating 25(OH) D3 significantly lower than the general population. It is important to note that NBCCS patients typically practice self-photoprotective measures to reduce UVB-induced skin cancer and this decreased sun exposure may play a strong contributory role in the decreased levels of circulating 25(OH)D3. Even so, these data provide a contrasting picture that does not definitively identify either overactivation or underactivation of the vitamin D signaling cascade as a causative mechanism for BCC development.

The established links between aberrant Hh signaling and BCC formation led researchers to explore the relationships between the two in VDR knockout mice and the analysis of these results has also proven complicated (Teichert, Elalieh, & Bikle, 2010; Teichert, Elalieh, Elias, Welsh, & Bikle, 2011). VDR knockout resulted in a 164-fold decrease in SHh mRNA in hair follicles, but this reduced expression was not observed until postnatal day 77 and comparable changes in expression were not seen for Ptch, Smo, Gli1, or Gli2 (Teichert et al., 2010). In addition, topical administration of a small molecule Smo agonist partially restored proper hair growth in the VDR knockout mice, suggesting the ability of enhanced Hh signaling to overcome the improper hair follicle cycling associated with the knockout mice (Teichert et al., 2010). By contrast, the overexpression of multiple Hh signaling components, albeit at modest levels, was seen in the epidermis of VDR knockout mice at the same postnatal timeframe (Teichert et al., 2011). SHh, Ptch, Smo, Gli1, and Gli2 expression levels were increased between 1.17- and 3.54-fold in this model. These Hh pathway components were also upregulated (1.53- to 3.65-fold) in DMBA- and UVB-induced skin tumors in the VDR knockout mice; however, there were a variety of tumors formed following carcinogenic insult, of which BCCs only constituted 11% (Teichert et al., 2011).

4. INHIBITION OF HH SIGNALING BY VITAMIN D-BASED SMALL MOLECULES

4.1 Natural Vitamin D Ligands

4.1.1 Vitamin D3/Cholecalciferol

Evidence that vitamin D ligands modulate Hh signaling was first established when studies designed to better understand the mechanisms through which Ptch exerts its inhibitory effects on Smo identified elevated levels of 3β-hydroxysteroids in conditioned media from Ptch1-transfected cells (Biljsma et al., 2006). These data suggested that Ptch-mediated efflux of a small molecule with the 3β-hydroxysteroid scaffold had direct implications on Hh activity. Follow-up studies to further characterize these findings identified vitamin D3 as a 3β-hydroxyl *seco*-steroid that exhibits modest inhibition of Hh signaling (IC_{50} values = 5–10 μM) in multiple Hh-dependent mouse embryonic fibroblast cell lines (MEFs; Banerjee, Ghosh, & Hadden, 2012; Biljsma et al., 2006). Surprisingly, even though VD3 does not bind to recombinant VDR, it has demonstrated the ability to activate canonical VDR signaling in cell culture, as measured by upregulation of Cyp24A1 mRNA, a well-characterized target gene associated with VDR activation (Banerjee et al., 2012). Most likely, metabolism of VD3 *in vitro* to 25(OH) D3 and/or calcitriol results in activation of VDR signaling; however, definitive evidence supporting this hypothesis has not been disclosed.

The ability of VD3 to reduce cellular proliferation and tumor growth via Hh inhibition has been explored in several *in vitro* and *in vivo* models of cancer. VD3 (5 and 10 μM) inhibited proliferation and reduced *Gli1* expression in a murine BCC cell line that demonstrates constitutive Hh signaling (ASZ) and an MB cell line that exhibits upregulated Hh signaling (DeBerardinis, Banerjee, & Hadden, 2013; Tang et al., 2011). By contrast, VD3 did not inhibit proliferation of a nontumorigenic keratinocyte cell line (C5N), highlighting the specific nature of its anti-Hh effects (Tang et al., 2011). Not surprisingly, VD3 also upregulated CYP24A1 expression in the ASZ cells, indicating it activates canonical VDR signaling in this model in a similar fashion to that previously demonstrated in MEFs. Direct topical administration of VD3 (1.3–2.6 mg/kg) to murine nodular BCCs significantly reduced *Gli1* expression and Ki67 staining, while concomitantly increasing CYP24A1 expression, highlighting its nonselective nature *in vivo* (Tang et al., 2011). Interestingly, BCCs treated with VD3 continued to express keratin14, but not keratin10, suggesting that VD3 does not induce differentiation of BCC cells *in vivo* (Tang et al., 2011).

Studies to evaluate whether inhibition of Hh signaling by VD3 is an effective therapeutic option against human cancers not dependent on Hh signaling have also been performed (Brüggemann et al., 2010; Dormoy et al., 2012). VD3 was a modest inhibitor (IC_{50} values ~5–10 μM) of *in vitro* cellular proliferation against a variety of pancreatic cancer cells lines that demonstrate overactive Hh signaling; however, it did not affect pancreatic adenocarcinoma growth *in vivo* as a monotherapy (5 μg/animal, IP) or in combination with nucleoside analogues or radiation (Brüggemann et al., 2010). Conflicting results were obtained when VD3 was evaluated against clear cell renal cell carcinoma (CCC; Dormoy et al., 2012). Vitamin D3 exhibited antiproliferative and proapoptotic properties in several human CCC cell lines at low concentrations (10–50 nM); however, it did not downregulate Gli1 mRNA expression in these cells. A significant decrease in Gli2 mRNA expression was observed in 786-O CCC cells following VD3 treatment (30 nM, 18–24 h). VD3 (6.25 μg/animal, IP) significantly reduced CCC tumor formation when administered prophylactically and decreased tumor size when administered after tumor onset (Dormoy et al., 2012). Similar to the *in vitro* results in CCC, decreased tumor growth correlated with significant reductions in Gli2 mRNA expression in the tumor tissues. Of note, hypercalcemic side effects were not observed in mice treated with VD3, suggesting its ability to inhibit Hh signaling and provide anticancer properties without the detrimental side effects commonly associated with calcitriol.

4.1.2 Calcitriol

Calcitriol has also been evaluated for its anti-Hh and anticancer properties in multiple Hh-dependent cellular model systems. Calcitriol significantly reduces *Gli1* expression in Hh-dependent MEFs at low nanomolar concentrations (100 nM) (Banerjee et al., 2012). Studies exploring the ability of calcitriol to inhibit Hh signaling in ASZ cells have proven contradictory (Tang et al., 2011; Uhmann et al., 2012a). In one report, neither cellular proliferation nor *Gli1* expression was significantly reduced in ASZ cells following administration of calcitriol (100 nM) (Tang et al., 2011). A second report demonstrated significant reductions in both *Gli1* expression and BrdU incorporation in ASZ cells treated with calcitriol (10 nM), suggesting it does maintain anti-Hh effects in this cellular model (Uhmann et al., 2012a). Gli1 expression and Ki67 staining were reduced in BCC-bearing skin punches from Hh-dependent murine BCC following administration of calcitriol (10 nM) (Uhmann et al., 2012a). Similar results for Gli1 downregulation and BrdU incorporation were seen in a primary cell culture model of embryonal rhabdomyosarcoma (ERMS; Uhmann et al., 2012b). In each of these cellular models, calcitriol significantly upregulated Cyp24A1 expression, further highlighting its nonselective effects on both the Hh and VDR signaling pathways.

Similar to the results seen in ASZ cells, the *in vivo* effects of calcitriol in murine models of Hh-dependent BCC have also proven contradictory (Tang et al., 2011; Uhmann et al., 2012a). Direct topical administration of calcipotriene cream (Dovonex, 0.0005% calcitriol, 4 days) to murine BCC tumors did not decrease Gli1 mRNA expression; however, it did reduce Ki67 staining, indicating inhibition of tumor cell proliferation (Tang et al., 2011). By contrast, daily IP administration of calcitriol (100 ng/kg, 90 days) resulted in decreased expression of Gli1 and Gli2 mRNA and a reduction in Ki67 positive cells in BCC tissue (Uhmann et al., 2012a). Most likely, these *in vivo* discrepancies are primarily a result of experimental differences, and prolonged topical administration with higher doses of calcitriol would result in effects similar to those seen with the IP studies. A significant increase in expression of the keratinocyte differentiation markers Tgm1 and K10 was also detected following IP dosing (Uhmann et al., 2012a). While IP dosing increased serum calcium levels, associated hypercalcemic side effects (weight loss, kidney damage) were not observed, suggesting a dose range that could deliver positive anticancer effects without the requisite side effects primarily associated with prolonged calcitriol treatment. IP administration of calcitriol (50 ng/kg/d for 8 weeks) in an Hh-dependent murine model of ERMS also reduced Gli1 expression, BrdU incorporation, and tumor volume (Uhmann et al., 2012b). Finally, both topical and IP administration of calcitriol in each of these *in vivo* models significantly upregulated CYP24A1 expression.

4.2 Synthetic Vitamin D Analogues

In order to improve the potency and selectively of VD3 with respect to inhibition of the Hh signaling pathway, extensive structure–activity relationship (SAR) studies on the scaffold have been reported (Banerjee, DeBerardinis, & Hadden, 2015; Banerjee et al., 2012; DeBerardinis, Banerjee, Miller, Lemieux, & Hadden, 2012; DeBerardinis, Raccuia, Maschinot, Thompson, & Hadden, 2015; DeBerardinis et al., 2013, 2014). Initial VD3 analogues synthesized and evaluated focused on truncating VD3 to determine the minimal pharmacophore of the scaffold required to retain potent inhibition against the Hh pathway. Grundmann's alcohol (**1**), also termed the "northern region" of VD3, retained all the activity of the parent VD3 structure, but did not upregulate CYP24A1, indicating its selectivity for the Hh pathway (DeBerardinis et al., 2012). These results demonstrated that the *seco* B-ring/triene and the cyclohexyl A-ring are not essential for targeting the Hh pathway. Seeking to improve the potency of Grundmann's alcohol, while maintaining the selective nature of its inhibition, an analogue which incorporated an aromatic A-ring mimic linked to the northern region of VD3 through an ester bond was prepared and evaluated (**2**, Fig. 4; DeBerardinis et al., 2013). Analogue **2** demonstrated potent inhibition of Hh signaling in both MEFs and ASZ cells with IC_{50} values between 0.74 and 5.2 μM. In addition, analogue **2** did not activate VDR, demonstrating its selectivity for the Hh signaling pathway. Based on the promising activity of **2**, SAR studies for an extensive series of ester-linked, aromatic A-ring containing VD3 analogues were pursued. Three analogue subclasses

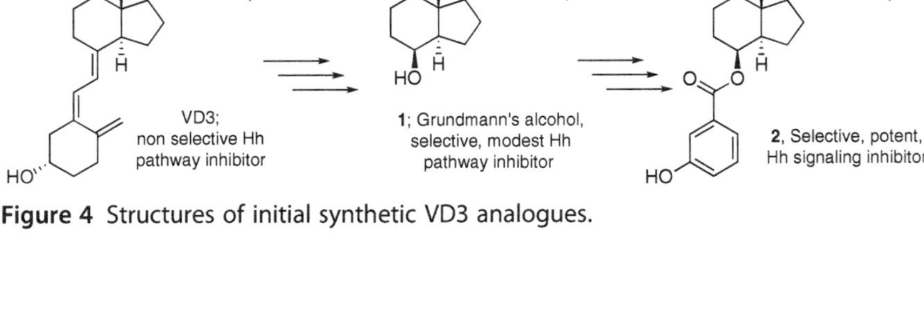

Figure 4 Structures of initial synthetic VD3 analogues.

containing either (1) a single substitution at the *ortho*, *meta*, or *para* position of the aromatic A-ring; (2) a heteroaryl or substituted biaryl A-ring moiety; or (3) multiple substituents on the aromatic A-ring were prepared and evaluated (DeBerardinis et al., 2013, 2014). The results of these studies demonstrated that a phenyl ring incorporating a hydrophilic moiety at the *meta* or *para* position retains potent, selective inhibition of Hh signaling, while hydrophobic substitutions and five- or seven-membered aromatic rings are inactive (DeBerardinis et al., 2013, 2014).

A second generation of VD3 analogues based on the scaffold of analogue **2** was recently reported (Banerjee et al., 2015; DeBerardinis et al., 2015). These compounds incorporated modifications to either the ester linker or alkyl side chain. For the side chain region, shorter alkyl groups were preferred compared to extended, bulky, or aryl side chains. The most potent compounds identified in this series (**3**–**5**, Fig. 5) inhibited Hh signaling with IC_{50} values between 1.1 and 6.6 μM (Banerjee et al., 2015). With respect to the ester region, flexible linkers that contain 1–3 atoms between the northern region and the aromatic A-ring were optimal (DeBerardinis et al., 2015). The most potent linker region analogues (**6**–**9**, shown in Fig. 5)

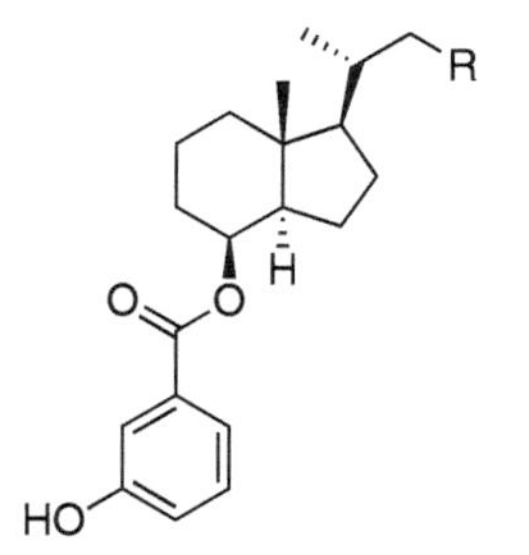

3; R = *i*-Butyl, IC_{50} values = 1.1–1.6 μM
4; R = *n*-Butyl, IC_{50} values = 1.1–6.6 μM
5; R = $CH(Et)_2$, IC_{50} values = 1.1–4.9 μM

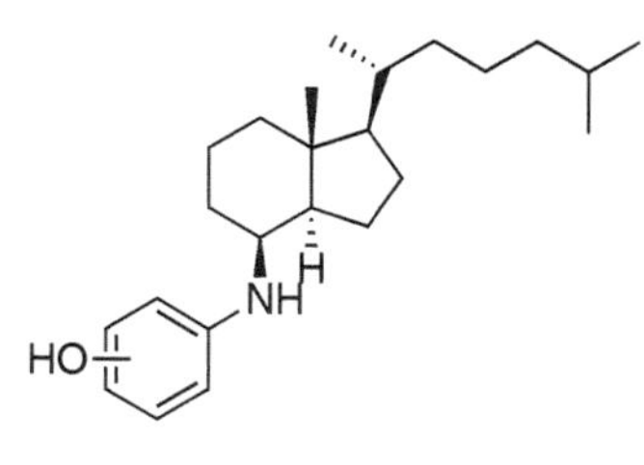

6; 3-OH, IC_{50} values = 1.0–1.7 μM
7; 4-OH, IC_{50} values = 0.4–0.67 μM

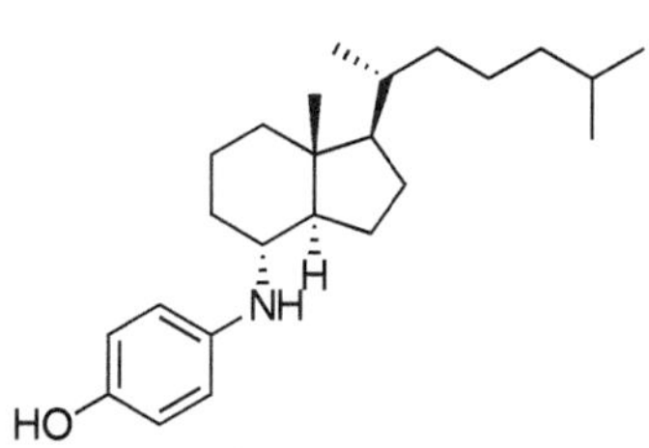

8; IC_{50} values = 0.32–0.57 μM

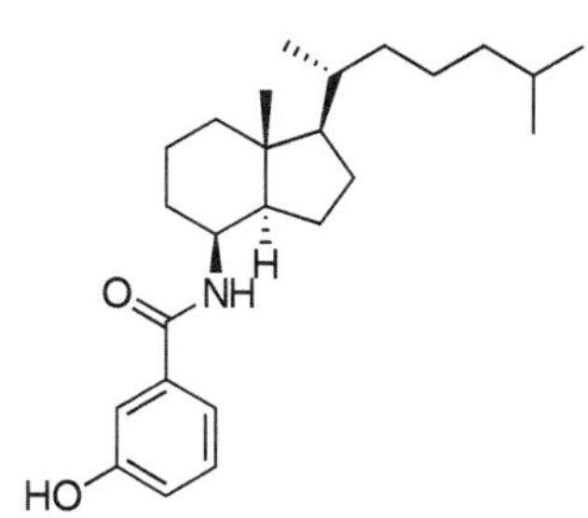

9; IC_{50} values = 0.98–1.8 μM

Figure 5 Second generation synthetic VD3 analogues that inhibit Hh signaling.

demonstrated potent and selective inhibition of the Hh pathway in both MEF and ASZ cell lines. Interestingly, analogues **7** and **8**, whose only structural difference is the stereochemical orientation of the amine attached to C-8 of the northern CD-ring core, were equipotent in their inhibition of Hh signaling (IC_{50} values in MEFs = 0.4 and 0.32, respectively) (DeBerardinis et al., 2015). None of the second generation analogues reported to date have bound to recombinant VDR or activated canonical VDR signaling in cell culture, clearly establishing the selective Hh inhibitory nature of this scaffold. To date, *in vivo* anti-Hh properties of the synthetic VD3 analogues have not been reported.

4.3 Cellular Mechanisms that Govern Hh Inhibition for Vitamin D-Based *Seco*-Steroids

Several approaches have been taken to determine the exact mechanisms through which VD3, calcitriol, and VD3 analogues inhibit Hh signaling. Initial studies focused on whether VD3 inhibits Hh signal transmission by direct binding to the key pathway component Smo, the target of multiple other small molecule Hh pathway inhibitors. Binding assays performed in a yeast-based system with inducible full-length human Smo demonstrated that VD3 binds to Smo in a Cyc-sensitive manner, suggesting that VD3 inhibits Hh signaling at the level of Smo and that it binds to the established Cyc-binding site on Smo (Biljsma et al., 2006). Subsequent studies designed to reproduce the yeast-based displacement studies in cell culture have raised questions as to whether vitamin D ligands bind Smo at the Cyc-binding site (DeBerardinis et al., 2014). Neither VD3 nor analogue **2** was able to displace BODIPY-Cyc from full-length human Smo overexpressed in HEK293T cells following two separate protocols: binding to intact cells via flow cytometry or fluorescence polarization on isolated membrane fractions (DeBerardinis et al., 2014). These binding study data certainly do not preclude vitamin D/Smo binding interactions; however, they provide strong evidence that *seco*-steroids and Cyc do not share a binding site on Smo. Additional cellular and *in vivo* studies provide further evidence that these compounds function at the level of Smo. VD3 maintained its ability to inhibit Hh signaling in *Ptch1*$^{-/-}$, but not *Smo*$^{-/-}$, MEFs, and VD3 treatment (1.2 mg/mL) mimicked the *Smo*$^{-/-}$ phenotype in zebrafish (Biljsma et al., 2006; DeBerardinis et al., 2014). Similar results suggesting calcitriol inhibits at the level of Smo have also been reported. Calcitriol did not downregulate Gli1 expression in *Ptch*$^{flox/flox}$ERT2$^{+/-}$ MEFs, a model system in which an aberrant, nonfunctional *Ptch* transcript is expressed, suggesting that calcitriol functions downstream of Ptch (Uhmann et al., 2012a). In addition, calcitriol did not reduce basal levels of Gli1 in *Smo*$^{-/-}$ MEFs and its ability to downregulate Gli1 was restored upon transfection of exogenous human Smo (Uhmann et al., 2012a). It should be noted that functional, active Smo is required for proper Hh pathway signaling and Gli1 expression; therefore, it is not surprising that Gli1 expression is negligible in *Smo*$^{-/-}$ MEFs and the inability of calcitriol (or other small molecule Hh pathway modulators) to affect Gli1 expression in this model system does not directly suggest calcitriol functions through direct binding to Smo.

Studies have also been performed to determine whether vitamin D ligand-mediated inhibition of Hh signaling is a downstream cellular consequence of VDR activation (Brüggemann et al., 2010; DeBerardinis et al., 2014; Uhmann et al., 2012a). Following knockdown of VDR expression with VDR-specific siRNA in Hh-dependent MEFs, the ability of VD3 to inhibit Hh signaling was completely abolished, providing evidence that VDR is required for inhibition of Hh signaling by VD3 (DeBerardinis et al., 2014). By contrast, calcitriol retained the ability to inhibit Hh signaling following VDR knockdown, suggesting that even though these compounds contain similar molecular scaffolds, they may inhibit Hh signaling through distinct mechanisms. A second possibility for the discrepancies in this data is the inherent problem that siRNA-mediated knockdown of VDR is transient and its expression is not completely abolished. As calcitriol binds VDR with higher affinity than VD3, it is possible that calcitriol is capable of activating the residual VDR and that Gli1 downregulation is downstream of this activation. By contrast, VD3 must be metabolized to 25(OH)D3 or calcitriol in order to activate VDR and the concentration of VD3 converted to one of these more active forms may not be enough to overcome the reduced VDR expression.

While the siRNA studies in Hh-dependent MEFs suggest VDR may be important for the Hh inhibitory properties of VD3, additional reports have seen no dependence on VDR for this anti-Hh activity of either VD3 or calcitriol (Brüggemann et al., 2010; Tang et al., 2011; Uhmann et al., 2012a). Both VD3 and calcitriol maintained their ability to inhibit Hh signaling in *VDR*$^{-/-}$ MEFs (Brüggemann et al., 2010; Uhmann et al., 2012a). In addition, siRNA-mediated knockdown of VDR in *Ptch*$^{-/-}$ and *Ptch*$^{flox/flox}$ERT2$^{+/-}$ MEFs had no effect on the ability of calcitriol to decrease Gli1 expression (Uhmann et al., 2012a). Finally, the anti-Hh effects of VD3 in ASZ cells were not affected by the addition of VDR-specific shRNA, indicating its inhibitory properties are independent of VDR in this cellular model (Tang et al., 2011). Taken together these data strongly support a model in which vitamin D ligands inhibit Hh signaling via interactions with a key protein downstream of Ptch, possibly Smo, in a VDR-independent fashion; however, definitive binding interactions between Smo or additional proteins must be identified before this model can be conclusively verified.

5. DISCUSSION AND CONCLUSIONS

Multiple studies have identified potential mechanisms through which the canonical signaling cascades of the vitamin D system and the Hh pathway interact to control cellular growth and differentiation. The first of these was the identification that vitamin D3 inhibits Hh signaling and Gli1 expression in Hh-dependent MEF cell lines (Biljsma et al., 2006). Additional studies that have demonstrated similar cellular activities for calcitriol provide strong support for these scaffolds as endogenous regulators of Hh signaling. In retrospect, this cross talk between the two pathways is not surprising as both signaling pathways play pivotal roles in the development of both normal skin and skin cancer. Vitamin D3 and calcitriol are synthesized in the skin and required for proper keratinocyte differentiation and protection against oncogenic transformation. The Hh pathway is essential for hair follicle development and mutations in pathway components account for ~90% of diagnosed BCCs, a form of cancer that arises from the oncogenic transformation of keratinocytes.

Even with these connections between the two signaling networks, there are several discrepancies that have not been fully explored or rationalized. First, if UVB light from the sun is required for maintaining appropriate levels of vitamin D ligands for tumor suppression in the skin, why does the same UVB irradiation lead to cancer formation? Most likely, the balance between chemoprevention and oncogenesis is under tight regulation and the mutagenesis associated with extended sun exposure overshadows the positive effects of increased vitamin D concentrations; however, an in-depth understanding of this balance has not been presented. Second, Hh signaling has not been implicated as a causative factor in other forms of skin cancer (squamous cell carcinoma or melanoma) that arise in VDR knockout mice or have been correlated with increased sunlight, implying that the relationships between VDR/Hh signaling may not be broadly applicable and is primarily related only to BCC. Finally, it is unclear what cellular protein/receptor mediates the anti-Hh activity of the vitamin D scaffold. Several studies suggest VD3 and calcitriol function at a level downstream of Ptch, potentially through direct binding interactions with Smo; however, the Hh inhibitory properties of these *seco*-steroids have not been conclusively identified as VDR-independent. Continued work in these areas will provide a better understanding of the relationships between vitamin D and Hh signaling and lay the foundation for identifying the mechanisms responsible for their cross talk.

REFERENCES

Amayke, D., Jagani, Z., & Dorsch, M. (2013). Unraveling the therapeutic potential of the hedgehog pathway in cancer. *Nature Medicine, 19*, 1410–1422.

Asgari, M. M., Tang, J., Warton, M. E., Chren, M.-M., Quesenberry, C. P., Bikle, D., et al. (2010). Association of prediagnostic serum vitamin D levels with the development of basal cell carcinoma. *Journal of Investigative Dermatology, 130*, 1438–1443.

Atwood, S. X., Sarin, K. Y., Whitson, R. J., Li, J. R., Kim, G., Rezaee, M., et al. (2015). Smoothened variants explain the majority of drug resistance in basal cell carcinoma. *Cancer Cell, 27*, 342–353.

Banerjee, U., DeBerardinis, A. M., & Hadden, M. K. (2015). Design, synthesis, and evaluation of hybrid vitamin D3 side chain analogues as hedgehog pathway inhibitors. *Bioorganic & Medicinal Chemistry, 23*, 548–555.

Banerjee, U., Ghosh, M., & Hadden, M. K. (2012). Evaluation of vitamin D3 A-ring analogues as hedgehog pathway inhibitors. *Bioorganic & Medicinal Chemistry Letters, 22*, 1330–1334.

Basit, S. (2013). Vitamin D in health and disease: A literature review. *British Journal of Biomedical Science, 70*, 161–172.

Berman, D. M., Karhadkar, S. S., Maitra, A., Montes De Oca, R., Gerstenblith, M. R., Briggs, K., et al. (2003). Widespread requirement for hedgehog ligand stimulation in growth of digestive tract tumours. *Nature, 425*, 846–851.

Bikle, D. D. (2004). Vitamin D regulated keratinocyte differentiation. *Journal of Cellular Biochemistry, 92*, 436–444.

Bikle, D. D. (2014). Vitamin D metabolism, mechanism of action, and clinical applications. *Chemistry & Biology, 21*, 319–329.

Biljsma, M. F., Spek, C. A., Zivkovic, D., van de Water, S., Rezaee, F., & Peppelenbosch, M. P. (2006). Repression of smoothened by patched-dependent (pro-)vitamin D3 secretion. *PLoS Biology, 4*, e232.

Brüggemann, L. W., Queiroz, K. C. S., Zamani, K., van Straaten, A., Spek, C. A., & Bijlsma, M. F. (2010). Assessing the efficacy of the hedgehog pathway inhibitor vitamin D3 in a murine xenograft model for pancreatic cancer. *Cancer Biology & Therapy, 10*, 79–88.

Buonamici, S., Williams, J., Morrissey, M., Wang, A., Guo, R., Vattay, A., et al. (2010). Interfering with resistance to smoothened antagonists by inhibition of the PI3K pathway in medulloblastoma. *Science Translational Medicine, 2*, 51ra70.

Das, S., Samant, R. S., & Shevde, L. A. (2013). Nonclassical activation of hedgehog signaling enhances multidrug resistance and makes cancer cells refractory to smoothened-targeting hedgehog inhibition. *Journal of Biological Chemistry, 288*, 11824–11833.

DeBerardinis, A. M., Banerjee, U., & Hadden, M. K. (2013). Identification of vitamin D3-based hedgehog pathway inhibitors that incorporate an aromatic A-ring isostere. *ACS Medicinal Chemistry Letters, 4*, 590–595.

DeBerardinis, A. M., Banerjee, U., Miller, M., Lemieux, S., & Hadden, M. K. (2012). Probing the structural requirements for vitamin D3 inhibition of hedgehog signaling. *Bioorganic & Medicinal Chemistry Letters*, *22*, 4859–4863.

DeBerardinis, A. M., Madden, D., Banerjee, U., Sail, V., De Carlo, D., Lemieux, S. M., et al. (2014). Structure-activity relationships for vitamin D3-based aromatic A-ring analogues as hedgehog pathway inhibitors. *Journal of Medicinal Chemistry*, *57*, 3724–3736.

DeBerardinis, A. M., Raccuia, D. S., Maschinot, C. A., Thompson, E., & Hadden, M. K. (2015). Vitamin D3 analogues that contain modified A- and seco-B-rings as hedgehog pathway inhibitors. *European Journal of Medicinal Chemistry*, *93*, 156–171.

DeLuca, G. C., Kimball, S. M., Kolasinski, J., Ramagopalan, S. V., & Ebers, G. C. (2013). Review: The role of vitamin D in nervous system health and disease. *Neuropathology and Applied Neurobiology*, *39*, 458–484.

Dierks, C., Grbic, J., Zirlik, K., Beigi, R., Englund, N. P., Guo, G. R., et al. (2007). Essential role of stromally induced hedgehog signaling in B-cell malignancy. *Nature Medicine*, *13*, 944–951.

Dijkgraaf, G. J., Alicke, B., Weinmann, L., Januario, T., West, K., Modrusan, Z., et al. (2011). Small molecule inhibition of GDC-0449 refractory mutants and downstream mechanisms of drug resistance. *Cancer Research*, *71*, 435–444.

Dixon, K. M., Deo, S. S., Wong, G., Slater, M., Norman, A. W., Bishop, J. E., et al. (2005). Skin cancer prevention: A possible role for 1,25dihydroxyvitamin D3 and its analogs. *Journal of Steroid Biochemistry and Molecular Biology*, *97*, 137–143.

Dlugosz, A., Agrawal, S., & Kirkpatrick, P. (2012). Pipeline pioneers: Vismodegib. *Nature Reviews. Drug Discovery*, *11*, 437–438.

Dormoy, V., Béraud, C., Lindner, V., Coquard, C., Barthelmebs, M., Brasse, D., et al. (2012). Vitamin D3 triggers antitumor activity through targeting hedgehog signaling in human renal cell carcinoma. *Carcinogenesis*, *33*, 2084–2093.

Ellison, T. I., Smith, M. K., Gilliam, A. C., & Macdonald, P. N. (2008). Inactivation of the vitamin D receptor enhances susceptibility of murine skin to UV-induced tumorigenesis. *Journal of Investigative Dermatology*, *128*, 2508–2517.

Gorlin, R. J. (1995). Nevoid basal cell carcinoma syndrome. *Dermatologic Clinics*, *13*, 113–125.

Hadden, M. K. (2014). Hedgehog pathway agonism: Therapeutic potential and small molecule development. *ChemMedChem*, *9*, 27–37.

Hahn, H., Wicking, C., Zaphiropoulous, P. G., Gailani, M. R., Shanley, S., Chidambaram, A., et al. (1996). Mutations in the human homolog ofDrosophila patched in the nevoid basal cell carcinoma syndrome. *Cell*, *85*, 841–851.

Haussler, M. R., Jurutka, P. W., Mizwicki, M., & Norman, A. W. (2011). Vitamin D receptor (VDR)-mediated actions of 1α,25(OH)$_2$vitamin D$_3$: Genomic and nongenomic mechanisms. *Best Practice & Research Clinical Endocrinology & Metabolism*, *25*, 543–559.

Hawker, N. P., Pennypacker, S. D., Chang, S. M., & Bikle, D. D. (2007). Regulation of human epidermal keratinocyte differentiation by the vitamin D receptor and its coactivators DRIP205, SRC2, and SRC3. *Journal of Investigative Dermatology*, *127*, 874–880.

Hegde, G. V., Peterson, K. J., Emanuel, K., Mittal, A. K., Joshi, A. D., Dickinson, J. D., et al. (2008). Hedgehog-induced survival of B-cell chronic lymphocytic leukemia cells in a stromal cell microenvironment: A potential new therapeutic target. *Molecular Cancer Research*, *6*, 1928–1936.

Hui, C.-C., & Angers, S. (2011). Gli proteins in development and disease. *Annual Review of Cell and Developmental Biology*, *27*, 513–537.

Jiang, Y. J., & Bikle, D. D. (2014). LncRNA: A new player in 1α,25(OH)(2)vitamin D (3)/VDR protection against skin cancer formation. *Experimental Dermatology*, *23*, 147–150.

Karhadkar, S. S., Bova, G. S., Abdallah, N., Dhara, S., Gardner, D., Maitra, A., et al. (2004). Hedgehog signaling in prostate regeneration, neoplasis, and metastasis. *Nature*, *431*, 707–712.

Köstner, K., Denzer, N., Koreng, M., Reichrath, S., Gräber, S., Klein, R., et al. (2012). Association of genetic variants of the vitamin D receptor (VDR) with cutaneous squamous cell carcinomas (SCC) and basal cell carcinomas (BCC): A pilot study in a German population. *Anticancer Research*, *32*, 327–333.

Kubo, M., Nakamura, M., Tasaki, A., Yamanaka, N., Nakashima, H., Nomura, M., et al. (2004). Hedgehog signaling pathway is a new therapeutic target for patients with breast cancer. *Cancer Research*, *64*, 6071–6074.

Lee, M. J., Hatton, B. A., Villavicencio, E. H., Khanna, P. C., Friedman, S. D., Ditzler, S., et al. (2012). Hedgehog pathway inhibitor saridegib (IPI-926) increases lifespan in a mouse medulloblastoma model. *Proceedings of the National Academy of Sciences of the United States of America*, *109*, 7859–7864.

Lesiak, A., Norval, M., Wodz-Naskiewick, K., Pawliczak, R., Rogowski-Tylman, M., Sysa-Jedrzejowska, A., et al. (2011). An enhanced risk of basal cell carcinoma is associated with particular polymorphisms in the VDR and MTHFR genes. *Experimental Dermatology*, *20*, 800–804.

Ling, G., Ahmadian, A., Persson, A., Unden, A. B., Afink, G., Williams, C., et al. (2001). PATCHED and p53 gene alterations in sporadic and hereditary basal cell cancer. *Oncogene*, *20*, 7770–7778.

Mitri, J., Muraru, M. D., & Pittas, A. G. (2011). Vitamin D and type 2 diabetes: A systematic review. *European Journal of Clinical Nutrition*, *65*, 1005–1015.

Mitschele, T., Diesel, B., Friedrich, M., Meineke, V., Maas, R. M., Gärtner, B. C., et al. (2004). Analysis of the vitamin D system in basal cell carcinomas (BCCs). *Laboratory Investigation*, *84*, 693–702.

Moll, P. R., Sander, V., Frischauf, A. M., & Richter, K. (2007). Expression profiling of vitamin D treated primary human keratinocytes. *Journal of Cellular Biochemistry*, *100*, 574–592.

Nagpal, S., Na, S., & Rathnachalam, R. (2005). Non-calcemic actions of vitamin D receptor ligands. *Endocrine Reviews*, *26*, 662–687.

Norman, P. E., & Powell, J. T. (2014). Vitamin D and cardiovascular disease. *Circulation Research*, *114*, 379–393.

Pike, J. W., & Meyer, M. B. (2010). The vitamin D receptor: New paradigms for the regulation of gene expression by 1,25-dihydroxyvitaminD(3). *Endocrinology and Metabolism Clinics of North America*, *39*, 255–269.

Pricl, S., Cartelazzi, B., Col, V. D., Marson, D., Laurini, E., Fermeglia, M., et al. (2015). Smoothened (SMO) receptor mutations dictate resistance to vismodegib in basal cell carcinoma. *Molecular Oncology*, *9*, 389–397.

Raffel, C., Jenkins, R. B., Frederick, L., Hebrink, D., Alderete, B., Fults, D. W., et al. (1997). Sporadic medulloblastomas contain PTCH mutations. *Cancer Research*, *57*, 842–845.

Reddy, K. K. (2013). Vitamin D level and basal cell carcinoma, squamous cell carcinoma, and melanoma risk. *Journal of Investigative Dermatology*, *133*, 589–592.

Reifenberger, J., Wolter, M., Knobbe, C. B., Köhler, B., Schönicke, A., Scharwächter, C., et al. (2005). Somatic mutations in the PTCH, SMOH, SUFUH and TP53 genes in sporadic basal cell carcinomas. *British Journal of Dermatology*, *152*, 43–51.

Revelli, A., Massobrio, M., & Tesarik, J. (1998). Non-genomic effects of 1α,25-dihydroxyvitamin D3. *Trends in Endocrinology and Metabolism, 9*, 419–427.
Riobo, N. A., Lu, K., Ai, X., Haines, G. M., & Emerson, C. P. (2006). Phosphoinositide 3-kinase and Akt are essential for sonic hedgehog signaling. *Proceedings of the National Academy of Sciences of the United States of America, 103*, 4505–4510.
Rudin, C. M., Hann, C. L., Laterra, J., Yauch, R. L., Callahan, C. A., Fu, L., et al. (2009). Treatment of medulloblastoma with hedgehog inhibitor GDC-0449. *New England Journal of Medicine, 361*, 1173–1178.
Sharpe, H. J., Pau, G., Dijkgraaf, G. J., Basses-Seguin, N., Modrusan, Z., Januario, T., et al. (2015). Genomic analysis of smoothened inhibitor resistance in basal cell carcinoma. *Cancer Cell, 27*, 327–341.
Sharpe, H. J., Wang, W., Hannoush, R. N., & de Sauvage, F. J. (2015). Regulation of the oncoprotein smoothened by small molecules. *Nature Chemical Biology, 11*, 246–255.
Tang, J. Y., Xiao, T. Z., Oda, Y., Chang, K. S., Shpall, E., Wu, A., et al. (2011). Vitamin D3 inhibits hedgehog signaling and proliferation in murine basal cell carcinomas. *Cancer Prevention Research, 4*, 744–751.
Taylor, M. D., Liu, L., Raffel, C., Hui, C., Mainprize, T. G., Zhang, X., et al. (2002). Mutations in SUFU predispose to medulloblastoma. *Nature Genetics, 31*, 306–310.
Teichert, A., Elalieh, H., & Bikle, D. (2010). Disruption of the hedgehog pathway contributes to the hair follicle cycling deficiency in Vdr knockout mice. *Journal of Cellular Physiology, 225*, 482–489.
Teichert, A. E., Elalieh, H., Elias, P. M., Welsh, J., & Bikle, D. D. (2011). Overexpression of hedgehog signaling is associated with epidermal tumor formation in vitamin D receptor-null mice. *Journal of Investigative Dermatology, 31*, 2289–2297.
Thayer, S. P., Pasca di Magliano, M., Heiser, P. W., Nielsen, C. M., Roberts, D. J., Lauwers, G. Y., et al. (2003). Hedgehog is an early and late mediator of pancreatic cancer tumorigenesis. *Nature, 425*, 851–856.
Tian, H., Callahan, C. A., DuPree, K. J., Darbonne, W. C., Ahn, C. P., Scales, S. J., et al. (2009). Hedgehog signaling is restricted to the stromal compartment during pancreatic carcinogenesis. *Proceedings of the National Academy of Sciences of the United States of America, 106*, 4254–4259.
Tostar, U., Malm, C. J., Meis-Kindblom, J. M., Kindblom, L., Toftgård, R., & Unden, A. B. (2006). Deregulation of the hedgehog signalling pathway: A possible role for the PTCH and SUFU genes in human rhabdomyoma and rhabdomyosarcoma development. *Journal of Pathology, 208*, 17–25.
Uhmann, A., Niemann, H., Lammering, B., Henkel, C., Hess, I., Nitzki, F., et al. (2012a). Antitumoral effects of calcitriol in basal cell carcinomas involve inhibition of hedgehog signaling and induction of vitamin D receptor signaling and differentiation. *Molecular Cancer Therapeutics, 10*, 2179–2188.
Uhmann, A., Niemann, H., Lammering, B., Henkel, C., Hess, I., Rosenberger, A., et al. (2012b). Calcitriol inhibits hedgehog signaling and induces vitamin D receptor signaling and differentiation in the patched mouse model of embryonal rhabdomyosarcoma. *Sarcoma, 2012*. 357040.
Van der Pols, J. C., Russell, A., Bauer, U., Neale, R. E., Kimlin, M. G., & Green, A. C. (2013). Vitamin D status and skin cancer risk independent of time outdoors: 11-year prospective study in an Australian community. *Journal of Investigative Dermatology, 133*, 637–641.
Vokes, S. A., Ji, H., McCuine, S., Tenzen, T., Giles, S., Zhong, S., et al. (2007). Genomic characterization of Gli-activator targets in sonic hedgehog-mediated neural patterning. *Development, 134*, 1977–1989.
Wang, Y., Ding, Q., Yen, C. J., Xia, W., Izzo, J. G., Lang, J. Y., et al. (2012). The crosstalk of mTOR/S6K1 and hedgehog pathways. *Cancer Cell, 21*, 374–387.

Watkins, D. N., Berman, D. M., Burkholder, S. G., Wang, B., Beachy, P. A., & Baylin, S. B. (2003). Hedgehog signaling within airway epithelial progenitors and in small-cell lung cancer. *Nature*, *422*, 313–317.

Wong, G., Gupta, R., Dixon, K. M., Deo, S. S., Choong, S. M., Halliday, G. M., et al. (2004). 1,25-Dihydroxyvitamin D and three low-calcemic analogs decrease UV-induced DNA damage via the rapid response pathway. *Journal of Steroid Biochemistry and Molecular Biology*, *89–90*, 567–570.

Yauch, R. L., Dijkgraaf, G. J., Alicke, B., Januario, T., Ahn, C. P., Holcomb, T., et al. (2009). Smoothened mutation confers resistance to a hedgehog pathway inhibitor in medulloblastoma. *Science*, *326*, 572–574.

Yauch, R. L., Gould, S. E., Scales, S. J., Tang, T., Tian, H., Ahn, C. P., et al. (2008). A paracrine requirement for hedgehog signaling in cancer. *Nature*, *455*, 406–410.

Yoon, J. W., Kita, Y., Frank, D. J., Majewski, R. R., Konicek, B. A., Nobrega, M. A., et al. (2002). Identification of GLI1-binding elements in target genes and a role for multiple downstream pathways in GLI1-induced cell transformation. *Journal of Biological Chemistry*, *277*, 5548–5555.

Zibat, A., Missiaglia, E., Rosenberger, A., Pritchard-Jones, K., Shipley, J., Hahn, H., et al. (2010). Activation of the hedgehog pathway confers a poor prognosis in embryonal and fusion gene-negative alveolar rhabdomyosarcoma. *Oncogene*, *29*, 6323–6330.

Zinser, G. M., Sundberg, J. P., & Welsh, J. (2002). Vitamin D(3) receptor ablation sensitizes skin to chemically induced tumorigenesis. *Carcinogenesis*, *23*, 2103–2109.

CHAPTER TEN

Molecular Approaches for Optimizing Vitamin D Supplementation

Carsten Carlberg[1]
Institute of Biomedicine, School of Medicine, University of Eastern Finland, Kuopio, Finland
[1]Corresponding author: e-mail address: carsten.carlberg@uef.fi

Contents

Abstract

Vitamin D can be synthesized endogenously within UV-B exposed human skin. However, avoidance of sufficient sun exposure via predominant indoor activities, textile coverage, dark skin at higher latitude, and seasonal variations makes the intake of vitamin D fortified food or direct vitamin D supplementation necessary. Vitamin D has via its biologically most active metabolite 1α,25-dihydroxyvitamin D and the transcription factor vitamin D receptor a direct effect on the epigenome and transcriptome of many human tissues and cell types. Different interpretation of results from observational studies with vitamin D led to some dispute in the field on the desired optimal vitamin D level and the recommended daily supplementation. This chapter will provide background on the epigenome- and transcriptome-wide functions of vitamin D and will outline how this insight may be used for determining of the optimal vitamin D status of human individuals. These reflections will lead to the concept of a personal vitamin D index that may be a better guideline for an optimized vitamin D supplementation than population-based recommendations.

Vitamins and Hormones, Volume 100
ISSN 0083-6729
http://dx.doi.org/10.1016/bs.vh.2015.10.001

ABBREVIATIONS

$1,25(OH)_2D$ 1α,25-dihydroxyvitamin D
25(OH)D 25-hydroxyvitamin D
CYP cytochrome P450
FAIRE-seq formaldehyde-assisted isolation of regulatory elements sequencing
OGTT oral glucose tolerance test
PBMC peripheral blood mononuclear cell
PTH parathyroid hormone
SNP single nucleotide polymorphism
VDR vitamin D receptor

1. INTRODUCTION

The natural way of obtaining vitamin D is its endogenous synthesis from the cholesterol precursor 7-dehydrocholesterol (Fig. 1). Thus the term "vitamin" is not used in its original sense. However, the essential need of UV-B for endogenous vitamin D synthesis restricts this pathway to sun exposed skin. Lifestyle changes during the last hundreds to thousand years, such as predominant indoor activities and textile coverage outdoors, as well as seasonal and climatic changes, often result in insufficient UV-B exposure and thus low endogenous vitamin D production. This can cause dependence on external vitamin D supply, i.e., under these conditions the compound is correctly termed a vitamin. Average human diet is rather low in vitamin D,

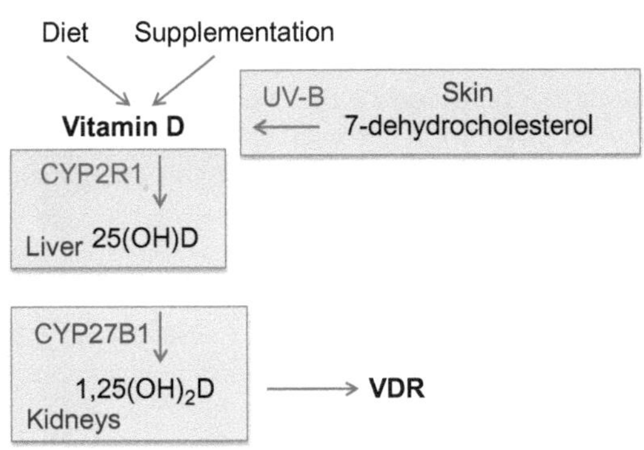

Figure 1 Vitamin D synthesis pathway. Vitamin D is taken up by diet or supplementation or synthesized endogenously in the skin under essential catalysis by UV-B. In the liver, vitamin D is converted by the enzyme CYP2R1 into the circulating form 25(OH)D and then in the kidneys by the enzyme CYP27B1 into the high-affinity VDR ligand $1,25(OH)_2D$. (See the color plate.)

since only fatty fish and some mushrooms contain reasonable amounts of the vitamin D isomers vitamin D_3 and vitamin D_2, respectively. Therefore, in some countries dietary products, such as milk, margarine, and juices, are fortified with vitamin D_3 or direct supplementation with the compound is recommended. This chapter discusses how for each human individual an optimal vitamin D level can be determined.

Vitamin D itself is an biologically inert molecule and has to be activated by hydroxylation first at position 25, leading to the prehormone 25-hydroxyvitamin D (25(OH)D), and then at position 1, creating the nuclear hormone 1α,25-dihydroxyvitamin D (1,25$(OH)_2$D) (Norman, 2008). The latter molecule is the only natural high-affinity ligand to the transcription factor vitamin D receptor (VDR) (Haussler et al., 1997). The first step of vitamin D synthesis is catalyzed in the liver by the cytochrome P450 (CYP) enzyme CYP2R1 and the second step by CYP27B1 in the kidneys (Fig. 1). The individual's vitamin D status is assessed best via the serum 25(OH)D concentration, since this vitamin D metabolite has a half-life of 15 days, while the half-life of 1,25$(OH)_2$D is with approximately 15 h far shorter. For both bone and overall health, the Institute of Medicine recommends 25(OH)D serum levels of $\geq$50 nM (Institute-of-Medicine, 2011). This translates to a daily vitamin D supplementation of 10–15 μg (400–600 IU) for children and 15–20 μg (600–800 IU) for adults. In contrast, the US Endocrine Society suggests a 25(OH)D serum concentration of 75 nM and daily vitamin D supplementations with 25 μg (1000 IU) or more (Holick et al., 2011).

The classical action of 1,25$(OH)_2$D is the promotion of dietary calcium and phosphorus absorption in the intestine, the facilitation of calcium reabsorption in the renal tubules, and the control of remodeling in the bones. Since CYP27B1 is not only expressed in the kidneys, there is also extra-renal production of 1,25$(OH)_2$D in tissues, such as in macrophages, the gastrointestinal tract, skin, vasculature, and placenta, in which the nuclear hormone has paracrine and autocrine functions. Furthermore, the rather ubiquitous expression of VDR supports the view that vitamin D and its receptor have far more functions than only the control of calcium homeostasis and bone remodeling. Accordingly, vitamin D deficiency does not only result in rickets in children and in a higher incidence of osteoporotic fractures in adults (Carlberg, 2014b), but it may also compromise the protective roles of vitamin D against cancer, cardiovascular diseases, diabetes, infections, and neuropsychiatric disorders (Holick, 2007), since vitamin D is involved in the regulation of cellular growth and differentiation (Feldman, Krishnan,

Swami, Giovannucci, & Feldman, 2014) as well as innate and adaptive immunity (Chun, Liu, Modlin, Adams, & Hewison, 2014).

In summary, it is undisputed that a sufficient vitamin D status is essential for bone health, such as the prevention of osteoporosis (Institute-of-Medicine, 2011). Tissue calcification is the main possible side effect that might be caused by overdosing with natural and synthetic vitamin D analogs (Cheskis, Freedman, & Nagpal, 2006). Therefore, there is some hesitation to recommend higher vitamin D doses for reaching nonskeletal effects of vitamin D. The latter are mainly based on observational studies and most of them lack proof of causal relations from randomized controlled trials.

2. A VIEW FROM EVOLUTION

Cholesterol synthesis is an evolutionary very old pathway, so that 7-dehydrocholesterol was already available in early marine organisms, such as phyto- and zooplankton (Holick, 2011). These species use the photochemical reaction resulting in vitamin D as a protection against UV-B-induced DNA damage, i.e., the historically first role of vitamin D was that of an inert molecule acting as a sunscreen. Since plankton is a major component in the marine food chain, vitamin D accumulates in the liver of many deep-water fish, such as cod. Vitamin D obtained endocrine functions when animals moved out of the water and needed to develop a stable skeleton based on calcium (Bouillon & Suda, 2014). Therefore, only vertebrates have a full vitamin D endocrine system, composed of plasma transport proteins, such as the vitamin D binding protein, metabolizing enzymes, such as CYP27B1 and CYP24A1, and a specific high-affinity nuclear receptor, such as VDR.

The evolutionary precursors of nuclear receptors were ligand-independent transcription factors being primarily involved in the control of cellular metabolism (Escriva, Bertrand, & Laudet, 2004). These ancestral nuclear receptors acquired in a multistep process the ability to bind and to be activated by metabolic compounds, such as bile acids in the case of the VDR precursor (Makishima et al., 2002). Further evolution of the ligand-binding domain created then a ligand-binding pocket that can harbor with high specificity and affinity 1,25(OH)$_2$D (Carlberg & Molnár, 2012). After the move from the calcium-rich ocean to the calcium-poor terrestrial renvironment, the control of calcium homeostasis became a speciality of the vitamin D endocrine system (Bouillon & Suda, 2014). However, the evolutionary background of vitamin D and the VDR implies that both are involved in a wider set of functions than only providing bones with calcium.

Anatomically modern humans developed some 200,000 years ago in East Africa. Already their ancestors had obtained dark skin following the loss of body hair more than 1 million years ago, in order to prevent UV-mediated degradation of the essential circulating methyl-group donor folate. The intensive sun exposure at the equator allowed, despite their dark skin, sufficient vitamin D_3 synthesis. In fact, the average circulating levels of 25(OH)D in members of the traditionally living Maasai tribe in East Africa is 119 nM (Luxwolda, Kuipers, Kema, Dijck-Brouwer, & Muskiet, 2012). Since modern humans lived some 150,000 years a similar lifestyle than the Maasai, at least concerning sun exposure, it can be assumed that human physiology and biochemistry has adapted well to this rather high vitamin D status. When some 50,000 years ago some modern humans started to move north toward Asia and Europe, the essential need of endogenous vitamin D production in less sunny regions at higher latitude caused an evolutionary pressure for skin lightening (Hochberg & Templeton, 2010). The gradual skin lightening process took 10,000–30,000 years and reflects the pace of human migration to nearly all regions of the planet. In an evolutionary scale, this is very fast, i.e., most other population-wide adaptations of the human genome took far more time. Nevertheless, the immigration of light skin Europeans to the Americas and Australia and the involuntary transfer of dark skin Africans to the Americas within the last 500 years were far too quick for any genetic adaption. In net effect, nowadays many humans live at a latitude, i.e., at an average yearly UV-B exposure, for which their skin color is not adapted. For example, dark skin persons living in Canada or in the UK do not have the chance of sufficient endogenous production of vitamin D. In addition, colder climate as well as cultural and religious traditions imposed to most humans a textile coverage of nearly their entire body being another reason for a low endogenous vitamin D synthesis in most human populations. Furthermore, genetic variations between human individuals are primarily found on the level of single nucleotide polymorphisms (SNPs). Some of these SNPs have been shown to be related to serum 25(OH)D levels, i.e., to the vitamin D status of the individuals (Wang et al., 2010). Thus, based on their genetics some persons are exposed throughout their entire life to either a higher or a lower serum 25(OH)D level than the average population.

The agricultural revolution that started some 10,000 years ago and in particular the industrial revolution that took place during the last

100–200 years drastically changed the lifestyle of nearly all human populations concerning the composition of diet, physical activity, and the intestinal microbiome. These changes are closely related and result in homeostatic imbalances that are the basis of many common non-communicative disorders, such as cardiovascular disease, diabetes, autoimmune disease, and cancer. Thus, today's rather low rate of endogenous vitamin D production parallels with many other potentially disease-promoting lifestyle changes of contemporary humans that are genetically still adapted to a large extend to the environmental conditions and lifestyle of their ancestors in the savannahs of East Africa. However, we should bear in mind that evolution selects for benefits that result in higher number of offspring reaching a reproductive age, i.e., getting and raising children, but not for aging-related diseases. Therefore, the vitamin D status was probably high, in order to protect against infectious diseases, such as tuberculosis, but has not been adapted for the protection against disorders that normally occur at higher age, such as cancer and cardiovascular disease.

Taken together, the genetic origin of all contemporary humans is an equatorial region (East Africa), where there are no seasonal changes in UV-B exposure. This means that the human genome is rather adapted to a constant vitamin D status than to level changes between summer and winter. The very high average 25(OH)D serum concentration in Maasai may represents an optimal vitamin D status but it could also reflect the maximal levels that evolution allowed in the presence at excessive sun exposure. Interestingly, the 25(OH)D serum levels of Maasai individuals ranged from 58 to 167 nM (Luxwolda et al., 2012). This may suggest that there is a wide personal range in the optimal or maximal vitamin D status that could be either genetically or epigenetically programmed.

3. VITAMIN D AND THE EPIGENOME

As a high-affinity ligand of the transcription factor VDR, $1,25(OH)_2D$ and its precursor vitamin D belong to a small group of natural compounds that have a direct effect on gene regulation (Carlberg & Dunlop, 2006). VDR is a member of the nuclear receptor superfamily (Evans & Mangelsdorf, 2014) that contains other endocrine receptors specifically binding lipophilic molecules in the size of cholesterol, such as estrogen, testosterone, and cortisol (Carlberg & Molnár, 2012). The nuclear receptor is essential for all molecular actions of $1,25(OH)_2D$, but in addition VDR bears also ligand-independent functions (Polly et al., 2000), i.e., the functional profile of the receptor exceeds that of its ligand. VDR acts as the mechanistic core of vitamin D signaling. It recognizes via its DNA-binding domain genomic target sequences, it uses its complete surface for the interaction with other nuclear proteins, and its inner surface, the ligand-binding pocket, serves as a cave specifically interacting with 1,25$(OH)_2D$ and its synthetic analogs (Carlberg, Molnár, & Mourino, 2012). The protein–protein interaction partners of VDR are either other transcription factors, such as the retinoid X receptor (blue in Fig. 2) that help the receptor to stabilize its contact with genomic DNA (Carlberg & Polly, 1998), or nuclear adaptor proteins, such as corepressors, coactivators, and the Mediator complex.

In all stages of a cell, genomic DNA forms a complex with nucleosomes (gray balls in Fig. 2), referred to as chromatin. As densely packed heterochromatin, it prevents the access to genomic DNA (Beisel & Paro, 2011). Chromatin has this intrinsic repressive potential in order to conserve the epigenetic landscape of a differentiated cell, i.e., by default it largely restricts the access of transcription factors to promoter and enhancer regions leaving only in the order of 50–100,000 accessible chromatin regions per cell type (ENCODE-Project-Consortium et al., 2012). The accessibility of chromatin is modulated by methylation of genomic DNA and by posttranslational modifications, such as acetylations and methylations, of nucleosome-forming histone proteins (Fig. 2) (Bell, Tiwari, Thoma, & Schubeler,

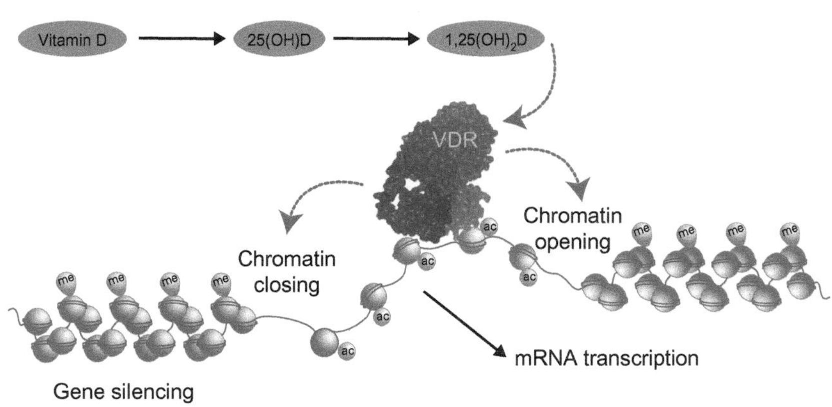

Figure 2 Chromatin model of vitamin D signaling. Gene regulation by VDR requires accessible genomic DNA, i.e., open chromatin. In turn, ligand-dependent actions of VDR result either in further opening of chromatin (i.e., in most cases, in upregulation of the transcription of the respective gene) or in closing of chromatin (and respective downregulation of transcription). (See the color plate.)

2011; Zhou, Goren, & Bernstein, 2011). These epigenomic modifications are controlled by chromatin modifying enzymes that read, write, or erase posttranslational marks on the histones.

Some of the protein–protein interaction partners of VDR attract chromatin remodeling complexes that rearrange the positioning of nucleosomes as well as chromatin modifying enzymes that locally open and close chromatin (Fig. 2). For example, VDR-mediated histone modifications can change the pattern of accessible chromatin regions. The interactions of VDR with nuclear proteins are modulated by the absence or presence of 1,25(OH)$_2$D within the receptor's ligand-binding domain, i.e., the most of them are ligand-dependent (Molnár, 2014). Experimentally, this can be monitored genome-wide by using the method formaldehyde-assisted isolation of regulatory elements sequencing (FAIRE-seq) (Giresi, Kim, McDaniell, Iyer, & Lieb, 2007). A detailed FAIRE-seq time course in a 1,25(OH)$_2$D-stimulated human monocytic cell line (THP-1) could demonstrate that some 87% of VDR binding sites colocalize with open chromatin (Seuter, Pehkonen, Heikkinen, & Carlberg, 2013). At approximately 20% of these loci, a stimulation with 1,25(OH)$_2$D leads to a significant increase in chromatin accessibility (Seuter et al., 2013). In this way, vitamin D has a direct effect on changes of the epigenome.

A gene can only be transcribed when the genomic regions of both its transcription start site and the binding sites of the transcription factors controlling the activity of RNA polymerase II, referred to as enhancers, are located within accessible chromatin (Carlberg & Campbell, 2013). Vitamin D-triggered epigenome changes are therefore the first step in the modulation of the transcriptome of a cell. Thus, these 1,25(OH)$_2$D-dependent epigenome modulations are the most direct functional readout of the molecular actions of vitamin D (Carlberg, 2014a). However, it should be kept in mind that not all changes within the epigenome result in a modulation of the transcriptome. Therefore, in most cases, changes in mRNA expression, such as measured by quantitative PCR or RNA-seq, act as molecular markers of the action of VDR and its ligand.

In summary, it can be assumed that the vitamin D status measured on the level of serum 25(OH)D concentrations is proportional to the availability of 1,25(OH)$_2$D in the nuclei of VDR expressing tissues and cell types. Since 1,25(OH)$_2$D has via the VDR a direct effect on chromatin accessibility, the vitamin D status of a human individual should have an impact on his/her epigenome and subsequently on the transcriptome.

4. MOLECULAR INSIGHT FROM VITAMIN D INTERVENTION TRIALS

Most vitamin D intervention studies have so far focused on the evaluation of the health status of the study participants via questionnaires, medical examination, or serum biochemistry. The 5-month vitamin D_3 intervention study VitDmet (NCT01479933) that investigated 71 elderly prediabetic persons, measured in addition at start and end of the trial the mRNA expression in peripheral blood mononuclear cells (PBMCs) and adipose tissue biopsies (Carlberg et al., 2013). This allowed the evaluation of vitamin D target genes as biomarkers for functional consequences of changes in the vitamin D status. The analysis of the VitDmet samples had been performed in an unconventional way, since the samples of this three-arm intervention (daily either 0, 1600, or 3200 IU vitamin D) were pooled, and the ratios (and not the differences) of the investigated parameters at end and start of the study were correlated with the respective ratios of the 25(OH)D levels (Carlberg et al., 2013). Nevertheless, this analysis approach was successful and demonstrated for all 24 investigated primary vitamin D target genes a significant correlation between mRNA expression changes and variations in the vitamin D status (Table 1) (Vukic et al., 2015). From the more than 100 clinical and biochemical parameters that had been determined in the VitDmet trial, only 12 displayed correlations with vitamin D status changes that were as significant as that of the 24 vitamin D target genes (Table 1) (Saksa et al., 2015). These 36 parameters were linked in a correlation network, the center of which was the serum parathyroid hormone (PTH) concentration change (Saksa et al., 2015; Vukic et al., 2015). PTH is a well-established marker of the vitamin D status (Bouillon et al., 2013) and an indication of the validity of the analysis. Accordingly, PTH is leading the relevance ranking of the 36 vitamin D-triggered parameters (Table 1). Interestingly, the following 12 parameters in the ranking are all expression changes of vitamin D target genes (*STS*, *BCL6*, *ITGAM*, *LRRC25*, *LPGAT1*, *TREM1*, *DUSP10*, *CD14*, *CD97*, *CD274*, *FUCA1*, and *NEF2*), which is technically easier to determine than most clinical and biochemical parameters. Interestingly, a recent meta-analysis (Standahl Olsen, Rylander, Brustad, Aksnes, & Lund, 2013) had already identified *CD14* as the most suited gene for describing the vitamin D status of primary blood samples.

Table 1 Parameters Relevant for Monitoring the Vitamin D Status of Human Individuals

Parameter	Significant Correlation with
PTH	25(OH)D, *STS*, *BCL6*, *ITGAM*, *LRRC25*, *LPGAT1*, *TREM1*, *CD274*, *FUCA1*, *CD38*, *FBP1*, *DUSP10*, *NRIP1*, *THBD*, *ASAP2*, *NINJ1*, *IL8*, *G0S2*, *CAMP*, *SLC37A2*, insulin resistance, HOMA-IR, fasting insulin, fasting FFAs TNFRSF1B, lymphocyte number, FFAs (120 min), IL6, ALAT, heart rate, adiponectin
STS	25(OH)D, *BCL6*, *ITGAM*, *LRRC25*, *LPGAT1*, *TREM1*, *CD274*, *NFE2*, *CD38*, *DUSP10*, *CD14*, *CD97*
BCL6	25(OH)D, *STS*, *ITGAM*, *LRRC25*, *LPGAT1*, *TREM1*, *FUCA1*, *NFE2*, *CD38*, *DUSP10*, *CD14*, *CD97*
ITGAM	25(OH)D, *STS*, *BCL6*, *LRRC25*, *LPGAT1*, *TREM1*, *FUCA1*, *NFE2*, *DUSP10*, *CD14*, *CD97*, *NRIP1*, *SLC37A2*
LRRC25	25(OH)D, *STS*, *BCL6*, *ITGAM*, *LPGAT1*, *TREM1*, *FUCA1*, *NFE2*, *FBP1*, *TMEM37*, *DUSP10*, *CD14*, lymphocyte number
LPGAT1	25(OH)D, *STS*, *BCL6*, *ITGAM*, *LRRC25*, *CD38*, *DUSP10*, *CD97*
TREM1	25(OH)D, *STS*, *BCL6*, *ITGAM*, *LRRC25*, *LPGAT1*, *FUCA1*, *NFE2*, *FBP1*, *DUSP10*, *CD14*
DUSP10	25(OH)D, *STS*, *BCL6*, *ITGAM*, *LRRC25*, *LPGAT1*, *CD274*, *NFE2*
CD14	25(OH)D, *STS*, *BCL6*, *ITGAM*, *LRRC25*, *LPGAT1*, *TREM1*, *FUCA1*, *NFE2*, *DUSP10*
CD97	25(OH)D, *STS*, *BCL6*, *ITGAM*, *LRRC25*, *LPGAT1*, *TREM1*, *NFE2*, *DUSP10*, *CD14*
CD274	*STS*, *BCL6*, *LPGAT1*, *CD38*, *DUSP10*, *CD14*, *CD97*, *ASAP2*, *LRRC8A*, insulin resistance
FUCA1	25(OH)D, *STS*, *BCL6*, *LRRC25*, *LPGAT1*, *TREM1*, *FBP1*
NEF2	25(OH)D, *STS*, *BCL6*, *ITGAM*, *TREM1*, *FUCA1*, *CD14*
Insulin sensitivity (OGTT)	25(OH)D, *CD38*, *TMEM37*, *CD14*, *CD97*, *ASAP2*, *NINJ1*, *G0S2*, PTH, HOMA-IR, fasting insulin, lymphocyte number
CD38	25(OH)D, *STS*, *ITGAM*, *CD274*, *DUSP10*, *NRIP1*, *LRRC8A*

Table 1 Parameters Relevant for Monitoring the Vitamin D Status of Human Individuals—cont'd

Parameter	Significant Correlation with
NRIP1	25(OH)D, *STS*, *BLC6*, *ITGAM*, *LPGAT1*, *FUCA1*, *NFE2*, *CD97*, *THBD*, *LRRC8A*, *SLC37A2*
FBP1	25(OH)D, *STS*, *BCL6*, *ITGAM*, *LRRC25*, *LPGAT1*, *TREM1*, *FUCA1*, *NFE2*, lymphocyte number
HOMA-IR (OGTT)	25(OH)D, *CD97*, *NRIP1*, *ASAP2*, *NINJ1*, insulin resistance, fasting insulin, TNFRSF1B, lymphocyte number
Fasting insulin (OGTT)	25(OH)D, *CD97*, *NRIP1*, *ASAP2*, *NINJ1*, insulin resistance, HOMA-IR, TNFRSF1B, FFAs (120 min)
THBD	25(OH)D, *STS*, *LRRC8A*, *LGPAT1*, *TREM1*, *FUCA1*, *FBP1*, *DUSP10*, *IL8*, lymphocyte number
ASAP2	25(OH)D, *STS*, *BCL6*, *ITGAM*, *LPGAT1*, *NFE2*
NINJ1	25(OH)D, *BCL6*, *LRRC25*, *TREM1*, *FUCA1*, *CD14*
Fasting FFAs (OGTT)	25(OH)D, *CD97*, *NRIP1*, *ASAP2*, *NINJ1*, insulin resistance, HOMA-IR, TNFRSF1B, FFAs (120 min)
IL8	25(OH)D, *CD274*, *FUCA1*, *CD38*, *CAMP*
TMEM37	25(OH)D, *G0S2*
LRRC8A	25(OH)D, *ITGAM*, *CD38*, *NRIP1*
G0S2	25(OH)D, *TMEM37*, insulin resistance
CAMP	25(OH)D, *CD14*, *NINJ1*, *IL8*, heart rate
TNFRSF1B	25(OH)D, *STS*, *BCL6*, *LPGAT1*, ALAT
Lymphocyte number	25(OH)D, *FBP1*
SLC37A2	25(OH)D
FFAs after 120 min (OGTT)	25(OH)D
IL6	25(OH)D, *TREM1*
ALAT	25(OH)D
Heart rate	25(OH)D
Adiponectin	25(OH)D

With serum and PBMC samples of the 71 VitDmet study participants, 24 vitamin D target genes (italic) and 12 biochemical/clinical parameters were found to be suitable for monitoring the vitamin D status of the investigated individuals. Some of the biochemical parameters were measured in the context of an oral glucose tolerance test (OGTT). For further details, please refer to the original publication (Vukic et al., 2015).

None of the VitDmet study participants was responsive to all 36 measured parameters (Vukic et al., 2015). However, the study subjects differed significantly by responding only to 10 (27.8%) up to 30 (83.3%) of all. This allowed a segregation of the study subjects into 26 high and 45 low responders and is an important demonstration how individual the response to vitamin D can be (Vukic et al., 2015). Comparable segregations had not yet been performed for other studies. To a minor extent, the observed differences may be based on genetic variations, such as SNPs affecting the vitamin D status (Wang et al., 2010). However, in analogy to most other traits underlying common aging-related disorders, such as type 2 diabetes, osteoporosis, or cardiovascular disease, a molecular explanation is found on the level of the epigenome rather than on the genome. Since the epigenome can respond to environmental changes, such as those imposed by the individual's lifestyle, it is far more dynamic than the genome. This implies that this responsiveness to vitamin D can also change during the life span of the human individual.

The time span of 5 months between the two measurements of the VitDmet participants suggests that the observed changes on the level of gene expression cannot be a direct consequence of transcriptional regulation but must be the result of vitamin D-triggered epigenomic changes. In order to observe direct transcriptional effects, a different type of intervention had to be performed. In the VitDbol (NCT02063334) intervention study, healthy human adults were treated once with high dose of vitamin D (2000 μg) and samples were taken already 1 and 2 days after onset of the intervention (Vukic et al., 2015). This study demonstrated that some (for example, *CAMP*, *TREM1*, *CD14*, and *ITGAM*) but not all of vitamin D target genes used in the VitDmet study can be used as markers of the rapid transcriptional response of human individuals to vitamin D.

Taken together, the observations made in the VitDmet and VitDbol trials suggest that the response to vitamin D is biphasic being composed of a fast direct transcriptional response and a more long-lasting epigenomic response. Epigenetic changes underlie also the fast transcriptional effects, but these may be only transient and different to the longer lasting effects of vitamin D on the epigenome.

5. CONSEQUENCES FOR VITAMIN D SUPPLEMENTATION

The results of molecular analysis of vitamin D supplementation studies, such as VitDmet (Carlberg et al., 2013; Ryynänen et al., 2014; Saksa et al., 2015; Wilfinger et al., 2014) and VitDbol (Vukic et al., 2015), indicate that each human individual displays a personal response to vitamin D. Importantly, this dynamic response to vitamin D, i.e., a comparison of vitamin D-triggered parameters at two or more time points, does not correlate with the static description of the vitamin D status based on one single measurement as performed in most observational studies. This finding implies that it is advisable to measure at least twice a person's vitamin D status together with a number of molecular and clinical parameters. These measurements can be done within the period of a few months, such as at begin and end of the winter season, or even at two consequent days. The first type of measurement describes the epigenomic response and is independent of any vitamin D supplementation protocol, while the second test monitors the transcriptional response and requires a vitamin D bolus, in order to observe significant effects. Obviously, the latter test provides quicker results that may be immediately implemented in an appropriate vitamin D supplementation protocol. However, the transcriptional and the epigenetic response are not necessarily identical. Therefore, a combination of both approaches, such as obtained by measurements at days 0, 1, and 30, may be most suited for computing a vitamin D index describing the personalized response of human individuals to vitamin D.

The vitamin D index analysis can indicate for each human individual a vitamin D supplementation protocol that will direct to a personal optimal vitamin D status. This concept may dissolve the scientific dispute about recommended 25(OH)D serum levels and vitamin D amounts of daily supplementation. The fact that in the United States, sales of vitamin D supplements increased by a factor of nearly 15 within the last decade indicates that vitamin D supplementation became very popular in the general population (Kupferschmidt, 2012). For these people, a smartphone app integrating the results of a vitamin D index measurement with their dietary vitamin D intake (fatty fish or fortified food), outdoor physical activity (correlating with sun exposure), and adiposity (decreasing 25(OH)D bioavailability) will rather accurately recommend a personalized vitamin D intake (which may change from day to day). Follow-up studies of individuals with a stable optimized vitamin D status will more likely prove or disprove claims about the impact of vitamin D in a variety of symptoms and disorders than traditional observational studies.

In summary, the inclusion of vitamin D index measurements in the stratification of study cohorts may be most appropriate, in order to challenge observational studies suggesting that high serum concentrations of vitamin D protect against cardiovascular disease, diabetes, colorectal cancer, and all-cause mortality.

6. CONCLUSION AND FUTURE DIRECTIONS

In the past, vitamin D and its metabolite 1,25(OH)$_2$D were known best for their role in calcium homeostasis and bone formation, but to date most genome-wide data are available for the actions vitamin D in cells of the hematopoietic system (Carlberg, 2014a). This emphasizes the impact of vitamin D and VDR for innate and adaptive immunity. The response to vitamin D supplementation varies considerably from individual to individual and depends on many factors, such as baseline level, adiposity, and genotype, i.e., a "one dose fits all" approach is not anymore appropriate. Thus, the dynamic vitamin D status of human individuals, which is introduced here as the vitamin D index, can be considered as a biomarker of the lifestyle of the person. This index is obtained from PBMCs that represent cells of both the adaptive and the innate immune system being in contact with most tissues of the human body. PBMCs are not only the most representative cell types for what is happening in the human body, but they are also the primary tissue that is easily obtained by a simple draw of blood. In future, the vitamin D index may be used in two rather different ways. On one hand, the test may be reduced to a routine measurement of a few biomarker-type genes, such as *CD14*, *ITGAM*, and *CAMP*, while on the other hand, next-generation sequencing type measurements may be applied that will allow an assessment far beyond the role of vitamin D. Since the vitamin D index has a large epigenetic component, measurements via epigenome- and/or transcriptome-wide methods, such as RNA-seq and FAIRE-seq, can integrate the environmental influences to the individual, such as physical activity, diet, and sun exposure. In this context, the vitamin D index exceeds its original scope describing the response to vitamin D but can monitor the health status of a human individual as a whole. Moreover, this will also better integrate confounding factors affecting the function of vitamin D, such as adiposity. This approach also implies that the status of a person in health as well as in disease cannot be reliably deduced from a single genotyping experiment, such as suggested by classical pharmacogenetics, but needs to be profiled on the level of the epigenome and transcriptome in a time series experiment (Carlberg & Raunio, 2014).

ACKNOWLEDGMENTS

C.C. thanks the Academy of Finland and the Juselius Foundation for support.

REFERENCES

Beisel, C., & Paro, R. (2011). Silencing chromatin: Comparing modes and mechanisms. *Nature Reviews. Genetics*, *12*, 123–135.

Bell, O., Tiwari, V. K., Thoma, N. H., & Schubeler, D. (2011). Determinants and dynamics of genome accessibility. *Nature Reviews. Genetics*, *12*, 554–564.

Bouillon, R., & Suda, T. (2014). Vitamin D: Calcium and bone homeostasis during evolution. *BoneKEy Reports*, *3*, 480.

Bouillon, R., Van Schoor, N. M., Gielen, E., Boonen, S., Mathieu, C., Vanderschueren, D., et al. (2013). Optimal vitamin D status: A critical analysis on the basis of evidence-based medicine. *The Journal of Clinical Endocrinology and Metabolism*, *98*, E1283–E1304.

Carlberg, C. (2014a). Genome-wide (over)view on the actions of vitamin D. *Frontiers in Physiology*, *5*, 167.

Carlberg, C. (2014b). The physiology of vitamin D—Far more than calcium and bone. *Frontiers in Physiology*, *5*, 335.

Carlberg, C., & Campbell, M. J. (2013). Vitamin D receptor signaling mechanisms: Integrated actions of a well-defined transcription factor. *Steroids*, *78*, 127–136.

Carlberg, C., & Dunlop, T. W. (2006). An integrated biological approach to nuclear receptor signaling in physiological control and disease. *Critical Reviews in Eukaryotic Gene Expression*, *16*, 1–22.

Carlberg, C., & Molnár, F. (2012). Current status of vitamin D signaling and its therapeutic applications. *Current Topics in Medicinal Chemistry*, *12*, 528–547.

Carlberg, C., Molnár, F., & Mourino, A. (2012). Vitamin D receptor ligands: The impact of crystal structures. *Expert Opinion on Therapeutic Patents*, *22*, 417–435.

Carlberg, C., & Polly, P. (1998). Gene regulation by vitamin D_3. *Critical Reviews in Eukaryotic Gene Expression*, *8*, 19–42.

Carlberg, C., & Raunio, H. (2014). From pharmacogenomics to integrated personal omics profiling: A gap in implementation into healthcare. *Personalized Medicine*, *11*, 625–629.

Carlberg, C., Seuter, S., de Mello, V. D., Schwab, U., Voutilainen, S., Pulkki, K., et al. (2013). Primary vitamin D target genes allow a categorization of possible benefits of vitamin D_3 supplementation. *PloS One*, *8*, e71042.

Cheskis, B. J., Freedman, L. P., & Nagpal, S. (2006). Vitamin D receptor ligands for osteoporosis. *Current Opinion in Investigational Drugs*, *7*, 906–911.

Chun, R. F., Liu, P. T., Modlin, R. L., Adams, J. S., & Hewison, M. (2014). Impact of vitamin D on immune function: Lessons learned from genome-wide analysis. *Frontiers in Physiology*, *5*, 151.

ENCODE-Project-Consortium, Bernstein, B. E., Birney, E., Dunham, I., Green, E. D., Gunter, C., et al. (2012). An integrated encyclopedia of DNA elements in the human genome. *Nature*, *489*, 57–74.

Escriva, H., Bertrand, S., & Laudet, V. (2004). The evolution of the nuclear receptor superfamily. *Essays in Biochemistry*, *40*, 11–26.

Evans, R., & Mangelsdorf, D. (2014). Nuclear receptors, RXR, and the Big Bang. *Cell*, *157*, 255–266.

Feldman, D., Krishnan, A. V., Swami, S., Giovannucci, E., & Feldman, B. J. (2014). The role of vitamin D in reducing cancer risk and progression. *Nature Reviews. Cancer*, *14*, 342–357.

Giresi, P. G., Kim, J., McDaniell, R. M., Iyer, V. R., & Lieb, J. D. (2007). FAIRE (formaldehyde-assisted isolation of regulatory elements) isolates active regulatory elements from human chromatin. *Genome Research*, *17*, 877–885.
Haussler, M. R., Haussler, C. A., Jurutka, P. W., Thompson, P. D., Hsieh, J. C., Remus, L. S., et al. (1997). The vitamin D hormone and its nuclear receptor: Molecular actions and disease states. *The Journal of Endocrinology*, *154*(Suppl.), S57–S73.
Hochberg, Z., & Templeton, A. R. (2010). Evolutionary perspective in skin color, vitamin D and its receptor. *Hormones (Athens, Greece)*, *9*, 307–311.
Holick, M. F. (2007). Vitamin D deficiency. *The New England Journal of Medicine*, *357*, 266–281.
Holick, M. F. (2011). Vitamin D: Evolutionary, physiological and health perspectives. *Current Drug Targets*, *12*, 4–18.
Holick, M. F., Binkley, N. C., Bischoff-Ferrari, H. A., Gordon, C. M., Hanley, D. A., Heaney, R. P., et al. (2011). Evaluation, treatment, and prevention of vitamin D deficiency: An Endocrine Society clinical practice guideline. *The Journal of Clinical Endocrinology and Metabolism*, *96*, 1911–1930.
Institute-of-Medicine (2011). *Dietary reference intakes for calcium and vitamin D*. Washington, DC: National Academies Press.
Kupferschmidt, K. (2012). Uncertain verdict as vitamin D goes on trial. *Science*, *337*, 1476–1478.
Luxwolda, M. F., Kuipers, R. S., Kema, I. P., Dijck-Brouwer, D. A., & Muskiet, F. A. (2012). Traditionally living populations in East Africa have a mean serum 25-hydroxyvitamin D concentration of 115 nmol/l. *The British Journal of Nutrition*, *108*, 1557–1561.
Makishima, M., Lu, T. T., Xie, W., Whitfield, G. K., Domoto, H., Evans, R. M., et al. (2002). Vitamin D receptor as an intestinal bile acid sensor. *Science*, *296*, 1313–1316.
Molnár, F. (2014). Structural considerations of vitamin D signaling. *Frontiers in Physiology*, *5*, 191.
Norman, A. W. (2008). From vitamin D to hormone D: Fundamentals of the vitamin D endocrine system essential for good health. *The American Journal of Clinical Nutrition*, *88*, 491S–499S.
Polly, P., Herdick, M., Moehren, U., Baniahmad, A., Heinzel, T., & Carlberg, C. (2000). VDR-Alien: A novel, DNA-selective vitamin D_3 receptor-corepressor partnership. *FASEB Journal*, *14*, 1455–1463.
Ryynänen, J., Neme, A., Tuomainen, T. P., Virtanen, J. K., Voutilainen, S., Nurmi, T., et al. (2014). Changes in vitamin D target gene expression in adipose tissue monitor the vitamin D response of human individuals. *Molecular Nutrition & Food Research*, *58*, 2036–2045.
Saksa, N., Neme, A., Ryynänen, J., Uusitupa, M., de Mello, V. D., Voutilainen, S., et al. (2015). Dissecting high from low responders in a vitamin D_3 intervention study. *The Journal of Steroid Biochemistry and Molecular Biology*, *148*, 275–282.
Seuter, S., Pehkonen, P., Heikkinen, S., & Carlberg, C. (2013). Dynamics of 1α,25-dihydroxyvitamin D-dependent chromatin accessibility of early vitamin D receptor target genes. *Biochimica et Biophysica Acta*, *1829*, 1266–1275.
Standahl Olsen, K., Rylander, C., Brustad, M., Aksnes, L., & Lund, E. (2013). Plasma 25 hydroxyvitamin D level and blood gene expression profiles: A cross-sectional study of the Norwegian Women and Cancer Post-genome Cohort. *European Journal of Clinical Nutrition*, *67*, 773–778.
Vukic, M., Neme, A., Seuter, S., Saksa, N., de Mello, V. D., Nurmi, T., et al. (2015). Relevance of vitamin D receptor target genes for monitoring the vitamin D responsiveness of primary human cells. *PloS One*, *10*, e0124339.

Wang, T. J., Zhang, F., Richards, J. B., Kestenbaum, B., van Meurs, J. B., Berry, D., et al. (2010). Common genetic determinants of vitamin D insufficiency: A genome-wide association study. *Lancet*, *376*, 180–188.

Wilfinger, J., Seuter, S., Tuomainen, T.-P., Virtanen, J. K., Voutilainen, S., Nurmi, T., et al. (2014). Primary vitamin D receptor target genes as biomarkers for the vitamin D_3 status in the hematopoietic system. *The Journal of Nutritional Biochemistry*, *25*, 875–884.

Zhou, V. W., Goren, A., & Bernstein, B. E. (2011). Charting histone modifications and the functional organization of mammalian genomes. *Nature Reviews. Genetics*, *12*, 7–18.

CHAPTER ELEVEN

The Role of Vitamin D_3 in the Development and Neuroprotection of Midbrain Dopamine Neurons

Rowan P. Orme*,†, Charlotte Middleditch*, Lauren Waite*, Rosemary A. Fricker*,1

*School of Medicine and Institute for Science and Technology in Medicine, Keele University, Keele, United Kingdom

†Cardiff Institute of Infection and Immunity, School of Medicine, Cardiff, United Kingdom

[1]Corresponding author: e-mail address: r.a.fricker@keele.ac.uk

Contents

Vitamins and Hormones, Volume 100
ISSN 0083-6729
http://dx.doi.org/10.1016/bs.vh.2015.10.007

Abstract

Vitamin D has long been synonymous with bone health. More recently, new health benefits are continually being associated with vitamin D, including a burgeoning field on neuroprotective properties. This has generated a huge explosion of interest in recent years in the potential for vitamin D to be used not only as a therapeutic in neurodegenerative disease, including Parkinson's disease, but also as biomarkers and for risk association.

With an emphasis on Parkinson's disease, this chapter will discuss recent evidence supporting the assertion that vitamin D can be a useful therapeutic agent used as an intervention therapy to be combined with existing treatments; and the case for further development of novel treatments utilizing the potential of vitamin D.

In addition, we present novel, previously unpublished evidence showing that in a unilateral model of Parkinson's disease, vitamin D can not only reduce the extent of denervation, but that this is also reflected in functional benefit to the animals.

The potential of vitamin D is slowly being realized; in the future, it will be widely associated with far more than just bone health and may even contribute to an elusive treatment of neurodegenerative illness.

1. INTRODUCTION

The generic term "vitamin D" is used to refer to a group of fat soluble steroid hormones that play key roles in normal development, homeostasis and in health and disease. Vitamin D has been synonymous with health since its discovery in the early twentieth century and was the result of a long search for a cure for rickets (Wolf, 2004). The classical role of vitamin D in maintaining good bone health is joined by a myriad of nonclassical health benefits.

"Vitamin D" is often used to refer to the biologically active form, calcitriol, but technically this is incorrect. In fact, vitamin D is not a vitamin as it is produced in the body from a biologically inactive precursor molecule. Additionally, there are two "vitamin D" molecules: vitamin D_2 is generally obtained from food and vitamin D_3 is produced by the action of sunlight on skin. For the sake of general discussion on the topic of vitamin D, this chapter will not discriminate between the generic term of vitamin D used to describe the active metabolite, and the specific terms used for many of the precursor molecules except where necessary.

2. BIOSYNTHESIS AND METABOLISM OF VITAMIN D_3

2.1 Source, Storage, and the Blood–Brain Barrier

There are two major sources of vitamin D: from the action of sunlight on skin where UVB causes changes to 7-dehydrocholesterol (vitamin D_3, cholecalciferol) and from dietary intake (vitamin D_2, ergocalciferol). Both of these require further metabolism, firstly in liver and then in kidneys, to produce the biologically active calcitriol. As vitamin D is a fat soluble vitamin, much of the total vitamin D in the body is retained in fat stores. It is only when it is transported to the liver to undergo the first hydroxylation reaction that it can circulate freely in blood plasma. Circulating vitamin D is also free to cross the blood–brain barrier and as such it is also found in cerebrospinal fluid.

2.2 Metabolism to Calcitriol and Other Active Metabolites

Much of the vitamin D circulating in serum begins as cholesterol, which is converted to 7-dehydrocholesterol by 7-dehydrocholesterol reductase. In the skin, 7-dehydrocholesterol is converted to cholecalciferol in a ring opening reaction. Ergocalciferol and cholecalciferol are subject to hydroxylation in the endoplasmic reticulum of hepatocytes by Liver-25-hydroxylase (CYP2R1) to produce 25-hdroxy vitamin D, calcidiol. Calcidiol then undergoes the second hydroxylation by 1-alpha-hydroxylase (CYP27B1) in the kidneys to finally produce the biologically active 1,25-dihydroxy vitamin D, calcitriol, which exerts its biological function through ligation with the nuclear vitamin D receptor (VDR). Excretion of vitamin D is through further hydroxylation by the CYP24A1 enzyme to produce 1,24,25-hydroxyvitamin D, followed by additional oxidation and cleavage of the side chain to generate calcitroic acid (Fig. 1; Brito Galvao, Nagode, Schenck, & Chew, 2013).

2.3 Physiological Activity/Levels

In the UK, the recommended daily intake (RDA) of vitamin D is met by the majority of the population. The RDA varies from 7 (280 IU) per day for children between 6 and 36 months of age to 8.5 μg (340 IU) per day for infants under 6 months old and up to 10 μg (400 IU) per day for adults (Pearce & Cheetham, 2010). Approximately 90% of this is suggested to come from biosynthesis in the skin; in the absence of sunlight exposure it is unlikely that sufficient vitamin D will be obtained from diet alone. It is important to note here, that even in cases of severe vitamin D deficiency, circulating calcitriol levels may be sufficient. Controversially, the RDA of vitamin D may be inadequate to maintain optimal levels for health (Kennel, Drake, & Hurley, 2010).

In normal health, vitamin D deficiency is defined as less than 30 ng/ml in serum. Normal levels are suggested to be between 30 and 50 ng/ml

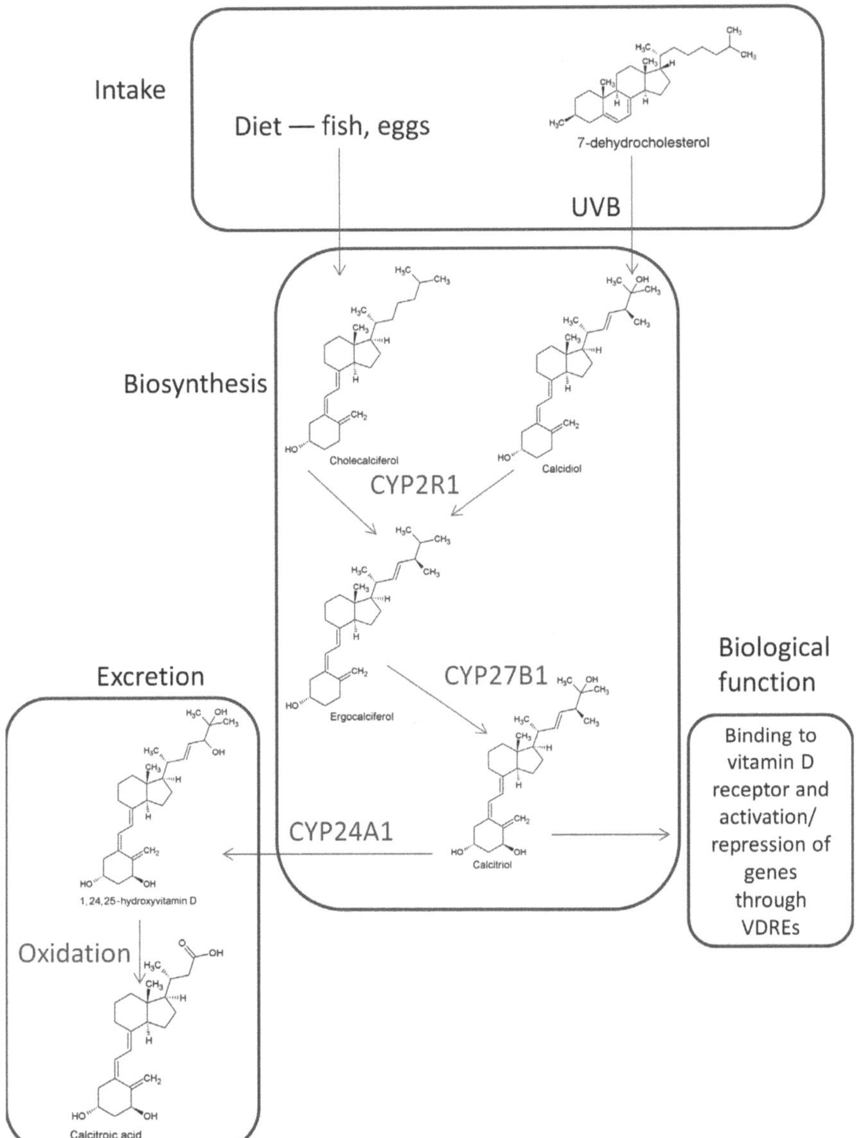

Figure 1 Biosynthesis of Vitamin D. Vitamin D is obtained from diet in the form of Ergocalciferol or from the action of UVB from the sun on 7-dehydroxycholesterol in the skin by way of cholecalciferol. These biologically inactive molecules undergo successive hydroxylation reactions in the liver and kidneys to produce calcidiol and finally the bioactive vitamin D molecule, calcitriol. Calcitriol binds to its nuclear receptor, vitamin D receptor, which once ligated binds to vitamin D response elements to activate or suppress genes, thus exerting its biological control. Excretion of vitamin requires further metabolism of calcitriol by further hydroxylation and finally oxidation and side chain cleavage to produce the highly water soluble calcitroic acid. (See the color plate.)

(Holick, 2009), with over 100 ng/ml being excessive and potentially harmful (Zittermann, 2003).

While discussing the levels of circulating vitamin D, it is important to consider what, exactly, is being measured. Although several different assays for measuring various different forms of vitamin D and its precursor molecules have been reported (Guo, Taylor, Singh, & Soldin, 2006; Hollis, 2004), the only measure used to determine vitamin D deficiency is 25-hydroxy vitamin D (Holick, 2009)—the product of the first hydroxylation reaction. This measure informs directly on neither stored vitamin D levels nor the biologically active calcitriol. However, owing to its long half-life and being a direct measure of circulating vitamin D culminated from both UV exposure and dietary intake, it is a sensible measure of vitamin D status.

3. VITAMIN D_3 AND HEALTH

The VDR is almost ubiquitously expressed, so it is not surprising that vitamin D exerts multiple effects in the human body. The effects come about through regulation of over 500 genes by the ligated VDR protein's binding to vitamin D response elements (VDREs; Zella et al., 2008).

3.1 Bone Health

The most well-known function of vitamin D is in bone health, stemming from its initial identification as a regulator of calcium homeostasis. This is achieved through facilitation of calcium into or out of bone, through the gut wall regulating the levels of calcium and phosphate in the blood stream (Lips & van Schoor, 2011). In vitamin D deficiency, calcium is inadequately absorbed from the gut and cannot be used for bone strengthening or remodeling, leading to osteoporosis and osteomalacia.

In addition to this classical role of vitamin D, there are numerous other examples of vitamin D affecting health, which are briefly discussed below; and increasing numbers of nonclassical vitamin functions including regulating the cell cycle. It is proposed that all of these functions are the effect of calcitriol binding to its nuclear receptor and subsequent binding to VDREs leading to gene activation and suppression.

3.2 Back Pain

Back pain has been linked with low levels of vitamin D. This is shown in one study where patients with lower back pain, which has not been caused by other factors, had significantly lower vitamin D levels in relation to control patients with no back pain (Lodh et al., 2015). This is further supported by another study, in which females with back pain lasting longer than 3 months had significantly lower Vitamin D_3 levels than control subjects (Lotfi, Abdel-Nasser, Hamdy, Omran, & El-Rehany, 2007).

3.3 Cancer

The links between vitamin D and antiproliferative effects have inevitably led to investigations of its potential use as a combination drug in treating cancers. A systematic review of vitamin D signaling pathways, includes its roles as an antiproliferative agent, activation of apoptotic pathways for clearance of cancer cells and the prevention of angiogenesis (Deeb, Trump, & Johnson, 2007). Chemotherapeutic potential seems to be linked directly to the circulating levels of vitamin D, with 1000 to 2000 IU per day required to reduce the risk of colorectal cancer and lower doses not having the same protective effect (Gorham et al., 2007).

There is some evidence that vitamin D levels provide a protective status to lower the risk of cancer. Analysis of 63 publications (colon (30), breast (13), prostate (26), ovarian (7)) revealed that in numerous cases vitamin D_3 levels correlated with reduced incidence of cancer (Garland et al., 2006) Conversely, other studies suggest no or only weak evidence for a link between vitamin D levels and cancer protection, and there are examples where high vitamin D levels may actually increase risk, for example of pancreatic cancer (Helzlsouer, 2010).

3.4 Immunity

An excellent overview of vitamin D and its requirements in immunity is provided by Aranow (2011). To summarize, vitamin D has been associated with immunity for a number of decades, with vitamin D deficient individuals more likely to suffer infection than those who have sufficient serum levels (Ginde, Mansbach, & Camargo, 2009). Indeed, there is a direct link between vitamin D levels and the ability to fight infection (Liu et al., 2006). This role of vitamin D in protective immunity may be a direct consequence of VDR expression on CD4+ T-cells (Mahon, Wittke, Weaver, & Cantorna, 2003).

3.5 Vitamin D Toxicity (Hypervitaminosis D)

Given the numerous studies linking increased vitamin D with improved health, one could be forgiven for thinking that high vitamin D intake should be recommended for everyone. This is not the case, as hypervitaminosis, or too much vitamin D, can cause problems as it increases the levels of calcium produced in the body. This can cause diseases such as calcinosis, the build-up of calcium and calcium deposits in bones and soft tissues, and hypercalcemia, which is a high blood level of calcium. For this reason, a better understanding of the mechanisms by which vitamin D exerts its actions, and the safe long and short term exposure limits is critical.

4. VITAMIN D_3 IN THE CENTRAL NERVOUS SYSTEM

Vitamin D metabolites naturally pass through the blood–brain barrier, giving them access to neuronal and glial cells. Hence, numerous roles for vitamin D have been implicated in various neurological/neuromuscular disorders. It has also been proposed that microglia within the CNS can generate calcitriol *in situ* and this might represent an antitumor response. Calcitriol can inhibit the synthesis of inducible nitric oxide synthase, leading to upregulation of glutathione, thus could play a role in neuroprotection or neuromodulation (for review see Garcion, Wion-Barbot, Montero-Menei, Berger, & Wion, 2002).

There is widespread expression of the VDR in the brains of adult rodents, with high levels found in sensory, motor and limbic systems, suggesting a role for vitamin D throughout life (Prufer, Veenstra, Jirikowski, & Kumar, 1999). Expression of functional VDRs within both neurons and glia of the adult hippocampus provide further evidence for vitamin D's importance in the adult CNS (Langub, Herman, Malluche, & Koszewski, 2001). In the human brain, both VDR and 1α-hydroxylase, the enzyme required for calcitriol production, have been observed in high levels in the substantia nigra, suggesting a potential link between this vitamin and the dopamine neuron population linked with Parkinson's disease (Eyles, Smith, Kinobe, Hewison, & McGrath, 2005).

4.1 Multiple Sclerosis

The neurological condition multiple sclerosis (MS) which effects the CNS has been subject to studies involving vitamin D. In white populations, higher vitamin D levels are associated with a reduced incidence of MS. Interestingly, there was no significant reduction in risk of MS in black or hispanic populations (Munger, Levin, Hollis, Howard, & Ascherio, 2006).

Vitamin D has been shown to improve reaction times in elderly people who repeatedly fall, helping to reduce the number of falls in these older patients. As increased vitamin D only improves reaction time but not muscle strength, researchers concluded that the increased vitamin D was improving neuromuscular function (Dhesi et al., 2004). A clinical trial combining vitamin D_3 with interferon β 1b treatment, showed lowered disease activity on MRI scans in patients receiving the dual therapy as compared to interferon alone (Soilu-Hanninen et al., 2012).

The mechanism of vitamin D's action in MS has been studied in animal models such as the cuprizone model of demyelination. In a study by Wergeland and colleagues, high doses of vitamin D_3 were able to reduce white matter demyelination, and were associated with reduced microglial activation (Wergeland et al., 2011).

4.2 Depression, Mental Health, and Cognitive Function

The areas of the brain that are associated with depression coincide with areas where VDR expression is high. Studies have shown that individuals with low serum vitamin D are more likely to be clinically depressed (Jozefowicz, Rabe-Jablonska, Wozniacka, & Strzelecki, 2014; Kerr et al., 2015). Low mood and impaired cognitive performance has also been associated with reduced vitamin D levels in Alzheimer's disease (AD) patients (Wilkins, Sheline, Roe, Birge, & Morris, 2006).

Lowered vitamin D levels have been associated with seasonal affective disorder, due to the vitamin's ability to alter serotonin levels. A recent study (Patrick & Ames, 2014) suggests that in its hormone form, vitamin D helps control the levels of serotonin in the brain by activating transcription of tryptophan hydroxylase 2, the gene responsible for the synthesis of serotonin. One proposed theory on the development of seasonal affective disorder is that calcitriol might inhibit the binding of melatonin to retinoid Z receptors, which would increase circulating serotonin levels (Partonen, 1998).

Other evidence links vitamin D with stress. Vitamin D_3 receptors are located in nucleus of adrenal medullary cells and these have been found to accumulate vitamin D in mice injected with radiolabeled vitamin D prehormone. In bovine adrenal medullary cells in culture, addition of vitamin D_3 caused an increase in mRNA levels of tyrosine hydroxylase the rate limiting enzyme involved in the synthesis of catecholamines (maximal at 10^{-8} M). It was suggested that this may be a mechanism to control response to stress (Puchacz, Stumpf, Stachowiak, & Stachowiak, 1996).

Vitamin D appears to influence many neuronal processes: combating oxidative stress, regulating neuronal calcium, enhancing nerve conduction, detoxification, and immunomodulation; this observation, linked with high expression of the VDR in hippocampal regions, has led researchers to propose its role in memory formation and cognitive function (reviewed in Buell & Dawson-Hughes, 2008). In support of this, low vitamin D levels have been linked with cognitive dysfunction. A recent study showed that serum levels of 25-hydroxy vitamin D correlated closely with mini mental state examination scores for older adults (Przybelski & Binkley, 2007).

There has been a lot of speculation on vitamin D's role in AD. Genetic studies have suggested that lower levels of VDR promoter activity are a risk factor for AD (Wang et al., 2012). Alzheimers patients have lower levels of 25-hydroxy vitamin D compared with age-matched healthy controls (Zhao, Sun, Ji, & Shen, 2013), and lower levels of 25-hydroxy vitamin D have been shown to increase the risk of developing AD (Afzal, Bojesen, & Nordestgaard, 2014). In cortical neuron cultures, a combination of vitamin D and memantine (an NMDA receptor antagonist) were found to be effective in reducing neuronal death following an insult with either glutamate or amyloid beta, providing evidence for vitamin D's potential as part of a pharmacological therapy regimen for AD (Annweiler, Brugg, Peyrin, Bartha, & Beauchet, 2014).

4.3 Vitamin D in CNS Development

There is a growing body of evidence that vitamin D is a key signaling factor required for normal brain development. The VDR is widely expressed in the developing CNS, in particular in neuroepithelial layers and regions undergoing active neuronal differentiation (Veenstra et al., 1998).

A recent study showed that rat mothers with low vitamin D levels during pregnancy produced offspring with changes in brain structure compared to offspring from mothers with normal vitamin D levels. There was a thinning of the cortex, along with enlargement of lateral ventricles, and an increase in the number of proliferating cells throughout the brain. Along with changes to brain structure, there were also differences in growth factor levels in the developing brain with vitamin D deficient offspring having less nerve growth factor and glial-derived neurotropic factor (GDNF; Eyles, Brown, Mackay-Sim, McGrath, & Feron, 2003; Eyles et al., 2005).

5. VITAMIN D AND NEUROPROTECTION IN PARKINSON'S DISEASE

Over the past 25 years, vitamin D has been increasingly associated with neuroprotection, including in Parkinson's disease. In this section, we will summarize the evidence responsible for generating such interest in the application of vitamin D for CNS protection and benefit to disease states, with a heavy emphasis on Parkinson's disease. The studies discussed in this section are those which are most relevant to PD, either through the use of Parkinson's models: mesencephalic cell cultures or lesions of the nigral-striatal dopamine system; or clinical studies and trials with an emphasis of vitamin D and PD. It should, however, be emphasized that this work is supported by a myriad of other studies using slightly different systems, many of which support the assertion that vitamin D may be relevant to PD. Discussion of this topic in its entirety is beyond the scope of this chapter and would require an entire volume in itself!

5.1 Population-Based Evidence Linking Vitamin D and Parkinson's Disease

Before assessing the scientific studies, both *in vitro* and *in vivo*, it is important to acknowledge the evidence directly linking vitamin D and Parkinson's disease. Maybe the most straightforward evidence is the direct link between circulating levels of vitamin D and PD. From a collection of over 7000 unique serum samples collected from 40 areas of Finland collected between 1978 and 1980, an inverse correlation between serum vitamin D levels and incidence of PD was observed. Individuals with 50 nmol/l or greater serum vitamin D had a 65% lower risk of developing PD than those with 25 nmol/l or lower serum vitamin D (Knekt et al., 2010). Given then, that the major source of circulating vitamin D is from the incidence of UV from sunlight on the skin, it is maybe less surprising that time spent outdoors and birthplace have both been linked to PD. From 3819 Danish males with PD and almost 20,000 age and sex matched controls, those with predominantly outdoor jobs had reduced risk of developing PD (Kenborg, Jorgensen, Budtz-Jorgensen, Knudsen, & Hansen, 2010). A second, unrelated study in the US, found a geographical gradient from East to West, with those in the East, where greater levels of UV are seen, having lower rates of PD (Betemps & Buncher, 1993). One caveat in these studies is the lack of distinction between the two major circulating forms of vitamin D: vitamin D_2, obtained only through diet; and vitamin D_3 which is majorly obtained through the action of sunlight on skin. To address this, Wang et al. moved away from using traditional immunoassays to measure total vitamin D, to modern mass spectrometric methods to determine levels of individual vitamin D forms. As well as observing an inverse relationship between total vitamin D and PD, both dietary vitamin D_2 and nondietary vitamin D_3 showed individual associations with PD. This suggests that total vitamin D levels are important, but the source of the vitamin is not crucial (Wang et al., 2015).

5.2 *In vitro* Evidence of Neuroprotective Properties of Vitamin D

A brief survey of current literature shows a substantial body of *in vitro* work relating vitamin D, or its active metabolite calcitriol, with neuroprotection, through a range of mechanisms. The majority of *in vitro* studies have focussed on the use of primary mesencephalic cultures of rodent dopamine neurons or of the human neuroblastoma cell line SH-SY5Y, which can be differentiated to dopamine neurons, using a variety of insults to induce toxicity in cells. The general findings of these studies are that vitamin D is able to protect neurons from cell death induced by reactive oxygen species or specific toxins.

In one of the first studies investigating the protective effects of calcitriol, the bioactive form of vitamin D, toward dopamine neurons, Ibi et al. cultured embryonic day 16 rat mesencephalic cells and used a variety of insults including hydrogen peroxide, 6-hydroxy dopamine, and glutamate to induce neuronal cell death (Ibi et al., 2001). Vitamin D was either coadministered with the toxic insult or used to pre-treat cultures prior to inducing cell death. Interestingly, coadministration of vitamin D had no protective effects on cultures; however, pretreatment with vitamin D provided significant protection. Neuroprotection was observed in a dose-dependent manner from nM concentrations. The protective effects were ameliorated by addition of the protein synthesis inhibitor, cycloheximide, leading to the conclusion that vitamin D exerts protective effects against a toxic insult by inducing synthesis of proteins that prevent the build-up of reactive oxygen species in cells. Additional work has shown two days' pretreatment with 100 pM of vitamin D is sufficient to protect embryonic day 14–15 mesencephalic cultures against 6-OHDA induced neurotoxicity, assessed by TH immunoreactivity (Wang et al., 2001).

Since these initial studies, several others have sought to replicate the findings using slightly different systems or with focus on other mechanisms of neuroprotection. Jang et al. have used SH-SY5Y cells to study protective benefits of calcitriol simultaneously added with rotenone to cultures (Jang et al., 2014). In contrast to the Ibi study (Ibi et al., 2001), pre-treatment of cells was not required to induce neuroprotection when concentrations of calcitriol upward of 300 nM were used. Concentrations below this were not reported and concentrations greater than 15 μM reduced cell viability even in the absence of rotenone. Neuroprotection was observed through use of an MTT assay and Annexin V and propidium iodide staining. In this study, protection was attributed to induction of autophagy, owing to observations that LC1 and beclin 1 were significantly upregulated in cells treated with calcitriol.

Our group have also observed a dose-dependent protective effect of calcitriol when added to cultures of embryonic day 12 rat mesencephalic dopamine neurons (Orme, Bhangal, & Fricker, 2013). When calcitriol was added to neuronal cultures, we observed a concurrent upregulation of the potent neurotrophic factor, GDNF, which is in line with previous literature reports (Naveilhan, Neveu, Wion, & Brachet, 1996). Calcitriol's effect was a protective effect, and not an artifact of increased cell division at the progenitor stage, shown by the observation of identical levels of BrdU incorporation in cultures either with or without calcitriol supplementation. Based on these results, we took the logical step of investigating whether or not calcitriol can protect against 6-OHDA induced striatal dopamine lesions in rats; see below.

The *in vitro* studies discussed above represent a small snapshot of the body of evidence showing neuroprotective effects of calcitriol. Many studies replicate these findings using a range of methods including MPP and L-buthionine sulfoximine induced toxicity of mesencephalic cultures (Shinpo, Kikuchi, Sasaki, Moriwaka, & Tashiro, 2000) and changes in calcium signaling induced by calcitriol. The effect on calcium signaling may be of particular interest as midbrain dopamine neurons are one of the few types of brain cell utilizing calcium to evoke a firing potential. It is well understood and acknowledged that high levels of intracellular calcium induce oxidative stress on cells and it has been hypothesized that this property of midbrain dopamine neurons is responsible for their extremely selective degeneration in PD (Cali, Ottolini, & Brini, 2014). Further work on the potential of calcitriol to modulate calcium signaling suggests that this can also be achieved through nongenomic effects, by altering intermediate filament protein phosphorylation and uptake of calcium through L-type voltage-dependent calcium channels (Zanatta et al., 2012).

5.3 Neuroprotection by Vitamin D *in vivo*

Following on from the *in vitro* assessments of calcitriol-induced neuroprotection, several studies have assessed its functioning *in vivo*. As with the *in vitro* studies, there have been a range of methods used with varying outcomes leading to a variety of hypotheses on vitamin D's role. *In vivo* experiments are a step closer to clinical applications so thorough use of *in vivo* models is warranted, especially in the light of promising *in vitro* work. Before embarking on such experiments, it is important to consider how the vitamin D hormone should be administered. Vitamin D itself is not biologically active and must be converted to the bioactive metabolite, calcitriol. As it is not clear if the CNS can convert vitamin D to calcitriol, direct administration of vitamin D intraparenchymally is not advised. Systemic administration of vitamin D is complicated by that fact that is fat soluble and often stored in fatty deposits, rather than being metabolized; thus circulating levels of calcitriol may not represent the levels of administered vitamin D. To circumvent these, many of the studies use intraperitoneal administration of calcitriol to ensure elevated circulating levels of bioactive vitamin D.

A summary of *in vivo* studies is presented in Table 1.

A brief glance at Table 1 shows that upregulation of striatal GDNF is a frequently hypothesized reason for calcitriol proffering protection to dopamine neurons against exogenous induced toxicity. This is an understandable hypothesis given the quantity of *in vitro* evidence showing upregulation of GDNF protein in mesencephalic cultures, described previously. Investigation into the regulation of GDNF protein or mRNA does not require chemical lesioning of the brain, and indeed some studies make use of this simpler but informative methodology. While it is often apparent that treatment of cell cultures with calcitriol can elicit very rapid changes in GDNF protein synthesis, it is less straightforward to assume the same to be true in a living animal. Rapid metabolism can also lead to reduced concentrations of administered compounds. It is for these reasons that many studies pretreat animals for seven or more days prior to assessing the effects of vitamin D. Two studies have used systemic administration of calcitriol to study directly the effects of brain GDNF levels (Cass et al., 2012; Sanchez et al., 2002). In both of these works, increased expression of GDNF protein was observed in the striatum of rats treated with calcitriol compared to control animals administered vehicle alone. The magnitude of the increase in striatal GDNF differs between the two studies, with one study showing approximately 80% increase in treated rats versus control animals at two doses of calcitriol (1 and

Table 1 Summary Table of *In Vivo* Studies of Vitamin D and Neuroprotection

Year	Animal	Toxin	Lesion Method	Vitamin D	Proposed Mechanism	Reference
2001	Rat	6-OHDA	Unilateral MFB	I.P., 8 day pretreatment	Reversal of free radical or ROS induced cell death	Wang et al. (2001)
2002	Rat	N/A	N/A	I.P., 7 day pretreatment	Upregulated striatal GDNF	Sanchez, Lopez-Martin, Segura, Labandeira-Garcia, and Perez-Fernandez (2002)
2003	Rat	Zinc	Unilateral SN	I.P., 7 day pretreatment	Reversal of zinc-induced oxidative injury	Dean et al. (2012)
2006	Rat/mouse	6-OHDA/MPTP	Unilateral MFB/I.P.	I.P., 7 day pretreatment	Inhibition of microglial activation	Kim et al. (2006)
2009	Rat	6-OHDA	Unilateral MFB	I.P., 7 day pretreatment	Upregulated striatal GDNF	Sanchez, Relova, Gallego, Ben-Batalla, and Perez-Fernandez (2009)
2012	Rat	N/A	N/A	S.C., 8 days	Upregulation of striatal GDNF, augmented DA release	Cass, Peters, Fletcher, and Yurek (2012)
2012	Mouse	MPTP	I.P.	Depletion, 35 days	N/A	Metz and Whishaw (2002)
2015	Rat	6-OHDA	Unilateral striatal	Oral, 14 day pretreatment, intrastriatal	Upregulation of striatal GDNF??	R.P. Orme, S.B. Dunnett, and R.A. Fricker (unpublished)

3 μg/kg/day) (Cass et al., 2012), while the other study showed a more modest increase at single dose of 1 μg/kg/day calcitriol (Sanchez et al., 2002). The former of these studies also shows an increase in nigral GDNF of over 100% at both calcitriol doses. This study also utilized HPLC to study striatal levels of dopamine and dopamine metabolites. While there was no significant difference in striatal dopamine between calcitriol-treated animals and control groups, there was an increase in DOPAC and HVA, two dopamine metabolites, in the treated group.

While GDNF is known to serve as a neuroprotectant, it is difficult to administer as it does not cross the blood–brain barrier and would require intracerebral administration. While this is technically possible, simple systemic drug administration through more traditional routes is far simpler and safer. The potential of vitamin D to upregulate CNS levels of GDNF is therefore of significant interest, especially if it is capable of halting degeneration. Building on evidence of increased expression of potent neurotrophic factors, further studies have sought to elucidate the neuroprotective effects of vitamin D against chemical toxins administered to the nigro-striatal system.

Initial studies using complete unilateral destruction of the nigro-striatal pathway by injection of 6-OHDA into the MFB of rats pretreated with calcitriol showed that calcitriol pretreatment reduced loss of motor function postlesion and led to elevated dopamine levels in the striatum of lesioned animals. It is not clear from this manuscript; however, whether there was an effect on total TH immunoreactive neurons in brains of lesioned animals (Wang et al., 2001). However, in a similar study which also describes increased GDNF expression in the striatum of 6-OHDA lesioned rats, immunostaining of nigral sections showed that calcitriol pretreatment served to protect dopamine neurons from 6-OHDA induced death. Crucially, this study also demonstrated that postlesion treatment with calcitriol can serve to regenerate the lesioned cells (Sanchez et al., 2009).

Another hypothesis for the mechanism of protection of dopamine neurons induced by vitamin D is the prevention of oxidative stress from reactive oxygen species. This is a conclusion drawn not only from 6-OHDA induced neurotoxicity (Sanchez et al., 2009; Wang et al., 2001) but also from oxidative stress induced by metals. It is well known that iron is a source of oxidative stress, indeed it has been postulated that iron may be a cause of neuronal cell death in numerous degenerative conditions, but there is also evidence from rodent studies that zinc-mediated oxidative stress may induce cell death (Lin, Fan, Yang, Hsu, & Yang, 2003). Zinc, infused intranigrally into the brains of rats, induced degeneration of the substantia nigra cells. When treated with calcitriol, lipid peroxidation was prevented and the reduction in striatal dopamine associated with zinc treatment was attenuated. This study represents a relatively crude destruction of brain tissue that is likely far less specific than toxins such as 6-OHDA and MPP, but the mechanism of neurotoxicity is far more simple, which leads to more conclusive evidence that calcitriol can act to protect neurons by reducing oxidative stress.

One final mode of protection proposed for vitamin D is via reduction in activated microglia. Activated microglia are present at sites of injury and inflammation and serve as the brain's immune cells. Activated microglia are immune reactive for the cell surface marker CD11b. Following MPTP or 6-OHDA lesions in mice and rats, respectively, CD11b positive cells were observed in the vicinity of the lesion. However, in animals pre-treated with calcitriol, there was a significant reduction in the numbers of activated microglia, indicating a reduction in inflammation and brain injury that led to an increase in the number of surviving TH positive cells (Kim et al., 2006).

Using an alternative strategy to investigate vitamin D's neuroprotective properties, Dean et al. depleted mice of vitamin D by removing it from the animals' diet for a period of 35 days prior to lesioning with MPTP toxin. In this case, no effect of depletion was observed on the extent of neuronal damage, suggested that reducing vitamin D from normal circulating levels does not affect the vulnerability of dopamine neurons (Dean et al., 2012).

To complement these studies, our group has carried out further work to investigate the use of vitamin D for protection of dopamine neurons in a more realistic rodent model of PD. Injection of 6-OHDA into the MFB, as used in the majority of existing studies, elicits a rapid and almost complete destruction of the dopamine pathway. In PD, this does not represent the gradual loss of neurons that gives rise to the characteristic slow decline in patients. We therefore used a terminal lesion model, injecting 6-OHDA unilaterally into the left striatum at six different coordinates to elicit an estimated 70% destruction that develops over days to weeks (Lee, Sauer, & Björklund, 1996). This may be more representative of late stage neuronal loss as it estimated that patients lose approximately 50% of neurons prior to symptomatic onset and presentation in the clinic. We concurrently injected calcitriol into neighboring striatal tissue to better understand the immediate impact of vitamin D-mediated neuroprotection. In addition, some animals were pretreated with oral vitamin D suspension, representing a high vitamin D systemic concentration, reported and described above to be effective in reducing the incidence of PD. Output measures of this study were both functional based behavioral testing and histological analysis of brain tissue.

From the battery of behavioral testing, we observed no functional improvement in animals receiving concurrent calcitriol administration direct to the striatum, in any trials, suggesting that vitamin D given at the same time as a toxic insult does not lead to any form of neuroprotection. Interestingly, in animals given vitamin D supplementation prior to lesioning, some functional benefit was observed (Fig. 2).

Unilateral lesions are of great use for functional testing of animals to assess lesion severity as they allow behavioral testing on the basis of drug induced rotational behavior. Apomorphine delivered to unilaterally lesioned animals induces rapid contralateral rotational behavior where substantial loss of innervation is present (Metz & Whishaw, 2002). Over a period of 40 min following subcutaneous administration of 0.05 mg/kg apomorphine, we observed greater rotational bias in control animals receiving vehicle alone than those which received systemic pretreatment with oral vitamin D (unpaired *t*-test, 5.3 ± 1.1 vs. 2.3 ± 0.5 mean net rotations per minute, $p < 0.01$; Fig. 2A).

The balance beam is a simple trial where animals are encouraged to walk across a short, narrow beam. Following training and customization to the "safe house," animals are allowed up to 60 s to navigate the narrowing beam. Animals with significant motor deficit have more difficulty in crossing the beam which manifests as an increased numbers of paw slips. Our work showed that systemically pretreated animals made significantly fewer errors than control animals (unpaired *t*-test, 3.3 ± 0.7 vs. 8.0 ± 1.5, $p < 0.001$; Fig. 2B). These data corroborate previous studies suggesting that systemic pretreatment with vitamin D to animals undergoing dopaminergic lesions provides some protection against neuronal loss.

In addition to behavioral testing, the brains of animals were histologically examined for TH immunoreactive innervation of the striatum by substantia nigra neurons. As expected the right hemisphere (lesioned) showed considerably less innervation than the intact hemisphere. There were no differences in TH innervation volume between control animals and those receiving neuroprotection; innervation of the left intact hemisphere was equal in all animals (15.3 ± 0.5 mm^3 for pretreated animals vs. 16.9 ± 0.5 mm^3 in control animals).When considering innervation of the right, lesioned, striatum, animals pretreated with vitamin D had on average 141% greater innervation volume than those receiving vehicle alone,

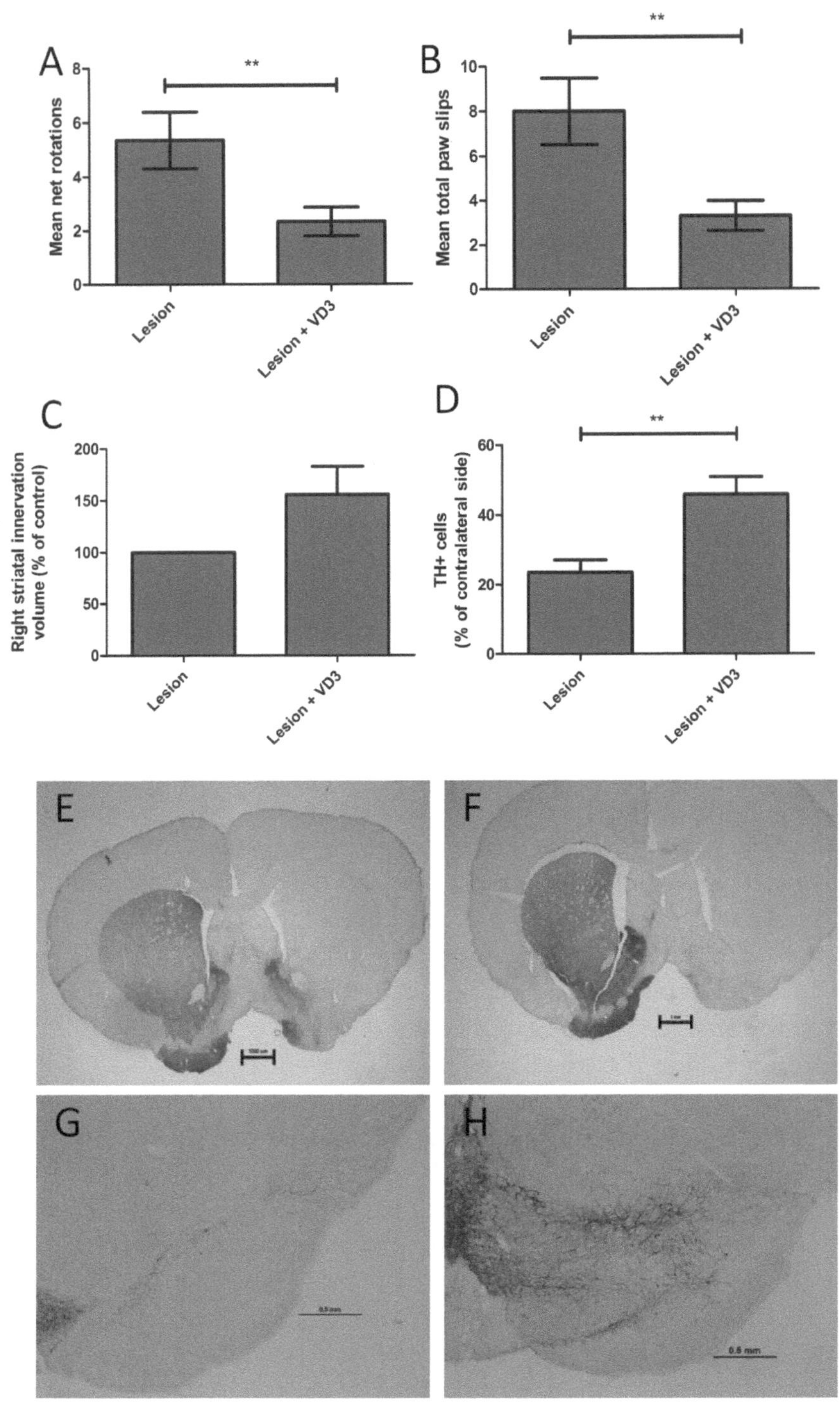
A
**
Mean net rotations
0
2
4
6
8
Lesion
Lesion + VD3
B
**
Mean total paw slips
0
2
4
6
8
10
Lesion
Lesion + VD3
C
Right striatal innervation volume (% of control)
0
50
100
150
200
Lesion
Lesion + VD3
D
**
TH+ cells (% of contralateral side)
0
20
40
60
Lesion
Lesion + VD3
E
F
G
H

although this did not reach significance (unpaired *t*-test $p=0.43$; Fig. 2C). Animals treated with calcitriol at the same time as lesioning displayed no difference in striatal innervation on the lesioned side compared with controls, again indicating that concurrent direct treatment alongside lesioning is not beneficial. In addition to measuring total innervation volume, we counted the numbers of surviving TH+ cells in the lesioned and contralateral SN. Animals pretreated with vitamin D had significantly more surviving TH+ neurons than those receiving vehicle alone on the lesioned side (Fig. 2D; $45.9 \pm 5.0\%$ of intact SN vs. $23.6 \pm 3.5\%$, unpaired *t*-test $p<0.01$). Combined with the functional data, this strongly supports the notion that systemic levels of vitamin D can protect neurons from death in the face of a toxic insult. Representative sections through the striatum of vitamin D pretreated and vehicle only animals are shown in Fig. 2E and F, respectively. The extent of striatal denervation can be clearly seen, along with the much smaller area of surviving innervation in both pretreated animals and controls. The lesioned substantia nigra from pretreated and control animals is shown in Fig. 2G and H, respectively, highlighting the increased survival of TH immunoreactive cells in pretreated animals.

At first glance, the results of intraparenchymal calcitriol administration alongside the toxin are disappointing; however, it should be considered that this hardly represents a realistic approach to neuroprotection where a small molecule alternative, such as vitamin D is available. It is clear that oral, systemic administration is a far better option than intracranial delivery, as a safe and far more feasible option.

Figure 2 Neuroprotective action of Vitamin D in a 6-OHDA model of Parkinson's disease. In functional testing of animals following lesion with 6-OHDA, those pretreated with vitamin D showed less severe drug induced rotational behavior when stimulated with apomorphine (A). The pretreated animals were also more successful at the balance beam test, having significantly fewer paw-slips than control animals (B). Brains were removed from animals and immunostained with antibodies specific to tyrosine hydroxylase to visualize striatal innervation. Striatal innervation volumes were not significantly different between pretreated animals and vehicle only controls following lesioning (C). However, in the substantia nigra of pretreated animals there was a significant increase in the number of TH+ cells when expressed as percent of the intact side (D), **$p < 0.01$. Representative images of sections through the striatum of pretreated (E) and control (F) animals clearly show the effect of lesioning on striatal innervation (scale bars = 1 mm). The lesioned substantia nigra shows increased loss of TH immunoreactive cells in the control animals (G) compared to vitamin D pretreated subjects (H); scale bars 0.5 mm. (See the color plate.)

5.4 Clinical Studies with Vitamin D and Parkinson's Disease

Whilst the body of research discussed above is of great academic interest, it is meaningless unless it translates to patient benefit. And given the body of evidence suggesting neuroprotective properties, it is of little surprise that clinical studies using vitamin D supplementation in treating PD are underway. In one published study conducted in Tokyo, 114 patients diagnosed with PD but with no family history of disease and who were not already using vitamin D supplementation were enrolled onto a double blind, randomized trial studying the effect of 1200 IU/day versus placebo on continued deterioration, assessed using the Hoehn and Yahr staging over a period of 12 months. Twelve patients were lost to follow-up or withdrew from the trial, leaving 55 and 57 patients in the test and placebo groups, respectively. Strikingly, patients assigned to the vitamin D supplementation group demonstrated significantly reduced decline in their Hoehn and Yahr scores when compared to the control placebo cohort (Suzuki et al., 2013).

6. FUTURE DEVELOPMENTS AND APPLICATIONS TO PARKINSON'S DISEASE

Clearly, vitamin D plays a role in neuroprotection, although the precise mechanisms by which it confers neuronal support are currently still under debate. Whatever the mechanism for vitamin D's role (GDNF upregulation, reducing oxidative stress, decreasing microglia activation), there are strong arguments for developing strategies to introduce this vitamin as another arm of treatment for Parkinson's disease, to potentially slow disease progression and improve patient's quality of life. Further preclinical and clinical trials are warranted to optimize dosage and to better understand vitamin D's effects across the length of the disease process.

In addition, the pro-neuronal survival effect of vitamin D makes it attractive when considering cell replacement therapies for PD. Addition of calcitriol to primary cell suspensions may enhance the survival of transplanted cells, and lead to the lowering of numbers of embryos required for tissue dissection. This would make clinical transplants more ethically and practically feasible, and thus increase their validity and availability to patients. Alternatively, as stem cell therapies are increasingly considered for PD, vitamin D may have a role to play in the production of dopamine neurons from pluripotent stem cells as an aid to enhance their survival, offering a simple, safe and cost-effective media supplement in the protocols currently being developed toward clinical application.

Clearly, whether as a direct neuroprotective agent or indirect supplement, vitamin D's future is compelling, in the search for therapies for PD.

REFERENCES

Afzal, S., Bojesen, S. E., & Nordestgaard, B. G. (2014). Reduced 25-hydroxyvitamin D and risk of Alzheimer's disease and vascular dementia. *Alzheimer's and Dementia*, *10*, 296–302. http://doi.org/10.1016/j.jalz.2013.05.1765.

Annweiler, C., Brugg, B., Peyrin, J.-M., Bartha, R., & Beauchet, O. (2014). Combination of memantine and vitamin D prevents axon degeneration induced by amyloid-beta and glutamate. *Neurobiology of Aging*, *35*(2), 331–335. http://doi.org/10.1016/j.neurobiolaging.2013.07.029.

Aranow, C. (2011). Vitamin D and the immune system. *Journal of Investigative Medicine: The Official Publication of the American Federation for Clinical Research*, *59*(6), 881–886. http://doi.org/10.231/JIM.0b013e31821b8755.

Betemps, E. J., & Buncher, C. R. (1993). Birthplace as a risk factor in motor neurone disease and Parkinson's disease. *International Journal of Epidemiology*, *22*(5), 898–904.

Brito Galvao, J. F., Nagode, L. A., Schenck, P. A., & Chew, D. J. (2013). Calcitriol, calcidiol, parathyroid hormone, and fibroblast growth factor-23 interactions in chronic kidney disease. *Journal of Veterinary Emergency and Critical Care*, *23*(2), 134–162. http://doi.org/10.1111/vec.12036.

Buell, J. S., & Dawson-Hughes, B. (2008). Vitamin D and neurocognitive dysfunction: Preventing "D"ecline? *Molecular Aspects of Medicine*, *29*(6), 415–422. http://doi.org/10.1016/j.mam.2008.05.001.

Cali, T., Ottolini, D., & Brini, M. (2014). Calcium signaling in Parkinson's disease. *Cell and Tissue Research*, *357*(2), 439–454. http://doi.org/10.1007/s00441-014-1866-0.

Cass, W. A., Peters, L. E., Fletcher, A. M., & Yurek, D. M. (2012). Evoked dopamine overflow is augmented in the striatum of calcitriol treated rats. *Neurochemistry International*, *60*(2), 186–191. http://doi.org/10.1016/j.neuint.2011.11.010.

Dean, E. D., Mexas, L. M., Capiro, N. L., McKeon, J. E., DeLong, M. R., Pennell, K. D., et al. (2012). 25-Hydroxyvitamin D depletion does not exacerbate MPTP-induced dopamine neuron damage in mice. *PLoS One*, *7*(7), e39227. http://doi.org/10.1371/journal.pone.0039227.

Deeb, K. K., Trump, D. L., & Johnson, C. S. (2007). Vitamin D signalling pathways in cancer: Potential for anticancer therapeutics. *Nature Reviews. Cancer*, *7*(9), 684–700. http://doi.org/10.1038/nrc2196.

Dhesi, J. K., Jackson, S. H. D., Bearne, L. M., Moniz, C., Hurley, M. V., Swift, C. G., et al. (2004). Vitamin D supplementation improves neuromuscular function in older people who fall. *Age and Ageing*, *33*(6), 589–595. http://doi.org/10.1093/ageing/afh209.

Eyles, D., Brown, J., Mackay-Sim, A., McGrath, J., & Feron, F. (2003). Vitamin D3 and brain development. *Neuroscience*, *118*(3), 641–653.

Eyles, D. W., Smith, S., Kinobe, R., Hewison, M., & McGrath, J. J. (2005). Distribution of the vitamin D receptor and 1 alpha-hydroxylase in human brain. *Journal of Chemical Neuroanatomy*, *29*(1), 21–30. http://doi.org/10.1016/j.jchemneu.2004.08.006.

Garcion, E., Wion-Barbot, N., Montero-Menei, C. N., Berger, F., & Wion, D. (2002). New clues about vitamin D functions in the nervous system. *Trends in Endocrinology and Metabolism*, *13*(3), 100–105.

Garland, C. F., Garland, F. C., Gorham, E. D., Lipkin, M., Newmark, H., Mohr, S. B., et al. (2006). The role of vitamin D in cancer prevention. *American Journal of Public Health, 96*(2), 252–261. http://doi.org/10.2105/AJPH.2004.045260.

Ginde, A. A., Mansbach, J. M., & Camargo, C. A. J. (2009). Association between serum 25-hydroxyvitamin D level and upper respiratory tract infection in the Third National Health and Nutrition Examination Survey. *Archives of Internal Medicine, 169*(4), 384–390. http://doi.org/10.1001/archinternmed.2008.560.

Gorham, E. D., Garland, C. F., Garland, F. C., Grant, W. B., Mohr, S. B., Lipkin, M., et al. (2007). Optimal vitamin D status for colorectal cancer prevention: A quantitative meta analysis. *American Journal of Preventive Medicine, 32*(3), 210–216. http://doi.org/10.1016/j.amepre.2006.11.004.

Guo, T., Taylor, R. L., Singh, R. J., & Soldin, S. J. (2006). Simultaneous determination of 12 steroids by isotope dilution liquid chromatography-photospray ionization tandem mass spectrometry. *Clinica Chimica Acta; International Journal of Clinical Chemistry, 372*(1–2), 76–82. http://doi.org/10.1016/j.cca.2006.03.034.

Helzlsouer, K. J. (2010). Overview of the cohort consortium vitamin D pooling project of rarer cancers. *American Journal of Epidemiology, 172*(1), 4–9. http://doi.org/10.1093/aje/kwq119.

Holick, M. F. (2009). Vitamin D status: Measurement, interpretation and clinical application. *Annals of Epidemiology, 19*(2), 73–78. http://doi.org/10.1016/j.annepidem.2007.12.001.

Hollis, B. W. (2004). Editorial: The determination of circulating 25-hydroxyvitamin D: No easy task. *Journal of Clinical Endocrinology and Metabolism, 89*(7), 3149–3151. http://doi.org/10.1210/jc.2004-0682.

Ibi, M., Sawada, H., Nakanishi, M., Kume, T., Katsuki, H., Kaneko, S., et al. (2001). Protective effects of 1 alpha,25-(OH)(2)D(3) against the neurotoxicity of glutamate and reactive oxygen species in mesencephalic culture. *Neuropharmacology, 40*(6), 761–771.

Jang, W., Kim, H. J., Li, H., Jo, K. D., Lee, M. K., Song, S. H., et al. (2014). 1,25-Dyhydroxyvitamin D(3) attenuates rotenone-induced neurotoxicity in SH-SY5Y cells through induction of autophagy. *Biochemical and Biophysical Research Communications, 451*(1), 142–147. http://doi.org/10.1016/j.bbrc.2014.07.081.

Jozefowicz, O., Rabe-Jablonska, J., Wozniacka, A., & Strzelecki, D. (2014). Analysis of vitamin D status in major depression. *Journal of Psychiatric Practice, 20*(5), 329–337. http://doi.org/10.1097/01.pra.0000454777.21810.15.

Kenborg, L., Jorgensen, A. D., Budtz-Jorgensen, E., Knudsen, L. E., & Hansen, J. (2010). Occupational exposure to the sun and risk of skin and lip cancer among male wage earners in Denmark: A population-based case–control study. *Cancer Causes & Control, 21*(8), 1347–1355. http://doi.org/10.1007/s10552-010-9562-1.

Kennel, K. A., Drake, M. T., & Hurley, D. L. (2010). Vitamin D deficiency in adults: When to test and how to treat. *Mayo Clinic Proceedings, 85*(8), 752–757. http://doi.org/10.4065/mcp.2010.0138.

Kerr, D. C. R., Zava, D. T., Piper, W. T., Saturn, S. R., Frei, B., & Gombart, A. F. (2015). Associations between vitamin D levels and depressive symptoms in healthy young adult women. *Psychiatry Research, 227*(1), 46–51. http://doi.org/10.1016/j.psychres.2015.02.016.

Kim, J.-S., Ryu, S.-Y., Yun, I., Kim, W.-J., Lee, K.-S., Park, J.-W., et al. (2006). 1alpha,25-Dihydroxyvitamin D(3) protects dopaminergic neurons in rodent models of Parkinson's disease through inhibition of microglial activation. *Journal of Clinical Neurology (Seoul, Korea), 2*(4), 252–257. http://doi.org/10.3988/jcn.2006.2.4.252.

Knekt, P., Kilkkinen, A., Rissanen, H., Marniemi, J., Saaksjarvi, K., & Heliovaara, M. (2010). Serum vitamin D and the risk of Parkinson disease. *Archives of Neurology*, *67*(7), 808–811. http://doi.org/10.1001/archneurol.2010.120.

Langub, M. C., Herman, J. P., Malluche, H. H., & Koszewski, N. J. (2001). Evidence of functional vitamin D receptors in rat hippocampus. *Neuroscience*, *104*(1), 49–56.

Lee, C. S., Sauer, H., & Björklund, A. (1996). Dopaminergic neuronal degeneration and motor impairments following axon terminal lesion by intrastriatal 6-hydroxydopamine in the rat. *Neuroscience*, *72*(3), 641–653. http://doi.org/http://dx.doi.org/10.1016/0306-4522(95)00571-4.

Lin, A. M. Y., Fan, S. F., Yang, D. M., Hsu, L. L., & Yang, C. H. J. (2003). Zinc-induced apoptosis in substantia nigra of rat brain: Neuroprotection by vitamin D3. *Free Radical Biology & Medicine*, *34*(11), 1416–1425.

Lips, P., & van Schoor, N. M. (2011). The effect of vitamin D on bone and osteoporosis. *Best Practice & Research Clinical Endocrinology & Metabolism*, *25*(4), 585–591. http://doi.org/10.1016/j.beem.2011.05.002.

Liu, P. T., Stenger, S., Li, H., Wenzel, L., Tan, B. H., Krutzik, S. R., et al. (2006). Toll-like receptor triggering of a vitamin D-mediated human antimicrobial response. *Science (New York, N.Y.)*, *311*(5768), 1770–1773. http://doi.org/10.1126/science.1123933.

Lodh, M., Goswami, B., Mahajan, R. D., Sen, D., Jajodia, N., & Roy, A. (2015). Assessment of vitamin D status in patients of chronic low back pain of unknown etiology. *Indian Journal of Clinical Biochemistry*, *30*(2), 174–179. http://doi.org/10.1007/s12291-014-0435-3.

Lotfi, A., Abdel-Nasser, A. M., Hamdy, A., Omran, A. A., & El-Rehany, M. A. (2007). Hypovitaminosis D in female patients with chronic low back pain. *Clinical Rheumatology*, *26*(11), 1895–1901. http://doi.org/10.1007/s10067-007-0603-4.

Mahon, B. D., Wittke, A., Weaver, V., & Cantorna, M. T. (2003). The targets of vitamin D depend on the differentiation and activation status of CD4 positive T cells. *Journal of Cellular Biochemistry*, *89*(5), 922–932. http://doi.org/10.1002/jcb.10580.

Metz, G. A., & Whishaw, I. Q. (2002). Drug-induced rotation intensity in unilateral dopamine-depleted rats is not correlated with end point or qualitative measures of forelimb or hindlimb motor performance. *Neuroscience*, *111*(2), 325–336.

Munger, K. L., Levin, L. I., Hollis, B. W., Howard, N. S., & Ascherio, A. (2006). Serum 25-hydroxyvitamin D levels and risk of multiple sclerosis. *JAMA*, *296*(23), 2832–2838. http://doi.org/10.1001/jama.296.23.2832.

Naveilhan, P., Neveu, I., Wion, D., & Brachet, P. (1996). 1,25-Dihydroxyvitamin D3, an inducer of glial cell line-derived neurotrophic factor. *Neuroreport*, 7(13), 2171–2175.

Orme, R. P., Bhangal, M. S., & Fricker, R. A. (2013). Calcitriol imparts neuroprotection in vitro to midbrain dopaminergic neurons by upregulating GDNF expression. *PLoS One*, *8*(4), e62040. http://doi.org/10.1371/journal.pone.0062040.

Partonen, T. (1998). Vitamin D and serotonin in winter. *Medical Hypotheses*, *51*(3), 267–268.

Patrick, R. P., & Ames, B. N. (2014). Vitamin D hormone regulates serotonin synthesis. Part 1: Relevance for autism. *FASEB Journal: Official Publication of the Federation of American Societies for Experimental Biology*, *28*(6), 2398–2413. http://doi.org/10.1096/fj.13-246546.

Pearce, S. H. S., & Cheetham, T. D. (2010). Diagnosis and management of vitamin D deficiency. *BMJ (Clinical Research Ed)*, *340*, b5664.

Prufer, K., Veenstra, T. D., Jirikowski, G. F., & Kumar, R. (1999). Distribution of 1,25-dihydroxyvitamin D3 receptor immunoreactivity in the rat brain and spinal cord. *Journal of Chemical Neuroanatomy*, *16*(2), 135–145.

Przybelski, R. J., & Binkley, N. C. (2007). Is vitamin D important for preserving cognition? A positive correlation of serum 25-hydroxyvitamin D concentration with cognitive function. *Archives of Biochemistry and Biophysics*, *460*(2), 202–205. http://doi.org/10.1016/j.abb.2006.12.018.

Puchacz, E., Stumpf, W. E., Stachowiak, E. K., & Stachowiak, M. K. (1996). Vitamin D increases expression of the tyrosine hydroxylase gene in adrenal medullary cells. *Brain Research. Molecular Brain Research*, *36*(1), 193–196.

Sanchez, B., Lopez-Martin, E., Segura, C., Labandeira-Garcia, J. L., & Perez-Fernandez, R. (2002). 1,25-Dihydroxyvitamin D(3) increases striatal GDNF mRNA and protein expression in adult rats. *Brain Research. Molecular Brain Research*, *108*(1–2), 143–146.

Sanchez, B., Relova, J. L., Gallego, R., Ben-Batalla, I., & Perez-Fernandez, R. (2009). 1,25-Dihydroxyvitamin D3 administration to 6-hydroxydopamine-lesioned rats increases glial cell line-derived neurotrophic factor and partially restores tyrosine hydroxylase expression in substantia nigra and striatum. *Journal of Neuroscience Research*, *87*(3), 723–732. http://doi.org/10.1002/jnr.21878.

Shinpo, K., Kikuchi, S., Sasaki, H., Moriwaka, F., & Tashiro, K. (2000). Effect of 1,25-dihydroxyvitamin D(3) on cultured mesencephalic dopaminergic neurons to the combined toxicity caused by L-buthionine sulfoximine and 1-methyl-4-phenylpyridine. *Journal of Neuroscience Research*, *62*(3), 374–382.

Soilu-Hanninen, M., Aivo, J., Lindstrom, B.-M., Elovaara, I., Sumelahti, M.-L., Farkkila, M., et al. (2012). A randomised, double blind, placebo controlled trial with vitamin D3 as an add on treatment to interferon beta-1b in patients with multiple sclerosis. *Journal of Neurology, Neurosurgery, and Psychiatry*, *83*(5), 565–571. http://doi.org/10.1136/jnnp-2011-301876.

Suzuki, M., Yoshioka, M., Hashimoto, M., Murakami, M., Noya, M., Takahashi, D., et al. (2013). Randomized, double-blind, placebo-controlled trial of vitamin D supplementation in Parkinson disease. *American Journal of Clinical Nutrition*, *97*(5), 1004–1013. http://doi.org/10.3945/ajcn.112.051664.

Veenstra, T. D., Prufer, K., Koenigsberger, C., Brimijoin, S. W., Grande, J. P., & Kumar, R. (1998). 1,25-Dihydroxyvitamin D3 receptors in the central nervous system of the rat embryo. *Brain Research*, *804*(2), 193–205.

Wang, L., Evatt, M. L., Maldonado, L. G., Perry, W. R., Ritchie, J. C., Beecham, G. W., et al. (2015). Vitamin D from different sources is inversely associated with Parkinson disease. *Movement Disorders: Official Journal of the Movement Disorder Society*, *30*(4), 560–566. http://doi.org/10.1002/mds.26117.

Wang, L., Hara, K., Van Baaren, J. M., Price, J. C., Beecham, G. W., Gallins, P. J., et al. (2012). Vitamin D receptor and Alzheimer's disease: A genetic and functional study. *Neurobiology of Aging*, *33*(8), 1844.e1–1844.e9. http://doi.org/10.1016/j.neurobiolaging.2011.12.038.

Wang, J. Y., Wu, J. N., Cherng, T. L., Hoffer, B. J., Chen, H. H., Borlongan, C. V., et al. (2001). Vitamin D(3) attenuates 6-hydroxydopamine-induced neurotoxicity in rats. *Brain Research*, *904*(1), 67–75.

Wergeland, S., Torkildsen, O., Myhr, K.-M., Aksnes, L., Mork, S. J., & Bo, L. (2011). Dietary vitamin D3 supplements reduce demyelination in the cuprizone model. *PLoS One*, *6*(10), e26262. http://doi.org/10.1371/journal.pone.0026262.

Wilkins, C. H., Sheline, Y. I., Roe, C. M., Birge, S. J., & Morris, J. C. (2006). Vitamin D deficiency is associated with low mood and worse cognitive performance in older adults. *American Journal of Geriatric Psychiatry: Official Journal of the American Association for Geriatric Psychiatry*, *14*(12), 1032–1040. http://doi.org/10.1097/01.JGP.0000240986.74642.7c.

Wolf, G. (2004). The discovery of vitamin D: The contribution of Adolf Windaus. *Journal of Nutrition*, *134*(6), 1299–1302. Retrieved from. http://www.ncbi.nlm.nih.gov/pubmed/15173387.

Zanatta, L., Goulart, P. B., Goncalves, R., Pierozan, P., Winkelmann-Duarte, E. C., Woehl, V. M., et al. (2012). 1alpha,25-dihydroxyvitamin D(3) mechanism of action:

Modulation of L-type calcium channels leading to calcium uptake and intermediate filament phosphorylation in cerebral cortex of young rats. *Biochimica et Biophysica Acta, 1823*(10), 1708–1719. http://doi.org/10.1016/j.bbamcr.2012.06.023.

Zella, L. A., Shevde, N. K., Hollis, B. W., Cooke, N. E., Pike, J. W., Zella, L. A., et al. (2008). Vitamin D-binding protein influences total circulating levels of 1,25-dihydroxyvitamin D(3) but does not directly modulate the bioactive levels of the hormone in vivo. *Endocrinology, 149*(7), 3656–3667. http://doi.org/10.1210/en.2008-0042.

Zhao, Y., Sun, Y., Ji, H.-F., & Shen, L. (2013). Vitamin D levels in Alzheimer's and Parkinson's diseases: A meta-analysis. *Nutrition (Burbank, Los Angeles County, Calif), 29*(6), 828–832. http://doi.org/10.1016/j.nut.2012.11.018.

Zittermann, A. (2003). Vitamin D in preventive medicine: Are we ignoring the evidence? *British Journal of Nutrition, 89*(05), 552–572.

CHAPTER TWELVE

Vitamin D and Cardiac Differentiation

Irene M. Kim*, Keith C. Norris†, Jorge N. Artaza*,†,1

*Department of Health & Life Sciences, Charles R. Drew University of Medicine and Science, Los Angeles, California, USA

†Department of Medicine, David Geffen School of Medicine at UCLA, Los Angeles, California, USA

[1]Corresponding author: e-mail address: joartaza@ucla.edu

Contents

Abstract

Calcitriol (1,25-dihydroxycholecalciferol or 1,25-D_3) is the hormonally active metabolite of vitamin D. Experimental studies of vitamin D receptors and 1,25-D_3 establish calcitriol to be a critical regulator of the structure and function of the heart. Clinical studies link vitamin D deficiency with cardiovascular disease (CVD). Emerging evidence demonstrates that calcitriol is highly involved in CVD-related signaling pathways, particularly the Wnt signaling pathway. Addition of 1,25-D_3 to cardiomyocyte cells and examination of its effects on cardiomyocytes and mainly Wnt11 signaling allowed the specific characterization of the role of calcitriol in cardiac differentiation. 1,25-D_3 is demonstrated to: (i) inhibit cell proliferation without promoting apoptosis; (ii) decrease

Vitamins and Hormones, Volume 100
ISSN 0083-6729
http://dx.doi.org/10.1016/bs.vh.2015.10.008

expression of genes related to the regulation of the cell cycle; (iii) promote formation of cardiomyotubes; (iv) induce expression of casein kinase-1-α1, a negative regulator of the canonical Wnt signaling pathway; and (v) increase expression of noncanonical Wnt11, which has been recognized to induce cardiac differentiation during embryonic development and in adult cells.

Thus, it appears that vitamin D promotes cardiac differentiation through negative modulation of the canonical Wnt signaling pathway and upregulation of noncanonical Wnt11 expression. Future work to elucidate the role(s) of vitamin D in cardiovascular disorders will hopefully lead to improvement and potentially prevention of CVD, including abnormal cardiac differentiation in settings such as postinfarction cardiac remodeling.

1. INTRODUCTION

Calcitriol, which is also referred to as 1,25-D_3, is the hormonally active metabolite of vitamin D. It plays a classical hormonal role in skeletal health to regulate calcium and phosphorus metabolism, by influencing intestinal absorption, bone resorption, and renal retention. It also has physiological functions in nonskeletal tissues (Norman & Powell, 2014). Experimental studies reveal the presence of vitamin D receptors (VDR) in all the major cardiovascular cell types and demonstrate the effects of 1,25-D_3 upon binding to VDR in cardiac cells (Gardner, Chen, & Glenn, 2013; Nibbelink, Tishkoff, Hershey, Rahman, & Simpson, 2007). These studies characterize vitamin D as a critical regulator of the structure and function of the heart.

Furthermore, clinical studies have generally established an association between vitamin D deficiency (hypovitaminosis D) and cardiovascular disease (CVD), the leading cause of mortality worldwide (Norman & Powell, 2014). Even more, Vitamin D deficiency was also associated with several cardiovascular-related diseases, including hypertension, coronary artery disease, cardiomyopathy, and diabetes (Vacek et al., 2012). However, little is known about the molecular mechanism by which 1,25-D_3 modulates heart development. A growing body of evidence indicates that 1,25-D_3 affects cardiac differentiation via signaling pathways related to CVD (Hliang et al., 2014). In particular, noncanonical Wnt11 signaling, which is sufficient to induce cardiomyogenic differentiation in unfractionated bone marrow mononuclear cells, is a key signaling pathway linked to 1,25-D_3 (Flaherty et al., 2008).

To further study, the molecular mechanism of calcitriol, 1,25-D_3 was added to cardiomyocyte cells and its effects on cardiomyocytes and Wnt11 signaling were examined. The role of calcitriol in cardiac differentiation was further elucidated. 1,25-D_3 caused cell proliferation inhibition without promoting apoptosis, decreased gene expression of cell cycle regulation, formation of cardiomyotubes, expression of casein kinase-1-α1 (a negative regulator of the canonical Wnt signaling pathway), and expression of noncanonical Wnt11, which has been demonstrated to induce cardiac differentiation during embryonic development and in adult cells (Hliang et al., 2014). This experiment and related studies provide evidence that vitamin D promotes cardiac differentiation via upregulation of noncanonical Wnt11 expression and negative modulation of the canonical Wnt signaling pathway. In this chapter, we review our current knowledge on calcitriol and its association with cardiovascular function and disease. We also describe the results of our recent analysis on the effect of calcitriol on cardiomyocyte cells, in order to advance understanding of its molecular mechanism in cardiac differentiation and explore potential methods through which vitamin D may enhance cardiac health.

2. CALCITRIOL AND CVD

Cardiovascular and related diseases are the most common cause of mortality and morbidity in the nation (Go et al., 2014). Observational studies strongly associate hypovitaminosis D with many cardiovascular disorders, including hypertension, peripheral vascular disease, diabetes mellitus, metabolic syndrome, coronary artery disease, and heart failure (Vanga, Good, Howard, & Vacek, 2010). Mechanistically, 1,25-D_3 binds to the specific VDR that regulates numerous genes involved in fundamental processes related to CVD, including cell proliferation and differentiation, apoptosis, oxidative stress, membrane transport, matrix homeostasis, and cell adhesion (Norman & Powell, 2014). Experimental studies have established an important role for vitamin D metabolites in pathways that are critical to cardiovascular function and disease, including inflammation, thrombosis, and the renin–angiotensin system (Norman & Powell, 2014). As portrayed in Table 1, there are multiple potential mechanisms by which calcitriol may affect cardiovascular health. The major pathways through which hypovitaminosis D may lead to CVD are illustrated in Fig. 1. Despite the uncertainty regarding its true mechanism within cardiovascular systems, calcitriol is recognized as crucial for cardiovascular health and hypovitaminosis D is recognized as a potential risk factor for several CVD processes (Artaza et al., 2011; Artaza, Mehrotra, & Norris, 2009).

Table 1 Potential Mechanisms Through Which Hypovitaminosis D May Affect Cardiovascular Health (Vanga et al., 2010)

Pathology	Proposed Mechanism of Action
Hypertensive vascular disease	• Increased intracellular calcium leading to decreased renin activity • Calcitriol suppression of renin promoter gene • Alteration of the sensitivity of vascular smooth muscle cells
Peripheral vascular disease	• Increased calcification
Diabetes mellitus	• Immunomodulatory effects by reducing tumor necrosis factor-α, parathyroid hormone, and interleukin-10 • Decreased insulin receptor expression, leading to peripheral resistance of insulin • Effect on intracellular calcium levels leading to decreased insulin secretion
Lipid metabolism	• Increase peripheral insulin resistance, contributing to high lipid profile • Statins may increase vitamin D levels by increasing 7-dehydrocholerterol • Increased vessel free radicals lead to oxidation of low-density lipoprotein and increased engulfment by macrophages, an early sign of atherosclerosis
Coronary artery disease	• Indirect effect through risk factor modification • Altering endothelial function • Increased coronary artery calcification
Heart failure	• Direct effect on myocardial contractility • Regulation of brain natriuretic peptide secretion • Reduction of left ventricular hypertrophy with effects on extracellular remodeling • Regulation of inflammatory cytokines • Secondary hyperparathyroidism, which leads to vasodilatation and positive inotropic stimulation
Arrhythmias	• Direct myocardial substrate modification • Indirectly via calcium levels and metabolism at a cellular level

Several epidemiologic and clinical studies support the association between hypovitaminosis D and CVD. In 1990, Scragg et al. reported an inverse relationship between serum 25-hydroxy vitamin D (25[OH] Vit D) and myocardial infarction, in a community setting. In 2007, Martins

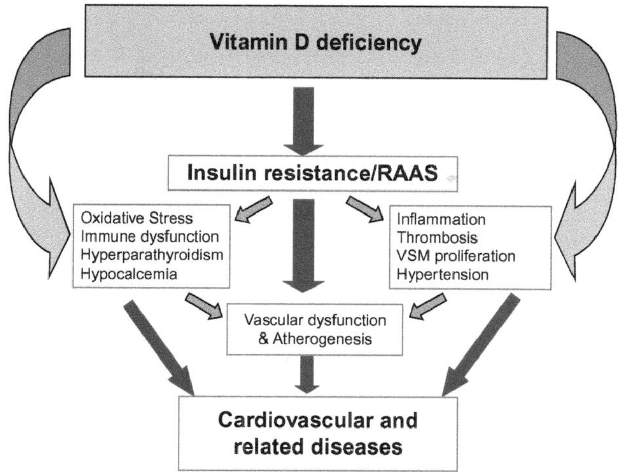

Figure 1 Conceptual model of major pathways through which hypovitaminosis D may lead to CVD (Artaza et al., 2009).

et al. analyzed over 13,000 adults in the Third National Health and Nutrition Examination Survey (NHANES III), and demonstrated a correlation between hypovitaminosis D and increased prevalence of CVD risk factors, independent of its association with all-cause mortality (Martins et al., 2007). Several CVD risk factors linked to hypovitaminosis D include obesity, insulin resistance, diabetes, systemic inflammation, and hypertriglyceridemia (Vanga et al., 2010). Powerful CVD risk factors include chronic kidney disease (CKD) and left ventricular hypertrophy (Artaza et al., 2009). Importantly, epidemiologic studies suggested that patients who had CKD and were treated with calcitriol had a survival advantage when compared with those who did not receive treatment with these agents (Artaza et al., 2009; Shoben, Rudser, de Boer, Young, & Kestenbaum, 2008; Wolf et al., 2008). Clinical studies have generally demonstrated an independent association between hypovitaminosis D and various manifestations of CVD including vascular calcification (Norman & Powell, 2014) and congestive heart failure (Lutsey et al., 2015) These findings reinforce the connection between calcitriol and cardiac function, and they provide hopeful evidence for the therapeutic success of calcitriol for CVD treatment. However, the role of vitamin D supplementation in the management of CVD has been met with variable results on both intermediate (Dong et al., 2010; Gepner et al., 2012; Martins et al., 2014; Yiu et al., 2013) and long-term outcomes (Donneyong et al., 2015; Vacek et al., 2012; Wood et al., 2012), and thus its role in a clinical setting remains unconfirmed (Manson & Bassuk, 2015). There is a need to advance understanding of biologic pathways through which vitamin D affects cardiovascular health and to conduct clinical interventions, in order to precisely define the cardio protective effects of vitamin D repletion (Artaza et al., 2009).

3. CALCITRIOL IS CRITICAL IN THE MODULATION AND MAINTENANCE OF HEART CELL STRUCTURE AND FUNCTION

Experimental studies reveal the presence of VDR in all the major cardiovascular cell types and demonstrate many of the effects of 1,25-D_3 upon binding to VDR in cardiac cells (Gardner et al., 2013; Nibbelink et al., 2007). These studies characterize vitamin D as a critical regulator of the structure and function of the heart. The metabolism and biologic actions of vitamin D, summarized in Fig. 2, involve the conversion of vitamin D, which is initially bound to a circulating glycoprotein called vitamin D-binding protein (DBP) (Vanga et al., 2010). The liver converts

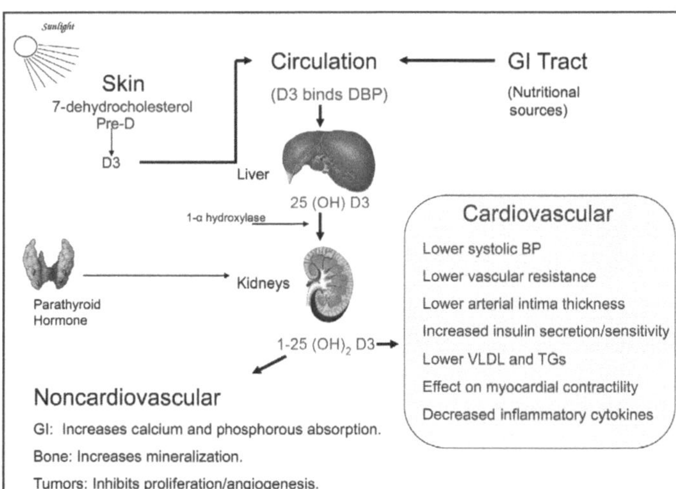

Figure 2 Metabolism and biologic actions of vitamin D. The major biologic form of vitamin D, vitamin D_3, is synthesized in the skin from the precursor (pre-D) under direct sunlight. Vitamin D from cutaneous synthesis and nutritional sources enters the circulation and is bound to DBP. A series of enzymatic hydroxylations in the liver and kidneys transform vitamin D to biologically active $1,25(OH)_2$ vitamin D. Parathyroid hormone regulates the hydroxylation in kidney. Activated vitamin D exerts multiple cardiovascular and noncardiovascular actions. BP, blood pressure; GI, gastrointestinal; TG, triglyceride; VLDL, very low-density lipoprotein (Vanga et al., 2010). (See the color plate.)

vitamin D to 25(OH) Vit D, and then, in a rate-limiting step, the kidneys convert 25(OH) Vit D to its active form 1,25$(OH)_2$ vitamin D (Vanga et al., 2010).

3.1 VDR Involved in Cardiovascular System

The active form of vitamin D binds to VDR, as illustrated in Fig. 3 (Norman & Powell, 2014). The conjugated vitamin D with its receptor forms a heterodimer complex with retinoid X receptor (RXR), and with several other factors and an activator, this complex attaches to vitamin D-responsive elements on deoxyribonucleic acid (DNA) and alters gene expression (Vanga et al., 2010).

VDR are prominently present in enterocytes, osteoblasts, parathyroid gland cells, and distal renal tubule cells, which supports the role of vitamin D in calcium absorption, parathyroid hormone suppression, and bone mineral density enhancement (Vanga et al., 2010). However, recent investigations have shown a significant presence of VDR in all of the major cardiovascular cell types, including cardiomyocytes, arterial wall cells, and immune cells (Gardner et al., 2013). This presence of VDR strongly implies that calcitriol plays a significant role in cardiac cell functions. Various experimental studies indicate that the liganded VDR is greatly involved in the

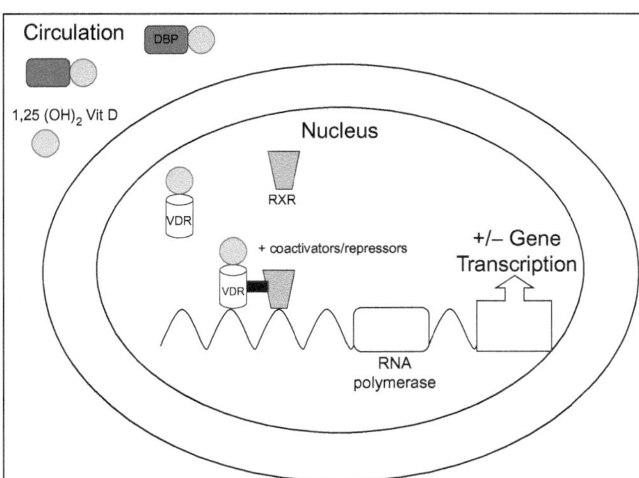

Figure 3 1,25$(OH)_2$ Vit D is bound to DBP in circulation, crosses the cell membrane, and binds to VDR. The conjugated vitamin D with its receptor forms a heterodimer complex with RXR and, with other factors, attaches to vitamin D-responsive elements on DNA and alters gene expression. RNA, ribonucleic acid (Vanga et al., 2010). (See the color plate.)

control of cardiac hypertrophy and fibrosis, regulation of blood pressure, and suppression of the development of atherosclerosis, and several clinical studies support these experimental findings, enhancing evidence that vitamin D or its analogs might have therapeutic effects, in the prevention or treatment of CVD (Artaza et al., 2009; Gardner et al., 2013).

3.2 1,25-D_3 Actions on HL-1 Cardiac Myocyte

A study by Nibbelink et al. (2007) provides evidence that VDR are present in murine cardiac myocytes (HL-1 cells). Additionally, it shows that 1,25-D_3 affects the growth, proliferation, morphology, and gene expression of these cells. In their research, murine HL-1 cardiac myocytes were grown and treated with 1,25-D_3 in culture. Growth and morphology were assessed with microscopic analysis, and cells were counted and protein levels were evaluated through western blot analysis. Subcellular localization of the VDR was determined using immunofluorescence and confocal microscopy. Their experimental results demonstrate that 1,25-D_3 decreases proliferation and alters cellular morphology of HL-1 cells, as shown in Fig. 4.

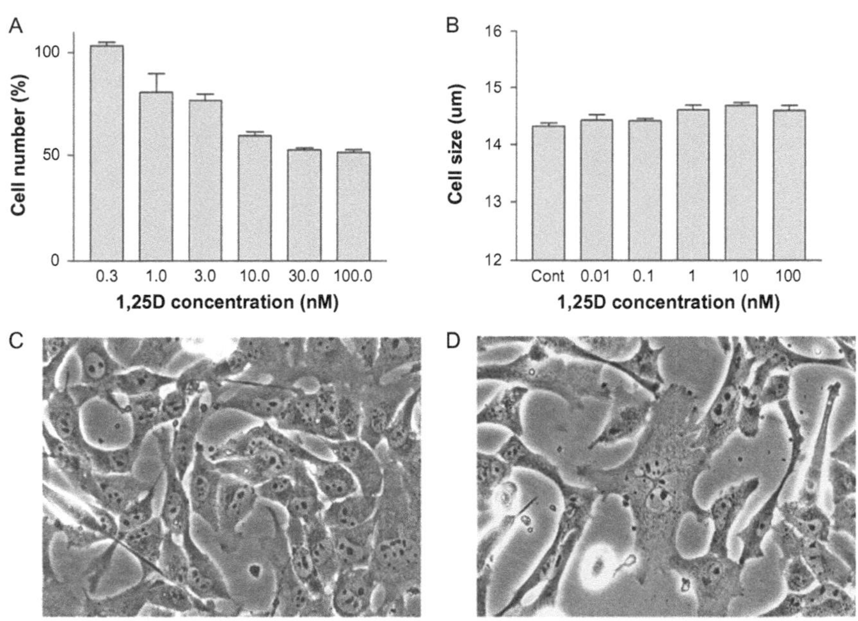

Figure 4 (A and B) Cells treated for 3 days with varying concentrations of 1,25-D_3 show reduced proliferation and slight hypertrophy relative to control. (C and D) Photomicrograph of cells treated for 2 days with (C) control versus (D) 100 nM 1,25-D_3 (Nibbelink et al., 2007).

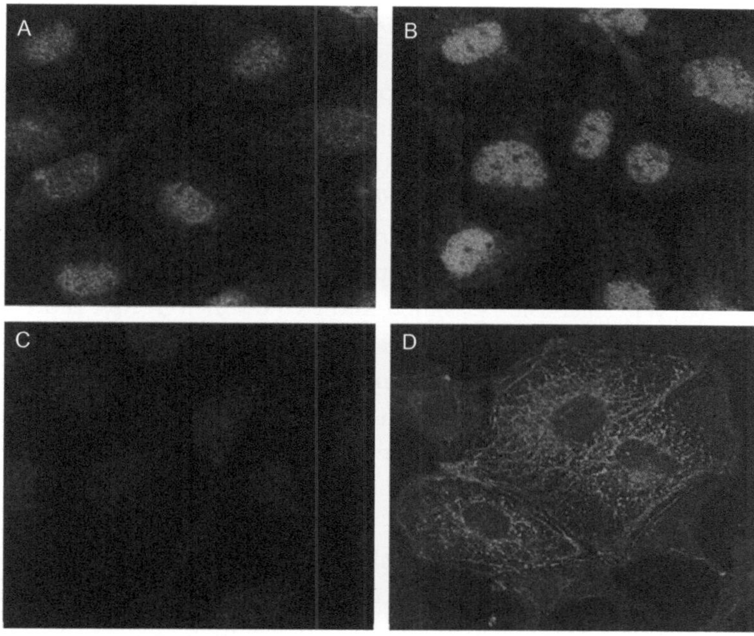

Figure 5 Immunofluorescent analysis of VDR. Confocal images of HL-1 cells treated with 100 nM 1,25-D_3 (B) versus control (A) for 24 h. To ensure specific staining a VDR specific peptide, SC1008P (Santa Cruz), was preincubated with primary VDR Antibodies (Ab) (C). Adequate Ab penetration and cell architecture preservation is evidenced by actinin staining (D). All digital images are 0.35 μm sections captured via Olympus FV500. VDR primary Ab is SC1008 with FITC conjugated secondary AP307F (Chemicon). Actinin primary Ab is A7811 with FITC conjugated secondary A8711 (both Sigma) (Nibbelink et al., 2007). (See the color plate.)

Treatment with 1,25-D_3 increases expression of myotrophin (which promotes growth of cardiomyocytes) and decreases expression of atrial natriuretic peptide and c-myc. 1,25-D_3 treatment also increases expression and nuclear localization of VDR, which is depicted in Fig. 5. Thus, 1,25-D_3 can be recognized as an important hormone for modulating and maintaining heart cell structure and function (Nibbelink et al., 2007).

4. NONCANONICAL WNT11 SIGNALING AND CARDIOGENESIS

The evolutionarily conserved Wnt family of glycoproteins, consisting at least of 19 known members in mammals, are secreted signaling proteins that serve as molecular signaling cues (Flaherty & Dawn, 2008). Wnt glycoproteins provoke and orchestrate various biological activities that include (i) determination of cellular polarity, proliferation, and migration; (ii) craniofacial and neural development; (iii) cardiac development; (iv) tumorigenesis and metastasis; and (v) stem cell differentiation and renewal (Eisenberg & Eisenberg, 2006; Logan & Nusse, 2004; Miller, Hocking, Brown, & Moon, 1999). Wnt pathways are primarily triggered by binding of Wnt to a seven-span transmembrane Frizzled (Fz) family of cell surface receptors, but their disparate subcellular signaling elements require that Wnt ligands be classified into two distinct groups: those using the canonical pathway for intracellular signaling and those using the noncanonical pathway (Flaherty & Dawn, 2008). The Wnt11 signaling pathway, which is the focus of vitamin D's link to cardiac differentiation, utilizes the noncanonical signaling network. Noncanonical, as opposed to canonical, signaling is widely represented as two distinct signaling modules, a calcium-dependent protein kinase C (PKC)-mediated Wnt/Ca^{2+} pathway and a Dsh-dependent c-jun-terminal kinase (JNK)-mediated planar cell polarity (PCP) pathway (Eisenberg & Eisenberg, 2006; Kuhl, Sheldahl, Malbon, & Moon, 2000; Miller et al., 1999). Figure 6 clarifies the distinctions between the canonical and noncanonical pathways.

Among all of the Wnt genes, only Wnt11 initially demonstrated a spatiotemporally regulated expression during gastrulation within or in close proximity to the precardiac mesoderm, which suggested a preeminent role of Wnt11 in cardiac specification (Eisenberg & Eisenberg, 1999; Eisenberg, Gourdie, & Eisenberg, 1997; Kispert, Vainio, Shen, Rowitch, & McMahon, 1996; Pandur, Lasche, Eisenberg, & Kuhl, 2002). Subsequent studies have demonstrated that noncanonical signaling via Wnt11 is sufficient to induce cardiomyocytic commitment in embryonic stem cell populations (Pandur et al., 2002; Terami, Hidaka, Katsumata, et al., 2004; Ueno, Weidinger, Osugi, et al., 2007), adult stem cell populations (Bedada et al., 2005; Flaherty et al., 2008), and unfractionated bone marrow mononuclear cell populations (Flaherty et al., 2008).

A review of several important studies involving Wnt11 signaling in cardiac cells illustrates the growing body of evidence for the mechanism behind cardiac differentiation. In 1999, Eisenberg and Eisenberg demonstrated that soluble Wnt11 was sufficient to promote cardiac tissue formation in posterior noncardiac mesoderm of quail embryonic explants, producing the first line of evidence implicating Wnt11 as an important molecular signal during vertebrate cardiogenesis. Later studies implicated the noncanonical signaling as the transducing apparatus for Wnt11 (Heisenberg et al., 2000;

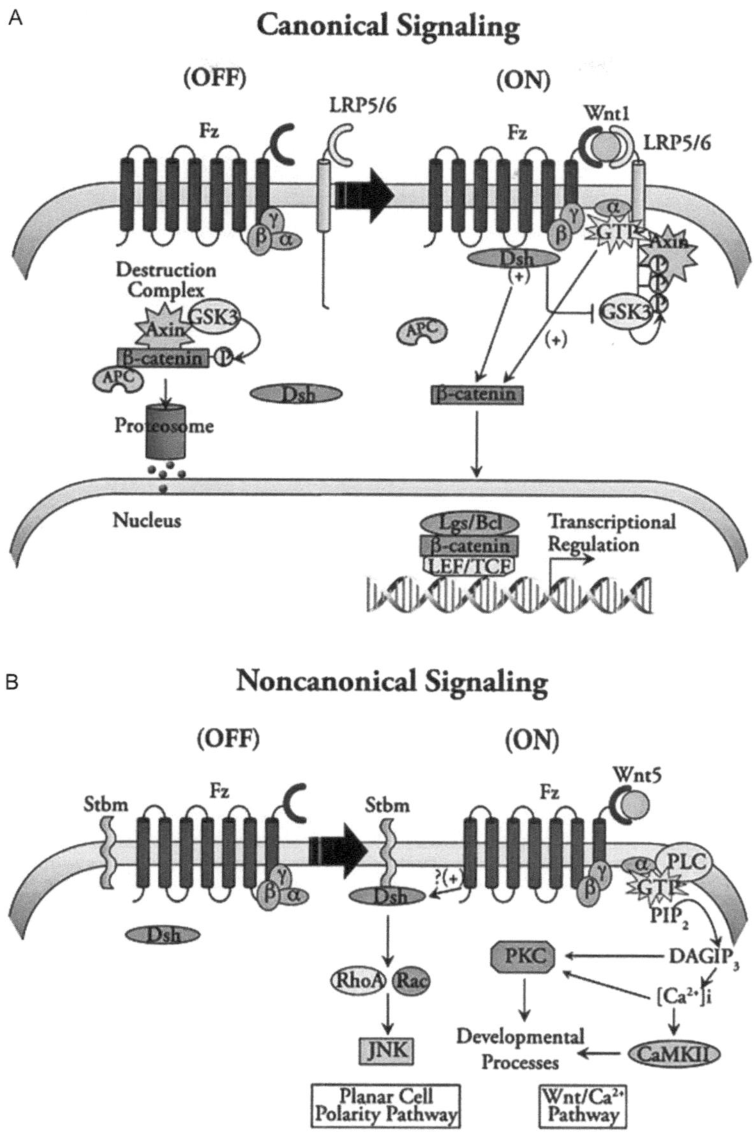

Figure 6 Schematic diagrams depicting Wnt signal transduction pathways. (A) Canonical signaling: (OFF) in the absence of Wnt ligand binding to a receptor complex consisting of Fz and low-density lipoprotein receptor-related protein (LRP, isoforms 5 or 6) coreceptors, cytoplasmic Dishevelled (Dsh) remains inactive and β-catenin, sequesetered within the destruction complex comprising adenomatosis polyposis coli and Axin, is phosphorylated by glycogen synthase kinase-3 (GSK-3), thereby targeting it for ubiquitination and proteosomal degradation. Note the presence of heterotrimeric G-proteins interacting with the Fz receptor. (ON) Wnt ligand binding to the intracellular domain of LRP5/6 where it binds axin with high affinity, sequestering it away from the

(*Continued*)

Kuhl et al., 2000). In 2002, Pandur et al. utilized elegant loss- and gain-of-function experiments in *Xenopus* embryos to demonstrate that Wnt11 is actually required for cardiogenesis and is sufficient to induce a contractile phenotype. This study also showed that noncanonical Wnt11 signaling transduction occurs in a JNK-dependent manner, requiring PKC as the upstream mediator. In 2008, Flaherty et al. experimented with adult density-gradient separated unfractionated bone marrow mononuclear cells and concluded that noncanonical signaling via Wnt11 is sufficient to induce robust cardiomyogenic differentiation in a protein kinase C- and c-Jun-N-terminal kinase-dependent manner. Collectively, these studies, as illustrated in Fig. 7, indicate that Wnt11 via the noncanonical signaling pathway is absolutely critical for vertebrate cardiogenesis (Flaherty & Dawn, 2008).

5. 1,25-VITAMIN D_3 PROMOTES CARDIAC DIFFERENTIATION THROUGH MODULATION OF THE WNT SIGNALING PATHWAY

To further study the molecular mechanism of calcitriol, 1,25-D_3 was added to H9c2 rat embryonic myocardium cells (ATCC, Manassas, VA, USA), and its effects were examined. Cardiomyocyte cell proliferation, cardiomyotube formation, and gene expression of cell cycle regulation were the primary focus points of the experiment (Hliang et al., 2014). The experimental results indicate that vitamin D may promote cardiac differentiation via upregulation of noncanonical Wnt11 expression and negative modulation of the canonical Wnt signaling pathway.

5.1 1,25-D_3 Inhibits Cell Proliferation of H9c2 Cardiomyocytes

Addition of 1,25-D_3 to H9c2 cardiomyocytes inhibits cell proliferation in a dose-dependent manner and downregulates the expression of *Mki67* and *Pcna*, two well known and broadly used cell proliferation markers (Hliang

Figure 6—Cont'd destruction complex, thereby enabling lymphoid enhancer factor-dependent transcription. The Gα subunit has also been implicated in the disassembly of the β-catenin destruction complex, linking it to the canonical signaling. (B) Noncanonical signaling represented as two distinct pathways: the Wnt/Ca^{2+} pathway and the planar cell polarity (PCP) pathway. Ca^{2+}/PKC signaling occurs via Wnt activation of Fz receptors causing hetertorimeric G-protein-dependent Ca^{2+} release, thus activating PKC and CaMKII. Rho/JNK signaling occurs via cooperative interaction between transmembrane receptor strabismus (Stbm) and Fz receptors, which activates Rho-Rac-mediated activation of JNK through Dsh (Flaherty & Dawn, 2008). (See the color plate.)

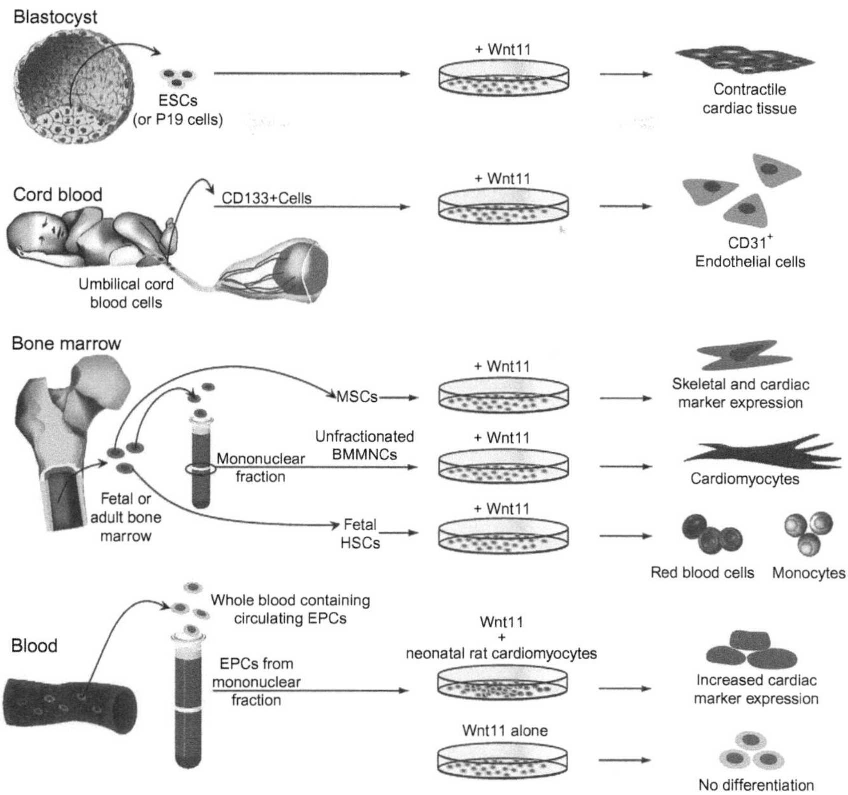

Figure 7 Differential impact of Wnt11 signaling on cell fate *in vitro*. Although blastocyst-derived pluripotent embryonic stem cells give rise to cardiac tissue with a contractile phenotype, Wnt11-treated, cord blood-derived CD133+ progenitors undergo endothelial lineage commitment. Wnt11-treated, bone marrow-derived cells show a remarkable degree of divergence with regard to their lineage commitment and the phenotypic characteristics they assume. In fetal hematopoietic progenitors (HSCs), Wnt11 promotes red blood cell commitment and favors monocyte precursor formation. Bone marrow-derived multipotent adult stem cells treated with Wnt11 differentiate along a cardiomyogenic pathway, but fail to adopt an adult phenotype, whereas Wnt11 is sufficient to produce a cardiomyocytic phenotype in bone marrow mononuclear cells (BMMNCs). Finally, in circulating progenitor cells (CPCs) that adopt a cardiac phenotype when cocultured with neonatal rat cardiomyocytes, Wnt11 enhances expression of cardiac markers, yet, by itself, is unable to affect the native phenotype. ESCs, embryonic stem cells; MSCs, multipotent stem cells; EPCs, endothelial progenitor cells (Flaherty & Dawn, 2008). (See the color plate.)

et al., 2014). Figures 8 and 9 summarize these experimental results. This inhibition was not accompanied by changes in apoptosis or, specifically, the expression of the proapoptotic caspase 3 or expression of the anti-apoptotic *Bc12*. These results were consistent with those of a previous study

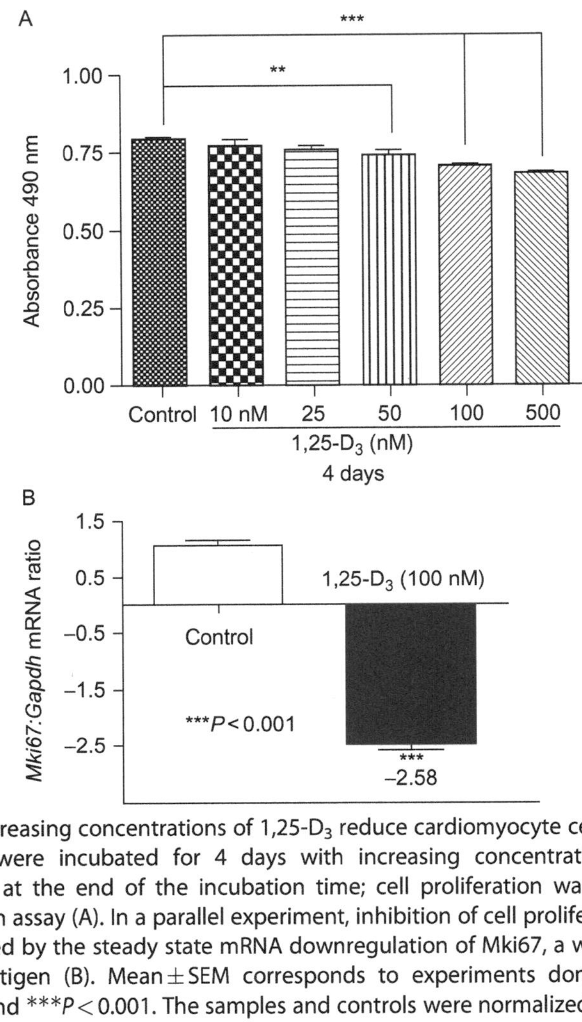

Figure 8 Increasing concentrations of 1,25-D_3 reduce cardiomyocyte cell proliferation. H9c2 cells were incubated for 4 days with increasing concentrations of 1,25-D (0–500 nM), at the end of the incubation time; cell proliferation was evaluated by the formazan assay (A). In a parallel experiment, inhibition of cell proliferation was also demonstrated by the steady state mRNA downregulation of Mki67, a well-known proliferation antigen (B). Mean $\pm$ SEM corresponds to experiments done in triplicate, $^{**}P<0.01$ and $^{***}P<0.001$. The samples and controls were normalized to the GAPDH housekeeping gene (Hliang et al., 2014).

utilizing mesenchymal multipotent cells (Artaza, Sirad, Ferrini, & Norris, 2010) and with those of a study that demonstrated hypovitaminosis D is not associated with increased apoptosis in cardiac tissue (Assalin et al., 2013).

5.2 1,25-D_3 Enhances Cardiomyotube Area and Promotes the Expression of Cardiac Troponin

Addition of 1,25-D_3 causes an increase in cardiomyotube cell area and expression of cardiac troponin, a major cardiac differentiation marker

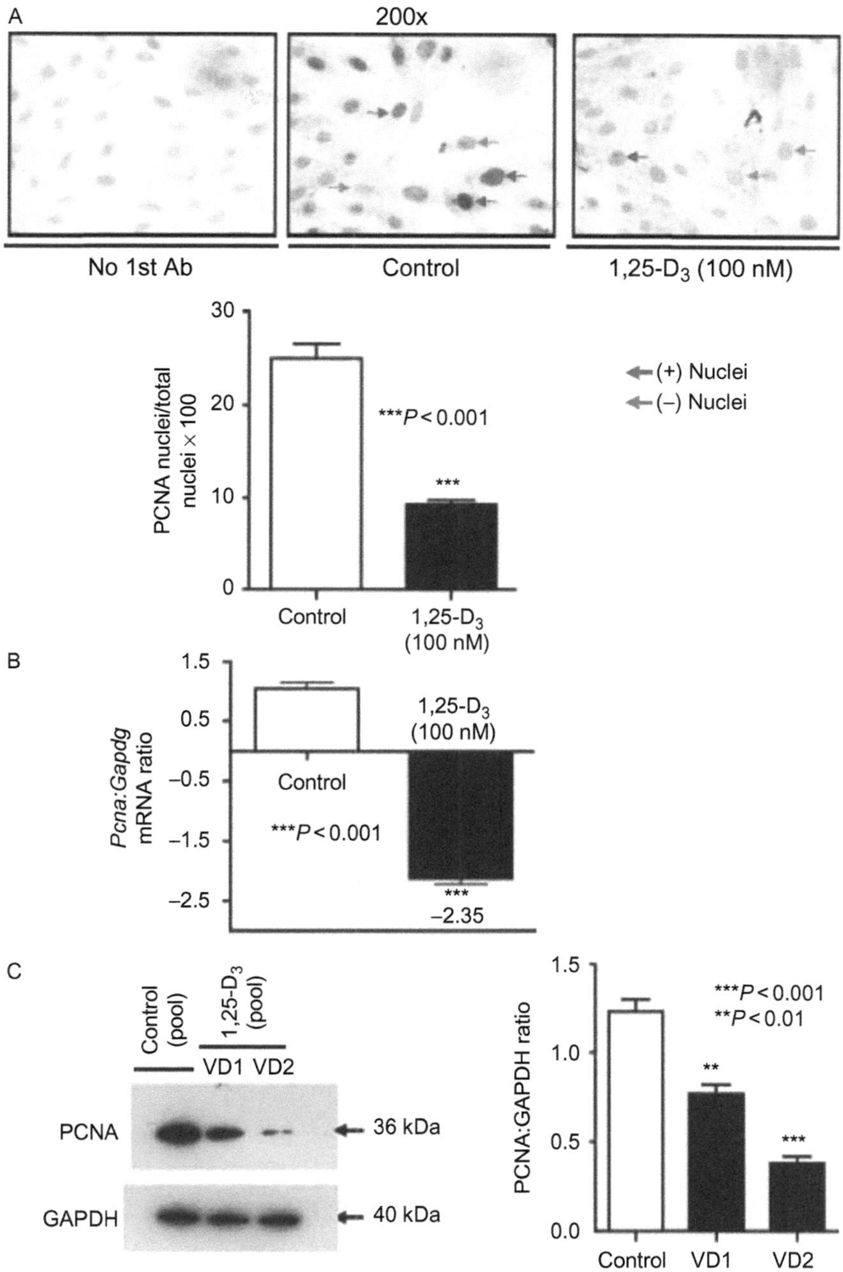

Figure 9 1,25-D_3 downregulates the expression of PCNA. Cultures of H9c2 cells were treated as illustrated in Fig. 8 for 4 days. Immunocytochemistry (ICC) reactions, real-time PCR, and western blotting were carried out at the end of the incubation period. Arrows indicate nuclei positive for PCNA. (A) Representative ICC pictures with the corresponding image analysis to determine the percentages of positive cells (brown nuclear staining) for experiments done in triplicate, $^{***}P < 0.001$. Magnification 200×.
(Continued)

(Hliang et al., 2014). Figure 10 clearly illustrates that the cell areas of cardiomyotubes treated with 1,25-D_3 are larger than those of the control, with increasing concentrations of 1,25-D_3 correlating with increasing cell areas. It also shows cardiomyotubes treated with 1,25-D_3 express more troponin than the control cells. These findings add to the growing evidence that addition of 1,25-D_3 promotes cardiac differentiation.

5.3 1,25-D_3 Induces Cell Cycle Exit of H9c2 Cardiac Cells, Without Inducing Apoptosis

1,25-D_3 blocks the transition of the cell cycle from G1 to S1-phase, and causes accumulation of cells in G1. It induces cell cycle exit by decreasing expression of genes related to the regulation of the cell cycle, such as cyclin A1, D1, D3, C, and E, as well as CDKs such as CDK2 and CDK4 (Hliang et al., 2014). Particularly in the H9c2 cardiac cell model, this cell cycle blocking effect is accompanied by increased expression of p21, a CDK inhibitor and prime candidate for the antiproliferative effect in H9c2 cardiac cells (Hliang et al., 2014). 1,25-D_3 also decreases expression of CHEK1, a protein required for checkpoint-mediated cell cycle arrest in response to DNA damage or presence of unreplicated DNA (Hliang et al., 2014). The cell cycle exit of cardiomyocytes is significant because it is a critical process in cardiac differentiation. The induction of cell cycle exit by 1,25-D_3 provides still more evidence of its role in cardiac differentiation.

5.4 1,25-D_3 Modulates the Expression of Key Components of the Canonical and Noncanonical Wnt Signaling Pathway

The mechanism of cardiac differentiation by 1,25-D_3 involves the modulation of the expression of critical members of the Wnt signaling pathway. Wnt signals are transduced to the canonical pathway for cell fate determination, and to the noncanonical pathway for control of cell movement and tissue polarity (Hliang et al., 2014). Canonical Wnt signaling inhibition has previously been established as essential for cardiogenic activity (Bergmann, 2010; Lanier et al., 2012). Inhibitors of the canonical Wnt

Figure 9—Cont'd (B) Steady-state mRNA downregulation of PCNA. Mean ± SEM corresponds to experiments done in triplicate, $***P < 0.001$. (C) Protein extracts were subjected to western blotting and the corresponding image analysis. Mean ± SEM corresponds to experiments done in triplicates, $**P < 0.01$ and $**P < 0.001$. Control, VD1, and VD2 are different pools of two samples each. In both cases, real-time PCR and western blotting, samples and controls were normalized to the GAPDH housekeeping gene (Hliang et al., 2014). (See the color plate.)

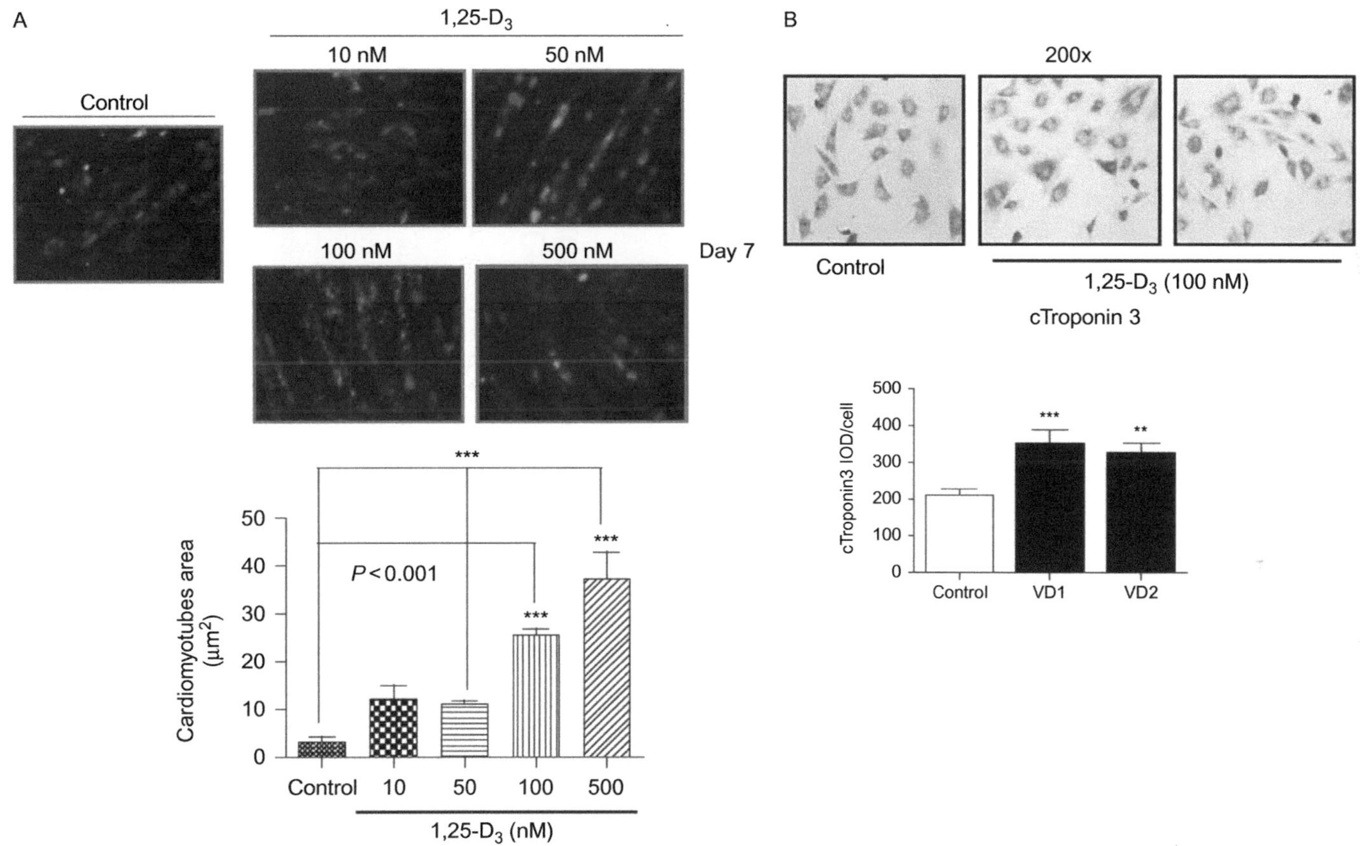

Figure 10 See figure legend on next page.

signaling pathway have potently promoted cardiomyocyte differentiation from human embryonic stem-cell-derived mesoderm (Willems et al., 2011).

Within the H9c2 cardiomyocytes, 1,25-D_3 induces expression of casein kinase-1-α1, a negative regulator of the canonical Wnt signaling pathway (Hliang et al., 2014). Inhibition of the canonical Wnt signaling pathway reverts the process of β-catenin accumulation, translocation, and activation of genes required for cell cycle progression (Hliang et al., 2014). Because Wnt/beta-catenin signaling negatively regulates cardiomyocyte differentiation, inhibition of canonical Wnt signaling at this stage decreases cell proliferation and, more importantly, increases cardiomyocyte differentiation.

Within the H9c2 cardiomyocytes, 1,25-D_3 also increases expression of noncanonical Wnt11, which has been demonstrated to induce cardiac differentiation during embryonic development, in adult cells, and in bone marrow mononuclear cells (Flaherty & Dawn, 2008; Flaherty, Kamerzell, & Dawn, 2012; Flaherty et al., 2008; Hliang et al., 2014; Zhang et al., 2012). Although initial cardiac specification requires balanced expression of both canonical and noncanonical Wnt signaling, committed cells, such as the H9c2 cardiomyocytes, require canonical inhibition for cardiac specification (Eisenberg & Eisenberg, 2006; Ueno et al., 2007).

6. REGULATION OF CARDIAC FUNCTION THROUGH INHIBITION OF WNT SIGNALING PATHWAY

Increasing evidence supports the importance of the Wnt signaling pathway in regulation of cardiac function, as well as cardiac differentiation, and there are studies documenting its pivotal role in adult cardiac hypertrophy and remodeling (Hliang et al., 2014). Normal adult cardiomocytes involve

Figure 10 1,25-D_3 increases cardiomyotubes area and the expression of cardiac troponin. (A) Cultures of H9c2 cells preincubated with PKH2 Green Fluorescence Cell Linker and counterstained with DAPI were treated as illustrated in Fig. 8 for 7 days. Representative pictures are presented for control (No 1,25-D_3 addition) and for increasing the concentrations (10–500 nM) of 1,25-D_3 with the corresponding image analysis of the cardiomyotube area. ***$P < 0.001$. Magnification 200 ×. In a parallel experiment, cultures of H9c2 cells incubated on eight-well chamber slides were treated as described in Fig. 8 for 7 days. At the end of the incubation period, the cells were fixed and subjected to ICC. (B) Representative ICC pictures of cardiac Troponin + cells with the corresponding image analysis expressing percentage IOD (area × intensity) for experiments done in triplicate. **$P < 0.01$ and ***$P < 0.001$. Magnification 200 × (Hliang et al., 2014). (See the color plate.)

quiescent Wnt/FZD signaling, but the pathway becomes reactivated in disease states, such as cardiac hypertrophy. Cardiac hypertrophy is characterized by an increase in cell size and involves protein synthesis, fibrosis, and upregulation of a fetal gene expression pattern. Later stages of hypertrophy can lead to maladaptive remodeling and heart failure (Hliang et al., 2014; Rohini, Agrawal, Koyani, & Singh, 2010). Inhibition of Wnt signaling by 1,25-D_3 attenuates the hypertrophic response. Additionally, inhibition of nuclear β-catenin signaling downstream of the canonical Wnt pathway significantly reduces post-infarct mortality and function decline of LV following chronic left anterior descending coronary artery ligation (Bergmann, 2010; Hliang et al., 2014). The Wnt pathway is also well established to be involved in the fibrosis of several other organ systems, such as lung, kidney, and liver (Hliang et al., 2014). Cardiac fibrosis, which can result from cardiac diseases like congestive heart failure or acute myocardiac infarction, can impair cardiac relaxation, causing diastolic dysfunction and potentially heart failure, and can also obstruct electrical wave propagation, causing arrhythmias (Hliang et al., 2014). Inhibition of the canonical Wnt signaling pathway by 1,25-D_3 could improve the fibrotic process.

7. CONCLUSION AND FUTURE DIRECTIONS

The supplementation of 1,25-D_3 to cardiomyocytes and examination of its effects have contributed a deeper understanding of the novel role of calcitriol in cardiac differentiation. Upon analysis of the experimental results, we conclude that calcitriol is demonstrated to promote cardiac differentiation, primarily through negative modulation of the canonical Wnt signaling pathway and upregulation of noncanonical Wnt11 expression. Increasing experimental and clinical studies continue to add to the growing body of evidence supporting this conclusion. Future work studying vitamin D repletion will hopefully lead to effective strategies for the treatment and potentially prevention of CVD, as well as other related cardiac conditions, in part by promoting effective cardiac differentiation. Targeting or modulating the Wnt signaling pathway with vitamin D may prove to be a suitable, natural, and cost-effective therapeutic method that deserves further investigation.

REFERENCES

Artaza, J. N., Contreras, S., Garcia, L. A., Mehrotra, R., Gibbons, G., Shohet, R., et al. (2011). Vitamin D and cardiovascular disease: Potential role in health disparities. *Journal of Health Care for the Poor and Underserved*, *22*(4 Suppl), 23–38.

Artaza, J. N., Mehrotra, R., & Norris, K. C. (2009). Vitamin D and the cardiovascular system. *Clinical Journal of the American Society of Nephrology*, *4*(9), 1515–1522.
Artaza, J. N., Sirad, F., Ferrini, M. G., & Norris, K. C. (2010). 1,25$(OH)_2$ vitamin D_3 inhibits cell proliferation by promoting cell cycle arrest without inducing apoptosis and modifies cell morphology of mesenchymal multipotent cells. *Journal of Steroid Biochemistry and Molecular Biology*, *119*, 73–83.
Assalin, H. B., Rafacho, B. P., dos Santos, P. P., Ardisson, L. P., Roscani, M. G., Chiuso-Minicucci, F., et al. (2013). Impact of the length of vitamin D deficiency on cardiac remodeling. *Circulation. Heart Failure*, *6*, 809–816.
Bedada, F. B., Technau, A., Ebelt, H., Schulze, M., & Braun, T. (2005). Activation of myogenic differentiation pathways in adult bone marrow-derived stem cells. *Molecular and Cellular Biology*, *25*, 9509–9519.
Bergmann, M. W. (2010). Wnt signaling in adult cardiac hypertrophy and remodeling: Lessons learned from cardiac development. *Circulation Research*, *107*, 1198–1208.
Dong, Y., Stallmann-Jorgensen, I. S., Pollock, N. K., Harris, R. A., Keeton, D., Huang, Y., et al. (2010). A 16-week randomized clinical trial of 2000 international units daily vitamin D3 supplementation in black youth: 25-hydroxyvitamin D, adiposity, and arterial stiffness. *Journal of Clinical Endocrinology and Metabolism*, *95*(10), 4584–4591.
Donneyong, M. M., Hornung, C. A., Taylor, K. C., Baumgartner, R. N., Myers, J. A., Eaton, C. B., et al. (2015). Risk of heart failure among postmenopausal women: A secondary analysis of the randomized trial of vitamin D plus calcium of the women's health initiative. *Circulation. Heart Failure*, *8*(1), 49–56.
Eisenberg, C. A., & Eisenberg, L. M. (1999). Wnt11 promotes cardiac tissue formation of early mesoderm. *Developmental Dynamics*, *216*, 45–58.
Eisenberg, L. M., & Eisenberg, C. A. (2006). Wnt signal transduction and the formation of the myocardium. *Developmental Biology*, *293*, 305–315.
Eisenberg, C. A., Gourdie, R. G., & Eisenberg, L. M. (1997). Wnt-11 is expressed in early avian mesoderm and required for the differentiation of the quail mesoderm cell line QCE-6. *Development*, *124*, 525–536.
Flaherty, M. P., Abdel-Latif, A., Li, Q., Hunt, G., Ranjan, S., Ou, Q., et al. (2008). Noncanonical Wnt11 signaling is sufficient to induce cardiomyogenic differentiation in unfractionated bone marrow mononuclear cells. *Circulation*, *117*, 2241–2252.
Flaherty, M. P., & Dawn, B. (2008). Noncanonical Wnt11 signaling and cardiomyogenic differentiation. *Trends in Cardiovascular Medicine*, *18*, 260–268.
Flaherty, M. P., Kamerzell, T. J., & Dawn, B. (2012). Wnt signaling and cardiac differentiation. *Progress in Molecular Biology and Translational Science*, *111*, 153–174.
Gardner, D. G., Chen, S., & Glenn, D. J. (2013). Vitamin D and the heart. *American Journal of Physiology. Regulatory, Integrative and Comparative Physiology*, *305*, R969–R977.
Gepner, A. D., Ramamurthy, R., Krueger, D. C., Korcarz, C. E., Binkley, N., & Stein, J. H. (2012). A prospective randomized controlled trial of the effects of vitamin D supplementation on cardiovascular disease risk. *PLoS One*, 7(5), e36617.
Go, A. S., Mozaffarian, D., Roger, V. L., Benjamin, E. J., Berry, J. D., Blaha, M. J., et al. (2014). Heart disease and stroke statistics—2014 update: A report from the American heart association. *Circulation*, *129*(3), e28–e292.
Heisenberg, C. P., Tada, M., Rauch, G. J., Saude, L., Concha, M. L., Geisler, R., et al. (2000). Silberblick/Wnt11 mediates convergent extension movements during zebrafish gastrulation. *Nature*, *405*, 76–81.
Hliang, S. M., Garcia, L. A., Contreras, J. R., Norris, K. C., Ferrini, M. G., & Artaza, J. N. (2014). 1,25-Vitamin D_3 promotes cardiac differentiation through modulation of the Wnt signaling pathway. *Journal of Molecular Endocrinology*, *53*, 303–317.

Kispert, A., Vainio, S., Shen, L., Rowitch, D. H., & McMahon, A. P. (1996). Proteoglycans are required for maintenance of Wnt-11 expression in the ureter tips. *Development*, *122*, 3627–3637.

Kuhl, M., Sheldahl, L. C., Malbon, C. C., & Moon, R. T. (2000). Ca^{2+}/calmodulin-dependent protein kinase II is stimulated by Wnt and Frizzled homologs and promotes ventral cell fates in Xenopus. *Journal of Biological Chemistry*, *275*, 12701–12711.

Lanier, M., Schade, D., Willems, E., Tsuda, M., Spiering, S., Kalisiak, J., et al. (2012). Wnt inhibition correlates with human embryonic stem cell cardiomyogenesis: A structure-activity relationship study based on inhibitors for the Wnt response. *Journal of Medicinal Chemistry*, *55*, 697–708.

Logan, C. Y., & Nusse, R. (2004). The Wnt signaling pathway in development and disease. *Annual Review of Cell and Developmental Biology*, *20*, 781–810.

Lutsey, P. L., Michos, E. D., Misialek, J. R., Pankow, J. S., Loehr, L., Selvin, E., et al. (2015). Race and vitamin D binding protein gene polymorphisms modify the association of 25-hydroxyvitamin D and incident heart failure: The ARIC (atherosclerosis risk in communities) study. *JACC. Heart Failure*, *3*(5), 347–356.

Manson, J. E., & Bassuk, S. S. (2015). Vitamin D research and clinical practice: At a crossroads. *Journal of the American Medical Association*, *313*(13), 1311–1312.

Martins, D., Meng, Y. X., Tareen, N., Artaza, J., Lee, J. E., Farodolu, C., et al. (2014). The effect of short term vitamin D supplementation on the inflammatory and oxidative mediators of arterial stiffness. *Health*, *6*(12), 1503–1511.

Martins, D., Wolf, M., Pan, D., Zadshir, A., Tareen, N., Thadhani, R., et al. (2007). Prevalence of cardiovascular risk factors and the serum levels of 25-hydroxyvitamin D in the United States: Data from the third national health and nutrition examination survey. *Archives of Internal Medicine*, *167*, 1159–1165.

Miller, J. R., Hocking, A. M., Brown, J. D., & Moon, R. T. (1999). Mechanism and function of signa transduction by the Wnt/beta-catenin and Wnt/Ca^{2+} pathways. *Oncogene*, *18*, 7860–7872.

Nibbelink, K. A., Tishkoff, D. X., Hershey, S. D., Rahman, A., & Simpson, R. U. (2007). $1{,}25(OH)_2$ Vitamin D_3 actions on cell proliferation, size, gene expression, and receptor localization, in the HL-1 cardiac myocyte. *Journal of Steroid Biochemistry and Molecular Biology*, *103*(3–5), 533–537.

Norman, P. E., & Powell, J. T. (2014). Vitamin D and cardiovascular disease. *Circulation Research*, *114*, 379–393.

Pandur, P., Lasche, M., Eisenberg, L. M., & Kuhl, M. (2002). Wnt-11 activation of a non-canonical Wnt signaling pathway is required for cardiogenesis. *Nature*, *418*, 636–641.

Rohini, A., Agrawal, N., Koyani, C. N., & Singh, R. (2010). Molecular targets and regulators of cardiac hypertrophy. *Pharmacological Research*, *61*, 269–280.

Scragg, R., Jackson, R., Holdaway, I. M., Lim, T., & Beaglehole, R. (1990). Myocardial infarction is inversely associated with plasma 25-hydroxyvitamin D3 levels: A community-based study. *International Journal of Epidemiology*, *19*, 559–563.

Shoben, A. B., Rudser, K. D., de Boer, I. H., Young, B., & Kestenbaum, B. (2008). Association of oral calcitriol with improved survival in nondialyzed CKD. *Journal of the American Society of Nephrology*, *19*, 1613–1619.

Terami, H., Hidaka, K., Katsumata, T., et al. (2004). Wnt11 facilitates embryonic stem cell differentiation to Nkx2,5-positive cardiomyocytes. *Biochemical and Biophysical Research Communications*, *325*, 968–975.

Ueno, S., Weidinger, G., Osugi, T., et al. (2007). Biphasic role for Wnt/beta-catenin signaling in cardiac specification in zebrafish and embryonic stem cells. *Proceedings of the National Academy of Sciences of the United States of America*, *104*, 9685–9690.

Vacek, J. L., Vanga, S. R., Good, M., Lai, S. M., Lakkireddy, D., & Howard, P. A. (2012). Vitamin D deficiency and supplementation and relation to cardiovascular health. *American Journal of Cardiology*, *109*(3), 359–363.

Vanga, S. R., Good, M., Howard, P. A., & Vacek, J. L. (2010). Role of vitamin D in cardiovascular health. *American Journal of Cardiology*, *106*, 798–805.

Willems, E., Spiering, S., Davidovics, H., Lanier, M., Xia, Z., Dawson, M., et al. (2011). Small-molecule inhibitors of the Wnt pathway potently promote cardiomyocytes from human embryonic stem cell-derived mesoderm. *Circulation Research*, *109*, 360–364.

Wolf, M., Betancourt, J., Chang, Y., Shah, A., Teng, M., Tamez, H., et al. (2008). Impact of activated vitamin D and race on survival among hemodialysis patients. *Journal of the American Society of Nephrology*, *19*, 1379–1388.

Wood, A. D., Secombes, K. R., Thies, F., Aucott, L., Black, A. J., Mavroeidi, A., et al. (2012). Vitamin D3 supplementation has no effect on conventional cardiovascular risk factors: A parallel-group, double-blind, placebo-controlled RCT. *Journal of Clinical Endocrinology and Metabolism*, *97*(10), 3557–3568.

Yiu, Y. F., Yiu, K. H., Siu, C. W., Chan, Y. H., Li, S. W., Wong, L. Y., et al. (2013). Randomized controlled trial of vitamin D supplement on endothelial function in patients with type 2 diabetes. *Atherosclerosis*, *227*(1), 140–146.

Zhang, Z., Li, H., Ma, Z., Feng, J., Gao, P., Dong, H., et al. (2012). Efficient cardiomyogenic differentiation of bone marrow mesenchymal stromal cells by combination of Wnt11 and bone morphogenetic protein 2. *Experimental Biology and Medicine*, *237*, 768–776.

CHAPTER THIRTEEN

Vitamin D in Prostate Cancer

Jungmi Ahn*, Sulgi Park*, Baltazar Zuniga*,†, Alakesh Bera*, Chung Seog Song*, Bandana Chatterjee*,‡,1

*Department of Molecular Medicine/Institute of Biotechnology, The University of Texas Health Science Center at San Antonio, Texas Research Park, San Antonio, Texas, USA
†The University of Texas at Austin, Austin, Texas, USA
‡South Texas Veterans Health Care System, Audie L Murphy VA Hospital, San Antonio, Texas, USA
[1]Corresponding author: e-mail address: chatterjee@uthscsa.edu

Contents

Abstract

Metastatic castration-resistant prostate cancer (mCRPC) is a progressive, noncurable disease induced by androgen receptor (AR) upon its activation by tumor tissue androgen, which is generated from adrenal steroid dehydroepiandrosterone (DHEA) through intracrine androgen biosynthesis. Inhibition of mCRPC and early-stage, androgen-dependent prostate cancer by calcitriol, the bioactive vitamin D_3 metabolite, is amply documented in cell culture and animal studies. However, clinical trials of calcitriol or synthetic analogs are inconclusive, although encouraging results have recently

Vitamins and Hormones, Volume 100
ISSN 0083-6729
http://dx.doi.org/10.1016/bs.vh.2015.10.012

emerged from pilot studies showing efficacy of a safe-dose vitamin D_3 supplementation in reducing tumor tissue inflammation and progression of low-grade prostate cancer. Vitamin D-mediated inhibition of normal and malignant prostate cells is caused by diverse mechanisms including G1/S cell cycle arrest, apoptosis, prodifferentiation gene expression changes, and suppressed angiogenesis and cell migration. Biological effects of vitamin D are mediated by altered expression of a gene network regulated by the vitamin D receptor (VDR), which is a multidomain, ligand-inducible transcription factor similar to AR and other nuclear receptors. AR-VDR cross talk modulates androgen metabolism in prostate cancer cells. Androgen inhibits vitamin D-mediated induction of CYP24A1, the calcitriol-degrading enzyme, while vitamin D promotes androgen inactivation by inducing phase I monooxygenases (e.g., CYP3A4) and phase II transferases (e.g., SULT2B1b, a DHEA-sulfotransferase). CYP3A4 and SULT2B1b levels are markedly reduced and CYP24A1 is overexpressed in advanced prostate cancer. In future trials, combining low-calcemic, potent next-generation calcitriol analogs with CYP24A1 inhibition or androgen supplementation, or cancer stem cell suppression by a phytonutrient such as sulfarophane, may prove fruitful in prostate cancer prevention and treatment.

ABBREVIATIONS

AR androgen receptor
CYP cytochrome P450, 1α,25-D_3, 1α,25-dihydroxy vitamin D_3
DBD DNA-binding domain
LBD ligand-binding domain
mCRPC metastatic castration-resistant prostate cancer
NR nuclear receptor
NTD N-terminal domain
VDR vitamin D receptor

1. INTRODUCTION

Biologically active vitamin D known as calcitriol (1α,25-dihydroxy vitamin D_3) is a secosteroid, best characterized for its essential endocrine role in bone mineralization, which is a consequence of the regulation of calcium and phosphate homeostasis by this hormone (DeLuca, 2014; Feldman, Krishnan, Swami, Giovannucci, & Feldman, 2014). Vitamin D deficiency causes softening of bone from insufficient mineralization, which manifests as rickets in children and osteomalacia in adults. Beyond bone health, a broad range of physiological processes including inflammation, angiogenesis, apoptosis, differentiation, and cell growth and proliferation are influenced by vitamin D's autocrine and paracrine actions on extraskeletal tissues. The antiproliferative effect of calcitriol has been demonstrated in cell culture and *in vivo* in animal models of various cancers including colon, breast, and prostate cancer (Deeb, Trump, & Johnson, 2007; Feldman et al., 2014). Cell growth inhibition of normal and malignant prostate cells by vitamin D has been linked to diverse mechanisms including G1 → S cell cycle arrest, DNA damage reduction, microRNA regulation, apoptosis induction, and prodifferentiation changes. Calcitriol inhibits primary prostate cancer cells isolated from clinical specimens, cell lines from prostate cancer, and tumor xenografts of prostate cancer. However, vitamin D's clinical benefit is uncertain, since a high serum vitamin D status is weakly linked to reduced prostate cancer risk in epidemiologic research, and clinical trials of vitamin D are inconclusive (Feldman et al., 2014). Reduction of tumor burden by vitamin D alone is untenable since calcitriol or its analog at a supraphysiologic clinical dose induces hypercalcemia, which elevates risks for cardiovascular disease and lethal prostate cancer (Datta & Schwartz, 2012; Okamoto et al., 2012). Nevertheless, a recent pilot study shows that progression of low-grade prostate cancer is prevented or even reversed by long-term, safe-dose vitamin D supplementation (Marshall et al., 2012).

Genomic action of vitamin D is mediated by the cognate nuclear vitamin D receptor (VDR), which is a ligand-inducible transcription factor (TF) (Carlberg, 2014; Haussler, Jurutka, Mizwicki, & Norman, 2011; Pike, Lee, & Meyer, 2014). The ligand-activated VDR, in association with coregulators, mediates transcriptional induction or repression of VDR/vitamin D target genes. Rapid, nongenomic vitamin D action (induced within several minutes) involving a membrane-bound VDR, which is unrelated to the nuclear VDR, has also been described (Haussler et al., 2011). Calcitriol can reduce DNA damage from thymine dimers caused by UV radiation, thus providing protection against sunlight-induced skin cancer. This DNA damage regulation involves nongenomic vitamin D action and requires interaction of the endoplasmic reticulum stress protein 57 (ERP57) with the membrane-bound VDR (Sequeira et al., 2012).

The androgen receptor (AR) plays a central role in prostate cancer development and progression. Adenocarcinoma of the prostate gland is a leading cause of cancer death in men worldwide, and other than skin cancer it is the most frequently diagnosed cancer in males of the Western society. American Cancer Society estimates that in 2015 prostate cancer will claim roughly 27,500 lives in the USA and more than 220,000 new cases will be diagnosed.

Prostate is also the primary site for sarcoma and carcinomas of neuroendocrine cells, small cells, transitional cells which, unlike adenocarcinoma, are extremely rare. Radiation and radical prostatectomy, individually or in combination, is the standard-of-care for gland-localized primary prostate cancer, and androgen deprivation therapy (ADT), which induces apoptosis for androgen-dependent prostate cancer cells, is used against locally invasive prostate cancer. ADT, which entails depletion of serum androgen to a castrate level via pharmacologic or surgical intervention, may be combined with AR antagonists (such as Casodex) for complete blockade of the androgen axis. A recent report on ~900 patients receiving ADT shows that for about 70% patients, cancer progressed within 20 months (Harshman et al., 2015). An overarching problem in prostate cancer management is that due to the lack of suitable biological and pathological markers, indolent tumors cannot be distinguished from aggressive tumors at initial diagnosis.

Metastatic castration-resistant prostate cancer (mCRPC), a noncurable terminal condition, is primarily driven by restored AR activity. Reactivation of AR occurs by several mechanisms—most prominently due to AR overexpression, ligand-independent activity of AR splice variants, and intracrine production of testosterone and 5α-dihydrotestosterone (androgens) *de novo* in tumor tissue (Mostaghel & Nelson, 2008; Scher, Buchanan, Gerald, Butlerand, & Tilley, 2004; Titus, Schell, Li, Tomer, & Mohler, 2005). Inhibition of reactivated AR accounts for the efficacy of second-generation AR antagonists (such as enzalutamide) and androgen biosynthesis blockers (such as abiraterone acetate) in inhibiting mCRPC progression for additional 4–5 months (de Bono et al., 2011; Mostaghel, 2014; Tran et al., 2009). Cross talk of the VDR and AR pathway contributes to prostate cancer inhibition by vitamin D in experimental models (Wang & Tenniswood, 2014). Detailed insights into the interplay of these two nuclear receptor (NR) pathways in castration-resistant prostate cancer are expected to identify novel approaches for controlling mCRPC.

This review covers an overview of (i) enzymes directing vitamin D biosynthesis and degradation and their relevance to prostate cancer; (ii) VDR as a ligand-inducible TF, its functional domains, its recognition of vitamin D response DNA element (VDRE), and domain-induced allostery; (iii) vitamin D-mediated regulation of cellular processes relevant to tumor growth inhibition; (iv) preclinical studies and clinical trials of calcitriol and analogs; (v) tumor tissue VDR levels and risks of lethal prostate cancer. Finally, vitamin D action in the context of VDR–AR cross talk and intracrine androgen metabolism in prostate cancer will be discussed.

2. VITAMIN D METABOLISM: SYNTHESIS, DEGRADATION, RELEVANCE TO PROSTATE CANCER

Vitamin D refers to molecules with a steroid-like structure (secosteroid) that serve as prohormones for the bioactive hormone calcitriol. Vitamin D_2 (ergocalciferol) produced in plants and vitamin D_3 (cholecalciferol), synthesized in the skin exposed to sun's ultraviolet radiation, are the two physiologically relevant vitamin D forms in humans, D_2 being a weaker prohormone than D_3. Besides natural production, dietary products (D_3-fortified dairy products, fatty fish, fish liver oil, and eggs) and D_3 supplement are the other two sources for D_3 intake. The vitamin D status of an individual is indicated by serum levels of the 25-hydroxy-D_2/D_3 (collectively 25(OH)D), which are produced predominantly in the liver by enzymatic hydroxylation of D_2 and D_3. Bioactivation of the circulating 25-hydroxy D to the potent hormone $1\alpha,25(OH)_2D_3$ (specified as calcitriol) entails additional hydroxylation at the carbon-1 position. Calcitriol biosynthesis occurs predominantly in the kidneys, although production of $1\alpha,25(OH)_2D_3$ can also occur in a number of extrarenal tissues including prostate (DeLuca, 2014; Feldman et al., 2014).

Vitamin D deficiency is prevalent among African Americans and higher incidences of aggressive prostate cancer and higher mortality rates from this malignancy are observed in men from this race group. Epidemiologic data in some examples have shown an inverse correlation between serum vitamin D and prostate cancer risks, although the data are inconsistent with several other population-based studies (Feldman et al., 2014; Li et al., 2007). Current thinking is that an intermediate range of vitamin D status may be optimal—both high and low serum vitamin D may be adversely linked to prostate cancer risks and prostate cancer progression (Albanes et al., 2011). The serum 25(OH) vitamin D level at 50–70 ng/ml range is recommended for the benefit of bone health; whether the same level of serum vitamin D would be beneficial against prostate cancer risks has not been settled (Feldman et al., 2014).

2.1 Enzymatic Machinery for Vitamin D Biosynthesis and Degradation

Vitamin D synthesis in skin starts from the conversion of 7-dehydrocholesterol to previtamin D_3 by ultraviolet B (UVB) at the 282–310 nm radiation of the solar energy spectrum. Previtamin D_3 undergoes temperature-dependent isomerization to vitamin D_3. As noted above, vitamin D_3 is sequentially hydroxylated first to 25-hydroxy D_3 in the liver by the microsomal CYP2R1 (Zhu, Ochalek, Kaufmann, Jones, & DeLuca, 2013) and next, in the kidney proximal tubule, to the active metabolite 1α,25-dihydroxy vitamin D_3 (abbreviated 1,25-D_3) by the mitochondrial CYP27B1 (DeLuca, 2014). Figure 1 schematically shows biosynthesis of functional vitamin D, its degradation by CYP24A1, and its autoregulation and regulation by other factors.

An inherited loss-of-function mutation of CYP2R1 is associated with low circulating vitamin D and classic symptoms of vitamin D deficiency (Al Mutair, Nasrat, & Russell, 2012). A critical role of CYP27B1 in

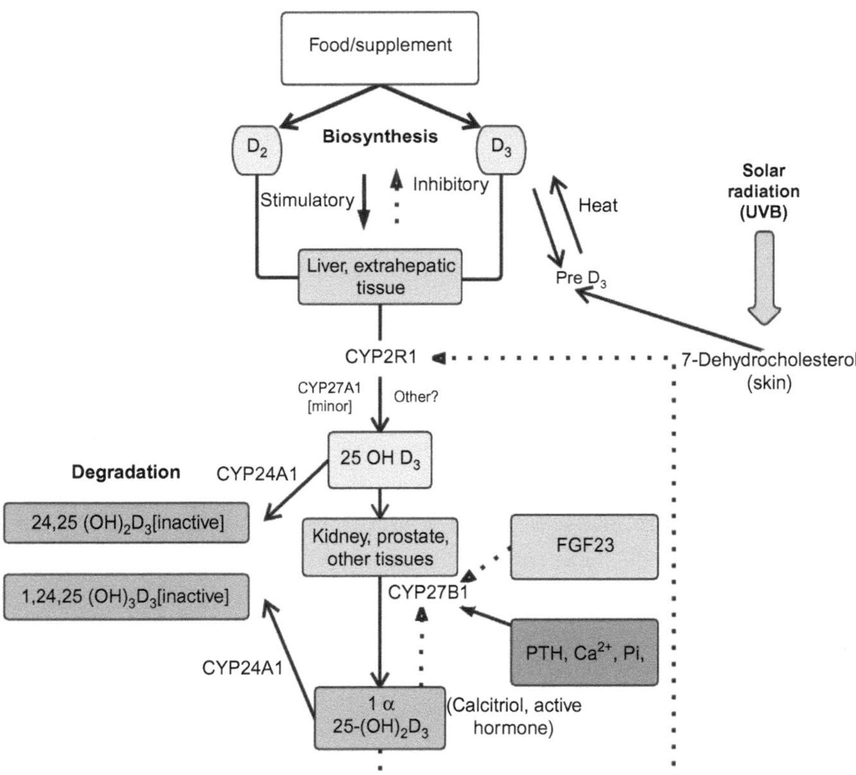

Figure 1 Schema showing enzyme-catalyzed pathways to the biosynthesis and degradation of calcitriol. Regulation of CYP2R1 and CYP27B1 by calcitriol (1α,25-dihydroxy vitamin D_3), parathyroid hormone (PTH), phosphorous (Pi), calcium (Ca^{2+}), and fibroblast growth factor-23 (FGF23) is shown. Stimulation is indicated by a solid arrow (⟶); inhibition is shown by a broken arrow (⇢). (See the color plate.)

vitamin D action is evident from patients carrying loss-of-function or reduced function mutations of this enzyme. Null inactivated or reduced CYP27B1 activity causes vitamin D-dependent rickets-type 1, an autosomal recessive disease exhibiting hypocalcemia, hypophosphatemia, and consequent development of fracture-prone soft, weak bone, and bowed legs (Babiker et al., 2014). CYP27B1 expression is transcriptionally upregulated by parathyroid hormone, calcium, phosphorous, and downregulated by 1,25-D_3. CYP27B1 expression is also suppressed by the bone-derived circulating peptide FGF23, which regulates parathyroid hormone secretion (Chanakul et al., 2013). CYP24A1, a mitochondrial 1, 25-dihydroxyvitamin D_3 24-hydroxylase (24-hydroxylase), inactivates calcitriol by converting it to a carbon-24 hydroxylated metabolite. Robust induction of CYP24A1 by calcitriol prevents excessive accumulation of this hormone in target tissues and helps maintain vitamin D homeostasis.

2.2 Vitamin D Metabolism in Prostate Cancer

Prostate cancer cells and prostate tumor tissue express the enzymatic machineries for supporting local synthesis of calcitriol (Chen, 2008; Chen & Holick, 2003; Chen, Sakai, Yamamoto, & Kittaka, 2012; Feldman et al., 2014). The activity of CYP27B1, the major mediator of 25(OH)D bioactivation to calcitriol, is repressed in prostate cancer cells in culture and in prostate tumor tissue, in contrast to the relative abundance of this enzyme in the nonmalignant prostate epithelium (Hsu, Feldman, McNeal, & Peehl, 2001). Epidermal growth factor can upregulate *CYP27B1* gene transcription in nonmalignant prostate epithelial cells, but not in prostate cancer cells (Jamieson, Holick, & Chen, 2004). It has been suggested that CYP27B1 plays a role in the normal regulation of prostate cell growth and dysregulation of CYP27B1 may, in part, contribute to the uncontrolled cell growth and proliferation in prostate cancer (Hsu et al., 2001). Nevertheless, a recent study concluded that cholecalciferol and calcitriol were equally effective in reducing tumor growth in a prostate cancer xenograft model (Swami et al., 2012). This observation indicates that 25-hydroxy-vitamin D_3-1α-hydroxylase activity is present in prostate cancer tissue.

Homeostasis of the vitamin D hormone in prostate is impacted by CYP24A1 expression. Since calcitriol strongly induces CYP24A1, resistance to vitamin D therapy may occur from elevated CYP24A1 expression. In fact, increased CYP24A1 expression correlated with advanced stages of prostate cancer (Tannour-Louet et al., 2014). Thus, combined targeting of prostate tumor by inhibiting CYP24A1 with a small-molecule inhibitor and activating VDR signaling by vitamin D_3 supplementation is potentially a fruitful avenue for therapeutic intervention. Chronic lymphocytic leukemia cells were inhibited when cotreated with a CYP24A1-specific small-molecule inhibitor (such as an imidazole styrylbenzamide) and calcitriol, which induced *GADD45α* (encoding growth arrest and DNA damage-inducible protein) and *CDKN1A* (encoding the cell cycle inhibitor p21; Ferla et al., 2014).

Homeostasis of active vitamin D is also impacted by its phase I oxidation mediated by CYP3A4 and by phase II glucuronidation mediated by 5′-diphosphoglucurosyltransferase (UGT). UGT1A4 and UGT1A3 are involved in the glucuronidation of $25(OH)D_3$. Glucuronide conjugates of $25(OH)D_3$ have been detected in human plasma and bile (Wang et al., 2014). Phase I/phase II modifications, occurring primarily in the liver and intestine, facilitate vitamin D clearance from the body. Since CYP3A4 and UGTs are induced by drug-activated nuclear receptors PXR (pregnane X receptor) and CAR (constitutive androstane receptor; Aleksunes & Klaassen, 2012), prescription drugs and other xenobiotic factors may reduce the serum vitamin D level, which in turn would reduce the availability of vitamin D as an anticancer agent. These confounders should be considered in order to correctly evaluate the efficacy of vitamin D supplementation for chemoprevention and its role as a cancer therapeutic.

3. VDR-REGULATED GENE TRANSCRIPTION: LIGAND SPECIFICITY, DNA RESPONSE ELEMENTS, DOMAIN-INDUCED ALLOSTERY

VDR is a ligand-inducible TF and a member of the NR superfamily, which in the case of humans includes 48 receptors. NR proteins, which regulate nearly all aspects of the physiological processes necessary to create and support life, are defined by a common structural organization containing multiple functional domains (Carlberg, 2014; Evans & Mangelsdorf, 2014; Helsen & Claessens, 2014). VDR (NR1I1) has a short (24 amino acids), unstructured N-terminal domain (NTD), a central DNA-binding domain (DBD) containing two helical zinc finger modules and a carboxyl-terminal ligand-binding domain (LBD). A mostly unstructured hinge domain connecting DBD and LBD allows for structural flexibility of the DNA-bound VDR and its partner retinoid X receptor (RXR). Structural determination of VDR and other NRs by X-ray crystallography and cryoelectron microscopy and in solution, by small-angle X-ray scattering (SAXS), small-angle neutrino scattering (SANS), and hydrogen–deuterium exchange revealed conformational details for each functional domain and for the full-length receptor in the ligand-bound, DNA-associated form (Carlberg & Campbell, 2013; Huang, Chandra, & Rastinejad, 2010; Watson et al., 2013; Zhang et al., 2011).

Nucleus-residing unliganded VDR remains in a chromatin-bound repressed state similar to what was first demonstrated for two other NRs, namely thyroid hormone receptor (TR) and retinoic acid receptor (RAR; Rosenfeld, Lunyak, & Glass, 2006). The unliganded VDR recruits corepressors (such as NCoR1, SMRT/NCoR2), which in turn recruit a histone deacetylase (HDAC) complex—the net effect being generation of a VDR-containing compact chromatin region and gene repression. As an example, our study showed that unliganded VDR suppressed the basal CYP24A1 level in breast cancer cells, and VDR silencing (by siRNA) elevated the basal level of CYP24A1 (Alimirah et al., 2010). The ligand-dependent activation phase of VDR is initiated when change in LBD conformation due to the binding of 1,25-D_3 at the ligand-binding pocket and repositioning of LBD helix-12 create an interaction surface for coregulator exchange, replacing corepressors with coactivators (VaÈisaÈnen, PeraÈkylaÈ, KaÈrkkaÈinen, Steinmeyer, & Carlberg, 2002). Binding of a pioneering factor is thought to open the chromatin region and facilitate VDR binding (Carlberg, 2014). The AF2 activation function at LBD for VDR (and all other NRs) arises from coactivator recruitment to LBD induced by repositioned helix-12. A coactivator of the p160 family (SRC-1/-2/-3) makes physical contact with VDR via an LXXLL motif and also recruits histone acetyltransferases (such as CBP/p300, p/CAF), which mediate acetylation at specific lysine and arginine residues of histone H3 and H4. Chromatin relaxation due to histone acetylation sets in motion the assembly of other classes of coregulator complex including additional histone modifiers (methyltransferase/demethylase, ubiquitin ligase/deubiquitinase, kinase/phosphatase), chromatin remodelers (such as switching defective (SWI)/sucrose nonfermentor (SNF) containing WSTF (Williams Syndrome Transcription Factor) Including Nucleosome Assembly Complex (WINAC) complex, Rosenfeld et al., 2006). DRIP, a VDR-interacting mediator complex, functionally couples VDR-associated coactivators to the regulatory machinery at the transcription start site (TSS), leading to increased RNA polymerase II (Pol II) activity and gene induction. VDREs in *CYP24A1* have been characterized (Carlberg, 2014; Luo et al., 2010). CYP24A1 mRNAs were induced more than 2000-fold when castration-resistant C4-2B prostate cancer cells were treated with EB1089 (Fig. 4).

VDR mediates gene repression through a direct mechanism known as transrepression (Rosenfeld et al., 2006). Reduced *CYP27B1* gene expression by 1,25-D_3 in the kidneys entails association of the ligand-bound VDR with several negative vitamin D response elements (nVDREs; Turunen, Dunlop, Carlberg, & Väisänen, 2007). The *CYP27B1* upstream promoter contains two types of nVDREs—one having sequence organization resembling a positive VDRE with specific affinity for the VDR/RXR dimer in a ligand-dependent manner, and the other with binding specificity for a specific TF, such as VDIR, whose activity is inhibited by tethered VDR or by nVDRE-bound VDR. The latter example of nVDRE bears no similarity to a classical VDRE (Turunen et al., 2007).

3.1 DNA Response Elements

VDR binds a VDRE as a dimer in association with a partner protein. RXR, the NR that binds to 9-*cis* retinoic acid is the partner for dimerization in majority of cases. Nevertheless, another DNA-bound TF may also be the dimer partner for VDR as noted above (Carlberg et al., 1993). Protein–protein interaction at the interface of VDR and RXR DBDs stabilizes the dimer, with further stabilization arising from the interaction of LBDs of VDR and RXR. Dimerization stabilizes VDR–DNA interaction by circumventing the otherwise low DNA-binding affinity of a VDR monomer (Helsen & Claessens, 2014). The classical VDRE for the VDR/RXR complex contains variations of the sequence 5′-AG(G/T)TCA (consensus half-site sequence), configured as a direct repeat (DR) with variable number of spacer nucleotides (DR-n) separating the AG(G/T)TCA repeat. VDR contacts the 6-base 5′ half-site within the major groove of DNA via the first zinc finger module of DBD (Carlberg & Campbell, 2013). The second zinc finger of VDR associates with the RXR DBD. A short C-terminal extension after the second zinc finger further stabilizes receptor–DNA interaction (Helsen & Claessens, 2014). Genome-wide analysis identified DR3 as the preferred response element for VDR in a chromatin context (Tuoresmäki, Väisänen, Neme, Heikkinen, & Carlberg, 2014). A DR3 element mediates robust VDR-mediated transactivation of *CYP24A1* by 1,25-D_3

(Carlberg & Campbell, 2013). Other DR configurations (most frequently DR4, DR5) are also involved in vitamin D-mediated induction of target genes. A DR4-type nVDRE mediates transrepression of *CCNC*, encoding cyclin C (Carlberg & Campbell, 2013). In our study, we reported the involvement of a DR7-type VDRE in the VDR-mediated induction of *SULT2B1* sulfotransferase in prostate cancer cells (Fig. 2A; Seo et al., 2013). SULT2B1b (hereon referred as SULT2B) is a prostate-expressed cholesterol- and DHEA-sulfotransferase encoded by *SULT2B1*. SULT2B levels are reduced in

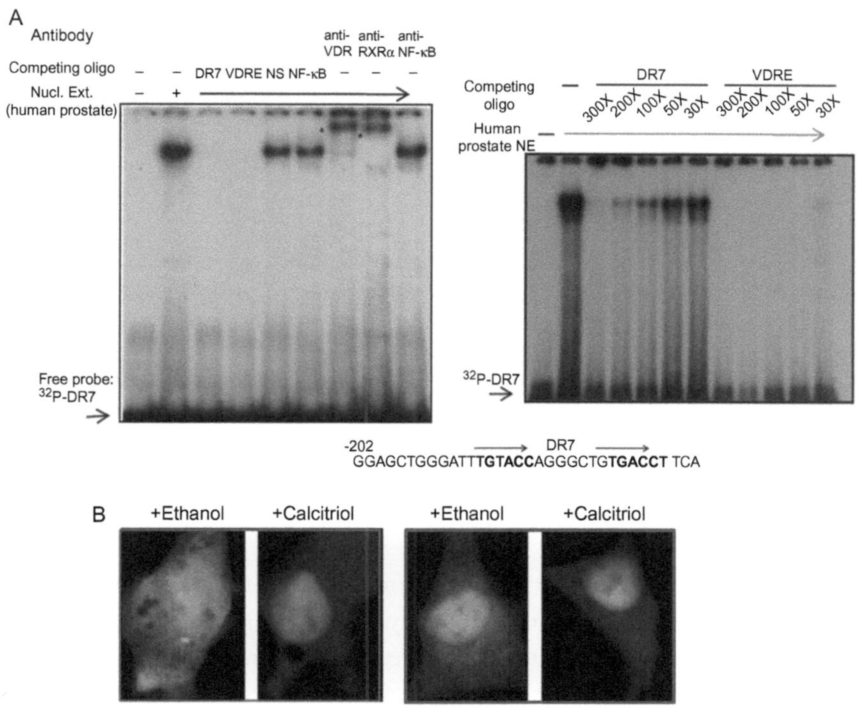

Figure 2 (A) Electrophoretic mobility shift assay (EMSA) showing binding of VDR and RXR-α (present in the nuclear extract of normal human prostate tissue) to the DR7-type element present in and around the -202 nucleotide position of the *SULT2B1* promoter. The promoter sequence containing the DR7-type element is shown. *Left panel*: oligonucleotide competition demonstrates specificity of the EMSA complex; antibody supershift assay demonstrates the presence of VDR and RXR-α within the EMSA complex. *Right panel*: Relative affinities of DR7 and a DR3-type VDRE (from the rat osteocalcin promoter) for the EMSA complex formed by ^{32}P-labeled DR7 element with VDR and RXR-α components of human prostate nuclear extract. These data from our experiments are taken from Seo et al. (2013). (B) Subcellular localization of transfected CFP-VDR in COS1 cells treated with ethanol or 10 nM calcitriol. Representative data from two separate transfection experiments are shown. CFP, cyan fluorescent protein. (See the color plate.)

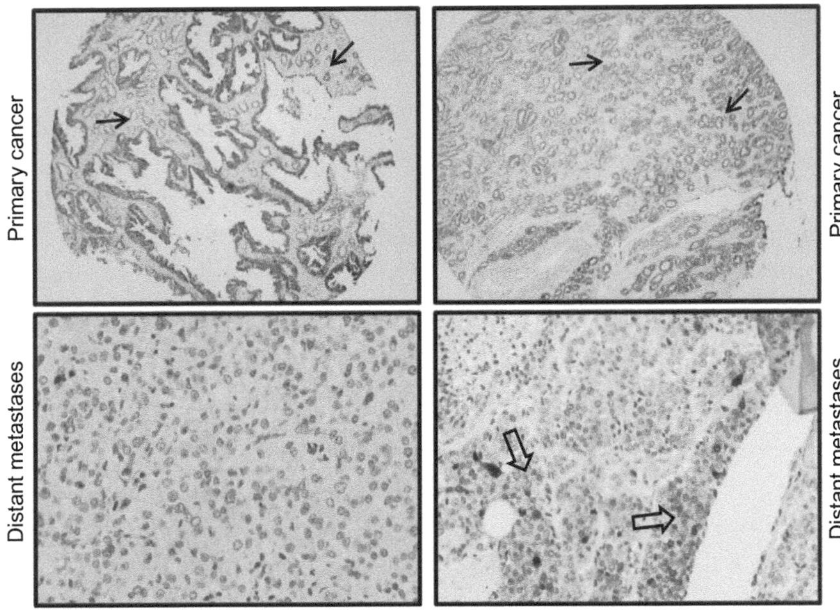

Figure 3 Photomicrographs of SULT2B-immunostained prostate cancer specimens from four cases. *Upper panel*: Primary prostate cancer from two patients; photomicrograph taken at 4 ×. The primary specimen at left shows nonmalignant acini, which stained strongly for SULT2B, as well as cancerous areas (solid arrows), which show markedly reduced SULT2B levels. The tissue core at the right shows only the malignant region of primary cancer (arrows), which stained weakly for SULT2B. *Lower panel*: distant metastases of prostate cancer from two patients; photomicrograph taken at 20 ×. Open arrows show remaining SULT2B expression in small percent of metastatic cancer cells for the second patient. (See the color plate.)

primary prostate cancer and its expression is nondetectable in >90% cases of distant metastasis that we have analyzed (Fig. 3). Consistent with its transactivation role, we found cyan fluorescent protein-labeled VDR (CFP-VDR) localized mostly in the nucleus of calcitriol-treated COS-1 cells (monkey kidney derived) that expressed transfected CFP-VDR. In the absence of the ligand, both nuclear and cytoplasmic compartment showed fluorescence indicating nucleocytoplasmic distribution of the receptor (Fig. 2B). The subcellular localization is tissue- and cell type-specific, since in MCF-7 breast cancer cells, unliganded VDR resides in the nucleus (Alimirah et al., 2010).

3.2 Domain-Induced Allostery

VDR-regulated gene activity can be influenced by the chemical structure of the VDR agonist, nucleotide sequence of the DNA response element, and surface topography of NTD—each specifying domain-dependent allostery.

Functional cross talk among various domains of VDR due to allosteric interaction was revealed by structural and biochemical analyses (Huang et al., 2010; Meijsing et al., 2009; Watson et al., 2013; Zhang et al., 2011). Relay of interdomain signal for the VDR/RXR complex was evident from hydrogen–deuterium exchange profiling of the heterodimer. Ligand binding to VDR increased solvent exchange at its DBD indicating that ligand occupancy and the ensuing conformational change of LBD impacted DBD conformation and thus, its affinity for the cognate element (Zhang et al., 2011). This conformational change would potentially alter gene expression.

Influence of the ligand structure on target gene expression is revealed from differential effect of the natural (1,25-D_3) versus synthetic (EB1089) ligand of VDR on the transactivation of androgen metabolism genes in prostate cancer cells (Doherty, Dvorkin, Rodriguez, & Thompson, 2014). The mRNAs of enzymes such as CYP3A4, CYP3A5, and AKR1C3 were induced strongly by EB1089 but not by calcitriol in androgen-dependent prostate cancer cells. On the other hand, calcitriol, not EB1089, induced mRNAs for 17β-hydroxysteroid dehydrogenase-2 and prostaglandin dehydrogenase (Doherty et al., 2014).

Signal transmission from DBD to LBD for DNA-bound VDR has been demonstrated as well (Zhang et al., 2011). Receptor binding to a DR3 element altered the LBD surface for coactivator contact that led to altered SRC1 interaction with VDR and with the VDR/RXR complex (Zhang et al., 2011). An example of NTD–DBD communication comes from our finding, which showed that a polymorphic site (FokI-FF) at the VDR NTD, which deletes three amino acids from the naturally occurring VDR variant, abolished repression of the *CYP24A1* promoter by unliganded VDR (Alimirah et al., 2010).

4. INHIBITION OF PROSTATE CANCER BY VITAMIN D: INSIGHTS FROM CELL CULTURE AND PRECLINICAL STUDIES, AND CLINICAL TRIALS

Growth-inhibitory effects of ligand-activated VDR on normal and malignant epithelial cells from diverse tissues including colon, breast, and prostate have been extensively investigated (Deeb et al., 2007; Feldman et al., 2014; Getzenberg et al., 1997; Salehi-Tabara et al., 2012). Impaired cell growth from vitamin D/VDR action is driven by several distinct mechanisms, such as cell accumulation at G0–G1 and consequent G1 → S cell cycle arrest; induction of differentiation, apoptosis, and inhibition of angiogenesis, cell adhesion, and cell migration (Deeb et al., 2007; Sung & Feldman, 2000). Calcitriol inhibits proliferation of normal prostate epithelial cell line, cells of primary culture from normal and cancerous prostate tissue and prostate cancer cell lines with an androgen-dependent or androgen-independent phenotype. Cells from normal prostate epithelium are more sensitive (approximately by two orders of magnitude) to 1,25-D_3-mediated proliferation inhibition than cells originating from prostate cancer. Higher corepressor levels (NCoR1, NCoR2/SMRT) and lower levels of VDR in prostate cancer cells compared to normal prostate epithelial cells may partly account for differential sensitivity of normal versus cancerous prostate cells to vitamin D-induced inhibition. Furthermore, less aggressive prostate cancer cells (such as androgen-dependent LNCaP cells) are more sensitive to the antiproliferative action of calcitriol than more aggressive prostate cancer cells such as PC3 and DU145. Mechanisms for growth inhibition are diverse and cell-type dependent. G1 → S arrest and apoptosis are induced in androgen-responsive, moderately differentiated LNCaP prostate cancer cells upon treatment with active vitamin D or analogs; but no G0–G1 accumulation occurs for calcitriol-treated PC3 cells and these cells were much less inhibited for proliferation by calcitriol than LNCaP cells (Zhuang & Burnstein, 1998). Calcitriol-treated PC3 and DU145 cells exhibited reduced cell adhesion, migration, and invasion (Sung & Feldman, 2000), and growth factor-stimulated proliferation and invasion of DU145 cells was inhibited by BXL-628, a calcitriol analog (Marchiani et al., 2006). Disparate expression/activity of CYP24A1 and resulting differences in the prostate level of the active vitamin D_3 metabolite also play a role in cell-type dependent variations of the antiproliferative response. Reduced metabolism of 1,25-D_3 in the presence of a CYP24A1 inhibitor elevated VDR signaling in DU145 cells (Yee, Campbell, & Simons, 2006). Clinical resistance to vitamin D therapy may result from augmented metabolism of 1,25-D_3 since *CYP24A1* gene amplification and its enhanced expression in clinical prostate cancer specimens have been reported (Tannour-Louet et al., 2014). Growth-promoting pathways that are known to be inhibited by VDR signaling in prostate cancer cells are described below (Deeb et al., 2007).

4.1 Mechanisms for Antiproliferative Actions

4.1.1 Cell Cycle Arrest

Inhibition of G1 → S cell cycle progression in VDR-activated nonmalignant and malignant prostate epithelial cells is associated with elevated expression of p21/Cip1, p27/Kip1, and p15/INK4b, which are cyclin-dependent kinase inhibitors (CKIs; Boyle, Zhao, Cohen, & Feldman, 2001; Campbell, Elstner, Holden, Uskokovic, & Koeffler, 1997; Moffatt, Johannes, Hedlund, & Miller, 2001; Yang & Burnstein, 2003; Zhuang & Burnstein, 1998). A well-documented consequence of p21 upregulation in calcitriol-treated prostate cancer cells is hypophosphorylation of the retinoblastoma protein pRb that results in the sequestration of E2F TFs and inhibition of E2F target genes that are involved in DNA synthesis, thereby blocking S-phase entry of cells. VDR-regulated overexpression of *CDKN1A* (encoding p21) in RWPE-1 nonmalignant prostate epithelial cells is epigenetically controlled by histone H3 acetylation at lysine-9 (Thorne et al., 2011). *CDKN1A* induction is also associated with stress-induced G1/S cell cycle arrest; in this case, trans-activation of *CDKN1A* is mediated by p53, which is a TF and tumor suppressor.

Induction of *CDKN2B* (encoding p15/INK4b, a CKI of the INK4 family) in normal prostate epithelial cells in response to calcitriol is an indirect effect caused by VDR-stimulated TGFβ1-Smad signaling (Robson, Gnanapragasam, Byrne, Cllins, & Neal, 1999). Activities of cyclin D bound CDK4 and CDK6 are inhibited by p15/INK4b (Sherr & Roberts, 1999), which leads to pRb accumulation in a hypophosphorylated form with a net result of G1 → S block due to E2F sequestration.

Reduction of the cMYC oncoprotein level, and consequent hypophosphorylated pRb and pRb-mediated E2F sequestration, is another mechanism for vitamin D/VDR-mediated G1 → S cell cycle block. Calcitriol-mediated reduction of the cMYC protein level has been reported for nonmalignant prostate cells (RWPE-1), androgen-dependent prostate cancer cells (LNCaP, LAPC4, VCaP), and for androgen-independent C4-2 prostate cancer cells (Rohan & Weigel, 2009; Salehi-Tabara et al., 2012; Washington & Weigel, 2010). cMYC downregulation in this case is a result of reduced cMYC mRNA expression along with destabilization of the cMYC protein (Rohan & Weigel, 2009).

In our study, we observed E2F1 mRNA and protein levels were reduced in calcitriol-treated LNCaP cells and a decline in E2F1 promoter activity accounts for this reduction. A calcitriol-responsive DNA element in the upstream E2F1 promoter binds several regulatory proteins (Ahn et al., manuscript in preparation). Reduction of E2F1 in LNCaP cells in response to calcitriol is consistent with what was reported earlier (Rao et al., 2004; Zhuang & Burnstein, 1998).

VDR-induced p27/Kip1 accumulation in LNCaP cells caused a reduced nuclear CDK2 level (due to extrusion of nuclear CDK2 to the cytoplasm)—the net consequence being inactivation of the cyclin E/CDK2 complex and calcitriol-mediated inhibition of the S-phase entry of LNCaP cells (Yang & Burnstein, 2003). Additionally, p27 was found to be stabilized due to inhibition of its proteasomal degradation caused by reduced p27 phosphorylation at threonine-187, and reduction of the level of Skp2, which is a component of the SKP1/Cullin1/F-box protein (SCF) ubiquitin ligase complex.

Calcitriol may also induce cell cycle arrest by increasing the expression of GADD45α, which is a stress-induced protein. DNA-damaging agents including docetaxel, a chemotherapeutic, induce *GADD45α*. GADD45α disrupts the CDK1/cyclinB1 complex by binding to CDK1 leading to G_2–M blockade. GADD45α also induces G1 → S arrest by binding p21/Cip1; however, the underlying mechanism for G1 → S block has not been determined. Overexpression of GADD45α was found to inhibit cell proliferation (Zhan et al., 1994). GADD45α is downregulated in prostate cancer.

4.1.2 Differentiation

VDR signaling can induce partial differentiation of prostate cancer cells since the gene expression signature of prostate cancer cells incubated with 1,25-D_3 was found to be similar in some respect to that specifying differentiated cells of the prostate epithelium (Campbell et al., 1997; Chen & Holick, 2003; Deeb et al., 2007; Feldman et al., 2014; Guzey, Luo, & Getzenberg, 2004). 1,25-D_3 treatment caused morphologic changes resembling induction of differentiation in the highly metastatic R3327-Mat-LyLu (MLL) Dunning rat prostate cancer cells (Getzenberg et al., 1997). Consistent with findings by others, in unpublished study, we observed that calcitriol treatment of LNCaP cells caused induction of androgen-inducible genes that are expressed in terminally differentiated prostate epithelial cells. Examples include the genes encoding the homeobox protein NKX3.1, prostate-expressed serine proteases of the kallikrein family (KLK2, KLK3 (*aka* PSA), KLK4, KLK15), TMPRSS2, and PMEPA1. We also observed calcitriol-mediated induction of *ID4* which encodes the inhibitor of DNA binding 4. ID4 is a dominant-negative basic helix–loop–helix (bHLH) protein, which promotes normal prostate development and its level is markedly reduced in castration-resistant prostate cancer (Patel et al., 2014). Calcitriol-mediated induction of E-cadherin, an epithelial marker and cell–cell adhesion molecule, was observed in LNCaP and PC3 cells (Campbell et al., 1997). E-cadherin induction by calcitriol caused differentiation of colon cancer cells due to suppression of β-catenin activation (Palmer et al., 2001).

4.1.3 Apoptosis

Several experimental systems demonstrated that 1,25-D_3-induced apoptosis can contribute to growth inhibition of prostate cancer cells (Blutt, McDonnell, Polek, & Weigel, 2000; Murthy, Agoulnik, & Weigel, 2005; Oades, Dredge, Kirby, & Colston, 2002; Saito et al., 2008; Washington & Weigel, 2010). In one study, LNCaP and ALVA-31 cells, which underwent apoptosis upon calcitriol treatment, showed reduced expression of a number of antiapoptotic proteins—Bcl-2, Bcl-XL, MCI-1, BAG1L, XIAP, cIAP1, and cIAP2; the proapoptotic proteins BAX and BAK did not show altered levels. Bcl-2 overexpression prevented calcitriol-induced apoptosis of these cells (Guzey, Kitada, & Reed, 2002). Caspase 3 and caspase 9, respectively the executor and initiator proteases in the mitochondria-dependent pathway to apoptosis, were induced by calcitriol, whereas caspase 8, which acts via the mitochondria-independent apoptotic pathway, was not induced.

Another mechanism for calcitriol-induced apoptosis of prostate cancer cells involves a role for the insulin-like growth factor binding protein-3 (IGFBP3), which is induced by calcitriol due to VDR-mediated transcriptional regulation of the IGFBP3 promoter (Boyle et al., 2001; Kojima et al., 2006). Calcitriol-mediated growth inhibition of LNCaP cells in serum-free media was found to depend on the induction of IGFBP3, which led to elevated expression of p21/Cip (Boyle et al., 2001). Knockdown of IGFBP3 by RNA interference abrogated growth inhibition and p21 accumulation in these cells. However, calcitriol-regulated growth inhibition of LNCaP cells in serum-containing media did not require IGFBP3 induction (Stewart & Weigel, 2005). IGFBP3 overexpression in doxycycline-inducible cells induced apoptosis of LNCaP cells but not in C4-2 cells, although calcitriol can induce IGFBP3 in both LNCaP and C4-2 cells (Kojima et al., 2006). Thus, under certain experimental conditions the proapoptotic activity of IGBP3 caused growth inhibition of prostate cancer cells by vitamin D/VDR action.

The prostaglandin pathway and ERK/MAP kinase pathway, which promote cell survival, are additional mechanisms that lead to prostate cancer inhibition by vitamin D (Krishnan et al., 2007).

4.1.4 Angiogenesis, Cell Migration, Metastasis

Invasive tumor growth and metastasis requires angiogenesis, i.e., formation of new blood vessels from existing vessels—a process that is essential for primary tumor growth, progression to locally invasive carcinoma, and then to metastasis at distant sites. Vitamin D signaling is thought to be active in both endothelial cells and smooth muscle cells of the vasculature since both cell types express VDR. 1,25-D_3 inhibited the angiogenic activity of endothelial cells. In a coculture experiment, interleukin-8 (IL-8), secreted from PC3 and DU145 cells, would normally stimulate endothelial cell migration and tube formation, which are the two critical steps in angiogenesis. Calcitriol inhibited these processes by reducing *IL-8* expression—a consequence of inhibition of NF-κB activity by calcitriol (Bao, Yeh, & Lee, 2006). A role for 1,25-D_3 in inhibiting the angiogenic activity of tumor-derived endothelial cells (TDECs) *in vivo* was demonstrated in an allograft prostate tumor model (Chung et al., 2009). In this study, allograft tumors were produced in VDR-intact and VDR-null mice from prostate cancer cells isolated from tumors of TRAMP (transgenic adenocarcinoma of mouse prostate) mice. Tumors were much larger in VDR-ablated mice at 30-day postinoculation of tumor cells. Furthermore, VDR signaling was detected for TDECs from wild-type but not VDR-null mice, and calcitriol inhibited TDECs originating from wild-type, VDR-expressing mice, not from VDR-null mice. Inhibition of angiogenesis by calcitriol was further supported by the findings that tumors from VDR-ablated mice had (i) larger vascular volume; (ii) enlarged vessels; (iii) less pericyte coverage; (iv) more vascular leakage; and (v) higher levels of hypoxia-inducible factor1α (HIF1α), vascular endothelial growth factor (VEGF), angiopoietin 1, and platelet-derived growth factor (PDGF-BB; Chung et al., 2009). In a second example, when calcitriol or a potent analog (inecalcitol) was administered to mice bearing LNCaP xenograft tumors, reduction of the tumor burden was accompanied by decreased vascularity surrounding the tumor mass and within the tumor tissue (Okamoto et al., 2012).

1,25-D_3 inhibited cell motility and cell invasiveness. Calcitriol reduced the highly invasive phenotype of DU145 cells by suppressing expression of MMP9 and cathepsins, the proteases which promote motility and invasion (Bao et al., 2006). Metastasis of prostate cancer was inhibited by 1,25-D_3 in the syngeneic Dunning prostate tumor model, where tumors were subcutaneously produced in Copenhagen rats from highly aggressive MLL Dunning rat prostate cancer cells. 1,25-D_3 inhibited the growth of MLL tumors and reduced the number and size of lung metastases (Getzenberg et al., 1997).

4.2 Preclinical Studies

A large number of preclinical studies involving prostate tumor xenografts and allografts and prostate cancer in genetically modified mice provided convincing evidence for the efficacy of calcitriol in reducing tumor burden and cancer metastases for androgen-dependent and castration-resistant prostate cancer (Deeb et al., 2007; Feldman et al., 2014). Furthermore, calcitriol acted synergistically with chemotherapeutics (Taxanes, platimum analogs) to reduce prostate tumor burden when administered either as a pretreatment or concurrently as a combined regimen (Hershberger et al., 2001). Several synthetic vitamin D analogs (such as EB1089, Inecalcitol) have proven more effective as anticancer agents than the natural hormone (calcitriol) in studies with animal models (Okamoto et al., 2012; Perez-Stable et al., 2002). Dietary vitamin D_3 supplementation raised blood levels of $25(OH)D_3$ and reduced tumor burden from xenografts of PC3 prostate cancer cells in athymic male nude mice (Swami et al., 2012). Interestingly, vitamin D_3 was as effective as calcitriol, which was injected intraperitoneally, in reducing tumor burden. In xenograft tumor tissue, dietary D_3 and injected calcitriol elevated to similar levels the mRNAs for CYP24A1, 15-PGDH, IGFBP-3, and p21/Cip, and also reduced mRNAs for cycloxygenase-2 and the prostaglandin receptor EP4 to similar levels. The mRNAs for Bcl-2 (antiapoptotic) and Bax (proapoptotic) did not change. These results highlight the potential benefit of a combined regimen of nonsteroidal anti-inflammatory drugs (NSAIDs) and calcitriol/calcitriol analog or dietary vitamin D_3 for the clinical management of prostate cancer, as proposed (Krishnan et al., 2007; Swami et al., 2012).

It was recently reported that for prostate tumors in TRAMP mice, early intervention with calcitriol elevated the tumor tissue E-cadherin level (a prodifferentiation marker) and inhibited the androgen-stimulated growth of primary tumors during the early phase of the disease; however, continued calcitriol treatment increased distant organ metastases (Ajibade et al., 2014). TRAMP mice develop aggressive, metastatic prostate cancer in a stepwise manner—advancing from precancerous lesions to low-grade neoplasia and then to highly invasive carcinoma leading to distant metastases. The results of Ajibade et al. (2014) indicate that long-term treatment of calcitriol may promote an aggressive disease phenotype. Whether increased miR-106-b levels and hence reduced expression of p21 (a miR-106-b target) would play a role for promoting disease progression in TRAMP mice under long-term calcitriol treatment should be assessed, since vitamin D can increase the microRNA miR-106-b level due to VDR-mediated induction of the DNA helicase *MCM7* gene (which harbors the miR-106 gene cluster within intron 13), and miR-106b targets p21/Cip for translational suppression (Thorne et al., 2011). In another study, a chemoprevention role for calcitriol has been implicated based on the finding that in *Nkx3.1*; *Pten* mutant mice, calcitriol prevented PIN (prostate intraepithelial neoplasia) lesions from progressing to cancerous lesions (Banach-Petrosky et al., 2007).

4.3 Tumor-Expressed VDR, Association with Lethal Cancer, Clinical Potential of Vitamin D

Malignant prostate tissue expresses VDR and supports VDR signaling, indicated by the tumor tissue expression of VDR target genes such as CYP24A1, IGFBP-3, COX-2, and CYP3A4. Based on the data from 841 prostate cancer patients, it was concluded that high tumor tissue VDR expression is significantly associated with reduced risks for lethal prostate cancer (Hendrickson et al., 2011). Since the highest quartile of VDR expression was associated with reduced risks for the lethal disease, it was suggested that a threshold level of VDR may be needed in tumor tissue before VDR activity can have an impact on prostate tumor biology. In this patient cohort, VDR expression levels in cancer tissues did not associate with prediagnostic levels of serum 25(OH)D_3 and 1,25-D_3 and with VDR polymorphism.

In another study, however, VDR *FokI* polymorphism was found to associate with total prostate cancer risk (Li et al., 2007). *FokI* polymorphism produces the *F* allele which encodes a 3-amino-acid shorter, but functionally more active VDR variant compared to the longer VDR arising from the *f* allele. Translation initiation from the second ATG codon of the VDR mRNA generates the shorter VDR encoded by the *F* allele. Men with a low serum vitamin D status and a VDR genotype corresponding to the less active *f* allele are at approximately twofold higher risk for prostate cancer than men who carry the *FF* or *Ff* allele and have high serum 5(OH)D_3 plus 1,25 $(OH)_2D_3$ (Li et al., 2007). Other VDR polymorphisms, i.e., *BsmI, ApaI, TaqI* polymorphisms—all located at the 3′ end of the VDR gene outside of the coding region, and a poly A repeat did not show association with prostate cancer risk. These association studies indicate the possibility that the VDR genotype and expression, coupled with serum vitamin D status at initial diagnosis, may predict lethal versus indolent prostate cancer. Vitamin D and Omeg*a*-3 *T*rial (VITAL) is examining whether daily vitamin D supplementation at 2000 international unit (IU) with or without ω-3 fatty acid supplementation (1 g fish oil daily) would reduce risks for prostate cancer and several other cancers, as well as risks for maladies like stroke and cardiovascular disease in healthy elderly men and women. Approximately, 20,000 participants are planned for the VITAL study; data gathering is scheduled to be completed by summer 2016 (Manson, Bassuk, et al., 2012).

Despite extensive validation of the antiproliferative and antitumor action of 1,25-D_3 in cell culture and animal studies, many clinical trials found no therapeutic benefit of calcitriol or its synthetic analogs as a single-agent antiprostate cancer therapeutic, except for limited cases when partial tumor growth response and PSA response was observed (Feldman et al., 2014). A major clinical concern with vitamin D therapy is hypercalcemia, which develops in patients receiving a supraphysiological vitamin D dose necessary for tumor growth inhibition. In randomized controlled trials designed to assess the efficacy of a combined regimen of α-calcidol and docetaxel in patients with mCRPC, no benefit was observed for overall survival or PSA response (Attia et al., 2008).

Nevertheless, results from two recent small-scale studies are generating renewed optimism for vitamin D's role in prostate cancer chemoprevention, especially for the low-grade disease (Hollis, 2015; Marshall et al., 2012). A pilot study evaluated the effect of enhancing serum vitamin D status by vitamin D_3 supplementation (4000 IU per day) on prostate cancer progression for 44 patients diagnosed with low-risk prostate cancer who chose active surveillance over radical prostatectomy. More than 50% of patients who took daily vitamin D_3 supplement for 1 year had decreased number of positive cores at repeat biopsy, decreased Gleason scores, and disappearance of some tumors (Marshall et al., 2012). 34% showed increase in the number of positive cores and increased Gleason scores. 11% showed no change in tumor pathology. Serum PSA levels did not change. No adverse health consequences were observed due to year-long vitamin D_3 supplementation. In another pilot study, 37 men awaiting elective prostatectomy were randomized into two groups—one received vitamin D_3 daily (4000 IU) for 2 months and the other group received placebo (Hollis, 2015). After two months, analysis of resected prostate tissues showed that for 60% in the vitamin D group, tumors shrank or disappeared and the tissue had dramatically reduced inflammation, judged from decreased levels of inflammation-related lipids and proteins and induction of the growth factor differentiation factor-15 (GDF-15). GDF-15, which counteracts inflammation, is markedly reduced in prostate cancer (Hollis, 2015). The placebo group showed either no change in tumor or tumor progressed. These results are consistent with the anti-inflammatory role of vitamin D (Krishnan & Feldman, 2011), and it should be noted that inflammation is strongly linked to various cancers including prostate cancer (DeMarzo et al., 2007). Since preclinical studies showed that 1,25-D_3 reduced the tumor tissue prostaglandin level by reducing expression of COX-2 and the prostaglandin receptor and inducing

15-prostaglandin dehydrogenase, it is important to know whether combining vitamin D_3 supplementation with an NSAID, as proposed by Feldman and colleagues (Moreno, Krishnan, Peehl, & Feldman, 2006), would inhibit slow-growing as well as aggressive prostate cancer.

5. FUNCTIONAL INTERPLAY OF AR AND VDR IN PROSTATE CANCER: IMPACT ON CELL GROWTH AND INTRACRINE ANDROGEN BIOSYNTHESIS

5.1 Impact on Cell Growth

Cross talk between VDR and AR signaling in prostate cancer cells was demonstrated in cell culture and animal model studies. Androgen-induced AR signaling was found to contribute to the 1,25-D_3-mediated inhibition of androgen-dependent (LNCaP, CWR22R) and castration-resistant (C4-2) prostate cancer cells, since inhibition of these AR-positive cells was blocked by Casodex (an AR antagonist) or by AR silencing (via RNA interference) or by targeted disruption of genomic AR (Bao, Hu, Ting, & Lee, 2004; Lau, Trump, & Johnson, 2004; Murthy et al., 2005; Zhao, Peehl, Navone, & Feldman, 2000). In calcitriol-treated LNCaP cells, androgen-induced hyperphosphorylation of pRb was prevented, despite elevation of the AR protein level (Feldman, Zhao, & Krishnan, 2000). However, an androgen-independent mechanism can also contribute to the inhibition, since Casodex did not prevent calcitriol-induced inhibition of other prostate cancer lines such as the MDA lines (PCa-2a, PCa-2b); LAPC4; CWR22RV1; ALVA-31; and LNCaP-104R1, even though functional AR is present in all of these cell lines (Weigel, 2007; Yang, Maiorino, Roos, Knight, & Burnstein, 2002; Zhao et al., 2000).

Cross talk of VDR with AR pathways *in vivo* is indicated by the finding that testosterone supplementation abolished VDR-dependent differential tumor growth and tumor cell proliferation in the LPB-Tag (long probasin promoter-driven SV40 T antigen) model of mouse prostate cancer (Mordan-McCombs et al., 2010). In this study, significantly higher tumor growth and tumor cell proliferation were observed in LPb-Tag x VDR-KO mice compared to LPb-Tag x VDR-intact mice, whereas testosterone administration obliterated these differences between the two groups. In epidemiologic research, interindividual differences in serum testosterone levels may have contributed to inconsistent results on the relationship between prostate cancer risk and serum vitamin D status (Mordan-McCombs et al., 2010). Further evidence for an interacting role of androgen and vitamin D signaling in the regulation of prostate growth comes from the finding that 1α,25-D_3 reduced the prostate size of testis-intact but not castrated Sprague Dawley rats (Leman, Ariotti, Dhir, & Getzenberg, 2003). Finally, tumor growth rate, enhanced by factors derived from endothelial cells within the prostate cancer microenvironment, may be modulated due to VDR–AR interaction, since serum testosterone and 25(OH)D_3 levels markedly influenced blood flow and vascularity of rat prostate, and both of these NRs are expressed in the endothelial cells of normal and malignant prostate (Godoy et al., 2013).

5.2 Calcitriol, Androgen, and Intracrine Androgen Metabolism in Prostate

Metabolic inactivation of AR ligands can potentially be a fruitful approach for targeting advanced prostate cancer, since androgen-induced AR activity is a key driver of prostate cancer progression. Increased intracrine androgen biosynthesis in cancer tissue from the adrenal dehydroepiandrosterone (DHEA) is an important mechanism for activating the AR axis in castration-resistant prostate cancer (Mostaghel & Nelson, 2008). Indeed, the androgen biosynthesis blocker Zytiga® (abiraterone acetate) can reduce tumor tissue AR activity and the drug is widely used in clinical practice to control castration-resistant prostate cancer (Mostaghel, 2014). It was reported that calcitriol can induce cytochrome P450 enzymes in androgen-dependent LNCaP and LAPC4 prostate cancer cells, and CYP3A4 induction in LNCaP cells was shown to induce oxidative inactivation of testosterone (to 6β-hydroxy-testosterone) and DHEA (to 16α-hyroxy-DHEA; Doherty et al., 2014; Maguire et al., 2012). Reduced CYP3A4 expression in specimens from aggressive prostate cancer and in circulating prostate cancer cells has been reported (Fujimura et al., 2009; Mitsiades et al., 2012).

Calcitriol and the analog EB1089 caused a two- to three-fold induction of SULT2B1b sulfotransferase mRNA and protein in prostate cancer cells and *in vivo* in mouse prostate, and a DR7-type VDRE is involved in the VDR-mediated induction of the corresponding gene (*SULT2B1*, Seo et al., 2013). The C-3 hydroxyl group of DHEA and cholesterol is targeted by SULT2B1b (aka SULT2B) for sulfoconjugation, which facilitates their metabolic clearance. Silencing of SULT2B by RNA interference caused accelerated proliferation of prostate cancer cells in culture (Seo et al., 2013) and in xenograft tumors (Park et al., in preparation). Abundant SULT2B expression in normal prostate epithelium is contrasted by its reduced level in primary prostate cancer and its almost negligible expression in metastatic CRPC (Fig. 3). No detectable SULT2B was observed in >90% cases of distant metastases that we have analyzed. Occasionally, weak SULT2B staining in metastatic specimens was detected—a representative example being the specimen at lower panel, right (Fig. 3).

The above findings make it likely that the prostatic androgen flux is regulated in part by SULT2B activity, and interference with this flux due to reduced SULT2B expression, as observed in primary cancer and in mCRPC, would promote prostate cancer growth by enhancing tumor tissue androgen synthesis. Confirmation of this possibility awaits experimental evidence.

Treatment of C4-2B castration-resistant cells concurrently with R1881 (a synthetic androgen) and EB1089 completely abrogated the induction of SULT2B1 mRNAs observed with EB1089 alone (Fig. 4B). Similarly, EB1089 induced CYP3A4 mRNA expression by more than 10-fold, while cotreatment with R1881 drastically reduced the level of CYP3A4 mRNAs, bringing it to the basal level or even lower (Fig. 4C). EB1089-induced robust expression of CYP24A1 (~2000-fold) was also drastically reduced (to less than 10-fold) by the combined action of R1881 and EB1089 (Fig. 4A). Dampened vitamin D/VDR-induced CYP24A1 mRNA expression in the presence of androgen is beneficial, as it would prevent ligand degradation, thereby enhancing VDR signaling.

An important example of the beneficial effect of androgen–vitamin D interaction is the inhibition of AR-regulated energy metabolism of LNCaP prostate cancer cells (Wang & Tenniswood, 2014; Wang, Welsh, & Tenniswood, 2013). Normally, prostate cancer cells generate energy using an aerobic mechanism that requires citrate oxidation by m-aconitase-2 activity and propagation of TCA cycle. Glucose usage stimulated by androgen-activated AR leads to the production of acetyl CoA, which feeds into the TCA cycle for citrate synthesis (Costello & Franklin, 1991a,1991b). Combined 1α,25-D_3 and testosterone treatment of LNCaP cells caused upregulation of zinc transporters (SLC39A1 and SLC39A11) that led to intracellular zinc accumulation, inactivation of m-aconitase-2 activity at the high zinc level, and secretion of citrate to the cytoplasm for lipid synthesis; also, reduced expression of the thiamine pyrophosphate (TPP) transporter (SLC25A19) in the presence of 1α,25-D_3 and testosterone together would lower the mitochondrial level of the coenzyme TPP, which in turn can decrease the activities of pyruvate dehydrogenase and α-ketoglutarate dehydrogenase and inhibit ATP production

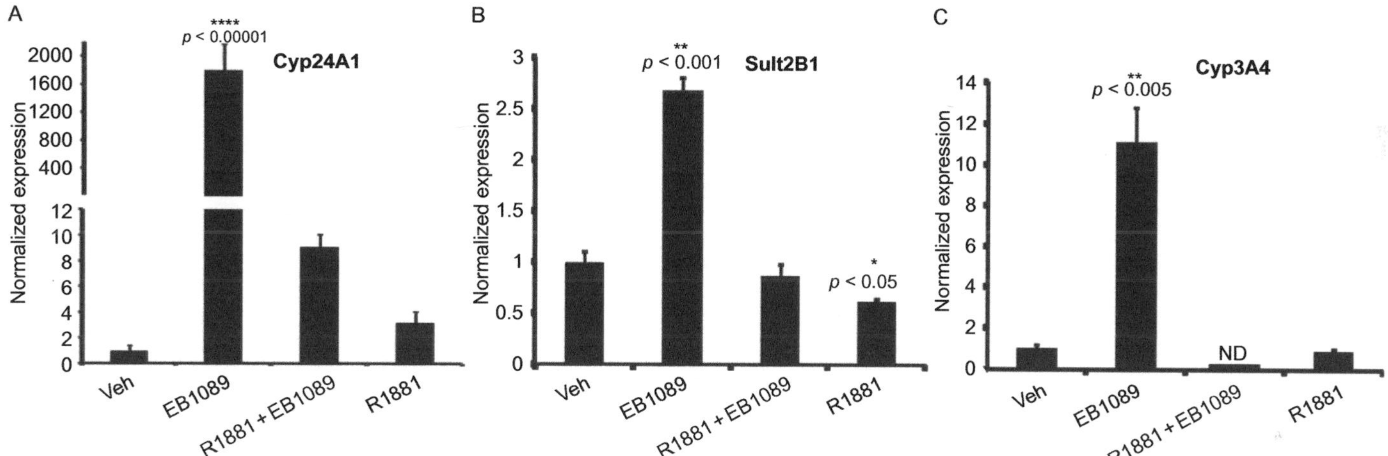

Figure 4 CYP24A1, SULT2B1, and CYP3A4 mRNAs in C4-2B castration-resistant human prostate cancer cells treated with vehicle; EB1089 (10 nM); R1881 (1 nM) + EB1089 (10 nM); R1881 (1 nM). (A) CYP24A1 mRNAs. (B) SULT2B 1b mRNAs. (C) CYP3A4 mRNAs. Cells were incubated with indicated hormones or vehicle for 20 h before RNAs were extracted from the treated cells and analyzed by qRT-PCR assay. Asterisks indicate supershifted EMSA complex in the presence of VDR and RXR antibodies.

(Costello & Franklin, 1991a, 1991b; Wang, Chatterjee, Chittur, Welsh, & Tenniswood, 2011; Wang & Tenniswood, 2014). Our observation that androgen dramatically reduced vitamin D-induced expression of CYP24A1 mRNAs in castration-resistant C4-2B cells implies that the vitamin D level in these cells would remain at a relatively high level. This result seems to predict that androgen supplementation, at least intermittently, has the potential to enhance the antiprostate tumor activity of calcitriol or analog. Human trials are needed to assess the potential of combined vitamin D and testosterone supplementation in prostate cancer prevention and treatment.

6. SUMMARY AND FUTURE POSSIBILITIES

Besides its essential role in bone mineralization, the bioactive vitamin D (i.e., calcitriol, 1α,25-D_3, or synthetic analogs) has wide impact on the physiology and pathophysiology of extraskeletal tissues, including growth inhibition of normal and malignant tissues of prostate. The efficacy of vitamin D for inhibiting prostate cancer cells and prostate tumor growth in cell culture and animal models is well established. However, clinical trials with calcitriol or its analog as a single agent are inconclusive, especially since the high dose of vitamin D needed for clinical efficacy is known to induce hypercalcemia. Nonsignificant association of a low serum vitamin D status with increased prostate cancer risk further adds to the uncertainty for the beneficial role of vitamin D as a single agent in prostate cancer therapy and chemoprevention. On the other hand, encouraging results were observed in two recent small-scale pilot studies. In one study, progression of low-grade prostate cancer was prevented or reversed for patients on active surveillance who received daily, safe-dose vitamin D supplementation, at 4000 IU each day for a 1-year period. In another study, after daily vitamin D supplementation (4000 IU/day) was given for 2 months to patients who elected surgery, cancer tissue in resected prostate samples showed markedly decreased inflammation, induction of the anti-inflammatory protein GDF-15 (which is reduced in prostate cancer), and lower Gleason score. These results (discussed in Section 4.3) indicate that vitamin D may inhibit prostate cancer progression in humans, at least for the low-grade, organ-confined disease.

The genomic action of vitamin D is mediated by the cognate VDR which is a ligand-inducible, DNA-binding TF, and a member of the NR superfamily. Nongenomic action of vitamin D through a membrane receptor plays a relatively minor role in the regulation of cell functions. VDR has a multidomain organization similar to other NRs. Due to interdomain allosteric interactions, the transcriptional activity of VDR is influenced by ligand structure and nucleotide organization of DNA response elements in target genes. Calcitriol is biosynthesized in human body from 7-dehydrocholesterol, which is converted in sunlight-exposed skin to the secosteroid vitamin D_3 (abbreviated D_3), which is then sequentially converted to 25-hydroxy D_3 in the liver, predominantly by CYP2R1, and then to 1α,25-dihydroxy D_3 (1,25-D_3) by the 1α-hydroxylase activity of CYP27B1 in the kidneys. Local synthesis of 1,25-D_3 in prostate and other extrarenal tissues from circulating 25-hydroxy D_3 is mediated by the 1α-hydroxylase activity of CYP27B1. Degradation of 1,25-D_3 is mediated by the 24-hydroxylase activity of the catabolic enzyme CYP24A1. Vitamin D homeostasis under normal physiology is maintained by a regulatory loop that entails 1,25-D_3-mediated induction of *CYP24A1* and repression of *CYP2R1* and *CYP27B1*. Overexpression of CYP24A1 in advanced prostate cancer has been reported (Tannour-Louet et al., 2014).

Functional interaction between the AR and VDR axis in prostate cancer cells has been demonstrated in experimental systems. This cross talk may facilitate vitamin D action against prostate cancer. AR, which promotes disease progression in mCRPC, is activated by tumor tissue androgen, which is produced by intracrine androgen biosynthesis from the adrenal steroid DHEA. The prostate-expressed SULT2B1b (aka SULT2B), a phase II sulfotransferase, may interfere with intracrine androgen production since conversion of DHEA to DHEA-sulfate by SULT2B would lower the precursor DHEA pool available for intracrine androgen synthesis. A role for SULT2B in prostate cancer progression is likely since its expression is significantly reduced in primary cancer and it is mostly nondetectable in cancer tissues of distant metastases. Vitamin D (calcitriol and EB1089) induced SULT2B expression in castration-resistant prostate cancer cells. The phase I enzyme CYP3A4, which inactivates DHEA and testosterone by oxidative modification, is induced by VDR in castration-resistant prostate cancer cells. Therefore, vitamin D may attenuate AR signaling in advanced prostate cancer by restricting the ligand source for AR. The observed induction of CYP24A1 by vitamin D in prostate cancer cells would antagonize vitamin D action, especially since CYP24A1 overexpression in advanced prostate cancer has been reported. We found that androgen suppressed VDR-mediated CYP24A1 induction in prostate cancer cells. Thus, VDR action may lower the ligand pool for AR (due to induction of CYP3A4, SULT2B) and AR activity may help protect the VDR ligand from CYP24A1-mediated degradation. However, VDR-mediated induction of SULT2B and CYP3A4 was also suppressed by androgen, which can potentially elevate AR signaling by preventing reduction of the tumor tissue androgen level. Elevated AR signaling, on the other hand, may further suppress VDR-induced CYP24A1 expression and provide additional protection against 1,25-D_3 degradation. Studies in animal models may reveal the net impact of androgen supplementation (continuously or intermittently) on the inhibition of prostate tumor growth by bioactive vitamin D.

Design of future clinical trials of vitamin D should take into consideration a number of results from *in vitro* and *in vivo* studies. For example, calcitriol induced the microRNA miR106-b as well as the cell cycle inhibitor p21 in prostate cancer cells (discussed in Section 4.1). Since p21 is a target of miR106-b, under certain conditions calcitriol-induced expression of miR106-b may prevent accumulation of p21 so that vitamin D may fail to cause inhibition of prostate tumor growth. In human trials, the combined action of calcitriol (or analog) and an oligonucleotide-based anti-miR, which targets miR106-b and specifically reduces its endogenous level, may be investigated. In another example, calcitriol inhibited the prostaglandin pathway by downregulating the expression of COX-2 and prostaglandin receptor and upregulating 15-prostaglandin dehydrogenase (discussed in Section 4.2). Therefore, calcitriol combined with a NSAID may be a potent regimen for inhibition of prostate cancer progression (Moreno, Krishnan, & Feldman, 2005). Vitamin D can induce growth arrest of prostate progenitor/stem cells (Li, Fleet, & Teegarden, 2009; Maund et al., 2011) and the natural compound sulfarophane, an organosulfur present in cruciferous vegetables, was shown to ablate pancreatic cancer stem cells (Hu & Fu, 2012). It will be of interest to assess the effect of the combined action of vitamin D and sulfarophane on prostate tumor growth in humans. Finally, a number of low calcemic alternatives to calcitriol are now available. Inecalcitol is one such example; it has higher antiproliferative activity and is 100-fold less hypercalcemic than calcitriol. Inecalcitol was reported to be highly effective in inhibiting the growth of LNCaP cells in culture and in xenograft tumors (Okamoto et al., 2012), and a phase I multicenter trial of inecalcitol combined with docetaxel has shown encouraging PSA response in naïve mCRPC patients (Medioni et al., 2014). With advances in the development of next-generation calcitriol analogs and recent encouraging pilot data showing efficacy of calcitriol in inhibiting low-grade prostate cancer at a safe, physiologically compatible dose, it is likely that calcitriol or synthetic analogs will be approved in the near future for use in prostate cancer prevention and treatment.

ACKNOWLEDGMENTS

This work was supported by a DOD-IDEA grant, a VA Merit-Review grant, Research Career Scientist award from VA (to B.C.) and a pilot grant from Morrison Trust Foundation, San Antonio. We are grateful to Colm Morrissey and Dr. Elahe Mostaghel, M.D. (Fred Hutchison Cancer Center at Seattle) for prostate cancer specimens, and Dr. Sherry Werner, M.D. (UTHSCSA) for her time and help in taking images of IHC-stained cancer tissues. Help from Ms. Debarati Mukherjee with manuscript reading and editing and for insightful comments is gratefully acknowledged.

REFERENCES

Ajibade, A. A., Kirk, J. S., Karasik, E., Gillard, B., Moser, M. T., Johnson, C. S., et al. (2014). Early growth inhibition is followed by increased metastatic disease with vitamin D (calcitriol) treatment in the TRAMP model of prostate cancer. *PLoS One*, *9*, e89555.

Albanes, D., Mondul, A. M., Yu, K., Parisi, D., Horst, R. L., Virtamo, J., et al. (2011). Serum 25-hydroxyvitamin D and prostate cancer risk in a large nested case-control study. *Cancer Epidemiology, Biomarkers and Prevention*, *20*(9), 1850–1860.

Aleksunes, L. M., & Klaassen, C. D. (2012). Coordinated regulation of hepatic phase I and II drug-metabolizing genes and transporters using AhR-, CAR-, PXR-, PPARα-, and Nrf2-null mice. *Drug Metabolism and Disposition*, *40*, 1366–1379.

Alimirah, F., Vaishnav, A., Mc Cormick, M., Echchgadda, I., Chatterjee, B., Mehta, R. G., et al. (2010). Functionality of unliganded VDR in breast cancer cells: Repressive action on CYP24 basal transcription. *Molecular and Cellular Biochemistry*, *342*, 143–150.

Al Mutair, A. N., Nasrat, G. H., & Russell, D. W. (2012). Mutation of the *CYP2R1* vitamin D 25-hydroxylase in a Saudi Arabian family with severe vitamin D deficiency. *The Journal of Clinical Endocrinology and Metabolism*, *97*, E2022–E2025.

Attia, S., Elckhoff, J., Wilding, G., McNeel, D., Blank, J., Ahuja, H., et al. (2008). Randomized double blinded phase II evaluation of docetaxel with or without doxercalciferol in patients with metastatic androgen-independent prostate cancer. *Clinical Cancer Research*, *14*, 2437–2443.

Babiker, A. M., Al Gadi, I., Al-Jurayyan, N. A., Al Nemri, A. M., Al Haboob, A. A., Al Boukai, A. A., et al. (2014). A novel pathogenic mutation of the CYP27B1 gene in a patient with vitamin D-dependent rickets type 1: A case report. *BMC Research Notes*, *7*, 783.

Banach-Petrosky, W., Jessen, W. J., Ouyang, X., Gao, H., Rao, J., Quinn, J., et al. (2007). Prolonged exposure to reduced levels of androgen accelerates prostate cancer progression in *Nkx3.1*; *Pten* mutant mice. *Cancer Research*, *67*, 9089–9096.

Bao, B.-Y., Hu, Y.-C., Ting, H.-J., & Lee, Y.-F. (2004). Androgen signaling is required for the vitamin D-mediated growth inhibition in human prostate cancer cells. *Oncogene*, *23*, 3350–3360.

Bao, B. Y., Yeh, S. D., & Lee, Y. F. (2006). 1α, 25 dihydroxy vitamin D3 inhibits prostate cancer cell invasion via modulation of selective proteases. *Carcinogenesis*, *27*, 32–42.

Blutt, S. E., McDonnell, T. J., Polek, T. C., & Weigel, N. L. (2000). Calcitriol-induced apoptosis in LNCaP cells is blocked by overexpression of Bcl-2. *Endocrinology*, *138*, 1491–1497.

Boyle, B. J., Zhao, X. Y., Cohen, P., & Feldman, D. (2001). Insulin-like growth factor binding protein3 mediates 1 alpha 25-dihydroxyvitamin d(3) growth inhibition in the LNCaP prostate cancer cell line through p21/WAF1. *The Journal of Urology*, *165*, 1319–1324.

Campbell, M. J., Elstner, E., Holden, S., Uskokovic, M., & Koeffler, H. P. (1997). Inhibition of proliferation of prostate cancer cells by a 19-nor-hexafluoride vitamin D3 analogue

involves induction of p21waf1, p27Kip1 and E-cadherin. *Journal of Molecular Endocrinology, 19*, 15–27.
Carlberg, C. (2014). Genome-wide(overview) on the actions of vitamin D. *Frontiers in Physiology, 5*, 1–10. http://dx.doi.org/10.3389/fphys.2014.00167 (article number: 167).
Carlberg, C., Bendik, I., Wyss, A., Meier, E., Sturzenbecker, L. A., Grippo, J. F., et al. (1993). Two nuclear signaling pathways for vitamin D. *Nature, 361*, 657–660.
Carlberg, C., & Campbell, M. J. (2013). Vitamin D receptor signaling mechanisms: Integrated actions of a well-defined transcription factor. *Steroids, 78*, 127–136.
Chanakul, A., Zhang, M. Y. H., Louw, A., Armbrecht, H. J., Miller, W. L., Portale, A. A., et al. (2013). FGF-23 regulates CYP27B1 transcription in the kidney and in extra-renal tissues. *PLoS One, 8*(9), e72816.
Chen, T. C. (2008). 25-Hydroxyvitamin D-1 alpha-hydroxylase (CYP27B1) is a new class of tumor suppressor in the prostate. *Anicancer Research, 28*(4A), 2015–2017.
Chen, T. C., & Holick, M. F. (2003). Vitamin D and prostate cancer prevention and treatment. *TRENDS in Endocrinology and Metabolism, 14*, 423–430.
Chen, T. C., Sakai, T., Yamamoto, K., & Kittaka, A. (2012). The roles of cytochrome P450 enzymes in prostate cancer development and treatment. *Anticancer Research, 32*, 291–298.
Chung, I., Han, G., Seshadri, M., Gillard, B. M., Yu, W.-d., Foster, B. A., et al. (2009). Role of VDR in anti-proliferative effects of calcitriol in tumor-derived endothelial cells and tumor angiogenesis in vivo. *Cancer Research, 69*, 967–975.
Costello, L. C., & Franklin, R. B. (1991a). Concept of citrate production and secretion by prostate 1. Metabolic relationships. *Prostate, 18*, 25–46.
Costello, L. C., & Franklin, R. B. (1991b). Concept of citrate production and secretion by prostate 2. Hormonal relationship in normal and neoplastic prostate. *Prostate, 19*, 181–205.
Datta, M., & Schwartz, G. G. (2012). Calcium and vitamin D supplementation during androgen deprivation therapy for prostate cancer: A critical review. *The Oncologist, 17*(9), 1171–1179.
de Bono, J. S., Logothetis, C. J., Molina, A., Fizazi, K., North, S., et al. (2011). Abiraterone and increased survival in metastatic prostate cancer. *The New England Journal of Medicine, 364*, 1995–2005.
Deeb, K. K., Trump, D. L., & Johnson, C. S. (2007). Vitamin D signalling pathways in cancer: Potential for anticancer therapeutics. *Nature Reviews. Cancer*, 7(9), 684–700.
DeLuca, H. F. (2014). History of the discovery of vitamin D and its active metabolites. *BoneKEy Reports, 3*, 1–8. http://dx.doi.org/10.1038/bonekey.2013.213 (article number: 479).
DeMarzo, A. M., Platz, E. A., Sutcliffe, S., Xu, J., Grönberg, H., Drake, C. G., et al. (2007). Inflammation in prostate carcinogenesis. *Nature Reviews. Cancer*, 7, 256–269.
Doherty, D., Dvorkin, S. A., Rodriguez, E. P., & Thompson, P. D. (2014). Vitamin D receptor agonist EB1089 is a potent regulator of prostatic intracrine metabolism. *Prostate, 74*, 273–285.
Evans, R. M., & Mangelsdorf, D. J. (2014). Nuclear receptors, RXR, and the big bang. *Cell, 157*, 255–266.
Feldman, D., Krishnan, A. V., Swami, S., Giovannucci, E., & Feldman, B. J. (2014). The role of vitamin D in reducing cancer risk and progression. *Nature Reviews. Cancer, 14*, 342–357.
Feldman, D., Zhao, X.-Y., & Krishnan, A. V. (2000). Vitamin D and prostate cancer. *Endocrinology, 141*(1), 5–9.
Ferla, S., Aboraia, A. S., Brancale, A., Pepper, C. J., Zhu, J., Ochalek, J. T., et al. (2014). Small-molecule inhibitors of 25-hydroxyvitamin D-24-hydroxylase (CYP24A1): Synthesis and biological evaluation. *Journal of Medicinal Chemistry, 57*, 7702–7715.
Fujimura, T., Takahashi, S., Uno, T., Kumagai, J., Murata, T., Takayama, K., et al. (2009). Expression of cytochrome P450 3A4 and its clinical significance in human prostate cancer. *Urology, 74*, 391–397.

Getzenberg, R. H., Light, B. W., Lapco, P. E., Konety, B. R., Nangia, A. K., Acierno, J. S., et al. (1997). Vitamin D inhibition of prostate adenocarcinoma growth and metastasis in the Dunning rat prostate model system. *Urology*, *50*, 999–1006.

Godoy, A. S., Chung, I., Montecinos, V. P., Buttyan, R., Johnson, C. S., & Smith, G. J. (2013). Role of androgen and vitamin D receptors in endothelial cells from benign and malignant human prostate. *American Journal of Physiology. Endocrinology and Metabolism*, *304*, E1131–E1139.

Guzey, M., Kitada, S., & Reed, J. C. (2002). Apoptosis induction by 1alpha,25-dihydroxyvitamin D3 in prostate cancer. *Molecular Cancer Therapeutics*, *1*, 667–677.

Guzey, M., Luo, J., & Getzenberg, R. H. (2004). Vitamin D_3 modulated gene expression patterns in human primary normal and cancer prostate cells. *Journal of Cellular Biochemistry*, *93*, 271–285.

Harshman, L. C., Wang, X., Nakabayashi, M., Xie, W., Valenca, L., Werner, L., et al. (2015). Statin use at the time of initiation of androgen deprivation therapy and time to progression in patients with hormone-sensitive prostate cancer. *JAMA Oncology*, *1*(4), 495–504. http://dx.doi.org/10.1001/jamaoncol.2015.0829.

Haussler, M. R., Jurutka, P. W., Mizwicki, M., & Norman, A. W. (2011). Vitamin D receptor (VDR)-mediated actions of 1α, $25(OH)_2$ vitamin D_3: Genomic and non-genomic mechanisms. *Best Practice & Research Clinical Endocrinology & Metabolism*, *25*, 543–559.

Helsen, C., & Claessens, F. (2014). Looking at nuclear receptors from a new angle. *Molecular and Cellular Endocrinology*, *382*, 97–106.

Hendrickson, W. K., Flavin, R., Kasperzyk, J. L., Fiorentino, M., Fang, F., Lis, R., et al. (2011). Vitamin D receptor protein expression in tumor tissue and prostate cancer progression. *Journal of Clinical Oncology*, *29*, 2378–2385.

Hershberger, P. A., Yu, W., Modzelewski, R. A., Rueger, R. M., Johnson, C. S., & Trump, D. L. (2001). Calcitriol (1,25-Dihydroxycholecalci-ferol) enhances paclitaxel antitumor activity in vitro and in vivo and accelerates paclitaxel-induced apoptosis. *Clinical Cancer Research*, 7, 1043–1051.

Hollis, B. W. (2015). Vitamin D in the prevention and treatment of cancer. In *Talk presented at American Chemical Society 249th national meeting and exposition; March 22–36, Denver, CO.*

Hsu, J. Y., Feldman, D., McNeal, J. E., & Peehl, D. M. (2001). Reduced 1a-hydroxylase activity in human prostate cancer cells correlates with decreased susceptibility to 25-hydroxyvitamin D3-induced growth inhibition. *Cancer Research*, *61*, 2852–2856.

Hu, Y., & Fu, L. (2012). Targeting cancer stem cells: A new therapy to cure cancer patients. *American Journal of Cancer Research*, *2*, 340–356.

Huang, P., Chandra, V., & Rastinejad, F. (2010). Structural overview of the nuclear receptor superfamily: Insights into physiology and therapeutics. *Annual Review of Physiology*, *72*, 247–272.

Jamieson, D., Holick, M. F., & Chen, T. C. (2004). Regulation of 25-hydroxyvitamin D-1 alpha-hydroxylase by epidermal growth factor in prostate cells. *The Journal of Steroid Biochemistry and Molecular Biology*, *89–90*, 127–130.

Kojima, S., Mulholland, D. J., Ettinger, S., Fazli, L., Nelson, C. C., & Gleave, M. E. (2006). Differential regulation of IGFBP-3 by the androgen receptor in the lineage-related androgen-dependent LNCaP and androgen-independent C4-2 prostate cancer models. *Prostate*, *66*, 971–986.

Krishnan, A. V., & Feldman, D. (2011). Mechanisms of the anti-cancer and anti-inflammatory actions of vitamin D. *Annual Review of Pharmacology and Toxicology*, *51*, 311–336.

Krishnan, A. V., Moreno, J., Nonn, L., Malloy, P., Swami, S., Peng, L., et al. (2007). Novel pathways that contribute to the anti-proliferative and chemopreventive activities of calcitriol in prostate cancer. *The Journal of Steroid Biochemistry and Molecular Biology*, *103*, 694–702.

Lau, Y. K., Trump, D. L., & Johnson, C. (2004). Interaction of vitamin D receptor and androgen receptor in human prostate cancer cell line. *Journal of Clinical Oncology*, *22*(14S) (2004 ASCO annual meeting proceedings (post-meeting edition); Abstract # 9550).

Leman, E. S., Ariotti, J. A., Dhir, R., & Getzenberg, R. H. (2003). Vitamin D and androgen regulation of prostatic growth. *Journal of Cellular Biochemistry*, *90*, 138–147.

Li, J., Fleet, J. C., & Teegarden, D. (2009). Activation of rapid signaling pathways does not contribute to 1alpha, 25-dihydroxy vitamin D3-induced growth inhibition of mouse prostate epithelial progenitor cells. *Journal of Cellular Biochemistry*, *107*, 1031–1036.

Li, H., Stampfer, M. J., Hollis, J. B. W., Mucci, L. A., Gaziano, J. M., Hunter, D., et al. (2007). A prospective study of plasma vitamin D metabolites, vitamin D receptor polymorphisms, and prostate cancer. *PLoS Medicine*, *4*(3), e103.

Luo, W., Karpf, A. R., Deeb, K. K., Muindi, J. R., Morrison, C. D., Johnson, C. S., et al. (2010). Epigenetic regulation of vitamin D 24-hydroxylase/*CYP24A1* in human prostate cancer. *Cancer Research*, *70*, 5953–5962.

Maguire, O., Pollock, C., Martin, P., Owen, A., Smyth, T., Doherty, D., et al. (2012). Regulation of CYP3A4 and CYP3A5 expression and modulation of intracrine metabolism of androgens in prostate cells by liganded vitamin D receptor. *Molecular and Cellular Endocrinology*, *364*, 54–64.

Manson, J. E., Bassuk, S., et al. (2012). The VITamin D and OmegA-3 TriaL (VITAL): Rationale and design of a large randomized controlled trial of vitamin D and marine omega-3 fatty acid supplements for the primary prevention of cancer and cardiovascular disease. *Contemporary Clinical Trials*, *33*, 159–171.

Marchiani, S., Bonaccorsi, L., Ferruzzi, P., Crescioli, C., Muratori, M., Adorini, L., et al. (2006). The vitamin D analogue BXL-628 inhibits growth factor-stimulated proliferation and invasion of DU145 prostate cancer cells. *Journal of Cancer Research and Clinical Oncology*, *132*, 408–416.

Marshall, D., Savage, S. J., Garrett-Mayer, E., Keane, T. E., Hollis, B. W., & Horst, R. L. (2012). Vitamin D3 supplementation at 4000 international units per day for one year results in a decrease of positive cores at repeat biopsy in subjects with low-risk prostate cancer under active surveillance. *The Journal of Clinical Endocrinology and Metabolism*, *97*, 2315–2324.

Maund, S. L., Barclay, W. W., Hover, L. D., Axanova, L. S., Sui, G., Hipp, J. D., et al. (2011). Interleukin-1 alpha mediates the anti-proliferative effects of 1,25 dihydroxyvitamin D3 in prostate progenitor/stem cells. *Cancer Research*, *71*(15), 5276–5286.

Medioni, J., Deplanque, G., Ferrero, J.-M., Maurina, T., Rodier, J.-M., Raymond, E., et al. (2014). Phase I safety and pharmacodynamic of inecalcitol, a novel VDR agonist with docetaxel in metastatic castration-resistant prostate cancer patients. *Clinical Cancer Research*, *20*(17), 4471–4477.

Meijsing, S. H., Pufall, M. A., So, A. Y., Bates, D. L., Chen, L., & Yamamoto, K. R. (2009). DNA binding site sequence directs glucocorticoid structure and activity. *Science*, *324*, 407–410.

Mitsiades, N., Sung, C. C., Schultz, N., Danilla, D. C., He, B., Eedunuri, V. K., et al. (2012). Distinct patterns of dysregulated expression of enzymes involved in androgen synthesis and metabolism in metastatic prostate cancer tumors. *Cancer Research*, *72*, 6142–6152.

Moffatt, K. A., Johannes, W. U., Hedlund, T. E., & Miller, G. J. (2001). Growth inhibitory effects of 1α, 25-dihydroxyvitamin D3 are mediated by increased levels of p21 in the prostatic carcinoma cell line ALVA-31. *Cancer Research*, *61*, 7122.

Mordan-McCombs, S., Brown, T., Wang, W. L., Gaupel, A. C., Welsh, J., & Tenniswood, M. (2010). Tumor progression in the LPB-Tag transgenic model of prostate cancer is altered by vitamin D receptor and serum testosterone status. *The Journal of Steroid Biochemistry and Molecular Biology*, *121*, 368–371.

Moreno, J., Krishnan, A. V., & Feldman, D. (2005). Molecular mechanisms mediating the anti-proliferative effects of Vitamin D in prostate cancer. *The Journal of Steroid Biochemistry and Molecular Biology, 97*, 31–36.

Moreno, J., Krishnan, A. V., Peehl, D. M., & Feldman, D. (2006). Mechanisms of vitamin D-mediated growth inhibition in prostate cancer cells: Inhibition of the prostaglandin pathway. *Anticancer Research, 26*, 2525–2530.

Mostaghel, E. A. (2014). Abiraterone in the treatment of metastatic castration-resistant prostate cancer. *Cancer Management and Research, 6*, 39–51.

Mostaghel, E. A., & Nelson, P. S. (2008). Intracrine androgen metabolism in prostate cancer progression: Mechanisms of castration resistance and therapeutic implications. *Best Practice & Research. Clinical Endocrinology & Metabolism, 22*, 243–258.

Murthy, S., Agoulnik, I. U., & Weigel, N. L. (2005). Androgen receptor signaling and vitamin D receptor action in prostate cancer cells. *Prostate, 64*, 362–372.

Oades, G. M., Dredge, K., Kirby, R. S., & Colston, K. W. (2002). Vitamin D receptor-dependent antitumour effects of 1,25-dihydroxyvitamin D3 and two synthetic analogues in three in vivo models of prostate cancer. *BJU International, 90*, 607–616.

Okamoto, R., Delansorne, R., Wakimoo, N., Doan, N. B., Akagi, T., Shen, M., et al. (2012). Inecalcitol, an analog of 1a,25(OH)2D3, induces growth arrest of androgen-dependent prostate cancer cells. *International Journal of Cancer, 130*, 2464–2473.

Palmer, H. G., Gonzalex-Sancho, J. M., Espada, J., Berciano, M. T., Puig, I., Baulida, J., et al. (2001). Vitamin D(3) promotes the differentiation of colon carcinoma cells by the induction of E-cadherin and the inhibition of β-catenin signaling. *The Journal of Cell Biology, 154*, 369–387.

Patel, D., Knowell, A. E., Korang-Yeboah, M., Sharma, P., Joshi, J., Glymph, S., et al. (2014). Inhibitor of differentiation 4 (ID4) inactivation promotes de novo steroidogenesis and castration-resistant prostate cancer. *Molecular Endocrinology, 28*, 1239–1253.

Perez-Stable, C. M., Schwartz, G. G., Farinas, A., Finegold, M., Binderup, L., Howard, G. A., et al. (2002). The Gγ/T-15 transgenic mouse model of androgen-independent prostate cancer: Target cells of carcinogenesis and the effect of the vitamin D analogue EB1089. *Cancer Epidemiology, Biomarkers and Prevention, 11*, 555–563.

Pike, J. W., Lee, S. M., & Meyer, M. B. (2014). Regulation of gene expression by 1,25-dihydroxyvitamin D3 in bone cells: Exploiting new approaches and defining new mechanisms. *BoneKEy Reports, 3*, 1–9. http://dx.doi.org/10.1038/bonekey.2013.216 (article number: 482).

Rao, A., Coan, A., Welsh, J. E., Barclay, W. W., Koumenis, C., Cramer, S. D., et al. (2004). Receptor and p21/WAF1 are targets of genistein and 1,25-dihydroxyvitamin D3 in human prostate cancer cells. *Cancer Research, 64*, 2143–2147.

Robson, C. N., Gnanapragasam, V., Byrne, R. L., Cllins, A. T., & Neal, D. E. (1999). Transforming growth factor-beta1 up-regulates p15, p21 and p27 and blocks cell cycling in G1 in human prostate epithelium. *The Journal of Endocrinology, 160*, 257–266.

Rohan, J. N., & Weigel, N. L. (2009). 1Alpha,25-dihydroxyvitamin D3 reduces c-Myc expression, inhibiting proliferation and causing G1 accumulation in C4-2 prostate cancer cells. *Endocrinology, 150*, 2046–2054.

Rosenfeld, M. G., Lunyak, V. V., & Glass, C. K. (2006). Sensors and signals: A coactivator/corepressor/epigenetic code for integrating signal-dependent programs of transcriptional response. *Genes & Development, 20*, 1405–1428.

Saito, T., Okamoto, R., Haritunians, T., O'Kelly, J., Uskokovic, M., Maehr, H., et al. (2008). Novel Gemini-vitamin D3 analogs have potent antitumor activity. *The Journal of Steroid Biochemistry and Molecular Biology, 112*, 151–156.

Salehi-Tabara, R., Nguyen-Yamamotoa, L., Tavera-Mendoza, L. E., Quail, T., Dimitrov, V., An, B. S., et al. (2012). Vitamin D receptor as a master regulator of the c-MYC/MXD1 network. *Proceedings of the National Academy of Sciences of the United States of America, 109*, 18827–18832.

Scher, H. I., Buchanan, G., Gerald, W., Butlerand, L. M., & Tilley, W. D. (2004). Targeting the androgen receptor: Improving outcomes for castration-resistant prostate cancer. *Endocrine-Related Cancer, 11*, 459–476.

Seo, Y. K., Mirkheshti, N., Song, C. S., Kim, S., Dodds, S., Ahn, S. C., et al. (2013). SULT2B1b sulfotransferase: Induction by vitamin D receptor and reduced expression in prostate cancer. *Molecular Endocrinology, 27*, 925–939.

Sequeira, V. B., Rybchyn, M. S., Tongkao-on, W., Gordon-Thomson, C., Malloy, P. J., Nemere, I., et al. (2012). The role of the vitamin D receptor and ERp57 in photoprotection by 1α,25-dihydroxyvitamin D3. *Molecular Endocrinology, 26*, 574–582.

Sherr, C. J., & Roberts, J. M. (1999). CDK inhibitors: Positive and negative regulators of G1-phase progression. *Genes & Development, 13*, 1501–1512.

Stewart, L. V., & Weigel, N. L. (2005). Role of insulin-like growth factor binding proteins in 1alpha,25-dihydroxy-vitamin D(3)-induced growth inhibition of human prostate cancer cells. *Prostate, 64*, 9–19.

Sung, V., & Feldman, D. (2000). 1,25-Dihydroxyvitamin D3 decreases human prostate cancer cell adhesion and migration. *Molecular and Cellular Endocrinology, 164*, 133–143.

Swami, S., Krishnan, A. V., Wang, J. Y., Jensen, K., Horst, R., Albertelli, A., et al. (2012). Dietary vitamin D3 and 1,25-dihydroxyvitamin D3 (calcitriol) exhibit equivalent anticancer activity in mouse xenograft models of breast and prostate cancer. *Endocrinology, 153*, 2576–2587.

Tannour-Louet, M., Lewis, S. K., Louet, J. F., Stewart, J., Addai, J. B., Sahin, A., et al. (2014). Increased expression of CYP24A1 correlates with advanced stages of prostate cancer and can cause resistance to vitamin D3-based therapies. *FASEB Journal, 28*, 364–372.

Thorne, J. L., Maguire, O., Doig, C. L., Battaglia, S., Fehr, L., Sucheston, L. E., et al. (2011). Epigenetic control of a VDR-governed feed-forward loop that regulates $p21^{(waf1/cip1)}$ expression and function in non-malignant prostate cells. *Nucleic Acids Research, 39*(6), 2045–2056.

Titus, M. A., Schell, M. J., Li, F. B., Tomer, K. B., & Mohler, J. L. (2005). Testosterone and dihydro-testosterone tissue levels in recurrent prostate cancer. *Clinical Cancer Research, 11*, 4653–4657.

Tran, C., Ouk, S., Clegg, N. J., Chen, Y., Watson, P. A., Arora, V., et al. (2009). Development of a second-generation antiandrogen for treatment of advanced prostate cancer. *Science, 324*, 787–790.

Tuoresmäki, P., Väisänen, S., Neme, A., Heikkinen, S., & Carlberg, C. (2014). Patterns of genome-wide VDR locations. *PLoS One, 9*(4), e96105.

Turunen, M. M., Dunlop, T. W., Carlberg, C., & Väisänen, S. (2007). Selective use of multiple vitamin D response elements underlies the 1α, 25-dihydroxyvitamin D3-mediated negative regulation of the human *CYP27B1* gene. *Nucleic Acids Research, 35*, 2734–2747.

VaÈisaÈnen, S., PeraÈkylaÈ, M., KaÈrkkaÈinen, J. I., Steinmeyer, A., & Carlberg, C. (2002). Critical role of helix 12 of the vitamin D3 receptor for the partial agonism of carboxylic ester antagonists. *Journal of Molecular Biology, 315*, 229–238.

Wang, W. W., Chatterjee, N., Chittur, S. V., Welsh, J., & Tenniswood, M. P. (2011). Effects of 1α, 25 dihydroxyvitamin D3 and testosterone on miRNA and mRNA expression in LNCaP cells. *Molecular Cancer, 10*, 58.

Wang, W. L., & Tenniswood, M. (2014). Vitamin D, intermediary metabolism and prostate cancer tumor progression. *Frontiers in Physiology, 5*, 1–9. http://dx.doi.org/10.3389/fphys.2014.00183 (article number: 183).

Wang, W. W., Welsh, J., & Tenniswood, M. (2013). 1, 25 Dihydroxy vitamin D3 modulates lipid metabolism in prostate cancer cells through miRNA mediated regulation of PPARα. *The Journal of Steroid Biochemistry and Molecular Biology, 136*, 247–251.

Wang, Z., Wong, T., Hashizume, T., Dickman, L. Z., Scian, M., Koszewski, N. J., et al. (2014). Human UGT1A4 and UGT1A3 conjugate 25-hydroxyvitamin D_3: Metabolite structure, kinetics, inducibility, and interindividual variability. *Endocrinology, 155*, 2052–2063.

Washington, M. N., & Weigel, N. L. (2010). 1α,25-Dihydroxyvitamin D3 inhibits growth of VCaP prostate cancer cells despite inducing the growth-promoting TMPRSS2:ERG gene fusion. *Endocrinology, 151*, 1409–1417.

Watson, L. C., Kuchenbecker, K. M., Schiller, B. J., Gross, J. D., Pufall, M. A., & Yamamoto, K. R. (2013). The glucocorticoid receptor dimer interface allosterically transmits sequence-specific DNA signals. *Nature Structural and Molecular Biology, 20*, 876–883.

Weigel, N. L. (2007). Interactions between vitamin D and androgen receptor signaling in prostate cancer cells. *Nutrition Reviews, 65*, S116–S117.

Yang, E. S., & Burnstein, K. L. (2003). Vitamin D inhibits G1 to S progression in LNCaP prostate cancer cells through p27Kip1 stabilization and Cdk2 mislocalization to the cytoplasm. *The Journal of Biological Chemistry, 278*, 46862–46868.

Yang, E. S., Maiorino, C. A., Roos, B. A., Knight, S. R., & Burnstein, K. L. (2002). Vitamin D-mediated growth inhibition of an androgen-ablated LNCaP cell line model of human prostate cancer. *Molecular and Cellular Endocrinology, 186*, 69–79.

Yee, S. W., Campbell, M. J., & Simons, C. (2006). Inhibition of vitamin D3 metabolism enhances VDR signalling in androgen-independent prostate cancer cells. *The Journal of Steroid Biochemistry and Molecular Biology, 98*, 228–235.

Zhan, Q., Lord, K. A., Alamo, I., Hollander, M. C., Carrier, F., Ron, D., et al. (1994). The gadd and MyD genes define a novel set of mammalian genes encoding acidic proteins that synergistically suppress cell growth. *Molecular and Cellular Biology, 14*, 2361–2371.

Zhang, J., Chalmers, M. J., Stayrook, K. R., Burris, L. L., Wang, Y., Busby, S. A., et al. (2011). DNA binding alters coactivator interaction surfaces of the intact VDR-RXR complex. *Nature Structural and Molecular Biology, 18*, 556–563.

Zhao, X.-Y., Peehl, D. M., Navone, N. M., & Feldman, D. (2000). 1α, 25-Dihydroxyvitamin D3 inhibits prostate cancer cell growth by androgen-dependent and androgen-independent mechanisms. *Endocrinology, 141*, 2548–2556.

Zhu, J. G., Ochalek, J. T., Kaufmann, M., Jones, G., & DeLuca, H. F. (2013). CYP2R1 is a major, but not exclusive, contributor to 25-hydroxyvitamin D production in vivo. *Proceedings of the National Academy of Sciences of the United States of America, 110*, 15650–15655.

Zhuang, S.-H., & Burnstein, K. L. (1998). Antiproliferative effect of 1α,25-dihydroxyvitamin D3 in human prostate cancer cell line LNCaP involves reduction of cyclin-dependent kinase 2 activity and persistent G1 accumulation. *Endocrinology, 139*, 1197–1207.

CHAPTER FOURTEEN

Metabolism and Action of 25-Hydroxy-19-nor-Vitamin D_3 in Human Prostate Cells

Eiji Munetsuna*,†, Atsushi Kittaka‡, Tai C. Chen§, Toshiyuki Sakaki*,1

*Department of Biotechnology, Faculty of Engineering, Toyama Prefectural University, Toyama, Japan
†Department of Biochemistry, Fujita Health University for Medical Science, Toyoake, Japan
‡Faculty of Pharmaceutical Sciences, Teikyo University, Tokyo, Japan
§Clinical Translational Science Institute and Department of Medicine, Boston University School of Medicine, Boston, Massachusetts, USA
[1]Corresponding author: e-mail address: tsakaki@pu-toyama.ac.jp

Contents

Abstract

Since the discovery of 1α,25(OH)$_2$D$_3$ in the early 1970s, it has been widely accepted that this metabolite is responsible for the biological actions of vitamin D. Likewise, we have assumed that 25(OH)-19-nor-D$_3$-dependent growth inhibition of human prostate PZ-HPV-7 cells was the result of its subsequent conversion to 1α,25(OH)$_2$-19-nor-D$_3$, catalyzed by CYP27B1 within the prostate cells. However, further *in vitro* studies in a reconstituted system using recombinant CYP27B1 revealed that 25(OH)-19-nor-D$_3$ was hardly converted to 1α,25(OH)$_2$-19-nor-D$_3$ by the enzyme. The kinetic analysis of 1α-hydroxylation of 25(OH)D$_3$ and 25(OH)-19-nor-D$_3$ demonstrated that the k_{cat}/K_m for 25(OH)-19-nor-D$_3$ is less than 0.1% of that for 25(OH)D$_3$. When 25(OH)-19-nor-D$_3$ was

Vitamins and Hormones, Volume 100
ISSN 0083-6729
http://dx.doi.org/10.1016/bs.vh.2015.10.009

added to cultured PZ-HPV-7 cells, eight metabolites were detected, while no 1α,25 $(OH)_2$-19-nor-D_3 was found. In addition, the time course of VDR translocation into the nucleus induced by 100 nM 25(OH)-19-nor-D_3, and the subsequent transactivation of *CYP24A1* gene were almost identical to those induced by 1 nM 1α,25$(OH)_2$-19-nor-D_3. These results strongly suggest that 25(OH)-19-nor-D_3 binds directly to VDR as a ligand to transport VDR into the nucleus to induce *CYP24A1* gene transactivation. Furthermore, knockdown of *CYP27B1* gene did not affect the antiproliferative activity of 25(OH)-19-nor-D_3, whereas *VDR* knockdown attenuated the effect, suggesting that the antiproliferative activity of 25(OH)-19-nor-D_3 is VDR dependent but CYP27B1 independent. Finally, our recent studies using the same cell line demonstrate that 25(OH)D_3 can act as a VDR agonist to induce gene transactivation. These findings suggest that vitamin D analogs without 1α-hydroxyl group could be developed as drugs for osteoporosis or cancer treatment.

1. INTRODUCTION

Vitamin D_3 can be synthesized in the skin or obtained from dietary sources (Holick, 2007). Since vitamin D_3 is inert and requires activation, it is generally believed that vitamin D_3 is first converted to 25-hydroxyvitamin D_3 (25(OH)D_3) in the liver by hepatic CYP2R1 and CYP27A1 (Sakaki, Kagawa, Yamamoto, & Inouye, 2005), followed by 1α-hydroxylation in the presence of renal 1α-hydroxylase (CYP27B1), leading to the synthesis of 1α,25$(OH)_2D_3$, the biological active form of the vitamin D_3 (Omdahl, Morris, & May, 2002; Sakaki et al., 2005). The serum level of 1α,25$(OH)_2D_3$ is maintained at a constant level because renal CYP27B1 expression is tightly regulated by various factors. For example, low blood calcium concentration will stimulate CYP27B1 expression, causing an increase in blood calcium level, which in turn downregulates the CYP27B1 activity (Jones & Prosser, 2011).

Genomic actions of 1α,25$(OH)_2D_3$ are mediated by the vitamin D receptor (VDR). Upon binding of the receptor to its agonist, VDR heterodimerizes with retinoid X receptor (RXR) and translocates into the nucleus (Prufer & Barsony, 2002; Racz & Barsony, 1999). The heterodimer then associates with vitamin D-responsive elements in the promoter of vitamin D target genes such as CYP24A1, which catalyzes the degradation of 1α,25$(OH)_2D_3$ and 25(OH)D_3 (Zierold, Darwish, & DeLuca, 1995).

Classically, 1α,25$(OH)_2D_3$ regulates bone formation, calcium, and phosphate homeostasis (Dusso, Brown, & Slatopolsky, 2005). However, during the past two decades, nonclassical effects of $1\alpha,25(OH)_2D_3$ have been demonstrated such as anticancer actions. The first evidence suggesting an association between vitamin D and prostate cancer incidence was reported in 1992 (Hanchette & Schwartz, 1992). The report demonstrated a link between reduced sunlight exposure and prostate cancer risk. The authors therefore hypothesized that vitamin D deficiency could be a risk factor for prostate cancer. Thereafter, many studies have demonstrated the antiproliferative activity of $1\alpha,25(OH)_2D_3$ in prostate cells derived from either normal or cancerous prostate tissues (Chen & Holick, 2003; Feldman, Krishnan, Swami, Giovannucci, & Feldman, 2014; Flanagan et al., 2009; Iglesias-Gato et al., 2011; Munetsuna et al., 2014). Furthermore, animal studies using xenograft models indicated an antitumor effects of dietary vitamin D_3 (Ray et al., 2012) and $1\alpha,25(OH)_2D_3$ (Blutt, Polek, Stewart, Kattan, & Weigel, 2000). Nowadays, the anticancer effects of vitamin D_3 have been well recognized. However, systemic administration of $1\alpha,25(OH)_2D_3$ in clinical trials can cause hypercalcemia and hypercalciuria. Therefore, $1\alpha,25(OH)_2D_3$ is not suitable as a therapeutic agent for cancer treatment (Chen & Holick, 2003; Leyssens, Verlinden, & Verstuyf, 2013). The analogs that are less calcemic but exhibit potent antiproliferative activity have potential as therapeutic agents. Among a large number of vitamin D_3 analogs, it has been reported that A-ring-modified 19-nor-vitamin D compounds have unique biological activity and can alter the VDR–coactivator interaction, resulting in selective potentiation of the transcriptional function (Arai et al., 2005; Kittaka et al., 2000; Konno et al., 2000; Ono et al., 2003; Suhara et al., 2001). In this chapter, we first briefly describe the chemistry and synthesis of 19-nor-vitamin D compounds and then review our current knowledge on the effects of 25-hydroxy-19-nor-vitamin D_3 (25(OH)-19-nor-D_3) on human prostate cells and intracellular metabolism of the analog. Also, based on our current and previous findings, we propose the possibility of a 1α-hydroxylation independent action of the vitamin D_3 and its analogs.

2. THE CHEMISTRY AND SYNTHETIC SCHEMES FOR 25-HYDROXY-19-NOR-VITAMIN D_3

To our knowledge, two approaches to synthesize 25-hydroxy-19-nor-vitamin D_3 [25(OH)-19-nor-D_3] have been reported, and they are summarized in Schemes 1 and 2. First, Mikami's group synthesized the A-ring part of the target molecule of 25(OH)-19-nor-D_3 starting from

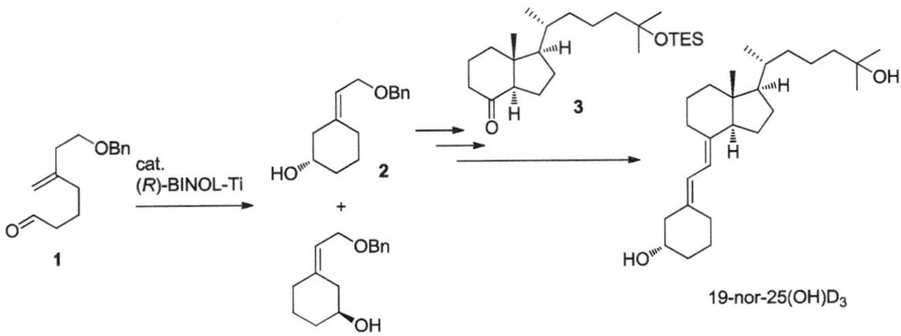

Scheme 1 Mikami's synthetic route to 25(OH)-19-nor-D_3.

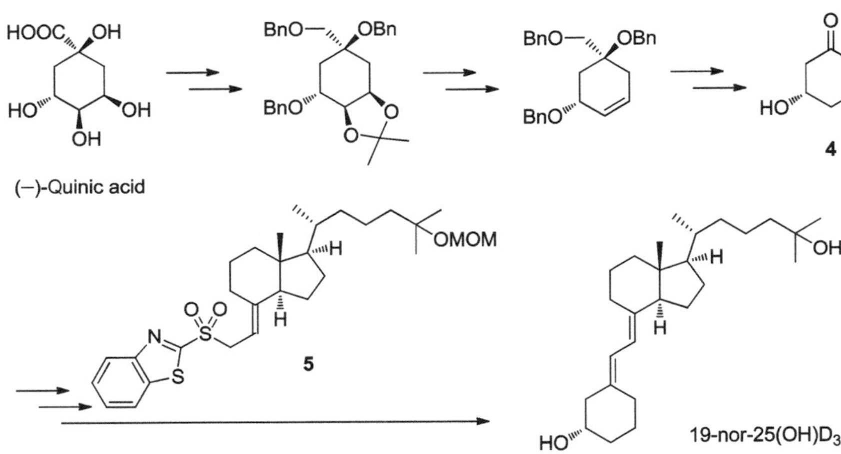

Scheme 2 Kittaka's synthetic route to 25(OH)-19-nor-D_3.

nonchiral aldehyde **1** using an asymmetric catalytic carbonyl-ene cyclization reaction with a binaphthol titanium complex (Okano et al., 2000). The resulting A-ring **2** was coupled with the Grundmann ketone derivative **3** to yield the 19-nor-vitamin D_3 skeleton. Using this synthetic method, 3-*epi*-25(OH)-19-nor-D_3, 3-deoxy-1α(OH)-19-nor-D_3, and 3-deoxy-1α (OH)-19-nor-D_3 were also obtained (Okano et al., 1998).

Second, Kittaka's group synthesized the A-ring precursor (**4**) from (−)-quinic acid (Scheme 2). The ketone **4** was coupled with the CD-ring sulfone **5** utilizing Julia olefination to obtain the target 25(OH)-19-nor-D_3 (Arai et al., 2005). The coupling reaction between the A-ring and the CD-ring parts was based on the 19-nor-2α-(3-hydroxypropyl)-1α,25-dihydroxyvitamin D_3 (MART-10) synthesis (Ono et al., 2003). This synthetic approach also gave 3-deoxy-1α(OH)-19-nor-D_3, and the structure was compared with that of 25(OH)-19-nor-D_3 and determined by NOE experiments of ^{1}H NMR.

3. 19-NOR-VITAMIN D_3 ANALOGS AS THERAPEUTIC AGENT FOR HUMAN PROSTATE CANCER

As mentioned in Section 1, systemic administration of 1α,25(OH)$_2$D$_3$ to cancer patients can cause hypercalcemia and hypercalciuria. Thus, 1α,25(OH)$_2$D$_3$ is not suitable as a therapeutic agent for cancer treatment (Chen & Holick, 2003; Leyssens et al., 2013). The analogs, that are less or non-calcemic but exhibit enhanced anticancer activity, have potential as therapeutic agents for treating cancers.

3.1 Vitamin D and Prostate Cancer

Prostate cancer is the most diagnosed and the second most fetal cancer among men in the US and Northern Europe (Siegel, Miller, & Jemal, 2015). Although the etiology of prostate cancer is not yet completely understood, ecological studies have demonstrated consistently the inversed relationship between sunlight or UVB exposure and the incidence of prostate cancer (Grant, 2012). This suggests that one cause of prostate cancer might be vitamin D insufficiency (Schwartz & Hulka, 1990). However, the epidemiological studies have failed to demonstrate conclusively the association between low serum 25(OH)D levels and prostate cancer incidence and/or mortality in clinical trials. Unlike the epidemiological studies, the biochemical evidence to support a role for vitamin D in prostate cancer is overwhelming (Chen & Holick, 2003; Miller, 1998). VDR has been demonstrated in cultured prostate cells derived from either cancerous or noncancerous prostate tissues (Chen & Holick, 2003; Chen & Kittaka, 2011; Chen, Sakaki, Yamamoto, & Kittaka, 2012). 1α,25(OH)$_2$D$_3$ and its analogs have been shown to inhibit cell proliferation and invasion, and stimulate cellular differentiation and apoptosis in prostate cancer cells and tumor progression in animal models (Bhatia et al., 2009; Blutt et al., 2000; Miller, 1998). These findings provide a strong base for the use of vitamin D compounds as therapeutic agents for prostate cancer under the condition that androgen deprivation therapy has failed. However, several early clinical trials using 1α,25(OH)$_2$D$_3$ revealed that the hormone caused serious hypercalcemia and hypercalceuria side effects when it was administered systemically (Gross, Stamey, Hancock, & Feldman, 1998; Osborn et al., 1995). To overcome these drawbacks, several thousand vitamin D analogs have been synthesized and their biological activities studied, attempting to attenuate calcemic effect, yet to maintain or enhance antitumor potency. For instance, one analog, EB1089 (seocalcitol), with extensive modification at the side chain of the $1\alpha,25(OH)_2D_3$ molecule has been investigated extensively *in vitro*, in animals and, in several cases, in clinical trials for the treatment of various cancers (Kittaka et al., 2012). This analog is less calcemic than $1\alpha,25(OH)_2D_3$ and is more potent than $1\alpha,25(OH)_2D_3$ in inhibiting the growth of human prostate cancer cells in nude mice (Bhatia et al., 2009; Blutt et al., 2000).

3.2 The Biological Activity of 19-nor-Vitamin D Analogs

A group of analogs known as 19-nor-vitamin D analogs, in which the ring A methylene group on C-19 is replaced with two hydrogen atoms, have been synthesized by Deluca's group (Perlman, Sicinski, Schnoes, & DeLuca, 1990) and Kittaka's group (Ono et al., 2003). One of those analogs, 19-nor-$1\alpha,25(OH)_2D_2$, also called Zemplar® or paricalcitol, is a FDA-approved drug for the treatment of secondary hyperparathyroidism. This analog has potency similar to $1\alpha,25(OH)_2D_3$ in inducing *CYP24A1* promoter activity, and in suppressing parathyroid hormone secretion in hemodialysis patients with secondary hyperparathyroidism, without inducing hypercalcemia or hyperphosphatemia (Llach et al., 1998; Perlman et al., 1990; Slatopolsky et al., 1995). The antiproliferative activity of 19-nor-$1\alpha,25(OH)_2D_2$ and 19-nor-$1\alpha,25(OH)_2D_3$ has been studied in LNCaP cells, an androgen-dependent cell line, and in the primary cultures of prostate cancer cells (Chen & Holick, 2003; Chen, Schwartz, Burnstein, Lokeshwar, & Holick, 2000). It was shown that these two 19-nor-vitamin D compounds have activity comparable to $1\alpha,25(OH)_2D_3$.

A substitution with 2α-methyl, 2α-(3-hydroxypropyl), or 2α-(3-hydroxypropoxy) group at the A-ring has been shown to increase their binding affinity for the VDR by two- to fourfold as compared to $1\alpha,25(OH)_2D_3$ (Saito, Honzawa, & Kittaka, 2006; Saito et al., 2004; Takahashi et al., 2006). Further X-ray co-crystallographic analysis of the VDR–ligand complexes using 2α-(3-hydroxypropyl) or 2α-(3-hydroxypropoxy) has revealed an enhanced C2α-effects on VDR binding affinity through the formation of additional hydrogen bonding with arginine-274 residue of the VDR molecule leading to the stabilization of the VDR–ligand complex (Hourai et al., 2006). Based on this X-ray crystallographic finding and the less-calcemic nature of 19-nor structure, the synthesis of a new analog with both features, MART-10, was attempted and was successfully synthesized in good yields for biological activity studies. The antiproliferative activity of MART-10 was first studied in PZ-HPV-7 prostate cells, an established cell line derived from the epithelial zone of a normal prostate, and in two prostate cancer cell lines, LNCaP and PC-3 (Chen et al., 2007; Flanagan et al., 2009; Iglesias-Gato et al., 2011). The results indicate that MART-10 is at least 500- to 1000-fold more active than $1\alpha,25(OH)_2D_3$ in these three cell lines. The higher potency exerted by MART-10 may be due to its higher affinity for VDR, lower affinity to vitamin D binding protein (Flanagan et al., 2009), and more resistance to CYP24A1 degradation (Iglesias-Gato et al., 2011). In addition to its higher activity in inhibiting prostate cancer cell proliferation, MART-10 is about 10-fold more active than $1\alpha,25(OH)_2D_3$ in inhibiting PC-3 cell invasion. More recently, a different analog of 19-nor-vitamin D_3 with additional modification at the side chain, 14-epi-23-yne-$1\alpha,25(OH)_2$-19-nor-vitamin D_3 (Inecalcitol), has been synthesized and its biological activity has been reported (Okamoto et al., 2012). It was demonstrated that this compound delayed tumor growth and decreased tumor weight and vascularity in prostate cancer xenograft mice (Leyssens, Verlinden, & Verstuyf, 2014; Okamoto et al., 2012).

4. THE BIOLOGICAL ACTIVITIES OF 25-HYDROXY-19-NOR-VITAMIN D_3

It has been shown that 19-demethylenated analogs of $1\alpha,25(OH)_2D_3$, i.e., $1\alpha,25(OH)_2$-19-nor-vitamin D_2 and $1\alpha,25(OH)_2$-19-nor-vitamin D_3, possess similar pro-differentiation and antiproliferative activities as $1\alpha,25(OH)_2D_3$ (Chen et al., 2000; Chen, Holick, Lokeshwar, Burnstein, & Schwartz, 2003) and are less calcemic than $1\alpha,25(OH)_2D_3$ when administered systemically (DeLuca et al., 2005; Sicinski et al., 2002). Since it is also known that prostate cells possess 1α-hydroxylase, (1α(OH)ase), and can activate $25(OH)D_3$ to $1\alpha,25(OH)_2D_3$ intracellularly to inhibit their proliferation (Krishnan, Peehl, & Feldman, 2003; Schwartz & Chen, 2005), we therefore hypothesized that 25-hydroxy-19-nor-vitamin D_3 (25(OH)-19-nor-D_3) could exert potent antiproliferative activity toward prostate cells which possess 1α(OH)ase activity through similar activation process and could be used as chemopreventive agents without causing hypercalcemic side effects. Based on the above hypothesis, 25(OH)-19-nor-D_3 was designed, synthesized (Arai et al., 2005), and its metabolism and mode of action were studied as described below.

4.1 The Translocation of VDR into the Nucleus

Most actions of vitamin D_3 are mediated through VDR. Racz and Barsony demonstrated that, following the binding of agonists to VDR, the activated VDR first heterodimerized with RXR and then translocated into the nucleus (Racz & Barsony, 1999). Thus, VDR translocation into the nucleus has been accepted as an indicator of the biological activities of vitamin D and its analogs. Although VDR is found in both cytoplasm and nucleus under normal conditions, $1\alpha,25(OH)_2D_3$ treatment can cause VDR tranlocation into the nucleus (Munetsuna et al., 2014).

Despite the fact that the binding affinity of $1\alpha,25(OH)_2$-19-nor-D_3 for VDR is approximately 4% of that of $1\alpha,25(OH)_2$-D_3 (Urushino et al., 2007), VDR translocation was observed after incubation of cells with the analog (1 nM) for 60–90 min (Munetsuna et al., 2011) (Fig. 1). Interestingly, we found that 25(OH)-19-nor-D_3 also caused VDR translocation (Munetsuna et al., 2011). Similar effect was observed in CYP27B1 knock-down PZ-HPV-7 cells, suggesting that 25(OH)-19-nor-D_3 itself could directly cause the translocation of VDR.

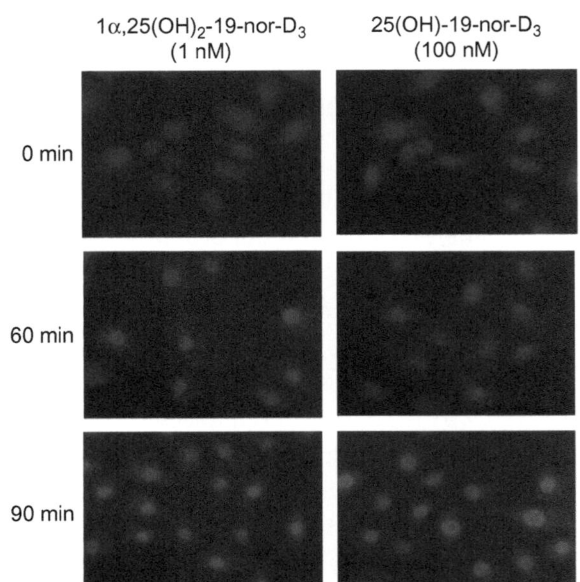

Figure 1 Translocation of VDR into the nucleus by $1\alpha,25(OH)_2$-19-nor-D_3 and 25(OH)-19-nor-D_3. VDR was detected by a fluorescence microscope with antihuman VDR rat monoclonal antibody, 9A7 (Abcam, Cambridge, UK; Munetsuna et al., 2011). (See the color plate.)

4.2 The Transcriptional Activation of *CYP24A1* Gene

Since the nuclear translocation of VDR has been shown to be closely linked to the transcriptional activation (Racz & Barsony, 1999) and both 1α,25$(OH)_2$-19-nor-D_3 and 25(OH)-19-nor-D_3 can induce nuclear translocation of VDR, transcriptional activation of CYP24A1 by VDR in the presence of either analog was examined. We first analyzed CYP24A1 mRNA level, a well-known indicator of the activated VDR (Flanagan et al., 2006; Zierold et al., 1995). Briefly, PZ-HPV-7 cells were incubated with 1α,25$(OH)_2$-19-nor-D_3 (1 and 10 nM) or 25(OH)-19-nor-D_3 (10 and 100 nM) for 2–20 h (Fig. 2). CYP24A1 mRNA was quantified by real-time RT-PCR. As expected, a significant increase in CYP24A1 mRNA level was observed. The CYP24A1 mRNA level was increased as high as 700× with 10 nM 1α,25$(OH)_2$-19-nor-D_3 at 5 h and then declined. The CYP24A1 mRNA level was also increased 150 and 800× with 10 and 100 nM 25(OH)-19-nor-D_3, respectively, in the same time-dependent manner as 1α,25$(OH)_2$-19-nor-D_3.

4.3 The Antiproliferative Effect of 25(OH)-19-nor-D_3 in the Human Prostate Cells

As discussed in Sections 3.1 and 3.2, vitamin D and its analogs have potent activity in suppressing cellular proliferation. Consistent with these findings, we also observed the antiproliferative effects of 25(OH)D_3 and 1α,25

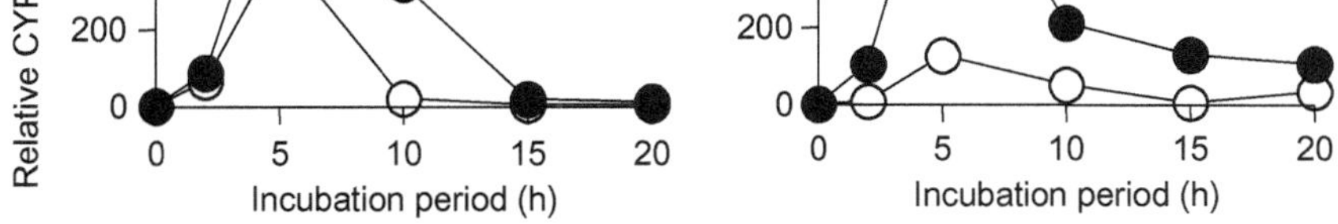

Figure 2 Effect of 1α,25$(OH)_2$-19-nor-D_3 and 25(OH)-19-nor-D_3 on CYP24A1 transcription (Munetsuna et al., 2011). After cells were incubating with 1α,25$(OH)_2$-19-nor-D_3 (1, 10 nM) (left) and 25(OH)-19-nor-D_3 (1, 10 nM) (right), CYP24A1 mRNA levels were quantified by RT-PCR.

$(OH)_2D_3$ in prostate cells (Munetsuna et al., 2014). In addition, we found that proliferation of PZ-HPV-7 cells was inhibited by 25(OH)-19-nor-D_3 (Munetsuna et al., 2011; Fig. 3). Originally, we assumed that the observed effect of this analog might be indirect and was likely the result of its conversion to 1α,25$(OH)_2$-19-nor-D_3. Indeed, 1α,25$(OH)_2$-19-nor-D_3 treatment greatly suppressed proliferation of PZ-HPV-7 cells and primary cultures of prostate cells (Chen et al., 2003; Munetsuna et al., 2011). However, as mentioned above, 25(OH)-19-nor-D_3, unlike 25(OH)D_3, is hardly subjected to 1α-hydroxylation. Thus, we conclude that this effect is more likely mediated by 25(OH)-19-nor-D_3 itself, and not through its conversion to 1α,25$(OH)_2$-19-nor-D_3 as we initially hypothesized. To further support this conclusion, we performed antiproliferative assay in CYP27B1 or VDR knockdown cells and found that knockdown of the enzyme did not affect the biological activities of 25(OH)-19-nor-D_3, while knockdown of the receptor eliminated the effect, indicating a CYP27B1-independent and VDR-dependent action of the analog (Fig. 3).

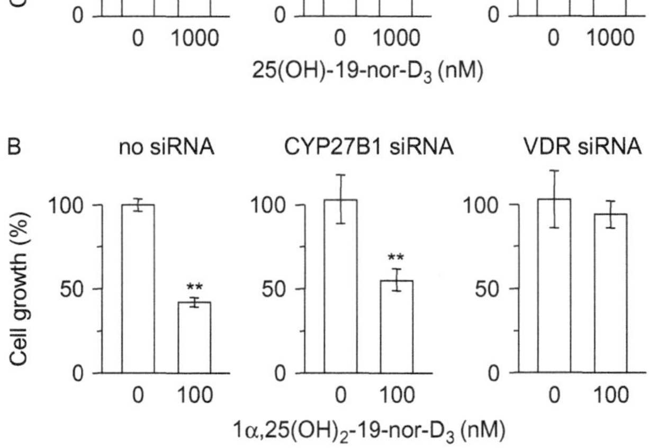

Figure 3 Effect of 25(OH)-19-nor-D_3 on cellular proliferation (Munetsuna et al., 2011). It should be noted that the effect of 25(OH)-19-nor-D_3 was not canceled by CYP27B1 knockdown. While, VDR knockdown ameliorated the effect. *$P < 0.05$, **$P < 0.01$, compared with control cells.

5. THE METABOLISM OF 25-HYDROXY-19-NOR-VITAMIN D_3 IN PROSTATE CELLS

As mentioned in Section 1, the biological active form of vitamin D_3 (1α,25$(OH)_2D_3$) is synthesized from 25(OH)D_3 by CYP27B1, while degradation of the hormone is mediated by CYP24A1. We have investigated extensively the enzymatic properties of CYP enzymes involved in the metabolism of vitamin D_3 and its analogs (Sakaki et al., 2000; Sakaki, Sawada, Nonaka, Ohyama, & Inouye, 1999; Sawada et al., 2004; Sawada, Sakaki, Ohta, & Inouye, 2000; Yasuda et al., 2013). In addition, PZ-HPV-7 prostate cells have been shown to express various enzymes involved in vitamin D metabolism (Chen et al., 2012), we therefore studied whether 25-hydroxy-19-nor-vitamin D_3 could be subject to 1α-hydroxylation first in a cell-free reconstituted system and then in PZ-HPV-7 cells.

5.1 25(OH)-19-nor-D_3 is Hardly Subjected to 1α-Hydroxylation

Using a reconstituted cell-free system containing recombinant CYP24A1, CYP27A1, or CYP27B1, we have identified the metabolites of 19-nor-vitamin D_3 compounds (Urushino et al., 2007). CYP24A1 catalyzes both C-23 and C-24 oxidation pathway and converts 1α,25$(OH)_2$-19-nor-D_3 into various metabolites such as 1α,24,25-trihydroxy-19-nor-vitamin D_3 (1α,24,25$(OH)_3$-19-nor-D_3) and 1α,23,25-trihydroxy-19-nor-vitamin D_3 (1α,23,25$(OH)_3$-19-nor-D_3). We have also revealed that CYP27A1 metabolizes 25(OH)-19-nor-D_3 into a single product, 25,26-dihydroxy-19-nor-vitamin D_3, which has almost no affinity for VDR (Urushino et al., 2007). The same product was also detected as a major metabolite by CYP27B1. To our surprise, 25(OH)-19-nor-D_3 was hardly converted to 1α,25$(OH)_2$-19-nor-D_3 by CYP27B1. The kinetic analysis of 1α-hydroxylation of 25(OH)D_3 and 25(OH)-19-nor-D_3 demonstrated that the k_{cat}/K_m value for 25(OH)-19-nor-D_3 is less than 0.1% of that for 25(OH)D_3 (Urushino et al., 2007).

The extremely low activity of CYP27B1 toward 1α-hydroxylation of 25(OH)-19-nor-D_3 could be explained by the docking model for CYP27B1. Although 25(OH)D_3 prefers the β-form, 25(OH)-19-nor-D_3 prefers the α-form. Distances between hydrogen at the C-1α position and heme iron for 25(OH)D_3 and 25(OH)-19-nor-D_3 were calculated to be 3.9 and 5.3 Å, respectively, resulting in the extremely low activity of CYP27B1 for 25(OH)-19-nor-D_3 (Urushino et al., 2007).

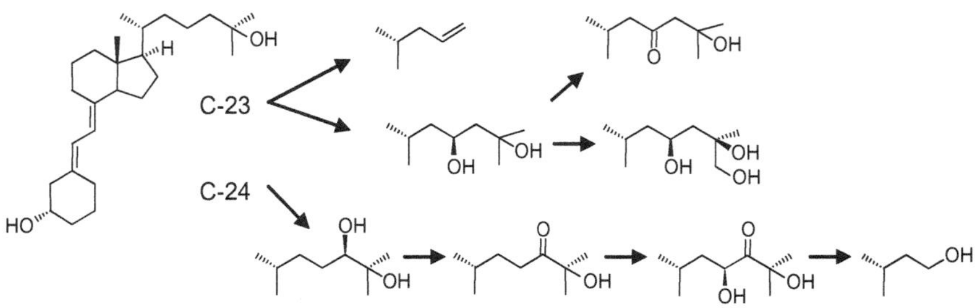

Figure 4 The proposed metabolic pathways of 25(OH)-19-nor-D_3 in PZ-HPV-7 cells (Munetsuna et al., 2011).

5.2 Cellular Metabolism of 25(OH)-19-nor-D_3

After the incubation of PZ-HPV-7 cells with 25(OH)-19-nor-vitamin D_3 (1 μM) for 24 h, at least eight metabolites were detected (Fig. 4). However, 1α,25$(OH)_2$-19-nor-D_3 was not among them, and all of the metabolites appeared to be CYP24A1-dependent metabolites. Judging from our findings that 25(OH)-19-nor-vitamin D_3 is a good substrate for CYP24A1 but not a good one for CYP27B1, these results are expected.

Our recent studies on the metabolism of 25(OH)D_3 in PZ-HPV-7 cells demonstrated that a very low level of 1α,25$(OH)_2D_3$ is produced in the cells (Munetsuna et al., 2014). In the presence of 10 or 100 nM 25(OH)D_3 in the culture medium, only 0.016 and 0.15 nM of 1α,25$(OH)_2D_3$ was detected, respectively. Major products were CYP24A1-dependent metabolites such as 24,25$(OH)_2D_3$ and 24-oxo-25(OH)D_3, and their maximal concentrations were 1–2 nM (Munetsuna et al., 2014). It is noted that even in the metabolism of 25(OH)D_3, which is a much better substrate for CYP27B1 than 25(OH)-19-nor-vitamin D_3, the 1α-hydroxylated product by CYP27B1 was quite low. Thus, no detectable conversion of 25(OH)-19-nor-vitamin D_3 to 1α,25$(OH)_2$-19-nor-D_3 in PZ-HPV-7 cells is reasonable.

6. NOVEL MECHANISM OF ACTION OF 25-HYDROXY-19-NOR-VITAMIN D_3

1α-Hydroxylation is believed to play a pivotal role for the biological effects of vitamin D. The affinity of 1α,25$(OH)_2D_3$ is much higher (about 500 ×) than that of 25(OH)D_3. Most vitamin D researchers may consider that endogenous ligand for VDR is only 1α,25$(OH)_2D_3$. However, based on the fact that vitamin D-dependent rickets type I patients who have no CYP27B1 activity were treated with vitamin D_3 before the identification of $1\alpha,25(OH)_2D_3$ (Fraser et al., 1973; Rhieu et al., 2014), high amounts of $25(OH)D_3$ might cause biological effects similar to $1\alpha,25(OH)_2D_3$. Consistent with this conception, Peng and colleagues demonstrated that $25(OH)D_3$ itself activated transcriptional activation of VDR target genes such as CYP24A1 (Peng, Hawthorne, Vaishnav, St-Arnaud, & Mehta, 2009). Our recent report also demonstrated the biological activity of $25(OH)D_3$ itself (Munetsuna et al., 2014).

So far, many biologically active vitamin D_3 analogs have been synthesized. Most of them are 1α-hydroxylated molecules. We found that the non-1α-hydroxylated analog 25(OH)-19-nor-D_3 has unique biological activity. Consistent with our results, Rhieu et al. (2014) also reported that a CYP27B1-independent action of a vitamin D_3 analog, 25-hydroxy-16-ene-23-yne-vitamin D_3. This analog has potent antiproliferative effects on PZ-HPV-7 cell, and this analog resists the catalytic activity of CYP27B1 and CYP24A1.

Based on our results, we propose a novel action model for 25(OH)-19-nor-D_3 as shown in Fig. 5. This model shows a direct action of 25(OH)-19-nor-D_3 as a ligand of VDR, and it is clearly distinct from a classical model

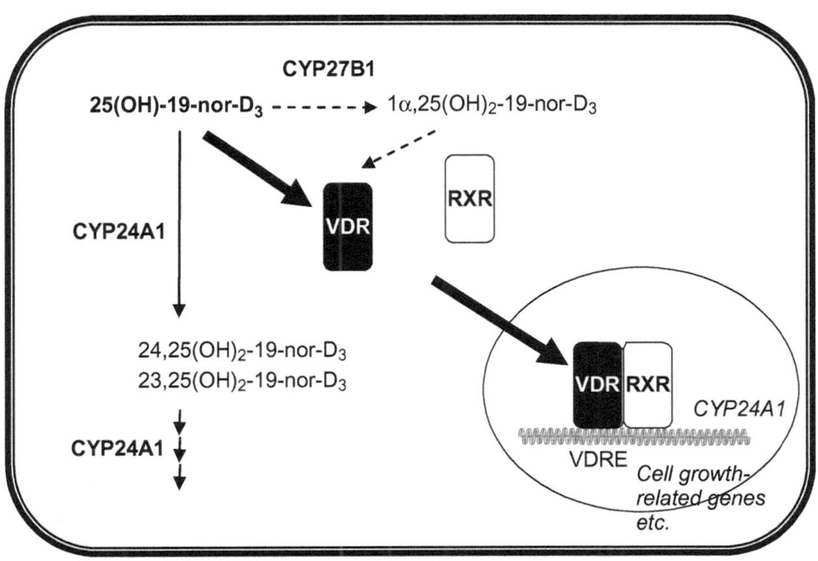

Figure 5 The proposed mode of 25(OH)-19-nor-D_3 action in PZ-HPV-7 cells based on our previous study (Munetsuna et al., 2011). 25(OH)-19-nor-D_3 directly binds to VDR as a ligand to induce transcription of many genes containing CYP24A1 and cell growth-related genes. 25(OH)-19-nor-D_3 was predominantly metabolized by CYP24A1 induced after the addition of 25(OH)-19-nor-D_3.

that $25(OH)D_3$ analogs exert their biological functions after 1α-hydroxylation by CYP27B1. It should be noted that the model in Fig. 5 resembles the action model proposed by Tuohimaa's group, suggesting a unique biological activity of $25(OH)D_3$ as a ligand for VDR in human prostate cells (Lou et al., 2004, 2010). Tuohimaa's group first reported the effect of non-1α-hydroxylated vitamin D_3 (Lou et al., 2004, 2010; Tuohimaa et al., 2005). Recently, DeLuca and associates also demonstrated a direct action of $25(OH)D_3$ as a VDR ligand in Cyp27b1 knockout mice (DeLuca, Prahl, & Plum, 2011). Consistent with those reports, we have confirmed the effects of $25(OH)D_3$ itself. The addition of 100 nM $25(OH)D_3$ decreased cell growth even in the CYP27B1 siRNA-transfected, CYP27B1-knowdown PZ-HPV-7 cells while VDR knockdown eliminated the antiproliferative activity of $25(OH)D_3$ (Munetsuna et al., 2014). These results strongly suggest a direct action of $25(OH)D_3$ itself as a VDR ligand. Our results may explain a significant correlation between the morbidity of prostate cancer and serum $25(OH)D_3$ levels (Ahonen, Tenkanen, Teppo, Hakama, & Tuohimaa, 2000; Polek & Weigel, 2002). The action model of 25(OH)-19-nor-D_3 shown in Fig. 5 suggests that $25(OH)D_3$ analogs, such as 25(OH)-19-nor-D_3, are attractive candidates for anticancer drugs with much less-calcemic activity.

7. CONCLUSIONS AND FUTURE DIRECTIONS

So far, no significant correlation between the morbidity of prostate cancer and genotypes of CYP27B1 has been observed (Hawkins et al., 2002). Barger-Lux et al. have suggested that 55–90% of the biological activities of vitamin D_3 metabolites are mediated by $25(OH)D_3$ in calcium absorption efficiency tests (Barger-Lux, Heaney, Lanspa, Healy, & DeLuca, 1995). Furthermore, Lou et al. have demonstrated that $25(OH)D_3$ but not $1\alpha,25(OH)_2D_3$ inhibits prostate stromal cell growth (Lou et al., 2004). In some normal prostate cell lines such as E-CZ-2 and E-PZ-8, and primary cultured prostate cells, cell growth has been shown to be inhibited by $25(OH)D_3$ and $1\alpha,25(OH)_2D_3$ (Chen et al., 2000). More recently, Tuohimaa's group reported the differential effect of vitamin D_3 metabolites such as $25(OH)D_3$ and $24,25(OH)_2D_3$ (Tuohimaa et al., 2013). They present a new conceptual insight into the vitamin D_3 endocrine system, which may guide the strategic use of vitamin D_3 in disease prevention and treatment.

Several reports demonstrated that antiproliferative effects of vitamin D_3 and its analogs are mediated by various mechanisms. For example, it has been reported that antiproliferative effect of 1α,25$(OH)_2D_3$ involves upregulation of cyclin-dependent kinase inhibitor p21 in human prostate cancer cells (Kittaka et al., 2012; Krishnan et al., 2003). Also, vitamin D_3 stimulates caspase-3 and enhances mitochondrial cytochrome *c* release in prostate cancer cell lines such as LNCaP and ALVA-31 (Guzey, Kitada, & Reed, 2002; Leyssens et al., 2013). Interestingly, prostate cellular proliferation appears to be regulated by semaphorin3B, cystatin D, and cystatin E/M according to our DNA microarray data with 25(OH)D_3. Thus, antiproliferative effects of 25(OH)-19-nor-D_3 could be mediated by these proteins (Munetsuna et al., 2014). These findings suggest that such non-1α-hydroxylated vitamin D analogs could be promising for prostate cancer treatment.

As shown in our previous report (Sakaki, Yasuda, Kittaka, Yamamoto, & Chen, 2014), vitamin D analogs which are resistant to CYP24A1-dependent catabolism could be useful for osteoporosis or cancer treatment. Actually, a CYP24A1-resistant analog eldecalcitol (ED-71), a 2β-hydroxylpropoxylated analog of 1α,25$(OH)_2D_3$, has been developed by Chugai Pharmaceutical Co., Ltd (Japan) and was approved as a new drug for the treatment of osteoporosis in Japan in 2011 (Matsumoto, 2012; Matsumoto & Kubodera, 2007; Matsumoto et al., 2005; Matsumoto, Takano, Yamakido, Takahashi, & Tsuji, 2010; Miyamoto, Murayama, Ochi, Watanabe, & Kubodera, 1993). Eldecalcitol shows strong inhibitory effects on bone resorption and increase in bone mineral density (BMD) to prevent osteoporotic fractures. On the other hand, another CYP24A1-resistant analog, 19-nor-2α-(3-hydroxypropyl)-1α,25$(OH)_2D_3$, (MART-10) has a potent antiproliferative activity with low calcemic effect, which is a suitable property as an anticancer drug. It is noted that both compounds have high affinity for VDR and resistance to CYP24A1-dependent catabolism. Using the same approach, we could substitute the carbon-2 of 25(OH)-19-nor-D_3 molecule with a 2α-(3-hydroxypropyl) moiety to enhance its VDR binding affinity and make it more resistant to CYP24A1 degradation. On the other hand, as mentioned above, 25(OH)-19-nor-D_3 is resistant to CYP27B1-dependent hydroxylation, it may be a useful compound to study the biological effects of 25-hydroxylated vitamin D compounds, instead of using 25(OH)D_3 itself. Recently, we have accumulated data on biological effects of 25(OH)D_3 itself using CYP27B1 knockout (KO) mice. BMD of the KO mice was dramatically increased by the administration of 25(OH)D_3 (data not shown). If the administration of 25(OH)-19-nor-D_3 or its C2 modified analogs to the osteoporosis animal models such as ovariectomized mice or rats can increase their BMD, vitamin D analogs without 1α-hydroxyl group, such as 19-nor-2α-(3-hydroxypropyl)-25(OH)D_3, could be developed as safe drugs for osteoporosis and cancer treatment.

REFERENCES

Ahonen, M. H., Tenkanen, L., Teppo, L., Hakama, M., & Tuohimaa, P. (2000). Prostate cancer risk and prediagnostic serum 25-hydroxyvitamin D levels (Finland). *Cancer Causes & Control, 11*(9), 847–852.

Arai, A., Tsutsumi, R., Hara, H., Chen, T. C., Sakaki, T., Urushino, N., et al. (2005). Synthesis of 25-hydroxy-19-norvitamin D3 analogs and their antiproliferative activities on prostate cells. *Heterocycles, 66*(3–5), 469–479.

Barger-Lux, M. J., Heaney, R. P., Lanspa, S. J., Healy, J. C., & DeLuca, H. F. (1995). An investigation of sources of variation in calcium absorption efficiency. *The Journal of Clinical Endocrinology and Metabolism, 80*(2), 406–411.

Bhatia, V., Saini, M. K., Shen, X., Bi, L. X., Qiu, S., Weigel, N. L., et al. (2009). EB1089 inhibits the parathyroid hormone-related protein-enhanced bone metastasis and xenograft growth of human prostate cancer cells. *Molecular Cancer Therapeutics, 8*(7), 1787–1798. http://dx.doi.org/10.1158/1535-7163.mct-09-0064.

Blutt, S. E., Polek, T. C., Stewart, L. V., Kattan, M. W., & Weigel, N. L. (2000). A calcitriol analogue, EB1089, inhibits the growth of LNCaP tumors in nude mice. *Cancer Research, 60*(4), 779–782.

Chen, T. C., & Holick, M. F. (2003). Vitamin D and prostate cancer prevention and treatment. *Trends in Endocrinology and Metabolism, 14*(9), 423–430.

Chen, T. C., Holick, M. F., Lokeshwar, B. L., Burnstein, K. L., & Schwartz, G. G. (2003). Evaluation of vitamin D analogs as therapeutic agents for prostate cancer. *Recent Results in Cancer Research, 164*, 273–288.

Chen, T. C., & Kittaka, A. (2011). Novel vitamin D analogs for prostate cancer therapy. *ISRN Urology, 2011*, 301490. http://dx.doi.org/10.5402/2011/301490.

Chen, T. C., Persons, K. S., Zheng, S., Mathieu, J., Holick, M. F., Lee, Y. F., et al. (2007). Evaluation of C-2-substituted 19-nor-1alpha,25-dihydroxyvitamin D3 analogs as therapeutic agents for prostate cancer. *The Journal of Steroid Biochemistry and Molecular Biology, 103*(3–5), 717–720. http://dx.doi.org/10.1016/j.jsbmb.2006.12.009.

Chen, T. C., Sakaki, T., Yamamoto, K., & Kittaka, A. (2012). The roles of cytochrome P450 enzymes in prostate cancer development and treatment. *Anticancer Research, 32*(1), 291–298.

Chen, T. C., Schwartz, G. G., Burnstein, K. L., Lokeshwar, B. L., & Holick, M. F. (2000). The in vitro evaluation of 25-hydroxyvitamin D3 and 19-nor-1alpha,25-dihydroxyvitamin D2 as therapeutic agents for prostate cancer. *Clinical Cancer Research, 6*(3), 901–908.

DeLuca, H. F., Plum, L. A., Clagett-Dame, M., Shevde, N. K., Pike, J. W., & Sicinski, R. R. (2005). 2-Carbon-modified analogs of 19-nor-1α,25-dihydroxyvittamin D_3. In D. Feldman, J. W. Pike, & F. H. Glorieux (Eds.), *Vitamin D* (pp. 1543–1555). New York, NY: Elsevier Academic Press.

DeLuca, H. F., Prahl, J. M., & Plum, L. A. (2011). 1,25-Dihydroxyvitamin D is not responsible for toxicity caused by vitamin D or 25-hydroxyvitamin D. *Archives of Biochemistry and Biophysics, 505*(2), 226–230. http://dx.doi.org/10.1016/j.abb.2010.10.012.

Dusso, A. S., Brown, A. J., & Slatopolsky, E. (2005). Vitamin D. *American Journal of Physiology. Renal Physiology*, *289*(1), F8–F28. http://dx.doi.org/10.1152/ajprenal.00336.2004.

Feldman, D., Krishnan, A. V., Swami, S., Giovannucci, E., & Feldman, B. J. (2014). The role of vitamin D in reducing cancer risk and progression. *Nature Reviews Cancer*, *14*(5), 342–357. http://dx.doi.org/10.1038/nrc3691.

Flanagan, J. N., Young, M. V., Persons, K. S., Wang, L., Mathieu, J. S., Whitlatch, L. W., et al. (2006). Vitamin D metabolism in human prostate cells: Implications for prostate cancer chemoprevention by vitamin D. *Anticancer Research*, *26*(4A), 2567–2572.

Flanagan, J. N., Zheng, S., Chiang, K. C., Kittaka, A., Sakaki, T., Nakabayashi, S., et al. (2009). Evaluation of 19-nor-2alpha-(3-hydroxypropyl)-1alpha,25-dihydroxyvitamin D3 as a therapeutic agent for androgen-dependent prostate cancer. *Anticancer Research*, *29*(9), 3547–3553.

Fraser, D., Kooh, S. W., Kind, H. P., Holick, M. F., Tanaka, Y., & DeLuca, H. F. (1973). Pathogenesis of hereditary vitamin-D-dependent rickets. An inborn error of vitamin D metabolism involving defective conversion of 25-hydroxyvitamin D to 1 alpha,25-dihydroxyvitamin D. *The New England Journal of Medicine*, *289*(16), 817–822. http://dx.doi.org/10.1056/nejm197310182891601.

Grant, W. B. (2012). Role of solar UVB irradiance and smoking in cancer as inferred from cancer incidence rates by occupation in Nordic countries. *Dermatoendocrinol*, *4*(2), 203–211. http://dx.doi.org/10.4161/derm.20965.

Gross, C., Stamey, T., Hancock, S., & Feldman, D. (1998). Treatment of early recurrent prostate cancer with 1,25-dihydroxyvitamin D3 (calcitriol). *The Journal of Urology*, *159*(6), 2035–2039. (discussion 2039–2040).

Guzey, M., Kitada, S., & Reed, J. C. (2002). Apoptosis induction by 1alpha,25-dihydroxyvitamin D3 in prostate cancer. *Molecular Cancer Therapeutics*, *1*(9), 667–677.

Hanchette, C. L., & Schwartz, G. G. (1992). Geographic patterns of prostate cancer mortality. Evidence for a protective effect of ultraviolet radiation. *Cancer*, *70*(12), 2861–2869.

Hawkins, G. A., Cramer, S. D., Zheng, S. L., Isaacs, S. D., Wiley, K. E., Chang, B. L., et al. (2002). Sequence variants in the human 25-hydroxyvitamin D3 1-alpha-hydroxylase (CYP27B1) gene are not associated with prostate cancer risk. *The Prostate*, *53*(3), 175–178. http://dx.doi.org/10.1002/pros.10144.

Holick, M. F. (2007). Vitamin D deficiency. *The New England Journal of Medicine*, *357*(3), 266–281. http://dx.doi.org/10.1056/NEJMra070553.

Hourai, S., Fujishima, T., Kittaka, A., Suhara, Y., Takayama, H., Rochel, N., et al. (2006). Probing a water channel near the A-ring of receptor-bound 1 alpha,25-dihydroxyvitamin D3 with selected 2 alpha-substituted analogues. *Journal of Medicinal Chemistry*, *49*(17), 5199–5205. http://dx.doi.org/10.1021/jm0604070.

Iglesias-Gato, D., Zheng, S., Flanagan, J. N., Jiang, L., Kittaka, A., Sakaki, T., et al. (2011). Substitution at carbon 2 of 19-nor-1alpha,25-dihydroxyvitamin D3 with 3-hydroxypropyl group generates an analogue with enhanced chemotherapeutic potency in PC-3 prostate cancer cells. *The Journal of Steroid Biochemistry and Molecular Biology*, *127*(3–5), 269–275. http://dx.doi.org/10.1016/j.jsbmb.2011.08.010.

Jones, G., & Prosser, D. E. (2011). *The activating enzymes of vitamin D metabolism (25- and 1α-Hydroxylase). Vitamin D* (3rd ed.). Amsterdam: Elsevier. (Chapter 3).

Kittaka, A., Suhara, Y., Takayanagi, H., Fujishima, T., Kurihara, M., & Takayama, H. (2000). A concise and efficient route to 2alpha-(omega-hydroxyalkoxy)-1alpha,25-dihydroxyvi tam in D3: Remarkably high affinity to vitamin D receptor. *Organic Letters*, *2*(17), 2619–2622.

Kittaka, A., Yoshida, A., Chiang, K. C., Takano, M., Sawada, D., Sakaki, T., et al. (2012). Potent 19-norvitamin D analogs for prostate and liver cancer therapy. *Future Medicinal Chemistry*, *4*(16), 2049–2065. http://dx.doi.org/10.4155/fmc.12.130.

Konno, K., Fujishima, T., Maki, S., Liu, Z., Miura, D., Chokki, M., et al. (2000). Synthesis, biological evaluation, and conformational analysis of A-ring diastereomers of 2-methyl-1,25-dihydroxyvitamin D(3) and their 20-epimers: Unique activity profiles depending on the stereochemistry of the A-ring and at C-20. *Journal of Medicinal Chemistry, 43*(22), 4247–4265.

Krishnan, A. V., Peehl, D. M., & Feldman, D. (2003). The role of vitamin D in prostate cancer. Recent results in cancer research. *Fortschritte der Krebsforschung. Progres Dans les Recherches Sur le Cancer, 164*, 205–221.

Leyssens, C., Verlinden, L., & Verstuyf, A. (2013). Antineoplastic effects of 1,25(OH)2D3 and its analogs in breast, prostate and colorectal cancer. *Endocrine-Related Cancer, 20*(2), R31–R47. http://dx.doi.org/10.1530/erc-12-0381.

Leyssens, C., Verlinden, L., & Verstuyf, A. (2014). The future of vitamin D analogs. *Frontiers in Physiology, 5*, 122. http://dx.doi.org/10.3389/fphys.2014.00122.

Llach, F., Keshav, G., Goldblat, M. V., Lindberg, J. S., Sadler, R., Delmez, J., et al. (1998). Suppression of parathyroid hormone secretion in hemodialysis patients by a novel vitamin D analogue: 19-nor-1,25-dihydroxyvitamin D_2. *American Journal of Kidney Diseases, 32*(2 Suppl. 2), S48–S54.

Lou, Y. R., Laaksi, I., Syvala, H., Blauer, M., Tammela, T. L., Ylikomi, T., et al. (2004). 25-Hydroxyvitamin D_3 is an active hormone in human primary prostatic stromal cells. *The FASEB Journal: Official Publication of the Federation of American Societies for Experimental Biology, 18*(2), 332–334. http://dx.doi.org/10.1096/fj.03-0140fje.

Lou, Y. R., Molnar, F., Perakyla, M., Qiao, S., Kalueff, A. V., St-Arnaud, R., et al. (2010). 25-Hydroxyvitamin D(3) is an agonistic vitamin D receptor ligand. *The Journal of Steroid Biochemistry and Molecular Biology, 118*(3), 162–170. http://dx.doi.org/10.1016/j.jsbmb.2009.11.011.

Matsumoto, T. (2012). Osteoporosis treatment by a new active vitamin D3 compound, eldecalcitol, in Japan. *Current Osteoporosis Reports, 10*(4), 248–250. http://dx.doi.org/10.1007/s11914-012-0116-1.

Matsumoto, T., & Kubodera, N. (2007). ED-71, a new active vitamin D3, increases bone mineral density regardless of serum 25(OH)D levels in osteoporotic subjects. *The Journal of Steroid Biochemistry and Molecular Biology, 103*(3–5), 584–586. http://dx.doi.org/10.1016/j.jsbmb.2006.12.088. S0960-0760(06)00454-7 [pii].

Matsumoto, T., Miki, T., Hagino, H., Sugimoto, T., Okamoto, S., Hirota, T., et al. (2005). A new active vitamin D, ED-71, increases bone mass in osteoporotic patients under vitamin D supplementation: A randomized, double-blind, placebo-controlled clinical trial. *The Journal of Clinical Endocrinology and Metabolism, 90*(9), 5031–5036. http://dx.doi.org/10.1210/jc.2004-2552. jc.2004-2552 [pii].

Matsumoto, T., Takano, T., Yamakido, S., Takahashi, F., & Tsuji, N. (2010). Comparison of the effects of eldecalcitol and alfacalcidol on bone and calcium metabolism. *The Journal of Steroid Biochemistry and Molecular Biology, 121*(1–2), 261–264. http://dx.doi.org/10.1016/j.jsbmb.2010.03.035. S0960-0760(10)00126-3 [pii].

Miller, G. J. (1998). Vitamin D and prostate cancer: Biologic interactions and clinical potentials. *Cancer Metastasis Reviews, 17*(4), 353–360.

Miyamoto, K., Murayama, E., Ochi, K., Watanabe, H., & Kubodera, N. (1993). Synthetic studies of vitamin D analogues. XIV. Synthesis and calcium regulating activity of vitamin D3 analogues bearing a hydroxyalkoxy group at the 2 beta-position. *Chemical & Pharmaceutical Bulletin (Tokyo), 41*(6), 1111–1113.

Munetsuna, E., Kawanami, R., Nishikawa, M., Ikeda, S., Nakabayashi, S., Yasuda, K., et al. (2014). Anti-proliferative activity of 25-hydroxyvitamin D3 in human prostate cells. *Molecular and Cellular Endocrinology, 382*(2), 960–970. http://dx.doi.org/10.1016/j.mce.2013.11.014.

Munetsuna, E., Nakabayashi, S., Kawanami, R., Yasuda, K., Ohta, M., Arai, M. A., et al. (2011). Mechanism of the anti-proliferative action of 25-hydroxy-19-nor-vitamin D3 in human prostate cells. *Journal of Molecular Endocrinology*. http://dx.doi.org/10.1530/JME-11-0008.

Okamoto, R., Delansorne, R., Wakimoto, N., Doan, N. B., Akagi, T., Shen, M., et al. (2012). Inecalcitol, an analog of 1alpha,25(OH)(2) D(3), induces growth arrest of androgen-dependent prostate cancer cells. *International Journal of Cancer*, *130*(10), 2464–2473. http://dx.doi.org/10.1002/ijc.26279.

Okano, T., Nakagawa, K., Kubodera, N., Ozono, K., Isaka, A., Osawa, A., et al. (2000). Catalytic asymmetric syntheses and biological activities of singly dehydroxylated 19-nor-1alpha,25-dihydroxyvitamin D(3) A-ring analogs in cancer cell differentiation and apoptosis. *Chemistry & Biology*, 7(3), 173–184.

Okano, T., Nakagawa, K., Tsugawa, N., Ozono, K., Kubodera, N., Osawa, A., et al. (1998). Singly dehydroxylated A-ring analogues of 19-nor-1alpha,25-dihydroxyvitamin D3 and 19-nor-22-oxa-1alpha,25-dihydroxyvitamin D3: Novel vitamin D3 analogues with potent transcriptional activity but extremely low affinity for vitamin D receptor. *Biological & Pharmaceutical Bulletin*, *21*(12), 1300–1305.

Omdahl, J. L., Morris, H. A., & May, B. K. (2002). Hydroxylase enzymes of the vitamin D pathway: Expression, function, and regulation. *Annual Review of Nutrition*, *22*, 139–166. http://dx.doi.org/10.1146/annurev.nutr.22.120501.150216.

Ono, K., Yoshida, A., Saito, N., Fujishima, T., Honzawa, S., Suhara, Y., et al. (2003). Efficient synthesis of 2-modified 1alpha,25-dihydroxy-19-norvitamin D3 with Julia olefination: High potency in induction of differentiation on HL-60 cells. *The Journal of Organic Chemistry*, *68*(19), 7407–7415. http://dx.doi.org/10.1021/jo034787y.

Osborn, J. L., Schwartz, G. G., Smith, D. C., Bahnson, R., Day, R., & Trump, D. L. (1995). Phase II trial of oral 1,25-dihydroxyvitamin D (calcitriol) in hormone refractory prostate cancer. *Urologic Oncology*, *1*(5), 195–198.

Peng, X., Hawthorne, M., Vaishnav, A., St-Arnaud, R., & Mehta, R. G. (2009). 25-Hydroxyvitamin D3 is a natural chemopreventive agent against carcinogen induced precancerous lesions in mouse mammary gland organ culture. *Breast Cancer Research and Treatment*, *113*(1), 31–41. http://dx.doi.org/10.1007/s10549-008-9900-0.

Perlman, K. L., Sicinski, R. R., Schnoes, H. K., & DeLuca, H. F. (1990). 1α,25-Dihydroxy-19-nor-vitamin D3, a novel vitamin D-related compound with potential therapeutic activity. *Tetrahedron Letters*, *31*(13), 1823–1824.

Polek, T. C., & Weigel, N. L. (2002). Vitamin D and prostate cancer. *Journal of Andrology*, *23*(1), 9–17.

Prufer, K., & Barsony, J. (2002). Retinoid X receptor dominates the nuclear import and export of the unliganded vitamin D receptor. *Molecular Endocrinology*, *16*(8), 1738–1751.

Racz, A., & Barsony, J. (1999). Hormone-dependent translocation of vitamin D receptors is linked to transactivation. *The Journal of Biological Chemistry*, *274*(27), 19352–19360.

Ray, R., Banks, M., Abuzahra, H., Eddy, V. J., Persons, K. S., Lucia, M. S., et al. (2012). Effect of dietary vitamin D and calcium on the growth of androgen-insensitive human prostate tumor in a murine model. *Anticancer Research*, *32*(3), 727–731.

Rhieu, S. Y., Annalora, A. J., LaPorta, E., Welsh, J., Itoh, T., Yamamoto, K., et al. (2014). Potent antiproliferative effects of 25-hydroxy-16-ene-23-yne-vitamin D(3) that resists the catalytic activity of both CYP27B1 and CYP24A1. *Journal of Cellular Biochemistry*, *115*(8), 1392–1402. http://dx.doi.org/10.1002/jcb.24789.

Saito, N., Honzawa, S., & Kittaka, A. (2006). Recent results on A-ring modification of 1alpha,25-dihydroxyvitamin D3: Design and synthesis of VDR-agonists and antagonists with high biological activity. *Current Topics in Medicinal Chemistry*, *6*(12), 1273–1288.

Saito, N., Suhara, Y., Kurihara, M., Fujishima, T., Honzawa, S., Takayanagi, H., et al. (2004). Design and efficient synthesis of 2 alpha-(omega-hydroxyalkoxy)-1 alpha,25-dihydroxyvitamin D3 analogues, including 2-epi-ED-71 and their 20-epimers with HL-60 cell differentiation activity. *The Journal of Organic Chemistry*, *69*(22), 7463–7471. http://dx.doi.org/10.1021/jo0491051.

Sakaki, T., Kagawa, N., Yamamoto, K., & Inouye, K. (2005). Metabolism of vitamin D3 by cytochromes P450. *Frontiers in Bioscience: A Journal and Virtual Library*, *10*, 119–134.

Sakaki, T., Sawada, N., Komai, K., Shiozawa, S., Yamada, S., Yamamoto, K., et al. (2000). Dual metabolic pathway of 25-hydroxyvitamin D3 catalyzed by human CYP24. *European Journal of Biochemistry/FEBS*, *267*(20), 6158–6165.

Sakaki, T., Sawada, N., Nonaka, Y., Ohyama, Y., & Inouye, K. (1999). Metabolic studies using recombinant *Escherichia coli* cells producing rat mitochondrial CYP24 CYP24 can convert 1alpha,25-dihydroxyvitamin D3 to calcitroic acid. *European Journal of Biochemistry/FEBS*, *262*(1), 43–48.

Sakaki, T., Yasuda, K., Kittaka, A., Yamamoto, K., & Chen, T. C. (2014). CYP24A1 as a potential target for cancer therapy. *Anti-Cancer Agents in Medicinal Chemistry*, *14*(1), 97–108.

Sawada, N., Kusudo, T., Sakaki, T., Hatakeyama, S., Hanada, M., Abe, D., et al. (2004). Novel metabolism of 1 alpha,25-dihydroxyvitamin D3 with C24-C25 bond cleavage catalyzed by human CYP24A1. *Biochemistry*, *43*(15), 4530–4537. http://dx.doi.org/10.1021/bi030207f.

Sawada, N., Sakaki, T., Ohta, M., & Inouye, K. (2000). Metabolism of vitamin D(3) by human CYP27A1. *Biochemical and Biophysical Research Communications*, *273*(3), 977–984. http://dx.doi.org/10.1006/bbrc.2000.3050.

Schwartz, G. G., & Chen, T. C. (2005). Vitamin D, sunlight, and the natural history of prostate cancer. In D. Feldman, J. W. Pike, & F. H. Glorieux (Eds.), *Vitamin D* (pp. 1599–1615). New York, NY: Elsevier Academic Press.

Schwartz, G. G., & Hulka, B. S. (1990). Is vitamin D deficiency a risk factor for prostate cancer? (Hypothesis). *Anticancer Research*, *10*(5A), 1307–1311.

Sicinski, R. R., Rotkiewics, P., Kolinski, A., Sicinska, W., Prahl, J. M., Smith, C. M., et al. (2002). 2-Ethyl and 2-ethylidene analogues of 1α,25-dihydroxy-19-norvitamin D3: Synthesis, conformational analysis, biological activities, and docking to the modeled rVDR ligand binding domain. *Journal of Medicinal Chemistry*, *45*, 3366–3380.

Siegel, R. L., Miller, K. D., & Jemal, A. (2015). Cancer statistics, 2015. *CA: A Cancer Journal for Clinicians*, *65*(1), 5–29.

Slatopolsky, E., Finch, J., Ritter, C., Denda, M., Morrissey, J., Brown, A., et al. (1995). A new analog of calcitriol, 19-nor-1,25-(OH)2D2, suppresses parathyroid hormone secretion in uremic rats in the absence of hypercalcemia. *American Journal of Kidney Diseases*, *26*(5), 852–860.

Suhara, Y., Nihei, K. I., Kurihara, M., Kittaka, A., Yamaguchi, K., Fujishima, T., et al. (2001). Efficient and versatile synthesis of novel 2alpha-substituted 1alpha,25-dihydroxyvitamin D(3) analogues and their docking to vitamin D receptors. *The Journal of Organic Chemistry*, *66*(26), 8760–8771.

Takahashi, E., Nakagawa, K., Suhara, Y., Kittaka, A., Nihei, K., Konno, K., et al. (2006). Biological activities of 2alpha-substituted analogues of 1alpha,25-dihydroxyvitamin D3 in transcriptional regulation and human promyelocytic leukemia (HL-60) cell proliferation and differentiation. *Biological & Pharmaceutical Bulletin*, *29*(11), 2246–2250.

Tuohimaa, P., Golovko, O., Kalueff, A., Nazarova, N., Qiao, S., Syvala, H., et al. (2005). Calcidiol and prostate cancer. *The Journal of Steroid Biochemistry and Molecular Biology*, *93*(2–5), 183–190. http://dx.doi.org/10.1016/j.jsbmb.2004.12.009.

Tuohimaa, P., Wang, J. H., Khan, S., Kuuslahti, M., Qian, K., Manninen, T., et al. (2013). Gene expression profiles in human and mouse primary cells provide new insights into the differential actions of vitamin D3 metabolites. *PLoS One*, *8*(10), e75338. http://dx.doi.org/10.1371/journal.pone.0075338.

Urushino, N., Nakabayashi, S., Arai, M. A., Kittaka, A., Chen, T. C., Yamamoto, K., et al. (2007). Kinetic studies of 25-hydroxy-19-nor-vitamin D3 and 1 alpha,25-dihydroxy-19-nor-vitamin D3 hydroxylation by CYP27B1 and CYP24A1. *Drug Metabolism and Disposition: The Biological Fate of Chemicals*, *35*(9), 1482–1488. http://dx.doi.org/10.1124/dmd.107.015602.

Yasuda, K., Ikushiro, S., Kamakura, M., Takano, M., Saito, N., Kittaka, A., et al. (2013). Human cytochrome P450-dependent differential metabolism among three 2alpha-substituted-1alpha,25-dihydroxyvitamin D(3) analogs. *The Journal of Steroid Biochemistry and Molecular Biology*, *133*, 84–92. http://dx.doi.org/10.1016/j.jsbmb.2012.09.006.

Zierold, C., Darwish, H. M., & DeLuca, H. F. (1995). Two vitamin D response elements function in the rat 1,25-dihydroxyvitamin D 24-hydroxylase promoter. *The Journal of Biological Chemistry*, *270*(4), 1675–1678.

CHAPTER FIFTEEN

Vitamin D Analogs with Nitrogen Atom at C2 Substitution and Effect on Bone Formation

Atsushi Kittaka*,1, Masashi Takano*, Hiroshi Saitoh†
*Faculty of Pharmaceutical Sciences, Teikyo University, Tokyo, Japan
†Teijin Institute for Bio-medical Research, Teijin Pharma Ltd., Tokyo, Japan
[1]Corresponding author: e-mail address: akittaka@pharm.teikyo-u.ac.jp

Contents

Abstract

The Arg274 residue of the ligand binding domain of human vitamin D receptor (hVDR) is important for 1α,25-dihydroxyvitamin D_3 (1α,25$(OH)_2D_3$) binding as a specific ligand through forming a hydrogen bond with the 1α-OH group of the active vitamin D_3, 1α,25$(OH)_2D_3$. An additional pincer-type hydrogen bond formation with Arg274 from a 2α-substituent of a synthetic 1α,25$(OH)_2D_3$ analog would enhance the binding affinity and biological activity. A series of 2α-[2-(heteroaryl)ethyl]-, 2α-(4-cyanobutyl)-, 2α-(ω-cyanoalkoxy)-, and 2β-(3-cyanopropoxy)-1α,25$(OH)_2D_3$ were designed and synthesized based on our original hVDR super agonists of 2α-(3-hydroxypropyl)- and 2α-(3-hydroxypropoxy)-1α,25$(OH)_2D_3$. Their potential biological activities, i.e., hVDR binding affinity, transactivation activity in HOS cells, and therapeutic effect on enhancing the bone mineral density of OVX rats, were studied.

Vitamins and Hormones, Volume 100
ISSN 0083-6729
http://dx.doi.org/10.1016/bs.vh.2015.10.010

1. INTRODUCTION

Vitamin D receptor (VDR) is a member of the nuclear receptor superfamily. 1α,25-Dihydroxyvitamin D_3 [1α,25$(OH)_2D_3$] is an active form of secosteroid hormone vitamin D_3 and acts as a ligand-dependent transcriptional regulator in the genomic pathway, which involves binding of ligand to VDR, formation of a heterodimer complex with retinoid X receptor, and subsequent modulation of gene expression (Mangelsdorf et al., 1995). It is well known that this natural hormone plays important roles for maintaining calcium and bone homeostasis, so not only 1α,25$(OH)_2D_3$ but also vitamin D derivatives such as 1α$(OH)D_3$ and eldecalcitol are used as therapeutic agents for osteoporosis. Besides, 1α,25$(OH)_2D_3$ acts as a regulator controlling cell growth and differentiation, immune function, embryonic development, and inflammatory reactions (Bikle, 2011; Bouillon et al., 2008; Mora, Iwata, & von Andrian, 2008). Therefore, vitamin D derivatives such as tacalcitol, maxacalcitol, and calcipotriol are used as therapeutic drugs for psoriasis (O'Neill & Feldman, 2010), and a new vitamin D derivative, inecalcitol, is being developed for cancers such as prostate cancer (Medioni et al., 2014) and chronic leukemia.

Up to the present, numerous vitamin D derivatives have been synthesized, but most of them are derivatives that are modified in side chains, and there are few publications describing modification of the A-ring (Bouillon, Okamura, & Norman, 1995). In the series of derivatives, it has been reported that some modifications at the C2 position enhance affinity to VDR (Saito, Honzawa, & Kittaka, 2006; Saito & Kittaka, 2006; Shevde et al., 2002; Sicinski, Prahl, Smith, & DeLuca, 1998).

When 1α,25$(OH)_2D_3$ binds to the human VDR (hVDR), six hydrogen bonds are formed between the three OH groups of the active ligand and the six amino acid residues of the hVDR. One of the most important hydrogen bonds connects between the 1α-OH group of 1α,25$(OH)_2D_3$ and the Arg274 residue of the receptor (Rochel, Wurtz, Mitschler, Klaholz, & Moras, 2000). The ligand binding domain (LBD) of the hVDR contains water molecules from the A-ring anchoring moiety to the surface of the protein to stabilize the VDR-[1α,25$(OH)_2D_3$] complex by forming a hydrogen bond network of these water molecules. This water molecule network is called a water channel (Hourai et al., 2006). Previously, we synthesized 2α-(3-hydroxypropyl)-1α,25$(OH)_2D_3$ (O1C3) and 2α-(3-hydroxypropoxy)-1α,25$(OH)_2D_3$

(O2C3), and they showed 3 and 1.8 times greater VDR binding affinity than 1α,25$(OH)_2D_3$, respectively (Saito et al., 2004; Suhara et al., 2001). X-Ray cocrystallographic analyses of both VDR-O1C3 and VDR-O2C3 complexes clearly demonstrated that the terminal hydroxy group of both synthetic ligands forms a hydrogen bond with the guanidino group of the Arg274 residue and replaces one of the water molecules originally forming an important hydrogen bond to the guanidino group in the LBD of the hVDR to stabilize the complex (Hourai et al., 2006). Therefore, O1C3 and O2C3 are able to make pincer-type complexes using two OH groups at the 1α-position and the ω-position of the 2α-side chain with hVDR through the guanidino group of Arg274.

Based on the above results, we thought about using nitrogen atoms, instead of the OH group on the introduced 2α-side chain, to create this type of additional hydrogen bond formation with the Arg274 residue. We studied the effects of a heteroaromatic ring including nitrogen atoms, such as tetrazole, triazole, and imidazole, as well as an ω-cyanoalkyl or ω-cyanoalkoxy group at the 2α-position of 1α,25$(OH)_2D_3$ on binding to the hVDR and on biological activities *in vitro* and *in vivo*.

First, 2α-[2-(tetrazol-2-yl)ethyl]-1α,25$(OH)_2D_3$ (**1a**) and the related compounds **1b**–**1f** were designed, since the number of atoms (as a linker) between the terminal OH group and the 2α-position of O1C3, which is a stronger binder to hVDR than O2C3, is three (*1*–*3*). Compounds **1a**–**1e** also consist of three atoms (as the linker, *1*–*3*) between the 2α-position and the nitrogen atom, i.e., *4* in **1a**, which could coordinate to Arg274, but only **1f** has four atoms (*1*–*4*) like O2C3 between the 2α-position and the heterocyclic nitrogen atom (*5*) (Fig. 1).

Next, 2α-(ω-cyanoalkoxy)-1α,25$(OH)_2D_3$ (**2**–**4** and **7**), 2α-(4-cyanobutyl)-1α,25$(OH)_2D_3$ (**5**), and 2β-(3-cyanopropoxy)-1α,25$(OH)_2D_3$ (**6**) were designed, since a linear structure of the cyano group having a carbon–nitrogen triple bond would act in a different way to stabilize a VDR–ligand complex from the heteroaryl group of **1a**–**1f** (Fig. 2).

The above compounds **1a**–**1f** and **2**–**7** were synthesized to test the effect of nitrogen atoms on VDR binding and osteocalcin promoter transactivation activity in human osteosarcoma (HOS) cells. Among them, the strong ligand **1a** was evaluated for its *in vivo* therapeutic effect using an ovariectomized (OVX) rat as an osteoporosis model animal. Lastly, we studied the crystal structures of the complexes between truncated hVDR LBD and **1a**, **1b**, **3**, as well as **7**.

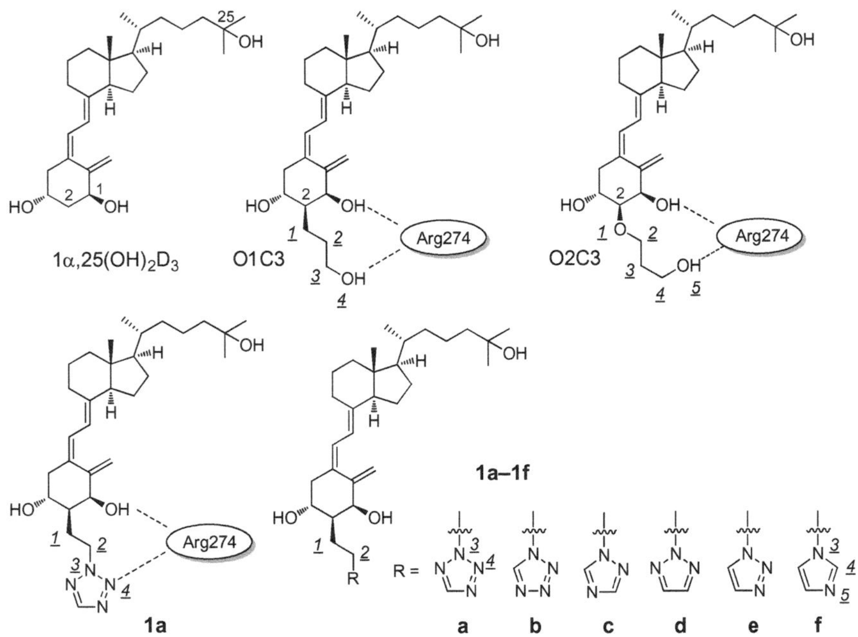

Figure 1 Structures of 1α,25(OH)$_2$D$_3$ and its 2α-functionalized analogs O1C3, O2C3, and **1a–1f**. *Adapted with permission from Matsuo et al. (2013).*

2. SYNTHESIS OF NEW VDR LIGANDS: (HETEROARYL) ETHYL GROUP AT C2α

2.1 Tetrazole

The A-ring precursors of enynes, N2-substituted tetrazole **10a** and N1-substituted tetrazole **10b**, were synthesized from methyl 3-deoxy-3-*C*-ethenylaltropyranoside **8** (Honzawa, Yamamoto, Hirasaka, Takayama, & Kittaka, 2003), which was available from methyl α-D-glucoside (Scheme 1). Compound **8** was converted to enyne alcohol **9**, and a Mitsunobu reaction between **9** and 1*H*-tetrazole gave the desired protected enynes **10a** and **10b** in yields of 81% and 19%, respectively. Each synthetic step proceeded smoothly, giving good to high yields (Matsuo et al., 2013). The chemical structures of these two regioisomers were determined by comparing the ^{1}H NMR and ^{13}C NMR chemical shifts of the correlated methylene and methyne CHs of **2a** and **2b** (Butler, 1984; Elguero, Marzin, & Roberts, 1974).

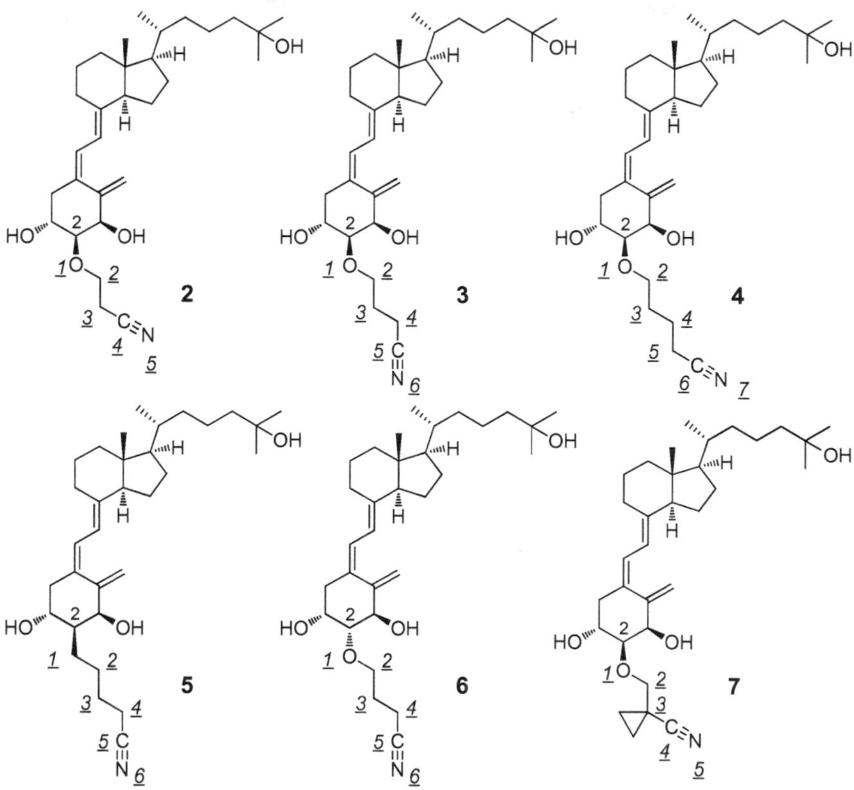

Figure 2 Structures of 2α-functionalized analogs with a ω-cyano group **2–7**. *Adapted with permission from Saitoh et al. (2015).*

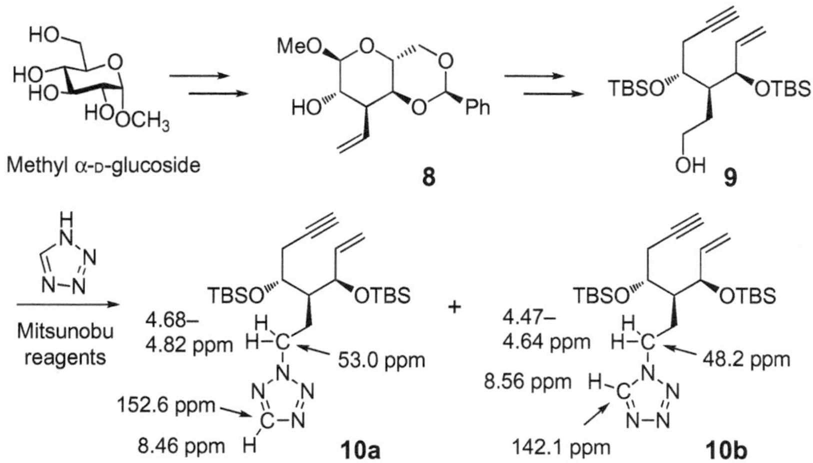

Scheme 1 Synthesis of A-ring precursors **10a** and **10b** with tetrazole. *Adapted with permission from Matsuo et al. (2013).*

2.2 Triazole

2.2.1 1,2,4-Triazole

1,2,4-Triazole reacted with enyne alcohol **9** using Mitsunobu reagents through S_N2 reaction to give a condensed product **10c** in 94% yield, and no N4-alkylated product was produced (Scheme 2).

2.2.2 1,2,3-Triazole

1,2,3-Triazol was also attached to enyne alcohol **9** using the Mitsunobu reaction, and in this case, two regioisomers were obtained: N2-alkylated product **10d** as a major product in 85% yield and N1-alkylated product **10e** as a minor one in 15% yield (Scheme 2).

2.3 Imidazole

The N-alkylation reaction of imidazole with alcohol **9** failed under the Mitsunobu reaction conditions. Imidazole anion was generated using a base,

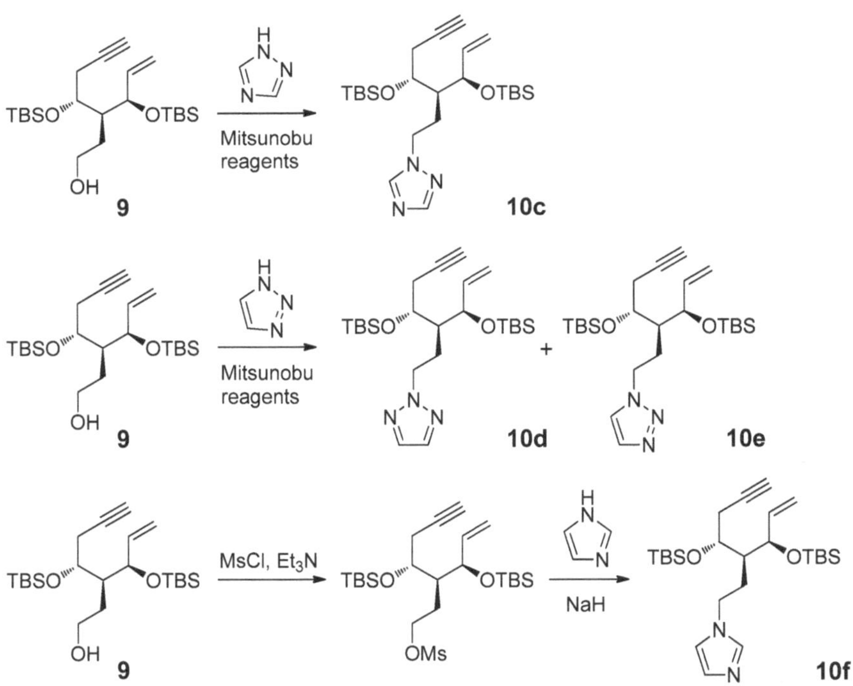

Scheme 2 Synthesis of A-ring precursors **10c–10f** with triazoles and imidazole. *Adapted with permission from Matsuo et al. (2013).*

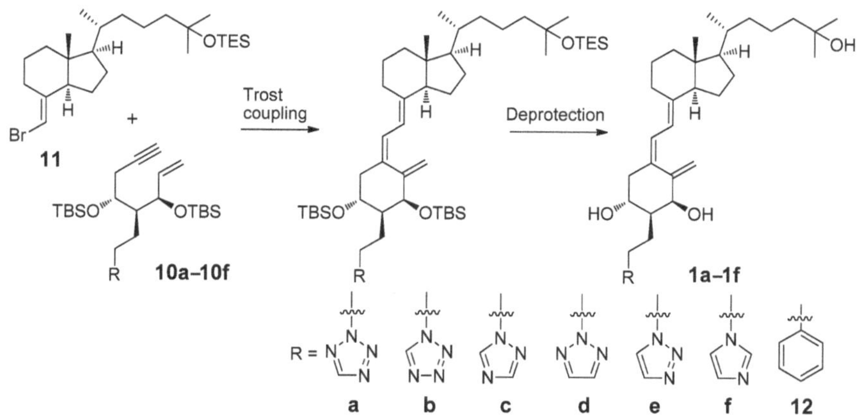

Scheme 3 Synthesis of **1a–1f** by Trost coupling. *Adapted with permission from Matsuo et al. (2013).*

NaH, and then this nitrogen anion was able to attack mesylate from alcohol **9** to give the connected product **10f** in 71% yield (Scheme 2).

The A-ring precursors **10a–10f** and the CD-ring bromoolefin **11** (Maeyama et al., 2006) were coupled under Trost coupling conditions (Trost, Dumas, & Villa, 1992). The products were deprotected, and the target 2α-heteroarylalkyl vitamin D_3 analogs **1a–1f** were purified with HPLC for biological evaluations, respectively (Scheme 3). For comparison, the structure of 2α-phenethyl analog **12** previously synthesized is also shown in Scheme 3 (Honzawa et al., 2005). Recently, four 2-[3-(tetrazolyl)propyl]-19-nor-1α,25$(OH)_2D_3$ analogs were synthesized, and they showed weak transactivation activity through hVDR in HOS cells (EC_{50} 7.3 nM, when 1α,25$(OH)_2D_3$ 0.23 nM) (Takano et al., 2015).

3. SYNTHESIS OF NEW VDR LIGANDS: CYANOALKYL OR CYANOALKOXY GROUP AT C2α AND C2β

Six kinds of A-ring precursors **19–24** were prepared from the enynes **13–18** (Hatakeyama et al., 2001; Saito et al., 2004; Saitoh et al., 2011, 2015), respectively, through selective deprotection of the primary alcohol, followed by a tosylation and cyanation process for **19–23** or oxidation, oxime formation, and dehydration to afford **24** (Scheme 4). Structural conversion from methyl D-glucoside to the enynes was studied in 2000 (Kittaka et al., 2000).

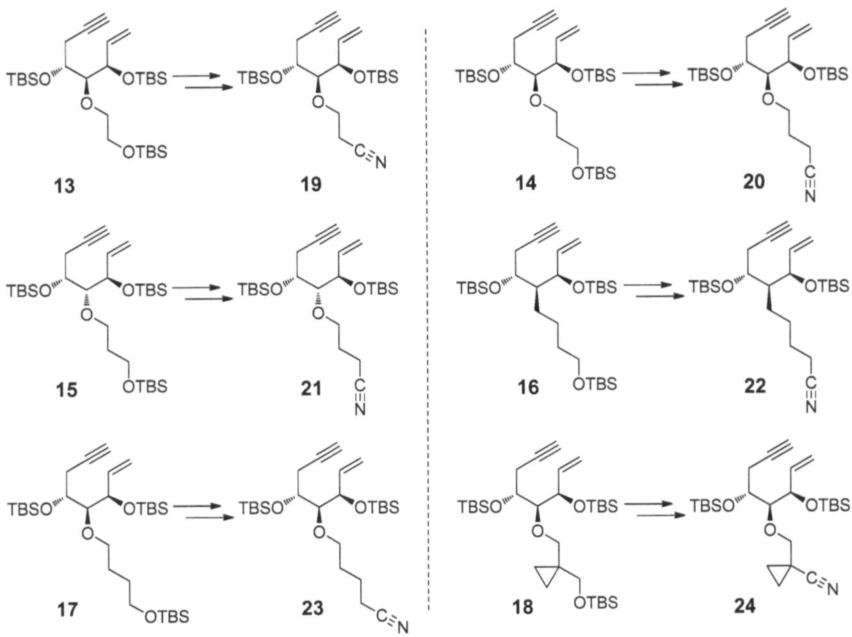

Scheme 4 Synthesis of A-ring precursors with the cyanoalkyl group at the C2 position. *Adapted with permission from Saitoh et al. (2015).*

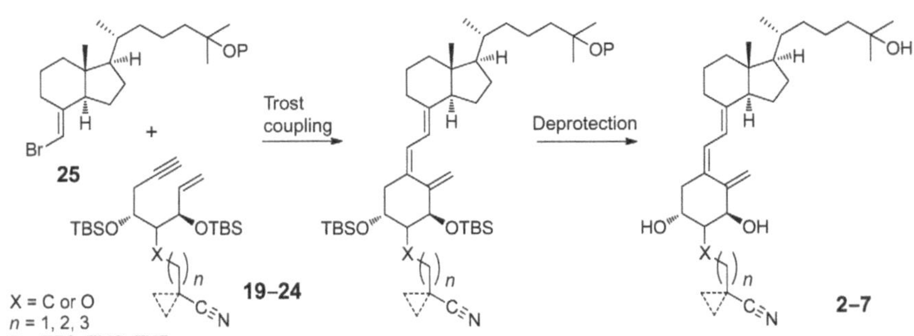

Scheme 5 Synthesis of **2–7** by Trost coupling. *Adapted with permission from Saitoh et al. (2015).*

The A-ring precursors **19–24** and the CD-ring bromoolefin **25** were connected under Trost coupling conditions (Trost et al., 1992). The coupled products were deprotected, and the target 2α-cyanobutyl (**5**) and 2-cyanoalkoxyvitamin D_3 analogs (**2**, **3**, **4**, **6**, and **7**) were purified with HPLC for biological evaluations, respectively (Scheme 5) (Saitoh et al., 2015). Compounds **3** and **6** were diastereoisomers based on the C2 chiral center.

4. BIOLOGICAL ACTIVITY OF NEW VDR LIGANDS

4.1 hVDR Binding Affinity and Transcriptional Activity

Binding affinity for hVDR and osteocalcin promoter transactivation activity in HOS cells of the new compounds **1a**–**1f** and **2**–**7** were evaluated (Tables 1 and 2). Among the compounds in Table 1, 2α-[2-(tetrazol-2-yl)ethyl]-1α,25$(OH)_2D_3$ (**1a**) showed potent binding affinity for hVDR and greater transactivation activity (EC_{50} 10 pM) than that of the natural hormone, 1α,25$(OH)_2D_3$. Analog **1d** showed moderate hVDR binding affinity and almost the same level of transactivation activity as 1α,25$(OH)_2D_3$. Compounds **1b**, **1e**, and **1f** showed no effective binding affinity for hVDR, as we previously found for 2α-phenethyl-1α,25$(OH)_2D_3$ (**12**, R=Ph in Scheme 3) (Honzawa et al., 2005). It seems that the aromatic ring of **1b**, **1e**, and **1f** was just a bulky group inhibiting hVDR binding.

As shown in Table 2, the binding affinity for hVDR of 2α-cyanobutyl or 2α-cyanoalkoxy-1α,25$(OH)_2D_3$ analogs decreased to 10–30% of that of active vitamin D_3, and the 2β-cyanopropoxy analog **6** hardly bound to hVDR. Interestingly, however, 2α-cyanoalkoxy-1α,25$(OH)_2D_3$ analogs **2**, **3**, **4**, and **7** possessed the same level of transactivation activity as 1α,25$(OH)_2D_3$.

Table 1 hVDR Binding Affinity and Osteocalcin Promoter Transactivation Activity in HOS Cells for Compounds **1a–1f**

Compound	Relative hVDR Binding Affinity (%)	Osteocalcin Transactivation Activity in HOS/SF Cells (EC_{50}, nM)
1α,25$(OH)_2D_3$	100	0.026
1a	64	0.010
1b	<1	0.29
1c	4	0.15
1d	12	0.044
1e	<1	0.21
1f	2	0.68
12	1	NT[a]

[a]Not tested.

Reproduced from Matsuo et al. (2013), with permission.

Table 2 hVDR Binding Affinity and Osteocalcin Promoter Transactivation Activity in HOS Cells for Compounds **2–7**

Compound	Relative hVDR Binding Affinity (%)	Osteocalcin Transactivation Activity in HOS/SF Cells (EC_{50}, nM)
1α,25$(OH)_2D_3$	100	0.0214–0.0430[a]
2	10	0.0157
3	12	0.0279
4	32	0.0383
5	7	NT[b]
6	<1	0.167
7	14	0.0305

[a]The range of EC_{50} values of 1α,25$(OH)_2D_3$ at the time of evaluation of each compound **2**–**7** except **5**.
[b]Not tested.
Reproduced from Saitoh et al. (2015), with permission.

Table 3 Therapeutic Effects of **1a** Using Ovariectomized (OVX) Rat Model

	Dose (μg/kg/day)	BMD (g/cm^2)	Serum Ca (mg/dL)
Control (sham)	–	0.231 ± 0.015	9.71 ± 0.27
Control (OVX)	–	0.209 ± 0.011	9.25 ± 0.11
1a (OVX)	0.007	0.214 ± 0.012	9.41 ± 0.20
1a (OVX)	0.02	0.223 ± 0.017	9.66 ± 0.25

Reproduced from Matsuo et al. (2013), with permission.

4.2 *In Vivo* Effects on Bone Mineral Density

Effects of the N2-alkylated tetrazole ring of **1a** on the *in vitro* biological results were highly potent, and the *in vivo* therapeutic effect of **1a** was tested using an OVX rat as an osteoporosis model animal. Twelve-week-old Sprague–Dawley female rats were OVX and fed a normal diet containing 1.0% Ca *ad libitum* for 4 weeks. The rats were then administered **1a** at doses of 0.007 and 0.02 μg/kg/day, 5 times a week for 4 weeks. Twenty-four hours after the final administration, bone mineral density (BMD) of the spine (L4–L5) bone mass was measured by dual X-ray absorption meter. The results of BMD and serum Ca density are shown in Table 3. The synthesized analog **1a** showed an increase of BMD at low doses of 0.007 and 0.02 μg/kg/day without a significant side effect of increased serum calcium, i.e., hypercalcemia, compared to 1α,25$(OH)_2D_3$ (Table 4).

Table 4 Therapeutic Effects of $1\alpha,25(OH)_2D_3$ Using Ovariectomized (OVX) Rat Model

	Dose (μg/kg/day)	BMD (g/cm^2)	Serum Ca (mg/dL)
Control (sham)	–	0.245 ± 0.010	9.72 ± 0.40
Control (OVX)	–	0.209 ± 0.019	9.14 ± 0.24
$1\alpha,25(OH)_2D_3$ (OVX)	0.025	0.218 ± 0.010	10.14 ± 0.37
$1\alpha,25(OH)_2D_3$ (OVX)	0.1	0.228 ± 0.013	10.35 ± 0.27

Reproduced from Matsuo et al. (2013), with permission.

5. NEW LIGANDS BOUND TO hVDR: X-RAY COCRYSTALLOGRAPHIC ANALYSES

The crystal structures of the complexes between truncated hVDR LBD and **1a**, **1b**, **3**, and **7** were studied for any new interactions between the tetrazole ring or the cyano group and amino acid residues of the LBD (Figs. 3–6). The hVDR LBD-**1a** complex demonstrated that one of the nitrogen atoms of the N2-alkylated tetrazole ring formed a direct hydrogen bond with Arg274 (2.99 Å distance), which was also shown in the hVDR LBD-O1C3 complex between the 2α-terminal OH group of O1C3 and Arg274 to stabilize the complex after missing two water molecules from the original hVDR LBD-$1\alpha,25(OH)_2D_3$ complex (Hourai et al., 2006). In addition, notable hydrogen bonding between the CH group of tetrazole ring **1a** and the carbonyl oxygen atom of Asp144 was seen in Fig. 3. It is known that CH has potential as a hydrogen bonding donor, and its distance of CH···O is close enough at 2.89 Å.

Although the binding affinity of **1b** for hVDR was low (Table 1), the crystal structure of the hVDR LBD-**1b** complex was successfully analyzed. Compound **1b** with the N1-alkylated tetrazole ring had a nitrogen atom as the hydrogen bonding acceptor at the same position of the CH group in **1a** shown above in the hVDR complex. It must possess unfavorable interaction energy with the carbonyl oxygen of Asp144 when compared with **1a**, and the nitrogen atom of the main chain of Asp144 makes bifurcated hydrogen bonds with tetrazole ring **1b** (Fig. 4). The bifurcated hydrogen bonds are not very strong as there is only a single hydrogen bond formation between the nitrogen of the main chain at the Asp144 residue and the tetrazole ring.

Surprisingly, the location of the terminal cyano group of **3** in the LBD of hVDR was very different from that of the terminal hydroxy group of the

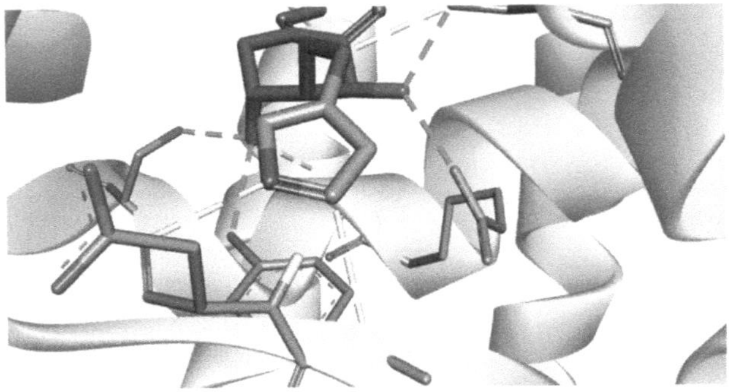

Figure 3 X-Ray studies on hVDR LBD-**1a** complex. The Protein Data Bank accession number for the coordinates of the structure of the VDR complex with **1a** is 4ITE. The A-ring part is magnified. *Adapted with permission from Matsuo et al. (2013, 2014).* (See the color plate.)

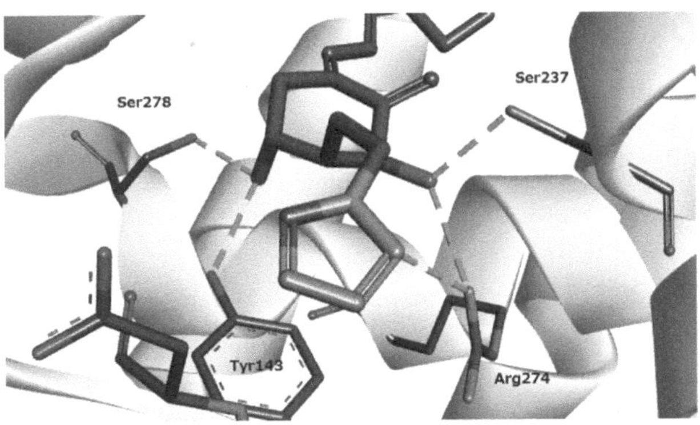

Figure 4 X-Ray studies on hVDR LBD-**1b** complex. The Protein Data Bank accession number for the coordinates of the structure of the VDR complex with **1b** is 4ITF. The A-ring part is magnified. *Adapted with permission from Matsuo et al. (2014).* (See the color plate.)

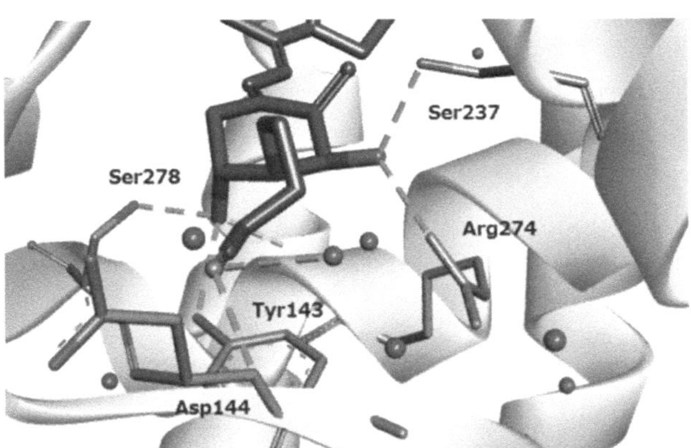

Figure 5 X-Ray studies on hVDR LBD-**3** complex. The Protein Data Bank accession number for the coordinates of the structure of the VDR complex with **3** is 4PA2. The A-ring part is magnified. *Adapted with permission from Saitoh et al. (2015).* (See the color plate.)

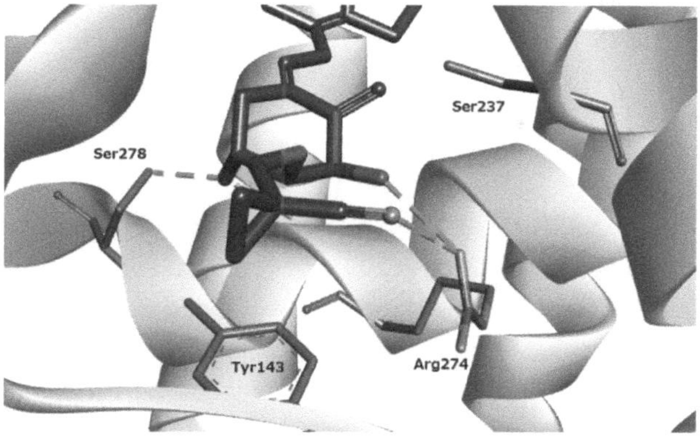

Figure 6 X-Ray studies on hVDR LBD-**7** complex. The Protein Data Bank accession number for the coordinates of the structure of the VDR complex with **7** is 3WWR. The A-ring part is magnified. *Adapted with permission from Saitoh et al. (2015).* (See the color plate.)

2α-substituted active vitamin D_3 analogs synthesized so far (Hourai et al., 2006), and the cyano group occupied the Tyr143 residue and formed a hydrogen bond with the backbone NH proton at the Asp144 residue (Fig. 5). The two hydroxy groups at C1 and C3 each formed two hydrogen bonds with Arg274 and Ser237 as well as Tyr143 and Ser278, respectively, as usually observed in the original hVDR LBD-1α,25$(OH)_2D_3$ and the hVDR LBD-synthetic analog complexes (Rochel et al., 2000).

Regarding compound **7**, the cyano group of **7** faced the Arg274 residue, and both cyano and C1-OH groups formed hydrogen bonds with Arg274. However, no hydrogen bonding between the C1-OH group of **7** and Ser237 was observed, because the distance between the hydroxy group and Ser278 was 3.58 Å, which is too long to form a hydrogen bond. Furthermore, the C3-OH group of **7** formed a single hydrogen bond only with Ser278 due to the long distance to Tyr143 (3.6 Å) (Fig. 6). These interactions were different from the original hVDR LBD-1α,25$(OH)_2D_3$ complex (Rochel et al., 2000).

6. SUMMARY

Twelve vitamin D analogs with nitrogen atom at the C2 position, heteroarylethyl **1a–1f** and cyanoalkyl or cyanoalkoxy **2–7**, were synthesized using the Trost coupling method and tested for biological activity. Among them, 2α-[2-(tetrazol-2-yl)ethyl]-1α,25$(OH)_2D_3$ (**1a**) showed higher osteocalcin promoter transactivation activity in HOS cells and a greater therapeutic effect *in vivo* in OVX rats on enhancing BMD without hypercalcemic side effects than those of 1α,25$(OH)_2D_3$. X-Ray cocrystallographic analysis of the hVDR LBD-**1a** complex confirmed new hydrogen bond formation between the nitrogen atom of the tetrazole ring and Arg274 as well as between the tetrazole CH group and the Asp144 carbonyl oxygen. 2α-(3-Cyanopropoxy)-1α,25$(OH)_2D_3$ (**3**) showed a novel hydrogen bond networks with the LBD of the hVDR; especially the terminal cyano group interacted with both Tyr143 and Asp144, and **3** possessed the same level of transactivation activity as that of the natural hormone, 1α,25$(OH)_2D_3$.

Introduction of nitrogen atom(s) to the 2α-side chain was able to stabilize the receptor–ligand complex using nitrogen coordination to amino acids of the LBD of hVDR and afforded potent biological activity. 2α-Substituted active vitamin D_3 analogs, O1C3, O2C3, and MART-10, for example, were CYP24A1-resistant and showed long half-life activity in the target cells (Yasuda et al., 2013). These effects would create an opportunity to discover novel vitamin D analogs for antiosteoporosis and/or anticancer chemotherapy (Kittaka et al., 2012; Matsumoto, Kittaka, & Chen, 2015).

ACKNOWLEDGMENTS

The authors are grateful for a grant-in-aid from the Ministry of Education, Culture, Sports, Science and Technology (No. 25860011 to M.T.), grants-in-aid from the Japan Society for the Promotion of Science (No. 15K08031 to M.T., No. 24590021 and No. 15K07869 to A.K.), and AMED-CREST, AMED for partial support in the preparation of this manuscript.

REFERENCES

Bikle, D. (2011). Vitamin D regulation of immune function. *Vitamins and Hormones*, *86*, 1–21.

Bouillon, R., Carmeliet, G., Verlinden, L., van Etten, E., Verstuyf, A., Luderer, H. F., et al. (2008). Vitamin D and human health: Lessons from vitamin D receptor null mice. *Endocrine Reviews*, *29*, 726–776.

Bouillon, R., Okamura, W., & Norman, W. H. (1995). Structure-function relationships in the vitamin D endocrine system. *Endocrine Reviews*, *16*, 200–257.

Butler, R. N. (1984). Tetrazoles. In A. R. Katritzky & W. R. Charles (Eds.), *Comprehensive heterocyclic chemistry: Vol. 5* (pp. 791–838). Oxford: Pergamon Press Ltd.

Elguero, J., Marzin, C., & Roberts, J. D. (1974). Carbon-13 magnetic resonance studies of azoles. Tautomerism, shift reagents, and solvent effects. *The Journal of Organic Chemistry*, *39*, 357–363.

Hatakeyama, S., Kawase, A., Uchiyama, Y., Maeyama, J., Iwabuchi, Y., & Kubodera, N. (2001). Synthesis and biological characterization of 1α,24,25-trihydroxy-2β-(3-hydroxypropoxy) vitamin D_3 (24-hydroxylated ED-71). *Steroids*, *66*, 267–276.

Honzawa, S., Hirasaka, K., Yamamoto, Y., Peleg, S., Fujishima, T., Kurihara, M., et al. (2005). Design, synthesis and biological evaluation of novel 1α,25-dihydroxyvitamin D_3 analogues possessing aromatic ring on 2α-position. *Tetrahedron*, *61*, 11253–11263.

Honzawa, S., Yamamoto, Y., Hirasaka, K., Takayama, H., & Kittaka, A. (2003). Synthesis of A-ring synthon of 2α-substituted vitamin D_3 analogues utilizing Grignard reaction towards methyl 2,3-anhydro-4,6-*O*-benzylidene-α-D-mannopyranoside. *Heterocycles, 61*, 327–338.

Hourai, S., Fujishima, T., Kittaka, A., Suhara, Y., Takayama, H., Rochel, N., et al. (2006). Probing a water channel near the A-ring of receptor-bound 1α,25-dihydroxyvitamin D_3 with selected 2α-substituted analogues. *Journal of Medicinal Chemistry, 49*, 5199–5205.

Kittaka, A., Suhara, Y., Takayanagi, H., Fujishima, T., Kurihara, M., & Takayama, H. (2000). A concise and efficient route to 2α-(ω-hydroxyalkoxy)-1α,25-dihydroxyvitamin D_3: Remarkably high affinity to vitamin D receptor. *Organic Letters, 2*, 2619–2622.

Kittaka, A., Yoshida, A., Chiang, K. C., Takano, M., Sawada, D., Sakaki, T., et al. (2012). Potent 19-norvitamin D analogs for prostate and liver cancer therapy. *Future Medicinal Chemistry, 4*, 2049–2065.

Maeyama, J., Hiyamizu, H., Takahashi, K., Ishihara, J., Hatakeyama, S., & Kubodera, N. (2006). Two convergent approaches to the synthesis of 1α,25-dihydroxy-2β-(3-hydroxypropoxy) vitamin D_3 (ED-71) by the Lythgoe and the Trost coupling reactions. *Heterocycles, 70*, 295–307.

Mangelsdorf, D. J., Thummel, C., Beato, M., Herrlich, P., Schütz, G., Umesono, K., et al. (1995). The nuclear receptor superfamily: The second decade. *Cell, 83*, 835–839.

Matsumoto, Y., Kittaka, A., & Chen, T. C. (2015). 19-Norvitamin D analogs for breast cancer therapy. *Canadian Journal of Physiology and Pharmacology, 93*, 333–348.

Matsuo, M., Hasegawa, A., Takano, M., Nakamura, Y., Saito, H., Kakuda, S., et al. (2013). Synthesis of 2α-heteroarylalkyl active vitamin D_3 with therapeutic effect on enhancing bone mineral density in vivo. *ACS Medicinal Chemistry Letters, 4*, 671–674.

Matsuo, M., Hasegawa, A., Takano, M., Saito, H., Kakuda, S., Takagi, K., et al. (2014). Design and synthesis of 2α-(tetrazolethyl)-1α,25-dihydroxyvitamin D_3 as a high affinity ligand for vitamin D receptor. *The Journal of Steroid Biochemistry and Molecular Biology, 144*, 201–203.

Medioni, J., Deplanque, G., Ferrero, J. M., Maurina, T., Rodier, J. M., Raymond, E., et al. (2014). Phase I safety and pharmacodynamic of inecalcitol, a novel VDR agonist with docetaxel in metastatic castration-resistant prostate cancer patients. *Clinical Cancer Research, 20*, 4471–4477.

Mora, J. R., Iwata, M., & von Andrian, U. H. (2008). Vitamin effects on the immune system: Vitamins A and D take centre stage. *Nature Review Immunology, 8*, 685–698.

O'Neill, J. L., & Feldman, S. R. (2010). Vitamin D analogue-based therapies for psoriasis. *Drugs Today (Barcelona Spain), 46*, 351–360.

Rochel, N., Wurtz, J. M., Mitschler, A., Klaholz, B., & Moras, D. (2000). The crystal structure of the nuclear receptor for vitamin D bound to its natural ligand. *Molecular Cell, 5*, 173–179.

Saito, N., Honzawa, S., & Kittaka, A. (2006). Recent results on A-ring modification of 1α,25-dihydroxyvitamin D_3: Design and synthesis of VDR-agonists and antagonists with high biological activity. *Current Topics in Medicinal Chemistry, 6*, 1273–1288.

Saito, N., & Kittaka, A. (2006). Highly potent vitamin D antagonists: Design, synthesis and biological evaluation. *ChemBioChem, 7*, 1478–1490.

Saito, N., Suhara, Y., Kurihara, M., Fujishima, T., Honzawa, S., Takayanagi, H., et al. (2004). Design and efficient synthesis of 2α-(ω-hydroxyalkoxy)-1α,25-dihydroxyvitamin D_3 including 2-*epi*-ED-71 and its 20-epimers with HL-60 cell differentiation activity. *The Journal of Organic Chemistry, 69*, 7463–7471.

Saitoh, H., Chida, T., Takagi, K., Horie, K., Sawai, Y., Nakamura, Y., et al. (2011). Synthesis of C-2 substituted vitamin D derivatives having ringed side chains and biological evaluation, especially biological effect on bone by modification at C-2 position. *Organic and Biomolecular Chemistry, 9*, 3954–3964.

Saitoh, H., Watanabe, H., Kakuda, S., Takimoto-Kamimura, M., Takagi, K., Takeuchi, A., et al. (2015). Synthesis and biological activities of vitamin D_3 derivatives with cyanoalkyl side chain at C-2 position. *The Journal of Steroid Biochemistry and Molecular Biology*, *148*, 27–30.

Shevde, N. K., Plum, L. A., Clagett-Dame, H., Yamamoto, H., Pike, J. W., & DeLuca, H. F. (2002). A potent analog of 1α,25-dihydroxyvitamin D_3 selectively induces bone formation. *Proceedings of the National Academy of Sciences of the United States of America*, *99*, 13487–13491.

Sicinski, R. R., Prahl, J. M., Smith, C. M., & DeLuca, H. F. (1998). New 1α,25-dihydroxy-19-norvitamin D_3 compounds of high biological activity: Synthesis and biological evaluation of 2-hydroxymethyl, 2-methyl, and 2-methylene analogues. *Journal of Medicinal Chemistry*, *41*, 4662–4674.

Suhara, Y., Nihei, K., Kurihara, M., Kittaka, A., Yamaguchi, K., Fujishima, T., et al. (2001). Efficient and versatile synthesis of novel 2α-substituted 1α,25-dihydroxyvitamin D_3 analogues and their docking to vitamin D receptors. *The Journal of Organic Chemistry*, *66*, 8760–8771.

Takano, M., Higuchi, E., Higashi, K., Hirano, K., Takeuchi, A., Sawada, D., et al. (2015). Synthesis and preliminary biological evaluation of 2-[3-(tetrazolyl)propyl]-1α,25-dihydroxy-19-norvitamin D_3. *Heterocycles*, *90*, 1274–1287.

Trost, B. M., Dumas, J., & Villa, M. (1992). New strategies for the synthesis of vitamin D metabolites via palladium-catalyzed reactions. *Journal of the American Chemical Society*, *114*, 9836–9845.

Yasuda, K., Ikushiro, S., Kamakura, M., Takano, M., Saito, N., Kittaka, A., et al. (2013). Human cytochrome P450-dependent differential metabolism among three 2α-substituted-1α,25-dihydroxyvitamin D_3 analogs. *The Journal of Steroid Biochemistry and Molecular Biology*, *133*, 84–92.

CHAPTER SIXTEEN

Mechanistic Insights of Vitamin D Anticancer Effects

Yingyu Ma*, Candace S. Johnson*, Donald L. Trump†,‡,1
*Department of Pharmacology and Therapeutics, Roswell Park Cancer Institute, Buffalo, New York, USA
†Department of Medicine, Roswell Park Cancer Institute, Buffalo, New York, USA
‡Inova Dwight and Martha Schar Cancer Institute, Falls Church, Virginia, USA
[1]Corresponding author: e-mail address: Donald.Trump@inova.org

Contents

Abstract

Vitamin D is a secosteroid hormone that regulates many biological functions in addition to its classical role in maintaining calcium homeostasis and bone metabolism. Vitamin D deficiency appears to predispose individuals to increased risk of developing a number of cancers. Compelling epidemiological and experimental evidence supports a role for vitamin D in cancer prevention and treatment in many types of cancers. Preclinical studies show that $1,25D_3$, the active metabolite of vitamin D, and its analogs have antitumor effects *in vitro* and *in vivo* through multiple mechanisms including the induction of cell cycle arrest, apoptosis, differentiation and the suppression of inflammation, angiogenesis, invasion, and metastasis. $1,25D_3$ also potentiates the effect of chemotherapeutic agents and other agents in the combination treatment. In this review, the antitumor effects of $1,25D_3$ and the potential underlying mechanisms will be discussed. The current findings support the application of $1,25D_3$ in cancer prevention and treatment.

Vitamins and Hormones, Volume 100
ISSN 0083-6729
http://dx.doi.org/10.1016/bs.vh.2015.11.003

1. OVERVIEW OF VITAMIN D ANTICANCER EFFECTS

Vitamin D is a steroid hormone that plays a major role in the regulation of calcium and phosphate homeostasis and is essential for bone metabolism. Beyond this classical function, vitamin D has important effects on many other physiological responses and may have protective effects against a variety of diseases including hypertension, diabetes, metabolic syndrome, and multiple sclerosis (Garland, Gorham, Mohr, & Garland, 2009). Increasing epidemiological and experimental studies support a role for vitamin D compounds in cancer prevention and treatment in a broad spectrum of cancers (Deeb, Trump, & Johnson, 2007; Garland et al., 2006; Giovannucci, Liu, Rimm, et al., 2006; Welsh, 2012). Serum 25(OH)D_3 concentration is currently accepted as the main indicator of vitamin D status and total body stores. Low levels of plasma 25(OH)D_3 are associated with higher cancer incidence and mortality in patients with colorectal, breast, lung, and prostate cancers (Giovannucci, 2009; Giovannucci, Liu, Stampfer, & Willett, 2006; Kilkkinen et al., 2008; Ng et al., 2008).

2. EPIDEMIOLOGICAL STUDIES

In 1980, Garland and Garland published their seminal hypothesis proposing ultraviolet-B (UV-B) exposure and consequent vitamin D synthesis is protective against colon cancer (Garland & Garland, 1980). This hypothesis is based on the observation that colon cancer mortality is inversely associated with exposure to natural light (Garland & Garland, 1980). Subsequently, this observation has been extended to additional cancer types in different countries through this ecological analysis approach (Boscoe & Schymura, 2006; Grant, 2002, 2007; Grant & Garland, 2006; Moan, Dahlback, Lagunova, Cicarma, & Porojnicu, 2009).

This hypothesis is supported by observational studies showing low serum 25(OH)D_3 levels are associated with higher incidence of breast, colorectal, and bladder cancers (Amaral et al., 2012; Bertone-Johnson et al., 2005; Garland et al., 1989). In line with these findings, a systematic review with meta-analysis suggests serum 25(OH)D_3 levels are inversely associated with total cancer incidence and mortality (Yin et al., 2013). Meta-analyses of five prospective cohort studies including 2330 colorectal cancer patients and five studies including 4413 breast cancer patients reveal an association between higher 25(OH)D_3 levels (>75 nmol/l) and lower mortality rates in colorectal and breast cancer patients (Maalmi, Ordonez-Mena, Schottker, & Brenner, 2014). Colorectal cancer patients who had higher 25(OH)D_3 levels had 29% and 35% reduced risk of all-cause and cancer-related mortality, respectively (Maalmi et al., 2014). Similarly, breast cancer patients with higher 25(OH)D_3 levels had 37% and 43% lower risk of all-cause and cancer mortality, respectively (Maalmi et al., 2014). In another meta-analysis of eight prospective cohort studies including 26,018 individuals, the lowest 25(OH)D_3 quintile is associated with increased all-cause mortality and cancer mortality, but only in individuals with cancer history (Schottker et al., 2014). The Australian Ovarian Cancer Study reports that higher serum 25(OH)D_3 levels at diagnosis are associated with longer survival in ovarian cancer patients (Webb et al., 2015). On the other hand, a systematic review of published case–control and cohort studies shows that among 14 prospective studies, 11 studies show no association between 25(OH)D_3 levels and prostate cancer risk (van der Rhee, Coebergh, & de Vries, 2013). One study shows no statistically significant trend in prostate cancer risk with 25(OH)D_3 levels standardized for season (Ahn et al., 2008). In addition, higher serum 25(OH)D_3 levels may be associated with a higher risk of aggressive disease (Gleason sum ≥ 7 or stage III or IV) (Ahn et al., 2008). Negative associations are found in two studies (van der Rhee et al., 2013). The functional single-nucleotide polymorphism in the group-specific component (GC) gene (vitamin D binding protein, T > G, rs2282679) is not found to be associated with recurrence, metastases, or overall survival in prostate cancer patients, despite the reported association of GC rs2282679 with 25(OH)D_3 and 1,25D_3 levels (Trummer et al., 2015).

Compared to the observational studies, few vitamin D replacement studies have been conducted to investigate its impact on cancer risk or mortality. The Women's Health Initiative (WHI) conducted a randomized placebo-controlled trial supplementing more than 36,000 postmenopausal women with 1000 mg of elemental calcium and 400 IU of vitamin D_3 daily for an average of 7 years with the incidence of pathologically confirmed colorectal cancer as a secondary outcome (Wactawski-Wende et al., 2006). No detectable effect on the incidence of colorectal cancer was observed in the supplementation group. However, the dose of vitamin D_3 was small and not adequate to raise serum 25(OH)D_3 levels (Wactawski-Wende et al., 2006). When this study was reanalyzed by excluding women taking personal calcium or vitamin D supplements (57%), the risks of total breast, and colorectal cancers were reduced in the intervention group (Bolland, Grey, Gamble, & Reid, 2011). Further analysis of this study suggests that concurrent estrogen therapy has a strong attenuating effect on the beneficial effects of calcium + vitamin D_3 supplementation on the risk for colorectal cancer. Among women in the placebo groups of the estrogen trials, calcium + vitamin D_3 appear to have beneficial effects (Ding, Mehta, Fawzi, & Giovannucci, 2008). Another analysis of the WHI study reveals no difference in the incidence of nonmelanoma skin cancer (NMSC) or melanoma between the supplementation group and placebo group (Tang et al., 2011). However, calcium + vitamin D_3 supplementation lowers the risk of melanoma in women who had a history of NMSC (Tang et al., 2011). Lappe et al. conducted a 4-year randomized trial of 1179 postmenopausal women to evaluate the value of calcium (1400–1500 mg/day) and/or vitamin D (1100 IU/day) on all-cancer risk (Lappe, Travers-Gustafson, Davies, Recker, & Heaney, 2007). These levels of supplementation dose raises serum $25(OH)D_3$ levels to >80 nmol/l. Although cancer occurrence was not the end point of this trial, total cancer incidence was significantly lower in the calcium + vitamin D_3 group compared to the placebo group (Lappe et al., 2007). Every 25 nmol/l increase of serum $25(OH)D_3$ was associated with a predicted 35% decreased risk of cancer (Lappe et al., 2007).

Taken together, the epidemiological evidence supports a positive effect of vitamin D on cancer prevention. Low serum $25D_3$ levels are related to an increased risk for a number of cancer types. With the strongest evidence observed in colorectal cancer, vitamin D supplementation studies show promising, but inconclusive results supporting the reduction of cancer incidence and/or mortality. The proper dose of vitamin D supplementation to achieve cancer preventive effect remains to be determined. Therefore, well-designed randomized trials such as the Vitamin D and Omega-3 Trial (VITAL) (Nicholas, 2011), led by Brigham and Women's Hospital at Harvard Medical School, will provide valuable information and improve our understanding of the role of vitamin D in cancer prevention and general health.

3. EXPERIMENTAL STUDIES

3.1 Antitumor Activity and Mechanisms of Vitamin D as a Single Agent

The broad-spectrum antitumor effects of the active metabolite of vitamin D, $1,25(OH)_2D_3$ ($1,25D_3$), have been extensively studied preclinically. The activities of $1,25D_3$ and its analogs are primarily achieved through genomic regulation of target genes and related pathways mediated by $1,25D_3$-vitamin D receptor (VDR) and DNA bindings. These involve the inhibition of cancer cell proliferation, induction of apoptosis and differentiation, inhibition of angiogenesis, invasion, and metastasis (Deeb et al., 2007).

3.2 Proliferation

One of the distinguishing characteristics of cancer cells compared to normal cells is their ability to sustain uncontrolled and unregulated proliferation. This can be achieved by several mechanisms. Cancer cells generate their own growth factors or promote the production of growth factors in stromal cells. The growth factor receptors on cancer cells may be modulated quantitatively or structurally to enhance their signaling. In addition, growth factor signaling pathways can be constitutively activated due to mutations or the loss of negative feedback regulation (Hanahan & Weinberg, 2011).

Inhibition of proliferation is a major mechanism of the antitumor effects of vitamin D. 1,25D$_3$ induces G_0/G_1 cell cycle arrest through the induction of p21$^{waf1/cip1}$ in breast and prostate cancer cells (Moffatt, Johannes, Hedlund, & Miller, 2001; Narvaez & Welsh, 1997). p21$^{waf1/cip1}$ is a direct target of VDR; there are multiple vitamin D response elements (VDREs) in the p21$^{waf1/cip1}$ promoter region (Liu, Lee, Cohen, Bommakanti, & Freedman, 1996; Saramaki, Banwell, Campbell, & Carlberg, 2006). 1,25D$_3$-mediated growth inhibition is associated with the induction of p21$^{waf1/cip1}$ and p27^{kip1} and the reduction of cyclins, cyclin-dependent kinases (CDKs) and CDK inhibitors in pancreatic cancer cells (Kawa et al., 1997). Interestingly, in squamous cell carcinoma (SCC) cells, 1,25D$_3$ induces G_0/G_1 cell cycle arrest via the induction of p27^{kip1} and reduction of p21$^{waf1/cip1}$ (Hershberger et al., 1999). In addition, 1,25D$_3$ may induce G_2/M arrest through p53-independent transcriptional induction of growth arrest and DNA damage-inducible 45 (GADD45) in ovarian cancer cells (Jiang, Li, Fornace, Nicosia, & Bai, 2003). In SCC cells, 1,25D$_3$-induced cell cycle arrest is accompanied by reduction of ERK1/2 and Akt activation, which can be further enhanced by the addition of dexamethasone (Bernardi et al., 2001). 1,25D$_3$ and its analog EB1089 exert growth inhibitory effects in thyroid carcinoma cells, at least in part through the induction of phosphatase and tensin homolog (PTEN), and the resulting inhibition of the Akt pathway (Liu, Asa, Fantus, Walfish, & Ezzat, 2002). 1,25D$_3$ (0.1 μg/kg body weight, 3 times/week intraperitoneal injection) suppresses cell proliferation through the inhibition of PI3K/Akt pathway in a rat thyroid carcinogenesis model *in vivo* (Kemmochi et al., 2011).

$1,25D_3$ and $1,25D_3$-bromoacetate ($1,25D_3$-BE) have antitumor effects *in vitro* and *in vivo* in kidney cancer model systems, which are associated with the inhibition of phosphorylation of Akt and its target caspase 9 (Lambert et al., 2010). The dose of $1,25D_3$ and $1,25D_3$-BE is 0.75 μg/kg body weight, injected intraperitoneally every third day during the study (Lambert et al., 2010). The antiproliferative effect of $1,25D_3$ on endothelial cells transformed by the viral G protein-coupled receptor (vGPCR) associated with Kaposi sarcoma is mediated by inhibition of NF-κB activation, which is independent of PI3K/Akt and MAPK pathways (Gonzalez-Pardo, D'Elia, Verstuyf, Boland, & Russo de Boland, 2012). NF-κB consists of a family of ubiquitously expressed transcription factors that play major roles in the regulation of important genes that are involved in immune responses, inflammation, and cancer development. $1,25D_3$ inhibits NF-κ B activation by reducing the transactivation potential of the NF-κB p65 subunit in a VDR-dependent manner in breast cancer MCF-7 cells (Tse et al., 2010). $1,25D_3$ also inhibits ERK1/2 phosphorylation by inhibiting Src tyrosine kinase activation and activating protein tyrosine phosphatases in MCF-7 cells (Capiati, Rossi, Picotto, Benassati, & Boland, 2004). $1,25D_3$ may also inhibit breast cancer cell growth by promoting the expression of insulin-like growth factor-binding protein 3 (IGFBP3) (Colston, Perks, Xie, & Holly, 1998), which has a functional VDRE in the promoter and may be a direct target of VDR (Peng, Malloy, & Feldman, 2004). IGFBP3 subsequently promotes the expression of $p21^{waf1/cip1}$ (Boyle, Zhao, Cohen, & Feldman, 2001). $1,25D_3$ upregulates the expression of transforming growth factor β, an epithelial cell growth inhibitor, in breast cancer cells (Koli & Keski-Oja, 1995; Wu, Fan, Li, Srinivas, & Brattain, 1998). In a progressive mouse model of lung SCC applying *N*-nitroso-tris-chloroethylurea as the carcinogen, SWR/J mice fed with a vitamin D_3-deficient diet develop more high-grade dysplastic (HGD) lesions, higher circulating white blood cells, and increased IL-6 levels in bronchial lavages compared with the mice fed with vitamin D_3-sufficient diet (Mazzilli et al., 2015). Treatment with $1,25D_3$ (80 mg/kg/week) reduces the HGD lesions in mice fed with a deficient diet (Mazzilli et al., 2015). These observations indicate that $1,25D_3$ has antiproliferative effects in many tumor types by regulating cell cycle-related molecules and growth factor signaling pathways.

miRNAs are identified as the targets of $1,25D_3$ and contribute to the antiproliferative effects. $1,25D_3$ has been reported to inhibit cell growth by suppressing human telomerase reverse transcriptase (hTERT) through the induction of miR-498 in ovarian cancer cells (Kasiappan et al., 2012). $1,25D_3$ induces the expression of miR-98 which contributes to G_2/M cell cycle arrest and growth inhibitory effect of $1,25D_3$ in LNCaP cells (Ting, Messing, Yasmin-Karim, & Lee, 2013). The regulation of miR-98 is through both direct mechanism involving VDRE and indirect mechanism involving the reduction of Lin28A and Lin28B proteins, which affect the synthesis of miRNA (Ting et al., 2013). $1,25D_3$ upregulates miR-22 expression in SW480-ADH colon cancer cells (Alvarez-Diaz et al., 2012). miR-22 inhibitor abrogated the antiproliferative and antimigratory activities of $1,25D_3$ in SW480-ADH or HCT116 colon cancer cells (Alvarez-Diaz et al., 2012), suggesting a contributory role of miR-22 in $1,25D_3$ antitumor activity. $1,25D_3$ increases let-7a-2 expression through a VDRE upstream of pre-let-7a-2 in A549 lung cancer cells, as demonstrated by EMSA, ChIP assay, and promoter-reporter assay results, which may be involved in the growth inhibitory effect of $1,25D_3$ (Guan et al., 2013). Consequently, $1,25D_3$ suppresses the growth of several types of tumor cells through the regulation of specific miRNAs.

In vitro exposure of many types of cancer cells to $1,25D_3$ compromises the ability of certain cancer cells to maintain their constitutive proliferative signaling. There are multiple mechanisms, which contribute to this antitumor activity of $1,25D_3$.

3.3 Apoptosis

Induction of apoptosis or programmed cell death is another important mechanism for the antitumor activity of $1,25D_3$. $1,25D_3$ induces apoptosis in a number of cancer types and the mechanisms appear to be cell type- and tumor type-dependent. $1,25D_3$ induces caspase-independent apoptosis in breast cancer MCF-7 cells by interrupting mitochondrial function, which involves Bax translocation and the production of reactive oxygen species (ROS) (Narvaez & Welsh, 2001). $1,25D_3$ and its analog EB1089 promote apoptosis in MCF-7 cells by increasing intracellular calcium, which activates the calcium-dependent cysteine protease μ-calpain (Mathiasen et al., 2002). Another study shows that EB1089 induces nuclear apoptosis through beclin 1-dependent autophagy in MCF-7 cells (Hoyer-Hansen, Bastholm, Mathiasen, Elling, & Jaattela, 2005). Inhibition of autophagy reduces EB1089-induced nuclear changes and cell death (Hoyer-Hansen et al., 2005). On the other hand, restoration of autophagy through beclin-1 promotes nuclear apoptosis (Hoyer-Hansen et al., 2005). $1,25D_3$ and EB1089 also induce apoptosis that is accompanied by increased expression of proapoptotic Bak in five colorectal adenoma and carcinoma cell lines (Diaz, Paraskeva, Thomas, Binderup, & Hague, 2000). In prostate cancer LNCaP and ALVA-31 cells, $1,25D_3$ treatment leads to apoptosis through the suppression of antiapoptotic Bcl-2 family member proteins and IAP proteins, as well as by the activation of the mitochondrial apoptotic pathway and caspases 9 and 3 (Guzey, Kitada, & Reed, 2002). In ovarian cancer OVCAR3 cells, $1,25D_3$ induces apoptosis via the downregulation of telomerase by decreasing the stability of the *hTERT* mRNA (Jiang, Bao, Li, Nicosia, & Bai, 2004). $1,25D_3$ induces apoptosis in SCC cells through the induction of mitogen-activated protein kinase kinase kinase (MEKK-1) expression and caspase-dependent MEK cleavage (McGuire, Trump, & Johnson, 2001). These findings support a proapoptotic role of $1,25D_3$ in many cancer types.

Contrary to the proapoptotic effects in many cancer cells, $1,25D_3$ has protective roles in normal cells. $1,25D_3$ inhibits UV-B-induced apoptosis via the inhibition of mitochondrial cytochrome *c* release in primary human keratinocytes (De Haes et al., 2003). $1,25D_3$ enhances Bcl-2 expression in normal human thyrocytes which protects them from apoptosis (Wang et al., 1999). $1,25D_3$ also protects human pancreatic islet cells from TNF/IL-1β/IFNγ-induced apoptosis by inducing the antiapoptotic *A20* gene and suppressing Fas expression (Riachy et al., 2002, 2006). $1,25D_3$ suppresses death receptor-mediated apoptosis in human osteoblasts by increasing the Bcl-2/Bax ratio and inhibiting caspase 8 activation (Duque, El Abdaimi, Henderson, Lomri, & Kremer, 2004). These observations suggest that $1,25D_3$ may selectively target cancer cells for apoptotic cell death.

Apoptosis is a major mechanism for $1,25D_3$-mediated cell killing. $1,25D_3$ differentially induces apoptosis in a number of cancers and mainly through intrinsic apoptotic pathways. The mechanisms appear to be cancer type- and cell type-specific. Taken together with the role of $1,25D_3$ against apoptosis in certain normal cells, $1,25D_3$ appears to be an appealing agent in chemotherapy.

3.4 Differentiation

In cancer, differentiation indicates how different the cancer cells/tissue are from the normal cells/tissue. Poorly or undifferentiated cancer cells tend to grow and spread faster than the well-differentiated cancer cells. As a generalization, cancer cells fail to differentiate. In 1981, $1,25D_3$ was first reported to promote the differentiation of mouse myeloid leukemia cells into macrophages *in vitro* (Abe et al., 1981). Analysis of the surface antigens reveals that the differentiation of acute myeloid leukemia (AML) HL-60 cells induced by $1{,}25D_3$ is associated with the development of the phenotype of mature monocytes (Brackman, Lund-Johansen, & Aarskog, 1995). $1{,}25D_3$ also induces the differentiation of retinoic acid-resistant acute promyelocytic leukemia (APL) and induces cell cycle arrest and increased expression of $p21^{waf1/cip1}$ and $p27^{kip1}$ (Muto et al., 1999).

Several different signaling pathways are reported to contribute to $1{,}25D_3$-induced differentiation of leukemia cells. Activation of PI3K contributes to the differentiation of human monocytic leukemia THP-1 cells (Hmama et al., 1999). Transient phosphorylation of ERK1/2 maintains cell proliferation with subsequent increased $p27^{kip1}$ to promote differentiation and cell cycle arrest occurs in HL-60 cells (Wang & Studzinski, 2001). Inhibition of p38 MAP kinase by specific inhibitors potentiates $1{,}25D_3$-mediated HL-60 differentiation (Wang, Rao, & Studzinski, 2000). Another MAP kinase, JNK, is also involved in $1{,}25D_3$-mediated differentiation of HL-60 and U937 cells as supported by enhanced phosphorylation of c-jun (Wang, Wang, & Studzinski, 2003). The activation of ERK1/2 and c-jun by $1{,}25D_3$ leads to the upregulation of C/EBPβ, which participates in the differentiation of HL-60 cells (Ji & Studzinski, 2004). Further examination of three C/EBPβ isoforms reveals that $1{,}25D_3$ treatment leads to the nuclear localization of C/EBPβ-2 and C/EBPβ-3, while C/EBPβ-1 mostly remains in the cytoplasma (Marcinkowska et al., 2006). Increased ceramide levels are induced by $1{,}25D_3$ and result in enhanced differentiation in HL-60 cells (Okazaki, Bielawska, Bell, & Hannun, 1990). $1{,}25D_3$ promotes the expression of kinase suppressor of Ras-1 (KSR-1), also known as ceramide-activated protein kinase, which contributes to the differentiation of monocytic leukemia HL-60 cells (Wang & Studzinski, 2004). Mechanistic studies show that the induction of KSR-1 by $1{,}25D_3$ is through direct transcriptional activation via a VDRE in the 5′ flanking region (Wang, Wang, White, & Studzinski, 2006). These studies indicate $1{,}25D_3$ plays important roles in the differentiation of leukemia cells.

$1{,}25D_3$ also promotes the differentiation of epithelial tumors. In colon cancer CaCo-2 cells, $1{,}25D_3$ induces the activity of the differentiation marker alkaline phosphatase (Chen, Davis, Bissonnette, Scaglione-Sewell, & Brasitus, 1999). This effect is mediated by the activation of AP-1 (activator protein-1) (Chen et al., 1999). In colon cancer SW480 cells, $1{,}25D_3$ promotes a differentiated phenotype which involves the induction of

E-cadherin and the adhesion proteins, occluding zonula occludens-1 (ZO-1), ZO-2, and vinculin, as well as the inhibition of β-catenin transcriptional activity (Palmer et al., 2001). The suppression of β-catenin activity is facilitated by the translocation of β-catenin to the plasma membrane and the binding of VDR to β-catenin, which subsequently blocks the binding of β-catenin to TCF-4 (Palmer et al., 2001). Among nonsmall cell lung cancer cells, those with high VDR levels and sensitive to vitamin D express epithelial markers (Upadhyay et al., 2013). In SKLU-1 cells, 1,25 induces the expression of E-cadherin and reduces the expression of Snail, ZEB1, and vimentin, which are associated with reduced cell migration and maintenance of epithelial morphology (Upadhyay et al., 2013). In breast cancer, $1,25D_3$ promotes the ability of certain cell lines to differentiate into a more epithelial phenotype (Pendas-Franco et al., 2007). In the responsive MDA-MB-453 cells, $1,25D_3$ induces the expression of claudin-7, occluding integrin αv and β5 and focal adhesion kinase. In this system, $1,25D_3$ suppresses the expression of mesenchymal marker N-cadherin, P-cadherin, integrin α6 and β4, and smooth muscle α-actin (Pendas-Franco et al., 2007). $1,25D_3$ and its analog $1\alpha\text{-(OH)}D_5$ promote the differentiation of breast cancer T-47D cells as assessed by the induction of casein expression and lipid production (Lazzaro et al., 2000). $1,25D_3$ promotes a more differentiated phenotype indicated by reduced ability to form anchorage-independent colonies in MCF-7 and T-47D breast cancer cells, which is accompanied by cell cycle arrest and increased expression of $p21^{waf1/cip1}$ and $p27^{kip1}$ (Wang et al., 2001). In MDA-MB-231 cells, $1,25D_3$ induces the expression of E-cadherin (Wang et al., 2001). The induction of E-cadherin by $1,25D_3$ may be mediated by a prodifferentiation gene *icb-1* (Haselberger et al., 2011). *Icb-1* and *E-cadherin* expression is positively correlated in 66 breast cancer tissue samples (Haselberger et al., 2011). E-cadherin is also induced by $1,25D_3$ in prostate cancer cells and associated with reduced cell rolling and adhesion to endothelial cells (Hsu et al., 2011). In prostate cancer LNCaP cells, $1,25D_3$ treatment results in the upregulation of androgen receptor and increased production of prostate-specific antigen, an indication of differentiation in prostate cells (Beer et al., 2006). In osteosarcoma cells, a vitamin D analog, $1,25D_3$-3β-bromoacetate, promotes the secretion of osteocalcin while suppressing the activity of alkaline phosphatase (Van Auken, Buckley, Ray, Holick, & Baran, 1996). In an orthotopic xenograft model of thyroid follicular carcinoma, treatment with $1,25D_3$ (0.75 μg/kg, i.p., 3 times/week for 21 days) reduces the tumor volume which is associated with increased $p27^{kip1}$ expression. Moreover, the tumors exhibit restored thyroglobulin staining, an indication of differentiation, as compared with tumors derived from vehicle-treated control mice (Dackiw, Ezzat, Huang, Liu, & Asa, 2004). 1,25D_3-treated mice also develop fewer lung metastases compared with control mice (Dackiw et al., 2004). Thus, 1,25D also has prodifferentiation effects in many solid tumors.

Taken together, these data indicate that 1,25D_3 promotes differentiation in leukemia and a number of epithelial tumors; this effect involves complex mechanisms. These indications of enhanced differentiation may be beneficial in therapy. In addition, differentiation is commonly associated with reduced proliferation in cancer cells. Therefore, induction of differentiation presents an important mechanism for the antitumor activity of 1,25D_3.

3.5 Angiogenesis

In 1971, based on the observation that solid tumors cannot grow beyond a size of approximately 2 mm in diameter without having their own blood supply, Folkman first proposed that tumor growth is dependent on angiogenesis (Folkman, 1971). In addition to supporting the growth of the primary tumor, angiogenesis is essential for tumor invasion and metastasis. Several studies demonstrate that endothelial cells express functional VDR (Bernardi, Johnson, Modzelewski, & Trump, 2002; Chung et al., 2006; Merke et al., 1989). 1α-Hydroxylase (CYP27B1), the enzyme that leads to local production of 1,25D_3 from its precursor 25(OH)D_3, is expressed and enzymatically active in endothelial cells isolated from human renal arteries, postcapillary venules from lymphoid tissue, and human umbilical vein endothelial cells (HUVECs) (Zehnder et al., 2002). Functionally, 1,25D_3 exerts inhibitory effects on endothelial cells. 1,25D_3 or its analog TX527 inhibits the growth of Simian virus 40 immortalized murine endothelial cells (SVECs) and SVECs transformed by the viral G protein-coupled receptor (SVEC-vGPCR). This inhibition is accompanied by reduction of cyclin D1 and increase of $p27^{kip1}$ in SVECs (Gonzalez-Pardo et al., 2010). 1,25D_3 suppresses vascular endothelial growth factor (VEGF)-induced endothelial cell sprouting, elongation, and proliferation and induces apoptosis in sprouting endothelial cells *in vitro* (Mantell, Owens, Bundred, Mawer, & Canfield, 2000). 1,25D_3 prevents retinal endothelial cells from forming capillary networks in Matrigel without affecting cell proliferation or migration (Albert et al., 2007). Further, 1,25D_3 (5 μg/kg/day, i.p. for 4 days) or TX527 (10 μg/kg/day, i.p. for 4 days) suppresses SVEC-vGPCR tumor progression *in vivo* (Gonzalez-Pardo et al., 2010). 1,25D_3 and analogs

7553, 6760, and EB1089 inhibit proliferation of tumor-derived endothelial cells (TDECs) isolated from SCC tumors (Bernardi et al., 2002). 1,25D_3 induces G_0/G_1 cell cycle arrest and apoptosis in TDEC but not Matrigel-derived endothelial cells (MDECs), which is accompanied by decreased $p21^{waf1/cip1}$ expression, increased $p27^{kip1}$ expression, and reduced phosphorylation of Akt and ERK1/2 (Chung et al., 2006). These differences may be explained, at least in part, by methylation-mediated silencing of the 1,25D_3-metabolizing enzyme 24-hydroxylase (CYP24A1) in TDEC but not MDEC, which potentiates 1,25D_3 effect in TDEC by decreasing its catabolism (Chung et al., 2007). These observations show that 1,25D_3 and certain analogs inhibit the growth of tumor-related endothelial cells.

The effect of vitamin D on angiogenesis was first reported in 1990, when 1,25D_3 and analog 22-oxa-1,25D_3 were described to inhibit embryonic angiogenesis in chorioallantoic membranes (Oikawa et al., 1990). Since then, increasing evidence supports an antiangiogenic role of vitamin D *in vivo* in various model systems. 1,25D_3 inhibits the proliferation of TDEC from $VDR^{+/+}$ but not from $VDR^{-/-}$ mice (Chung et al., 2009). Tumors from $VDR^{-/-}$ mice contain enlarged blood vessels, increased vascular volume, less pericyte coverage of vessels, and more vascular leakage compared to those from $VDR^{+/+}$ mice (Chung et al., 2009). In addition, hypoxia-inducible factor-1α, VEGF, Ang1, and PDGF-BB expression is higher in tumors from $VDR^{-/-}$ mice (Chung et al., 2009). In an MCF-7 tumor xenograft model overexpressing VEGF, 1,25D_3 treatment (12.5 pmol/day for 8 weeks) results in less vascularized tumors in comparison with control-treated tumors (Mantell et al., 2000). 1,25D_3 (0.025 or 0.05 μg/mouse, i.p., 5 times/week for 5 weeks) reduces the mean vessel count in retinoblastoma in a dose-dependent manner in a transgenic murine retinoblastoma model system (Shokravi et al., 1995). 1,25D_3 or 1(OH)D_3 (2.5 or 5.0 nmol/kg, i.p., every other day starting on day 1, 3, or 7 and ending on day 9, 11, or 15) inhibits angiogenesis, tumor growth, and metastasis, thus prolonging survival in a murine renal cell carcinoma model (Fujioka et al., 1998). 1,25D_3 (0.03 or 0.06 μg/kg, i.p., every other day for 45 weeks) or 1(OH)D_3 (0.06 or 0.12 μg/kg, i.p., every other day for 45 weeks) also inhibits tumorigenesis, angiogenesis, and VEGF expression in an azoxymethane-induced rat colon cancer model (Iseki et al., 1999). 1,25D_3 and 22-oxa-1,25D_3 inhibit angiogenesis in a mouse dorsal air sac model and an *in vivo* chamber angiogenesis model, using Lewis lung carcinoma (LLC) tumor cells and bFGF as angiogenesis activators, respectively (Nakagawa, Sasaki, Kato, Kubodera, & Okano, 2005). Topical application of 1,25D_3 (10^{-7}, 10^{-8}, 10^{-9} M eye drops) has antiangiogenic effects in a suture-induced cornea inflammation mouse model (Suzuki, Sano, & Kinoshita, 2000) and in an oxygen-induced ischemic retinopathy mouse model (Albert et al., 2007). 1,25D_3 and retinoids (all-*trans* retinoic acid, 13-*cis* retinoic and 9-*cis* retinoic acid) synergistically inhibit tumor cell-induced angiogenesis *in vivo* in mouse xenograft models (Majewski et al., 1996; Majewski, Szmurlo, Marczak, Jablonska, & Bollag, 1993). These studies using animal models indicate that 1,25D and analogs have anti-angiogenic activities in cancer.

Mechanisms underlying the antiangiogenic effects of vitamin D remain to be fully elucidated. Reduction of the expression of MMP-2, MMP-9, and VEGF by 1,25D_3 or 22-oxa-1,25D_3 contributes to their antiangiogenic activity in LLC model *in vivo* (Nakagawa, Sasaki, et al., 2005). 1,25D_3 inhibits prostate cancer cell-conditioned media-induced HUVEC tube formation, migration, and MMP-9 expression by reducing the secretion of IL-8 in the prostate cancer cell lines (Bao, Yao, & Lee, 2006).

Interestingly, vitamin D promotes angiogenesis under physiological conditions. 1,25D_3 (0.3 ml of 10(−9) M, i.p.) enhances the vascularization of growth plate cartilage and induces VEGF expression in growth plate chondrocytes and osteoblasts *in vivo* (Lin et al., 2002). 1,25D_3 promotes the proangiogenic properties, including the formation of capillary-like structures and cell proliferation, of endothelial colony-forming cells (ECFCs) isolated from cord blood (Grundmann et al., 2012). These effects are mediated by the promotion of VEGF expression and pro-MMP2 activity by 1,25D_3 (Grundmann et al., 2012). 1,25D_3 analog ED-71 promotes blood vessel formation in bone marrow cavity following bone marrow ablation in mice; this effect is associated with enhanced VEGF120 expression in bone marrow cells (Okuda et al., 2007). It appears that 1,25D_3 differentially regulates angiogenesis in normal and tumor microenvironments.

Vitamin D has a significant impact on the vasculature and angiogenesis. 1,25D_3 inhibits the growth of endothelial cells derived from tumor environment through the induction of cell cycle arrest and apoptosis. Further, 1,25D_3 has *in vivo* antiangiogenic effects in several tumor model systems. These findings provide additional support for the use of 1,25D_3 in cancer therapy. Further studies on the mechanisms for vitamin D antiangiogenic effects are needed to enhance our understanding of its role in vasculature.

3.6 Invasion and Metastasis

Metastasis causes 90% of cancer-related deaths (Sporn, 1996). Metastasis is a complex process that involves multiple steps and several cellular processes such as cell adhesion, migration, invasion, proliferation, and angiogenesis. $1,25D_3$ and its analogs have been demonstrated to modulate these crucial steps in various cancer types.

$1,25D_3$ and its analog EB1089 restore the "adhesiveness" of thyroid cancer cells to collagen type I matrix through a PTEN-dependent induction of fibronectin (Liu, Asa, & Ezzat, 2005). By contrast, $1,25D_3$ inhibits adhesion and migration by downregulating integrins in prostate cancer cells (Sung & Feldman, 2000). $1,25D_3$ inhibits SCC cell migration through the promotion of E-cadherin expression (Ma et al., 2013). E-cadherin induction is VDR dependent since $1,25D_3$ has no effect on E-cadherin level in $1,25D_3$-resistant SCC-DR cells which have defective VDR signaling (Ma et al., 2013). $1,25D_3$ also induces the expression of E-cadherin in colon and prostate cancer cells (Hsu et al., 2011; Palmer et al., 2001). E-cadherin induction reduces prostate cancer cell rolling and adhesion to the endothelium and may lower their metastasis potential (Hsu et al., 2011). $1,25D_3$ inhibits SCC cell invasion, which is accompanied by reduced expression and secretion of MMP-2 and MMP-9 (Ma et al., 2013). $1,25D_3$ suppresses prostate cancer cell invasion through the reduction of the expression of MMP-9 and cathepsins and increased tissue inhibitors of metalloproteinase-1 activity (Bao, Yeh, & Lee, 2006). $1,25D_3$ reduces migration (Young et al., 1993) and invasion (Young & Lozano, 1997) of LLC cell line LLC-LN7, which is accompanied by decreased GM-CSF production (Young et al., 1993) and protein kinase A activity (Young & Lozano, 1997), respectively. $1,25D_3$ suppresses colon cancer cell migration and invasion through the inhibition of *DICKKOPF-4 (DKK-4)* gene, a downstream target of Wnt/β-catenin and a Wnt pathway antagonist (Pendas-Franco et al., 2008). The expression of *VDR* and *DKK-4* mRNA is inversely correlated in colon tumor tissues (Pendas-Franco et al., 2008). In pancreatic cancer BxPC-3 and PANC cells, $1,25D_3$ and its analog MART-10 inhibit epithelial–mesenchymal transition by reducing the expression of Snail, Slug, and vimentin, which results in reduced migration and invasion (Chiang et al., 2014). Taken together, 1,25D suppresses cancer cell adhesion, migration, and invasion *in vitro* in several tumor types.

$1,25D_3$ inhibits *in vivo* metastatic growth in several tumor models. $1,25D_3$ (0.5 μg/kg daily for 28 days) suppresses spontaneous and experimental pulmonary metastasis in a B16 melanoma mouse model (Yudoh, Matsuno, & Kimura, 1999). $1,25D_3$ (0.3 or 0.6 μg/mouse daily for 3 days) inhibits SCC lung metastasis in an experimental mouse model through the tail vein injection of SCC cells (Ma et al., 2013). 22-Oxa-$1,25D_3$ reduces angiogenesis in the basic fibroblast growth factor-induced angiogenesis model and lung colony formation in an intravenous injection model (Nakagawa, Sasaki, et al., 2005). $1,25D_3$ or 22-oxa-$1,25D_3$ suppresses lung metastasis of LCC cells in an experimental metastasis model (Nakagawa, Sasaki, et al., 2005). EB1089 suppresses bone metastasis in a breast cancer metastasis model (El Abdaimi et al., 2000). EB1089 inhibits parathyroid hormone-related protein-mediated xenograft tumor growth and bone metastasis in LNCaP-C4-2 prostate cancer mouse model (Bhatia et al., 2009).

Conversely, vitamin D deficiency promotes growth in favored sites of metastasis in several animal models. In a diet-induced vitamin D-deficient mouse model, enhanced growth of human breast cancer cells injected into the tibia and larger osteolytic lesions are observed in comparison to vitamin D-sufficient mice (Ooi, Zheng, et al., 2010; Ooi, Zhou, et al., 2010). Likewise, prostate cancer cell growth in bone is increased in vitamin D-deficient mice (Zheng et al., 2011). Metastatic growth of LLC cells is markedly reduced in $VDR^{-/-}$ mice which have high levels of serum $1,25D_3$ when compared to $VDR^{+/+}$ mice (Nakagawa, Kawaura, Kato, Takeda, & Okano, 2005). In the aggressive autochthonous transgenic adenocarcinoma of mouse prostate model, early treatment with $1,25D_3$ (20 μg/kg, i.p., 3 times/week) slows androgen-stimulated prostate tumor growth (Ajibade et al., 2014). However, long-term treatment with $1,25D_3$, from 4 weeks of age until palpable tumors developed at 20–25 weeks of age, leads to an increased number of distant organ metastases including in the liver, kidney, and lungs (Ajibade et al., 2014). This contrary result may be due to the animal model and the regimen used. Overall, the preclinical studies support an antimetastatic role for $1,25D_3$, but further studies are needed to assess clinical relevance.

Interestingly, $1,25D_3$ may play a facilitating role in regulating cell motility in normal cells. Enhanced cell motility of benign vascular smooth muscle cells is observed with $1,25D_3$ treatment and is associated with phosphatidylinositol 3-kinase activation (Rebsamen, Sun, Norman, & Liao, 2002). This differential regulation of cell motility is beneficial in cancer treatment because it allows the normal cells to be spared from the inhibitory effect.

In summary, compelling evidence supports an inhibitory role of $1,25D_3$ in cancer cell migration, invasion, and metastasis in a number of *in vitro* and *in vivo* cancer model systems. However, the molecular mechanisms for the antimetastatic activity of $1,25D_3$ remain to be fully understood. Further investigations are warranted to provide more insight into the design of $1,25D_3$-based therapy to prevent or treat metastasis.

3.7 Inflammation

Inflammation is one of the hallmarks of cancer (Hanahan & Weinberg, 2011). Chronic inflammation has been shown to increase the risk of a number of cancers including colorectal, lung, liver, bladder, and gastric cancers (Atsumi et al., 2014; Gagliani, Hu, Huber, Elinav, & Flavell, 2014). Inflammatory cells, cytokines, and other inflammatory mediators exist in the tumor microenvironment and regulate tumor progression. Both local inflammation and systemic inflammation, consisting of circulating immune cells, circulating cytokines, small inflammatory mediators such as prostaglandins (PGs), ROS, and acute phase proteins, may contribute to tumor growth and influence response to treatment (Diakos, Charles, McMillan, & Clarke, 2014). Cancer-related inflammation is involved in many aspects of tumor development and progression, such as tumor cell growth and survival, angiogenesis, invasion and metastasis, tumor immunity, and response to therapeutics (Diakos et al., 2014).

The VDR is expressed in many types of immune cells including activated CD4+ and CD8+ T cells, B cells, macrophages, and dendritic cells (Veldman, Cantorna, & DeLuca, 2000). In addition, T cells, B cells, macrophages, and dendritic cells express CYP27B1 ($25(OH)D_3$ 1-α-hydroxylase) and are able to synthesize the active metabolite $1,25D_3$ from $25(OH)D_3$ to regulate local vitamin D availability through intracrine, autocrine, or paracrine mechanisms (Adams & Hewison, 2008; Kundu, Chain, Coussens, Khoo, & Noursadeghi, 2014). The regulation of immune cells has an important impact on the inflammation in tumor microenvironments.

$1,25D_3$ has an anti-inflammatory effect in certain types of cancer. $1,25D_3$ directly inhibits the expression of NF-κB family member p50 and c-Rel as well as NF-κB transcriptional activity in activated lymphocytes (Yu, Bellido, & Manolagas, 1995). $1,25D_3$ inhibits the nuclear translocation of the NF-κB subunit p65 and subsequent NF-κB DNA binding, which results in decreased production of the proinflammatory cytokine IL-8 in human prostate cancer cells (Bao, Yao, et al., 2006). On the other hand, the VDR antagonist ZK 191732 enhances basal NF-κB activity by reducing the level of IκB in colon cancer cells HT-29 (Schwab et al., 2007). Similarly, the endogenous level of NF-κB inhibitor α (IκBα) is significantly lower in mouse embryonic fibroblasts (MEFs) derived from $VDR^{-/-}$ mice compared with $VDR^{+/-}$ MEFs (Sun et al., 2006). Increased induction of IL-6 by TNFα or IL-1β treatment is observed in $VDR^{-/-}$ MEFs than in $VDR^{+/-}$ MEFs, indicating enhanced inflammation in the absence of VDR (Sun et al., 2006).

PGs promote cancer growth and metastasis in a number of cancer types through the induction of proliferation and angiogenesis and the inhibition of apoptosis (Hawk, Viner, Dannenberg, & DuBois, 2002). $1,25D_3$ suppresses the production of PGs through multiple pathways (Krishnan & Feldman, 2011). In prostate and breast cancer cells, $1,25D_3$ treatment results in decreased expression of cyclooxygenase (COX)-2, which is the enzyme responsible for PG synthesis (Krishnan et al., 2010; Moreno et al., 2005). In addition, $1,25D_3$ increases the expression of 15-hydroxyprostaglandin dehydrogenase, which catalyzes the degradation of PGs (Krishnan et al., 2010; Moreno et al., 2005). The levels of active PGs which suppress PG-induced proliferation and inflammation in culture supernatant are reduced. $1,25D_3$ further suppresses the expression of PG receptors EP2 and FP at the mRNA level in prostate cancer cells, which limits the functional activity of PGs (Moreno et al., 2005). The combination of $1,25D_3$ with nonsteroidal anti-inflammatory drugs (NSAIDs), which inhibit PG synthesis through COX-2 inhibition, synergistically inhibits cell proliferation compared with single agent treatment in prostate cancer cells (Moreno et al., 2005).

$1,25D_3$ promotes the expression of mitogen-activated protein kinase phosphatase 5, which dephosphorylates and thereby inactivates p38 MAPK, in normal human prostate epithelial cells and primary prostate adenocarcinoma cells (Nonn, Peng, Feldman, & Peehl, 2006). Pretreatment with $1,25D_3$ results in reduced production of IL-6 upon UV or TNFα stimulation in these cells, which is mediated by the inhibition of p38 (Nonn et al., 2006).

In summary, $1,25D_3$ exerts anti-inflammatory activities in a number of cancer model systems through multiple mechanisms. Considering the important role of inflammation in carcinogenesis and tumor progression, these compelling findings suggest that $1,25D_3$, as a therapeutic agent, may have beneficial effects on cancer prevention and cancer treatment (Fig. 1).

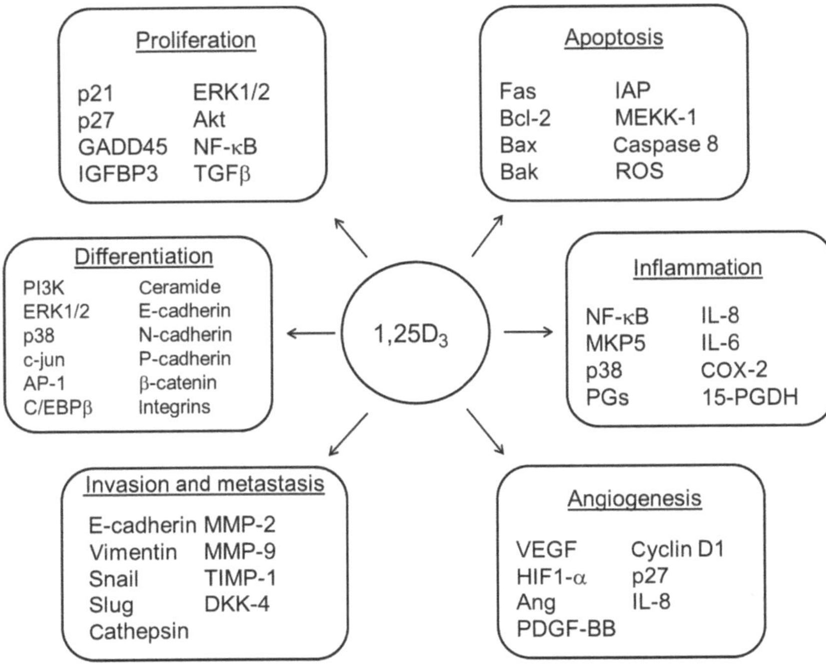

Figure 1 Key molecules involved in the antitumor effects of 1,25D$_3$ and its analogs.

3.8 1,25D$_3$ in Combinational Treatment

In addition to its direct antitumor effects, 1,25D$_3$ has additive or synergistic effects in many combinations with chemotherapeutic agents or other compounds. Cisplatin and its analog carboplatin are widely used DNA-damaging drugs. 1,25D$_3$ potentiates the growth inhibitory effect of carboplatin and cisplatin in breast cancer MCF-7 cells and prostate cancer LNCaP and DU145 cells (Cho et al., 1991; Moffatt, Johannes, & Miller, 1999). 1,25D$_3$ enhances cisplatin antitumor effect in a Y-79 human retinoblastoma xenograft model (Kulkarni, van Ginkel, Darjatmoko, Lindstrom, & Albert, 2009) and canine breast cancer, osteosarcoma, and mastocytoma cells (Rassnick et al., 2008). 1,25D$_3$ promotes MEKK-1 expression and caspase 3 cleavage when used in combination with cisplatin in SCC cells (Hershberger et al., 2002). In line with these findings, pretreatment with 1,25D$_3$ sensitizes SCC cells to cisplatin-induced growth inhibition through the enhanced expression of p73, a p53 family member (Ma et al., 2008). 1,25D$_3$ enhances caspase-dependent apoptosis and synergistically promotes the antiproliferative effects of gemcitabine and cisplatin; the current standard chemotherapy regimen for locally advanced and metastatic bladder cancer, in human bladder cancer cells T24 and UMUC3. This finding is associated with the increased level of p73 (Ma, Yu, Trump, & Johnson, 2010). A 4-week short-term treatment with $1,25D_3$ also augmented growth inhibition and apoptosis induced by gemcitabine in human pancreatic cancer Capan-1 cells (Ma et al., 2004). A 4-week short-term treatment with 1,25D and the EGFR inhibitor erlotinib results in stronger reduction of tumor incidence and volume than any single agent treatment in the 4-nitroquinoline-1-oxide-induced model of head and neck SCC, as well as the tumor growth in patient-derived xenografts in mice (Bothwell et al., 2015). The combination treatment also leads to the reduction of phospho-EGFR and phosphor-Akt levels in the whole tongue extracts (Bothwell et al., 2015).

$1,25D_3$ promotes tumor cell sensitivity to several antimetabolites, which interfere with the synthesis of RNA and DNA. $1,25D_3$ enhances cellular sensitivity of human colon cancer cells to 5-fluorouracil through a calcium-sensing receptor (Liu, Hu, & Chakrabarty, 2010). $1,25D_3$ promotes the accumulation of DNA fragments and cytotoxicity of ara-C (cytarabine) (Studzinski, Reddy, Hill, & Bhandal, 1991). The combination of $1,25D_3$ and cytarabine prolongs remission in elderly patients with AML and myelodysplastic syndrome in clinic (Ferrero et al., 2004; Slapak, Desforges, Fogaren, & Miller, 1992). $1,25D_3$ sensitizes breast cancer cells to DNA-damaging agent doxorubicin by inhibiting the expression and activity of cytoplasmic antioxidant enzyme Cu/Zn superoxide dismutase (Ravid et al., 1999). Tamoxifen and $1,25D_3$ or its analog EB1089, KH1060, CB966, or OCT combined lead to enhanced growth inhibition in breast cancer cells MCF-7 than either agent alone (Vink-van Wijngaarden et al., 1994).

$1,25D_3$ potentiates antitumor activity of microtubule-disrupting agents such as paclitaxel (Hershberger et al., 2001; Wang, Yang, Uytingco, Christakos, & Wieder, 2000) and docetaxel (Ting, Hsu, Bao, & Lee, 2007). This effect is associated with reduced expression level of p21 in prostate cancer cells (Hershberger et al., 2001) or increased Bcl-2 phosphorylation in breast cancer cells (Wang, Yang, et al., 2000) and multidrug resistance-associated protein 1 (Ting et al., 2007). $1,25D_3$ analog $1,25(OH)_2$-16-ene-23-yne-19-nor-26,27-F6-D_3 (LH) or EB1089 also potentiates antitumor activity of paclitaxel in breast cancer model systems (Koshizuka et al., 1999).

In EGFR and HER2 positive breast cancer cells, which are associated with poor prognosis and high metastasis rate, the combination of a tyrosine kinase inhibitor gefitinib and $1,25D_3$ or the analogs calcipotriol or EB1089 result in greater growth inhibitory effect than either agent alone (Segovia-Mendoza et al., 2014). This effect is associated with reduced ERK1/2 phosphorylation and increased apoptosis through upregulation of Bim and caspase 3 activation (Segovia-Mendoza et al., 2014).

The $1,25D_3$ analog ILX 23-7553 additively enhances the antitumor effects of both doxorubicin and ionizing irradiation in breast tumor cells MCF-7 through growth inhibition and apoptosis induction (Chaudhry, Sundaram, Gennings, Carter, & Gewirtz, 2001). EB1089 promotes the sensitivity of MCF-7 cells to irradiation through the increase of autophagy (Demasters, Di, Newsham, Shiu, & Gewirtz, 2006). In anaplastic thyroid cancer cells, $1,25D_3$ or its analog CD578 enhances the antiproliferative effect of paclitaxel or suberoylanilide hydroxamic acid, a potent histone deacetylase inhibitor, additively or synergistically (Clinckspoor et al., 2011). The adenosine deaminase-resistant analog fludarabine synergistically enhances $1,25D_3$-induced differentiation of human monoblastic leukemia U937 cells (Niitsu, Umeda, & Honma, 2000). $1,25D_3$ acts synergistically with hydroxyurea, cytarabine, or camptothecin to inhibit human monoblastic leukemia U937 cell growth (Makishima, Okabe-Kado, & Honma, 1998). Hydroxyurea also promotes $1,25D_3$-mediated U937 cell differentiation (Makishima et al., 1998). These findings indicate that $1,25D_3$ enhances the activity of multiple chemotherapy agents.

Besides chemotherapy, vitamin D has also been examined in combination with other types of cancer treatment. $1,25D_3$ or its analog 19-nor-1α,25-$(OH)_2D_2$ and ionizing radiation synergistically inhibit the proliferation of LNCaP prostate cancer cells and primary human tumor cells (Dunlap et al., 2003). Breast cancer cells overexpress one of the NF-κB subunits RelB, which promotes cancer cell survival. $1,25D_3$ treatment results in reduced mRNA and protein levels of RelB and its target genes *survivin*, *Bcl-2,* and *MnSOD*, and sensitizes the breast cancer cell lines, Hs578T and NF639, to gamma irradiation (Mineva et al., 2009). $1,25D_3$ also enhances the phototoxic response of human SCC A431 cells to methyl aminolevulinate-based photodynamic therapy (Cicarma, Tuorkey, Juzeniene, Ma, & Moan, 2009).

$1,25D_3$ in combination with nonspecific COX inhibitors acetyl salicylic acid or indomethacin markedly induces the differentiation of leukemia cell lines into monocytes and G_1 cell cycle arrest (Jamshidi, Zhang, Harrison, Wang, & Studzinski, 2008). This differentiation is dependent on the phosphorylation of Raf1 (Jamshidi et al., 2008). The combination treatment of ibuprofen, a NSAID, with $1,25D_3$ leads to enhanced growth inhibition and

G1 cell cycle arrest in human prostate cancer LNCaP cells compared to single agent treatment (Gavrilov, Steiner, & Shany, 2005). A $1,25D_3$ analog 22-oxa-1α,25-$(OH)_2D_3$, when used in combination with vitamin K_2, promotes the differentiation of HL-60 leukemia cells into monocytes in a synergistic nature as examined by morphology and cell surface CD14 expression (Funato, Miyazawa, Yaguchi, Gotoh, & Ohyashiki, 2002). This combination also leads to cell cycle arrest at G_0/G_1 phase and reduced apoptosis compared to vitamin K_2 alone (Funato et al., 2002). Carnosic acid, a plant-derived polyphenolic antioxidant, potentiates the monocytic differentiation effects of $1,25D_3$ in HL-60 cells (Danilenko et al., 2003). The combination treatment results in decreased intracellular ROS, increased intracellular glutathione, and the activation of Raf-1/MEK1/ERK1/2 pathway (Danilenko et al., 2003). A marine bryozoan-derived natural compound bryostatin-1 exerts antitumor activities in both solid and lymphoid tumors (Clark, Konyer, & Meckling, 2004). Bryostatin-1 synergizes with $1,25D_3$ to induce monocytic differentiation of NB4 cells (Clark et al., 2004; Song & Norman, 1999), which involves G_1 phase cell cycle arrest, decreased cell growth, and increased plastic adhesion (Clark et al., 2004). $25(OH)D_3$, when used together with iron deprivation agents including iron chelators or transferrin receptor antibody A24, induces the differentiation of myeloid leukemia cell lines and primary myeloblasts from AML patients into monocytes/macrophages (Callens et al., 2010). These effects are dependent on the induction of ROS and the activation of JNK MAPK pathway (Callens et al., 2010).

Treatment of $1,25D_3$ in combination with the retinoid X receptor (RXR) ligand 9-*cis* retinoic acid (9-*cis*-RA), delays tumor progression in PC3 prostate tumor xenograft model (Ikeda et al., 2003). This combination treatment leads to direct binding of VDR/RXR heterodimer to the promoter region of *hTERT* and the inhibition of *hTERT* expression, which subsequently results in decreased telomerase activity (Ikeda et al., 2003). $1,25D_3$ analog 20-*epi*-22oxa-24a,26a,27a-tri-homo-1α,25$(OH)_2D_3$ (KH1060) synergizes with 9-*cis*-RA to inhibit proliferation and promote differentiation of APL NB4 cells (Elstner et al., 1997) and myeloblastic HL-60 cells (Elstner et al., 1996). The combination treatment induces apoptosis which is associated with decreased Bcl-2 expression and increased Bax expression (Elstner et al., 1997). $1,25D_3$ antitumor effect may also involve histone deacetylation. Combining histone deacetylase inhibitor sodium butyrate or trichostatin A with $1,25D_3$ or its analog LH or 1α,25-$(OH)_2$-16,23*E*-diene-26,27-hexafluoride-D_3 (LT) synergistically suppresses the growth inhibitory effect of 1,25D_3 or its analog in prostate cancer cell lines LNCaP, PC3, and DU145 (Rashid et al., 2001), which is associated with apoptosis but not cell cycle arrest (Rashid et al., 2001).

In addition, the combination treatment of 1,25D_3 with other agents may potentiate the antitumor activity of 1,25D_3. Pretreatment with dexamethasone, which prevents 1,25D_3-induced hypercalcemia, followed by 1,25D_3 further inhibits cell proliferation in SCC cells compared with cells treated with 1,25D_3 alone (Yu et al., 1998). The enhanced antitumor activity by the combination treatment with dexamethasone and 1,25D_3 is also shown *in vivo* in a SCC xenograft mouse model. These findings are associated with the observations that dexamethasone enhances VDR expression in SCC cells and VDR ligand binding activities in tumor cell extracts and kidneys but decreases that in the intestinal mucosa (Yu et al., 1998). The 1,25D_3 and dexamethasone combination induces apoptosis and G_0/G_1 cell cycle arrest and inhibits the activation of Akt and ERK1/2 pathways in SCC cells (Bernardi et al., 2001). Further studies on the underlying mechanisms reveal that dexamethasone-induced VDR expression depends on glucocorticoid receptor binding to a glucocorticoid response element upstream of the transcription start site of Vdr gene (Hidalgo, Deeb, Pike, Johnson, & Trump, 2011). The upregulation of VDR expression leads to increased VDR transcription activity supported by increased $p27^{kip1}$ and reduced expression of $p21^{waf1/cip1}$ and cyclin D1 in SCC cells (Hidalgo et al., 2011).

The metabolizing enzyme 24-hydroxylase (CYP24A1), a mitochondrial cytochrome P450 enzyme, reduces the activity of 1,25D_3 by inducing the degradation of 1,25D_3. A broad-spectrum cytochrome P450 enzyme inhibitor ketoconazole (KTZ) or a CYP24A1-specific inhibitor RC2204 effectively inhibits CYP24A1 expression and its enzyme activity in prostate cancer PC3 cells and mouse kidney tissue. The combination treatment with KTZ or RC2204 with 1,25D_3 synergistically enhances the antiproliferative effect of 1,25D_3 in human prostate PC3 cells (Muindi et al., 2009). Dexamethasone is applied with KTZ since steroids are required when KTZ is administered in patients. The apoptosis observed is independent of caspase 3 activation, but involves nuclear translocation of apoptosis-inducing factor. 1,25D_3 combined with KTZ/dexamethasone enhances the growth inhibitory effect of 1,25D_3 alone in the PC3 xenograft mouse model (Muindi et al., 2009). Likewise, KTZ is shown to inhibit the induction of CYP24A1 and potentiates the antiproliferative effect of 1,25D_3 or analog EB1089 in a preclinical prostate cancer model (Peehl, Seto, Hsu, & Feldman, 2002). KTZ or a more potent CYP24A1 inhibitor, tetralone derivative 2-(4-hydroxybenzyl)-6-methoxy-3,4-dihydro-2*H*-naphthalen-1-one, enhances the antiproliferative effect of $1,25D_3$ in DU-145 and PC3 cells (Yee, Campbell, & Simons, 2006). The tetralone derivative in combination with $1,25D_3$ increases the expression of VDR target genes $p21^{waf/cip1}$ and *GADD45a* in DU-145 cells (Yee et al., 2006). An imidazole derivative liarozole inhibits the enzyme activity of CYP24A1 and thus sensitizes prostate cancer DU145 cells to $1,25D_3$-mediated growth inhibition, which is accompanied by increased VDR expression (Ly, Zhao, Holloway, & Feldman, 1999). CYP24A1 expression is frequently detected in NSCLC cells but not in the nontumorigenic bronchial epithelial Beas2B cells (Parise et al., 2006). In patient-derived tissue samples, CYP24A1 is detected in 10/18 of the primary lung tumors versus 1/11 of the normal lung tissue samples (Parise et al., 2006). In lung cancer H292 cells, a highly selective CYP24A1 inhibitor CTA091, which is a 24-sulfoximine analog of $1,25D_3$ binding to the substrate binding pocket of CYP24, enhances the growth inhibitory effect of $1,25D_3$ (Zhang et al., 2012). $1,25D_3$ induces G0/G1 arrest, which is accompanied by decreased levels of cyclin E2. In combination with CTA091, cell cycle arrest is further induced and cyclins E2, D1, and A are decreased (Zhang et al., 2012).

Together, $1,25D_3$ in combination of various chemotherapeutic agents results in enhanced growth inhibitory effects in a number of tumors. The effect of $1,25D_3$ is also potentiated by agents such as dexamethasone and CYP24A1 inhibitors. Further understanding of the effect and mechanisms of $1,25D_3$ combination treatment will identify potential therapeutic targets and facilitate the development of new and more effective treatment regimens in cancer.

4. CONCLUSIONS

Beyond its classical roles in maintaining calcium homeostasis and bone metabolism, increasing evidence has recognized important roles of vitamin D in many biological functions and diseases including cancer. Epidemiological as well as preclinical studies show $1,25D_3$ and its analogs have antitumor effects in many tumor types. The mechanisms involve the induction of apoptosis and differentiation, as well as the inhibition of cell proliferation, inflammation, angiogenesis, invasion, and metastasis. Multiple pathways and molecular targets contribute to these effects in a cancer type- and cell type-dependent manner. The reasons for the cancer-type specificity are still unclear. $1,25D_3$ differentially affects apoptosis, angiogenesis, and cell motility in normal noncancer cells, which is beneficial in cancer treatment. However, the mechanisms underlying these observations remain to be investigated. When used in combination treatment with chemotherapeutic agents or other agents, $1,25D_3$ additively or synergistically enhances the antitumor activities of these agents. On the other hand, the antitumor activities of $1,25D_3$ are regulated by agents such as dexamethasone or its metabolizing enzyme CYP24A1. Although data such as reviewed here support the use of $1,25D_3$ in cancer prevention and treatment, our understanding on the effects and mechanisms of $1,25D_3$ in cancer is still limited. Further randomized supplementation trials in general population and clinical trials in cancer patients are needed to evaluate the impact of $1,25D_3$ in cancer prevention and treatment, and guide the proper application of $1,25D_3$ for these purposes.

ACKNOWLEDGMENTS

We thank Drs. Pamela A. Hershberger, David W. Goodrich, Mukund Seshadri, Xinjiang Wang, Dhyan Chandra, Leigh Ellis, Wendy J. Huss, Neelu Yadav, Sebastiano Battaglia, Wendy Swetzig and Ms. Victoria N. Cranwell for their critical review of the manuscript.

REFERENCES

Abe, E., Miyaura, C., Sakagami, H., Takeda, M., Konno, K., Yamazaki, T., et al. (1981). Differentiation of mouse myeloid leukemia cells induced by 1 alpha,25-dihydroxyvitamin D3. *Proceedings of the National Academy of Sciences of the United States of America*, *78*(8), 4990–4994. Epub 1981/08/01.

Adams, J. S., & Hewison, M. (2008). Unexpected actions of vitamin D: New perspectives on the regulation of innate and adaptive immunity. *Nature Clinical Practice. Endocrinology & Metabolism*, *4*(2), 80–90. Epub 2008/01/24.

Ahn, J., Peters, U., Albanes, D., Purdue, M. P., Abnet, C. C., Chatterjee, N., et al. (2008). Serum vitamin D concentration and prostate cancer risk: A nested case-control study. *Journal of the National Cancer Institute*, *100*(11), 796–804. Epub 2008/05/29.

Ajibade, A. A., Kirk, J. S., Karasik, E., Gillard, B., Moser, M. T., Johnson, C. S., et al. (2014). Early growth inhibition is followed by increased metastatic disease with vitamin D (calcitriol) treatment in the TRAMP model of prostate cancer. *PLoS One*, *9*(2), e89555. Epub 2014/03/04.

Albert, D. M., Scheef, E. A., Wang, S., Mehraein, F., Darjatmoko, S. R., Sorenson, C. M., et al. (2007). Calcitriol is a potent inhibitor of retinal neovascularization. *Investigative Ophthalmology & Visual Science*, *48*(5), 2327–2334.

Alvarez-Diaz, S., Valle, N., Ferrer-Mayorga, G., Lombardia, L., Herrera, M., Dominguez, O., et al. (2012). MicroRNA-22 is induced by vitamin D and contributes to its antiproliferative, antimigratory and gene regulatory effects in colon cancer cells. *Human Molecular Genetics*, *21*(10), 2157–2165. Epub 2012/02/14.

Amaral, A. F., Mendez-Pertuz, M., Munoz, A., Silverman, D. T., Allory, Y., Kogevinas, M., et al. (2012). Plasma 25-hydroxyvitamin D3 and bladder cancer risk according to tumor stage and FGFR3 status: A mechanism-based epidemiological study. *Journal of the National Cancer Institute*, *104*(24), 1897–1904. Epub 2012/10/31.

Atsumi, T., Singh, R., Sabharwal, L., Bando, H., Meng, J., Arima, Y., et al. (2014). Inflammation amplifier, a new paradigm in cancer biology. *Cancer Research*, *74*(1), 8–14. Epub 2013/12/24.

Bao, B. Y., Yao, J., & Lee, Y. F. (2006). 1Alpha, 25-dihydroxyvitamin D3 suppresses interleukin-8-mediated prostate cancer cell angiogenesis. *Carcinogenesis*, *27*(9), 1883–1893.

Bao, B. Y., Yeh, S. D., & Lee, Y. F. (2006). 1Alpha,25-dihydroxyvitamin D3 inhibits prostate cancer cell invasion via modulation of selective proteases. *Carcinogenesis*, *27*(1), 32–42.

Beer, T. M., Garzotto, M., Park, B., Mori, M., Myrthue, A., Janeba, N., et al. (2006). Effect of calcitriol on prostate-specific antigen in vitro and in humans. *Clinical Cancer Research: An Official Journal of the American Association for Cancer Research*, *12*(9), 2812–2816. Epub 2006/05/06.

Bernardi, R. J., Johnson, C. S., Modzelewski, R. A., & Trump, D. L. (2002). Antiproliferative effects of 1alpha,25-dihydroxyvitamin D(3) and vitamin D analogs on tumor-derived endothelial cells. *Endocrinology*, *143*(7), 2508–2514.

Bernardi, R. J., Trump, D. L., Yu, W. D., McGuire, T. F., Hershberger, P. A., & Johnson, C. S. (2001). Combination of 1alpha,25-dihydroxyvitamin D(3) with dexamethasone enhances cell cycle arrest and apoptosis: Role of nuclear receptor cross-talk and Erk/Akt signaling. *Clinical Cancer Research: An Official Journal of the American Association for Cancer Research*, 7(12), 4164–4173.

Bertone-Johnson, E. R., Chen, W. Y., Holick, M. F., Hollis, B. W., Colditz, G. A., Willett, W. C., et al. (2005). Plasma 25-hydroxyvitamin D and 1,25-dihydroxyvitamin D and risk of breast cancer. *Cancer Epidemiology, Biomarkers & Prevention: A Publication of the American Association for Cancer Research, Cosponsored by the American Society of Preventive Oncology*, *14*(8), 1991–1997. Epub 2005/08/17.

Bhatia, V., Saini, M. K., Shen, X., Bi, L. X., Qiu, S., Weigel, N. L., et al. (2009). EB1089 inhibits the parathyroid hormone-related protein-enhanced bone metastasis and xenograft growth of human prostate cancer cells. *Molecular Cancer Therapeutics*, *8*(7), 1787–1798.

Bolland, M. J., Grey, A., Gamble, G. D., & Reid, I. R. (2011). Calcium and vitamin D supplements and health outcomes: A reanalysis of the Women's Health Initiative (WHI) limited-access data set. *The American Journal of Clinical Nutrition*, *94*(4), 1144–1149. Epub 2011/09/02.

Boscoe, F. P., & Schymura, M. J. (2006). Solar ultraviolet-B exposure and cancer incidence and mortality in the United States, 1993–2002. *BMC Cancer*, *6*, 264. Epub 2006/11/14.

Bothwell, K. D., Shaurova, T., Merzianu, M., Suresh, A., Kuriakose, M. A., Johnson, C. S., et al. (2015). Impact of short-term 1,25-dihydroxyvitamin D3 on the chemopreventive efficacy of erlotinib against oral cancer. *Cancer Prevention Research (Philadelphia, PA)*, *8*(9), 765–776. Epub 2015/06/24.

Boyle, B. J., Zhao, X. Y., Cohen, P., & Feldman, D. (2001). Insulin-like growth factor binding protein-3 mediates 1 alpha,25-dihydroxyvitamin d(3) growth inhibition in the LNCaP prostate cancer cell line through p21/WAF1. *The Journal of Urology*, *165*(4), 1319–1324. Epub 2001/03/21.

Brackman, D., Lund-Johansen, F., & Aarskog, D. (1995). Expression of cell surface antigens during the differentiation of HL-60 cells induced by 1,25-dihydroxyvitamin D3, retinoic acid and DMSO. *Leukemia Research*, *19*(1), 57–64. Epub 1995/01/01.

Callens, C., Coulon, S., Naudin, J., Radford-Weiss, I., Boissel, N., Raffoux, E., et al. (2010). Targeting iron homeostasis induces cellular differentiation and synergizes with differentiating agents in acute myeloid leukemia. *The Journal of Experimental Medicine*, *207*(4), 731–750.

Capiati, D. A., Rossi, A. M., Picotto, G., Benassati, S., & Boland, R. L. (2004). Inhibition of serum-stimulated mitogen activated protein kinase by 1alpha,25(OH)2-vitamin D3 in MCF-7 breast cancer cells. *Journal of Cellular Biochemistry*, *93*(2), 384–397. Epub 2004/09/16.

Chaudhry, M., Sundaram, S., Gennings, C., Carter, H., & Gewirtz, D. A. (2001). The vitamin D3 analog, ILX-23-7553, enhances the response to adriamycin and irradiation in MCF-7 breast tumor cells. *Cancer Chemotherapy and Pharmacology*, *47*(5), 429–436.

Chen, A., Davis, B. H., Bissonnette, M., Scaglione-Sewell, B., & Brasitus, T. A. (1999). 1,25-Dihydroxyvitamin D(3) stimulates activator protein-1-dependent Caco-2 cell differentiation. *The Journal of Biological Chemistry*, *274*(50), 35505–35513. Epub 1999/12/10.

Chiang, K. C., Yeh, C. N., Hsu, J. T., Jan, Y. Y., Chen, L. W., Kuo, S. F., et al. (2014). The vitamin D analog, MART-10, represses metastasis potential via downregulation of epithelial-mesenchymal transition in pancreatic cancer cells. *Cancer Letters*, *354*(2), 235–244. Epub 2014/08/26.

Cho, Y. L., Christensen, C., Saunders, D. E., Lawrence, W. D., Deppe, G., Malviya, V. K., et al. (1991). Combined effects of 1,25-dihydroxyvitamin D3 and platinum drugs on the growth of MCF-7 cells. *Cancer Research*, *51*(11), 2848–2853.

Chung, I., Han, G., Seshadri, M., Gillard, B. M., Yu, W. D., Foster, B. A., et al. (2009). Role of vitamin D receptor in the antiproliferative effects of calcitriol in tumor-derived endothelial cells and tumor angiogenesis in vivo. *Cancer Research*, *69*(3), 967–975.

Chung, I., Karpf, A. R., Muindi, J. R., Conroy, J. M., Nowak, N. J., Johnson, C. S., et al. (2007). Epigenetic silencing of CYP24 in tumor-derived endothelial cells contributes to selective growth inhibition by calcitriol. *The Journal of Biological Chemistry*, *282*(12), 8704–8714.

Chung, I., Wong, M. K., Flynn, G., Yu, W. D., Johnson, C. S., & Trump, D. L. (2006). Differential antiproliferative effects of calcitriol on tumor-derived and matrigel-derived endothelial cells. *Cancer Research*, *66*(17), 8565–8573.

Cicarma, E., Tuorkey, M., Juzeniene, A., Ma, L. W., & Moan, J. (2009). Calcitriol treatment improves methyl aminolaevulinate-based photodynamic therapy in human squamous cell carcinoma A431 cells. *The British Journal of Dermatology*, *161*(2), 413–418.

Clark, C. S., Konyer, J. E., & Meckling, K. A. (2004). 1Alpha,25-dihydroxyvitamin D3 and bryostatin-1 synergize to induce monocytic differentiation of NB4 acute promyelocytic leukemia cells by modulating cell cycle progression. *Experimental Cell Research*, *294*(1), 301–311.

Clinckspoor, I., Verlinden, L., Overbergh, L., Korch, C., Bouillon, R., Mathieu, C., et al. (2011). 1,25-Dihydroxyvitamin D3 and a superagonistic analog in combination with paclitaxel or suberoylanilide hydroxamic acid have potent antiproliferative effects on anaplastic thyroid cancer. *The Journal of Steroid Biochemistry and Molecular Biology*, *124*(1–2), 1–9. Epub 2010/12/25.

Colston, K. W., Perks, C. M., Xie, S. P., & Holly, J. M. (1998). Growth inhibition of both MCF-7 and Hs578T human breast cancer cell lines by vitamin D analogues is associated with increased expression of insulin-like growth factor binding protein-3. *Journal of Molecular Endocrinology*, *20*(1), 157–162. Epub 1998/03/26.

Dackiw, A. P., Ezzat, S., Huang, P., Liu, W., & Asa, S. L. (2004). Vitamin D3 administration induces nuclear p27 accumulation, restores differentiation, and reduces tumor burden in a mouse model of metastatic follicular thyroid cancer. *Endocrinology*, *145*(12), 5840–5846. Epub 2004/08/21.

Danilenko, M., Wang, Q., Wang, X., Levy, J., Sharoni, Y., & Studzinski, G. P. (2003). Carnosic acid potentiates the antioxidant and prodifferentiation effects of 1alpha,25-dihydroxyvitamin D3 in leukemia cells but does not promote elevation of basal levels of intracellular calcium. *Cancer Research*, *63*(6), 1325–1332.

Deeb, K. K., Trump, D. L., & Johnson, C. S. (2007). Vitamin D signalling pathways in cancer: Potential for anticancer therapeutics. *Nature Reviews. Cancer*, 7(9), 684–700.

De Haes, P., Garmyn, M., Degreef, H., Vantieghem, K., Bouillon, R., & Segaert, S. (2003). 1,25-Dihydroxyvitamin D3 inhibits ultraviolet B-induced apoptosis, Jun kinase activation, and interleukin-6 production in primary human keratinocytes. *Journal of Cellular Biochemistry*, *89*(4), 663–673.

Demasters, G., Di, X., Newsham, I., Shiu, R., & Gewirtz, D. A. (2006). Potentiation of radiation sensitivity in breast tumor cells by the vitamin D3 analogue, EB 1089, through promotion of autophagy and interference with proliferative recovery. *Molecular Cancer Therapeutics*, *5*(11), 2786–2797. Epub 2006/11/24.

Diakos, C. I., Charles, K. A., McMillan, D. C., & Clarke, S. J. (2014). Cancer-related inflammation and treatment effectiveness. *The Lancet Oncology*, *15*(11), e493–e503. Epub 2014/10/05.

Diaz, G. D., Paraskeva, C., Thomas, M. G., Binderup, L., & Hague, A. (2000). Apoptosis is induced by the active metabolite of vitamin D3 and its analogue EB1089 in colorectal adenoma and carcinoma cells: Possible implications for prevention and therapy. *Cancer Research*, *60*(8), 2304–2312.

Ding, E. L., Mehta, S., Fawzi, W. W., & Giovannucci, E. L. (2008). Interaction of estrogen therapy with calcium and vitamin D supplementation on colorectal cancer risk: Reanalysis of Women's Health Initiative randomized trial. *International Journal of Cancer*, *122*(8), 1690–1694. Epub 2007/12/20.

Dunlap, N., Schwartz, G. G., Eads, D., Cramer, S. D., Sherk, A. B., John, V., et al. (2003). 1Alpha,25-dihydroxyvitamin D(3) (calcitriol) and its analogue, 19-nor-1alpha,25(OH)(2)D(2), potentiate the effects of ionising radiation on human prostate cancer cells. *British Journal of Cancer*, *89*(4), 746–753.

Duque, G., El Abdaimi, K., Henderson, J. E., Lomri, A., & Kremer, R. (2004). Vitamin D inhibits Fas ligand-induced apoptosis in human osteoblasts by regulating components of both the mitochondrial and Fas-related pathways. *Bone*, *35*(1), 57–64.

El Abdaimi, K., Dion, N., Papavasiliou, V., Cardinal, P. E., Binderup, L., Goltzman, D., et al. (2000). The vitamin D analogue EB 1089 prevents skeletal metastasis and prolongs survival time in nude mice transplanted with human breast cancer cells. *Cancer Research*, *60*(16), 4412–4418.

Elstner, E., Linker-Israeli, M., Le, J., Umiel, T., Michl, P., Said, J. W., et al. (1997). Synergistic decrease of clonal proliferation, induction of differentiation, and apoptosis of acute promyelocytic leukemia cells after combined treatment with novel 20-epi vitamin D3 analogs and 9-cis retinoic acid. *The Journal of Clinical Investigation*, *99*(2), 349–360.

Elstner, E., Linker-Israeli, M., Umiel, T., Le, J., Grillier, I., Said, J., et al. (1996). Combination of a potent 20-epi-vitamin D3 analogue (KH 1060) with 9-cis-retinoic acid irreversibly inhibits clonal growth, decreases bcl-2 expression, and induces apoptosis in HL-60 leukemic cells. *Cancer Research*, *56*(15), 3570–3576.

Ferrero, D., Campa, E., Dellacasa, C., Campana, S., Foli, C., & Boccadoro, M. (2004). Differentiating agents + low-dose chemotherapy in the management of old/poor prognosis patients with acute myeloid leukemia or myelodysplastic syndrome. *Haematologica*, *89*(5), 619–620.

Folkman, J. (1971). Tumor angiogenesis: Therapeutic implications. *The New England Journal of Medicine*, *285*(21), 1182–1186.

Fujioka, T., Hasegawa, M., Ishikura, K., Matsushita, Y., Sato, M., & Tanji, S. (1998). Inhibition of tumor growth and angiogenesis by vitamin D3 agents in murine renal cell carcinoma. *The Journal of Urology*, *160*(1), 247–251.

Funato, K., Miyazawa, K., Yaguchi, M., Gotoh, A., & Ohyashiki, K. (2002). Combination of 22-oxa-1,25-dihydroxyvitamin D(3), a vitamin D(3) derivative, with vitamin K(2) (VK2) synergistically enhances cell differentiation but suppresses VK2-inducing apoptosis in HL-60 cells. *Leukemia*, *16*(8), 1519–1527.

Gagliani, N., Hu, B., Huber, S., Elinav, E., & Flavell, R. A. (2014). The fire within: Microbes inflame tumors. *Cell*, *157*(4), 776–783. Epub 2014/05/13.

Garland, C. F., Comstock, G. W., Garland, F. C., Helsing, K. J., Shaw, E. K., & Gorham, E. D. (1989). Serum 25-hydroxyvitamin D and colon cancer: Eight-year prospective study. *Lancet*, *2*(8673), 1176–1178. Epub 1989/11/18.
Garland, C. F., & Garland, F. C. (1980). Do sunlight and vitamin D reduce the likelihood of colon cancer? *International Journal of Epidemiology*, *9*(3), 227–231. Epub 1980/09/01.
Garland, C. F., Garland, F. C., Gorham, E. D., Lipkin, M., Newmark, H., Mohr, S. B., et al. (2006). The role of vitamin D in cancer prevention. *American Journal of Public Health*, *96*(2), 252–261.
Garland, C. F., Gorham, E. D., Mohr, S. B., & Garland, F. C. (2009). Vitamin D for cancer prevention: Global perspective. *Annals of Epidemiology*, *19*(7), 468–483. Epub 2009/06/16.
Gavrilov, V., Steiner, M., & Shany, S. (2005). The combined treatment of 1,25-dihydroxyvitamin D3 and a non-steroid anti-inflammatory drug is highly effective in suppressing prostate cancer cell line (LNCaP) growth. *Anticancer Research*, *25*(5), 3425–3429.
Giovannucci, E. (2009). Vitamin D, and cancer incidence in the Harvard cohorts. *Annals of Epidemiology*, *19*(2), 84–88.
Giovannucci, E., Liu, Y., Rimm, E. B., Hollis, B. W., Fuchs, C. S., Stampfer, M. J., et al. (2006). Prospective study of predictors of vitamin D status and cancer incidence and mortality in men. *Journal of the National Cancer Institute*, *98*(7), 451–459.
Giovannucci, E., Liu, Y., Stampfer, M. J., & Willett, W. C. (2006). A prospective study of calcium intake and incident and fatal prostate cancer. *Cancer Epidemiology, Biomarkers & Prevention: A Publication of the American Association for Cancer Research, Cosponsored by the American Society of Preventive Oncology*, *15*(2), 203–210.
Gonzalez-Pardo, V., D'Elia, N., Verstuyf, A., Boland, R., & Russo de Boland, A. (2012). NFkappaB pathway is down-regulated by 1alpha,25(OH)(2)-vitamin D(3) in endothelial cells transformed by Kaposi sarcoma-associated herpes virus G protein coupled receptor. *Steroids*, 77(11), 1025–1032. Epub 2012/06/12.
Gonzalez-Pardo, V., Martin, D., Gutkind, J. S., Verstuyf, A., Bouillon, R., de Boland, A. R., et al. (2010). 1 Alpha,25-dihydroxyvitamin D3 and its TX527 analog inhibit the growth of endothelial cells transformed by Kaposi sarcoma-associated herpes virus G protein-coupled receptor in vitro and in vivo. *Endocrinology*, *151*(1), 23–31. Epub 2009/11/17.
Grant, W. B. (2002). An estimate of premature cancer mortality in the U.S. due to inadequate doses of solar ultraviolet-B radiation. *Cancer*, *94*(6), 1867–1875. Epub 2002/03/29.
Grant, W. B. (2007). An ecologic study of cancer mortality rates in Spain with respect to indices of solar UVB irradiance and smoking. *International Journal of Cancer*, *120*(5), 1123–1128. Epub 2006/12/07.
Grant, W. B., & Garland, C. F. (2006). The association of solar ultraviolet B (UVB) with reducing risk of cancer: Multifactorial ecologic analysis of geographic variation in age-adjusted cancer mortality rates. *Anticancer Research*, *26*(4A), 2687–2699. Epub 2006/08/05.
Grundmann, M., Haidar, M., Placzko, S., Niendorf, R., Darashchonak, N., Hubel, C. A., et al. (2012). Vitamin D improves the angiogenic properties of endothelial progenitor cells. *American Journal of Physiology. Cell Physiology*, *303*(9), C954–C962. Epub 2012/08/31.
Guan, H., Liu, C., Chen, Z., Wang, L., Li, C., Zhao, J., et al. (2013). 1,25-Dihydroxyvitamin D3 up-regulates expression of hsa-let-7a-2 through the interaction of VDR/VDRE in human lung cancer A549 cells. *Gene*, *522*(2), 142–146. Epub 2013/04/10.
Guzey, M., Kitada, S., & Reed, J. C. (2002). Apoptosis induction by 1alpha,25-dihydroxyvitamin D3 in prostate cancer. *Molecular Cancer Therapeutics*, *1*(9), 667–677.
Hanahan, D., & Weinberg, R. A. (2011). Hallmarks of cancer: The next generation. *Cell*, *144*(5), 646–674. Epub 2011/03/08.

Haselberger, M., Springwald, A., Konwisorz, A., Lattrich, C., Goerse, R., Ortmann, O., et al. (2011). Silencing of the icb-1 gene inhibits the induction of differentiation-associated genes by vitamin D3 and all-trans retinoic acid in gynecological cancer cells. *International Journal of Molecular Medicine, 28*(1), 121–127. Epub 2011/04/02.

Hawk, E. T., Viner, J. L., Dannenberg, A., & DuBois, R. N. (2002). COX-2 in cancer—A player that's defining the rules. *Journal of the National Cancer Institute, 94*(8), 545–546. Epub 2002/04/18.

Hershberger, P. A., McGuire, T. F., Yu, W. D., Zuhowski, E. G., Schellens, J. H., Egorin, M. J., et al. (2002). Cisplatin potentiates 1,25-dihydroxyvitamin D3-induced apoptosis in association with increased mitogen-activated protein kinase kinase kinase 1 (MEKK-1) expression. *Molecular Cancer Therapeutics, 1*(10), 821–829.

Hershberger, P. A., Modzelewski, R. A., Shurin, Z. R., Rueger, R. M., Trump, D. L., & Johnson, C. S. (1999). 1,25-Dihydroxycholecalciferol (1,25-D3) inhibits the growth of squamous cell carcinoma and down-modulates p21(Waf1/Cip1) in vitro and in vivo. *Cancer Research, 59*(11), 2644–2649.

Hershberger, P. A., Yu, W. D., Modzelewski, R. A., Rueger, R. M., Johnson, C. S., & Trump, D. L. (2001). Calcitriol (1,25-dihydroxycholecalciferol) enhances paclitaxel antitumor activity in vitro and in vivo and accelerates paclitaxel-induced apoptosis. *Clinical Cancer Research: An Official Journal of the American Association for Cancer Research,* 7(4), 1043–1051.

Hidalgo, A. A., Deeb, K. K., Pike, J. W., Johnson, C. S., & Trump, D. L. (2011). Dexamethasone enhances 1alpha,25-dihydroxyvitamin D3 effects by increasing vitamin D receptor transcription. *The Journal of Biological Chemistry, 286*(42), 36228–36237.

Hmama, Z., Nandan, D., Sly, L., Knutson, K. L., Herrera-Velit, P., & Reiner, N. E. (1999). 1Alpha,25-dihydroxyvitamin D(3)-induced myeloid cell differentiation is regulated by a vitamin D receptor-phosphatidylinositol 3-kinase signaling complex. *The Journal of Experimental Medicine, 190*(11), 1583–1594. Epub 1999/12/10.

Hoyer-Hansen, M., Bastholm, L., Mathiasen, I. S., Elling, F., & Jaattela, M. (2005). Vitamin D analog EB1089 triggers dramatic lysosomal changes and Beclin 1-mediated autophagic cell death. *Cell Death and Differentiation, 12*(10), 1297–1309. Epub 2005/05/21.

Hsu, J. W., Yasmin-Karim, S., King, M. R., Wojciechowski, J. C., Mickelsen, D., Blair, M. L., et al. (2011). Suppression of prostate cancer cell rolling and adhesion to endothelium by 1alpha,25-dihydroxyvitamin D3. *The American Journal of Pathology, 178*(2), 872–880. Epub 2011/02/02.

Ikeda, N., Uemura, H., Ishiguro, H., Hori, M., Hosaka, M., Kyo, S., et al. (2003). Combination treatment with 1alpha,25-dihydroxyvitamin D3 and 9-cis-retinoic acid directly inhibits human telomerase reverse transcriptase transcription in prostate cancer cells. *Molecular Cancer Therapeutics, 2*(8), 739–746.

Iseki, K., Tatsuta, M., Uehara, H., Iishi, H., Yano, H., Sakai, N., et al. (1999). Inhibition of angiogenesis as a mechanism for inhibition by 1alpha-hydroxyvitamin D3 and 1,25-dihydroxyvitamin D3 of colon carcinogenesis induced by azoxymethane in Wistar rats. *International Journal of Cancer, 81*(5), 730–733.

Jamshidi, F., Zhang, J., Harrison, J. S., Wang, X., & Studzinski, G. P. (2008). Induction of differentiation of human leukemia cells by combinations of COX inhibitors and 1,25-dihydroxyvitamin D3 involves Raf1 but not Erk 1/2 signaling. *Cell Cycle,* 7(7), 917–924.

Ji, Y., & Studzinski, G. P. (2004). Retinoblastoma protein and CCAAT/enhancer-binding protein beta are required for 1,25-dihydroxyvitamin D3-induced monocytic differentiation of HL60 cells. *Cancer Research, 64*(1), 370–377. Epub 2004/01/20.

Jiang, F., Bao, J., Li, P., Nicosia, S. V., & Bai, W. (2004). Induction of ovarian cancer cell apoptosis by 1,25-dihydroxyvitamin D3 through the down-regulation of telomerase. *The Journal of Biological Chemistry, 279*(51), 53213–53221.

Jiang, F., Li, P., Fornace, A. J., Jr., Nicosia, S. V., & Bai, W. (2003). G2/M arrest by 1,25-dihydroxyvitamin D3 in ovarian cancer cells mediated through the induction of GADD45 via an exonic enhancer. *The Journal of Biological Chemistry*, *278*(48), 48030–48040. Epub 2003/09/25.

Kasiappan, R., Shen, Z., Tse, A. K., Jinwal, U., Tang, J., Lungchukiet, P., et al. (2012). 1,25-Dihydroxyvitamin D3 suppresses telomerase expression and human cancer growth through microRNA-498. *The Journal of Biological Chemistry*, *287*(49), 41297–41309. Epub 2012/10/12.

Kawa, S., Nikaido, T., Aoki, Y., Zhai, Y., Kumagai, T., Furihata, K., et al. (1997). Vitamin D analogues up-regulate p21 and p27 during growth inhibition of pancreatic cancer cell lines. *British Journal of Cancer*, *76*(7), 884–889. Epub 1997/01/01.

Kemmochi, S., Fujimoto, H., Woo, G. H., Hirose, M., Nishikawa, A., Mitsumori, K., et al. (2011). Preventive effects of calcitriol on the development of capsular invasive carcinomas in a rat two-stage thyroid carcinogenesis model. *The Journal of Veterinary Medical Science/The Japanese Society of Veterinary Science*, *73*(5), 655–664. Epub 2011/01/06.

Kilkkinen, A., Knekt, P., Heliovaara, M., Rissanen, H., Marniemi, J., Hakulinen, T., et al. (2008). Vitamin D status and the risk of lung cancer: A cohort study in Finland. *Cancer Epidemiology, Biomarkers & Prevention: A Publication of the American Association for Cancer Research, Cosponsored by the American Society of Preventive Oncology*, *17*(11), 3274–3278.

Koli, K., & Keski-Oja, J. (1995). 1,25-Dihydroxyvitamin D3 enhances the expression of transforming growth factor beta 1 and its latent form binding protein in cultured breast carcinoma cells. *Cancer Research*, *55*(7), 1540–1546. Epub 1995/04/01.

Koshizuka, K., Koike, M., Asou, H., Cho, S. K., Stephen, T., Rude, R. K., et al. (1999). Combined effect of vitamin D3 analogs and paclitaxel on the growth of MCF-7 breast cancer cells in vivo. *Breast Cancer Research and Treatment*, *53*(2), 113–120.

Krishnan, A. V., & Feldman, D. (2011). Mechanisms of the anti-cancer and anti-inflammatory actions of vitamin D. *Annual Review of Pharmacology and Toxicology*, *51*, 311–336. Epub 2010/10/13.

Krishnan, A. V., Swami, S., Peng, L., Wang, J., Moreno, J., & Feldman, D. (2010). Tissue-selective regulation of aromatase expression by calcitriol: Implications for breast cancer therapy. *Endocrinology*, *151*(1), 32–42. Epub 2009/11/13.

Kulkarni, A. D., van Ginkel, P. R., Darjatmoko, S. R., Lindstrom, M. J., & Albert, D. M. (2009). Use of combination therapy with cisplatin and calcitriol in the treatment of Y-79 human retinoblastoma xenograft model. *The British Journal of Ophthalmology*, *93*(8), 1105–1108.

Kundu, R., Chain, B. M., Coussens, A. K., Khoo, B., & Noursadeghi, M. (2014). Regulation of CYP27B1 and CYP24A1 hydroxylases limits cell-autonomous activation of vitamin D in dendritic cells. *European Journal of Immunology*, *44*(6), 1781–1790. Epub 2014/03/20.

Lambert, J. R., Eddy, V. J., Young, C. D., Persons, K. S., Sarkar, S., Kelly, J. A., et al. (2010). A vitamin D receptor-alkylating derivative of 1alpha,25-dihydroxyvitamin D3 inhibits growth of human kidney cancer cells and suppresses tumor growth. *Cancer Prevention Research (Philadelphia, PA)*, *3*(12), 1596–1607. Epub 2010/12/15.

Lappe, J. M., Travers-Gustafson, D., Davies, K. M., Recker, R. R., & Heaney, R. P. (2007). Vitamin D and calcium supplementation reduces cancer risk: Results of a randomized trial. *The American Journal of Clinical Nutrition*, *85*(6), 1586–1591. Epub 2007/06/09.

Lazzaro, G., Agadir, A., Qing, W., Poria, M., Mehta, R. R., Moriarty, R. M., et al. (2000). Induction of differentiation by 1alpha-hydroxyvitamin D(5) in T47D human breast cancer cells and its interaction with vitamin D receptors. *European Journal of Cancer*, *36*(6), 780–786. Epub 2000/04/14.

Lin, R., Amizuka, N., Sasaki, T., Aarts, M. M., Ozawa, H., Goltzman, D., et al. (2002). 1Alpha,25-dihydroxyvitamin D3 promotes vascularization of the chondro-osseous junction by stimulating expression of vascular endothelial growth factor and matrix metalloproteinase 9. *Journal of Bone and Mineral Research, 17*(9), 1604–1612.

Liu, W., Asa, S. L., & Ezzat, S. (2005). 1Alpha,25-dihydroxyvitamin D3 targets PTEN-dependent fibronectin expression to restore thyroid cancer cell adhesiveness. *Molecular Endocrinology, 19*(9), 2349–2357.

Liu, W., Asa, S. L., Fantus, I. G., Walfish, P. G., & Ezzat, S. (2002). Vitamin D arrests thyroid carcinoma cell growth and induces p27 dephosphorylation and accumulation through PTEN/akt-dependent and -independent pathways. *The American Journal of Pathology, 160*(2), 511–519. Epub 2002/02/13.

Liu, G., Hu, X., & Chakrabarty, S. (2010). Vitamin D mediates its action in human colon carcinoma cells in a calcium-sensing receptor-dependent manner: Downregulates malignant cell behavior and the expression of thymidylate synthase and survivin and promotes cellular sensitivity to 5-FU. *International Journal of Cancer, 126*(3), 631–639.

Liu, M., Lee, M. H., Cohen, M., Bommakanti, M., & Freedman, L. P. (1996). Transcriptional activation of the Cdk inhibitor p21 by vitamin D3 leads to the induced differentiation of the myelomonocytic cell line U937. *Genes & Development, 10*(2), 142–153. Epub 1996/01/15.

Ly, L. H., Zhao, X. Y., Holloway, L., & Feldman, D. (1999). Liarozole acts synergistically with 1alpha,25-dihydroxyvitamin D3 to inhibit growth of DU 145 human prostate cancer cells by blocking 24-hydroxylase activity. *Endocrinology, 140*(5), 2071–2076.

Ma, Y., Liu, H., Tu-Rapp, H., Thiesen, H. J., Ibrahim, S. M., Cole, S. M., et al. (2004). Fas ligation on macrophages enhances IL-1R1-Toll-like receptor 4 signaling and promotes chronic inflammation. *Nature Immunology, 5*(4), 380–387.

Ma, Y., Yu, W. D., Hershberger, P. A., Flynn, G., Kong, R. X., Trump, D. L., et al. (2008). 1Alpha,25-dihydroxyvitamin D3 potentiates cisplatin antitumor activity by p73 induction in a squamous cell carcinoma model. *Molecular Cancer Therapeutics, 7*(9), 3047–3055.

Ma, Y., Yu, W. D., Su, B., Seshadri, M., Luo, W., Trump, D. L., et al. (2013). Regulation of motility, invasion, and metastatic potential of squamous cell carcinoma by 1alpha,25-dihydroxycholecalciferol. *Cancer, 119*(3), 563–574. Epub 2012/07/27.

Ma, Y., Yu, W. D., Trump, D. L., & Johnson, C. S. (2010). 1,25D(3) enhances antitumor activity of gemcitabine and cisplatin in human bladder cancer models. *Cancer, 116*(13), 3294–3303.

Maalmi, H., Ordonez-Mena, J. M., Schottker, B., & Brenner, H. (2014). Serum 25-hydroxyvitamin D levels and survival in colorectal and breast cancer patients: Systematic review and meta-analysis of prospective cohort studies. *European Journal of Cancer, 50*(8), 1510–1521. Epub 2014/03/04.

Majewski, S., Skopinska, M., Marczak, M., Szmurlo, A., Bollag, W., & Jablonska, S. (1996). Vitamin D3 is a potent inhibitor of tumor cell-induced angiogenesis. *The Journal of Investigative Dermatology. Symposium Proceedings, 1*(1), 97–101.

Majewski, S., Szmurlo, A., Marczak, M., Jablonska, S., & Bollag, W. (1993). Inhibition of tumor cell-induced angiogenesis by retinoids, 1,25-dihydroxyvitamin D3 and their combination. *Cancer Letters, 75*(1), 35–39.

Makishima, M., Okabe-Kado, J., & Honma, Y. (1998). Growth inhibition and differentiation induction in human monoblastic leukaemia cells by 1alpha-hydroxyvitamin D derivatives and their enhancement by combination with hydroxyurea. *British Journal of Cancer, 77*(1), 33–39.

Mantell, D. J., Owens, P. E., Bundred, N. J., Mawer, E. B., & Canfield, A. E. (2000). 1 Alpha,25-dihydroxyvitamin D(3) inhibits angiogenesis in vitro and in vivo. *Circulation Research, 87*(3), 214–220.

Marcinkowska, E., Garay, E., Gocek, E., Chrobak, A., Wang, X., & Studzinski, G. P. (2006). Regulation of C/EBPbeta isoforms by MAPK pathways in HL60 cells induced to differentiate by 1,25-dihydroxyvitamin D3. *Experimental Cell Research, 312*(11), 2054–2065. Epub 2006/04/21.

Mathiasen, I. S., Sergeev, I. N., Bastholm, L., Elling, F., Norman, A. W., & Jaattela, M. (2002). Calcium and calpain as key mediators of apoptosis-like death induced by vitamin D compounds in breast cancer cells. *The Journal of Biological Chemistry, 277*(34), 30738–30745.

Mazzilli, S. A., Hershberger, P. A., Reid, M. E., Bogner, P. N., Atwood, K., Trump, D. L., et al. (2015). Vitamin D repletion reduces the progression of premalignant squamous lesions in the NTCU lung squamous cell carcinoma mouse model. *Cancer Prevention Research (Philadelphia, PA), 8*(10), 895–904. Epub 2015/08/16.

McGuire, T. F., Trump, D. L., & Johnson, C. S. (2001). Vitamin D(3)-induced apoptosis of murine squamous cell carcinoma cells. Selective induction of caspase-dependent MEK cleavage and up-regulation of MEKK-1. *The Journal of Biological Chemistry, 276*(28), 26365–26373.

Merke, J., Milde, P., Lewicka, S., Hugel, U., Klaus, G., Mangelsdorf, D. J., et al. (1989). Identification and regulation of 1,25-dihydroxyvitamin D3 receptor activity and biosynthesis of 1,25-dihydroxyvitamin D3. Studies in cultured bovine aortic endothelial cells and human dermal capillaries. *The Journal of Clinical Investigation, 83*(6), 1903–1915.

Mineva, N. D., Wang, X., Yang, S., Ying, H., Xiao, Z. X., Holick, M. F., et al. (2009). Inhibition of RelB by 1,25-dihydroxyvitamin D3 promotes sensitivity of breast cancer cells to radiation. *Journal of Cellular Physiology, 220*(3), 593–599.

Moan, J., Dahlback, A., Lagunova, Z., Cicarma, E., & Porojnicu, A. C. (2009). Solar radiation, vitamin D and cancer incidence and mortality in Norway. *Anticancer Research, 29*(9), 3501–3509. Epub 2009/08/12.

Moffatt, K. A., Johannes, W. U., Hedlund, T. E., & Miller, G. J. (2001). Growth inhibitory effects of 1alpha, 25-dihydroxyvitamin D(3) are mediated by increased levels of p21 in the prostatic carcinoma cell line ALVA-31. *Cancer Research, 61*(19), 7122–7129. Epub 2001/10/05.

Moffatt, K. A., Johannes, W. U., & Miller, G. J. (1999). 1Alpha,25dihydroxyvitamin D3 and platinum drugs act synergistically to inhibit the growth of prostate cancer cell lines. *Clinical Cancer Research: An Official Journal of the American Association for Cancer Research, 5*(3), 695–703.

Moreno, J., Krishnan, A. V., Swami, S., Nonn, L., Peehl, D. M., & Feldman, D. (2005). Regulation of prostaglandin metabolism by calcitriol attenuates growth stimulation in prostate cancer cells. *Cancer Research, 65*(17), 7917–7925. Epub 2005/09/06.

Muindi, J. R., Johnson, C. S., Trump, D. L., Christy, R., Engler, K. L., & Fakih, M. G. (2009). A phase I and pharmacokinetics study of intravenous calcitriol in combination with oral dexamethasone and gefitinib in patients with advanced solid tumors. *Cancer Chemotherapy and Pharmacology, 65*(1), 33–40.

Muto, A., Kizaki, M., Yamato, K., Kawai, Y., Kamata-Matsushita, M., Ueno, H., et al. (1999). 1,25-Dihydroxyvitamin D3 induces differentiation of a retinoic acid-resistant acute promyelocytic leukemia cell line (UF-1) associated with expression of p21(WAF1/CIP1) and p27(KIP1). *Blood, 93*(7), 2225–2233.

Nakagawa, K., Kawaura, A., Kato, S., Takeda, E., & Okano, T. (2005). 1 Alpha,25-dihydroxyvitamin D(3) is a preventive factor in the metastasis of lung cancer. *Carcinogenesis, 26*(2), 429–440.

Nakagawa, K., Sasaki, Y., Kato, S., Kubodera, N., & Okano, T. (2005). 22-Oxa-1alpha,25-dihydroxyvitamin D3 inhibits metastasis and angiogenesis in lung cancer. *Carcinogenesis, 26*(6), 1044–1054.

Narvaez, C. J., & Welsh, J. (1997). Differential effects of 1,25-dihydroxyvitamin D3 and tetradecanoylphorbol acetate on cell cycle and apoptosis of MCF-7 cells and a vitamin D3-resistant variant. *Endocrinology, 138*(11), 4690–4698. Epub 1997/11/05.

Narvaez, C. J., & Welsh, J. (2001). Role of mitochondria and caspases in vitamin D-mediated apoptosis of MCF-7 breast cancer cells. *The Journal of Biological Chemistry*, *276*(12), 9101–9107.

Ng, K., Meyerhardt, J. A., Wu, K., Feskanich, D., Hollis, B. W., Giovannucci, E. L., et al. (2008). Circulating 25-hydroxyvitamin D levels and survival in patients with colorectal cancer. *Journal of Clinical Oncology: Official Journal of the American Society of Clinical Oncology*, *26*(18), 2984–2991.

Nicholas, J. (2011). Vitamin D, and cancer: Uncertainty persists; research continues. *Journal of the National Cancer Institute*, *103*(11), 851–852. Epub 2011/05/25.

Niitsu, N., Umeda, M., & Honma, Y. (2000). Myeloid and monocytoid leukemia cells have different sensitivity to differentiation-inducing activity of deoxyadenosine analogs. *Leukemia Research*, *24*(1), 1–9.

Nonn, L., Peng, L., Feldman, D., & Peehl, D. M. (2006). Inhibition of p38 by vitamin D reduces interleukin-6 production in normal prostate cells via mitogen-activated protein kinase phosphatase 5: Implications for prostate cancer prevention by vitamin D. *Cancer Research*, *66*(8), 4516–4524.

Oikawa, T., Hirotani, K., Ogasawara, H., Katayama, T., Nakamura, O., Iwaguchi, T., et al. (1990). Inhibition of angiogenesis by vitamin D3 analogues. *European Journal of Pharmacology*, *178*(2), 247–250.

Okazaki, T., Bielawska, A., Bell, R. M., & Hannun, Y. A. (1990). Role of ceramide as a lipid mediator of 1 alpha,25-dihydroxyvitamin D3-induced HL-60 cell differentiation. *The Journal of Biological Chemistry*, *265*(26), 15823–15831. Epub 1990/09/15.

Okuda, N., Takeda, S., Shinomiya, K., Muneta, T., Itoh, S., Noda, M., et al. (2007). ED-71, a novel vitamin D analog, promotes bone formation and angiogenesis and inhibits bone resorption after bone marrow ablation. *Bone*, *40*(2), 281–292.

Ooi, L. L., Zheng, Y., Zhou, H., Trivedi, T., Conigrave, A. D., Seibel, M. J., et al. (2010). Vitamin D deficiency promotes growth of MCF-7 human breast cancer in a rodent model of osteosclerotic bone metastasis. *Bone*, *47*(4), 795–803. Epub 2010/07/20.

Ooi, L. L., Zhou, H., Kalak, R., Zheng, Y., Conigrave, A. D., Seibel, M. J., et al. (2010). Vitamin D deficiency promotes human breast cancer growth in a murine model of bone metastasis. *Cancer Research*, *70*(5), 1835–1844. Epub 2010/02/18.

Palmer, H. G., Gonzalez-Sancho, J. M., Espada, J., Berciano, M. T., Puig, I., Baulida, J., et al. (2001). Vitamin D(3) promotes the differentiation of colon carcinoma cells by the induction of E-cadherin and the inhibition of beta-catenin signaling. *The Journal of Cell Biology*, *154*(2), 369–387.

Parise, R. A., Egorin, M. J., Kanterewicz, B., Taimi, M., Petkovich, M., Lew, A. M., et al. (2006). CYP24, the enzyme that catabolizes the antiproliferative agent vitamin D, is increased in lung cancer. *International Journal of Cancer*, *119*(8), 1819–1828. Epub 2006/05/19.

Peehl, D. M., Seto, E., Hsu, J. Y., & Feldman, D. (2002). Preclinical activity of ketoconazole in combination with calcitriol or the vitamin D analogue EB 1089 in prostate cancer cells. *The Journal of Urology*, *168*(4 Pt. 1), 1583–1588.

Pendas-Franco, N., Garcia, J. M., Pena, C., Valle, N., Palmer, H. G., Heinaniemi, M., et al. (2008). DICKKOPF-4 is induced by TCF/beta-catenin and upregulated in human colon cancer, promotes tumour cell invasion and angiogenesis and is repressed by 1alpha,25-dihydroxyvitamin D3. *Oncogene*, *27*(32), 4467–4477.

Pendas-Franco, N., Gonzalez-Sancho, J. M., Suarez, Y., Aguilera, O., Steinmeyer, A., Gamallo, C., et al. (2007). Vitamin D regulates the phenotype of human breast cancer cells. *Differentiation*, *75*(3), 193–207.

Peng, L., Malloy, P. J., & Feldman, D. (2004). Identification of a functional vitamin D response element in the human insulin-like growth factor binding protein-3 promoter. *Molecular Endocrinology*, *18*(5), 1109–1119. Epub 2004/02/14.

Rashid, S. F., Moore, J. S., Walker, E., Driver, P. M., Engel, J., Edwards, C. E., et al. (2001). Synergistic growth inhibition of prostate cancer cells by 1 alpha,25 dihydroxyvitamin D(3) and its 19-nor-hexafluoride analogs in combination with either sodium butyrate or trichostatin A. *Oncogene, 20*(15), 1860–1872.

Rassnick, K. M., Muindi, J. R., Johnson, C. S., Balkman, C. E., Ramnath, N., Yu, W. D., et al. (2008). In vitro and in vivo evaluation of combined calcitriol and cisplatin in dogs with spontaneously occurring tumors. *Cancer Chemotherapy and Pharmacology, 62*(5), 881–891.

Ravid, A., Rocker, D., Machlenkin, A., Rotem, C., Hochman, A., Kessler-Icekson, G., et al. (1999). 1,25-Dihydroxyvitamin D3 enhances the susceptibility of breast cancer cells to doxorubicin-induced oxidative damage. *Cancer Research, 59*(4), 862–867.

Rebsamen, M. C., Sun, J., Norman, A. W., & Liao, J. K. (2002). 1Alpha,25-dihydroxyvitamin D3 induces vascular smooth muscle cell migration via activation of phosphatidylinositol 3-kinase. *Circulation Research, 91*(1), 17–24.

Riachy, R., Vandewalle, B., Kerr Conte, J., Moerman, E., Sacchetti, P., Lukowiak, B., et al. (2002). 1,25-dihydroxyvitamin D3 protects RINm5F and human islet cells against cytokine-induced apoptosis: Implication of the antiapoptotic protein A20. *Endocrinology, 143*(12), 4809–4819.

Riachy, R., Vandewalle, B., Moerman, E., Belaich, S., Lukowiak, B., Gmyr, V., et al. (2006). 1,25-Dihydroxyvitamin D(3) protects human pancreatic islets against cytokine-induced apoptosis via down-regulation of the fas receptor. *Apoptosis, 11*(2), 151–159.

Saramaki, A., Banwell, C. M., Campbell, M. J., & Carlberg, C. (2006). Regulation of the human p21(waf1/cip1) gene promoter via multiple binding sites for p53 and the vitamin D3 receptor. *Nucleic Acids Research, 34*(2), 543–554. Epub 2006/01/26.

Schottker, B., Jorde, R., Peasey, A., Thorand, B., Jansen, E. H., Groot, L., et al. (2014). Vitamin D and mortality: Meta-analysis of individual participant data from a large consortium of cohort studies from Europe and the United States. *BMJ, 348*, g3656. Epub 2014/06/19.

Schwab, M., Reynders, V., Loitsch, S., Steinhilber, D., Stein, J., & Schroder, O. (2007). Involvement of different nuclear hormone receptors in butyrate-mediated inhibition of inducible NF kappa B signalling. *Molecular Immunology, 44*(15), 3625–3632. Epub 2007/05/25.

Segovia-Mendoza, M., Diaz, L., Gonzalez-Gonzalez, M. E., Martinez-Reza, I., Garcia-Quiroz, J., Prado-Garcia, H., et al. (2014). Calcitriol and its analogues enhance the antiproliferative activity of gefitinib in breast cancer cells. *The Journal of Steroid Biochemistry and Molecular Biology, 148*, 122–131. Epub 2014/12/17.

Shokravi, M. T., Marcus, D. M., Alroy, J., Egan, K., Saornil, M. A., & Albert, D. M. (1995). Vitamin D inhibits angiogenesis in transgenic murine retinoblastoma. *Investigative Ophthalmology & Visual Science, 36*(1), 83–87.

Slapak, C. A., Desforges, J. F., Fogaren, T., & Miller, K. B. (1992). Treatment of acute myeloid leukemia in the elderly with low-dose cytarabine, hydroxyurea, and calcitriol. *American Journal of Hematology, 41*(3), 178–183.

Song, X. D., & Norman, A. W. (1999). Bryostatin-1 and 1alpha,25-dihydroxyvitamin D3 synergistically stimulate the differentiation of NB4 acute promyelocytic leukemia cells. *Leukemia, 13*(2), 275–281.

Sporn, M. B. (1996). The war on cancer. *Lancet, 347*(9012), 1377–1381. Epub 1996/05/18.

Studzinski, G. P., Reddy, K. B., Hill, H. Z., & Bhandal, A. K. (1991). Potentiation of 1-beta-D-arabinofuranosylcytosine cytotoxicity to HL-60 cells by 1,25-dihydroxyvitamin D3 correlates with reduced rate of maturation of DNA replication intermediates. *Cancer Research, 51*(13), 3451–3455.

Sun, J., Kong, J., Duan, Y., Szeto, F. L., Liao, A., Madara, J. L., et al. (2006). Increased NF-kappaB activity in fibroblasts lacking the vitamin D receptor. *American Journal of Physiology. Endocrinology and Metabolism*, *291*(2), E315–E322. Epub 2006/03/02.

Sung, V., & Feldman, D. (2000). 1,25-Dihydroxyvitamin D3 decreases human prostate cancer cell adhesion and migration. *Molecular and Cellular Endocrinology*, *164*(1–2), 133–143.

Suzuki, T., Sano, Y., & Kinoshita, S. (2000). Effects of 1alpha,25-dihydroxyvitamin D3 on Langerhans cell migration and corneal neovascularization in mice. *Investigative Ophthalmology & Visual Science*, *41*(1), 154–158.

Tang, J. Y., Fu, T., Leblanc, E., Manson, J. E., Feldman, D., Linos, E., et al. (2011). Calcium plus vitamin D supplementation and the risk of nonmelanoma and melanoma skin cancer: Post hoc analyses of the Women's Health Initiative Randomized Controlled Trial. *Journal of Clinical Oncology: Official Journal of the American Society of Clinical Oncology*, *29*(22), 3078–3084. Epub 2011/06/29.

Ting, H. J., Hsu, J., Bao, B. Y., & Lee, Y. F. (2007). Docetaxel-induced growth inhibition and apoptosis in androgen independent prostate cancer cells are enhanced by 1alpha,25-dihydroxyvitamin D3. *Cancer Letters*, *247*(1), 122–129.

Ting, H. J., Messing, J., Yasmin-Karim, S., & Lee, Y. F. (2013). Identification of microRNA-98 as a therapeutic target inhibiting prostate cancer growth and a biomarker induced by vitamin D. *The Journal of Biological Chemistry*, *288*(1), 1–9. Epub 2012/11/29.

Trummer, O., Langsenlehner, U., Krenn-Pilko, S., Pieber, T. R., Obermayer-Pietsch, B., Gerger, A., et al. (2015). Vitamin D and prostate cancer prognosis: A Mendelian randomization study. *World Journal of Urology* . Epub 2015/07/26. PMID:26209090.

Tse, A. K., Zhu, G. Y., Wan, C. K., Shen, X. L., Yu, Z. L., & Fong, W. F. (2010). 1Alpha,25-dihydroxyvitamin D3 inhibits transcriptional potential of nuclear factor kappa B in breast cancer cells. *Molecular Immunology*, *47*(9), 1728–1738. Epub 2010/04/08.

Upadhyay, S. K., Verone, A., Shoemaker, S., Qin, M., Liu, S., Campbell, M., et al. (2013). 1,25-Dihydroxyvitamin D3 (1,25(OH)2D3) signaling capacity and the epithelial-mesenchymal transition in non-small cell lung cancer (NSCLC): Implications for use of 1,25(OH)2D3 in NSCLC treatment. *Cancers*, *5*(4), 1504–1521. Epub 2013/11/13.

Van Auken, M., Buckley, D., Ray, R., Holick, M. F., & Baran, D. T. (1996). Effects of the vitamin D3 analog 1 alpha, 25-dihydroxyvitamin D3-3 beta-bromoacetate on rat osteosarcoma cells: Comparison with 1 alpha, 25-dihydroxyvitamin D3. *Journal of Cellular Biochemistry*, *63*(3), 302–310.

van der Rhee, H., Coebergh, J. W., & de Vries, E. (2013). Is prevention of cancer by sun exposure more than just the effect of vitamin D? A systematic review of epidemiological studies. *European Journal of Cancer*, *49*(6), 1422–1436. Epub 2012/12/15.

Veldman, C. M., Cantorna, M. T., & DeLuca, H. F. (2000). Expression of 1,25-dihydroxyvitamin D(3) receptor in the immune system. *Archives of Biochemistry and Biophysics*, *374*(2), 334–338. Epub 2000/02/10.

Vink-van Wijngaarden, T., Pols, H. A., Buurman, C. J., van den Bemd, G. J., Dorssers, L. C., Birkenhager, J. C., et al. (1994). Inhibition of breast cancer cell growth by combined treatment with vitamin D3 analogues and tamoxifen. *Cancer Research*, *54*(21), 5711–5717.

Wactawski-Wende, J., Kotchen, J. M., Anderson, G. L., Assaf, A. R., Brunner, R. L., O'Sullivan, M. J., et al. (2006). Calcium plus vitamin D supplementation and the risk of colorectal cancer. *The New England Journal of Medicine*, *354*(7), 684–696. Epub 2006/02/17.

Wang, S. H., Koenig, R. J., Giordano, T. J., Myc, A., Thompson, N. W., & Baker, J. R., Jr. (1999). 1Alpha,25-dihydroxyvitamin D3 up-regulates Bcl-2 expression and protects normal human thyrocytes from programmed cell death. *Endocrinology, 140*(4), 1649–1656.

Wang, Q., Lee, D., Sysounthone, V., Chandraratna, R. A. S., Christakos, S., Korah, R., et al. (2001). 1,25-Dihydroxyvitamin D3 and retonic acid analogues induce differentiation in breast cancer cells with function- and cell-specific additive effects. *Breast Cancer Research and Treatment, 67*(2), 157–168.

Wang, X., Rao, J., & Studzinski, G. P. (2000). Inhibition of p38 MAP kinase activity up-regulates multiple MAP kinase pathways and potentiates 1,25-dihydroxyvitamin D(3)-induced differentiation of human leukemia HL60 cells. *Experimental Cell Research, 258*(2), 425–437. Epub 2000/07/18.

Wang, X., & Studzinski, G. P. (2001). Activation of extracellular signal-regulated kinases (ERKs) defines the first phase of 1,25-dihydroxyvitamin D3-induced differentiation of HL60 cells. *Journal of Cellular Biochemistry, 80*(4), 471–482. Epub 2001/02/13.

Wang, X., & Studzinski, G. P. (2004). Kinase suppressor of RAS (KSR) amplifies the differentiation signal provided by low concentrations 1,25-dihydroxyvitamin D3. *Journal of Cellular Physiology, 198*(3), 333–342. Epub 2004/02/03.

Wang, Q., Wang, X., & Studzinski, G. P. (2003). Jun N-terminal kinase pathway enhances signaling of monocytic differentiation of human leukemia cells induced by 1,25-dihydroxyvitamin D3. *Journal of Cellular Biochemistry, 89*(6), 1087–1101. Epub 2003/08/05.

Wang, X., Wang, T. T., White, J. H., & Studzinski, G. P. (2006). Induction of kinase suppressor of RAS-1 (KSR-1) gene by 1, alpha25-dihydroxyvitamin D3 in human leukemia HL60 cells through a vitamin D response element in the 5′-flanking region. *Oncogene, 25*(53), 7078–7085. Epub 2006/05/30.

Wang, Q., Yang, W., Uytingco, M. S., Christakos, S., & Wieder, R. (2000). 1,25-Dihydroxyvitamin D3 and all-trans-retinoic acid sensitize breast cancer cells to chemotherapy-induced cell death. *Cancer Research, 60*(7), 2040–2048.

Webb, P. M., de Fazio, A., Protani, M. M., Ibiebele, T. I., Nagle, C. M., Brand, A. H., et al. (2015). Circulating 25-hydroxyvitamin D and survival in women with ovarian cancer. *The American Journal of Clinical Nutrition, 102*(1), 109–114. Epub 2015/05/15.

Welsh, J. (2012). Cellular and molecular effects of vitamin D on carcinogenesis. *Archives of Biochemistry and Biophysics, 523*(1), 107–114. Epub 2011/11/17.

Wu, G., Fan, R. S., Li, W., Srinivas, V., & Brattain, M. G. (1998). Regulation of transforming growth factor-beta type II receptor expression in human breast cancer MCF-7 cells by vitamin D3 and its analogues. *The Journal of Biological Chemistry, 273*(13), 7749–7756. Epub 1998/04/29.

Yee, S. W., Campbell, M. J., & Simons, C. (2006). Inhibition of vitamin D3 metabolism enhances VDR signalling in androgen-independent prostate cancer cells. *The Journal of Steroid Biochemistry and Molecular Biology, 98*(4–5), 228–235. Epub 2006/02/18.

Yin, L., Ordonez-Mena, J. M., Chen, T., Schottker, B., Arndt, V., & Brenner, H. (2013). Circulating 25-hydroxyvitamin D serum concentration and total cancer incidence and mortality: A systematic review and meta-analysis. *Preventive Medicine, 57*(6), 753–764. Epub 2013/09/17.

Young, M. R., Halpin, J., Hussain, R., Lozano, Y., Djordjevic, A., Devata, S., et al. (1993). Inhibition of tumor production of granulocyte-macrophage colony-stimulating factor by 1 alpha, 25-dihydroxyvitamin D3 reduces tumor motility and metastasis. *Invasion & Metastasis, 13*(4), 169–177.

Young, M. R., & Lozano, Y. (1997). Inhibition of tumor invasiveness by 1alpha,25-dihydroxyvitamin D3 coupled to a decline in protein kinase A activity and an increase in cytoskeletal organization. *Clinical & Experimental Metastasis, 15*(2), 102–110.

Yu, X. P., Bellido, T., & Manolagas, S. C. (1995). Down-regulation of NF-kappa B protein levels in activated human lymphocytes by 1,25-dihydroxyvitamin D3. *Proceedings of the*

National Academy of Sciences of the United States of America, 92(24), 10990–10994. Epub 1995/11/21.

Yu, W. D., McElwain, M. C., Modzelewski, R. A., Russell, D. M., Smith, D. C., Trump, D. L., et al. (1998). Enhancement of 1,25-dihydroxyvitamin D3-mediated antitumor activity with dexamethasone. *Journal of the National Cancer Institute, 90*(2), 134–141.

Yudoh, K., Matsuno, H., & Kimura, T. (1999). 1Alpha,25-dihydroxyvitamin D3 inhibits in vitro invasiveness through the extracellular matrix and in vivo pulmonary metastasis of B16 mouse melanoma. *The Journal of Laboratory and Clinical Medicine, 133*(2), 120–128.

Zehnder, D., Bland, R., Chana, R. S., Wheeler, D. C., Howie, A. J., Williams, M. C., et al. (2002). Synthesis of 1,25-dihydroxyvitamin D(3) by human endothelial cells is regulated by inflammatory cytokines: A novel autocrine determinant of vascular cell adhesion. *Journal of the American Society of Nephrology, 13*(3), 621–629.

Zhang, Q., Kanterewicz, B., Buch, S., Petkovich, M., Parise, R., Beumer, J., et al. (2012). CYP24 inhibition preserves 1alpha,25-dihydroxyvitamin D(3) anti-proliferative signaling in lung cancer cells. *Molecular and Cellular Endocrinology, 355*(1), 153–161. Epub 2012/03/06.

Zheng, Y., Zhou, H., Ooi, L. L., Snir, A. D., Dunstan, C. R., & Seibel, M. J. (2011). Vitamin D deficiency promotes prostate cancer growth in bone. *Prostate, 71*(9), 1012–1021. Epub 2011/05/05.

CHAPTER SEVENTEEN

Vitamin D Signaling Modulators in Cancer Therapy

Wei Luo*, Candace S. Johnson*, Donald L. Trump[†,‡,1]
*Department of Pharmacology and Therapeutics, Roswell Park Cancer Institute, Buffalo, New York, USA
[†]Department of Medicine, Roswell Park Cancer Institute, Buffalo, New York, USA
[‡]Inova Dwight and Martha Schar Cancer Institute, Falls Church, Virginia, USA
[1]Corresponding author: e-mail address: Donald.Trump@inova.org

Contents

Abstract

The antiproliferative and pro-apoptotic effects of 1α,25-dihydroxycholecalciferol (1,25 $(OH)_2D_3$, $1,25D_3$, calcitriol) have been demonstrated in various tumor model systems *in vitro* and *in vivo*. However, limited antitumor effects of $1,25D_3$ have been observed in clinical trials. This may be attributed to a variety of factors including overexpression of the primary $1,25D_3$ degrading enzyme, CYP24A1, in tumors, which would lead to rapid local inactivation of $1,25D_3$. An alternative strategy for improving the antitumor activity of $1,25D_3$ involves the combination with a selective CYP24A1 inhibitor. The validity of this approach is supported by numerous preclinical investigations, which demonstrate that CYP24A1 inhibitors suppress $1,25D_3$ catabolism in tumor cells and increase the effects of $1,25D_3$ on gene expression and cell growth. Studies are now required to determine whether selective CYP24A1 inhibitors + $1,25D_3$ can be used safely and effectively in patients. CYP24A1 inhibitors plus $1,25D_3$ can cause dose-limiting toxicity of vitamin D (hypercalcemia) in some patients. Dexamethasone significantly reduces $1,25D_3$-mediated hypercalcemia and enhances the antitumor activity of $1,25D_3$, increases VDR–ligand binding, and increases VDR protein expression. Efforts to dissect the mechanisms responsible for CYP24A1 overexpression and combinational effect of $1,25D_3$/dexamethasone in tumors are underway. Understanding the cross talk between vitamin D receptor (VDR) and glucocorticoid receptor (GR) signaling axes is of crucial

Vitamins and Hormones, Volume 100
ISSN 0083-6729
http://dx.doi.org/10.1016/bs.vh.2015.11.004

importance to the design of new therapies that include $1{,}25D_3$ and dexamethasone. Insights gained from these studies are expected to yield novel strategies to improve the efficacy of $1{,}25D_3$ treatment.

1. INTRODUCTION

The active form of vitamin D, 1α,25-dihydroxycholecalciferol ($1{,}25(OH)_2D_3$, $1{,}25D_3$, calcitriol), regulates calcium and phosphate homeostasis and bone mineralization (Reichel, Koeffler, & Norman, 1989). The serum $1{,}25D_3$ level is tightly controlled through feedback regulation of its synthesis and catabolism by 1α-hydroxylase encoded by *CYP27B1* and 24-hydroxylase encoded by *CYP24A1*, respectively. Both CYP27B1 and CYP24A1 are highly regulated enzymes and respond to modulating factors such as parathyroid hormone, calcitonin, calcium, phosphorus as well as $1{,}25D_3$. Vitamin D_3 (cholecalciferol), the major precursor of $1{,}25D_3$, is available in the diet, but is primarily supplied through synthesis from 7-dehydrocholesterol in the skin after exposure to ultraviolet B light. Vitamin D_3 is carried as a complex with the plasma vitamin D binding protein (DBP) to the liver and hydroxylated in the liver by 25-hydroxylase encoded by *CYP2R1* to $25(OH)D_3$ ($25D_3$, calcidiol), which is the major circulating form of vitamin D_3 (Omdahl, Morris, & May, 2002; White & Cooke, 2000). $25D_3$ is 1α-hydroxylated to $1{,}25D_3$ by *CYP27B1* mainly in kidney. $1{,}25D_3$ is also synthesized in many extrarenal sites in addition to the kidney (Bareis, Bises, Bischof, Cross, & Peterlik, 2001; Chen, Wang, Whitlatch, Flanagan, & Holick, 2003; Cross, 2007; Schwartz, Whitlatch, Chen, Lokeshwar, & Holick, 1998; Townsend et al., 2005). CYP24A1 is the major enzyme which inactivates $1{,}25D_3$ (Omdahl, Bobrovnikova, Choe, Dwivedi, & May, 2001; Omdahl & Brian, 2005; Omdahl et al., 2002; Prosser & Jones, 2004).

In addition to its classic role in bone and mineral metabolism, $1{,}25D_3$ has antiproliferative activity in a variety of murine, rat, and human tumor model systems (Ahmed, Johnson, Rueger, & Trump, 2002; Getzenberg et al., 1997; Hershberger et al., 1999, 2001; McElwain, Modzelewski, Yu, Russell, & Johnson, 1997; Trump et al., 2004) and in multiple tumor types including squamous cell carcinoma (SCC), prostate, breast, colon, lung, and other cancers (Colston, Chander, Mackay, & Coombes, 1992; Deeb, Trump, & Johnson, 2007; Eisman, Barkla, & Tutton, 1987; Getzenberg et al., 1997; Light et al., 1997; Ma, Yu, Trump, & Johnson, 2010; Mangelsdorf,

Koeffler, Donaldson, Pike, & Haussler, 1984; Peehl et al., 1994; Shabahang et al., 1994; Zhou et al., 1990). 1,25D_3 induces significant cell cycle arrest, induces and modulates apoptotic markers and decreases survival signals both *in vitro* and *in vivo* in a number of model systems (Ahmed et al., 2002; Getzenberg et al., 1997; Hershberger et al., 2002, 1999, 2001; McElwain et al., 1997; McGuire, Trump, & Johnson, 2001; Trump et al., 2004). Antitumor effects of 1,25D_3 have been observed in patients with leukemia and lymphoma (Cunningham et al., 1985). In patients with prostate cancer, 1,25D_3 can reduce the rate of rising or absolute serum content of prostate-specific antigen (PSA), a biochemical marker of progression of prostate cancer (Krishnan, Peehl, & Feldman, 2003; Peehl & Feldman, 2004).

However, clinical vitamin D-mediated antitumor activity could be limited by abnormal 1,25D_3 signaling in cancer cells and the occurrence of dose-limiting toxicity of 1,25D_3. Frequent alterations of vitamin D signaling and metabolism enzymes, including CYP24A1, are observed in cancer (Albertson et al., 2000; Anderson, Nakane, Ruan, Kroeger, & Wu-Wong, 2006; Mimori et al., 2004; Zhang et al., 2010). CYP24A1 is constitutively expressed in kidney, gastrointestinal mucosa, and in most other tissues and is transcriptionally induced by 1,25D_3 (Omdahl et al., 2001, 2002; Prosser & Jones, 2004). Increased expression of CYP24A1 has been reported in several human cancer tissues (Albertson et al., 2000; Anderson et al., 2006; Mimori et al., 2004). In comparison to normal cells, clearing of 1,25D_3 is enhanced in malignant cells due to differences in transcriptional regulation of *CYP24A1* (Matilainen, Malinen, Turunen, Carlberg, & Vaisanen, 2010). By stimulating 1,25D_3 degradation, overexpression of CYP24A1 limits 1,25D_3 biologic activity and may reduce the efficacy of 1,25D_3-mediated antitumor actions. CYP24A1 inhibition can increase the bioavailability of endogenous and administered 1,25D_3, thereby enhancing its anticancer effects and increasing risk of hypercalcemia (Wang, Swami, Krishnan, & Feldman, 2012). High level of vitamin D-produced hypercalcemia is observed in some patients (Cunningham et al., 1985; Koeffler, Hirji, & Itri, 1985; Rolla et al., 1993). To overcome this dose-limiting toxicity of 1,25D_3, glucocorticoids have been utilized to ameliorate hypercalcemic effects and to enhance the antitumor activity of 1,25D_3 (Lee, Choi, & Jeung, 2006; Trump, Potter, Muindi, Brufsky, & Johnson, 2006).

This review summarizes available information about the abnormal expression of CYP24A1 and potential dose-limiting toxicity of 1,25D_3, and available strategies for overcoming these obstacles in vitamin D anticancer therapy.

2. VITAMIN D RECEPTOR AND 1,25D$_3$ SIGNALING

The majority of the biologic effects of steroid hormones are mediated by the hormone's specific intracellular receptors. The effects of 1,25D$_3$ are primarily mediated by the vitamin D receptor (VDR) which is a member of the superfamily of nuclear receptors (Mangelsdorf et al., 1995). VDR is expressed in many cell types and tissues with the highest expression in tissues which are involved in the maintenance of calcium homeostasis (e.g., kidneys, bone, parathyroid, and intestine) (Wang, Zhu, & DeLuca, 2012). The activation process of VDR includes a translocation of VDR from the cytoplasm to the nucleus (Racz & Barsony, 1999). VDR acts as a ligand-dependent transcription factor. The binding of 1,25D$_3$ to VDR causes conformational structure changes in VDR that promote the formation of heterodimers with its coreceptor, another member of the nuclear receptor superfamily, the retinoid X receptor (RXR) and its ligand 9-*cis*-retinoic acid (9cRA) (Cheskis & Freedman, 1996; Jaaskelainen, Ryhanen, & Maenpaa, 2003; Sanchez-Martinez, Castillo, Steinmeyer, & Aranda, 2006). A 1,25D$_3$-liganded VDR–RXR–9cRA heterocomplex then binds to vitamin D responsive elements (VDREs) to enhance or repress the transcriptional activity of 1,25D$_3$ (Carlberg, 1995; Cheskis & Freedman, 1996; Jaaskelainen et al., 2003; Lemon, Fondell, & Freedman, 1997; Norman, 2006; Sanchez-Martinez et al., 2006). VDREs are generally found in the promoter region of 1,25D$_3$-regulated genes. In some cases, multiple copies of the VDREs are positioned not only in the proximal promoter but dispersed up to several thousand kilobases of 5′ of the transcription start site in vitamin D responsive genes (Vaisanen, Dunlop, Sinkkonen, Frank, & Carlberg, 2005). Transcription activation through 1,25D$_3$ and VDR is enhanced by nuclear receptor coactivator proteins such as steroid receptor coactivators, vitamin D receptor interacting protein complexes (Barletta, Freedman, & Christakos, 2002; Herdick & Carlberg, 2000; MacDonald, Baudino, Tokumaru, Dowd, & Zhang, 2001; Zhang et al., 2003). In addition, 1,25D$_3$ may act independent of VDR/DNA binding through nongenomic pathways (Norman et al., 1992). 1,25D$_3$ can activate a variety of rapid signal transduction pathways that may include kinases, phosphatases, or ion channels. The binding of 1,25D$_3$ to the VDR also induces VDR phosphorylation and stimulates transcription (Weigel, 1996). 1,25D$_3$ can activate protein kinase C (PKC), Raf, and mitogen-activated protein kinase phosphatase (MAPK; ERK1/2) (Gniadecki, 1996; Slater et al., 1995), modulate phospholipid metabolism (de Boland, Morelli, & Boland, 1994), stimulate the formation of cyclic nucleotides (Khare et al., 1994), and trigger calcium transport (de Boland & Norman, 1990). Most likely, nongenomic response is also mediated by VDR (Chen et al., 2013; Doroudi, Chen, Boyan, & Schwartz, 2014; Schaefer, Bonor, Joglekar, van Golen, & Nohe, 2013). Caveolae are the source of VDR-mediated signal transduction pathways. When $1,25D_3$–VDR is associated with plasma membrane caveolae, these signaling cascades can regulate gene expression through activation of VDR or their cognate promoter element (Chen et al., 2013; Doroudi et al., 2014; Schaefer et al., 2013). Membrane-associated protein disulfide isomerase, family A, member 3 (Pdia3) which is present in caveolae is identified as another receptor to mediate the $1,25D_3$ rapid response (Chen et al., 2013; Doroudi et al., 2014; Nemere et al., 2004). Interacting with Pdia3, $1,25D_3$ can initiate rapid signaling via phospholipase A2 $Ca(2+)$/calmodulin-dependent kinase II, leading to activation of PKC and PKC-dependent responses (Doroudi, Schwartz, & Boyan, 2015).

The presence of VDR in normal and cancer cells is essential for $1,25D_3$ and vitamin D activity. VDR may be overexpressed or suppressed in human cancers (Bikle, 2014; Deeb et al., 2007; Feldman, Krishnan, Swami, Giovannucci, & Feldman, 2014). VDR levels have been reported to decline with the aggression from nevus to melanoma in pigmented skin lesions (Brozyna, Jozwicki, Janjetovic, & Slominski, 2011) and development of invasive tumor from *in situ* cancer in breast and ovarian cancer (Lopes et al., 2010; Thill et al., 2010). High VDR levels are associated with improved prognosis and overall survival in breast, prostate cancer, and non-small-cell lung carcinoma (Berger et al., 1991; Ditsch et al., 2012; Hendrickson et al., 2011; Srinivasan, Parwani, Hershberger, Lenzner, & Weissfeld, 2011). A study examining the expression of VDR protein by immunohistochemistry in 841 patients with prostate cancer demonstrated that high level of VDR in prostate tumors was associated with a reduced risk of lethal cancer (Hendrickson et al., 2011). This study also suggests that the inhibition of prostate cancer progression is dependent upon VDR content and not just the serum level of $25D_3$ (Hendrickson et al., 2011). In colon cancer, VDR protein level declines as a function of colon tumor dedifferentiation (Matusiak, Murillo, Carroll, Mehta, & Benya, 2005). However, a study reports that VDR is overexpressed in colorectal cancer tumors with K-Ras mutations (Kure et al., 2009).

VDR expression is tightly regulated at the transcriptional level by several transcription factors including p63, p53, Sp1, SNAIL, and VDR itself

(Jehan & DeLuca, 2000; Larriba & Munoz, 2005; Maruyama et al., 2006; Pike et al., 2010). Modulation of the concentration of VDR in target cells alters the magnitude of the response to 1,25D_3 (Zhao, Eccleshall, Krishnan, Gross, & Feldman, 1997). Many factors have been proposed to influence VDR expression in cancer. SNAIL1 is absent from normal tissue and expressed in colon tumors, whereas SNAIL2 is expressed in normal tissue but its levels are increased in colon tumors (Larriba et al., 2009; Pena et al., 2005). Overexpression of SNAIL transcription factors can reduce *VDR* gene expression and suppress 1,25D_3-mediated cell growth arrest in colon cancer cell lines (Larriba et al., 2009, 2007; Palmer et al., 2004). SNAIL1 and 2 can bind to E-boxes in the proximal promoter of the *VDR* gene and downregulate *VDR* gene expression by increasing the recruitment of corepressors in colon and breast cancer cells (Larriba et al., 2009; Mittal, Myers, Misra, Bailey, & Chaudhuri, 2008). p53 is also involved in regulating *VDR* gene expression. Overexpression of wild-type p53 in p53-null cell lines Saos-2 and H1299 increases VDR protein levels by enhancing p53 association with p53-binding sites within the *VDR* gene (Maruyama et al., 2006).

VDR gene expression can be affected by tumor suppressor gene and oncogene mutations which contribute to tumorigenesis. Reduction of *VDR* mRNA and protein expression is observed in the small intestine and colon tissue of Apcmin$^{min/+}$ mice which have truncated mutation in the Apc gene (Xu, Posner, Stevenson, & Campbell, 2010). Ras activation that causes cancer can lead to the reduction of VDR expression. Reduced VDR expression is observed in *H-Ras* transformed HC-11 mammary cells and *K-Ras* transformated RWPE-2 human prostate epithelial cells (Escaleira & Brentani, 1999; Rozenchan, Folgueira, Katayama, Snitcovsky, & Brentani, 2004; Zhang et al., 2010). In addition to the reduction of VDR levels, Ras activation can impair vitamin D-mediated transcriptional activity. *H-Ras*-transformed keratinocytes have reduced *VDR* transcriptional activity due to phosphorylation of the VDR heterodimeric partner RXRα on Ser^{260} (Solomon, White, & Kremer, 1999). The mutant p53 can interact with VDR and modulate the expression of VDR-regulated genes in breast and lung cancer cell lines (Stambolsky et al., 2010). The effect of tumor suppressor gene and oncogene mutations on *VDR* gene expression may vary in different type of tumors. Increased *VDR* expression is associated with *K-Ras* mutation in colorectal cancer (Kure et al., 2009).

VDR expression is also regulated by epigenetic mechanisms in cancer. Reduction of VDR protein levels is associated with increased CpG island methylation in the *VDR* gene in colonic tissue from ovariectomized rats treated with dimethylhydrazine (Smirnoff, Liel, Gnainsky, Shany, & Schwartz, 1999). The association of lower *VDR* expression with the *VDR* promoter CpG island methylation is also observed in breast cancer tissues and breast cancer cell lines (Marik et al., 2010). Demethylation of the VDR promoter with 5-aza-2′-deoxycytidine, an inhibitor of DNA methyltransferase, restores the *VDR* expression, increases VDR-responsive gene expression, and potentiates $1,25D_3$-mediated the growth inhibitory response in breast cancer cell lines (Marik et al., 2010).

Decreased tissue sensitivity to $1,25D_3$ could be caused by the epigenetic silencing of the *VDR* gene in cancer (Marik et al., 2010). *VDR* promoter methylation may result in the expression of truncated VDR protein in cancer cells. *VDR* transcripts detected in primary breast cancers are predominantly 5′-truncated, while normal breast tissue expressed the full-length variants. Hypermethylation of *VDR* promoter CpG islands in primary tumors was observed in breast primary tumor, not in normal breast tissues. Treatment with 5-aza-2′-deoxycytidine restored expression of the active full-length transcript variant of VDR in breast cancer cell lines, indicating that *VDR* methylation affects expression of the active forms of VDR and *VDR* promoter methylation could be a cause of truncated protein expression in breast cancer cells (Marik et al., 2010).

The histone deacetylase HDAC3 is one of the most frequently upregulated genes in cancer and appears to suppress *VDR* expression. Knockdown of HDAC3 increases *VDR* expression and restores sensitivity of cancer cells to $1,25D_3$ in colorectal cancer cell lines HCT116 and SW480 (Godman et al., 2008). The polycomb group protein enhancer of zeste homologue 2 (EZH2), which is a histone-lysine N-methyltransferase enzyme, is overexpressed in many malignant tumors (Kondo, 2014). EZH2 regulates the histone trimethylation of lysine 27 (H3K27me3) in the *VDR* promoter (Lin et al., 2013). EZH2-induced cell invasion in colorectal cancer is proposed to be dependent on the downregulation of VDR (Lin et al., 2013).

Posttranslational regulation is also involved in VDR expression (Mohri, Nakajima, Takagi, Komagata, & Yokoi, 2009). Overexpression of miR-125b reduces VDR protein levels and attenuates $1,25D_3$-induced expression of CYP24A1 in MCF-7 cells (Mohri et al., 2009). Similarly, melanoma cell lines with low miR-125b levels have higher VDR levels and are more sensitive to $1,25D_3$-induced growth suppression than cells with high miR-125b levels (Essa et al., 2010).

3. GLUCOCORTICOID, $1,25D_3$-MEDIATED ANTITUMOR EFFECT AND HYPERCALCEMIA

The dose-limiting toxicities of $1,25D_3$ (hypercalcemia and hypercalcuria) can limit the usefulness of $1,25D_3$ as a cancer therapy (Gross, Stamey, Hancock, & Feldman, 1998; Osborn et al., 1995; Smith et al., 1999). $1,25D_3$ induces hypercalcemia through alterations in intestinal calcium absorption and renal calcium excretion (Klein, Arnaud, Gallagher, Deluca, & Riggs, 1977; Suzuki, Ichikawa, Saito, & Homma, 1983). Glucocorticoids, especially dexamethasone, have been extensively used to blunt/reverse many causes of hypercalcemia including $1,25D_3$-induced hypercalcemia.

Glucocorticoids are widely used in the treatment of inflammation and various immune-mediated diseases, such as rheumatoid arthritis, polymyalgia rheumatica, systemic lupus erythematosus, vasculitis, inflammatory bowel disease, immune liver, and skin diseases, and to suppress rejection in organ transplant recipients (Maalouf, Battistella, & Bouaziz, 2015; Malnick, Duek, Melzer, & Basevitz, 2012; Triadafilopoulos, 2014; van der Goes, Jacobs, & Bijlsma, 2014; van Sandwijk, Bemelman, & Ten Berge, 2013). $1,25D_3$ has significant additive effects on dexamethasone-mediated inhibition of lymphocyte proliferation and TH1 cytokine production (Jirapongsananuruk, Melamed, & Leung, 2000). The combination of $1,25D_3$ and dexamethasone further decreases lymphocyte IFN-γ production compared with either agent alone (Jirapongsananuruk et al., 2000). Furthermore, $1,25D_3$ enhances glucocorticoid-induced MAPK-1 and glucocorticoid-mediated inhibition of lipopolysaccharide-induced IL-6 production in human peripheral blood mononuclear cells (Zhang, Leung, & Goleva, 2013). These data suggest substantial actions of $1,25D_3$ on glucocorticoid-mediated anti-inflammatory effects in human monocytes. In addition to inflammation, glucocorticoids have antiproliferative activity in certain lymphoproliferative diseases and breast cancers as well as palliative effects in patients with prostate cancer and have been used widely in combination with other treatment for patients with cancer (Hardin et al., 1994; Herr & Pfitzenmaier, 2006; Inaba et al., 2011; Jakubowiak et al., 2009; Rutz, 2002; Tannock et al., 1989; Walsh & Avashia, 1992). A combination of $1,25D_3$ with glucocorticoids has demonstrated to be both effective and relatively nontoxic and exerts synergistic anti-inflammation and antiproliferative effects (Kerley, Elnazir, Faul, & Cormican, 2015; Kupfer, Maranville, Baxter, Huang, & Di Rienzo, 2013; Litonjua, 2013; Yu et al., 1998).

The combination of $1,25D_3$ and dexamethasone enhances $1,25D_3$-mediated antitumor effect and reduces the hypercalcemic effects of $1,25D_3$. Dexamethasone increases transcription of *VDR* resulting in increased VDR protein expression and potentiates VDR-mediated transcription of *VDR* (Hidalgo, Deeb, Pike, Johnson, & Trump, 2011). Induction of VDR expression by glucocorticoids has been shown in normal and cancer cells (Hidalgo, Trump, & Johnson, 2010; Kang et al., 2015). Treatment with dexamethasone increases *VDR* transcripts in a time- and a dose-dependent manner (Hidalgo et al., 2011). Dexamethasone regulates *VDR* expression at transcriptional level since dexamethasone-mediated VDR mRNA synthesis could be inhibited by actinomycin D, an antitumor antibiotic that inhibits transcription by binding DNA at the transcription initiation complex (Hidalgo et al., 2011; Sobell, 1985). The increased VDR expression by dexamethasone is likely GR-dependent since RU486, a glucocorticoid receptor (GR) antagonist, inhibits dexamethasone-induced VDR expression. In addition, the silencing of *GR* by siRNA abolishes the induction of *VDR* mRNA and protein expression by dexamethasone (Hidalgo et al., 2011).

Besides increasing the VDR expression, dexamethasone significantly enhances the antitumor activity of $1,25D_3$ *in vitro* and *in vivo* mouse models including the murine SCCVII/SF SCC, human prostate cancer PC3 xenograft (Bernardi et al., 2001; Hidalgo et al., 2011; Muindi et al., 2010; Yu et al., 1998). Dexamethasone increases $1,25D_3$–VDR complex binding to VDR-responsive gene DNA promoter, promotes the expression of these genes, such as p21, p27, cyclin D1, and calbindin D9K, and increases cell sensitivity to $1,25D_3$ (Bernardi et al., 2001; Hidalgo et al., 2011; Yu et al., 1998). Treatment with $1,25D_3$ and dexamethasone results in a greater decrease of prosurvival proteins p-ERK and p-AKT than treatment with either $1,25D_3$ or dexamethasone alone. $1,25D_3$ and dexamethasone combination treatment further increases cell cycle arrest (a greater G0/G1 accumulation) and the activation of apoptosis (increased cleavage of caspase-3 and poly(ADP-ribose) polymerase) compared with treatment with either $1,25D_3$ or dexamethasone alone (Bernardi et al., 2001). Dexamethasone also enhances the antitumor effect of $1,25D_3$ in combination with a number of cytotoxic drugs (Ahmed et al., 2002; Hershberger et al., 1999). Results from preliminary clinical trials are inconsistent. Preliminary results from clinical trials of $1,25D_3$ (12 g qd three times/week) with dexamethasone (4 mg qd four times/week) in castration resistant prostate cancer (CRPC) demonstrated a 50% reduction in PSA in 5 of 24 evaluable patients (24%) and a decrease in PSA velocity in 19 patients (79%)

(Muindi et al., 2002; Trump et al., 2006). Meanwhile, mild and transient hypercalcemia was observed in only two patients. In another phase II trial of 34 patients with hormone-refractory prostate cancer, combination of dexamethasone, calcitriol, and carboplatin produced a PSA response in 13 of 34 patients and had an acceptable side-effect profile (Flaig et al., 2006). However, in another phase 2 trial of weekly intravenous $1{,}25D_3$ in combination with dexamethasone for 18 evaluable patients with CRPC, high dose (74 μg/weekly) in combination with dexamethasone was well tolerated, but failed to produce a clinical or PSA response in men with CRPC (Chadha et al., 2010).

The reduction of vitamin D-induced hypercalcemia by glucocorticoids involves increased urinary excretion of calcium, decreased calcium absorption in the intestine, and potentially other mechanisms (Ragavan, Smith, & Bilezikian, 1982). Both *in vitro* and *in vivo*, dexamethasone increases the binding of $1{,}25D_3$ to VDR in the tumor while decreasing binding in intestinal mucosa, the site of calcium absorption (Klein et al., 1977; Yu et al., 1998). Reduction of the expression of $1{,}25D_3$-mediated duodenal calcium absorption regulatory genes may serve as a molecular mechanism linking $1{,}25D_3$ signaling and decreased intestinal calcium absorption. The reduction of intestinal calcium absorption with glucocorticoid treatment is associated with suppression of the expression of transient receptor potential cation channel subfamily V member 6 (*TRPV6*, influx) and calbindin-D9k (intracellular transfer) in intestines (Huybers, Naber, Bindels, & Hoenderop, 2007). In a mouse model, long-term use of large doses of dexamethasone results in impaired gastrointestinal absorption of calcium and decreased expression of active duodenal calcium absorption regulatory genes including duodenal $1{,}25D_3$ regulating genes *TRPV6*, calbindin D9k (*S100G*), *NCX1* (Na^+/Ca^{2+} exchanger 1), and plasma membrane Ca^{2+}-ATPase 1b (*PMCA 1b*, extrusion) (Kim, Lee, Jung, Choi, & Jeung, 2009). The negative effect of glucocorticoids on $1{,}25D_3$ target genes in intestine and bone is consistent with the reported correlation between glucocorticoid treatment and decreased bone mineral density (Reid, 1997). However, it cannot be excluded that the effect of glucocorticoids on duodenal calcium transporters may be independent of $1{,}25D_3$ (Feher & Wasserman, 1979; Huybers et al., 2007; Tohmon, Fukase, Kishihara, Kadowaki, & Fujita, 1988).

In addition, glucocorticoids affect vitamin D metabolism by inducing $1{,}25D_3$ catabolism after tissue localization (Carre, Ayigbede, Miravet, & Rasmussen, 1974). Dexamethasone enhances *CYP24A1* mRNA and activity in osteoblastic and renal cells treated with 1,25D_3 (Akeno, Matsunuma, Maeda, Kawane, & Horiuchi, 2000; Kurahashi, Matsunuma, Kawane, Abe, & Horiuchi, 2002; Van Cromphaut et al., 2007). Elevated 1,25D_3 levels at specific sites in bone are observed in *CYP24A1* null mutant mice (St-Arnaud et al., 2000). Increased *CYP24A1* mRNA as well as CYP24A1 activity in kidney is noted in mice treated with dexamethasone (Akeno et al., 2000; Van Cromphaut et al., 2007). These findings suggest that increased local catabolism of 1,25D_3 by induction of CYP24A1 expression serves as an additional mechanism of the glucocorticoid-mediated abrogation of 1,25D_3-mediated hypercalcemia.

Although there is increased CYP24A1 expression in certain tissues, such as kidney, It is observed that serum 1,25D_3 levels in dexamethasone-treated animals compared to control may be reduced or may be no change in animal models (Akeno et al., 2000; Van Cromphaut et al., 2007). 25D_3 and 1,25D_3 serum levels in human are varied after glucocorticoid treatment (Chesney, Mazess, Hamstra, DeLuca, & O'Reagan, 1978; Cosman, Nieves, Herbert, Shen, & Lindsay, 1994; Klein et al., 1977; Prummel, Wiersinga, Lips, Sanders, & Sauerwein, 1991). The variations have been suggested to be due to different glucocorticoid doses or different schedules of dosing or different methods to measure 25D_3 and 1,25D_3 serum levels (Klein et al., 1977).

The primary mechanism by which glucocorticoids enhance 1,25D_3-induced transcription of *CYP24A1* is GR dependent (Hidalgo et al., 2011). In addition, the enhancement of 1,25D_3-induced transcription of *CYP24A1* by glucocorticoids is through functionally cooperating of GR with CCAAT-enhancer-binding protein (C/EBP)β at the C/EBP binding site of the *CYP24A1* promoter (Dhawan & Christakos, 2010; Van Cromphaut et al., 2007). C/EBPs are a family of transcription factors composed of six members. C/EBPβ is known to act as an enhancer of *CYP24A1* transcription (Dhawan et al., 2005). They promote the expression of certain genes through interaction with their promoters. C/EBPβ can be induced by 1,25D_3 in kidney and osteoblastic cells (Dhawan et al., 2005; Gutierrez et al., 2002) and can be regulated by glucocorticoids in different tissues (Balazs, Schweizer, Frey, Rohner-Jeanrenaud, & Odermatt, 2008; Delany, Durant, & Canalis, 2001; McCarthy, Ji, Chen, Kim, & Centrella, 2000; Roesler, 2001). The induction of C/EBPβ by both 1,25D_3 and glucocorticoids as well as the cooperative effects between GR and C/EBPβ on transcription suggests combinational effects of 1,25D_3 and glucocorticoids on *CYP24A1* transcription.

4. GR AND VDR SIGNALING

The mechanism by which glucocorticoids enhances $1,25D_3$ action remains to be elucidated. Increased expression of VDR by dexamethasone may enhance sensitivity to $1,25D_3$ and potentiate $1,25D_3$ antitumor action. Both $1,25D_3$ and dexamethasone are known to bind members of the nuclear receptor superfamily (Mangelsdorf et al., 1995). Binding of $1,25D_3$ to the VDR promotes the binding of this $1,25D_3$–VDR complex to VDRE and activation of transcription of vitamin D responsive genes. Binding of glucocorticoid–GR to glucocorticoid-responsive elements (GRE) on vitamin D responsive genes further promotes $1,25D_3$-VDR-mediated activation of transcription of vitamin D responsive genes (Fig. 1).

The action of glucocorticoids is GR-dependent since the effect of glucocorticoids can be reversed by a specific GR antagonist (Lee et al., 2006). The activation of GR by ligand binding involves a conformational change, increased phosphorylation of GR, and release of associated regulatory proteins. Activated GR complex translocates to the nucleus, binds as a homodimer to specific DNA sequences GREs, and regulates the transcription of genes associated with the GREs (Lu et al., 2006). GR also regulates the transcription of genes through cross talk between the GR and other transcription factors (TFs), such as AP-1 and NF-κB (Kassel & Herrlich, 2007). These actions are thought to be the result of GR interfering with transduction pathways upstream of other TFs or GR and other TFs modulating each other's activity when bound to the promoters of their target genes (Kassel & Herrlich, 2007).

By using ChIP-on-chip, Zella et al. (2010) identified four DNA regions that bind GR–ligand complex within the vicinity of the VDR gene. pTK-U1 luciferase reporter activity assays in MC3T3-E1 cells showed that of the four regions, only the U1 region is able to induce GR-mediated transcription with dexamethasone (Zella et al., 2010). Hidalgo et al. analyzed a fragment located 5.2 kb upstream of VDR transcription start site containing two putative GREs using a luciferase-based reporter assay. A distal GRE responsible for GR-mediated transcription is identified in the U1 regulatory regions of mouse VDR (Hidalgo et al., 2011). Deletion of the distal GRE is sufficient to abolish dexamethasone-mediated induction of luciferase activity (Hidalgo et al., 2011). ChIP with a GR-specific antibody reveals the recruitment of GR to distal GRE on the VDR gene with dexamethasone treatment. The active GRE in the regulatory region of the VDR gene

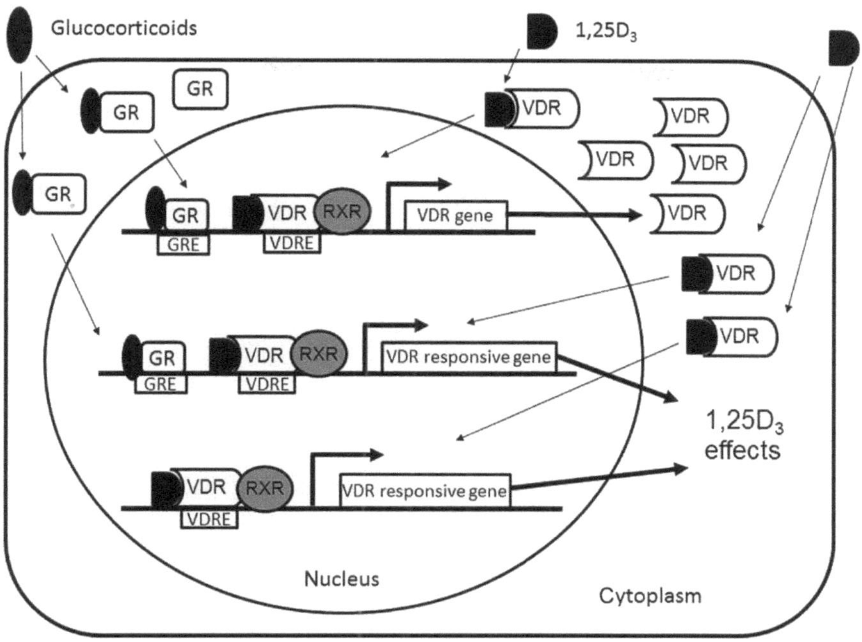

Figure 1 Interactions between vitamin D receptor (VDR) and glucocorticoid receptor (GR) signaling axes. Proposed model illustrates how $1{,}25D_3$ and glucocorticoids (GCs) function together to increase VDR expression and regulate the expression of VDR-responsive genes. In the presence of $1{,}25D_3$, VDR transcriptionally regulates VDR target genes including *VDR* gene itself. Binding of the complex of GCs–GRs to glucocorticoid-responsive elements (GRE) on *VDR* gene, together with the complex of $1{,}25D_3$–VDR/RXR, increases *VDR* gene expression. Interacting with $1{,}25D_3$–VDR/RXR complex, the GC–GR further promotes the expression of the VDR-responsive genes. This cross talk between GR and VDR signaling axes implicates synergistic effects on $1{,}25D_3$ and glucocorticoid function. *Modified from Hidalgo et al. (2011) and Cutolo et al. (2014).*

contains the precise consensus sequence TGTTCT (Hidalgo et al., 2011; Merkulov & Merkulova, 2009).

Although most of the activity of nuclear receptors is regulated by the binding of ligand, cross talk between nuclear receptors can affect receptor-mediated signaling through the binding of the ligand to another receptor. Cross talk between nuclear receptors can modulate expression of nuclear receptors at both the transcriptional or posttranscriptional levels in a unilateral or reciprocal manner (Bagamasbad & Denver, 2011). A line of evidence supports the cross talk between VDR and GR signaling axes in a variety of tissues. It has been reported that $1{,}25D_3$ increases glucocorticoid production in human adipocytes by increasing 11β-hydroxysteroid dehydrogenase type 1, which promotes the conversion of the glucocorticoid precursor cortisone to cortisol (Morris & Zemel, 2005). On the other hand, upregulation or downregulation of VDR by glucocorticoids is observed depending on the tissue and species examined (Chen, Cone, Morey-Holton, & Feldman, 1982). An increasing body of evidence now convincingly demonstrates that enhancement of $1,25D_3$ effect by glucocorticoids is primarily through increasing VDR transcription in a GR-dependent manner (Bernardi et al., 2001; Flynn et al., 2006; Hidalgo et al., 2011; Kallay et al., 2001; Kim et al., 2009; Sun & Zemel, 2008; Yu et al., 1998). Furthermore, studies reveal that dexamethasone significantly increases $1,25D_3$ ligand binding in SCC cells (Yu et al., 1998). Thus, dexamethasone may enhance the antiproliferative effects of $1,25D_3$ by increasing the expression of VDR and other vitamin D target genes which are critical for the antiproliferative effects. This hypothesis is supported by the finding that dexamethasone alone has little antiproliferative effect in the SCC model (Yu et al., 1998). Taken together, glucocorticoids trigger GR-responsive genes and together with the complex of VDR increase the expression of the VDR gene and other VDR-responsive genes. This cross talk between GR and VDR signaling axes implicates synergistic effects on $1,25D_3$ and glucocorticoid function (Fig. 1).

5. REGULATION OF CYP24A1 EXPRESSION

The human *CYP24A1* gene is located on chromosome 20q 13.2 (Kallioniemi et al., 1994). CYP24A1 protein is located in the mitochondria and belongs to a member of cytochrome class I P450 superfamily of enzymes. CYP24A1 is the major enzyme responsible for the conversion of $1,25D_3$ and $25D_3$, into less active metabolites $1,24,25D_3$ and $24,25D_3$, respectively. CYP24A1 is capable of further catalyzing a sequential monooxygenation of $1,24,25D_3$ and $24,25D_3$ in an *in vitro* reconstituted system (Sakaki, Kagawa, Yamamoto, & Inouye, 2005; Sakaki, Sawada, Nonaka, Ohyama, & Inouye, 1999). CYP24A1 has both 24-hydoxylase and 23-hydroxylase activity which is species-dependent (Jones, Prosser, & Kaufmann, 2012). A high basal level of CYP24A1 expression is present in the kidney and gastrointestinal mucosa (Iwata et al., 1995; Omdahl et al., 2002; Prosser & Jones, 2004). In most other normal tissues, *CYP24A1* gene is expressed at very low basal levels but is highly inducible by $1,25D_3$ through transcriptional regulation (Meyer, Zella, Nerenz, & Pike, 2007; Omdahl et al., 2001; Prosser & Jones, 2004; Roff & Wilson, 2008;

St-Arnaud, 2011). In turn, 1,25D_3 is rapidly inactivated by the induced catabolic enzyme CYP24A1 to prevent the accumulation of toxic level of 1,25D_3.

Expression of the CYP24A1 by 1,25D_3 is a VDR signaling dependent process (Endres, Kato, & DeLuca, 2000; Haussler et al., 1998; Zhu, Malloy, Delvin, Chabot, & Feldman, 1998). In VDR null mice, there are lower levels of CYP24A1 and 24-hydroxylated metabolites, demonstrating that CYP24A1 expression is dependent on VDR (Yoshizawa et al., 1997). CYP24A1 expression is regulated through the binding of 1,25D_3 to VDR and the subsequent interaction of the VDR–1,25D_3 complex with its heterodimer partner RXR–9cRA (Pike & Meyer, 2010). The binding of 1,25D_3 to VDR leads to the replacement of the repressor by a coactivator complex such as nuclear coactivators (Leo & Chen, 2000; Rachez & Freedman, 2000, 2001; Zhang et al., 2001). These coactivators link the ligand-activated VDR to enzymes displaying histone acetyltransferase activity that causes chromatin relaxation and activate the target genes and triggers transcription (Carlberg & Polly, 1998; Castillo, Jimenez-Lara, Tolon, & Aranda, 1999; Rachez et al., 1999). A cluster of VDREs has been reported in the proximal promoter of both the human *CYP24A1* gene (Chen & DeLuca, 1995) and the rat *CYP24a1* gene (Hahn, Kerry, Omdahl, & May, 1994; Kahlen & Carlberg, 1994; Zierold, Darwish, & DeLuca, 1995). SNPs in the human *CYP24A1* promoter containing VDRE could cause a lower affinity for VDR–RXR heterodimer, decrease transactivation, and reduce the expression of CYP24A1 (Roff & Wilson, 2008).

CYP24A1 expression can be induced by other endogenous compounds with structures very different from vitamin D, such as lithocholic acid (Makishima et al., 2002), RXR or RAR ligands (Zou, Elgort, & Allegretto, 1997). Chronic ethanol consumption and benzopyrene produced by cigarette combustion induce CYP24A1 expression leading to the disruption of vitamin D_3 homeostasis (Matsunawa et al., 2009; Shankar et al., 2008).

The regulation of *CYP24A1* expression may also be involved in nongenomic action of 1,25D_3 by stimulation of intracellular pathways such as PKC and MAPK, through a plasma membrane-associated VDR (Cui, Zhao, et al., 2009; Norman, Song, Zanello, Bula, & Okamura, 1999; Prosser & Jones, 2004). The signaling mechanisms that regulate transcriptional activation of the *CYP24A1* promoter are poorly defined and further investigation is needed. Some transcription factors that enhance 1,25D_3-mediated induction of the *CYP24A1* expression by interacting with VDR and intracellular signaling pathways have been identified.

For example, it has been shown that Ets-1 induces *CYP24A1* promoter activity in enterocyte-like Caco-2 cells, human embryonic kidney 293T (HEK-293T) cells, and monkey kidney epithelial COS-1 cells (Cui, Klopot, Jiang, & Fleet, 2009; Dwivedi et al., 2002; Nutchey et al., 2005). Ets-1 can be activated by the Ras/Raf signaling pathway through phosphorylation of threonine residue (Yang et al., 1996). Furthermore, 1,25D_3 stimulates MAP kinase and ERK5 to activate Ets-1 through phosphorylation in COS-1 monkey kidney fibroblast cells (Nutchey et al., 2005). The interaction of VDR and Ets-1 causes a significant induction of *CYP24A1* promoter activity in COS-1 cells (Cui, Zhao, et al., 2009; Dwivedi et al., 2002). The direct interaction between VDR and Ets-1 protein was demonstrated by an *in vitro* GST-protein pull-down study (Tolon, Castillo, Jimenez-Lara, & Aranda, 2000). intracellular signal transduction-regulated *CYP24A1* expression could be cell type specific. COS-1 cells treated with inhibitors for ERK1/2 and ERK5 abolished 1,25D_3 induction of the *CYP24A1* promoter (Dhawan & Christakos, 2010). However, this was not found to be the case in HEK-293T cells (Nutchey et al., 2005). In HEK-293T cells, induction of *CYP24A1* by 1,25D_3 requires JNK (c-Jun N-terminal kinase) but not the ERK1/2 (extracellular signal-regulated kinase 1/2) (Nutchey et al., 2005). In addition, cellular modulators, such as glucocorticoids, also regulate *CYP24A1* expression. Increased VDR gene expression by glucocorticoids and the cooperation between C/EBP and GR result in enhancement of 1,25D_3-induced *CYP24A1* expression (Dhawan & Christakos, 2010; Hidalgo et al., 2011). *CYP24A1* transcriptional activity is also regulated by ATP-dependent chromatin remodeling enzymes. SWI/SNF multisubunit complex, an ATP-dependent chromatin remodeling factor, contributes to transcriptional regulation of *CYP24A1* by cooperating with C/EBPβ, BRG1, and VDR (Seth-Vollenweider, Joshi, Dhawan, Sif, & Christakos, 2014). Brahma-related gene 1 (BRG1), an ATPase that is a component of the SWI/SNF complex, plays an important role in 1,25D_3-induced transcription of *CYP24A1*. PRMT5, a member of the protein arginine methyltransferases (PRMTs), is a histone-modifying enzyme. PRMT5 has been found to negatively regulate 1,25D_3-induced *CYP24A1* transcription by interacting with BRG1 via its methylation of H3R8 and H4R3 (Seth-Vollenweider et al., 2014).

In addition, it has been suggested that *CYP24A1* gene polymorphisms is associated with the enzyme activity. *CYP24A1* mutations have been identified in infants who develop severe hypercalcemia after high-dose prophylaxis with vitamin D. Functional characterization reveals a complete loss of function in all *CYP24A1* mutations (Schlingmann et al., 2011). A recent genome-wide study associated in more than 30,000 individuals indicates that genetic variation of *CYP24A1* is responsible for individuals who have substantially elevated risk of vitamin D insufficiency (Wang et al., 2010).

6. CYP24A1 EXPRESSION IN CANCER

Aberrant CYP24A1 expression is observed in cancer. Overexpression of CYP24A1 has been reported in breast tumors, ovarian, cervical, lung, cutaneous basal cell carcinoma, colorectal cancer and SCC (Albertson et al., 2000; Anderson et al., 2006; Friedrich et al., 2003; Mitschele et al., 2004). This suggests that increased $1,25D_3$ catabolism may be a common feature of advanced cancer. Besides the overexpression of CYP24A1 in cancer, studies showed that location of CYP24A1 shifts from the nuclear compartment to the cytoplasm in tumors and metastatic colon cancer (Matusiak & Benya, 2007).

Studies in several cancer cell lines demonstrate that the differences in $1,25D_3$-mediated growth inhibition among various cancer cell lines correlate inversely to levels of CYP24A1 expression in these cells (Anderson et al., 2006; Chen et al., 2011; Miller, Stapleton, Hedlund, & Moffat, 1995). Among the well-studied human prostate cancer cell lines DU145, PC3, and LNCaP (Miller et al., 1995; Skowronski, Peehl, & Feldman, 1993), DU145 cells have the highest level of CYP24A1 expression and are the least responsive to $1,25D_3$. LNCaP cells exhibit very low CYP24A1 expression of and are the most sensitive one to $1,25D_3$. PC3 cells have intermediate level of CYP24A1 and an intermediate level of growth inhibition (Miller et al., 1995; Skowronski et al., 1993). This suggests that overexpression of CYP24A1 may promote the rapid degradation of $1,25D_3$, thereby reducing its ability to regulate genes that control cellular proliferation and differentiation.

The mechanisms underlying the aberrant expression of CYP24A1 in tumors are not well defined. It seems that CYP24A1 overexpression in tumor is VDR-independent or does not correlate with VDR expression level in several types of tumors, since the latter are actually reduced or unchanged. This suggests that high CYP24A1 levels are not the result of the normal physiological transcriptional process observed when binding of $1,25D_3$-VDR complex to VDREs on the promoter of *CYP24A1* gene activates CYP24A1.

Accumulated evidence suggests that multiple factors might be involved in dysregulation of CYP24A1 expression in cancer (Cross, 2007; Khorchide, Lechner, & Cross, 2005; Komagata et al., 2009; Matilainen et al., 2010). One possibility is amplification at the *CYP24A1* locus. Overexpression of CYP24A1 in breast tumors are may be associated with amplification at the *CYP24A1* locus, and *CYP24A1* gene has been proposed to be a candidate oncogene in breast cancer (Albertson et al., 2000; Collins et al., 1998; Kallioniemi et al., 1994). Similar amplification is also seen in lung (Wong et al., 2003), colon (Hidaka et al., 2000), and esophageal (Rygiel et al., 2008) cancers. It is noteworthy that the gene amplification in the colon is only found in the malignant but not in benign tumors (Weiss, Hermsen, Meijer, & van Diest, 2002). These results suggest that the amplification at *CYP24A1* locus may lead to the transition of benign tumors to carcinoma.

Alternatively, CYP24A1 regulation in tumors may be related to miRNAs that target the *CYP24A1* gene. miR-125b binds at the 3′ UTR of *CYP24A1* mRNA to posttranscriptionally repress CYP24A1 translation (Komagata et al., 2009). In MCF-7 cells, CYP24A1 protein levels are increased by the inhibition of miR-125b and are decreased by miR-125b overexpression (Komagata et al., 2009). Furthermore, CYP24A1 levels in breast cancer tissues are inversely associated with miR-125b levels (Komagata et al., 2009). The miR-17 to -92 cluster regulates CYP24A1 expression in p53-depleted lung cancer cells (Borkowski et al., 2015). This is evident by that the miR-92a inhibitor, which reduces levels of the miR-17 to -92 primary transcripts, reduces the level of the CYP24A1 expression in p53-depleted lung cancer cells (Borkowski et al., 2015).

Additionally, the number of functional VDR binding sites could be responsible for a difference in *CYP24A1* gene expression level induced by vitamin D in malignant cells (Matilainen et al., 2010). The number of functional VDREs is higher in malignant breast MCF-7 cells with high CYP24A1 expression than in normal breast MCF-10A cells (Matilainen et al., 2010). ChIP analysis using anti-VDR antibody demonstrates that VDR is associated with both proximal and distal VDREs in MCF-7 cells, but in MCF-10A cells only with the proximal VDRE (Matilainen et al., 2010). Furthermore, CYP24A1 expression in cancer cells may be modified via epigenetic processes. In normal human tissues, *CYP24A1* promoter methylation is tissue specific. No methylation is detected in most normal human tissues including kidney, skeletal muscle, skin fibroblasts, brain (prefrontal cortex), sperm, whole blood, buccal mucosa, endometrial stroma, decidualized stroma, bone marrow, umbilical cord tissue, and peripheral blood lymphocytes (Novakovic et al., 2009). However, *CYP24A1* promoter methylation is seen in human placenta. The placenta-specific epigenetic repression of the *CYP24A1* is in concert with elevated maternal circulating 1,25D_3 during pregnancy (Kovacs & Kronenberg, 1997; Novakovic et al., 2009), suggesting vitamin D feedback catabolism plays an important role in maximizing active vitamin D bioavailability at the fetomaternal interface (Novakovic et al., 2009). CYP24A1 expression can be lower in some tumor-derived endothelial cells and cancers (Chung et al., 2007; Deeb et al., 2011; Luo et al., 2010; Ramnath et al., 2014). Studies demonstrate that the reduction of CYP24A1 expression is inversely correlated with *CYP24A1* promoter methylation in lung and prostate cancer (Khorchide et al., 2005; Luo et al., 2010; Ramnath et al., 2014). Bisulfite DNA sequencing and pyrosequencing analysis reveal that *CYP24A1* gene is heterogeneously methylated in prostate cancer and lung adenocarcinoma (Luo et al., 2010; Ramnath et al., 2014). Treatment with methyltransferase (DNMT) inhibitor 5-aza-2′-deoxycytidine (5-Aza) increases CYP24A1 expression in prostate cancer and lung adenocarcinoma (Luo et al., 2010; Ramnath et al., 2014). *CYP24A1* DNA promoter hypermethylation was also observed in human choriocarcinoma cell lines (Novakovic et al., 2009). The tumor microenvironment also affects the methylation status of the *CYP24A1* promoter. Differential methylation of the *CYP24A1* gene promoter was also observed in endothelium from benign and malignant human prostate (Deeb et al., 2011). In addition, studies indicate that repression of *CYP24A1* gene expression in human prostate cancer cells was mediated in part by repressive histone modifications. ChIP-qPCR studies reveal that specific histone modifications are associated with the *CYP24A1* promoter region. Treatment with deacetylase inhibitor (HDAC) trichostatin A (TSA) increases active histone mark H3K9ac and H3K4me2 and simultaneously decreases repressive histone mark H3K9me2 at the *CYP24A1* promoter. Treatment with DAC and TSA increases the recruitment of VDR to the *CYP24A1* promoter (Luo et al., 2010; Ramnath et al., 2014).

Lastly, studies demonstrate that some protein kinases may serve as a mechanism for controlling CYP24A1 expression in human cancers (Luo et al., 2013; Maier et al., 2012). Overexpression of CK2 has been noted in a variety of human cancers including prostate cancer and correlates with poor clinical outcomes (Gapany et al., 1995; Landesman-Bollag et al., 2001; Laramas et al., 2007; Oc et al., 2004). CK2 positively regulates CYP24A1 expression. The increase of CK2 expression is significantly associated with increased CYP24A1 expression in prostate cancer (Luo et al., 2013).

Taking together, increased CYP24A1 expression in malignant cells in comparison to normal cells is due to differences in the transcriptional and posttranscriptional regulation of *CYP24A1*.

7. CYP24A1 INHIBITORS

As a key enzyme responsible for vitamin D catabolism, the overexpression of CYP24A1 in tumors may reduce the biologic effects and therapeutic utility of vitamin D compounds. Furthermore, the induction of CYP24A1 expression through the negative feedback mechanisms following administration of $1,25D_3$ also increases degradation of $1,25D_3$ in tumor cells. *In vitro* studies reveal that differences in $1,25D_3$-mediated growth inhibition among various cancer cell lines correlate inversely to the levels of CYP24A1 expression in these cells (Anderson et al., 2006; Chen et al., 2011; Miller et al., 1995). Data obtained using CYP24A1 inhibitors suggest that increased CYP24A1 expression in tumors restricts $1,25D_3$ antitumor activity (Ly, Zhao, Holloway, & Feldman, 1999). Inhibition of CYP24A1 slows the loss of $1,25D_3$, leading to longer sustained $1,25D_3$ locally in tumor and thereby increasing $1,25D_3$ exposure and antiproliferative effects (Beumer et al., 2012; Ly et al., 1999; Parise et al., 2006; Schuster et al., 2006; Zhang et al., 2012).

Cytochrome P450 inhibitors have been developed to inhibit CYP24A1 activity. The tetralone derivative (2-(4-hydroxybenzyl)-6-methoxy-3,4-dihydro-2*H*-naphthalen-1-one), a nonazole CYP24A1 inhibitor, can enhance $1,25D_3$ antiproliferative activity in DU145 cells and increase the expression of vitamin D target genes, p21waf1/cip1 and GADD45a (Yee, Campbell, & Simons, 2006). Ketoconazole and liarozole (Ly et al., 1999; Peehl, Seto, Hsu, & Feldman, 2002; Reinhardt & Horst, 1989; Schuster et al., 2006) are nonselective azole CYP24A1 inhibitors and target the active sites of many P450 enzymes (Yee & Simons, 2004). Combination of $1,25D_3$ with liarozole or ketoconazole increases the half-life of $1,25D_3$ and potentiates its antiproliferative effects in human cancer cells (Peehl et al., 2002; Reinhardt & Horst, 1989; Yee et al., 2006). Plant derived isoflavones, such as genistein, are nonspecific inhibitors of the CYP enzymes (Shanmugam, Kannaiyan, & Sethi, 2011). Genistein inhibits the expression of CYP24A1, leading to an increase in the half-life of $1,25D_3$ (Farhan & Cross, 2002). Combination of $1,25D_3$ with genistein result in synergistic growth inhibition of prostate cancer cells by enhancing vitamin D signaling and $1,25D_3$-mediated functional responses including upregulation of mRNA levels of p21/WAF1 and IGFBP-3 (Rao et al., 2004; Swami, Krishnan, Peehl, & Feldman, 2005). These nonselective CYP24A1 inhibitors can also inhibit other cytochrome P450 enzymes including vitamin D activating enzymes.

The broad inhibitory effects of nonselective inhibitors, such as ketoconazole on other P450 enzymes, limit their clinical utility as CYP24A1 inhibitors. Effort has been made to develop CYP24A1 selective inhibitors. Schuster, Egger, Nussbaumer, and Kroemer (2003) synthesized and screened a large number of azole compounds using primary human keratinocyte cultures as a source of CYP24A1 and CYP27B1, with $25D_3$ as the substrate to analyze the complex metabolite profiles. Among them, VID400 ((*R*)-*N*-(2-(1*H*-imidazol-1-yl)-2-phenylethyl)-4-chlorobiphenyl-4-carboxamide) exhibits strong and selective CYP24A1 inhibitory activity (IC_{50}: 15 nM), with little effect on CYP27B1 activity (IC_{50}: 616 nM). Combination of $1,25D_3$ with VID400 shows a synergistic antiproliferative effect in human keratinocytes (Schuster, Egger, Reddy, & Vorisek, 2003). Posner et al. have developed two promising $1,25D_3$ analogs: the sulfone GHP-GH-16,23-diene-25S02-1 and sulfoximine MK-24(S)-S(O)(NH)-Ph-1, designated CTA018 and CTA091, respectively (Posner et al., 2010). These analogs specifically target the substrate binding pocket of CYP24A1. CTA091 significantly increases $1,25D_3$-mediated gene expression and antiproliferative activity in tumor cells and also significantly increases tumor exposure to coadministered $1,25D_3$ in lung tumor xenograft models (Beumer et al., 2012; Zhang et al., 2012).

8. VITAMIN D ANALOGS

In order to induce the nonclassical effects of vitamin D signaling such as pro-differentiating, antiproliferative, and immunomodulatory effects on cancer cells, supra-physiological levels of $1,25D_3$ (nanomolar range) are necessary, while physiological $1,25D_3$ serum concentrations are in the picomolar range. The preclinical data very strongly suggest that high concentrations of $1,25D_3$ are achievable strong antitumor effects in *in vitro* and *in vivo* models (Deeb et al., 2007; Feldman et al., 2014). In addition, serum concentrations of $1,25D_3$ comparable to those associated with distinct antitumor effects can be achieved safely in patients (Beer, Eilers, et al., 2003; Beer, Lemmon, Lowe, & Henner, 2003; Beer et al., 2007; Blanke et al.,

2009; Muindi et al., 2002; Trump et al., 2006). However, high level of vitamin D-produced hypercalcemia is observed in some patients (Cunningham et al., 1985; Koeffler et al., 1985; Rolla et al., 1993). Administration of high dose of $1,25D_3$ even as an intermittent schedule causes transient, but significant hypercalcemia (Smith et al., 1999). To minimize the calcemic side effects while preserving or augmenting the beneficial effects of $1,25D_3$, effort has been made to develop vitamin D analogs with tissue-specific effects and low calcemic side effects.

Large number of vitamin D analogs have been developed during recent years (Leyssens, Verlinden, & Verstuyf, 2014). Some have been approved for clinical treatment of hyperparathyroidism, psoriasis, and osteoporosis. These vitamin D analogs demonstrate several characteristics which are distinct from $1,25D_3$. MART-10 (19-nor-2α-(3-hydroxypropyl)-1α,$25(OH)_2D_3$) is a A-ring-modified $1,25D_3$ analog and exhibits CYP24A1-resistant biological activity (Kittaka et al., 2000; Saito et al., 2004; Suhara et al., 2000). MART-10 shows more potent than $1,25D_3$ in inhibiting cell proliferation in prostate cancer and breast cancer (Chiang et al., 2014, 2012; Flanagan et al., 2009). Seocalcitol (EB1089) and 20-epi-$1,25D_3$ with modifications of the side chain of $1,25D_3$ are degraded more slowly than $1,25D_3$, leading to more prolong exposure of these analogs to the tissues (Hansen & Maenpaa, 1997; Kissmeyer et al., 1997; Shankar et al., 1997; Zella, Meyer, Nerenz, & Pike, 2009). Lexicalcitol (KH1060) is more effective in stabilizing VDR against degradation than $1,25D_3$ (Jaaskelainen, Ryhanen, Mahonen, DeLuca, & Maenpaa, 2000; van den Bemd, Pols, Birkenhager, & van Leeuwen, 1996). The affinity for the DBP also plays a role in the activity of vitamin D analogs. Eldecalcitol has a higher DBP affinity compared to the parent compound, leading to sustaining longer in the circulation and is thus more suitable for the treatment of osteoporosis (Okano et al., 1989). 19-nor-14-epi-23-yne-$1,25(OH)_2 D_3$ (inecalcitol) was 11-fold more potent than $1,25D_3$ in causing 50% clonal growth inhibition of androgen-sensitive human prostate cancer LNCaP cells. Inecalcitol reduced in a dose-dependent manner the expression levels of the transcription factor ETS variant 1 and the serine/threonine protein kinase Pim-1, both of which are upregulated in prostate cancer (Okamoto et al., 2012). Inecalcitol suppressed SCC cell proliferation in a dose-dependent manner with an IC_{50} value 30× lower than that of $1,25D_3$. Both inecalcitol and $1,25D_3$ induced a comparable level of $G0/G_1$ cell cycle arrest in SCC cells. The level of apoptosis induced by inecalcitol was markedly higher than that of $1,25D_3$ (Ma et al., 2013). The exact mechanisms underlying the differences between

$1,25D_3$ and vitamin D analogs in terms of action and capacity need to be elucidated. The different catabolism of vitamin D analogs, interaction with the VDR, coactivators, and VDREs may contribute to super-agonistic actions of vitamin D analogs. For example, inecalcitol triggers stronger VDR-coactivator interactions than $1,25D_3$ (Eelen et al., 2005). Analogs which are more resistant to the catabolic degradation of CYP24A1 will have longer half-life and potentially more bioavailablity than $1,25D_3$. MART-10 is more resistant to the CYP24A1-mediated degradation and more potent than $1,25D_3$ (Chen & Kittaka, 2011; Kittaka et al., 2012).

9. CONCLUSIONS

Despite of promising preclinical data of vitamin D compounds-mediated antitumor action in cancer models, the role of vitamin D compounds in the therapy of cancer is still considerable uncertainty. Frequent alterations of vitamin D signaling and metabolism enzymes, including CYP24A1, and potential dose-limiting toxicity of $1,25D_3$, such as hypercalcemia, limit the usefulness of vitamin D in anticancer therapy. Mechanisms underlying the alterations of vitamin D signaling and metabolism of enzymes in cancer remain largely unexplored. To continue, understanding and exploring the molecular mechanisms of vitamin D and its analogs in cancer, such as cross talk of VDR and other nuclear receptors, is necessary. Development of more potent CYP24A1-resistant vitamin D analogs with less hypercalcemic effect and specific CYP24A1 inhibitors is expected to significantly enhance the antitumor activity of vitamin D compounds. Furthermore, combination of vitamin D and its analogs with standard cancer therapies is probably more effective than vitamin D monotherapy.

REFERENCES

Ahmed, S., Johnson, C. S., Rueger, R. M., & Trump, D. L. (2002). Calcitriol (1,25-dihydroxycholecalciferol) potentiates activity of mitoxantrone/dexamethasone in an androgen independent prostate cancer model. *The Journal of Urology*, *168*, 756–761.

Akeno, N., Matsunuma, A., Maeda, T., Kawane, T., & Horiuchi, N. (2000). Regulation of vitamin D-1alpha-hydroxylase and -24-hydroxylase expression by dexamethasone in mouse kidney. *The Journal of Endocrinology*, *164*, 339–348.

Albertson, D. G., Ylstra, B., Segraves, R., Collins, C., Dairkee, S. H., Kowbel, D., et al. (2000). Quantitative mapping of amplicon structure by array CGH identifies CYP24 as a candidate oncogene. *Nature Genetics*, *25*, 144–146.

Anderson, M. G., Nakane, M., Ruan, X., Kroeger, P. E., & Wu-Wong, J. R. (2006). Expression of VDR and CYP24A1 mRNA in human tumors. *Cancer Chemotherapy and Pharmacology*, *57*, 234–240.

Bagamasbad, P., & Denver, R. J. (2011). Mechanisms and significance of nuclear receptor auto- and cross-regulation. *General and Comparative Endocrinology, 170*, 3–17.

Balazs, Z., Schweizer, R. A., Frey, F. J., Rohner-Jeanrenaud, F., & Odermatt, A. (2008). DHEA induces 11–HSD2 by acting on CCAAT/enhancer-binding proteins. *Journal of the American Society of Nephrology: JASN, 19*, 92–101.

Bareis, P., Bises, G., Bischof, M. G., Cross, H. S., & Peterlik, M. (2001). 25-Hydroxy-vitamin D metabolism in human colon cancer cells during tumor progression. *Biochemical and Biophysical Research Communications, 285*, 1012–1017.

Barletta, F., Freedman, L. P., & Christakos, S. (2002). Enhancement of VDR-mediated transcription by phosphorylation: Correlation with increased interaction between the VDR and DRIP205, a subunit of the VDR-interacting protein coactivator complex. *Molecular Endocrinology, 16*, 301–314.

Beer, T. M., Eilers, K. M., Garzotto, M., Egorin, M. J., Lowe, B. A., & Henner, W. D. (2003). Weekly high-dose calcitriol and docetaxel in metastatic androgen-independent prostate cancer. *Journal of Clinical Oncology: Official Journal of the American Society of Clinical Oncology, 21*, 123–128.

Beer, T. M., Lemmon, D., Lowe, B. A., & Henner, W. D. (2003). High-dose weekly oral calcitriol in patients with a rising PSA after prostatectomy or radiation for prostate carcinoma. *Cancer, 97*, 1217–1224.

Beer, T. M., Ryan, C. W., Venner, P. M., Petrylak, D. P., Chatta, G. S., Ruether, J. D., et al. (2007). Double-blinded randomized study of high-dose calcitriol plus docetaxel compared with placebo plus docetaxel in androgen-independent prostate cancer: A report from the ASCENT investigators. *Journal of Clinical Oncology: Official Journal of the American Society of Clinical Oncology, 25*, 669–674.

Berger, U., McClelland, R. A., Wilson, P., Greene, G. L., Haussler, M. R., Pike, J. W., et al. (1991). Immunocytochemical determination of estrogen receptor, progesterone receptor, and 1,25-dihydroxyvitamin D3 receptor in breast cancer and relationship to prognosis. *Cancer Research, 51*, 239–244.

Bernardi, R. J., Trump, D. L., Yu, W. D., McGuire, T. F., Hershberger, P. A., & Johnson, C. S. (2001). Combination of 1alpha,25-dihydroxyvitamin D(3) with dexamethasone enhances cell cycle arrest and apoptosis: Role of nuclear receptor cross-talk and Erk/Akt signaling. *Clinical Cancer Research: An Official Journal of the American Association for Cancer Research, 7*, 4164–4173.

Beumer, J. H., Parise, R. A., Kanterewicz, B., Petkovich, M., D'Argenio, D. Z., & Hershberger, P. A. (2012). A local effect of CYP24 inhibition on lung tumor xenograft exposure to 1,25-dihydroxyvitamin D(3) is revealed using a novel LC-MS/MS assay. *Steroids, 77*, 477–483.

Bikle, D. D. (2014). The vitamin D receptor: A tumor suppressor in skin. *Advances in Experimental Medicine and Biology, 810*, 282–302.

Blanke, C. D., Beer, T. M., Todd, K., Mori, M., Stone, M., & Lopez, C. (2009). Phase II study of calcitriol-enhanced docetaxel in patients with previously untreated metastatic or locally advanced pancreatic cancer. *Investigational New Drugs, 27*, 374–378.

Borkowski, R., Du, L., Zhao, Z., McMillan, E., Kosti, A., Yang, C. R., et al. (2015). Genetic mutation of p53 and suppression of the miR-17 approximately 92 cluster are synthetic lethal in non-small cell lung cancer due to upregulation of vitamin D signaling. *Cancer Research, 75*, 666–675.

Brozyna, A. A., Jozwicki, W., Janjetovic, Z., & Slominski, A. T. (2011). Expression of vitamin D receptor decreases during progression of pigmented skin lesions. *Human Pathology, 42*, 618–631.

Carlberg, C. (1995). Mechanisms of nuclear signalling by vitamin D3. Interplay with retinoid and thyroid hormone signalling. *European Journal of Biochemistry/FEBS, 231*, 517–527.

Carlberg, C., & Polly, P. (1998). Gene regulation by vitamin D3. *Critical Reviews in Eukaryotic Gene Expression*, *8*, 19–42.

Carre, M., Ayigbede, O., Miravet, L., & Rasmussen, H. (1974). The effect of prednisolone upon the metabolism and action of25-hydroxy-and 1,25-dihydroxyvitamin D3. *Proceedings of the National Academy of Sciences of the United States of America*, *71*, 2996–3000.

Castillo, A. I., Jimenez-Lara, A. M., Tolon, R. M., & Aranda, A. (1999). Synergistic activation of the prolactin promoter by vitamin D receptor and GHF-1: Role of the coactivators, CREB-binding protein and steroid hormone receptor coactivator-1 (SRC-1). *Molecular Endocrinology*, *13*, 1141–1154.

Chadha, M. K., Tian, L., Mashtare, T., Payne, V., Silliman, C., Levine, E., et al. (2010). Phase 2 trial of weekly intravenous 1,25 dihydroxy cholecalciferol (calcitriol) in combination with dexamethasone for castration-resistant prostate cancer. *Cancer*, *116*, 2132–2139.

Chen, T. L., Cone, C. M., Morey-Holton, E., & Feldman, D. (1982). Glucocorticoid regulation of 1,25(OH)2-vitamin D3 receptors in cultured mouse bone cells. *The Journal of Biological Chemistry*, *257*, 13564–13569.

Chen, K. S., & DeLuca, H. F. (1995). Cloning of the human 1 alpha,25-dihydroxyvitamin D-3 24-hydroxylase gene promoter and identification of two vitamin D-responsive elements. *Biochimica et Biophysica Acta*, *1263*, 1–9.

Chen, J., Doroudi, M., Cheung, J., Grozier, A. L., Schwartz, Z., & Boyan, B. D. (2013). Plasma membrane Pdia3 and VDR interact to elicit rapid responses to 1alpha,25(OH)(2)D(3). *Cellular Signalling*, *25*, 2362–2373.

Chen, G., Kim, S. H., King, A. N., Zhao, L., Simpson, R. U., Christensen, P. J., et al. (2011). CYP24A1 is an independent prognostic marker of survival in patients with lung adenocarcinoma. *Clinical Cancer Research: An Official Journal of the American Association for Cancer Research*, *17*, 817–826.

Chen, T. C., & Kittaka, A. (2011). Novel vitamin D analogs for prostate cancer therapy. *ISRN Urology*, *2011*, 301490.

Chen, T. C., Wang, L., Whitlatch, L. W., Flanagan, J. N., & Holick, M. F. (2003). Prostatic 25-hydroxyvitamin D-1alpha-hydroxylase and its implication in prostate cancer. *Journal of Cellular Biochemistry*, *88*, 315–322.

Cheskis, B., & Freedman, L. P. (1996). Modulation of nuclear receptor interactions by ligands: Kinetic analysis using surface plasmon resonance. *Biochemistry*, *35*, 3309–3318.

Chesney, R. W., Mazess, R. B., Hamstra, A. J., DeLuca, H. F., & O'Reagan, S. (1978). Reduction of serum-1, 25-dihydroxyvitamin-D3 in children receiving glucocorticoids. *Lancet*, *2*, 1123–1125.

Chiang, K. C., Chen, S. C., Yeh, C. N., Pang, J. H., Shen, S. C., Hsu, J. T., et al. (2014). MART-10, a less calcemic vitamin D analog, is more potent than 1alpha,25-dihydroxyvitamin D3 in inhibiting the metastatic potential of MCF-7 breast cancer cells in vitro. *The Journal of Steroid Biochemistry and Molecular Biology*, *139*, 54–60.

Chiang, K. C., Yeh, C. N., Chen, S. C., Shen, S. C., Hsu, J. T., Yeh, T. S., et al. (2012). MART-10, a new generation of vitamin D analog, is more potent than 1alpha,25-dihydroxyvitamin D(3) in inhibiting cell proliferation and inducing apoptosis in ER+ MCF-7 breast cancer cells. *Evidence-Based Complementary and Alternative Medicine: eCAM*, *2012*, 310872.

Chung, I., Karpf, A. R., Muindi, J. R., Conroy, J. M., Nowak, N. J., Johnson, C. S., et al. (2007). Epigenetic silencing of CYP24 in tumor-derived endothelial cells contributes to selective growth inhibition by calcitriol. *The Journal of Biological Chemistry*, *282*, 8704–8714.

Collins, C., Rommens, J. M., Kowbel, D., Godfrey, T., Tanner, M., Hwang, S. I., et al. (1998). Positional cloning of ZNF217 and NABC1: Genes amplified at 20q13.2

and overexpressed in breast carcinoma. *Proceedings of the National Academy of Sciences of the United States of America, 95*, 8703–8708.

Colston, K. W., Chander, S. K., Mackay, A. G., & Coombes, R. C. (1992). Effects of synthetic vitamin D analogues on breast cancer cell proliferation in vivo and in vitro. *Biochemical Pharmacology, 44*, 693–702.

Cosman, F., Nieves, J., Herbert, J., Shen, V., & Lindsay, R. (1994). High-dose glucocorticoids in multiple sclerosis patients exert direct effects on the kidney and skeleton. *Journal of Bone and Mineral Research: The Official Journal of the American Society for Bone and Mineral Research, 9*, 1097–1105.

Cross, H. S. (2007). Extrarenal vitamin D hydroxylase expression and activity in normal and malignant cells: Modification of expression by epigenetic mechanisms and dietary substances. *Nutrition Reviews, 65*, S108–S112.

Cui, M., Klopot, A., Jiang, Y., & Fleet, J. C. (2009). The effect of differentiation on 1,25 dihydroxyvitamin D-mediated gene expression in the enterocyte-like cell line, Caco-2. *Journal of Cellular Physiology, 218*, 113–121.

Cui, M., Zhao, Y., Hance, K. W., Shao, A., Wood, R. J., & Fleet, J. C. (2009). Effects of MAPK signaling on 1,25-dihydroxyvitamin D-mediated CYP24 gene expression in the enterocyte-like cell line, Caco-2. *Journal of Cellular Physiology, 219*, 132–142.

Cunningham, D., Gilchrist, N. L., Cowan, R. A., Forrest, G. J., McArdle, C. S., & Soukop, M. (1985). Alfacalcidol as a modulator of growth of low grade non-Hodgkin's lymphomas. *British Medical Journal (Clinical Research Ed.), 291*, 1153–1155.

Cutolo, M., Paolino, S., Sulli, A., Smith, V., Pizzorni, C., & Seriolo, B. (2014). Vitamin D, steroid hormones, and autoimmunity. *Annals of the New York Academy of Sciences, 1317*, 39–46.

de Boland, A. R., Morelli, S., & Boland, R. (1994). 1,25(OH)2-vitamin D3 signal transduction in chick myoblasts involves phosphatidylcholine hydrolysis. *The Journal of Biological Chemistry, 269*, 8675–8679.

de Boland, A. R., & Norman, A. (1990). Evidence for involvement of protein kinase C and cyclic adenosine 3′,5′ monophosphate-dependent protein kinase in the 1,25-dihydroxyvitamin D3-mediated rapid stimulation of intestinal calcium transport, (transcaltachia). *Endocrinology, 127*, 39–45.

Deeb, K. K., Luo, W., Karpf, A. R., Omilian, A. R., Bshara, W., Tian, L., et al. (2011). Differential vitamin D 24-hydroxylase/CYP24A1 gene promoter methylation in endothelium from benign and malignant human prostate. *Epigenetics: Official Journal of the DNA Methylation Society, 6*, 994–1000.

Deeb, K. K., Trump, D. L., & Johnson, C. S. (2007). Vitamin D signalling pathways in cancer: Potential for anticancer therapeutics. *Nature Reviews. Cancer, 7*, 684–700.

Delany, A. M., Durant, D., & Canalis, E. (2001). Glucocorticoid suppression of IGF I transcription in osteoblasts. *Molecular Endocrinology, 15*, 1781–1789.

Dhawan, P., & Christakos, S. (2010). Novel regulation of 25-hydroxyvitamin D3 24-hydroxylase (24(OH)ase) transcription by glucocorticoids: Cooperative effects of the glucocorticoid receptor, C/EBP beta, and the Vitamin D receptor in 24(OH)ase transcription. *Journal of Cellular Biochemistry, 110*, 1314–1323.

Dhawan, P., Peng, X., Sutton, A. L., MacDonald, P. N., Croniger, C. M., Trautwein, C., et al. (2005). Functional cooperation between CCAAT/enhancer-binding proteins and the vitamin D receptor in regulation of 25-hydroxyvitamin D3 24-hydroxylase. *Molecular and Cellular Biology, 25*, 472–487.

Ditsch, N., Toth, B., Mayr, D., Lenhard, M., Gallwas, J., Weissenbacher, T., et al. (2012). The association between vitamin D receptor expression and prolonged overall survival in breast cancer. *The Journal of Histochemistry and Cytochemistry: Official Journal of the Histochemistry Society, 60*, 121–129.

Doroudi, M., Chen, J., Boyan, B. D., & Schwartz, Z. (2014). New insights on membrane mediated effects of 1alpha,25-dihydroxy vitamin D3 signaling in the musculoskeletal system. *Steroids*, *81*, 81–87.

Doroudi, M., Schwartz, Z., & Boyan, B. D. (2015). Membrane-mediated actions of 1,25-dihydroxy vitamin D3: A review of the roles of phospholipase A2 activating protein and Ca(2+)/calmodulin-dependent protein kinase II. *The Journal of Steroid Biochemistry and Molecular Biology*, *147*, 81–84.

Dwivedi, P. P., Hii, C. S., Ferrante, A., Tan, J., Der, C. J., Omdahl, J. L., et al. (2002). Role of MAP kinases in the 1,25-dihydroxyvitamin D3-induced transactivation of the rat cytochrome P450C24 (CYP24) promoter. Specific functions for ERK1/ERK2 and ERK5. *The Journal of Biological Chemistry*, *277*, 29643–29653.

Eelen, G., Verlinden, L., Rochel, N., Claessens, F., De Clercq, P., Vandewalle, M., et al. (2005). Superagonistic action of 14-epi-analogs of 1,25-dihydroxyvitamin D explained by vitamin D receptor-coactivator interaction. *Molecular Pharmacology*, *67*, 1566–1573.

Eisman, J. A., Barkla, D. H., & Tutton, P. J. (1987). Suppression of in vivo growth of human cancer solid tumor xenografts by 1,25-dihydroxyvitamin D3. *Cancer Research*, *47*, 21–25.

Endres, B., Kato, S., & DeLuca, H. F. (2000). Metabolism of 1alpha,25-dihydroxyvitamin D(3) in vitamin D receptor-ablated mice in vivo. *Biochemistry*, *39*, 2123–2129.

Escaleira, M. T., & Brentani, M. M. (1999). Vitamin D3 receptor (VDR) expression in HC-11 mammary cells: Regulation by growth-modulatory agents, differentiation, and Ha-ras transformation. *Breast Cancer Research and Treatment*, *54*, 123–133.

Essa, S., Denzer, N., Mahlknecht, U., Klein, R., Collnot, E. M., Tilgen, W., et al. (2010). VDR microRNA expression and epigenetic silencing of vitamin D signaling in melanoma cells. *The Journal of Steroid Biochemistry and Molecular Biology*, *121*, 110–113.

Farhan, H., & Cross, H. S. (2002). Transcriptional inhibition of CYP24 by genistein. *Annals of the New York Academy of Sciences*, *973*, 459–462.

Feher, J. J., & Wasserman, R. H. (1979). Intestinal calcium-binding protein and calcium absorption in cortisol-treated chicks: Effects of vitamin D3 and 1,25-dihydroxyvitamin D3. *Endocrinology*, *104*, 547–551.

Feldman, D., Krishnan, A. V., Swami, S., Giovannucci, E., & Feldman, B. J. (2014). The role of vitamin D in reducing cancer risk and progression. *Nature Reviews. Cancer*, *14*, 342–357.

Flaig, T. W., Barqawi, A., Miller, G., Kane, M., Zeng, C., Crawford, E. D., et al. (2006). A phase II trial of dexamethasone, vitamin D, and carboplatin in patients with hormone-refractory prostate cancer. *Cancer*, *107*, 266–274.

Flanagan, J. N., Zheng, S., Chiang, K. C., Kittaka, A., Sakaki, T., Nakabayashi, S., et al. (2009). Evaluation of 19-nor-2alpha-(3-hydroxypropyl)-1alpha,25-dihydroxyvitamin D3 as a therapeutic agent for androgen-dependent prostate cancer. *Anticancer Research*, *29*, 3547–3553.

Flynn, G., Chung, I., Yu, W. D., Romano, M., Modzelewski, R. A., Johnson, C. S., et al. (2006). Calcitriol (1,25-dihydroxycholecalciferol) selectively inhibits proliferation of freshly isolated tumor-derived endothelial cells and induces apoptosis. *Oncology*, *70*, 447–457.

Friedrich, M., Rafi, L., Mitschele, T., Tilgen, W., Schmidt, W., & Reichrath, J. (2003). Analysis of the vitamin D system in cervical carcinomas, breast cancer and ovarian cancer. *Recent Results in Cancer Research*, *164*, 239–246.

Gapany, M., Faust, R. A., Tawfic, S., Davis, A., Adams, G. L., & Ahmed, K. (1995). Association of elevated protein kinase CK2 activity with aggressive behavior of squamous cell carcinoma of the head and neck. *Molecular Medicine*, *1*, 659–666.

Getzenberg, R. H., Light, B. W., Lapco, P. E., Konety, B. R., Nangia, A. K., Acierno, J. S., et al. (1997). Vitamin D inhibition of prostate adenocarcinoma growth and metastasis in the Dunning rat prostate model system. *Urology, 50*, 999–1006.

Gniadecki, R. (1996). Activation of Raf-mitogen-activated protein kinase signaling pathway by 1,25-dihydroxyvitamin D3 in normal human keratinocytes. *The Journal of Investigative Dermatology, 106*, 1212–1217.

Godman, C. A., Joshi, R., Tierney, B. R., Greenspan, E., Rasmussen, T. P., Wang, H. W., et al. (2008). HDAC3 impacts multiple oncogenic pathways in colon cancer cells with effects on Wnt and vitamin D signaling. *Cancer Biology & Therapy, 7*, 1570–1580.

Gross, C., Stamey, T., Hancock, S., & Feldman, D. (1998). Treatment of early recurrent prostate cancer with 1,25-dihydroxyvitamin D3 (calcitriol). *The Journal of Urology, 159*, 2035–2039 (discussion 9–40).

Gutierrez, S., Javed, A., Tennant, D. K., van Rees, M., Montecino, M., Stein, G. S., et al. (2002). CCAAT/enhancer-binding proteins (C/EBP) beta and delta activate osteocalcin gene transcription and synergize with Runx2 at the C/EBP element to regulate bone-specific expression. *The Journal of Biological Chemistry, 277*, 1316–1323.

Hahn, C. N., Kerry, D. M., Omdahl, J. L., & May, B. K. (1994). Identification of a vitamin D responsive element in the promoter of the rat cytochrome P450(24) gene. *Nucleic Acids Research, 22*, 2410–2416.

Hansen, C. M., & Maenpaa, P. H. (1997). EB 1089, a novel vitamin D analog with strong antiproliferative and differentiation-inducing effects on target cells. *Biochemical Pharmacology, 54*, 1173–1179.

Hardin, J., MacLeod, S., Grigorieva, I., Chang, R., Barlogie, B., Xiao, H., et al. (1994). Interleukin-6 prevents dexamethasone-induced myeloma cell death. *Blood, 84*, 3063–3070.

Haussler, M. R., Whitfield, G. K., Haussler, C. A., Hsieh, J. C., Thompson, P. D., Selznick, S. H., et al. (1998). The nuclear vitamin D receptor: Biological and molecular regulatory properties revealed. *Journal of Bone and Mineral Research: The Official Journal of the American Society for Bone and Mineral Research, 13*, 325–349.

Hendrickson, W. K., Flavin, R., Kasperzyk, J. L., Fiorentino, M., Fang, F., Lis, R., et al. (2011). Vitamin D receptor protein expression in tumor tissue and prostate cancer progression. *Journal of Clinical Oncology: Official Journal of the American Society of Clinical Oncology, 29*, 2378–2385.

Herdick, M., & Carlberg, C. (2000). Agonist-triggered modulation of the activated and silent state of the vitamin D(3) receptor by interaction with co-repressors and co-activators. *Journal of Molecular Biology, 304*, 793–801.

Herr, I., & Pfitzenmaier, J. (2006). Glucocorticoid use in prostate cancer and other solid tumours: Implications for effectiveness of cytotoxic treatment and metastases. *The Lancet Oncology, 7*, 425–430.

Hershberger, P. A., McGuire, T. F., Yu, W. D., Zuhowski, E. G., Schellens, J. H., Egorin, M. J., et al. (2002). Cisplatin potentiates 1,25-dihydroxyvitamin D3-induced apoptosis in association with increased mitogen-activated protein kinase kinase kinase 1 (MEKK-1) expression. *Molecular Cancer Therapeutics, 1*, 821–829.

Hershberger, P. A., Modzelewski, R. A., Shurin, Z. R., Rueger, R. M., Trump, D. L., & Johnson, C. S. (1999). 1,25-Dihydroxycholecalciferol (1,25-D3) inhibits the growth of squamous cell carcinoma and down-modulates p21(Waf1/Cip1) in vitro and in vivo. *Cancer Research, 59*, 2644–2649.

Hershberger, P. A., Yu, W. D., Modzelewski, R. A., Rueger, R. M., Johnson, C. S., & Trump, D. L. (2001). Calcitriol (1,25-dihydroxycholecalciferol) enhances paclitaxel antitumor activity in vitro and in vivo and accelerates paclitaxel-induced apoptosis. *Clinical Cancer Research: An Official Journal of the American Association for Cancer Research, 7*, 1043–1051.

Hidaka, S., Yasutake, T., Takeshita, H., Kondo, M., Tsuji, T., Nanashima, A., et al. (2000). Differences in 20q13.2 copy number between colorectal cancers with and without liver metastasis. *Clinical Cancer Research: An Official Journal of the American Association for Cancer Research, 6*, 2712–2717.

Hidalgo, A. A., Deeb, K. K., Pike, J. W., Johnson, C. S., & Trump, D. L. (2011). Dexamethasone enhances 1alpha,25-dihydroxyvitamin D3 effects by increasing vitamin D receptor transcription. *The Journal of Biological Chemistry, 286*, 36228–36237.

Hidalgo, A. A., Trump, D. L., & Johnson, C. S. (2010). Glucocorticoid regulation of the vitamin D receptor. *The Journal of Steroid Biochemistry and Molecular Biology, 121*, 372–375.

Huybers, S., Naber, T. H., Bindels, R. J., & Hoenderop, J. G. (2007). Prednisolone-induced Ca2+ malabsorption is caused by diminished expression of the epithelial Ca2+ channel TRPV6. *American Journal of Physiology. Gastrointestinal and Liver Physiology, 292*, G92–G97.

Inaba, H., Cao, X., Pounds, S., Pui, C. H., Rubnitz, J. E., Ribeiro, R. C., et al. (2011). Randomized trial of 2 dosages of prophylactic granulocyte-colony-stimulating factor after induction chemotherapy in pediatric acute myeloid leukemia. *Cancer, 117*, 1313–1320.

Iwata, K., Yamamoto, A., Satoh, S., Ohyama, Y., Tashiro, Y., & Setoguchi, T. (1995). Quantitative immunoelectron microscopic analysis of the localization and induction of 25-hydroxyvitamin D3 24-hydroxylase in rat kidney. *The Journal of Histochemistry and Cytochemistry: Official Journal of the Histochemistry Society, 43*, 255–262.

Jaaskelainen, T., Ryhanen, S., & Maenpaa, P. H. (2003). 9-cis retinoic acid accelerates calcitriol-induced osteocalcin production and promotes degradation ofboth vitamin D receptor and retinoid X receptor in human osteoblastic cells. *Journal of Cellular Biochemistry, 89*, 1164–1176.

Jaaskelainen, T., Ryhanen, S., Mahonen, A., DeLuca, H. F., & Maenpaa, P. H. (2000). Mechanism of action of superactive vitamin D analogs through regulated receptor degradation. *Journal of Cellular Biochemistry, 76*, 548–558.

Jakubowiak, A. J., Kendall, T., Al-Zoubi, A., Khaled, Y., Mineishi, S., Ahmed, A., et al. (2009). Phase II trial of combination therapy with bortezomib, pegylated liposomal doxorubicin, and dexamethasone in patients with newly diagnosed myeloma. *Journal of Clinical Oncology: Official Journal of the American Society of Clinical Oncology, 27*, 5015–5022.

Jehan, F., & DeLuca, H. F. (2000). The mouse vitamin D receptor is mainly expressed through an Sp1-driven promoter in vivo. *Archives of Biochemistry and Biophysics, 377*, 273–283.

Jirapongsananuruk, O., Melamed, I., & Leung, D. Y. (2000). Additive immunosuppressive effects of 1,25-dihydroxyvitamin D3 and corticosteroids on TH1, but not TH2, responses. *The Journal of Allergy and Clinical Immunology, 106*, 981–985.

Jones, G., Prosser, D. E., & Kaufmann, M. (2012). 25-Hydroxyvitamin D-24-hydroxylase (CYP24A1): Its important role in the degradation of vitamin D. *Archives of Biochemistry and Biophysics, 523*, 9–18.

Kahlen, J. P., & Carlberg, C. (1994). Identification of a vitamin D receptor homodimer-type response element in the rat calcitriol 24-hydroxylase gene promoter. *Biochemical and Biophysical Research Communications, 202*, 1366–1372.

Kallay, E., Pietschmann, P., Toyokuni, S., Bajna, E., Hahn, P., Mazzucco, K., et al. (2001). Characterization of a vitamin D receptor knockout mouse as a model of colorectal hyperproliferation and DNA damage. *Carcinogenesis, 22*, 1429–1435.

Kallioniemi, A., Kallioniemi, O. P., Piper, J., Tanner, M., Stokke, T., Chen, L., et al. (1994). Detection and mapping of amplified DNA sequences in breast cancer by comparative genomic hybridization. *Proceedings of the National Academy of Sciences of the United States of America, 91*, 2156–2160.

Kang, S., Tsai, L. T., Zhou, Y., Evertts, A., Xu, S., Griffin, M. J., et al. (2015). Identification of nuclear hormone receptor pathways causing insulin resistance by transcriptional and epigenomic analysis. *Nature Cell Biology*, *17*, 44–56.

Kassel, O., & Herrlich, P. (2007). Crosstalk between the glucocorticoid receptor and other transcription factors: Molecular aspects. *Molecular and Cellular Endocrinology*, *275*, 13–29.

Kerley, C. P., Elnazir, B., Faul, J., & Cormican, L. (2015). Vitamin D as an adjunctive therapy in asthma. Part 1: A review of potential mechanisms. *Pulmonary Pharmacology & Therapeutics*, *32*, 60–74.

Khare, S., Tien, X. Y., Wilson, D., Wali, R. K., Bissonnette, B. M., Scaglione-Sewell, B., et al. (1994). The role of protein kinase-C alpha in the activation of particulate guanylate cyclase by 1 alpha,25-dihydroxyvitamin D3 in CaCo-2 cells. *Endocrinology*, *135*, 277–283.

Khorchide, M., Lechner, D., & Cross, H. S. (2005). Epigenetic regulation of vitamin D hydroxylase expression and activity in normal and malignant human prostate cells. *The Journal of Steroid Biochemistry and Molecular Biology*, *93*, 167–172.

Kim, M. H., Lee, G. S., Jung, E. M., Choi, K. C., & Jeung, E. B. (2009). The negative effect of dexamethasone on calcium-processing gene expressions is associated with a glucocorticoid-induced calcium-absorbing disorder. *Life Sciences*, *85*, 146–152.

Kissmeyer, A. M., Binderup, E., Binderup, L., Mork Hansen, C., Andersen, N. R., Makin, H. L., et al. (1997). Metabolism of the vitamin D analog EB 1089: Identification of in vivo and in vitro liver metabolites and their biological activities. *Biochemical Pharmacology*, *53*, 1087–1097.

Kittaka, A., Suhara, Y., Takayanagi, H., Fujishima, T., Kurihara, M., & Takayama, H. (2000). A concise and efficient route to 2alpha-(omega-hydroxyalkoxy)-1alpha,25-dihydroxyvi tam in D3: Remarkably high affinity to vitamin D receptor. *Organic Letters*, *2*, 2619–2622.

Kittaka, A., Yoshida, A., Chiang, K. C., Takano, M., Sawada, D., Sakaki, T., et al. (2012). Potent 19-norvitamin D analogs for prostate and liver cancer therapy. *Future Medicinal Chemistry*, *4*, 2049–2065.

Klein, R. G., Arnaud, S. B., Gallagher, J. C., Deluca, H. F., & Riggs, B. L. (1977). Intestinal calcium absorption in exogenous hypercortisonism. Role of 25-hydroxyvitamin D and corticosteroid dose. *The Journal of Clinical Investigation*, *60*, 253–259.

Koeffler, H. P., Hirji, K., & Itri, L. (1985). 1,25-Dihydroxyvitamin D3: In vivo and in vitro effects on human preleukemic and leukemic cells. *Cancer Treatment Reports*, *69*, 1399–1407.

Komagata, S., Nakajima, M., Takagi, S., Mohri, T., Taniya, T., & Yokoi, T. (2009). Human CYP24 catalyzing the inactivation of calcitriol is post-transcriptionally regulated by miR-125b. *Molecular Pharmacology*, *76*, 702–709.

Kondo, Y. (2014). Targeting histone methyltransferase EZH2 as cancer treatment. *Journal of Biochemistry*, *156*, 249–257.

Kovacs, C. S., & Kronenberg, H. M. (1997). Maternal-fetal calcium and bone metabolism during pregnancy, puerperium, and lactation. *Endocrine Reviews*, *18*, 832–872.

Krishnan, A. V., Peehl, D. M., & Feldman, D. (2003). Inhibition of prostate cancer growth by vitamin D: Regulation of target gene expression. *Journal of Cellular Biochemistry*, *88*, 363–371.

Kupfer, S. S., Maranville, J. C., Baxter, S. S., Huang, Y., & Di Rienzo, A. (2013). Comparison of cellular and transcriptional responses to 1,25-dihydroxyvitamin D3 and glucocorticoids in peripheral blood mononuclear cells. *PloS One*, *8*, e76643.

Kurahashi, I., Matsunuma, A., Kawane, T., Abe, M., & Horiuchi, N. (2002). Dexamethasone enhances vitamin D-24-hydroxylase expression in osteoblastic (UMR-106) and renal (LLC-PK1) cells treated with 1alpha,25-dihydroxyvitamin D3. *Endocrine*, *17*, 109–118.

Kure, S., Nosho, K., Baba, Y., Irahara, N., Shima, K., Ng, K., et al. (2009). Vitamin D receptor expression is associated with PIK3CA and KRAS mutations in colorectal cancer. *Cancer Epidemiology, Biomarkers & Prevention: A Publication of the American Association for Cancer Research, Cosponsored by the American Society of Preventive Oncology, 18*, 2765–2772.

Landesman-Bollag, E., Romieu-Mourez, R., Song, D. H., Sonenshein, G. E., Cardiff, R. D., & Seldin, D. C. (2001). Protein kinase CK2 in mammary gland tumorigenesis. *Oncogene, 20*, 3247–3257.

Laramas, M., Pasquier, D., Filhol, O., Ringeisen, F., Descotes, J. L., & Cochet, C. (2007). Nuclear localization of protein kinase CK2 catalytic subunit (CK2alpha) is associated with poor prognostic factors in human prostate cancer. *European Journal of Cancer, 43*, 928–934.

Larriba, M. J., Martin-Villar, E., Garcia, J. M., Pereira, F., Pena, C., de Herreros, A. G., et al. (2009). Snail2 cooperates with Snail1 in the repression of vitamin D receptor in colon cancer. *Carcinogenesis, 30*, 1459–1468.

Larriba, M. J., & Munoz, A. (2005). SNAIL vs vitamin D receptor expression in colon cancer: Therapeutics implications. *British Journal of Cancer, 92*, 985–989.

Larriba, M. J., Valle, N., Palmer, H. G., Ordonez-Moran, P., Alvarez-Diaz, S., Becker, K. F., et al. (2007). The inhibition of Wnt/beta-catenin signalling by 1alpha,25-dihydroxyvitamin D3 is abrogated by Snail1 in human colon cancer cells. *Endocrine-Related Cancer, 14*, 141–151.

Lee, G. S., Choi, K. C., & Jeung, E. B. (2006). Glucocorticoids differentially regulate expression of duodenal and renal calbindin-D9k through glucocorticoid receptor-mediated pathway in mouse model. *American Journal of Physiology. Endocrinology and Metabolism, 290*, E299–E307.

Lemon, B. D., Fondell, J. D., & Freedman, L. P. (1997). Retinoid X receptor:vitamin D3 receptor heterodimers promote stable preinitiation complex formation and direct 1,25-dihydroxyvitamin D3-dependent cell-free transcription. *Molecular and Cellular Biology, 17*, 1923–1937.

Leo, C., & Chen, J. D. (2000). The SRC family of nuclear receptor coactivators. *Gene, 245*, 1–11.

Leyssens, C., Verlinden, L., & Verstuyf, A. (2014). The future of vitamin D analogs. *Frontiers in Physiology, 5*, 122.

Light, B. W., Yu, W. D., McElwain, M. C., Russell, D. M., Trump, D. L., & Johnson, C. S. (1997). Potentiation of cisplatin antitumor activity using a vitamin D analogue in a murine squamous cell carcinoma model system. *Cancer Research, 57*, 3759–3764.

Lin, Y. W., Ren, L. L., Xiong, H., Du, W., Yu, Y. N., Sun, T. T., et al. (2013). Role of STAT3 and vitamin D receptor in EZH2-mediated invasion of human colorectal cancer. *The Journal of Pathology, 230*, 277–290.

Litonjua, A. A. (2013). Vitamin D, and corticosteroids in asthma: Synergy, interaction and potential therapeutic effects. *Expert Review of Respiratory Medicine*, 7, 101–104.

Lopes, N., Sousa, B., Martins, D., Gomes, M., Vieira, D., Veronese, L. A., et al. (2010). Alterations in vitamin D signalling and metabolic pathways in breast cancer progression: A study of VDR, CYP27B1 and CYP24A1 expression in benign and malignant breast lesions. *BMC Cancer, 10*, 483.

Lu, N. Z., Wardell, S. E., Burnstein, K. L., Defranco, D., Fuller, P. J., Giguere, V., et al. (2006). International Union of Pharmacology. LXV. The pharmacology and classification of the nuclear receptor superfamily: Glucocorticoid, mineralocorticoid, progesterone, and androgen receptors. *Pharmacological Reviews, 58*, 782–797.

Luo, W., Karpf, A. R., Deeb, K. K., Muindi, J. R., Morrison, C. D., Johnson, C. S., et al. (2010). Epigenetic regulation of vitamin D 24-hydroxylase/CYP24A1 in human prostate cancer. *Cancer Research, 70*, 5953–5962.

Luo, W., Yu, W. D., Ma, Y., Chernov, M., Trump, D. L., & Johnson, C. S. (2013). Inhibition of protein kinase CK2 reduces Cyp24a1 expression and enhances 1,25-dihydroxyvitamin D(3) antitumor activity in human prostate cancer cells. *Cancer Research*, *73*, 2289–2297.

Ly, L. H., Zhao, X. Y., Holloway, L., & Feldman, D. (1999). Liarozole acts synergistically with 1alpha,25-dihydroxyvitamin D3 to inhibit growth of DU 145 human prostate cancer cells by blocking 24-hydroxylase activity. *Endocrinology*, *140*, 2071–2076.

Ma, Y., Yu, W. D., Hidalgo, A. A., Luo, W., Delansorne, R., Johnson, C. S., et al. (2013). Inecalcitol, an analog of 1,25D3, displays enhanced antitumor activity through the induction of apoptosis in a squamous cell carcinoma model system. *Cell Cycle*, *12*, 743–752.

Ma, Y., Yu, W. D., Trump, D. L., & Johnson, C. S. (2010). 1,25D3 enhances antitumor activity of gemcitabine and cisplatin in human bladder cancer models. *Cancer*, *116*, 3294–3303.

Maalouf, D., Battistella, M., & Bouaziz, J. D. (2015). Neutrophilic dermatosis: Disease mechanism and treatment. *Current Opinion in Hematology*, *22*, 23–29.

MacDonald, P. N., Baudino, T. A., Tokumaru, H., Dowd, D. R., & Zhang, C. (2001). Vitamin D receptor and nuclear receptor coactivators: Crucial interactions in vitamin D-mediated transcription. *Steroids*, *66*, 171–176.

Maier, C. J., Maier, R. H., Rid, R., Trost, A., Hundsberger, H., Eger, A., et al. (2012). PIM-1 kinase interacts with the DNA binding domain of the vitamin D receptor: A further kinase implicated in 1,25-(OH)2D3 signaling. *BMC Molecular Biology*, *13*, 18.

Makishima, M., Lu, T. T., Xie, W., Whitfield, G. K., Domoto, H., Evans, R. M., et al. (2002). Vitamin D receptor as an intestinal bile acid sensor. *Science*, *296*, 1313–1316.

Malnick, S., Duek, G., Melzer, E., & Basevitz, A. (2012). The treatment of autoimmune hepatitis. *Current Clinical Pharmacology*, 7, 318–327.

Mangelsdorf, D. J., Koeffler, H. P., Donaldson, C. A., Pike, J. W., & Haussler, M. R. (1984). 1,25-Dihydroxyvitamin D3-induced differentiation in a human promyelocytic leukemia cell line (HL-60): Receptor-mediated maturation to macrophage-like cells. *The Journal of Cell Biology*, *98*, 391–398.

Mangelsdorf, D. J., Thummel, C., Beato, M., Herrlich, P., Schutz, G., Umesono, K., et al. (1995). The nuclear receptor superfamily: The second decade. *Cell*, *83*, 835–839.

Marik, R., Fackler, M., Gabrielson, E., Zeiger, M. A., Sukumar, S., Stearns, V., et al. (2010). DNA methylation-related vitamin D receptor insensitivity in breast cancer. *Cancer Biology & Therapy*, *10*, 44–53.

Maruyama, R., Aoki, F., Toyota, M., Sasaki, Y., Akashi, H., Mita, H., et al. (2006). Comparative genome analysis identifies the vitamin D receptor gene as a direct target of p53-mediated transcriptional activation. *Cancer Research*, *66*, 4574–4583.

Matilainen, J. M., Malinen, M., Turunen, M. M., Carlberg, C., & Vaisanen, S. (2010). The number of vitamin D receptor binding sites defines the different vitamin D responsiveness of the CYP24 gene in malignant and normal mammary cells. *The Journal of Biological Chemistry*, *285*, 24174–24183.

Matsunawa, M., Amano, Y., Endo, K., Uno, S., Sakaki, T., Yamada, S., et al. (2009). The aryl hydrocarbon receptor activator benzo[a]pyrene enhances vitamin D3 catabolism in macrophages. *Toxicological Sciences: An Official Journal of the Society of Toxicology*, *109*, 50–58.

Matusiak, D., & Benya, R. V. (2007). CYP27A1 and CYP24 expression as a function of malignant transformation in the colon. *The Journal of Histochemistry and Cytochemistry: Official Journal of the Histochemistry Society*, *55*, 1257–1264.

Matusiak, D., Murillo, G., Carroll, R. E., Mehta, R. G., & Benya, R. V. (2005). Expression of vitamin D receptor and 25-hydroxyvitamin D3-1{alpha}-hydroxylase in normal and malignant human colon. *Cancer Epidemiology, Biomarkers & Prevention: A Publication of the*

American Association for Cancer Research, Cosponsored by the American Society of Preventive Oncology, 14, 2370–2376.

McCarthy, T. L., Ji, C., Chen, Y., Kim, K., & Centrella, M. (2000). Time- and dose-related interactions between glucocorticoid and cyclic adenosine 3′,5′-monophosphate on CCAAT/enhancer-binding protein-dependent insulin-like growth factor I expression by osteoblasts. *Endocrinology, 141*, 127–137.

McElwain, M. C., Modzelewski, R. A., Yu, W. D., Russell, D. M., & Johnson, C. S. (1997). Vitamin D: An antiproliferative agent with potential for therapy of squamous cell carcinoma. *American Journal of Otolaryngology, 18*, 293–298.

McGuire, T. F., Trump, D. L., & Johnson, C. S. (2001). Vitamin D(3)-induced apoptosis of murine squamous cell carcinoma cells. Selective induction of caspase-dependent MEK cleavage and up-regulation of MEKK-1. *The Journal of Biological Chemistry, 276*, 26365–26373.

Merkulov, V. M., & Merkulova, T. I. (2009). Structural variants of glucocorticoid receptor binding sites and different versions of positive glucocorticoid responsive elements: Analysis of GR-TRRD database. *The Journal of Steroid Biochemistry and Molecular Biology, 115*, 1–8.

Meyer, M. B., Zella, L. A., Nerenz, R. D., & Pike, J. W. (2007). Characterizing early events associated with the activation of target genes by 1,25-dihydroxyvitamin D3 in mouse kidney and intestine in vivo. *The Journal of Biological Chemistry, 282*, 22344–22352.

Miller, G. J., Stapleton, G. E., Hedlund, T. E., & Moffat, K. A. (1995). Vitamin D receptor expression, 24-hydroxylase activity, and inhibition of growth by 1alpha,25-dihydroxyvitamin D3 in seven human prostatic carcinoma cell lines. *Clinical Cancer Research: An Official Journal of the American Association for Cancer Research, 1*, 997–1003.

Mimori, K., Tanaka, Y., Yoshinaga, K., Masuda, T., Yamashita, K., Okamoto, M., et al. (2004). Clinical significance of the overexpression of the candidate oncogene CYP24 in esophageal cancer. *Annals of Oncology: Official Journal of the European Society for Medical Oncology/ESMO, 15*, 236–241.

Mitschele, T., Diesel, B., Friedrich, M., Meineke, V., Maas, R. M., Gartner, B. C., et al. (2004). Analysis of the vitamin D system in basal cell carcinomas (BCCs). *Laboratory Investigation, 84*, 693–702.

Mittal, M. K., Myers, J. N., Misra, S., Bailey, C. K., & Chaudhuri, G. (2008). In vivo binding to and functional repression of the VDR gene promoter by SLUG in human breast cells. *Biochemical and Biophysical Research Communications, 372*, 30–34.

Mohri, T., Nakajima, M., Takagi, S., Komagata, S., & Yokoi, T. (2009). MicroRNA regulates human vitamin D receptor. *International Journal of Cancer, 125*, 1328–1333.

Morris, K. L., & Zemel, M. B. (2005). 1,25-Dihydroxyvitamin D3 modulation of adipocyte glucocorticoid function. *Obesity Research, 13*, 670–677.

Muindi, J. R., Peng, Y., Potter, D. M., Hershberger, P. A., Tauch, J. S., Capozzoli, M. J., et al. (2002). Pharmacokinetics of high-dose oral calcitriol: Results from a phase 1 trial of calcitriol and paclitaxel. *Clinical Pharmacology and Therapeutics, 72*, 648–659.

Muindi, J. R., Yu, W. D., Ma, Y., Engler, K. L., Kong, R. X., Trump, D. L., et al. (2010). CYP24A1 inhibition enhances the antitumor activity of calcitriol. *Endocrinology, 151*, 4301–4312.

Nemere, I., Farach-Carson, M. C., Rohe, B., Sterling, T. M., Norman, A. W., Boyan, B. D., et al. (2004). Ribozyme knockdown functionally links a 1,25(OH)2D3 membrane binding protein (1,25D3-MARRS) and phosphate uptake in intestinal cells. *Proceedings of the National Academy of Sciences of the United States of America, 101*, 7392–7397.

Norman, A. W. (2006). Minireview: Vitamin D receptor: New assignments for an already busy receptor. *Endocrinology, 147*, 5542–5548.

Norman, A. W., Nemere, I., Zhou, L. X., Bishop, J. E., Lowe, K. E., Maiyar, A. C., et al. (1992). 1,25(OH)2-vitamin D3, a steroid hormone that produces biologic effects via both genomic and nongenomic pathways. *The Journal of Steroid Biochemistry and Molecular Biology*, *41*, 231–240.

Norman, A. W., Song, X., Zanello, L., Bula, C., & Okamura, W. H. (1999). Rapid and genomic biological responses are mediated by different shapes of the agonist steroid hormone, 1alpha,25(OH)2 vitamin D3. *Steroids*, *64*, 120–128.

Novakovic, B., Sibson, M., Ng, H. K., Manuelpillai, U., Rakyan, V., Down, T., et al. (2009). Placenta-specific methylation of the vitamin D 24-hydroxylase gene: Implications for feedback autoregulation of active vitamin D levels at the fetomaternal interface. *The Journal of Biological Chemistry*, *284*, 14838–14848.

Nutchey, B. K., Kaplan, J. S., Dwivedi, P. P., Omdahl, J. L., Ferrante, A., May, B. K., et al. (2005). Molecular action of 1,25-dihydroxyvitamin D3 and phorbol ester on the activation of the rat cytochrome P450C24 (CYP24) promoter: Role of MAP kinase activities and identification of an important transcription factor binding site. *The Biochemical Journal*, *389*, 753–762.

Oc, P., Rusch, V., Talbot, S. G., Sarkaria, I., Viale, A., Socci, N., et al. (2004). Casein kinase II alpha subunit and C1-inhibitor are independent predictors of outcome in patients with squamous cell carcinoma of the lung. *Clinical Cancer Research: An Official Journal of the American Association for Cancer Research*, *10*, 5792–5803.

Okamoto, R., Delansorne, R., Wakimoto, N., Doan, N. B., Akagi, T., Shen, M., et al. (2012). Inecalcitol, an analog of 1alpha,25(OH)(2) D(3), induces growth arrest of androgen-dependent prostate cancer cells. *International Journal of Cancer*, *130*, 2464–2473.

Okano, T., Tsugawa, N., Masuda, S., Takeuchi, A., Kobayashi, T., Takita, Y., et al. (1989). Regulatory activities of 2 beta-(3-hydroxypropoxy)-1 alpha, 25-dihydroxyvitamin D3, a novel synthetic vitamin D3 derivative, on calcium metabolism. *Biochemical and Biophysical Research Communications*, *163*, 1444–1449.

Omdahl, J. L., Bobrovnikova, E. A., Choe, S., Dwivedi, P. P., & May, B. K. (2001). Overview of regulatory cytochrome P450 enzymes of the vitamin D pathway. *Steroids*, *66*, 381–389.

Omdahl, J., & Brian, M. (2005). The 25-hydroxyvitamin D 24-hydroxylase. In D. Feldman, W. J. Pike, & F. Glorieux (Eds.), *Vitamin D* (2nd ed., pp. 85–104). San Diego, CA: Elsevier.

Omdahl, J. L., Morris, H. A., & May, B. K. (2002). Hydroxylase enzymes of the vitamin D pathway: Expression, function, and regulation. *Annual Review of Nutrition*, *22*, 139–166.

Osborn, J. L., Schwartz, G. G., Smith, D. C., Bahnson, R., Day, R., & Trump, D. L. (1995). Phase II trial of oral 1,25-dihydroxyvitamin D (calcitriol) in hormone refractory prostate cancer. *Urologic Oncology*, *1*, 195–198.

Palmer, H. G., Larriba, M. J., Garcia, J. M., Ordonez-Moran, P., Pena, C., Peiro, S., et al. (2004). The transcription factor SNAIL represses vitamin D receptor expression and responsiveness in human colon cancer. *Nature Medicine*, *10*, 917–919.

Parise, R. A., Egorin, M. J., Kanterewicz, B., Taimi, M., Petkovich, M., Lew, A. M., et al. (2006). CYP24, the enzyme that catabolizes the antiproliferative agent vitamin D, is increased in lung cancer. *International Journal of Cancer*, *119*, 1819–1828.

Peehl, D. M., & Feldman, D. (2004). Interaction of nuclear receptor ligands with the vitamin D signaling pathway in prostate cancer. *The Journal of Steroid Biochemistry and Molecular Biology*, *92*, 307–315.

Peehl, D. M., Seto, E., Hsu, J. Y., & Feldman, D. (2002). Preclinical activity of ketoconazole in combination with calcitriol or the vitamin D analogue EB 1089 in prostate cancer cells. *The Journal of Urology*, *168*, 1583–1588.

Peehl, D. M., Skowronski, R. J., Leung, G. K., Wong, S. T., Stamey, T. A., & Feldman, D. (1994). Antiproliferative effects of 1,25-dihydroxyvitamin D3 on primary cultures of human prostatic cells. *Cancer Research*, *54*, 805–810.

Pena, C., Garcia, J. M., Silva, J., Garcia, V., Rodriguez, R., Alonso, I., et al. (2005). E-cadherin and vitamin D receptor regulation by SNAIL and ZEB1 in colon cancer: Clinicopathological correlations. *Human Molecular Genetics*, *14*, 3361–3370.

Pike, J. W., & Meyer, M. B. (2010). The vitamin D receptor: New paradigms for the regulation of gene expression by 1,25-dihydroxyvitamin D(3). *Endocrinology and Metabolism Clinics of North America*, *39*, 255–269 (table of contents).

Pike, J. W., Meyer, M. B., Martowicz, M. L., Bishop, K. A., Lee, S. M., Nerenz, R. D., et al. (2010). Emerging regulatory paradigms for control of gene expression by 1,25-dihydroxyvitamin D3. *The Journal of Steroid Biochemistry and Molecular Biology*, *121*, 130–135.

Posner, G. H., Helvig, C., Cuerrier, D., Collop, D., Kharebov, A., Ryder, K., et al. (2010). Vitamin D analogues targeting CYP24 in chronic kidney disease. *The Journal of Steroid Biochemistry and Molecular Biology*, *121*, 13–19.

Prosser, D. E., & Jones, G. (2004). Enzymes involved in the activation and inactivation of vitamin D. *Trends in Biochemical Sciences*, *29*, 664–673.

Prummel, M. F., Wiersinga, W. M., Lips, P., Sanders, G. T., & Sauerwein, H. P. (1991). The course of biochemical parameters of bone turnover during treatment with corticosteroids. *The Journal of Clinical Endocrinology and Metabolism*, *72*, 382–386.

Rachez, C., & Freedman, L. P. (2000). Mechanisms of gene regulation by vitamin D(3) receptor: A network of coactivator interactions. *Gene*, *246*, 9–21.

Rachez, C., & Freedman, L. P. (2001). Mediator complexes and transcription. *Current Opinion in Cell Biology*, *13*, 274–280.

Rachez, C., Lemon, B. D., Suldan, Z., Bromleigh, V., Gamble, M., Naar, A. M., et al. (1999). Ligand-dependent transcription activation by nuclear receptors requires the DRIP complex. *Nature*, *398*, 824–828.

Racz, A., & Barsony, J. (1999). Hormone-dependent translocation of vitamin D receptors is linked to transactivation. *The Journal of Biological Chemistry*, *274*, 19352–19360.

Ragavan, V. V., Smith, J. E., & Bilezikian, J. P. (1982). Vitamin A toxicity and hypercalcemia. *The American Journal of the Medical Sciences*, *283*, 161–164.

Ramnath, N., Nadal, E., Jeon, C. K., Sandoval, J., Colacino, J., Rozek, L. S., et al. (2014). Epigenetic regulation of vitamin D metabolism in human lung adenocarcinoma. *Journal of Thoracic Oncology: Official Publication of the International Association for the Study of Lung Cancer*, *9*, 473–482.

Rao, A., Coan, A., Welsh, J. E., Barclay, W. W., Koumenis, C., & Cramer, S. D. (2004). Vitamin D receptor and p21/WAF1 are targets of genistein and 1,25-dihydroxyvitamin D3 in human prostate cancer cells. *Cancer Research*, *64*, 2143–2147.

Reichel, H., Koeffler, H. P., & Norman, A. W. (1989). The role of the vitamin D endocrine system in health and disease. *The New England Journal of Medicine*, *320*, 980–991.

Reid, I. R. (1997). Glucocorticoid osteoporosis—Mechanisms and management. *European Journal of Endocrinology/European Federation of Endocrine Societies*, *137*, 209–217.

Reinhardt, T. A., & Horst, R. L. (1989). Ketoconazole inhibits self-induced metabolism of 1,25-dihydroxyvitamin D3 and amplifies 1,25-dihydroxyvitamin D3 receptor up-regulation in rat osteosarcoma cells. *Archives of Biochemistry and Biophysics*, *272*, 459–465.

Roesler, W. J. (2001). The role of C/EBP in nutrient and hormonal regulation of gene expression. *Annual Review of Nutrition*, *21*, 141–165.

Roff, A., & Wilson, R. T. (2008). A novel SNP in a vitamin D response element of the CYP24A1 promoter reduces protein binding, transactivation, and gene expression. *The Journal of Steroid Biochemistry and Molecular Biology*, *112*, 47–54.

Rolla, D., Paoletti, E., Marsano, L., Mulas, D., Peloso, G., & Cannella, G. (1993). Effects of subcutaneous calcitriol administration on plasma calcium and parathyroid hormone concentrations in continuous ambulatory peritoneal dialysis uremic patients. *Peritoneal Dialysis International: Journal of the International Society for Peritoneal Dialysis*, *13*, 118–121.

Rozenchan, P. B., Folgueira, M. A., Katayama, M. L., Snitcovsky, I. M., & Brentani, M. M. (2004). Ras activation is associated with vitamin D receptor mRNA instability in HC11 mammary cells. *The Journal of Steroid Biochemistry and Molecular Biology, 92*, 89–95.

Rutz, H. P. (2002). Effects of corticosteroid use on treatment of solid tumours. *Lancet, 360*, 1969–1970.

Rygiel, A. M., Milano, F., Ten Kate, F. J., Schaap, A., Wang, K. K., Peppelenbosch, M. P., et al. (2008). Gains and amplifications of c-myc, EGFR, and 20.q13 loci in the no dysplasia-dysplasia-adenocarcinoma sequence of Barrett's esophagus. *Cancer Epidemiology, Biomarkers & Prevention: A Publication of the American Association for Cancer Research, Cosponsored by the American Society of Preventive Oncology, 17*, 1380–1385.

Saito, N., Suhara, Y., Kurihara, M., Fujishima, T., Honzawa, S., Takayanagi, H., et al. (2004). Design and efficient synthesis of 2 alpha-(omega-hydroxyalkoxy)-1 alpha,25-dihydroxyvitamin D3 analogues, including 2-epi-ED-71 and their 20-epimers with HL-60 cell differentiation activity. *The Journal of Organic Chemistry, 69*, 7463–7471.

Sakaki, T., Kagawa, N., Yamamoto, K., & Inouye, K. (2005). Metabolism of vitamin D3 by cytochromes P450. *Frontiers in Bioscience: A Journal and Virtual Library, 10*, 119–134.

Sakaki, T., Sawada, N., Nonaka, Y., Ohyama, Y., & Inouye, K. (1999). Metabolic studies using recombinant *Escherichia coli* cells producing rat mitochondrial CYP24 CYP24 can convert 1alpha,25-dihydroxyvitamin D3 to calcitroic acid. *European Journal of Biochemistry/FEBS, 262*, 43–48.

Sanchez-Martinez, R., Castillo, A. I., Steinmeyer, A., & Aranda, A. (2006). The retinoid X receptor ligand restores defective signalling by the vitamin D receptor. *EMBO Reports, 7*, 1030–1034.

Schaefer, R. J., Bonor, J. C., Joglekar, M. S., van Golen, K. L., & Nohe, A. G. (2013). 1,25 Dihydroxyvitamin D3 uptake is localized at caveolae and requires caveolar function. *Journal of Biomedical Nanotechnology, 9*, 1707–1715.

Schlingmann, K. P., Kaufmann, M., Weber, S., Irwin, A., Goos, C., John, U., et al. (2011). Mutations in CYP24A1 and idiopathic infantile hypercalcemia. *The New England Journal of Medicine, 365*, 410–421.

Schuster, I., Egger, H., Herzig, G., Reddy, G. S., Schmid, J. A., Schussler, M., et al. (2006). Selective inhibitors of vitamin D metabolism—New concepts and perspectives. *Anticancer Research, 26*, 2653–2668.

Schuster, I., Egger, H., Nussbaumer, P., & Kroemer, R. T. (2003). Inhibitors of vitamin D hydroxylases: Structure-activity relationships. *Journal of Cellular Biochemistry, 88*, 372–380.

Schuster, I., Egger, H., Reddy, G. S., & Vorisek, G. (2003). Combination of vitamin D metabolites with selective inhibitors of vitamin D metabolism. *Recent Results in Cancer Research, 164*, 169–188.

Schwartz, G. G., Whitlatch, L. W., Chen, T. C., Lokeshwar, B. L., & Holick, M. F. (1998). Human prostate cells synthesize 1,25-dihydroxyvitamin D3 from 25-hydroxyvitamin D3. *Cancer Epidemiology, Biomarkers & Prevention: A Publication of the American Association for Cancer Research, Cosponsored by the American Society of Preventive Oncology*, 7, 391–395.

Seth-Vollenweider, T., Joshi, S., Dhawan, P., Sif, S., & Christakos, S. (2014). Novel mechanism of negative regulation of 1,25-dihydroxyvitamin D3-induced 25-hydroxyvitamin D3 24-hydroxylase (Cyp24a1) transcription: Epigenetic modification involving cross-talk between protein-arginine methyltransferase 5 and the SWI/SNF complex. *The Journal of Biological Chemistry, 289*, 33958–33970.

Shabahang, M., Buras, R. R., Davoodi, F., Schumaker, L. M., Nauta, R. J., Uskokovic, M. R., et al. (1994). Growth inhibition of HT-29 human colon cancer cells by analogues of 1,25-dihydroxyvitamin D3. *Cancer Research, 54*, 4057–4064.

Shankar, V. N., Dilworth, F. J., Makin, H. L., Schroeder, N. J., Trafford, D. J., Kissmeyer, A. M., et al. (1997). Metabolism of the vitamin D analog EB1089 by cultured

human cells: Redirection of hydroxylation site to distal carbons of the side-chain. *Biochemical Pharmacology, 53*, 783–793.
Shankar, K., Liu, X., Singhal, R., Chen, J. R., Nagarajan, S., Badger, T. M., et al. (2008). Chronic ethanol consumption leads to disruption of vitamin D3 homeostasis associated with induction of renal 1,25 dihydroxyvitamin D3-24-hydroxylase (CYP24A1). *Endocrinology, 149*, 1748–1756.
Shanmugam, M. K., Kannaiyan, R., & Sethi, G. (2011). Targeting cell signaling and apoptotic pathways by dietary agents: Role in the prevention and treatment of cancer. *Nutrition and Cancer, 63*, 161–173.
Skowronski, R. J., Peehl, D. M., & Feldman, D. (1993). Vitamin D and prostate cancer: 1,25 Dihydroxyvitamin D3 receptors and actions in human prostate cancer cell lines. *Endocrinology, 132*, 1952–1960.
Slater, S. J., Kelly, M. B., Taddeo, F. J., Larkin, J. D., Yeager, M. D., McLane, J. A., et al. (1995). Direct activation of protein kinase C by 1 alpha,25-dihydroxyvitamin D3. *The Journal of Biological Chemistry, 270*, 6639–6643.
Smirnoff, P., Liel, Y., Gnainsky, J., Shany, S., & Schwartz, B. (1999). The protective effect of estrogen against chemically induced murine colon carcinogenesis is associated with decreased CpG island methylation and increased mRNA and protein expression of the colonic vitamin D receptor. *Oncology Research, 11*, 255–264.
Smith, D. C., Johnson, C. S., Freeman, C. C., Muindi, J., Wilson, J. W., & Trump, D. L. (1999). A Phase I trial of calcitriol (1,25-dihydroxycholecalciferol) in patients with advanced malignancy. *Clinical Cancer Research: An Official Journal of the American Association for Cancer Research, 5*, 1339–1345.
Sobell, H. M. (1985). Actinomycin and DNA transcription. *Proceedings of the National Academy of Sciences of the United States of America, 82*, 5328–5331.
Solomon, C., White, J. H., & Kremer, R. (1999). Mitogen-activated protein kinase inhibits 1,25-dihydroxyvitamin D3-dependent signal transduction by phosphorylating human retinoid X receptor alpha. *The Journal of Clinical Investigation, 103*, 1729–1735.
Srinivasan, M., Parwani, A. V., Hershberger, P. A., Lenzner, D. E., & Weissfeld, J. L. (2011). Nuclear vitamin D receptor expression is associated with improved survival in non-small cell lung cancer. *The Journal of Steroid Biochemistry and Molecular Biology, 123*, 30–36.
Stambolsky, P., Tabach, Y., Fontemaggi, G., Weisz, L., Maor-Aloni, R., Siegfried, Z., et al. (2010). Modulation of the vitamin D3 response by cancer-associated mutant p53. *Cancer Cell, 17*, 273–285.
St-Arnaud, R. (2011). CYP24A1: Structure, function, and physiological role. In D. Feldman, P. J. Wesley, & J. S. Adams (Eds.), *Vitamin D* (3rd ed., pp. 43–56). San Diego, CA: Elsevier.
St-Arnaud, R., Arabian, A., Travers, R., Barletta, F., Raval-Pandya, M., Chapin, K., et al. (2000). Deficient mineralization of intramembranous bone in vitamin D-24-hydroxylase-ablated mice is due to elevated 1,25-dihydroxyvitamin D and not to the absence of 24,25-dihydroxyvitamin D. *Endocrinology, 141*, 2658–2666.
Suhara, Y., Nihei, K. I., Tanigawa, H., Fujishima, T., Konno, K., Nakagawa, K., et al. (2000). Syntheses and biological evaluation of novel 2alpha-substituted 1alpha,25-dihydroxyvitamin D3 analogues. *Bioorganic & Medicinal Chemistry Letters, 10*, 1129–1132.
Sun, X., & Zemel, M. B. (2008). 1Alpha, 25-dihydroxyvitamin D and corticosteroid regulate adipocyte nuclear vitamin D receptor. *International Journal of Obesity, 32*, 1305–1311.
Suzuki, Y., Ichikawa, Y., Saito, E., & Homma, M. (1983). Importance of increased urinary calcium excretion in the development of secondary hyperparathyroidism of patients under glucocorticoid therapy. *Metabolism, Clinical and Experimental, 32*, 151–156.
Swami, S., Krishnan, A. V., Peehl, D. M., & Feldman, D. (2005). Genistein potentiates the growth inhibitory effects of 1,25-dihydroxyvitamin D3 in DU145 human prostate

cancer cells: Role of the direct inhibition of CYP24 enzyme activity. *Molecular and Cellular Endocrinology*, *241*, 49–61.

Tannock, I., Gospodarowicz, M., Meakin, W., Panzarella, T., Stewart, L., & Rider, W. (1989). Treatment of metastatic prostatic cancer with low-dose prednisone: Evaluation of pain and quality of life as pragmatic indices of response. *Journal of Clinical Oncology: Official Journal of the American Society of Clinical Oncology*, *7*, 590–597.

Thill, M., Fischer, D., Kelling, K., Hoellen, F., Dittmer, C., Hornemann, A., et al. (2010). Expression of vitamin D receptor (VDR), cyclooxygenase-2 (COX-2) and 15-hydroxyprostaglandin dehydrogenase (15-PGDH) in benign and malignant ovarian tissue and 25-hydroxycholecalciferol (25(OH2)D3) and prostaglandin E2 (PGE2) serum level in ovarian cancer patients. *The Journal of Steroid Biochemistry and Molecular Biology*, *121*, 387–390.

Tohmon, M., Fukase, M., Kishihara, M., Kadowaki, S., & Fujita, T. (1988). Effect of glucocorticoid administration on intestinal, renal, and cerebellar calbindin-D28K in chicks. *Journal of Bone and Mineral Research: The Official Journal of the American Society for Bone and Mineral Research*, *3*, 325–331.

Tolon, R. M., Castillo, A. I., Jimenez-Lara, A. M., & Aranda, A. (2000). Association with Ets-1 causes ligand- and AF2-independent activation of nuclear receptors. *Molecular and Cellular Biology*, *20*, 8793–8802.

Townsend, K., Evans, K. N., Campbell, M. J., Colston, K. W., Adams, J. S., & Hewison, M. (2005). Biological actions of extra-renal 25-hydroxyvitamin D-1alpha-hydroxylase and implications for chemoprevention and treatment. *The Journal of Steroid Biochemistry and Molecular Biology*, *97*, 103–109.

Triadafilopoulos, G. (2014). Glucocorticoid therapy for gastrointestinal diseases. *Expert Opinion on Drug Safety*, *13*, 563–572.

Trump, D. L., Hershberger, P. A., Bernardi, R. J., Ahmed, S., Muindi, J., Fakih, M., et al. (2004). Anti-tumor activity of calcitriol: Pre-clinical and clinical studies. *The Journal of Steroid Biochemistry and Molecular Biology*, *89-90*, 519–526.

Trump, D. L., Potter, D. M., Muindi, J., Brufsky, A., & Johnson, C. S. (2006). Phase II trial of high-dose, intermittent calcitriol (1,25 dihydroxyvitamin D3) and dexamethasone in androgen-independent prostate cancer. *Cancer*, *106*, 2136–2142.

Vaisanen, S., Dunlop, T. W., Sinkkonen, L., Frank, C., & Carlberg, C. (2005). Spatio-temporal activation of chromatin on the human CYP24 gene promoter in the presence of 1alpha,25-dihydroxyvitamin D3. *Journal of Molecular Biology*, *350*, 65–77.

Van Cromphaut, S. J., Stockmans, I., Torrekens, S., Van Herck, E., Carmeliet, G., & Bouillon, R. (2007). Duodenal calcium absorption in dexamethasone-treated mice: Functional and molecular aspects. *Archives of Biochemistry and Biophysics*, *460*, 300–305.

van den Bemd, G. C., Pols, H. A., Birkenhager, J. C., & van Leeuwen, J. P. (1996). Conformational change and enhanced stabilization of the vitamin D receptor by the 1,25-dihydroxyvitamin D3 analog KH1060. *Proceedings of the National Academy of Sciences of the United States of America*, *93*, 10685–10690.

van der Goes, M. C., Jacobs, J. W., & Bijlsma, J. W. (2014). The value of glucocorticoid co-therapy in different rheumatic diseases—Positive and adverse effects. *Arthritis Research & Therapy*, *16*(Suppl. 2), S2.

van Sandwijk, M. S., Bemelman, F. J., & Ten Berge, I. J. (2013). Immunosuppressive drugs after solid organ transplantation. *The Netherlands Journal of Medicine*, *71*, 281–289.

Walsh, D., & Avashia, J. (1992). Glucocorticoids in clinical oncology. *Cleveland Clinic Journal of Medicine*, *59*, 505–515.

Wang, J. Y., Swami, S., Krishnan, A. V., & Feldman, D. (2012). Combination of calcitriol and dietary soy exhibits enhanced anticancer activity and increased hypercalcemic toxicity in a mouse xenograft model of prostate cancer. *Prostate*, *72*, 1628–1637.

Wang, T. J., Zhang, F., Richards, J. B., Kestenbaum, B., van Meurs, J. B., Berry, D., et al. (2010). Common genetic determinants of vitamin D insufficiency: A genome-wide association study. *Lancet*, *376*, 180–188.

Wang, Y., Zhu, J., & DeLuca, H. F. (2012). Where is the vitamin D receptor? *Archives of Biochemistry and Biophysics*, *523*, 123–133.

Weigel, N. L. (1996). Steroid hormone receptors and their regulation by phosphorylation. *The Biochemical Journal*, *319*(Pt. 3), 657–667.

Weiss, M. M., Hermsen, M. A., Meijer, G. A., & van Diest, P. J. (2002). Comparative genomic hybridization in cancer investigations. *Methods in Molecular Biology*, *204*, 369–378.

White, P., & Cooke, N. (2000). The multifunctional properties and characteristics of vitamin D-binding protein. *Trends in Endocrinology and Metabolism*, *11*, 320–327.

Wong, M. P., Fung, L. F., Wang, E., Chow, W. S., Chiu, S. W., Lam, W. K., et al. (2003). Chromosomal aberrations of primary lung adenocarcinomas in nonsmokers. *Cancer*, *97*, 1263–1270.

Xu, H., Posner, G. H., Stevenson, M., & Campbell, F. C. (2010). Apc(MIN) modulation of vitamin D secosteroid growth control. *Carcinogenesis*, *31*, 1434–1441.

Yang, B. S., Hauser, C. A., Henkel, G., Colman, M. S., Van Beveren, C., Stacey, K. J., et al. (1996). Ras-mediated phosphorylation of a conserved threonine residue enhances the transactivation activities of c-Ets1 and c-Ets2. *Molecular and Cellular Biology*, *16*, 538–547.

Yee, S. W., Campbell, M. J., & Simons, C. (2006). Inhibition of Vitamin D3 metabolism enhances VDR signalling in androgen-independent prostate cancer cells. *The Journal of Steroid Biochemistry and Molecular Biology*, *98*, 228–235.

Yee, S. W., & Simons, C. (2004). Synthesis and CYP24 inhibitory activity of 2-substituted-3,4-dihydro-2H-naphthalen-1-one (tetralone) derivatives. *Bioorganic & Medicinal Chemistry Letters*, *14*, 5651–5654.

Yoshizawa, T., Handa, Y., Uematsu, Y., Takeda, S., Sekine, K., Yoshihara, Y., et al. (1997). Mice lacking the vitamin D receptor exhibit impaired bone formation, uterine hypoplasia and growth retardation after weaning. *Nature Genetics*, *16*, 391–396.

Yu, W. D., McElwain, M. C., Modzelewski, R. A., Russell, D. M., Smith, D. C., Trump, D. L., et al. (1998). Enhancement of 1,25-dihydroxyvitamin D3-mediated antitumor activity with dexamethasone. *Journal of the National Cancer Institute*, *90*, 134–141.

Zella, L. A., Meyer, M. B., Nerenz, R. D., Lee, S. M., Martowicz, M. L., & Pike, J. W. (2010). Multifunctional enhancers regulate mouse and human vitamin D receptor gene transcription. *Molecular Endocrinology*, *24*, 128–147.

Zella, L. A., Meyer, M. B., Nerenz, R. D., & Pike, J. W. (2009). The enhanced hypercalcemic response to 20-epi-1,25-dihydroxyvitamin D3 results from a selective and prolonged induction of intestinal calcium-regulating genes. *Endocrinology*, *150*, 3448–3456.

Zhang, C., Baudino, T. A., Dowd, D. R., Tokumaru, H., Wang, W., & MacDonald, P. N. (2001). Ternary complexes and cooperative interplay between NCoA-62/Ski-interacting protein and steroid receptor coactivators in vitamin D receptor-mediated transcription. *The Journal of Biological Chemistry*, *276*, 40614–40620.

Zhang, C., Dowd, D. R., Staal, A., Gu, C., Lian, J. B., van Wijnen, A. J., et al. (2003). Nuclear coactivator-62 kDa/Ski-interacting protein is a nuclear matrix-associated coactivator that may couple vitamin D receptor-mediated transcription and RNA splicing. *The Journal of Biological Chemistry*, *278*, 35325–35336.

Zhang, Q., Kanterewicz, B., Buch, S., Petkovich, M., Parise, R., Beumer, J., et al. (2012). CYP24 inhibition preserves 1alpha,25-dihydroxyvitamin D(3) anti-proliferative signaling in lung cancer cells. *Molecular and Cellular Endocrinology*, *355*, 153–161.

Zhang, Z., Kovalenko, P., Cui, M., Desmet, M., Clinton, S. K., & Fleet, J. C. (2010). Constitutive activation of the mitogen-activated protein kinase pathway impairs vitamin D signaling in human prostate epithelial cells. *Journal of Cellular Physiology*, *224*, 433–442.

Zhang, Y., Leung, D. Y., & Goleva, E. (2013). Vitamin D enhances glucocorticoid action in human monocytes: Involvement of granulocyte-macrophage colony-stimulating factor and mediator complex subunit 14. *The Journal of Biological Chemistry*, *288*, 14544–14553.
Zhao, X. Y., Eccleshall, T. R., Krishnan, A. V., Gross, C., & Feldman, D. (1997). Analysis of vitamin D analog-induced heterodimerization of vitamin D receptor with retinoid X receptor using the yeast two-hybrid system. *Molecular Endocrinology*, *11*, 366–378.
Zhou, J. Y., Norman, A. W., Chen, D. L., Sun, G. W., Uskokovic, M., & Koeffler, H. P. (1990). 1,25-Dihydroxy-16-ene-23-yne-vitamin D3 prolongs survival time of leukemic mice. *Proceedings of the National Academy of Sciences of the United States of America*, *87*, 3929–3932.
Zhu, W., Malloy, P. J., Delvin, E., Chabot, G., & Feldman, D. (1998). Hereditary 1,25-dihydroxyvitamin D-resistant rickets due to an opal mutation causing premature termination of the vitamin D receptor. *Journal of Bone and Mineral Research*, *13*, 259–264.
Zierold, C., Darwish, H. M., & DeLuca, H. F. (1995). Two vitamin D response elements function in the rat 1,25-dihydroxyvitamin D 24-hydroxylase promoter. *The Journal of Biological Chemistry*, *270*, 1675–1678.
Zou, A., Elgort, M. G., & Allegretto, E. A. (1997). Retinoid X receptor (RXR) ligands activate the human 25-hydroxyvitamin D3-24-hydroxylase promoter via RXR heterodimer binding to two vitamin D-responsive elements and elicit additive effects with 1,25-dihydroxyvitamin D3. *The Journal of Biological Chemistry*, *272*, 19027–19034.

INDEX

Note: Page numbers followed by "*f*" indicate figures, "*t*" indicate tables, and "*s*" indicate schemes.

P

R

S

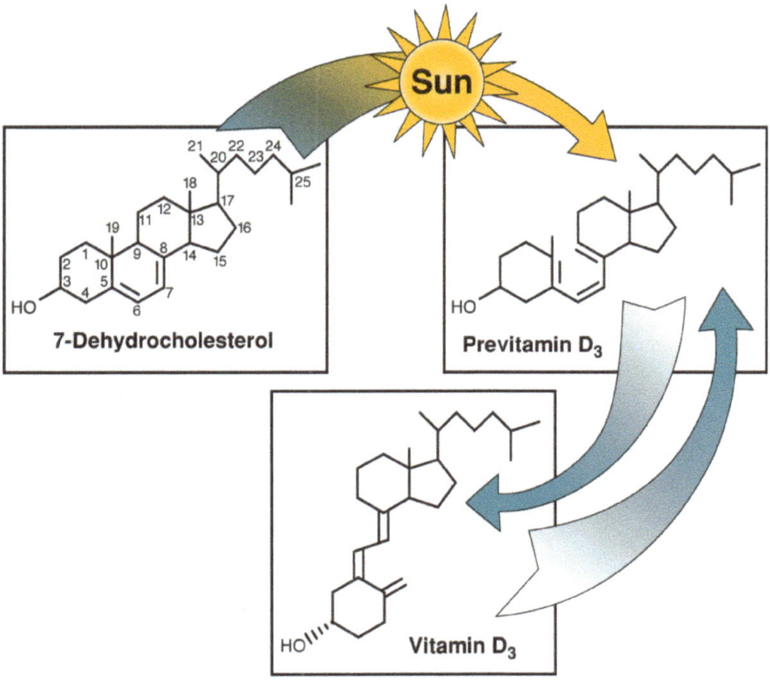

Hector F. DeLuca, Figure 2 Conversion of 7-dehydrocholesterol to previtamin D_3 by ultraviolet light (280–310 nm) from sunlight or artificial light. Previtamin D_3 isomerizes to vitamin D_3 and is always in a equilibrium mixture of 5.94% previtamin D_3 and 94.06% vitamin D_3 at body temperature.

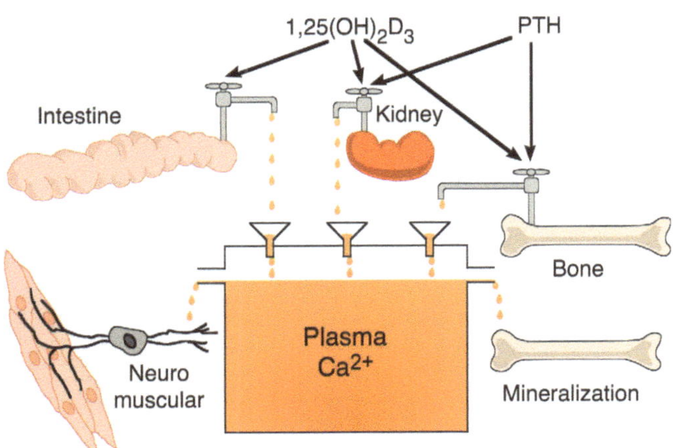

Hector F. DeLuca, Figure 3 A diagrammatic representation of the mineralization of bone brought about by vitamin D after its conversion to the hormonal form, 1,25-$(OH)_2D_3$. Note that hypocalcemic tetany is also prevented by the action of vitamin D and parathyroid hormone (PTH).

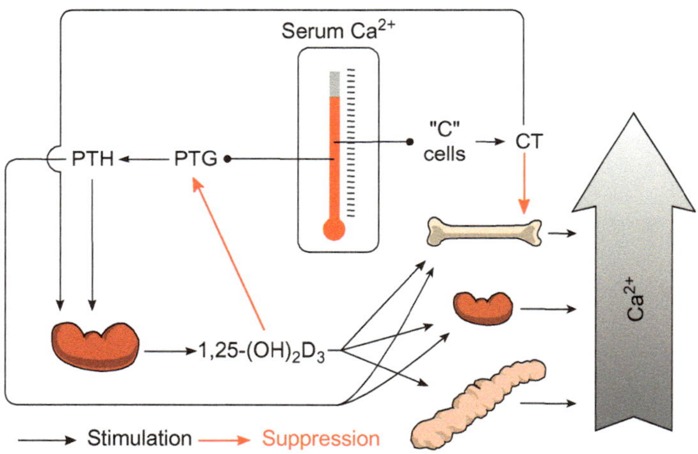

Hector F. DeLuca, Figure 5 A diagrammatic representation of the regulation of serum calcium concentration by the actions of 1,25-$(OH)_2D_3$, PTH, and calcitonin (CT).

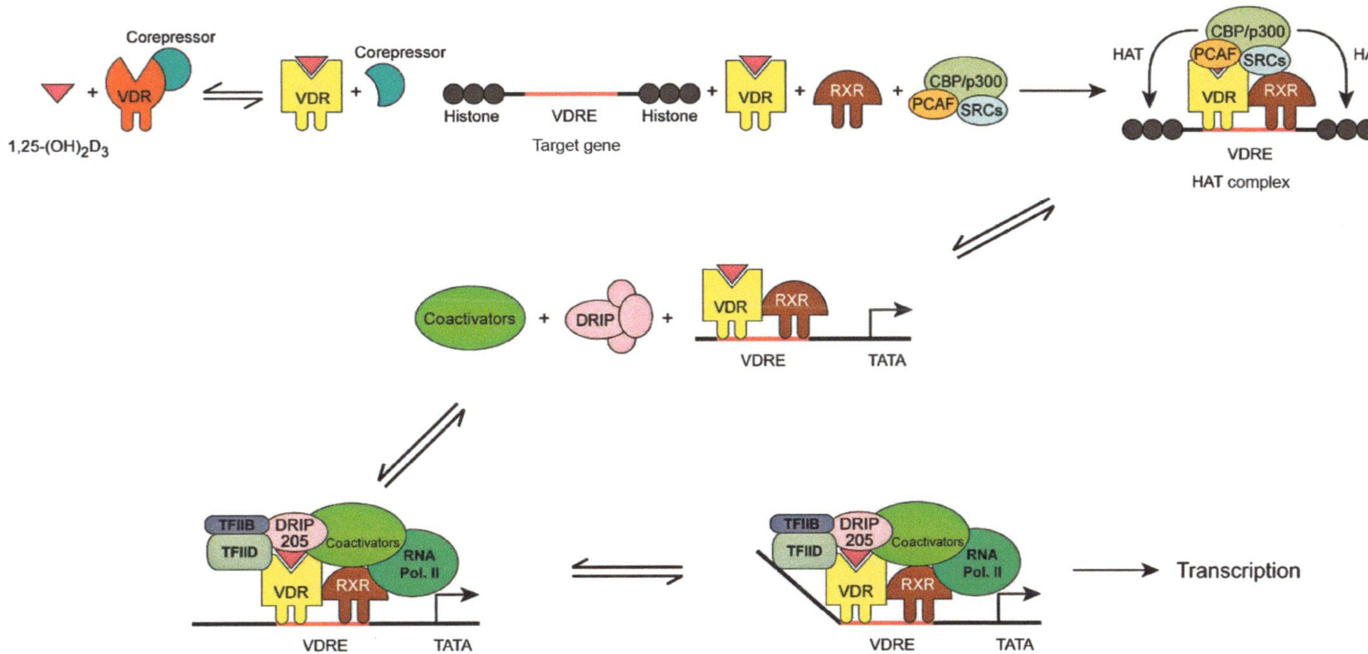

Hector F. DeLuca, Figure 6 A current model of gene expression regulated by the vitamin D hormone interacting with the vitamin D receptor (VDR) that requires a participation of the retinoid X receptor (RXR). These protein factors bind to the vitamin D responsive elements throughout the target gene including those in the promoter region. These elements are imperfect hexamers separated by three unspecified nucleotides. The factors involved are many including histone acetylase (HAT), Drip protein, CBP/p300, PCAF, and SRCs.

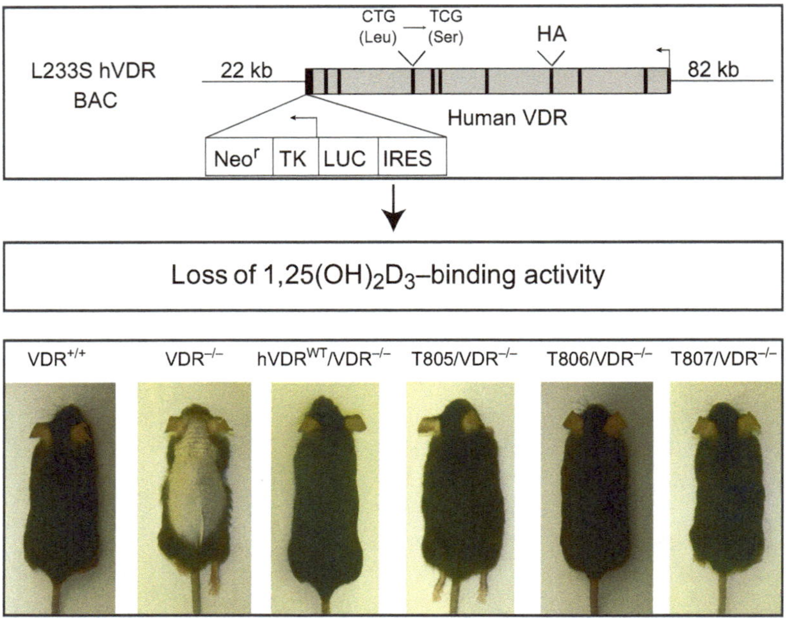

J. Wesley Pike *et al.*, Figure 1 A mutant human VDR deficient in $1,25(OH)_2D_3$-binding activity restores hair follicle cycling in VDR-null mice and prevents alopecia. Minigenes containing the human VDR gene locus were introduced into a VDR-null mouse background to create humanized mice expressing either wild-type (hVDR) or mutant (hVDR-L233S) VDR proteins. Both hVDR-WT and hVDR-L233S rescue the alopecia observed in VDR-null mice. VDR+/+, unmodified normal mice; hVDR-L233S 805, 806, and 807 represent three rescued mouse strains that express increasing concentrations of VDR protein in all appropriate tissues. See Lee, Goellner, et al. (2014).

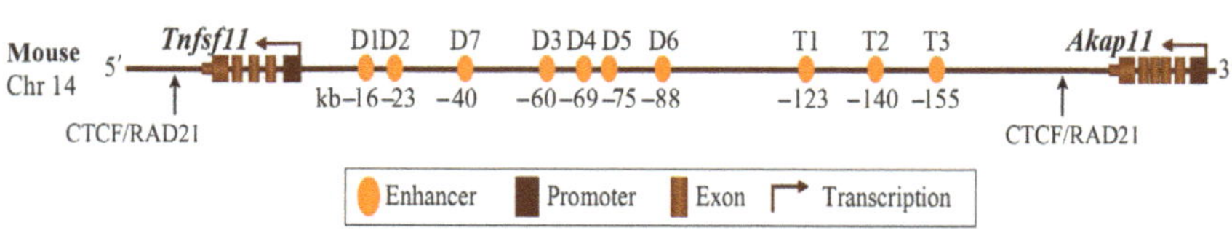

J. Wesley Pike *et al.*, Figure 2 Schematic linear structure of the mouse *Tnfsf11* gene locus with associated regulatory enhancers located on chromosome 14. The gene spans over 220 kb and is defined by two CTCF/RAD21 sites that serve as boundary elements for the gene. The *Tnfsf11* exons are defined as brown rectangles, and enhancers are defined by orange ovals. Arrows define the direction of transcription. Both the *Tnfsf11* and neighboring *Akap11* genes are located on the reverse strand. D1–D7 represent enhancers that regulate *Tnfsf11* expression in osteoblast-lineage cells, whereas T1–T3 represent enhancers that regulate *Tnfsf11* expression in hematopoietic B and T cells. See Onal et al. (2014).

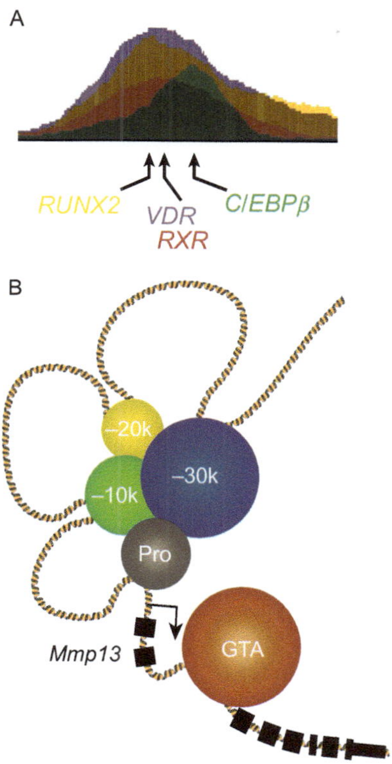

J. Wesley Pike *et al.*, Figure 3 Organization of osteoblast enhancer complexes (OEC) that bind RUNX2, C/EBPβ, and the VDR. (A) Consolidated OEC-binding arrangement for RUNX2, C/EBPβ, and VDR at the *Spp1* (osteopontin) gene as defined by ChIP-seq analysis. (B) Dispersed OEC-binding arrangement for RUNX2, C/EBPβ, and VDR at the *Mmp13* gene locus as defined by ChIP-seq and other functional analyses. Schematic depiction of the individual RUNX2 (blue), C/EBPβ (yellow), and VDR (green) enhancers shown clustered and interacting upstream of the *Mmp13* gene locus with the promoter proximal region (gray) and the general transcription assembly (brown) located near the transcriptional start site. See Meyer, Benkusky, and Pike (2015).

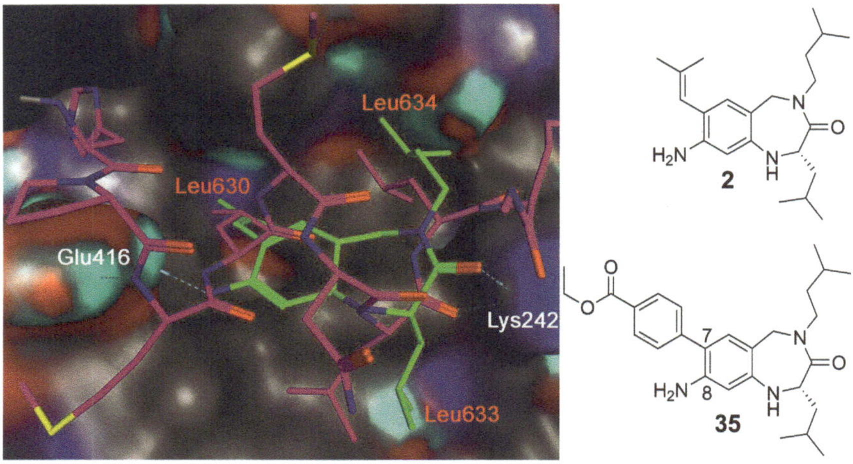

Kelly A. Teske *et al.*, Figure 4 Overlay between a crystal structure of VDR and coactivator peptide DRIP205 and docked conformation of compound 2 as well as structures of compounds 2 and 35.

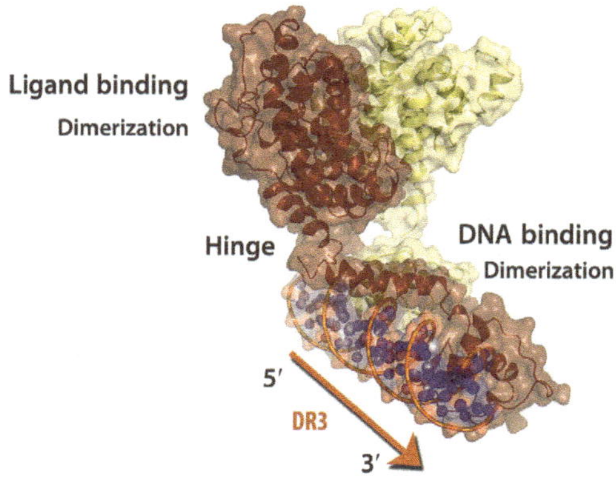

Anna Y. Belorusova and Natacha Rochel, Figure 1 Structural organization of the VDR–RXRα heterodimer bound to canonical DR3 response element as suggested by solution studies.

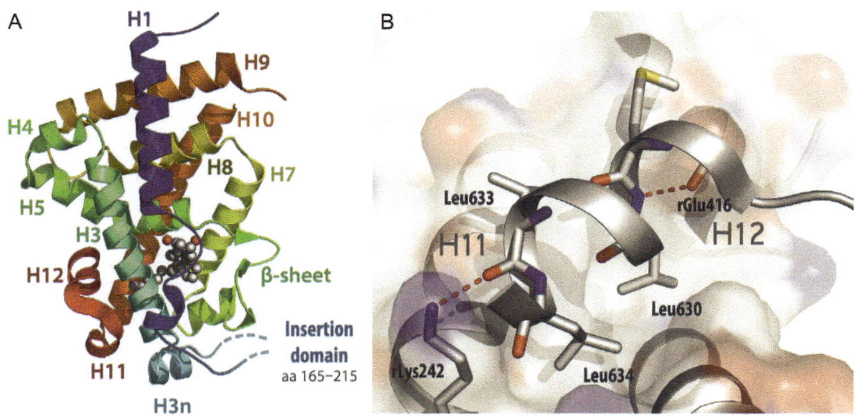

Anna Y. Belorusova and Natacha Rochel, Figure 2 Overall structure of the VDR LBD. (A) The VDR LBD bound to 1,25$(OH)_2D_3$ is composed of 13 helices (H1, 2, 2n, 3–12) (PDB ID: 1DB1). VDR members present an additional insertion domain (ID) between helices H1 and H2n. (B) Binding mode of the DRIP205 coactivator peptide LXXLL (PDB ID: 1RK3).

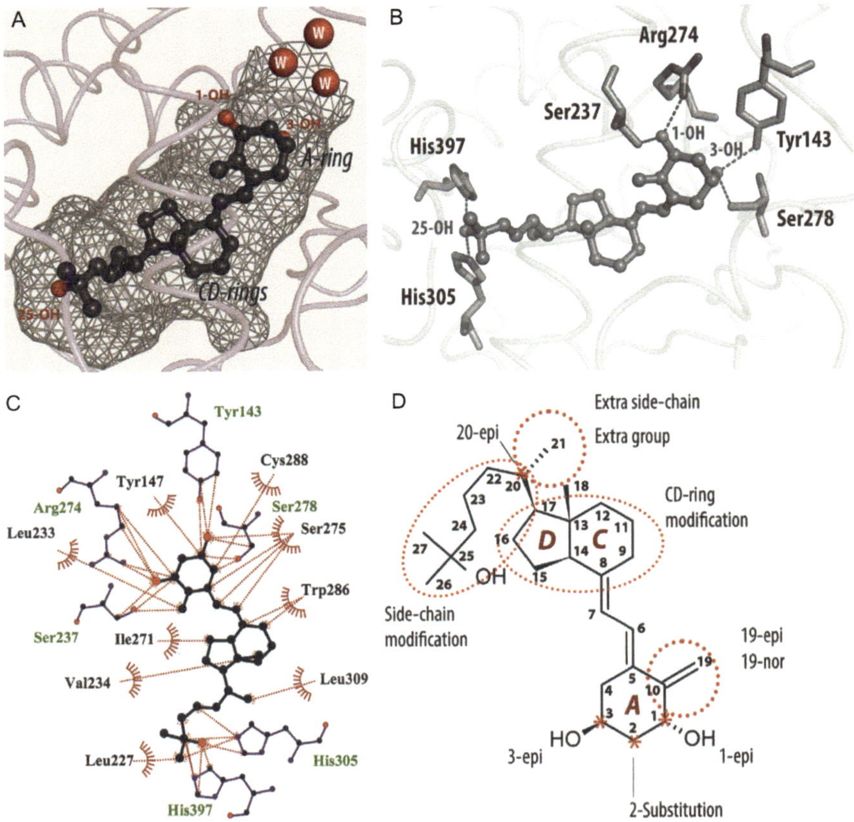

Anna Y. Belorusova and Natacha Rochel, Figure 3 Atomic details of the $1,25(OH)_2D_3$ interactions to the human VDR LBD (PDB ID: 1DB1). (A) Binding mode of $1,25(OH)_2D_3$ in the VDR LBP. The volume of the LBP is shown as a gray surface. (B) Specific H-bonds anchoring the three hydroxyl groups of the $1,25(OH)_2D_3$. (C) Electrostatic and hydrophobic interactions between $1,25(OH)_2D_3$ and LBP residues presented as 2D diagram. Obtained using LigPlot+ (Laskowski & Swindells, 2011). (D) Chemical structure and regions of $1,25(OH)_2D_3$ subjected to chemical modifications.

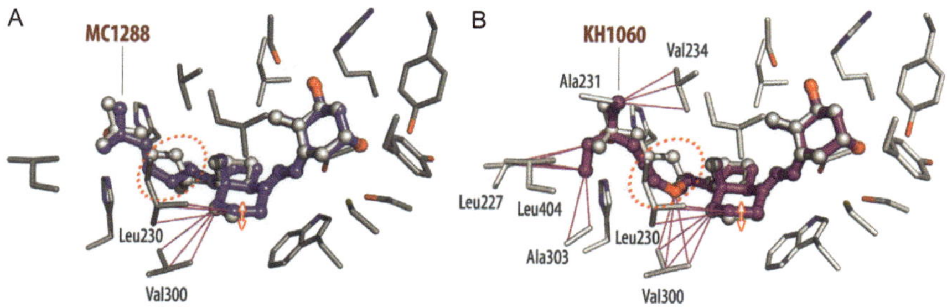

Anna Y. Belorusova and Natacha Rochel, Figure 4 Crystal structures of hVDR LBD in complex with 20-epi-analogs. (A) hVDR–LBD complex with KH1060 (PDB ID: 1IE8). (B) hVDR–LBD complex with MC1288 (PDB ID: 1IE9). Specific hydrophobic interactions are presented as lines. Different positioning of the aliphatic chain in comparison with 1,25$(OH)_2D_3$ is highlighted.

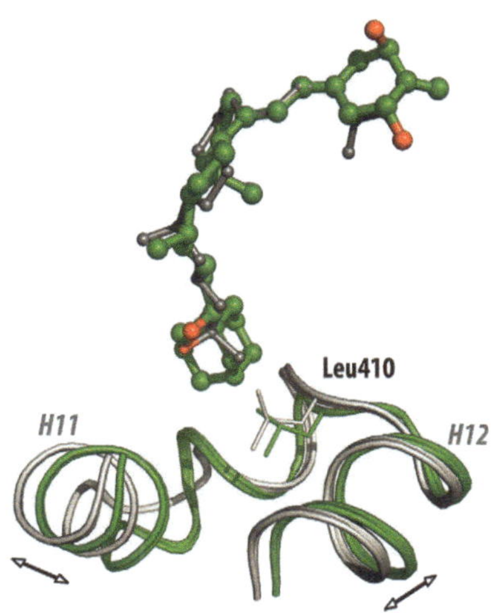

Anna Y. Belorusova and Natacha Rochel, Figure 9 Crystal structure of rVDR–LBD complex with adamantyl compound ADTT. Overlay of rVDR–LBD complex with ADTT (PDB ID: 2ZMI) and hVDR–LBD complex with 1,25$(OH)_2D_3$ (PDB ID: 1DB1; light gray). Repositioning of helices 11 and 12 is shown by arrows.

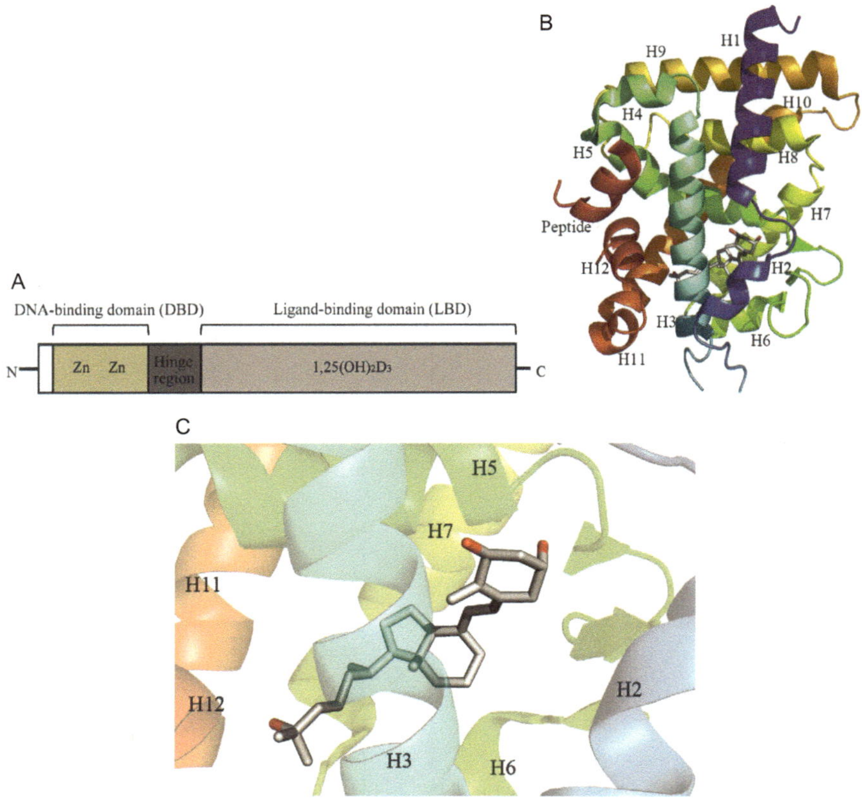

Teikichi Ikura and Nobutoshi Ito, Figure 1 Structure of rVDR in complex with the ligand $1,25(OH)_2D_3$. (A) Schematic of domain structures of rVDR. Two zinc molecules bind to the N-terminal DNA-binding domain (DBD), while the ligand binds to the C-terminal ligand-binding domain (LBD). (B) Tertiary structure of the LBD of rVDR in ternary complex with the ligand $1,25(OH)_2D_3$ and the peptide derived from the coactivator MED1 (PDB code: 2ZLC). Twelve helices of the LDB are numbered after those of the human RXR. (C) Close-up view of the ligand $1,25(OH)_2D_3$ in the ternary complex. The helices are drawn transparently.

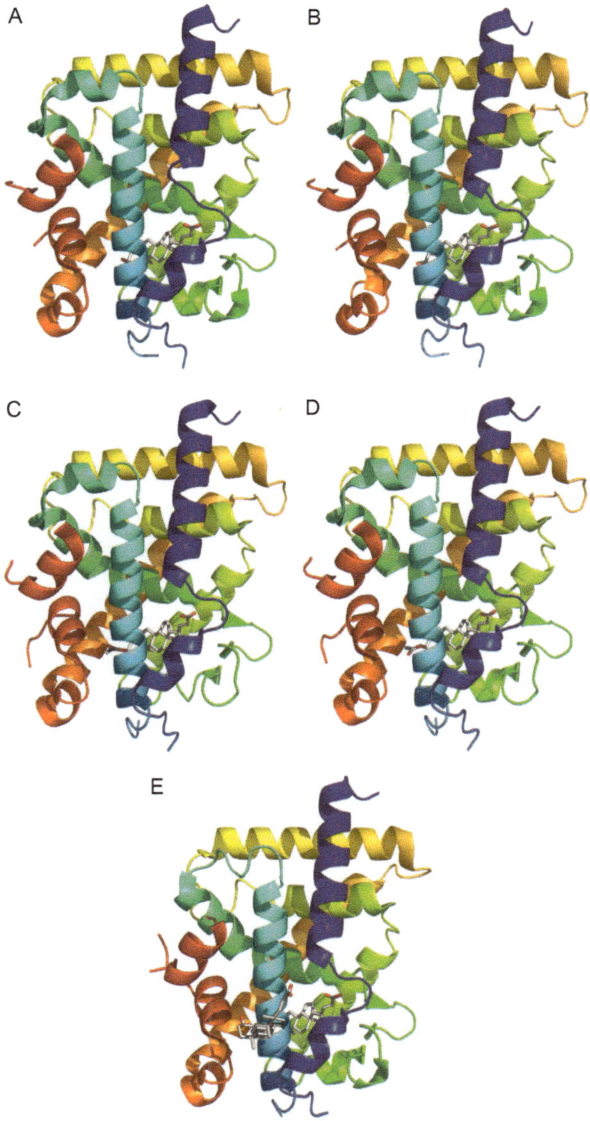

Teikichi Ikura and Nobutoshi Ito, Figure 3 Overall structures of the ternary complexes of VDR-LBD with LCA and its derivatives: (A) rVDR-LBD with LCA (PDB code: 3W5P), (B) rVDR-LBD with 3-keto LCA (PDB code: 3W5Q), (C) rVDR-LBD with LCA acetate (PDB code: 3W5R), (D) rVDR-LBD with LCA propionate (PDB code: 3W5T), and (E) zVDR-LBD with LCA (PDB code: 4Q0A).

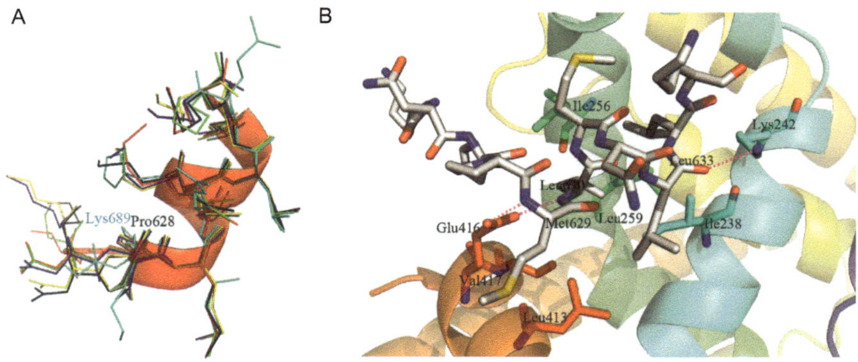

Teikichi Ikura and Nobutoshi Ito, Figure 4 Structures of the coactivator peptides in VDR complexes. (A) Superposition of MED1 peptides in rVDR complexes with 1,25$(OH)_2D_3$ (black), LCA (red), 3-keto LCA (green), LCA acetate (blue), LCA propionate (yellow), and SRC-2 peptide in zVDR complex with LCA (cyan). Pro628 of MED1 peptide corresponds to Lys689 of SRC-2 peptide. Secondary structure of the peptide in rVDR complex with LCA is also drawn with red. (B) Interactions between MED1 and rVDR-LBD complex with LCA. Hydrogen bonds are drawn with magenta dashed lines.

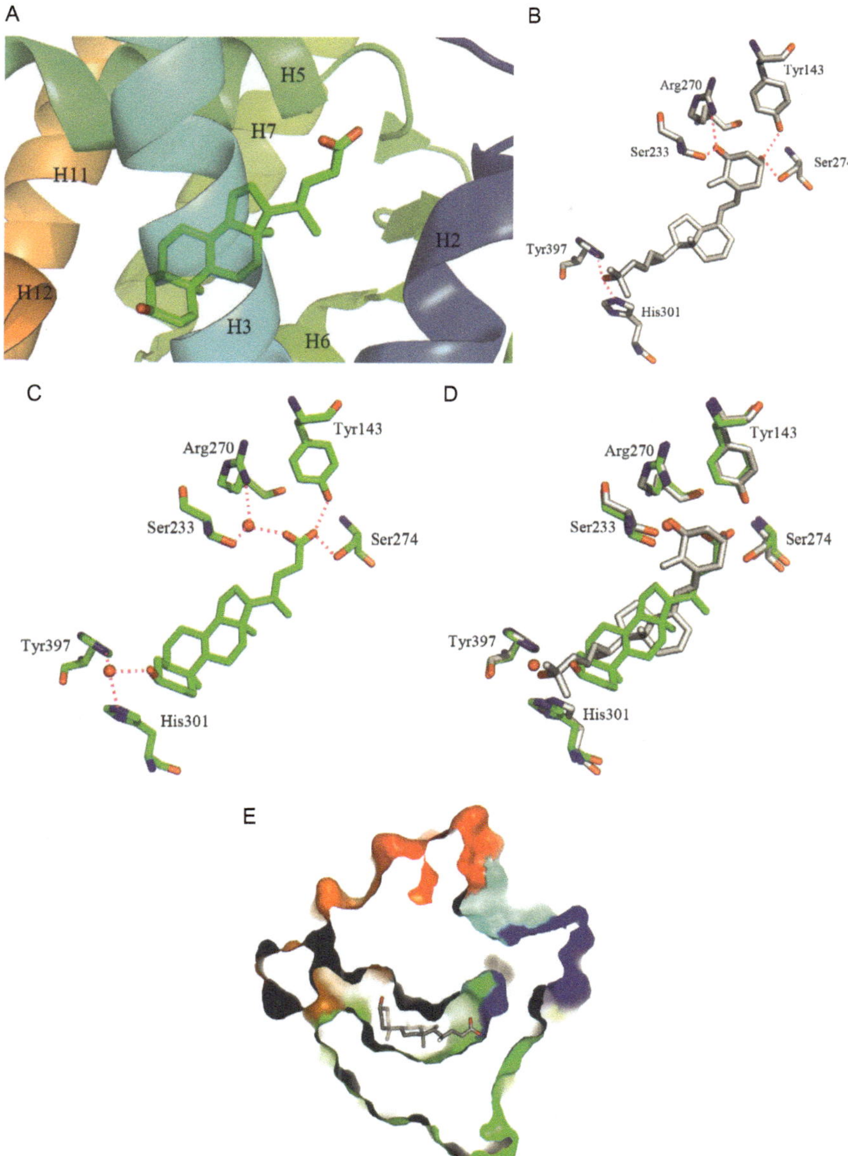

Teikichi Ikura and Nobutoshi Ito, Figure 5 Structures of the ligands in the ternary complexes and interactions between them and rVDR-LBD. (A) Structure of LCA in the LBP of rVDR-LBD. Its location is comparable with that of 1,25$(OH)_2D_3$ shown in Fig. 1C, but its orientation is opposite. (B) Interactions between 1,25$(OH)_2D_3$ and rVDR-LBD shown in Fig. 1C. (C) Interactions between LCA and rVDR-LBD. (D) Superposition of 1,25$(OH)_2D_3$ (light gray) and LCA (green) in the ternary complexes of rVDR-LDB. (E) Location of LCA in the LBP of rVDR-LBD, which shows its hydrophobic interactions with the alkyl and aromatic groups of the surrounding residues. Hydrogen bonds are drawn with magenta dashed lines in panels (B) and (C). Red spheres stand for water molecules in panels (C) and (D).

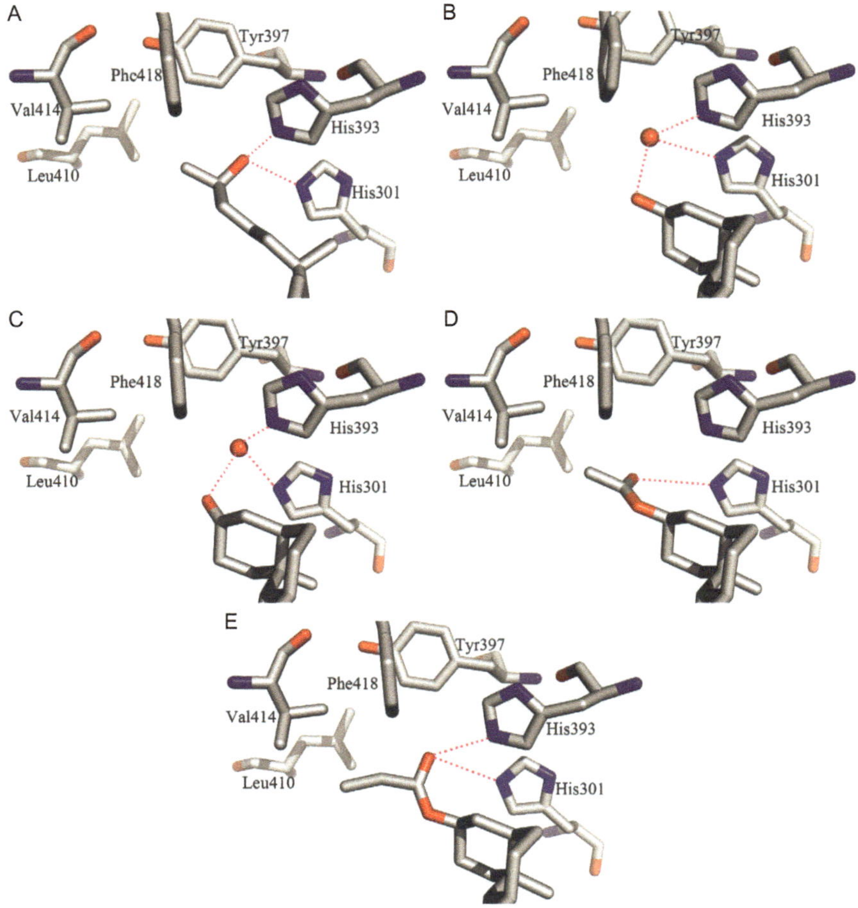

Teikichi Ikura and Nobutoshi Ito, Figure 6 Interactions of the ligands with residues in helix 6, loop 6–7, and helix 11 of rVDR-LDB. (A) Interactions between $1,25(OH)_2D_3$ and its surrounding residues. (B) Interactions between LCA and its surrounding residues. (C) Interactions between 3-keto LCA and its surrounding residues. (D) Interactions between LCA acetate and its surrounding residues. (E) Interactions between LCA propionate and its surrounding residues. Hydrogen bonds are drawn with magenta dashed lines in all panels. Red spheres stand for water molecules in panels (B) and (C).

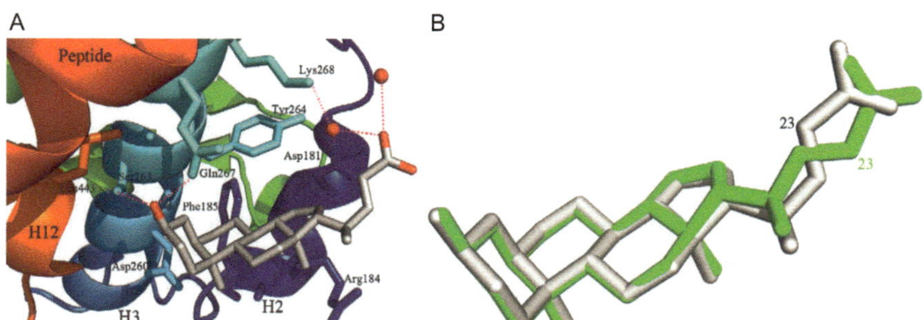

Teikichi Ikura and Nobutoshi Ito, Figure 7 Feature of the second LCA-binding site. (A) Interactions between LCA and its surrounding residues in the second LCA-binding site of zVDR-LBD. Hydrogen bonds are drawn with magenta dashed lines. Red spheres stand for water molecules. (B) Superposition of LCA molecules in the first canonical site (green) and the second site (gray). The orientation of the C-23 carbon of the second LCA is different from that of the first one.

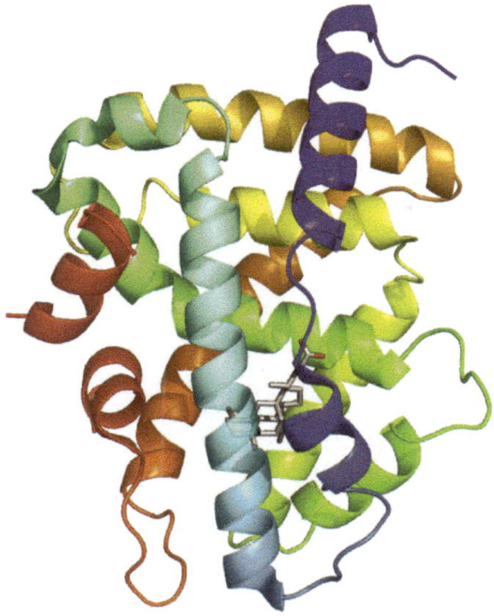

Teikichi Ikura and Nobutoshi Ito, Figure 8 Tertiary structure of the LBD of rat FXR in ternary complex with 6α-ethyl-CDCA and the peptide derived from the coactivator GRIP-1 (PDB code: 1OSV). The orientation of the ligand is the same as that of LCA bound to the VDR.

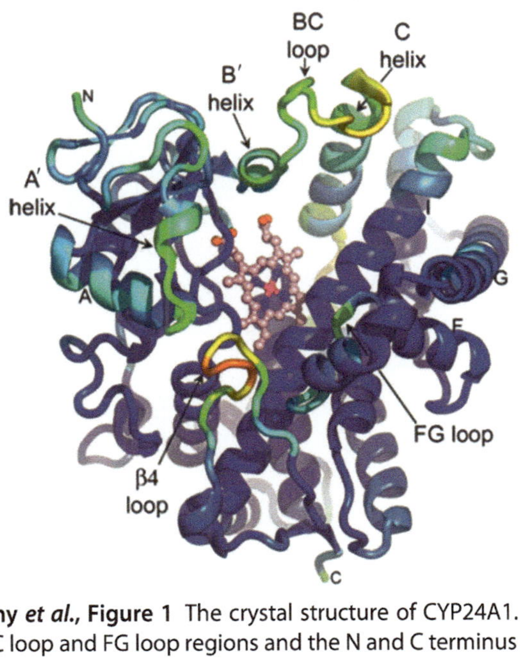

Vaishali Veldurthy *et al.*, Figure 1 The crystal structure of CYP24A1. α-Helices, the β4 loop as well as BC loop and FG loop regions and the N and C terminus are labeled. *From Andrew J. Annalora and G. David Stout with permission.*

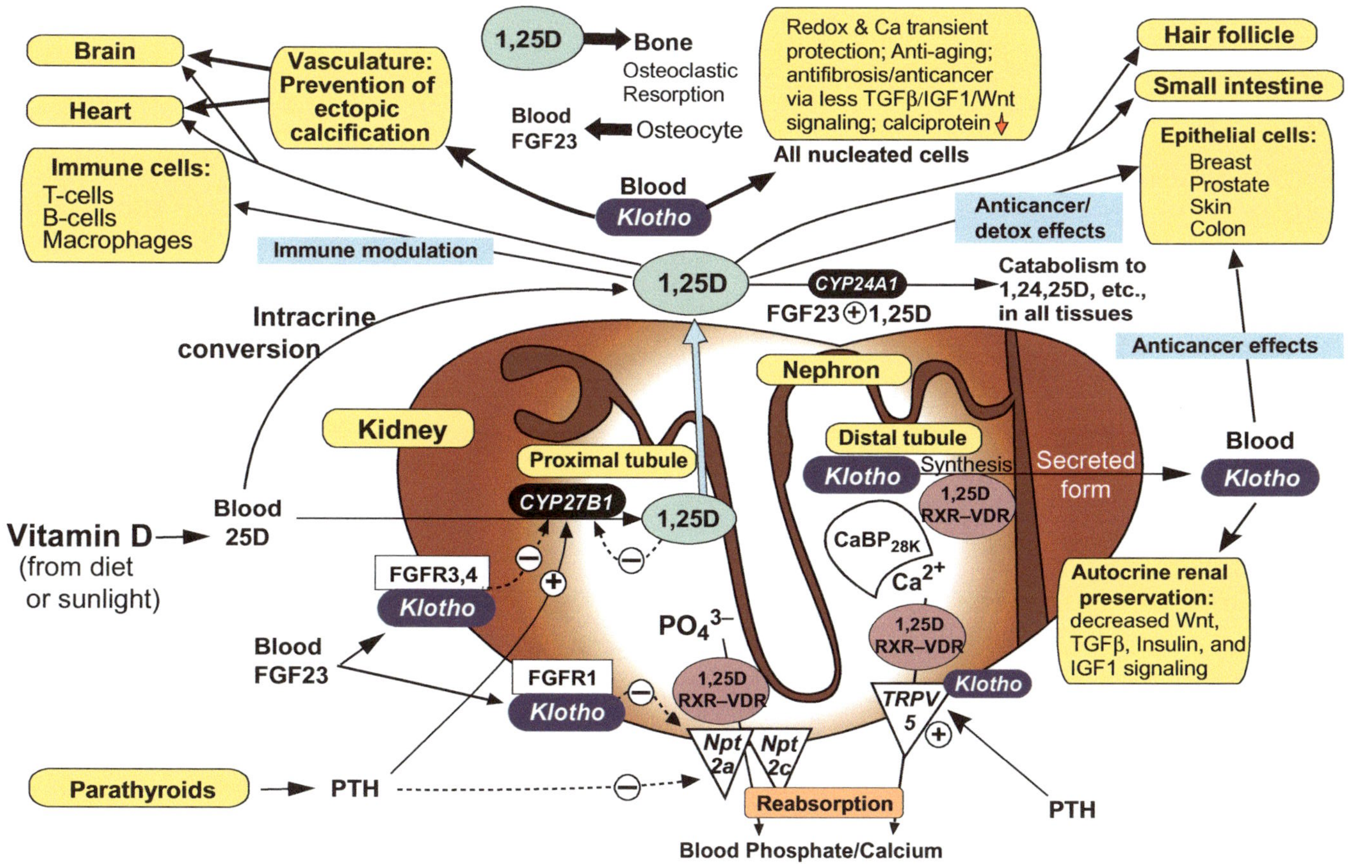

Mark R. Haussler *et al.*, Figure 1 See figure legend on next page

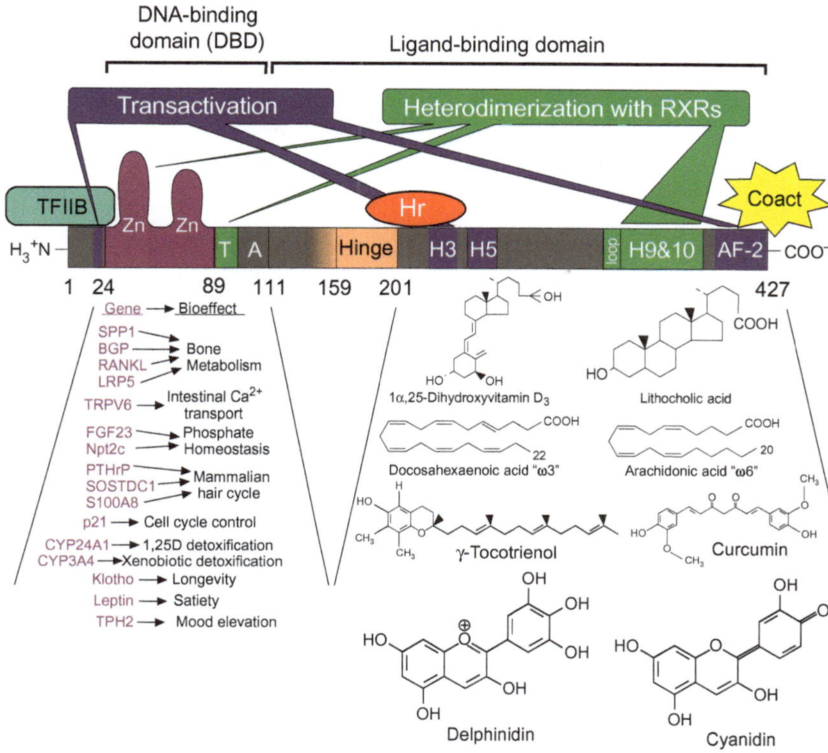

Mark R. Haussler *et al.*, Figure 2 Functional domains in human VDR. Highlighted at the left is the human VDR zinc finger DNA-binding domain that, in cooperation with the corresponding domain in the RXR heteropartner, mediates direct association with the target genes listed at the lower left, leading to the indicated physiological effects. The official gene symbol for bone Gla protein (BGP) is *BGLAP*, for RANKL is *TNFSF11*, for Npt2c is *SLC34A3*, for PTHrP is *PTHLH*, and for klotho is *KL*. Below the ligand-binding domain (at the right) are illustrated selected VDR ligands, including several novel ligands discussed in the text. Also shown above the schematic are the proposed interactions with TFIIB, hairless (Hr), and coactivators (coact), the latter of which associate with the activation function-2 (AF-2) domain.

Mark R. Haussler *et al.*, Figure 1 The kidney is the nexus of healthful aging. The kidney responds to 1,25D, FGF23, and PTH to regulate vitamin D bioactivation and calcium/phosphate reabsorption, and serves as an endocrine source of 1,25D and klotho. Thus, the kidney is the endocrine nexus of health by conserving calcium, eliminating phosphate, and producing 1,25D and klotho "fountain of youth" hormones. Renal hormones 1,25D (shaded in light blue) and klotho (shaded in dark blue) play crucial roles in bone mineral homeostasis to prevent osteomalacia and osteoporosis, but reach beyond these traditional roles to delay chronic disorders of aging such as ectopic calcification, fibrosis, vascular stiffening, heart and kidney function decline, epithelial cell cancers, autoimmune disease, hair loss, and neuropsychiatric conditions.

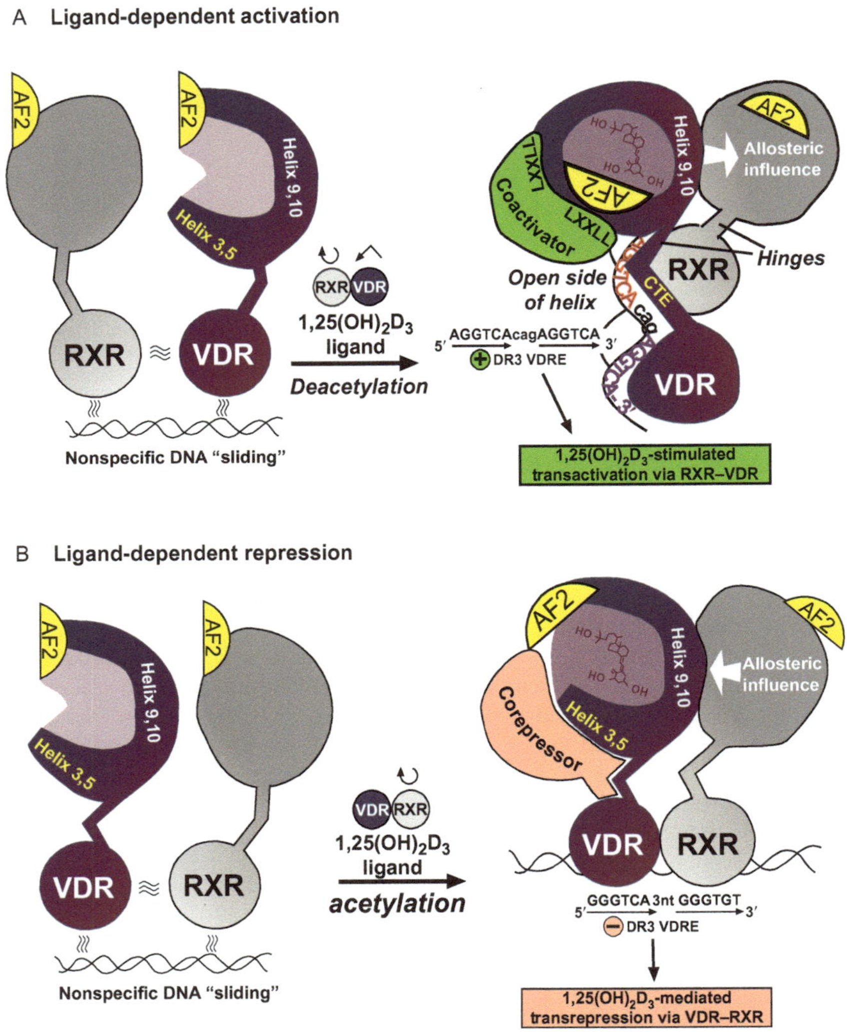

Mark R. Haussler *et al.*, Figure 3 Proposed mechanisms of gene induction and repression by VDR. (A) Allosteric model of RXR–VDR activation after binding 1,25D and coactivator, deacetylation, and docking on a generic positive VDRE with the sequence acAGGTCAcagAGGTCActc (Orlov et al., 2012). See text for explanation. (B) Allosteric model for VDR–RXR inactivation after binding 1,25D and corepressor, acetylation, and docking in reverse polarity on a high-affinity negative VDRE (chicken PTH). See text for explanation.

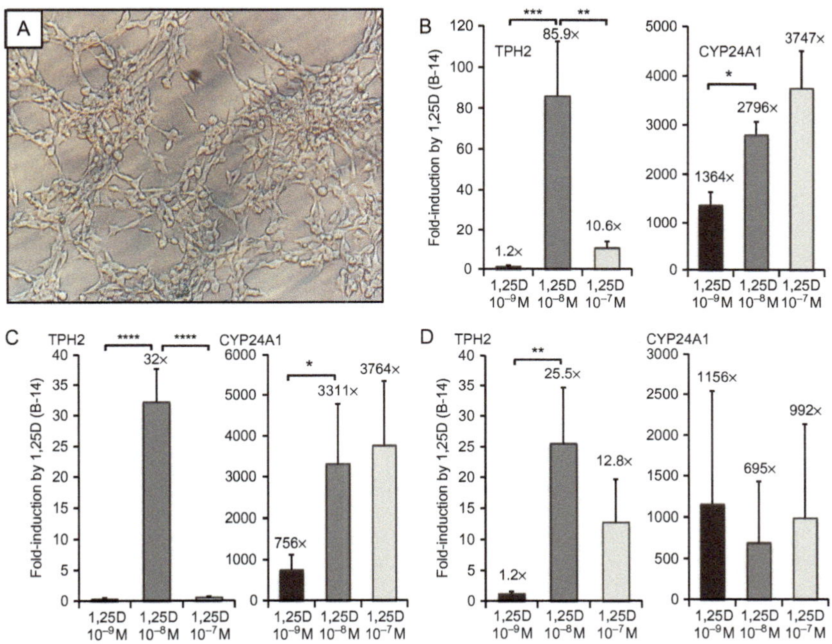

Mark R. Haussler *et al.*, Figure 5 The rat *TPH2* gene is induced biphasically by 1,25D in embryonic medullary raphe neuronal (RN46A-B14) cells. (A) RN46A-B14 cells exhibit a fusiform stellate morphology with neurite interconnections. (B–D) RN46A-B14 cells were treated with 1,25D for 24 h at the indicated concentrations, and mRNA levels determined by real-time PCR as described (Kaneko, Sabir, et al., 2015). Passage numbers are 3 for panel (B), 8 for panel C, and 30 for panel (D). Fold induction by 1,25D is shown on the ordinate of each panel for *TPH2* or *CYP24A1* as indicated. Statistical significance is shown in the body of each panel: $^{*}p < 0.05$; $^{**}p < 0.01$; $^{***}p < 0.001$; $^{****}p < 0.00001$. *Kaneko, Saini, et al. (2015), Copyright (2015) Federation of Societies for Experimental Biology, USA.*

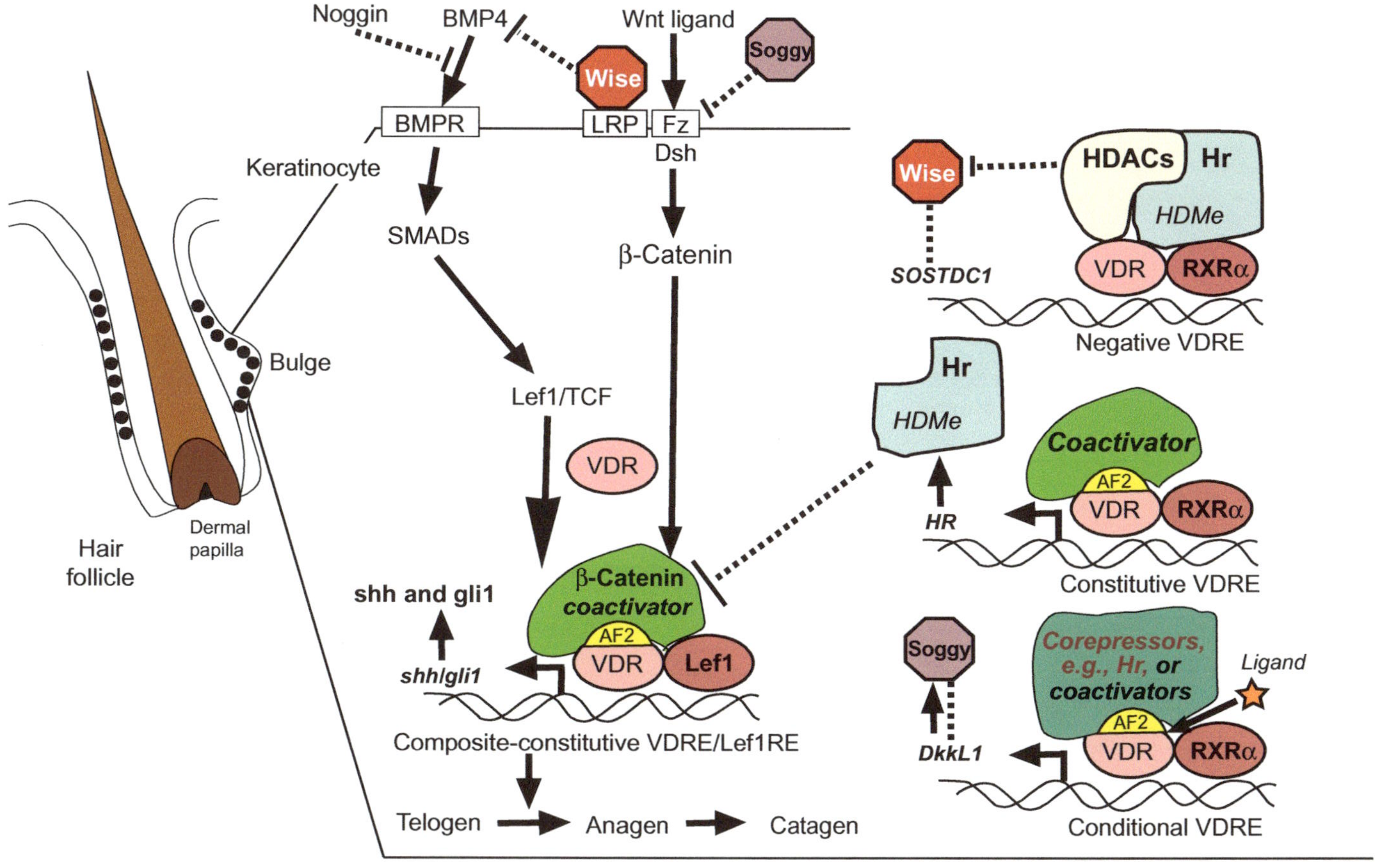

Mark R. Haussler *et al.*, Figure 6 See figure legend on next page

Mark R. Haussler *et al.*, Figure 6 Model for control of the hair cycle by unliganded VDR and β-catenin. Regulation of *HR, SOSTDC1*, and *DkkL1* expression by VDR may contribute to epidermal cell functions and hair cycling by upregulating *HR,* and in turn repressing *SOSTDC1* (Wise) and *DkkL1* (Soggy) expression. Signaling in the mammalian hair cycling is complex, consisting of the convergence of two signaling pathways, BMP and Wnt. Noggin from the dermal papilla initially antagonizes BMP4 signaling in bulge keratinocytes, allowing for the accumulation of Lef1/TCF, a DNA-binding protein that targets genes and controls their expression via transcriptional coactivator partners such as β-catenin. Cessation of Noggin signaling reinstates BMP signal transduction via SMADs provided that Wise (encoded by *SOSTDC1*), which antagonizes both BMP and Wnt pathways (Lintern, Guidato, Rowe, Saldanha, & Itasaki, 2009), has also been repressed, either directly by VDR (upper right), indirectly through VDR induction of Hr (right center), or by a combination thereof. Wnt ligand (e.g., Wnt 10b) signaling leads to accumulation of β-catenin, which cooperates with unliganded VDR and Lef1/TCF to induce genes encoding factors such as sonic hedgehog (*shh*), and *gli1* that trigger the hair cycle to transition from telogen (resting) to anagen (growth). See text for additional discussion. "Constitutive" VDREs refers to those *cis*-elements that dock unliganded VDR (shown as a pink oval), often in combination with RXR or potentially another transcription factor such as Lef1/TCF. "Conditional" VDREs signifies that the direction of gene control (positive vs. negative) may be cell context specific as well as differentiation stage selective. Abbreviations not defined in the text are: HDMe, histone demethylase; Wnt, ortholog of *Drosophila* wingless and mouse int-1; Lef1, lymphoid enhancer factor-1; TCF, T cell-specific factor; and Fz, frizzled. Factors that are membrane receptors or transporters are boxed. Solid arrows indicate activation and dotted lines ending in a solid perpendicular line denote inhibition.

```
shh:
aacccatttccagctccagtcatacgtgcatggagtttcagaaagtctcg
atttggctgggagattggcagcctggaaatctcaaaggaggtgggatggg
aagagaggctcgtgctttgctttgccgtctgtccatccaccctcgtcgcc
gaatatttattcgcttttaattcttatgcaagcaggttaaaaattaaagc
gattgcaaagccagcaagttccaagtccctcccctaaggtaccgcgggct
ctggagaaatgaggagcatccttaaagaaatatcaatacattctctgacc
```

```
gli1:
gaggaggtcatagagtaaggtcagagcctctgaaagtggatttatgagag
atcgtgagccatcatgtggttgctgggaattgagctcaggacctctggaa
gagcagtccttgctcttagccactgagccatcttgccagcccctgattgg
atgattgcttttgaaacagagtttctccatgttggtcctggttgacctgg
aactttctatgtagatcaggctggccttgaacttacagagatgcccttgc
ttctgtcttcccagtgctgggattaaaggcatgtactattatgcacatct
```

Mark R. Haussler *et al.*, Figure 7 *Cis*-regulatory modules (CRMs) in the *shh* and *gli1* genes. CRMs are demarcated by green (forward) and red (reverse) highlighted primers employed by Luderer et al. (2011) in ChIP analysis to localize VDR and Lef1/TCF binding regions. The *shh* CRM, located in the first intron of the mouse *shh* gene, contains 3 Lef1/TCF sites (purple highlight, italics) and 6 VDRE half-elements (yellow highlight, except when antisense (pale blue)). The *gli1* CRM, located approximately 2 kb upstream of the mouse *gli1* transcriptional start site, contains 3 Lef1/TCF sites (pale blue or purple/italics highlight) and 5 VDRE half-elements (yellow highlight), with two of the latter configured in a DR4 motif.

```
-28141
AGGGCAAATGAGTCAGCAGTAGGTTTAAGCAATGAACAGTTGCACCACGGGGTGTTTCTC
AAATTCTTGTCTACTTTTAGAAAGTACGCAATGCCATTCAGTGGGCACCGTGATTCTGAA
ATAGCTGCGAAAATGTTAAAAATGCTGTGGGCAAAGATGTGCCCTCATTGACCTCTTTCT
GAAAACAATAGTCCACAATTGTGTGGCGCGCCCTTCAGCAGCCTGGCTGCTCCCTCTGAT
CCTTGTTGGTGACATGAAGGGAATGCATGAAACTCTCTCTCTGGAGGCTCCTTCCAGCCT
CGAATTCTGATACCTGACACTAGGCCCACATTCCTGATACTCCGGCTGCATCTCTACAGG
ACCAAGACCACCAAGCCAATCGGGTGAGTGTGGACAAAATGTTCTCCCCAATCGATTCTT
TGGCACAGGGTAGTTGAAACAATTAGTTAGCTGGCATTGGTTTTCATTTCACAACTGGCG
GCTCTGGCATCATCGTGACCAGCGGTTTTTCCCACAGCGGTGTCCTAACATGCTGCATGC
CTGGAGCAGTTTTGAGGTGTGTCATAGGCAAGTGGCTACCTAATAACAAACTGTCGAGGC
TTTCGAATGATTGATGATCTTCCTGGTTTTAATCAATTAGAACAGATTCCACATAACCCC
GTAATGATGCCCAGACGGTTGCTTGCATTCTGACATGCCTCTTTGTATGGAAGACGTTTT
GCAACTGGCAGACCAGGTCTGCTTGGAGCTCCAATTCTGCTGGATGCCACGGGAAGAGTG
TCACGAGCCATACGCATGGATATCTGTCATCGGAAGCAGCCGCAGGCTGGCCATCGGAAG
CCCCTGATGCATCCTTAAACATTAAACCACTAAAGATCCAAGCTCTAGAATAACCCGGCC
TAAACAGCCTGCTACCGGCATAGGCTGAGGTTTAGCAAGACAGCTGGTAAACAAGGCAGT
TCAGAAGAACATGGAACAAAGGGCCTGCTCATTATGCTGATCCTTGCTCTAAAAAGAAAC
ATTCCACCATGATCTGTTCATCGATGAGGAAGGGAATGGGGTGTGATGCAGTATAGATTG
AGTATTTGGGTGTCCATTCCTGTACCAATCAATCATGGCCAGTTTAATACTGTTCATGAG
AGACTTTGGGTACTCAGCCTGCGAGCTCGTATTTCCTCCCTCCTGTGGTG
                                                      -26951
```

Mark R. Haussler *et al.*, Figure 8 Sequence of a proposed 1,25D/VDR–RXR CRM in the mouse leptin gene. The positioning of the 1190 bp CRM (depicted in sense orientation) in relation to the start of transcription is indicated by numbers bordering the sequence. DR3 VDREs are highlighted yellow when they are sense, and green (italicized typeface) when they occur antisense. The C/EBP site (antisense) is highlighted pale blue and the RUNX2 site (sense) is highlighted light purple. *Kaneko, Saini, et al. (2015), Copyright (2015) Federation of Societies for Experimental Biology, USA.*

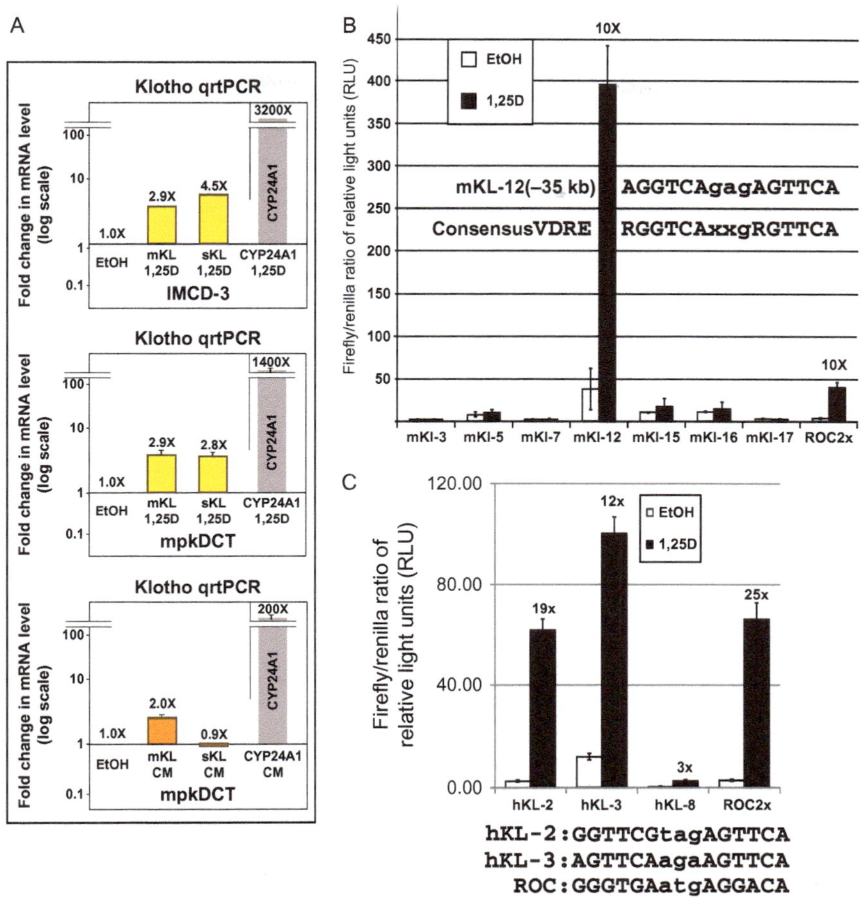

Mark R. Haussler *et al.*, Figure 9 Regulation of membrane or secreted klotho forms (abbreviated mKL or sKL) by 1,25 and activity of candidate mouse and human klotho VDREs. (A) Upregulation of mKL and sKL in mouse intermedullary collecting duct (IMCD-3) cells (upper panel) or mouse distal convoluted (mpkDCT) cells (lower two panels) by either 1,25D (upper two panels), or curcumin (CM, lower panel). Cells were treated with ligand for 24 h prior to RNA isolation and real-time PCR (Forster et al., 2011). (B) Candidate mouse klotho VDREs were cloned into a pLUC-MCS reporter vector, cotransfected into HK-2 human kidney cells along with a pSG5-VDR cDNA expression plasmid and treated with 1,25D (10^{-8} M) for 24 h. Firefly luciferase values were normalized to expression of Renilla luciferase. Data are depicted as a fold effect of 1,25D. Results reveal that only the mouse klotho VDRE located at −35 kb (mKl-12) displays transactivation ability. (C) Transfection of candidate human klotho VDREs demonstrates a striking (>10-fold) 1,25D responsiveness of VDREs corresponding to sequences at −46 kb (hKL-2) and −31 kb (hKL-3), but not +3.2 kb (hKL-8). The sequences of the active VDREs in the mouse and human *Klotho* genes are listed in panels (B) and below (C); they are very similar to proven VDREs, with the mouse VDRE at −35 kb (mKL-12) conforming exactly to the consensus VDRE.

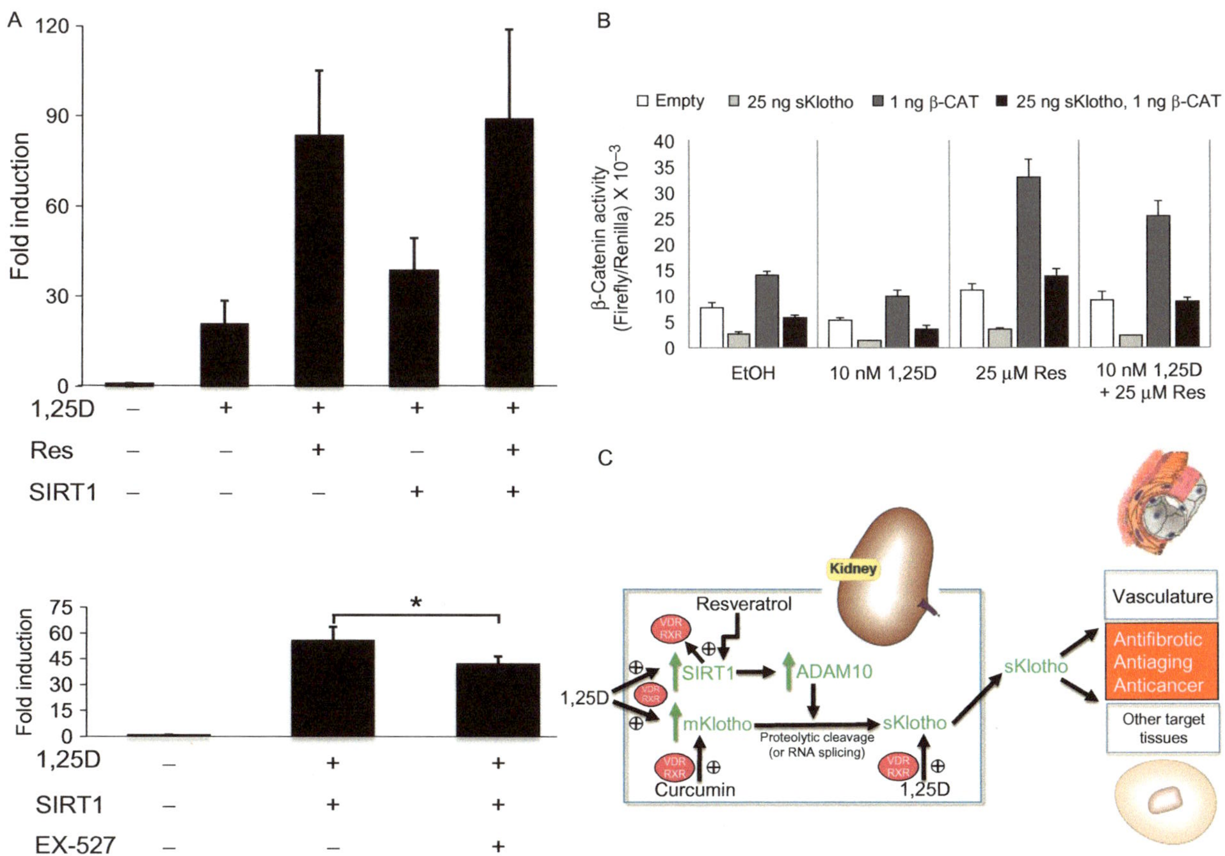

Mark R. Haussler *et al.*, Figure 10 See figure legend on next page

Mark R. Haussler *et al.*, Figure 10 Resveratrol and SIRT1 cooperate with 1,25D to enhance VDR signaling and the production of α-klotho. (A) Upper panel: HEK-293 cells were treated with 1,25D, resveratrol and/or SIRT1, and endogenous *CYP24A1* expression in human embryonic kidney cells was measured by real-time PCR. Lower panel: As in the upper panel, but treatments included the selective SIRT1 inhibitor, EX-527. (B) HEK-293 cells were cotransfected with a Firefly luciferase plasmid containing a β-catenin responsive element along with the indicated expression plasmids encoding soluble klotho (sKlotho) or β-catenin (β-CAT), with Firefly luciferase results normalized to Renilla luciferase. (C) A hypothetical model for resveratrol activation of VDR via stimulation of SIRT1 (Baur, 2010). SIRT1 catalyzes deacetylation of VDR (to increase the capacity of 1,25D binding), RXR, or comodulators. SIRT1 activation also leads to ADAM10 stimulation (Donmez, Wang, Cohen, & Guarente, 2010) to produce soluble klotho (sKlotho) via ADAM10-mediated cleavage of membrane klotho (mKlotho). Curcumin selectively induces mKlotho, while 1,25D stimulates SIRT1 activity (An et al., 2010) as well as expression of both mKlotho and sKlotho. The integration of these regulatory circuits, which are controlled by the levels of nutritionally derived "healthy" lipids (1,25D, curcumin, and resveratrol), culminates in the elaboration of sKlotho from the kidney to exert proposed endocrine anti-aging effects consisting of antifibrogenic and antineoplastic actions in the vasculature and other target tissues.

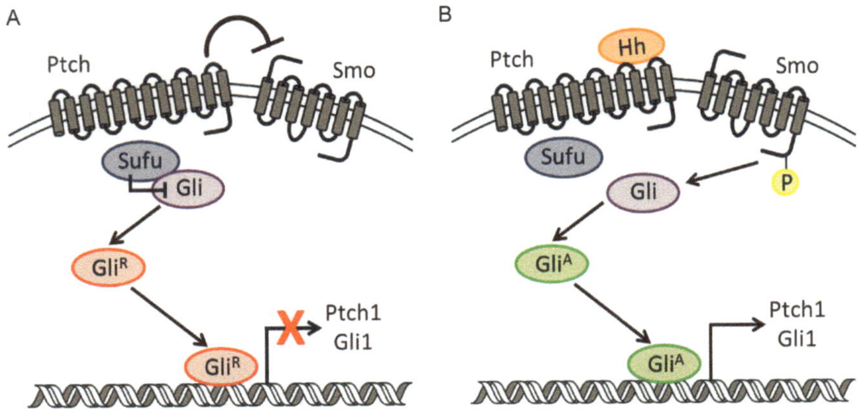

M. Kyle Hadden, Figure 1 The Hh signaling pathway. *Adapted from Hadden (2014).*

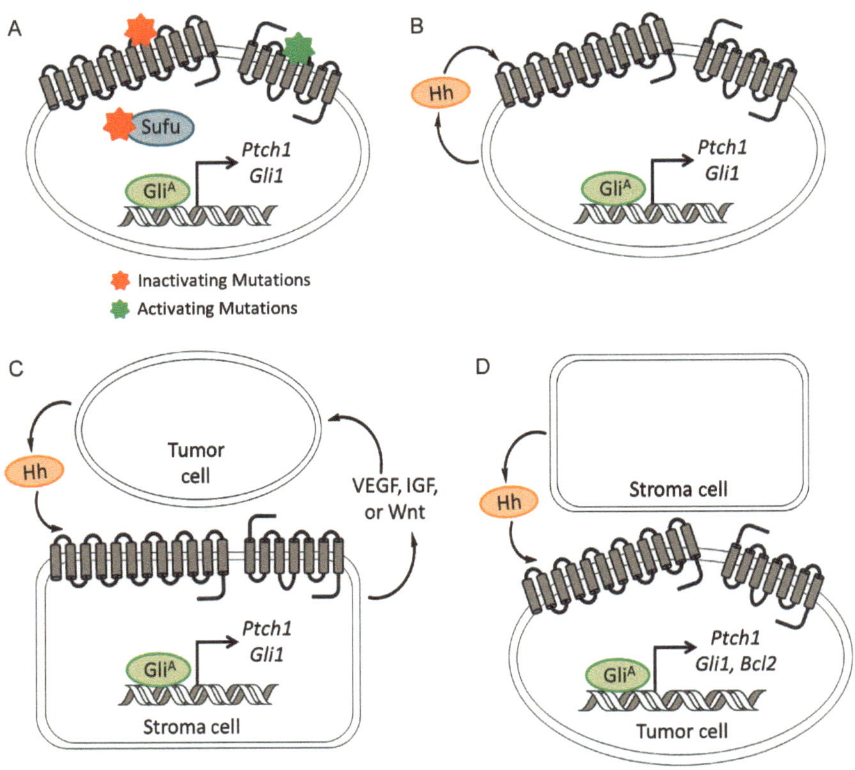

M. Kyle Hadden, Figure 2 Mechanisms of dysregulated Hh pathway signaling that have been implicated in the development of human cancer. (A) Mutations in key pathway components, (B) autocrine signaling, (C) paracrine signaling, and (D) reverse paracrine signaling.

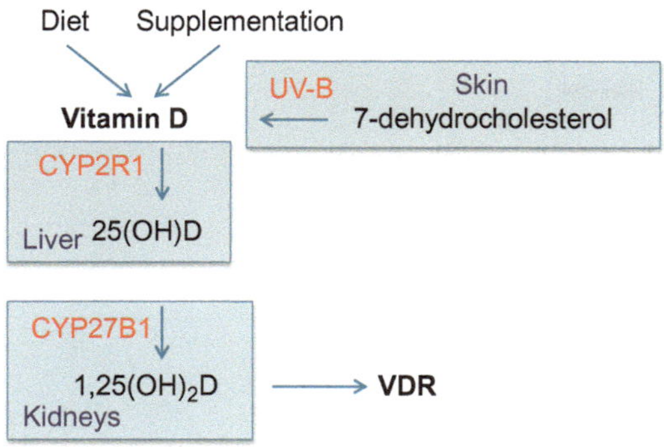

Carsten Carlberg, Figure 1 Vitamin D synthesis pathway. Vitamin D is taken up by diet or supplementation or synthesized endogenously in the skin under essential catalysis by UV-B. In the liver, vitamin D is converted by the enzyme CYP2R1 into the circulating form 25(OH)D and then in the kidneys by the enzyme CYP27B1 into the high-affinity VDR ligand 1,25(OH)$_2$D.

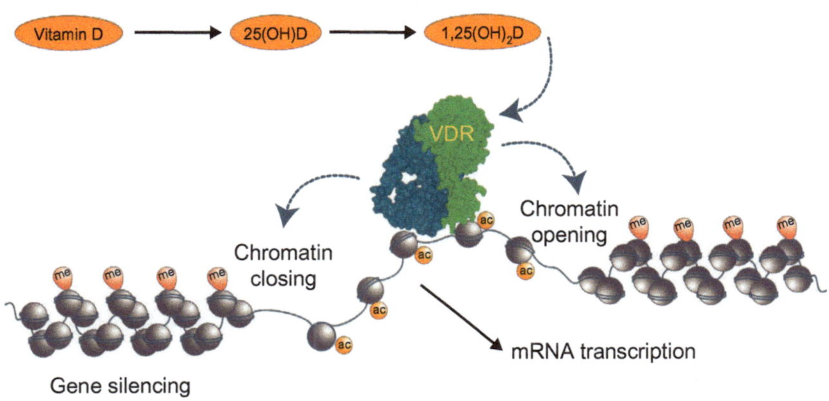

Carsten Carlberg, Figure 2 Chromatin model of vitamin D signaling. Gene regulation by VDR requires accessible genomic DNA, i.e., open chromatin. In turn, ligand-dependent actions of VDR result either in further opening of chromatin (i.e., in most cases, in upregulation of the transcription of the respective gene) or in closing of chromatin (and respective downregulation of transcription).

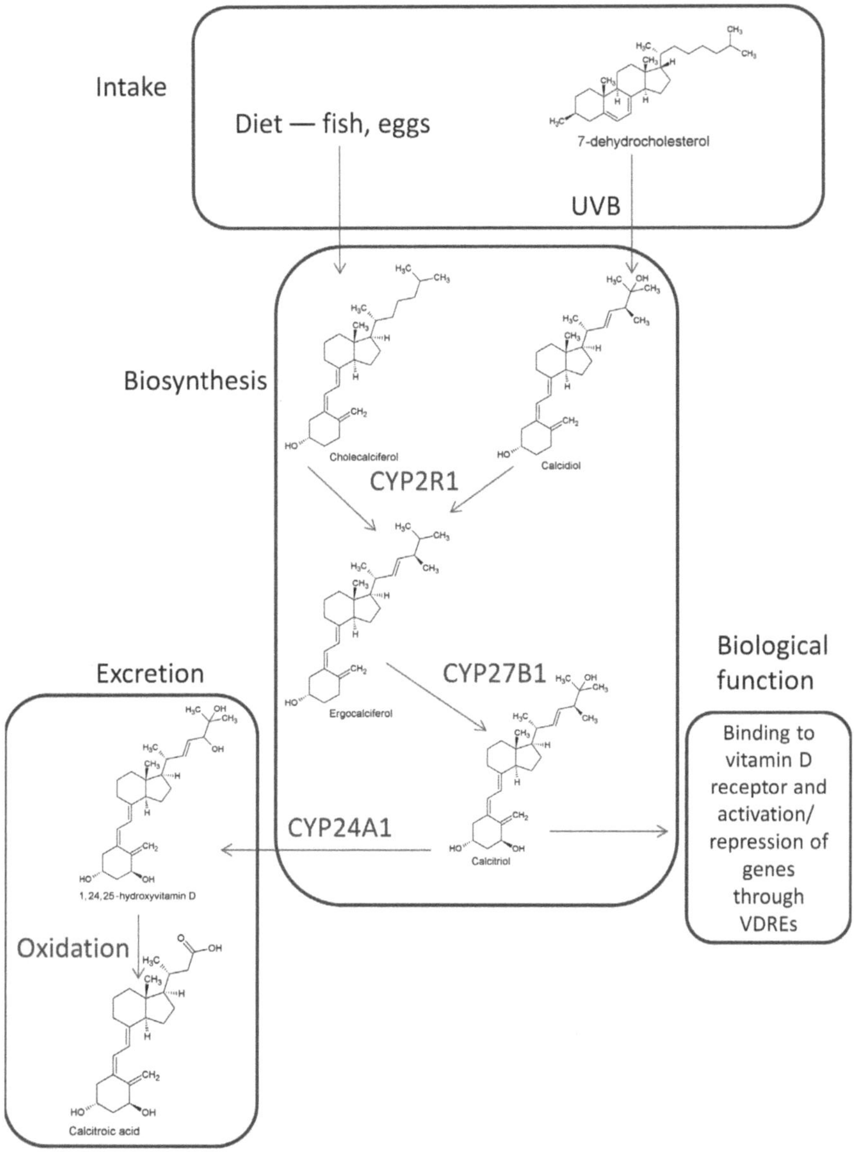

Rowan P. Orme *et al.*, Figure 1 Biosynthesis of Vitamin D. Vitamin D is obtained from diet in the form of Ergocalciferol or from the action of UVB from the sun on 7-dehydroxycholesterol in the skin by way of cholecalciferol. These biologically inactive molecules undergo successive hydroxylation reactions in the liver and kidneys to produce calcidiol and finally the bioactive vitamin D molecule, calcitriol. Calcitriol binds to its nuclear receptor, vitamin D receptor, which once ligated binds to vitamin D response elements to activate or suppress genes, thus exerting its biological control. Excretion of vitamin requires further metabolism of calcitriol by further hydroxylation and finally oxidation and side chain cleavage to produce the highly water soluble calcitroic acid.

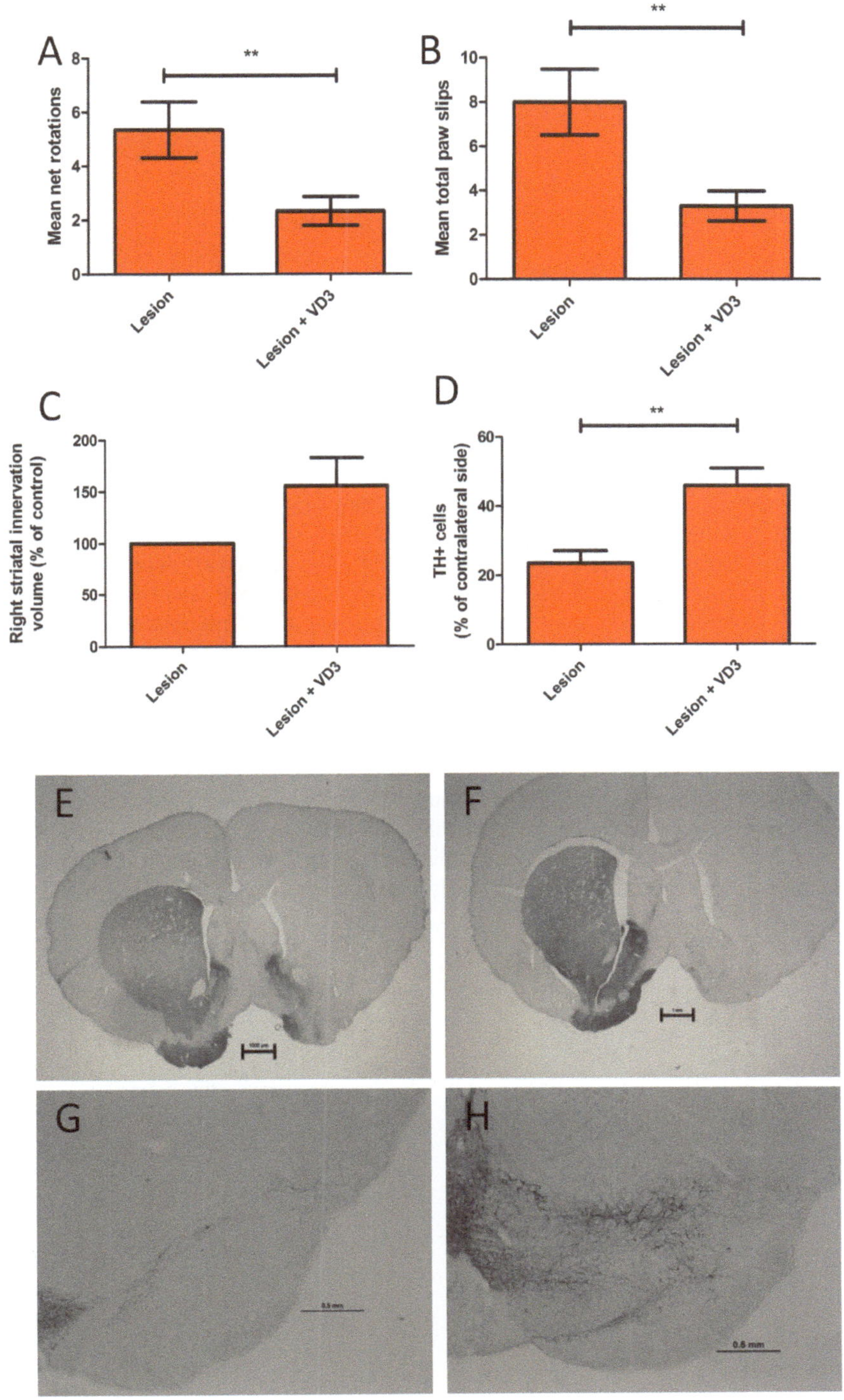

Rowan P. Orme *et al.*, Figure 2 See figure legend on opposite page.

Rowan P. Orme *et al.*, Figure 2 Neuroprotective action of Vitamin D in a 6-OHDA model of Parkinson's disease. In functional testing of animals following lesion with 6-OHDA, those pretreated with vitamin D showed less severe drug induced rotational behavior when stimulated with apomorphine (A). The pretreated animals were also more successful at the balance beam test, having significantly fewer paw-slips than control animals (B). Brains were removed from animals and immunostained with antibodies specific to tyrosine hydroxylase to visualize striatal innervation. Striatal innervation volumes were not significantly different between pretreated animals and vehicle only controls following lesioning (C). However, in the substantia nigra of pretreated animals there was a significant increase in the number of TH+ cells when expressed as percent of the intact side (D), $^{**}p < 0.01$. Representative images of sections through the striatum of pretreated (E) and control (F) animals clearly show the effect of lesioning on striatal innervation (scale bars = 1 mm). The lesioned substantia nigra shows increased loss of TH immunoreactive cells in the control animals (G) compared to vitamin D pretreated subjects (H); scale bars 0.5 mm.

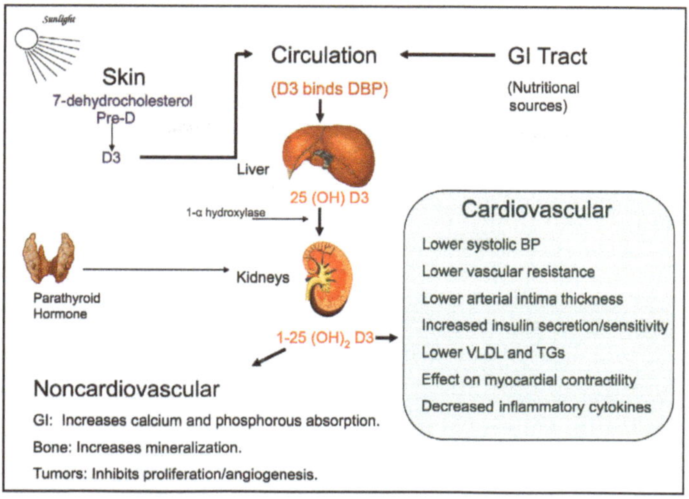

Irene M. Kim *et al.*, Figure 2 Metabolism and biologic actions of vitamin D. The major biologic form of vitamin D, vitamin D_3, is synthesized in the skin from the precursor (pre-D) under direct sunlight. Vitamin D from cutaneous synthesis and nutritional sources enters the circulation and is bound to DBP. A series of enzymatic hydroxylations in the liver and kidneys transform vitamin D to biologically active $1,25(OH)_2$ vitamin D. Parathyroid hormone regulates the hydroxylation in kidney. Activated vitamin D exerts multiple cardiovascular and noncardiovascular actions. BP, blood pressure; GI, gastrointestinal; TG, triglyceride; VLDL, very low-density lipoprotein (Vanga et al., 2010).

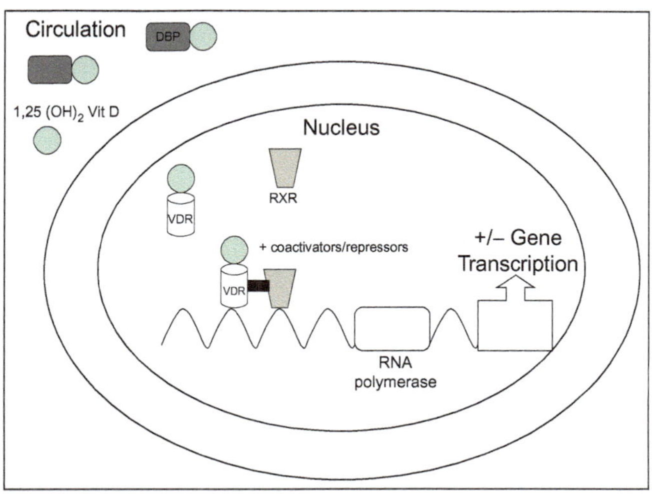

Irene M. Kim *et al.*, Figure 3 $1,25(OH)_2$ Vit D is bound to DBP in circulation, crosses the cell membrane, and binds to VDR. The conjugated vitamin D with its receptor forms a heterodimer complex with RXR and, with other factors, attaches to vitamin D-responsive elements on DNA and alters gene expression. RNA, ribonucleic acid (Vanga et al., 2010).

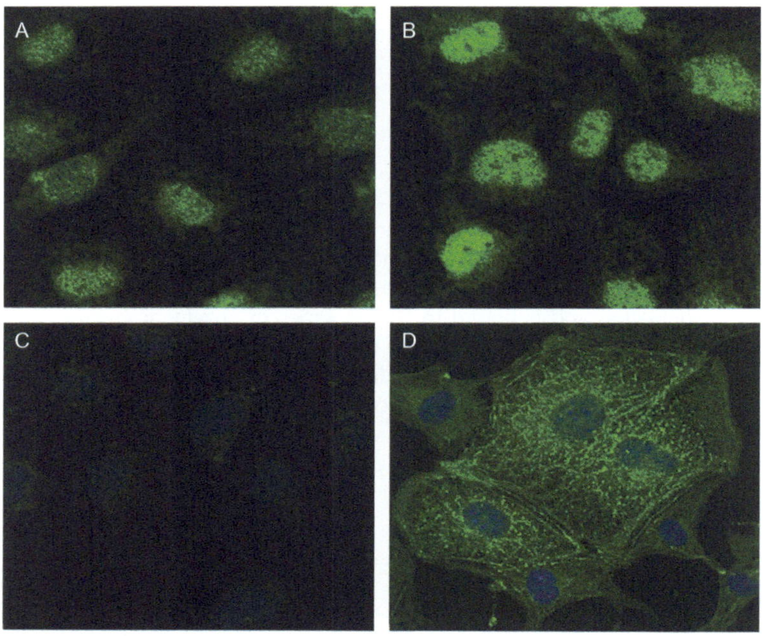

Irene M. Kim *et al.*, Figure 5 Immunofluorescent analysis of VDR. Confocal images of HL-1 cells treated with 100 nM 1,25-D_3 (B) versus control (A) for 24 h. To ensure specific staining a VDR specific peptide, SC1008P (Santa Cruz), was preincubated with primary VDR Antibodies (Ab) (C). Adequate Ab penetration and cell architecture preservation is evidenced by actinin staining (D). All digital images are 0.35 μm sections captured via Olympus FV500. VDR primary Ab is SC1008 with FITC conjugated secondary AP307F (Chemicon). Actinin primary Ab is A7811 with FITC conjugated secondary A8711 (both Sigma) (Nibbelink et al., 2007).

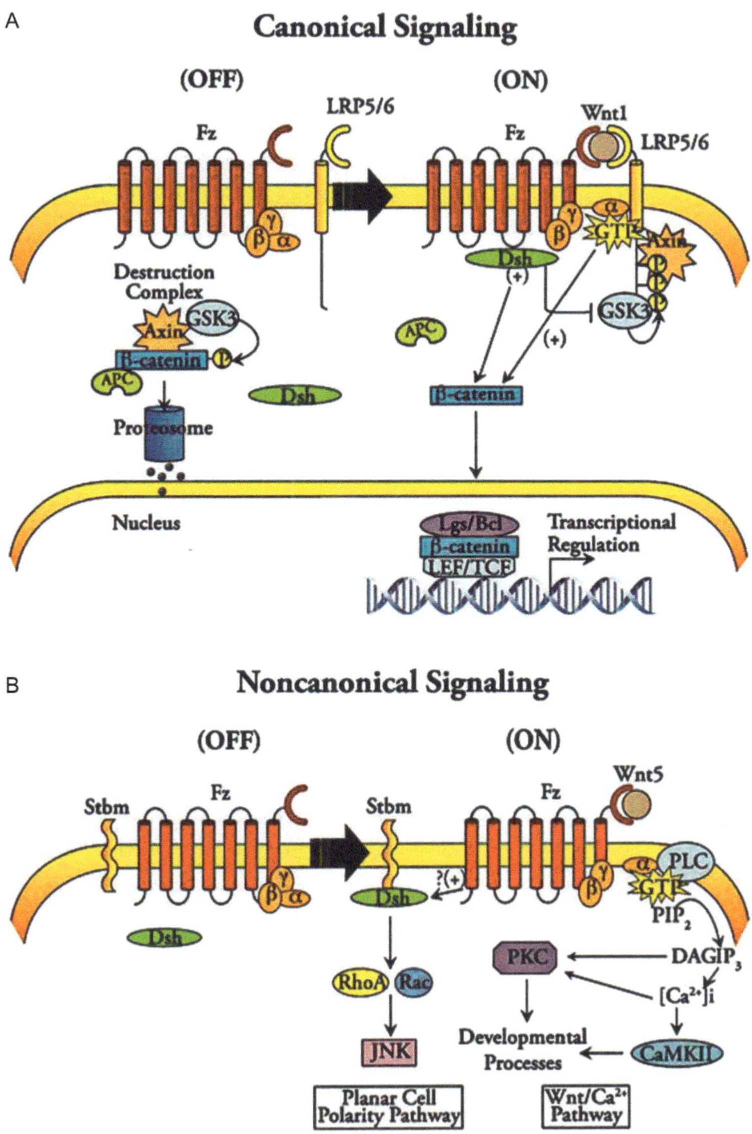

Irene M. Kim *et al.*, Figure 6 Schematic diagrams depicting Wnt signal transduction pathways. (A) Canonical signaling: (OFF) in the absence of Wnt ligand binding to a receptor complex consisting of Fz and low-density lipoprotein receptor-related protein (LRP, isoforms 5 or 6) coreceptors, cytoplasmic Dishevelled (Dsh) remains inactive and β-catenin, sequesetered within the destruction complex comprising adenomatosis polyposis coli and Axin, is phosphorylated by glycogen synthase kinase-3 (GSK-3), thereby targeting it for ubiquitination and proteosomal degradation. Note the presence of heterotrimeric G-proteins interacting with the Fz receptor. (ON) Wnt ligand binding to the intracellular domain of LRP5/6 where it binds axin with high affinity, sequestering it away from the destruction complex, thereby enabling lymphoid enhancer factor-dependent transcription. The Gα subunit has also been implicated in the disassembly of the β-catenin destruction complex, linking it to the canonical signaling. (B) Noncanonical signaling represented as two distinct pathways: the Wnt/Ca^{2+} pathway and the planar cell polarity (PCP) pathway. Ca^{2+}/PKC signaling occurs via Wnt activation of Fz receptors causing hetertorimeric G-protein-dependent Ca^{2+} release, thus activating PKC and CaMKII. Rho/JNK signaling occurs via cooperative interaction between transmembrane receptor strabismus (Stbm) and Fz receptors, which activates Rho-Rac-mediated activation of JNK through Dsh (Flaherty & Dawn, 2008).

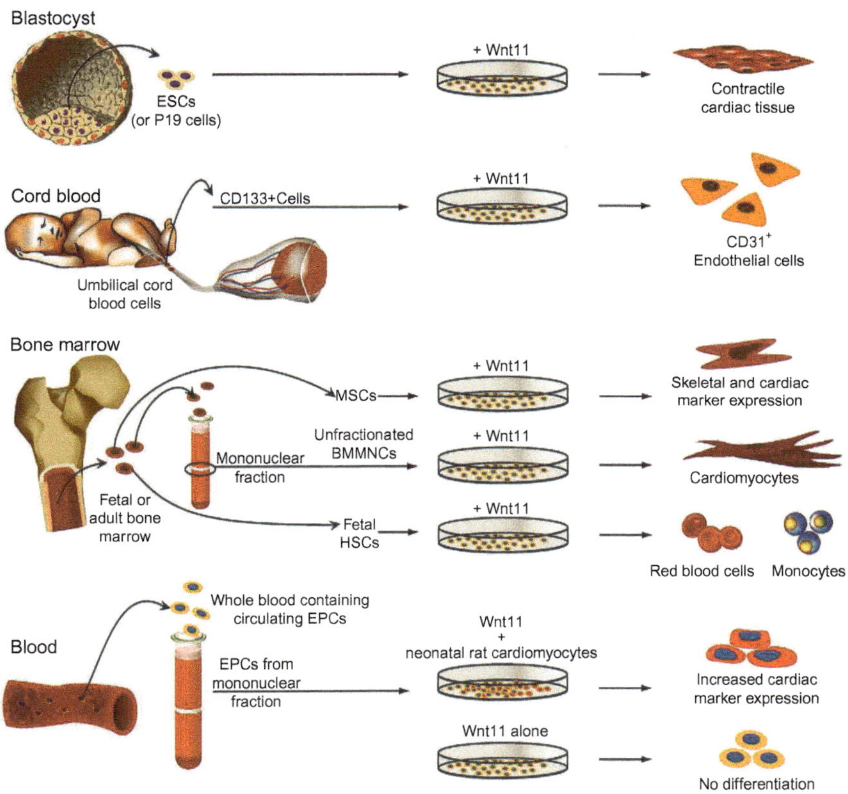

Irene M. Kim *et al.*, Figure 7 Differential impact of Wnt11 signaling on cell fate *in vitro*. Although blastocyst-derived pluripotent embryonic stem cells give rise to cardiac tissue with a contractile phenotype, Wnt11-treated, cord blood-derived CD133+ progenitors undergo endothelial lineage commitment. Wnt11-treated, bone marrow-derived cells show a remarkable degree of divergence with regard to their lineage commitment and the phenotypic characteristics they assume. In fetal hematopoietic progenitors (HSCs), Wnt11 promotes red blood cell commitment and favors monocyte precursor formation. Bone marrow-derived multipotent adult stem cells treated with Wnt11 differentiate along a cardiomyogenic pathway, but fail to adopt an adult phenotype, whereas Wnt11 is sufficient to produce a cardiomyocytic phenotype in bone marrow mononuclear cells (BMMNCs). Finally, in circulating progenitor cells (CPCs) that adopt a cardiac phenotype when cocultured with neonatal rat cardiomyocytes, Wnt11 enhances expression of cardiac markers, yet, by itself, is unable to affect the native phenotype. ESCs, embryonic stem cells; MSCs, multipotent stem cells; EPCs, endothelial progenitor cells (Flaherty & Dawn, 2008).

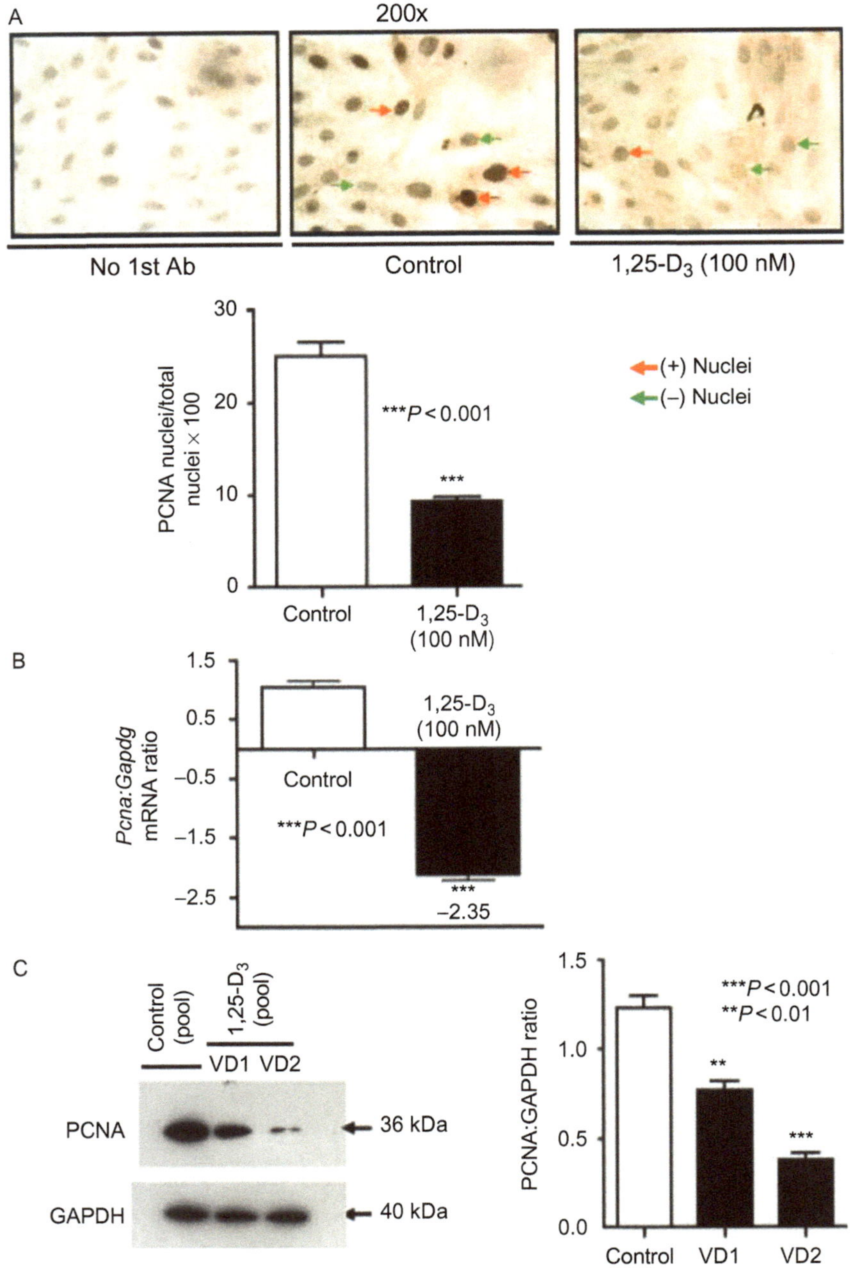

Irene M. Kim *et al.*, Figure 9 See figure legend on opposite page.

Irene M. Kim *et al.*, Figure 9 1,25-D_3 downregulates the expression of PCNA. Cultures of H9c2 cells were treated as illustrated in Fig. 8 for 4 days. Immunocytochemistry (ICC) reactions, real-time PCR, and western blotting were carried out at the end of the incubation period. Arrows indicate nuclei positive for PCNA. (A) Representative ICC pictures with the corresponding image analysis to determine the percentages of positive cells (brown nuclear staining) for experiments done in triplicate, $^{***}P < 0.001$. Magnification 200 ×. (B) Steady-state mRNA downregulation of PCNA. Mean ± SEM corresponds to experiments done in triplicate, $^{***}P < 0.001$. (C) Protein extracts were subjected to western blotting and the corresponding image analysis. Mean ± SEM corresponds to experiments done in triplicates, $^{**}P < 0.01$ and $^{**}P < 0.001$. Control, VD1, and VD2 are different pools of two samples each. In both cases, real-time PCR and western blotting, samples and controls were normalized to the GAPDH housekeeping gene (Hliang et al., 2014).

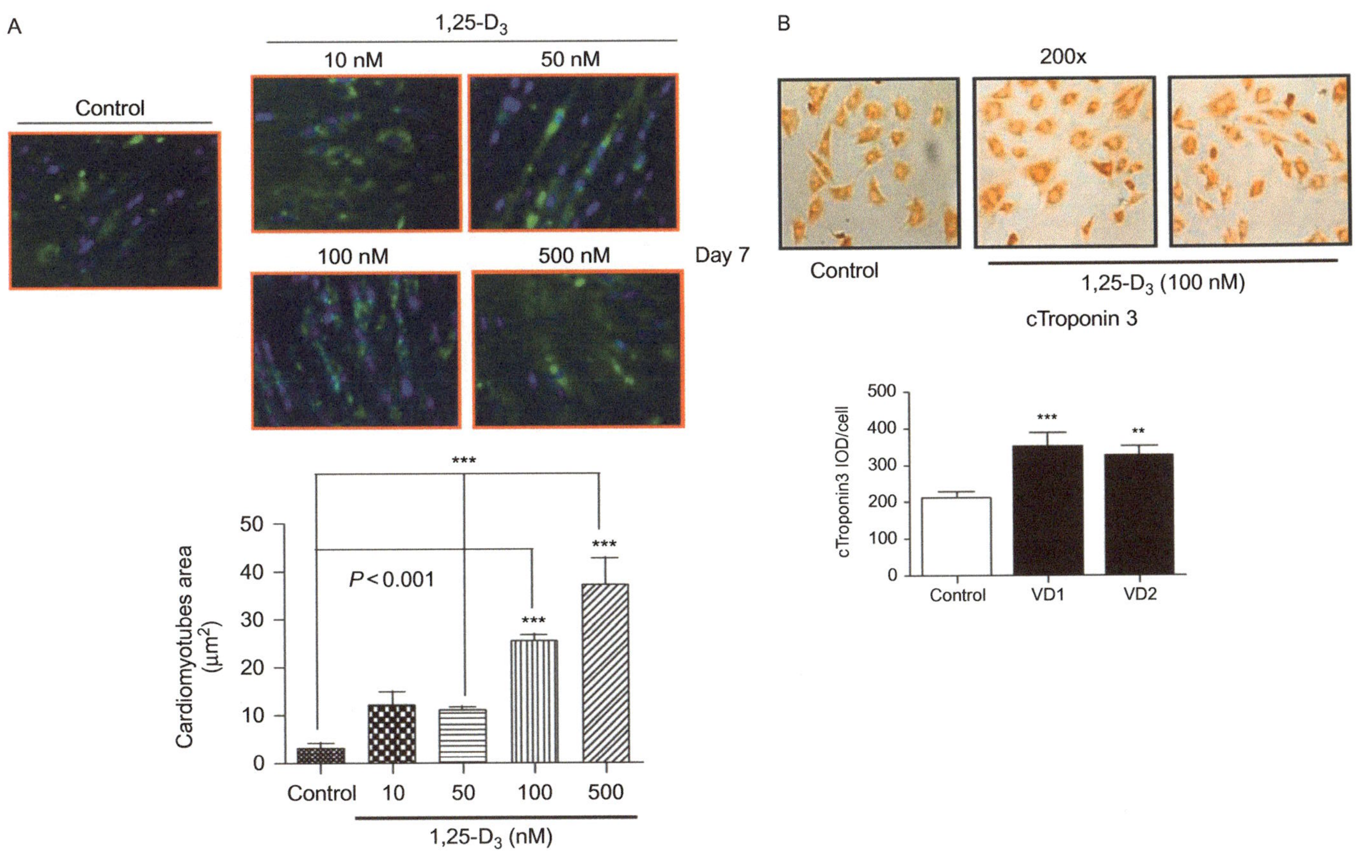

Irene M. Kim *et al.*, Figure 10 See figure legend on next page.

Irene M. Kim *et al.*, Figure 10 1,25-D_3 increases cardiomyotubes area and the expression of cardiac troponin. (A) Cultures of H9c2 cells preincubated with PKH2 Green Fluorescence Cell Linker and counterstained with DAPI were treated as illustrated in Fig. 8 for 7 days. Representative pictures are presented for control (No 1,25-D_3 addition) and for increasing the concentrations (10–500 nM) of 1,25-D_3 with the corresponding image analysis of the cardiomyotube area. $^{***}P<0.001$. Magnification 200×. In a parallel experiment, cultures of H9c2 cells incubated on eight-well chamber slides were treated as described in Fig. 8 for 7 days. At the end of the incubation period, the cells were fixed and subjected to ICC. (B) Representative ICC pictures of cardiac Troponin+cells with the corresponding image analysis expressing percentage IOD (area × intensity) for experiments done in triplicate. $^{**}P<0.01$ and $^{***}P<0.001$. Magnification 200× (Hliang et al., 2014).

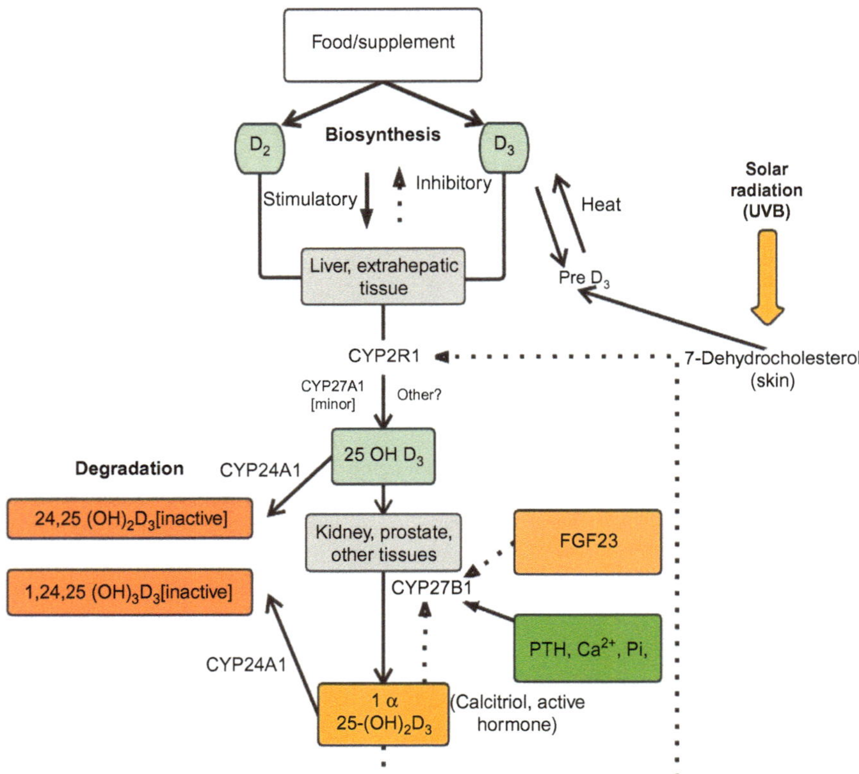

Jungmi Ahn *et al.*, Figure 1 Schema showing enzyme-catalyzed pathways to the biosynthesis and degradation of calcitriol. Regulation of CYP2R1 and CYP27B1 by calcitriol (1α,25-dihydroxy vitamin D_3), parathyroid hormone (PTH), phosphorous (Pi), calcium (Ca^{2+}), and fibroblast growth factor-23 (FGF23) is shown. Stimulation is indicated by a solid arrow (⟶); inhibition is shown by a broken arrow (⇢).

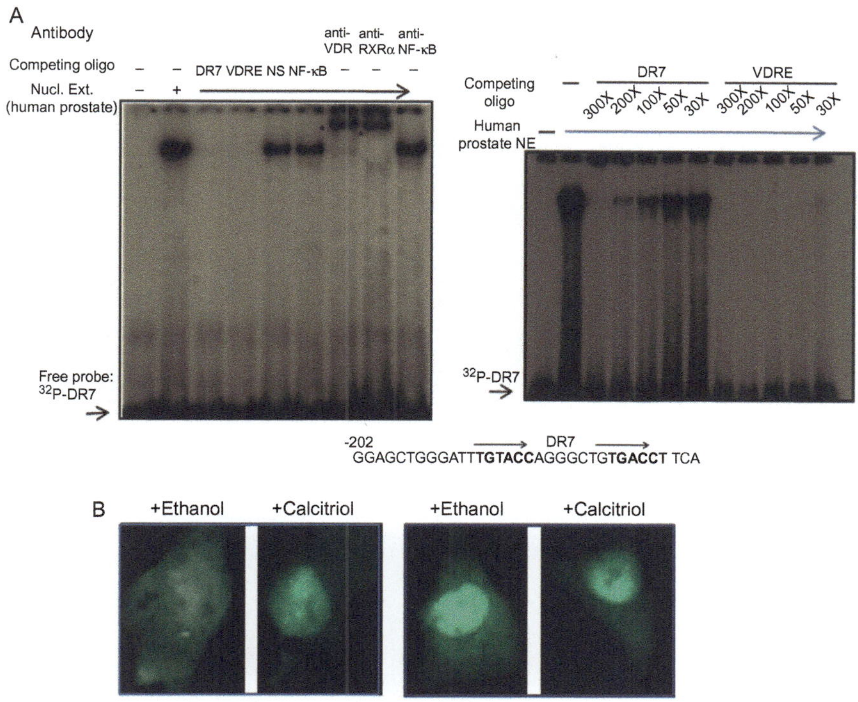

Jungmi Ahn *et al.*, Figure 2 (A) Electrophoretic mobility shift assay (EMSA) showing binding of VDR and RXR-α (present in the nuclear extract of normal human prostate tissue) to the DR7-type element present in and around the -202 nucleotide position of the *SULT2B1* promoter. The promoter sequence containing the DR7-type element is shown. *Left panel*: oligonucleotide competition demonstrates specificity of the EMSA complex; antibody supershift assay demonstrates the presence of VDR and RXR-α within the EMSA complex. *Right panel*: Relative affinities of DR7 and a DR3-type VDRE (from the rat osteocalcin promoter) for the EMSA complex formed by ^{32}P-labeled DR7 element with VDR and RXR-α components of human prostate nuclear extract. These data from our experiments are taken from Seo et al. (2013). (B) Subcellular localization of transfected CFP-VDR in COS1 cells treated with ethanol or 10 nM calcitriol. Representative data from two separate transfection experiments are shown. CFP, cyan fluorescent protein.

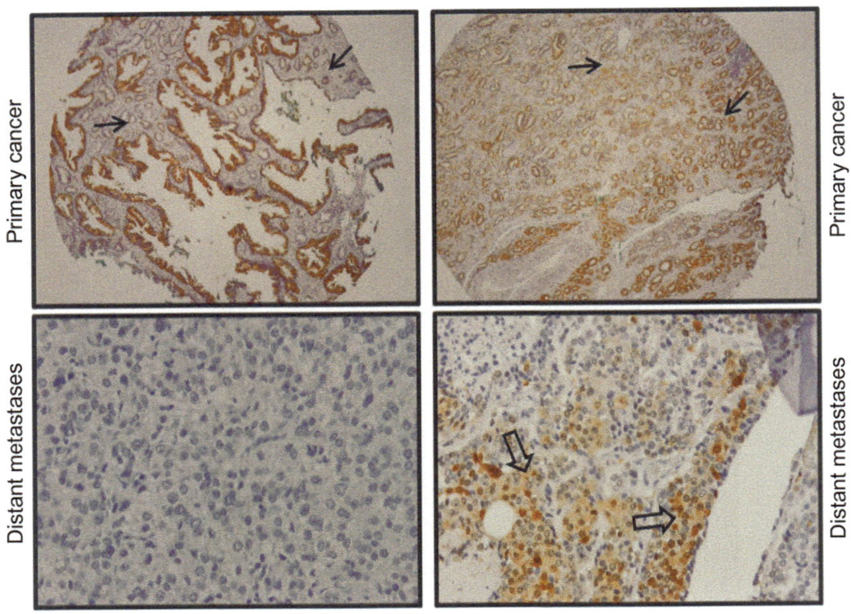

Jungmi Ahn *et al.*, Figure 3 Photomicrographs of SULT2B-immunostained prostate cancer specimens from four cases. *Upper panel*: Primary prostate cancer from two patients; photomicrograph taken at 4 ×. The primary specimen at left shows non-malignant acini, which stained strongly for SULT2B, as well as cancerous areas (solid arrows), which show markedly reduced SULT2B levels. The tissue core at the right shows only the malignant region of primary cancer (arrows), which stained weakly for SULT2B. *Lower panel*: distant metastases of prostate cancer from two patients; photomicrograph taken at 20 ×. Open arrows show remaining SULT2B expression in small percent of metastatic cancer cells for the second patient.

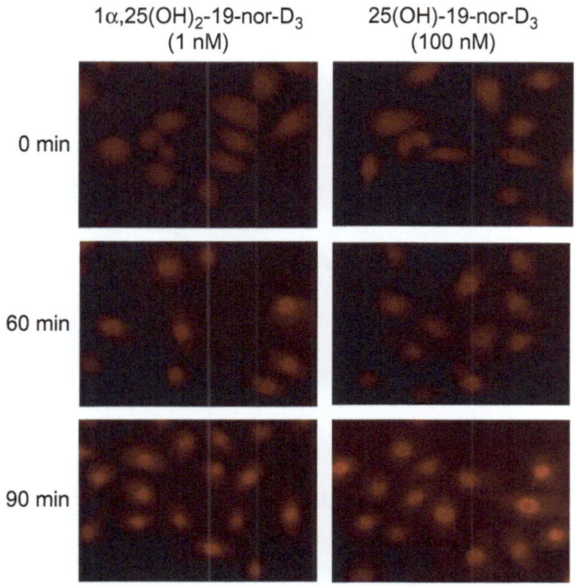

Eiji Munetsuna *et al.*, Figure 1 Translocation of VDR into the nucleus by 1α,25$(OH)_2$-19-nor-D_3 and 25(OH)-19-nor-D_3. VDR was detected by a fluorescence microscope with antihuman VDR rat monoclonal antibody, 9A7 (Abcam, Cambridge, UK; Munetsuna et al., 2011).

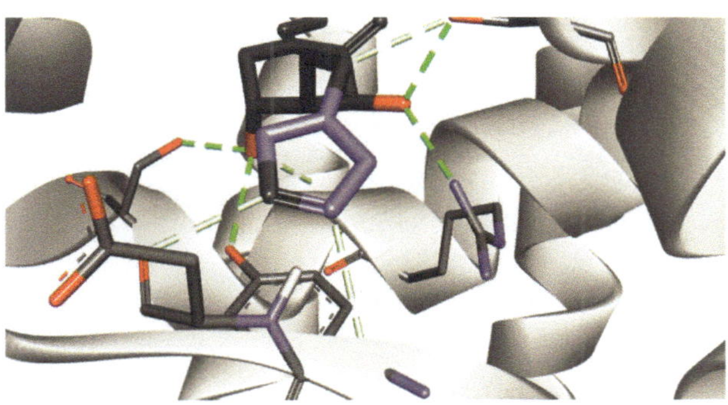

Atsushi Kittaka *et al.*, Figure 3 X-Ray studies on hVDR LBD-**1a** complex. The Protein Data Bank accession number for the coordinates of the structure of the VDR complex with **1a** is 4ITE. The A-ring part is magnified. *Adapted with permission from Matsuo et al. (2013, 2014).*

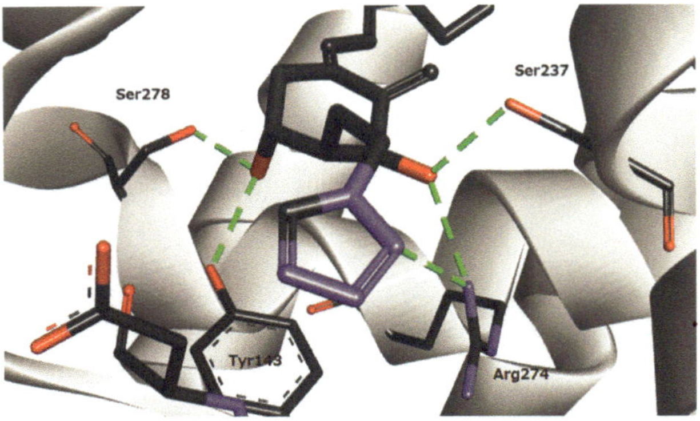

Atsushi Kittaka *et al.*, Figure 4 X-Ray studies on hVDR LBD-**1b** complex. The Protein Data Bank accession number for the coordinates of the structure of the VDR complex with **1b** is 4ITF. The A-ring part is magnified. *Adapted with permission from Matsuo et al. (2014).*

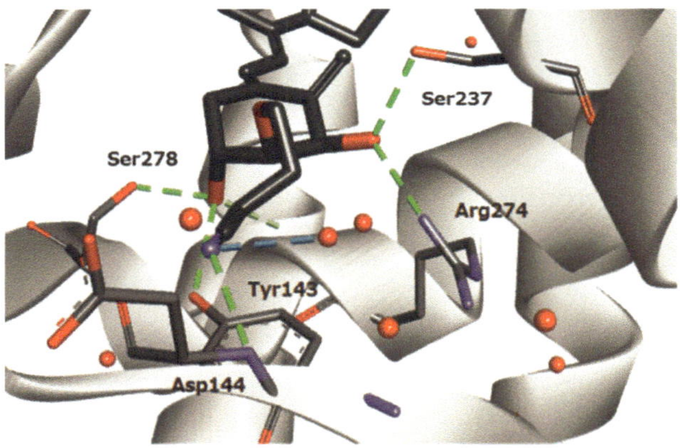

Atsushi Kittaka *et al.*, Figure 5 X-Ray studies on hVDR LBD-**3** complex. The Protein Data Bank accession number for the coordinates of the structure of the VDR complex with **3** is 4PA2. The A-ring part is magnified. *Adapted with permission from Saitoh et al. (2015).*

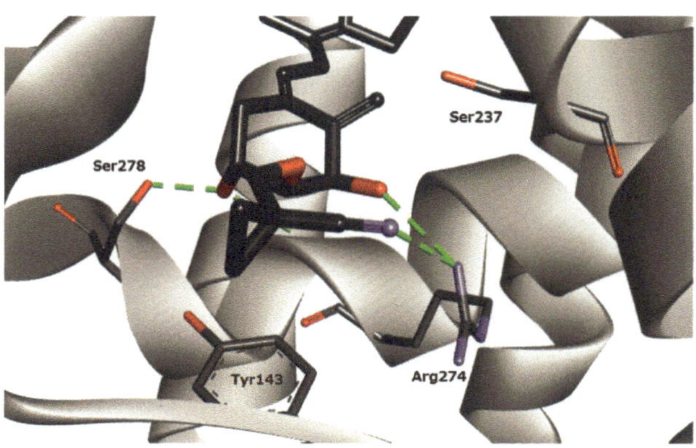

Atsushi Kittaka *et al.*, Figure 6 X-Ray studies on hVDR LBD-**7** complex. The Protein Data Bank accession number for the coordinates of the structure of the VDR complex with **7** is 3WWR. The A-ring part is magnified. *Adapted with permission from Saitoh et al. (2015).*

MIX
Papier | Fördert
gute Waldnutzung
FSC® C083411

Kolibri 360 GmbH
Ferdinand-Jühlke-Straße 7
99095 Erfurt, Deutschland
produktsicherheit@kolibri360.de